2nd International Conference on Innovative Materials, Structures and Technologies

IOP Conference Series: Materials Science and Engineering
Volume 96

Riga, Latvia
30 September – 2 October 2015

Editor:

Sandris Rucevskis

ISBN: 978-1-5108-1467-7
ISSN: 1757-8981

Printed from e-media with permission by:

Curran Associates, Inc.
57 Morehouse Lane
Red Hook, NY 12571

Some format issues inherent in the e-media version may also appear in this print version.

Printed by Curran Associates, Inc. (2016)

Institute of Physics
Dirac House, Temple Back
Bristol BS1 6BE UK

Phone: 44 1 17 929 7481
Fax: 44 1 17 920 0979

techtracking@iop.org

Additional copies of this publication are available from:

Curran Associates, Inc.
57 Morehouse Lane
Red Hook, NY 12571 USA
Phone: 845-758-0400
Fax: 845-758-2634
Email: curran@proceedings.com
Web: www.proceedings.com

Table of contents

Volume 96

2nd International Conference on Innovative Materials, Structures and Technologies
30 September to 2 October 2015, Riga, Latvia

Accepted papers received: 8 October 2015
Published online: 2 November 2015

Preface

Papers

Alkali Activation of Minerals and Waste

Composite Materials

Geodesy and Surveying, Geomatics, Mapping and GIS

High Performance Concrete

Innovative Timber, Concrete and Steel Structures

Inspection, Assessment and Monitoring

IOP Conf. Series: Materials Science and Engineering **96** (2015) 011001 doi:10.1088/1757-899X/96/1/011001

Preface

The 2nd International Conference on Innovative Materials, Structures and Technologies (IMST 2015) took place in Riga, Latvia from 30th September – 2nd October, 2015. The first event of the conference series, dedicated to the 150th anniversary of the Faculty of Civil Engineering of Riga Technical University, was held in 2013. Following the established tradition, the aim of the conference was to promote and discuss the latest results of industrial and academic research carried out in the following engineering fields: analysis and design of advanced structures and buildings; innovative, ecological and energy efficient building materials; maintenance, inspection and monitoring methods; construction technologies; structural management; sustainable and safe transport infrastructure; and geomatics and geotechnics. The conference provided an excellent opportunity for leading researchers, representatives of the industrial community, engineers, managers and students to share the latest achievements, discuss recent advances and highlight the current challenges.

IMST 2015 attracted over 120 scientists from 24 countries. After rigorous reviewing, over 80 technical papers were accepted for publication in the conference proceedings. On behalf of the organizing committee I would like to thank all the speakers, authors, session chairs and reviewers for their efficient and timely effort.

The 2nd International Conference on Innovative Materials, Structures and Technologies was organized by the Faculty of Civil Engineering of Riga Technical University with the support of the Latvia State Research Programme under the grant agreement "INNOVATIVE MATERIALS AND SMART TECHNOLOGIES FOR ENVIRONMENTAL SAFETY, IMATEH". I would like to express sincere gratitude to Juris Smirnovs, Dean of the Faculty of Civil Engineering, and Andris Chate, manager of the Latvia State Research Programme.

Finally, I would like to thank all those who helped to make this event happen. Special thanks go to Diana Bajare, Laura Sele, Liga Radina and Jana Galilejeva for their major contribution to organizing the conference and to the literary editor Tatjana Smirnova and technical editor Daira Erdmane for their hard work on the conference proceedings.

Sincerely,
Sandris Ručevskis

Conference Chairman
Faculty of Civil Engineering
Riga Technical University

IOP Conf. Series: Materials Science and Engineering **96** (2015) 011001 doi:10.1088/1757-899X/96/1/011001

Peer review statement

All papers published in this volume of *IOP Conference Series: Materials Science and Engineering* have been peer reviewed through processes administered by the proceedings Editors. Reviews were conducted by expert referees to the professional and scientific standards expected of a proceedings journal published by IOP Publishing.

IOP Conf. Series: Materials Science and Engineering **96** (2015) 012001 doi:10.1088/1757-899X/96/1/012001

The Effects of Aggressive Environments on the Properties of Fly Ash based Geopolymers

Z Baščarević[1,2], M Komljenović[1], V Nikolić[1] and N Marjanović[1]
[1]University of Belgrade, Institute for Multidisciplinary Research, Department of Materials Science, Serbia

E-mail: zvezdana@imsi.bg.ac.rs

Abstract. This paper analyzes the effects of two different aggressive environments, concentrated ammonium nitrate solution (480 g/dm^3) and sodium sulphate solution (50 g/dm^3), on the structure and mechanical strength of fly ash based geopolymers. Geopolymer samples were subjected to the aggressive solutions over a period of 365 days. It was found that exposure to the NH_4NO_3 and Na_2SO_4 solutions caused small decrease in geopolymer strength (10-20 %). The most valuable insight into the structural changes caused by testing of the geopolymer samples in the aggressive solutions was provided by means of ^{29}Si MAS NMR. It was found that the immersion of geopolymer samples in the NH_4NO_3 solution caused breaking of Si-O-Al bonds in the aluminosilicate geopolymer gel structure. On the other hand, treatment of the geopolymer samples with the Na_2SO_4 solution resulted in breaking of Si-O-Si bonds in geopolymer gel structure and leaching of Si. It was concluded that the major changes in the geopolymer structure were associated with the changes in the pH values of aggressive solutions during the testing.

1. Introduction

Geopolymers are a relatively new group of binder materials that are formed by alkali activation of aluminosilicate materials such as fly ash (FA). Alkali activation reaction transforms glassy phase from starting material into the main reaction product, aluminosilicate geopolymer gel. Due to highly cross-linked three dimensional structure of the aluminosilicate geopolymer gel, geopolymers show good mechanical properties. In addition, these binders usually perform better in aggressive aqueous solutions than Portland cement [1].

Ammonium salts solutions are very aggressive to Portland cement structure [2]. NH_4NO_3 is often used as a fertilizer and concrete structures in contact with surface water from the processed agricultural soil can suffer serious damage. The concentrated NH_4NO_3 solution is frequently used for

[2] To whom any correspondence should be addressed.

the accelerated decalcification test of Portland cement [2]. Decalcification process, which involves leaching of calcium from portlandite and calcium silicate hydrate, is completed after several weeks in the concentrated NH_4NO_3 solution and is accompanied by a loss of ~90 % of the initial strength of the Portland cement samples.

Sulphate ions present in groundwater or soils surrounding a concrete structure also represent a threat to long-term durability of concrete. Sulphate attack on Portland cement based concrete involves diffusion of sulphate ions into the structure and reactions with some of the phases present in the structure, thus forming new phases which can cause stress, expansion and strength loss [1, 2]. Sulphate attack on binders is most often studied in Na_2SO_4 solutions in concentrations of 44 – 50 g/dm^3 [2].

This paper investigates the effects of two aggressive solutions, concentrated NH_4NO_3 and Na_2SO_4 (50 g/dm^3) on the properties of FA based geopolymer samples.

2. Materials and methods

2.1. Materials

Two FA samples from Serbian thermal power plants were used as starting raw materials: FA Kolubara from thermal power plant (TPP) "Kolubara" and FA Svilajnac, from TPP "Morava". In order to increase reactivity of the FA samples, fractions of the samples smaller than 63 μm were used [3, 4].

Alkaline activator was sodium silicate solution with modulus (SiO_2/Na_2O mass ratio) of 3.04 ("Galenika-Magmasil", Serbia, 8.72% Na_2O, 26.5% SiO_2, 64.78% H_2O). Silicate modulus of the activating solution was adjusted by adding NaOH solution ("ZorkaPharm", Serbia, p.a. 98%).

Aggressive testing solutions were prepared using NH_4NO_3 and Na_2SO_4 of 99 % purity (Superlab, Serbia).

2.2. Preparation of geopolymer samples

Sodium silicate solution with modulus of 1.5 was used as the activating solution. The activator solution/FA ratio was the same for all geopolymer samples (10 % Na_2O content with respect to the FA mass). Geopolymer mortars were prepared by adding the silicate solution to water and then mixing the solution with FA and standard sand. Different amounts of water were required in order to obtain equal workability of the mortars based on the two FA samples (mortar flow measured on a flow table was 125 ± 5 mm), so water/FA ratio was 0.76 for the mortars based on FA Kolubara and 0.56 for the mortars based on FA Svilajnac (water in water/FA ratio was calculated as water from sodium silicate solution + water added for consistency). Sand/FA ratio was 3/1. The moulds with three mortar prisms (40 × 40 × 160 mm) were placed in plastic bags to avoid moisture loss during curing.

Geopolymer pastes were prepared by mixing the FA with the alkaline activator solution in the same proportion as in the preparation of the geopolymer mortars. Water/FA ratio was 0.69 for the pastes prepared with FA Kolubara and 0.49 for the pastes based on FA Svilajnac. Dimensions of the paste samples were 25 × 25 × 30 mm. Geopolymer mortar and paste samples were cured at 95 °C for 24 hours.

2.3. Experimental design

Testing solutions, concentrated (480g/dm^3) NH_4NO_3 solution and 50 g/dm^3 Na_2SO_4 solution, were prepared by dissolving NH_4NO_3 and Na_2SO_4 in distilled water.

Three geopolymer mortar prisms and corresponding paste sample were immersed in plastic containers with the solutions and kept at 20 ± 2 °C for a period of up to 365 days. Geopolymer samples were placed on plastic nettings to ensure that the entire sample surface was in contact with the aggressive solution. Solution/geopolymer mass ratio was 4/1. At the same time, reference samples were cured in the humid chamber (temperature 20 ± 2 °C, humidity ~98%).

2.4. Analytical methods

Compressive strength of the geopolymer mortars was tested using Tony Technik (Toninorm) testing machine.

Nitrogen adsorption-desorption isotherms were determined using Micrometrics ASAP 2020 instrument and powdered mortar samples.

The pH values of the aggressive solutions before and after the testing of the geopolymer samples were measured by pH-meter (pH Testr30, Eutech Instruments).

Inductively coupled plasma optical emission spectrometer (ICP-OES, SpectroGenesis EOP II, Spectro Analytical Instruments GmbH, Kleve) was used to determine the ion exchange between the geopolymers and the aggressive solutions.

Mineralogical characterization was performed by Philips PW 1710, with CuKα X-rays operating at 30 mA and 40 kV, with 0.02° 2θ steps, 0.5 s $step^{-1}$ in 5 – 50° 2θ range.

Scanning electron microscope (SEM, VEGA TS 5130 MM, Tescan) was used for morphological characterization of the samples.

^{29}Si MAS NMR were obtained using Bruker MSL 400 system, Apollo console upgraded (Tecmag) at a Larmor frequency of 79.49 MHz. Single pulse sequence was used for data acquisition with pulse width of 4 μs (corresponding to π/2 flip angle). Recycle delays were typically 20 s, the number of scans was 4,096 and the spectral width was 20,000 Hz for all recorded spectra. Chemical shifts were externally referenced to 2.2-dimethyl-2-silapentane-5-sulfonate (DSS) standard. Prior to ^{29}Si MAS NMR analysis, iron content of the geopolymer samples was reduced by exposing the samples to a strong magnetic field.

3. Results and discussion

3.1. Effects of the aggressive solutions on mechanical properties of the geopolymer samples

Initial compressive strengths of the geopolymer mortars based on two different FA samples were significantly different (figure 1). After 24 hours of curing, geopolymer mortar based on FA Kolubara showed compressive strength of about 20 MPa, while the strength of the geopolymer mortar based on FA Svilajnac was about two times higher. Differences in the compressive strength of the geopolymer mortars based on the two FA samples were the consequences of different chemical and mineralogical composition of the FA samples and different particle size distribution [3, 4]. FA Kolubara had high quartz content [3], and significantly lower amount of fine, more reactive, FA particles compared to FA Svilajnac [4]. Due to different fine particle content in the FA samples, different quantities of water had to be added in preparation of geopolymer mortars (Section 2.2). Different water/FA ratios affected porosity of the geopolymer samples based on the two FA samples: the porosity of the geopolymer mortar based on FA Kolubara was 0.160 cm^3/g, while the porosity of the mortar based on FA Svilajnac was lower (0.055 cm^3/g) [4].

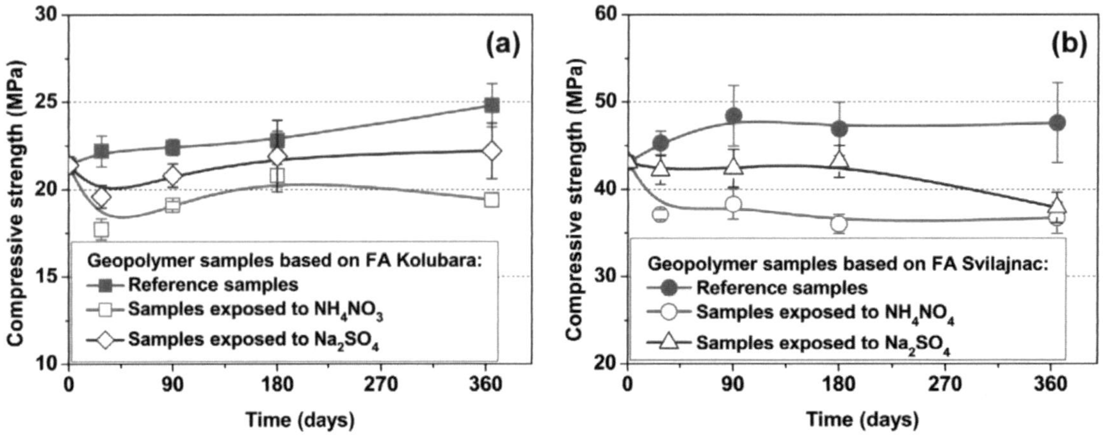

Figure 1. Compressive strength of the geopolymer samples: (a) FA Kolubara geopolymers, (b) FA Svilajnac geopolymers.

Compressive strength of both groups of geopolymer samples (based on FA Kolubara and FA Svilajnac) decreased due to the exposure to the aggressive solutions (figure 1). The strength decrease was more pronounced in the samples exposed to the effects of the NH_4NO_3 solution. The greatest decrease in strength (~20 %) was observed after the first 28 days in the NH_4NO_3 solution (figure 1).

After the first 28 days in the Na_2SO_4 solution, compressive strength of the geopolymer samples based on FA Kolubara decreased about 10 %. After the initial decrease in strength, the subsequent investigation showed there was a strength increase (figure 1a). Compressive strength of the geopolymer samples based on FA Svilajnac immersed into the Na_2SO_4 solution did not change significantly during the first 180 days of testing. A more noticeable decrease in strength of these samples was noticed only after 365 days in the Na_2SO_4 solution (figure 1b). Differences in strength change of the geopolymer samples based on FA Kolubara and FA Svilajnac exposed to the effects of the Na_2SO_4 solution were probably due to differences in the starting compressive strength (figure 1) and porosity of the samples [4].

It must be noted that testing of Portland cement samples in the same aggressive solutions usually results in much higher decrease in mechanical properties [1, 2].

3.2. Analyses of the aggressive solutions

The initial pH value of the NH_4NO_3 solution was ~4 while the initial pH value of the Na_2SO_4 solution was ~6 (Table 1). During the testing of the geopolymer samples, pH values of both solutions increased due to the leaching of pore solution alkalis [3, 4]. While the pH value of the NH_4NO_3 was lower than 9.25 throughout the whole testing period [3], the pH value of the Na_2SO_4 solution reached the value of ~12 already after the first 28 days of testing.

Table 1. pH values of the aggressive solutions.

	NH_4NO_3 solution Geopolymer samples		Na_2SO_4 solution Geopolymer samples	
Time (days)	FA Kolubara	FA Svilajnac	FA Kolubara	FA Svilajnac
0	4.17	4.17	5.64	5.90
28	8.27	8.31	11.90	11.90
90	8.32	8.40	12.11	12.13
180	8.34	8.44	11.86	12.06
365	7.98	8.47	11.68	11.67

ICP-OES analysis of the NH_4NO_3 solutions showed that alkali (Na, K) and alkaline earth (Ca, Mg) elements were leached from the geopolymers (data shown elsewhere [3]). The main elements of the geopolymer structure (Si and Al) were not detected. Analyses of the Na_2SO_4 solutions indicated that exposure of the geopolymer samples to the effects of the sulphate solution caused leaching of Si (Table 2). The maximum concentrations of Si in the Na_2SO_4 solutions were detected after the same testing periods, in which the maximum strength decrease of the geopolymer mortars was observed (figure1), thus suggesting that Si was probably leached from the aluminosilicate geopolymer gel.

Table 2. Concentrations of Si in the Na_2SO_4 solutions (mg/dm^3).

	Geopolymer samples	
Time (days)	FA Kolubara	FA Svilajnac
0	0	0
28	157.6	44.88
365	154.3	163.5

3.3. Effects of the aggressive solutions on the structure of the geopolymer samples
XRD analysis of the geopolymer samples did not indicate formation of any new crystalline phases formed due to a reaction with the aggressive testing solutions (figures 2 and 3). Besides the phases typical for FA [3], only two new phases were observed: calcite, formed as a result of carbonation of the samples, and NH_4NO_3 in the samples that were exposed to the NH_4NO_3 solution.

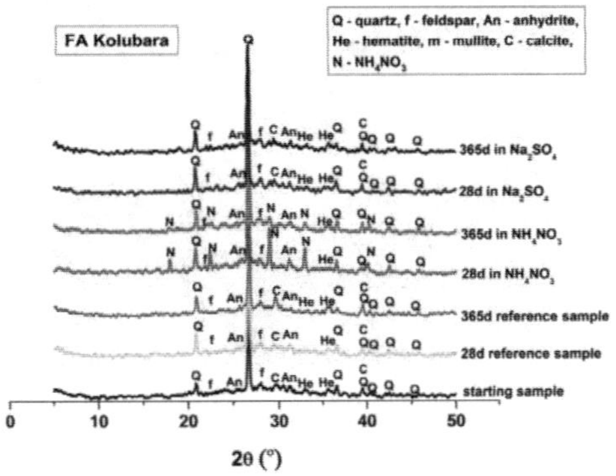

Figure 2. XRD patterns of the FA Kolubara geopolymer samples.

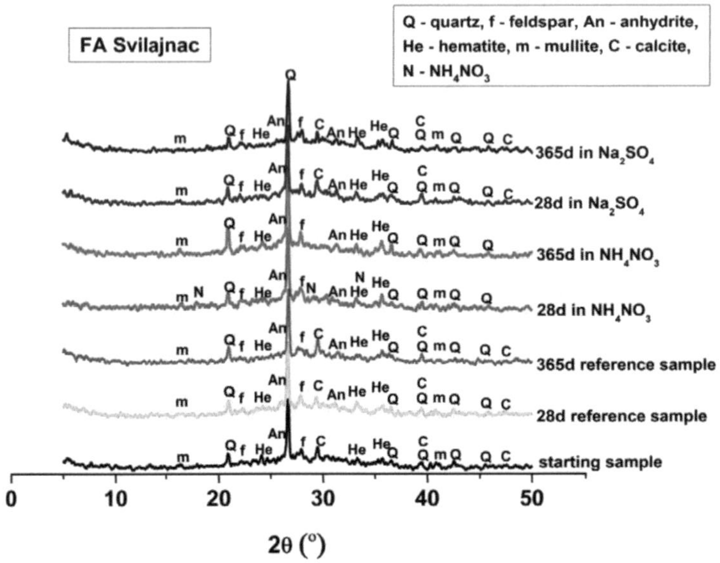

Figure 3. XRD patterns of the FA Svilajnac geopolymer samples.

No significant changes in the structure of the geopolymer samples exposed to the aggressive solutions were detected by SEM analysis of the samples (figures 4 and 5). Both reference and geopolymer samples treated with the NH_4NO_3 and Na_2SO_4 solutions showed microstructure typical for FA based geopolymers, with remnant FA particles embedded in matrix (aluminosilicate geopolymer gel).

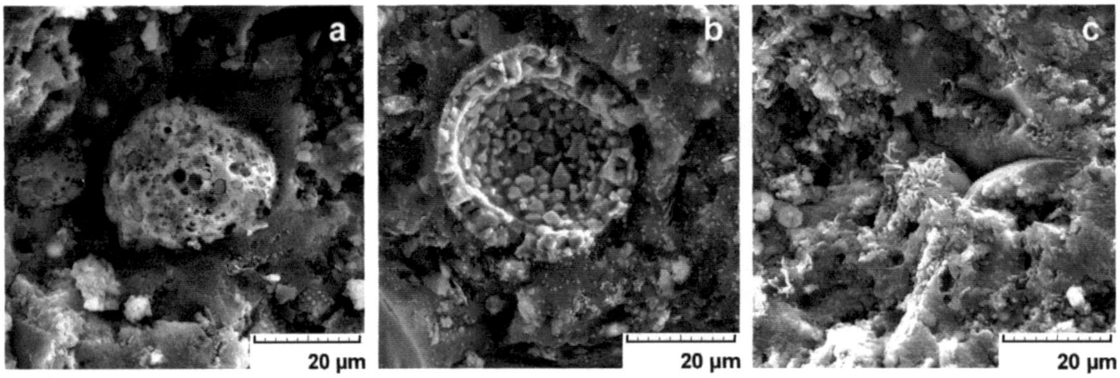

Figure 4. Microstructure of the FA Kolubara geopolymer samples: (a) 365 days reference sample, (b) 365 days in NH_4NO_3, (c) 365 days in Na_2SO_4.

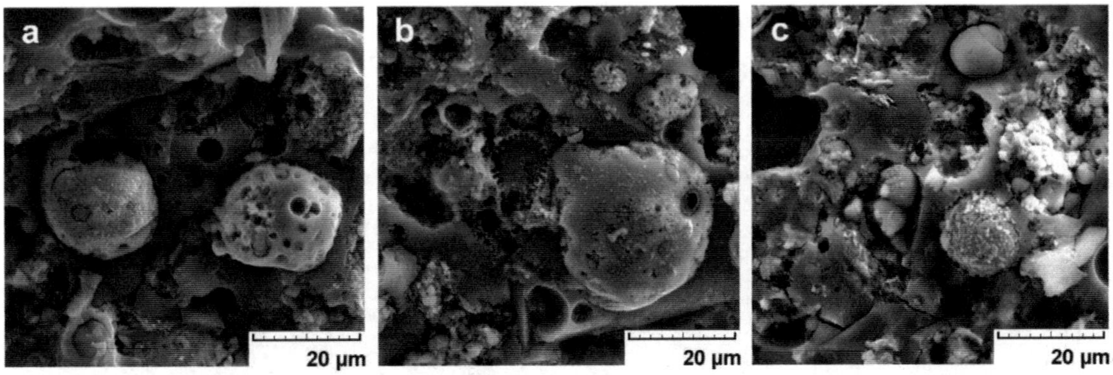

Figure 5. Microstructure of the FA Svilajnac geopolymer samples: (a) 365 days reference sample, (b) 365 days in NH_4NO_3, (c) 365 days in Na_2SO_4.

^{29}Si MAS NMR analysis of the geopolymer samples based on FA Kolubara and FA Svilajnac showed broad resonances, centred at -90 to -100 ppm, typical for geopolymers (figures 6 and 7) [3, 4]. Broad ^{29}Si MAS NMR spectra indicate that silicon is surrounded by a wide range of environments. A ^{29}Si MAS NMR spectrum of FA based geopolymer consists of overlapping resonances attributed to silicon sites in different silicate and aluminosilicate phases present in the material: aluminosilicate geopolymer gel, un-reacted glassy and crystalline phases (such as quartz and mullite) from the starting FA [3].

The chemical shift at which maximum peak intensity of the ^{29}Si MAS NMR spectrum occurs depends on the Si/Al atomic ratio of the material [5]. By comparing the ^{29}Si MAS NMR spectra of the reference samples and the geopolymer samples exposed to the aggressive environments, it is clear that the treatment of the geopolymer samples with the NH_4NO_3 and Na_2SO_4 solutions caused changes in the Si/Al atomic ratio of the samples, i.e. changes in the structure of the aluminosilicate geopolymer gel.

The ^{29}Si MAS NMR spectra of the starting geopolymer sample based on FA Kolubara and the reference sample after 28 days of curing were centred at -94 ppm (figure 6). After curing of the reference sample in the humid chamber for 365 days, the maximum peak intensity was centred at -93 ppm. The maximum peak intensity of the starting geopolymer sample based on FA Svilajnac and the reference samples cured for 28 and 365 days was centred at -93 ppm (figure 7).

Shifting of the maximum peak intensities of the geopolymer samples treated with the NH_4NO_3 solution to more negative chemical shifts (figures 6 and 7) indicated that the structure of these samples consisted of species richer in silicon. Since none of the main elements of the aluminosilicate geopolymer gel structure (Si and Al) were detected by the ICP-OES analysis of the solution, it is more likely that exposure of the geopolymer samples led to breaking of -Si-O-Al- bonds in the aluminosilicate gel structure. Breaking of -Si-O-Al- bonds caused the strength decrease of the samples (figure 1) and changed the local environment of the silicon in the aluminosilicate structure, as indicated by the ^{29}Si MAS NMR analysis. The observed degradation mechanism represents the first step of the dealumination reaction, the process that is usually observed in the case of acid attack on geopolymer binder [1, 3].

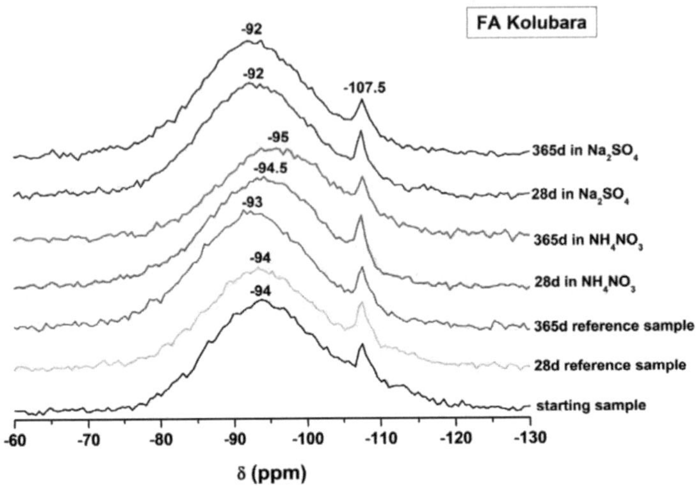

Figure 6. ^{29}Si MAS NMR spectra of the FA Kolubara geopolymer samples.

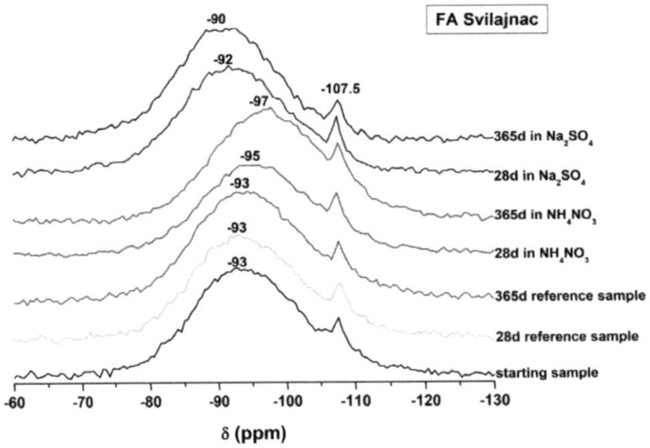

Figure 7. ^{29}Si MAS NMR spectra of the FA Svilajnac geopolymer samples.

^{29}Si MAS NMR analysis of the geopolymer samples treated with the Na_2SO_4 solution indicated a decrease in the Si/Al atomic ratio of these samples, i.e. confirmed leaching of Si (Table 2). Decrease in the Si/Al atomic ratio of the samples exposed to the Na_2SO_4 solution was indicated by shifting of the maximum peak intensities of the ^{29}Si MAS NMR spectra to less negative chemical shifts (figures 6 and 7). The observed small decrease in strength of the geopolymer samples treated with the Na_2SO_4 solution was caused by breaking of the -Si-O-Si- bonds and leaching of Si. This type of degradation mechanism has already been observed in geopolymer samples exposed to moderately alkaline solutions [1]. Subsequent increase in strength of the geopolymer samples based on FA Kolubara (figure 1) was most probably due to continuing of the alkali activation reaction [1, 4]. Conditions to which the geopolymer samples were subjected throughout the testing in the Na_2SO_4 solution, high concentration of Na in the solution and high pH values of the solution (Table 1) were favourable for continuing the alkali activation reaction [1].

4. Conclusions

In this paper, effects of two different aggressive environments, NH_4NO_3 solution (480 g/dm^3) and Na_2SO_4 solution (50 g/dm^3), on mechanical strength and structure of FA based geopolymers were investigated over a period of 365 days. It was found that testing of the geopolymer samples in the aggressive solutions caused small decrease in compressive strength (10-20 %). No new phases were formed in the geopolymer structure due to a reaction with either of the aggressive solutions. Based on the ^{29}Si MAS NMR analyses, it was found that the exposure of the geopolymer samples to the aggressive solutions caused alterations in the aluminosilicate gel. Small strength decrease of the geopolymer samples exposed to the NH_4NO_3 solution was due to breaking of -Si-O-Al- bonds in the aluminosilicate geopolymer gel structure. On the other hand, exposure of the geopolymer samples to the Na_2SO_4 solution caused breaking of the -Si-O-Si- bonds and leaching of Si.

The observed degradation mechanisms can be associated with changes in pH values of the aggressive solutions. It was found that leaching of pore solution alkalis caused increase in pH values of both testing solutions. The pH value of the initially acidic (pH = ~4) NH_4NO_3 solution increased to ~8.3, while the pH value of the Na_2SO_4 solution increased from the starting ~6 to ~12 after the first 28 days of testing. Changes in the geopolymer structure observed after the treatment with the NH_4NO_3 solution represent the first step of the degradation mechanism observed in the case of acid attack on geopolymers. Conversely, high pH value of the Na_2SO_4 solution caused breaking of the -Si-O-Si- bonds and leaching of Si, degradation process usually observed when geopolymer samples are exposed to moderately alkaline conditions.

The results obtained in this work suggest that, compared to traditional Portland cement, FA based geopolymers show better resistance to some types of aggressive environments.

5. References

[1] Baščarević Z 2015 The resistance of alkali-activated cement-based binders to chemical attack *Handbook of Alkali-activated Cements, Mortars and Concretes* ed F Pacheco-Torgal, J Labrincha, C Leonelli, A Palomo and P Chindaprasit (Cambridge, UK: Woodhead Publishing) chapter 14 pp 373 – 396

[2] Alexander M, Bertron A and De Belie N (ed) 2013 *Performance of Cement-Based Materials in Aggressive Aqueous Environment* (Berlin: Springer)

[3] Baščarević Z, Komljenović M, Miladinović Z, Nikolić V, Marjanović N, Žujović Z and Petrović R 2013 *Constr. Build. Mater.* **41** 570 – 579

[4] Baščarević Z, Komljenović M, Miladinović Z, Nikolić V, Marjanović N and Petrović R 2015 *Mater. Struct.* **48** 683 – 697

[5] Engelhardt G and Michel D 1987 *High resolution solid state NMR of silicates and zeolites* (New York: John Wiley & Sons)

Acknowledgement

This work was carried out within the Project TR34026 funded by the Ministry of Education, Science and Technological Development, Republic of Serbia.

IOP Conf. Series: Materials Science and Engineering **96** (2015) 012002 doi:10.1088/1757-899X/96/1/012002

Structural Investigation of Alkali Activated Clay Minerals for Application in Water Treatment Systems

G Bumanis[1,2], **D Bajare**[1] **and L Dembovska**[1]
[1]Riga Technical University, Department of Building Materials and Products, Institute of Materials and Structures, Kalku Str. 1, Riga, LV-1658, Latvia

E-mail: girts.bumanis@rtu.lv

Abstract. Alkali activation technology can be applied for a wide range of alumo-silicates to produce innovative materials with various areas of application. Most researches focuse on the application of alumo-silicate materials in building industry as cement binder replacement to produce mortar and concrete [1]. However, alkali activation technology offers high potential also in biotechnologies [2]. In the processes where certain pH level, especially alkaline environment, must be ensured, alkali activated materials can be applied. One of such fields is water treatment systems where high level pH (up to pH 10.5) ensures efficient removal of water pollutants such as manganese [3]. Previous investigations had shown that alkali activation technology can be applied to calcined clay powder and aluminium scrap recycling waste as a foam forming agent to create porous alkali activated materials. This investigation focuses on the structural investigation of calcined kaolin and illite clay alkali activation processes. Chemical and mineralogical composition of both clays were determined and structural investigation of alkali activated materials was made by using XRD, DTA, FTIR analysis; the microstructure of hardened specimens was observed by SEM. Physical properties of the obtained material were determined. Investigation indicates the essential role of chemical composition of the clay used in the alkali activation process, and potential use of the obtained material in water treatment systems.

1. Introduction

Inorganic binders can be divided into fundamentally different groups according to the properties of their chemical composition. The main parameters which identify mineral binders are the content of Ca, Al and Na oxides in the chemical composition of the binder. The chemical composition of alkali activated binders (AAB) is substantially different compared to the traditional cementitious binders. AAB could be divided in high calcium (Ca) binder systems which are characterised by high Ca

[2] To whom any correspondence should be addressed.

content in raw materials of AAB and low Ca systems AAB. The typical phases for AAB systems with high Ca content are C-S-H (calcium silicate hydrates) and C-A-S-H (calcium aluminate silicate hydrates) while for low Ca AAB systems N-A-S-H (sodium aluminate silicate hydrates) is the predominantly binding phase. Raw materials such as fly ash or calcined kaolin clay together with highly concentrated alkaline base (Na or K base from 6 to 10 M) or sodium silicate solution (activators) could be used to create low calcium AAB [4].

The properties of AAB are defined by the chemical composition of raw materials and the type of alkali activation solution used for activation together with their proportions in mixture composition of AAB. According to the scientific literature, a wide range of AAB could be obtained by using raw material compositions with SiO_2/Al_2O_3 (Si/Al) ratio which could vary from $0.5 < Si/Al < 300$ [5]. However, in order to obtain AAB with high durability and good mechanical property performance, it is recommended to provide the chemical composition in the following range: $1<Si/Al<5$ and Na_2O/Al_2O_3 (Na/Al) ~1 [4]. AAB created outside this range is characterised by low strength and chemical stability (durability) and it could be unstable in the water environment. If Na/Al ratio is high in mixture composition, free alkalis can remain in the structure of AAB and form carbonate crystals and efflorescence could be observed. Therefore, using different raw materials, AAB with a wide range of properties can be obtained. There is no common guidance for prediction of the properties of AAB by defining chemical compound ratio in the mixture composition; therefore potential properties and application should be tested experimentally.

The main compound ratios can be determined from mixture composition of AAB; therefore it has essential effect on leaching of alkalis from the structure of AAB. By reducing the Al content in the mixture composition, the amount of free alkalis increases pore amount in the material. [6]. It is concluded that low Al and high Si content in mixture composition of AAB promotes alkali leaching from the material structure and this could be the main factor that affects potential buffer capacity of water solution provided by AAB. The buffer capacity could be described as diffusion of free alkalis from material structure in the water environment. Na^+ and K^+ ions ensure highly concentrated alkaline environment, which enhances the dissolution of amorphous phase being an essential part of alkali activation process and binding phase development; however, these alkali ions are weakly bonded to the (N,K)-A-S-H phase and are available for leaching. Alkali ions can diffuse to AAB surface and react with CO_2 forming crystalline salts like $Na_2CO_3{\cdot}nH_2O$, $NaHCO_3$, K_2CO_3, $KHCO_3$[7]. The content of alkalis in pore solution increases if a highly concentrated alkaline activation solution is used in the alkali activation processes [6]. Ali has reported that pore solution transfer to the surrounding environment will occur only in direct surface contact between the material surface and environment (i.e. water). Therefore the leaching rate of alkalis in highly porous AAB will be more intense because of the high specific surface available for leaching [8].

In water treatment systems, where continuous ensuring of stable pH is vital for purification of water, this leaching effect from the structure of AAB could be regarded as a positive factor for passive pH or buffer capacity controlling systems. To enhance leaching from AAB structure, low calcium AAB with low Al and high Si content would be an advantage while for AAB designed for construction industry it would be an advantage to use raw materials with high amount of Al. However, the strength of AAB must be considered also in water treatment systems due to high water pressure during filtration through material.

2. Materials and methods

2.1. Materials

Two types of clay were used to obtain low calcium alkali activated binders (AAB): metakaolin (MK) obtained from kaolin clay from "Surin-Kaolin" Ltd (Poland) and calcined local illite clay (IC) obtained from raw illite clay from "Lode" Ltd (Latvia). The additional secondary raw materials used in this investigation were aluminium scrap recycling waste (ASRW) received from aluminium scrap recycling industry facility "Dilers" Ltd (Latvia), glass waste (G) received from the glass recycling factory "Lampu Demerkurizacijas Centrs" Ltd (Latvia). In order to increase material homogeneity and fineness, raw materials were ground for 30 minutes by using planetary ball mill Retsch PM 400. Chemical composition of raw materials is given in table 1.

Table 1. Chemical composition of raw materials (wt. %).

Chemical component	ASRW	IC	MK	G
Al_2O_3	63.2	14.6	34.2	1.0
SiO_2	7.9	73.8	51.8	68.1
CaO	2.6	0.9	0.1	1.4
SO_3	0.4	-	-	-
TiO_2	0.5	0.6	0.6	-
MgO	4.4	1.1	0.1	-
Fe_2O_3	4.5	4.1	0.5	0.2
PbO	-	-	-	20.0
Na_2O	3.8	0.1	0.6	8.0
K_2O	3.8	2.8	-	1.2
$CaCO_3$*$MgCO_3$	-	-	-	-
Others	2.6	1.1	0.5	0.1
LOI, 1000 °C	6.21	0.98	0.2	-

According to XRD analysis, calcined IC has crystalline structure and, in addition to quartz (SiO_2) and illite ($K(AlFe)_2AlSiO_3O_{10}(OH)_2 \cdot H_2O$), it also contains microcline ($KAlSiO_8$) while MK is largely amorphous with the amorphous phase in 2θ region from 18 to 33°. The content of reactive SiO_2 was 1.3% for MK and 1.1% - for IC, but total amount of R_2O_3 was 55.9% for MK and 19.5% for IC respectively. The ASRW is the final waste product in the aluminium scrap recycling factories. According to the XRD analysis data, the ASRW contains metallic aluminium (Al), iron sulphite ($FeSO_3$), aluminium nitride (AlN), corundum (Al_2O_3), aluminium iron oxide ($FeAlO_3$), magnesium dialuminium ($MgAl_2O_4$), quartz (SiO_2), aluminium chloride ($AlCl_3$) and aluminium hydroxide ($Al(OH)_3$). Glass powder is used as an additive to AAB mixture to change SiO_2 amount in the composition. The results of grading analysis showed that the additionally ground glass contains a wide range of particle sizes: from 8 μm to 30 μm with the average grain size of 26.3 μm. The specific surface area of G ground for 30 minutes was 1.126 m^2/g. The chemical compositions of ASRW, IC, IM and G are given in table 1.

The alkali activation solution was prepared by using commercially available sodium silicate solution from "Vincents Polyline" Ltd characterized by the silica modulus M_s=3.22. To obtain alkali activation solution with the required chemical composition, the modification with an addition of sodium hydroxide was done. Commercially available sodium hydroxide flakes from "Tianye

Chemicals" (China) with 99% purity were used. By the change of NaOH content, sodium silicate solution with silica modulus from M_s=1.3 to M_s=3.22 was obtained.

2.2. Methods

Mineralogical composition of raw materials and AAB was determined by X-ray diffraction (XRD) (PAN analytical X'Pert PRO) and chemical composition for raw materials was determined according to LVS EN-196-2 with sensibility ±0.5w%. Elements were analysed by EDX (energy dispersive X-ray spectrometry - EDS, Oxford instruments 7378). Reactive SiO_2 and R_2O_3 for calcined clays were determined according to LVS EN 196-2 and gravimetric analysis was used to determine content of SiO_2 and Al_2O_3 (Florentine method). Structural characterization of different functional groups of raw materials and AAB was performed by using Fourier transform infrared (FTIR) spectrometer (VARIAN 800 FT-IR) between 400 and 4000 cm^{-1}. The particle size distribution for powdered raw material was determined by laser granulometer Analysette 22 Nano Tec. Specific surface area was detected by BET method ("Nova 1200 E-Series, Quantachrome Instruments"). The thermoanalysis of AAB was performed with DTA/DSC Baehr DTA 703. Scanning electron microscope (SEM) (Tescan Mira/LMU) was used for microstructural investigation of AAB and for description of raw materials. The XRD, FTIR, SEM and DTA analyses were performed for 28 days old AAB specimens.

2.3. Sample preparation

AAB specimens based on MK or IC with ASRW and with or without glass (G) additive were created. Sodium silicate solution with M_s=1.63 was used. The mixture proportion of AAB is given in table 2.

Table 2. Mixture composition of alkali activated binders, mass ratio.

Mixture composition	MK	IC	ASRW	G	Sodium silicate solution to solid content ratio
MK-A1-G0	1	-	1	-	0.75
MK-A1-G1	1	-	1	1	0.75
IC-A1-G0	-	1	1	-	0.75
IC-A1-G1	-	1	1	1	0.75

The solid raw materials were mixed together and alkali activation solution was added according to mixture composition. After mixing, pastes were immediately poured into sealed prismatic moulds measuring 4×4×16 cm. The amount of paste poured into the mould was calculated according to the expansion of each mixture to avoid limited expansion of the paste. Moulds were covered and samples were cured at 80°C for 24 h. Afterwards the demoulded samples were cured at room environment until planned tests were performed.

3. Results and discussion

According to the XRD diffractograms, AABs are typically "X-ray amorphous"[9]. The broad peak observed from 20 to 30° 2θ in MK has been broadened up to 40° 2θ after the reaction with the activation solution M_s=1.67 and shifted higher. This shift demonstrates the formation of new amorphous phases, usually described as geopolymeric gels (MK-A1-G0 and MK-A1-G1) (figure 1) [10].

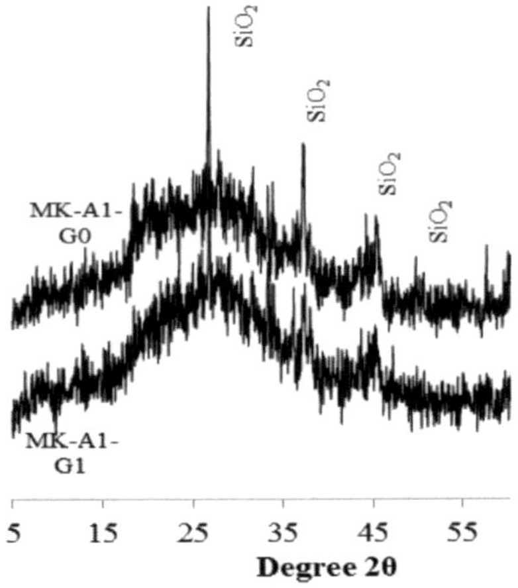

Figure 1. Diffractograms of AAB for 28 day old specimens made of metakaolin.

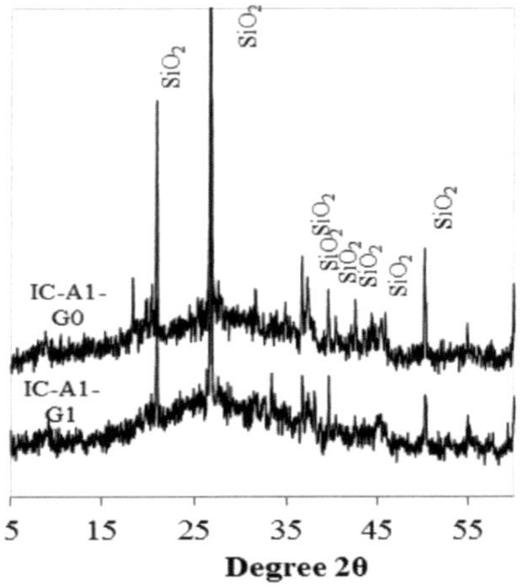

Figure 2. Diffractograms of AAB for 28 day old specimens made of illite clay.

XRD diffractograms showed that the mineralogical composition of AAB made with IC was not affected significantly by the amount of glass in the composition; however, the structure is not completely X-ray amorphous, comparing with AAB made with MK (figure 2). The halo of IC-A1-G0 and IC-A1-G1 is not very expressive in the diffractograms in comparison with MK. It could be explained by the mineralogical composition of IC calcined at temperature 800°C and low amount of Al_2O_3. The only crystalline phase is clearly identified by XRD analysis of AAB made with IC, which is quartz. Comparing the results obtained by XRD, it was concluded that alkali activation of IC creates weak amorphous structure of AAB; however, IC could be used to create AAB but the weak reactions could define mechanical properties of AAB.

Restructuration of mineral compounds in raw materials and formation of new compounds occurs during alkali activation and new bonds could be detected by FTIR analysis. FTIR spectroscopy technique was used for raw materials and the results compared to those obtained for AAB (figures 3 and 4). All AAB samples based on MK or IC have a vibration mode at 446 cm^{-1} which corresponds to Si-O-Si or Al-O-Si bending vibration [11]. This absorbtion mode was not characteristic for raw materials and indicates the formation of new bonds. Mode at 540 cm^{-1} is typical for MK and it is associated to Si-O-Al bonds, where Al is present in octahedral coordination. After alkali activation this bond vibration disappears in the structure of AAB. The small modes at 668 cm^{-1} could also be attributed to O-Si-O (zeolite species) bond band (figure 3) and was characteristic for calcined clay. The absorbtion spectra become less intense after alkali activation.

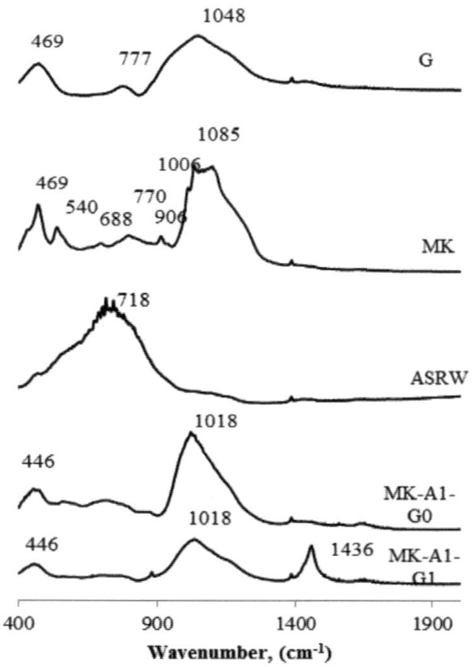

Figure 3. FTIR patterns of raw materials and 28 days old AAB based on MK.

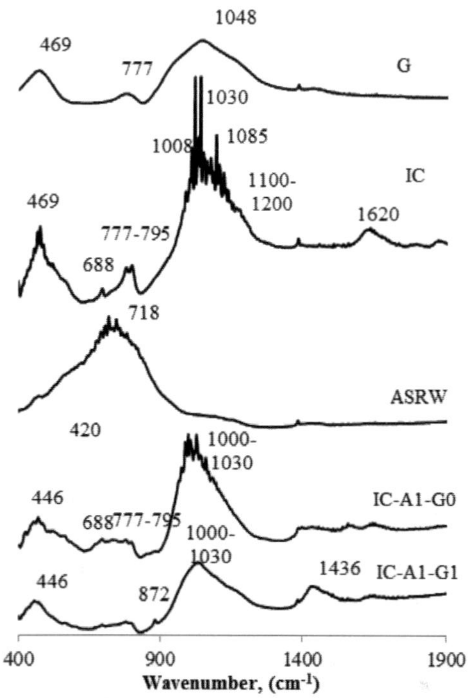

Figure 4. FTIR patterns of raw materials and 28 days old AAB based on IC.

Vibration spectra with the maximum at 1085 cm^{-1}, which is characteristic for raw materials (MK and IC), moves to higher frequencies after alkali activation, while vibration mode at 770 cm^{-1} in MK was not detected in AAB. The shift of vibration spectra between raw materials and AAB indicates the structural changes under process of alkali activation. The shoulder at 1100 and 1200 cm^{-1} is assigned to amorphous SiO_2, which indicates that unreacted amorphous SiO_2 could remain after alkali activation. This phenomenon was observed for samples with glass additive (MK-A1-G1 and IC-A1-G1) which contained more amorphous SiO_2 in the mixture composition. Vibration spectra with the maximum at 1085 cm^{-1}, which is characteristic for raw materials (MK and IC), moves to higher frequencies (1018 to 1030 cm^{-1}) after alkali activation, while vibration mode at 770 cm^{-1} in MK was not detected in AAB.

The results from DTA provide limited information on the phase composition of the system. It is evidently the consequence of the amorphous or near-amorphous character of the ingredients for the studied materials, thus a use of any method is rather restricted. DTA histograms are characterized by the occurrence of first endoeffect in the region about 110–150°C due to the loss of evaporable water (figure 5) [12]. The endoeffect in the region of 250–300°C most likely characterizes decomposition of pore solution, which probably is a mixture of NaOH (melting point 318°C) and $NaHCO_3$ (melting point 50°C) [13]. The loss of chemically bonded water was observed during the heat treatment of AAB. Glass additive in the mixture composition increased the intensity of endoeffects of AAB under heat treatment, which indicates that pore solution was more saturated with chemically bonded water.

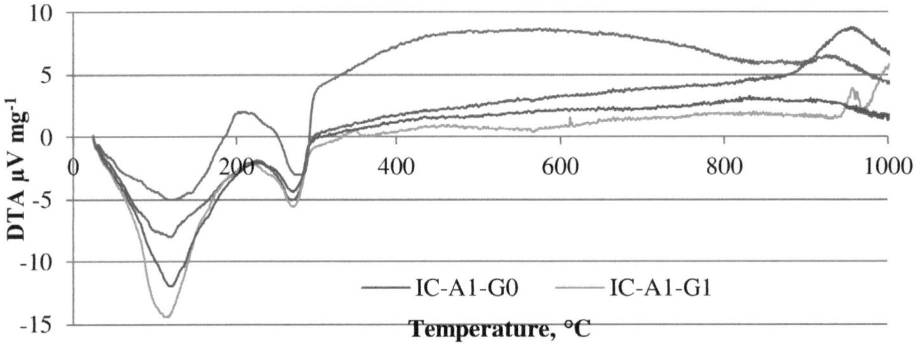

Figure 5. DTA curves of AAB based on IC and MK.

The density of obtained materials was from 319 to 382 kg/m^3 and the density of mixture composition made without glass additive was slightly lower in comparison with the mixture composition made with glass. The open porosity was from 16 to 29%, but the total porosity of the obtained materials increased from 84 to 87% (table 3). AAB mainly have close pore system. The BET specific surface area of MK-A1-G1 and MK-A1-G0 was from 5.1 to 6.1 m^2/g, while for IC-A1-G1 and IC-A1-G0 specific surface increased to 45 and 60 m^2/g. SEM images reveal the structure of AAB, where crystalline substances are concentrated on the pore surface. Crystals on the material surface were well built for AAB based on IC, while for AAB based on MK crystals were enclosed in the material structure (figure 6 and figure 7).

Table 3. Mass ratio of main compounds in mixture composition and physical properties of alkali activated material.

Mixture	Mass ratio			Density	Porosity, %	
composition	SiO_2/Al_2O_3	SiO_2/Na_2O	Na_2O/Al_2O_3	(kg/m^3)	Open	Total
IC-A1-G0	1.6	2.9	0.6	319	26	84
IC-A1-G1	2.5	3.7	0.7	382	24	82
MK-A1-G0	1.1	2.8	0.4	327	29	84
MK-A1-G1	1.8	3.7	0.5	355	16	84

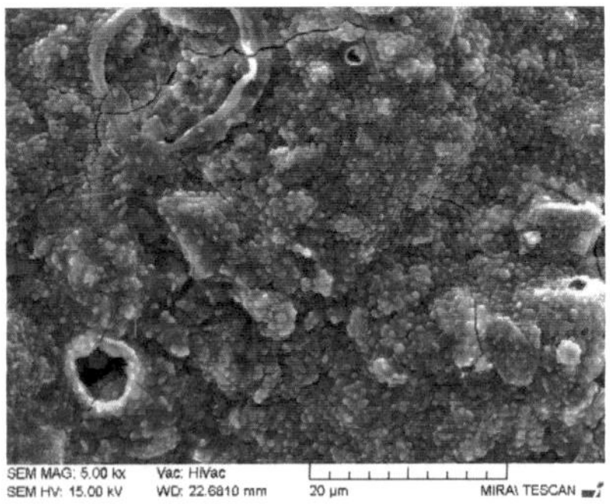

Figure 6. SEM images of AAB for 28 day old specimens made with metakaolin.

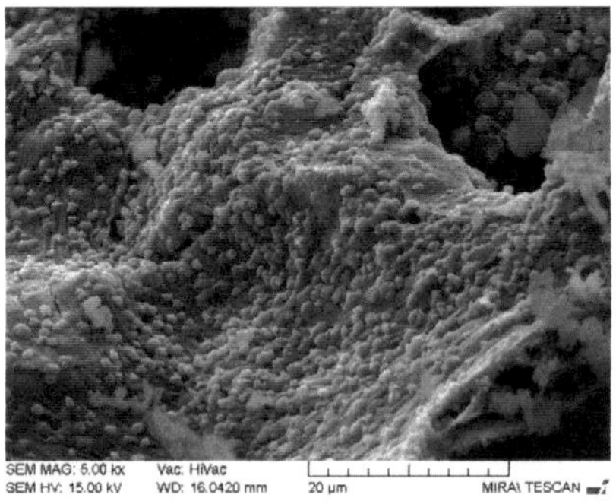

Figure 7. SEM images of AAB for 28 day old specimens made with illite clay.

4. Conclusions

Alkali activated binders based on metakaolin (MK), which contain reactive components (total R_2O_3) up to 55.9%, have expressive amorphous structure that characterizes changes (restructuration) of the MK minerals in the highly concentrated alkaline environment. Whereas binders based on illite clay (IC) (reactive components R_2O_3 - 19.5%) have less explicit amorphous structure according to the results of XRD. Restructuration of the existing chemical bonds and formation of new bonds of calcined kaolinite and illite clay takes place in the highly concentrated alkaline environment indicating the formation of a new binder structure. The shift of vibration spectra between raw materials and AAB (1085 cm^{-1} to 1018 cm^{-1}) indicates the structural changes under process of alkali activation.

Glass additive in the composition of alkali activated binder increases presence of semi-crystalline sodium carbonate and sodium bicarbonate in the pore solution, which is not possible to detect by XRD, but they are clearly identifiable in FTIR spectra (1436 cm^{-1}) and SEM. The pH level of such compounds could be from pH 8.4 to 11.6 and up to 14.0 for NaOH depending on the concentration in the water solution. This pH range ensured by salts and basis could be effectively used in the waste water treatment systems where optimal pH range is from 9.5 to 11.5. Therefore, it is possible to create AAB with high OH^- ion leaching rate by using glass additive in the mixture composition and AAB could be used as the material for passive pH stabilization in waste water treatment systems.

5. References

[1] Bajare D, Bumanis G, Korjakins A and Sele L 2014 *Constr. Mat. and Struct.* 136–144
[2] Rugele K, Bumanis G, Bajare D, Lakevics V and Rubulis J 2014 *Key Eng. Mater.* **604** 223–226
[3] Bajare D and Bumanis G 2014 *NTCC 2014: Int. Conf. on Non-Traditional Cem. and Con.* 1–9
[4] Provis J L, Lukey G C and van Deventer J S J 2005 *Chem. Mater.* **17** 3075–85
[5] Fletcher R A, MacKenzie K J D, Nicholson C L and Shimada S 2005 *J. Eur. Ceram. Soc.* **25** 1471–77
[6] Lloyd R, Provis J and van Deventer J 2010 *Cem. Concr. Res.* **40** 1386–92
[7] Škvára F, Šmilauer V, Hlaváček P, Kopecký L and Cílová Z 2012 *Ceram.-Silikaty* **56** 374–382
[8] Aly Z, Vance E R and Perera D S J 2012 *Nucl. Mater.* **424** 164–170
[9] Zhang Z, Wang H, Provis J L, Bullen F, Reid A and Zhu Y 2012 *Thermochim. Acta* **539** 23–33
[10] Gougazeh M and Tafila P O B Jordan 2013 *J. Civ. Eng.* **7** 236–257
[11] Criado M, Palomo A and Fernandezjimenez A 2005 *Fuel* **84** 2048–54
[12] Hounsi A D, Lecomte-Nana G L, Djétéli G and Blanchart P 2013 *Constr. Build. Mater.* **42** 105–113
[13] Souza A E, Teixeira S R, Santos G T A and Longo E 2013 *Cerâmica* **59** 147–155

Acknowledgments

The research work was carried out in the framework of the Latvian Council of Science Project: "Smart biofilters" (No.Z12.0491).

Alkali-Activated Aluminium-Silicate Composites as Insulation Materials for Industrial Application

L Dembovska[1,3], D Bajare[1], I Pundiene[2] and G Bumanis[1]

[1]Riga Technical University, Department of Building Materials and Products, Institute of Materials and Structures, Kalku Str. 1, LV-1658, Riga, Latvia
[2]Vilnius Gediminas Technical University, Sauletekio Str. 11, 10223, Vilnius, Lithuania

E-mail: laura.dembovska@rtu.lv

Abstract. The article reports on the study of thermal stability of alkali-activated aluminium-silicate composites (ASC) at temperature 800–1100°C. ASC were prepared by using calcined kaolinite clay, aluminium scrap recycling waste, lead-silicate glass waste and quartz sand. As alkali activator, commercial sodium silicate solution modified with an addition of sodium hydroxide was used. The obtained alkali activation solution had silica modulus M_s=1.67. Components of aluminium scrap recycling waste (aluminium nitride (AlN) and iron sulphite ($FeSO_3$)) react in the alkali media and create gases – ammonia and sulphur dioxide, which provide the porous structure of the material [1].

Changes in the chemical composition of ASC during heating were identified and quantitatively analysed by using DTA/TG, dimension changes during the heating process were determined by using HTOM, pore microstructure was examined by SEM, and mineralogical composition of ASC was determined by XRD. The density of ASC was measured in accordance with EN 1097-7.

ASC with density around 560 kg/m^3 and heat resistance up to 1100°C with shrinkage less than 5% were obtained. The intended use of this material is the application as an insulation material for industrial purposes at elevated temperatures.

1. Introduction

Heat-resistant concrete is a widely used material in high temperature conditions for industrial machinery. It is constantly subjected to operating and deterioration under combined thermo-chemo-hydro-mechanical effects, which often cause chemical and physical alteration of the concrete and result in excessive cracking, spalling and loss of strength [2]. In the industry, there is a need for

[3] To whom any correspondence should be addressed.

insulation materials stable at the elevated temperatures in order to preserve equipment from heating and cooling processes, and to satisfy the safety requirements, such as protecting personnel from hot or cold surfaces.

Studies have shown that scientists all over the world are searching to upgrade heat-resistant concretes and are looking for new materials, which can be applied at the elevated temperatures. The proposed alkali activated materials could also be used as an alternative to concrete; they have emerged as solutions to overcome the problem of massive usage of virgin raw materials and reduce CO_2 emissions [3]. Useful utilisation of the industrial waste is a much more attractive alternative than depositing it on the waste sites. Recycling of industrial waste into new materials reduces possible pollution of the environment and decreases the costs of waste deposits, as well as saves resources for manufacturing new materials and building units.

Alkali activation is a type of geosynthesis that involves naturally occurring aluminium-silicates; it is based on the chemistry of alkali-activated inorganic binders. These amorphous, three-dimensional aluminium-silicate binder materials, the formation process of which was first explained by Glukhovsky, were first introduced to the world of inorganic cementations materials by Davidovits [3]. Alkali activated materials may be synthesized at room temperature or at slightly elevated temperatures by the alkaline activation of aluminium-silicates obtained from industrial wastes, calcined clays, natural minerals or mixtures of two or more of these materials as well as different types of inert fillers. In a strong alkaline solution, aluminium-silicate reactive materials rapidly dissolve to form free SiO_4 and AlO_4 tetrahedral units [4,5]. During the polycondensation reaction, the tetrahedral units are linked in an alternating manner to yield amorphous structure.

A key attribute of alkali activation technology is the robustness and versatility of the manufacturing process; it enables products to be tailor-made from a range of aluminium-silicate raw materials so that they have specific properties for a given application at a competitive cost [6]. It is clear that alkali activated materials can exhibit performance properties such as high compressive strength [7]; high level of resistance to a range of different acids and salt solutions and high freeze–thaw resistance [8]; not subject to deleterious alkali–aggregate reactions [8]; low shrinkage [9]; low thermal conductivity [10]; good adhesion to steel [11]; inherent protection of steel reinforcing due to high residual pH and low chloride diffusion rates [12]; high stability when exposed to high temperatures [8]; effective solution for the stabilization of industrial by-products and the immobilization of heavy metals [13]. Although several researchers demonstrated an excellent durability of alkali activated materials in their experiments, the major unsolved problem in the development and application of alkali activation technology is still related to the durability, because it seems to depend strongly on the adequate curing regimes and composition.

This study attempts to respond to both issues: industrial waste and by-product usage in the production of alkali-activated aluminium-silicate composites (ASC), thus minimizing the usage of non-renewable natural resource and obtaining improved heat resistance of materials, while adjusting dimensional changes by adding quartz sand fillers, and secondly, porous aluminium-silicate composite suitability as thermal insulation material for industrial production purposes.

2. Materials and test methods

2.1. Test methods

Chemical composition of raw materials – calcined kaolinite clay (MK), waste from aluminium scrap recycling factory (ASRW) and lead-silicate glass waste (LSG) from glass recycling factory – was determined according to LVS EN 196-2 with precision +/- 0.5%. The analysis of ASRW elements was carried out with the help of inductively coupled plasma optical emission spectrometry (ICP-OES), atomic absorption spectroscopy (AAS) and potentiometer titration analyses. Particle size distribution in powder materials was determined by laser diffraction. Specific surface area was detected by BET method ('Nova 1200 E-Series, Quantachrome Instruments'). Effective diameter was detected by Zeta potential ('90 Plus' and 'MAS Zeta PALS Brookhaven Instr').

Mostly instrumental methods were used for investigation of aluminium-silicate binders without quartz filler to avoid dominance of quartz. Thermogravimetrical research was carried out by using High temperature DTA / DSC Baehr DTA 703. The mineralogical composition by X-ray diffraction (XRD) (PAN analytical X'Pert PRO) was determined for raw materials and ASC. Thermal conductivity test was performed with heat flow meter instrument LaserComp FOX. The upper and lower plate temperatures were 0°C and 20°C with average temperature of 10ºC. High temperature microscopy (HTOM) EM201, HT163 was used to determinate heat resistance and shrinkage of ASC in temperatures up to 1400ºC. The sample has been tested by heating rate 80°C/min up to the temperature of 500°C and then switched to 15ºC/min while reaching the temperature of 1400°C. Alkali-activated aluminium-silicate binders (ASB) and composites (ASC) with quartz sand filler were observed using scanning electron microscope ('TESCAN Mira / LMU Field-Emission-Gun').

2.2. Raw materials

Raw materials used in this study were calcined kaolinite clay (MK), aluminium scrap recycling waste (ASRW), lead-silicate glass waste (LSG) received from recycling industry and quartz sand (Q). Alkali-activation of raw material composition was done by modified sodium silicate solution with an addition of sodium hydroxide. MK, ASRW and LSG were ground in the laboratory planetary ball mill Retsch PM 400 for 30 min with speed of 300 rpm to obtain homogenous powder raw material.

In the current study, MK was obtained in the laboratory furnace by sintering commercially available kaolinite clay at the temperature of 800°C. Sintering process of calcined kaolinite clay lasted approximately 3 hours with the temperature increase of 15ºC/min. After additional grinding specific surface area of MK was 15.86 m^2/g, whereby effective diameter of MK particles was 743.1 nm.
Chemical composition of MK is given in table 1. According to XRD analysis, MK used is largely amorphous, with a small quantity of quartz (SiO_2) as an impurity phase, and minor kaolinite.

Crystalline layer structure, which is a common characteristic of clay minerals, leads to a fine particle size (1-3 μm) and plate like morphology in MK, it was determined by SEM. ASRW is the final waste in the aluminium scrap recycling process (chemical composition is given in table 1). The content analysis of ASRW shows the following results: aluminium (Al) – 34.40%, silicon (Si) – 4.4%, magnesium (Mg) – 2.4%, calcium (Ca) – 1.3%, sodium (Na) – 1.7%, potassium (K) – 2.3%, sulphur (S) – 0.1%, chlorine (Cl) – 4.2%, iron (Fe) – 3.6%, copper (Cu) – 1.0%, lead (Pb) – 0.1% and zinc (Zn) – 0.6%. These data correspond to the chemical composition of ASRW. In terms of chemical composition, the analysed ASRW contained also aluminium nitride (AlN) – on average 5 wt.%, aluminium chloride ($AlCl_3$) – on average 3 wt.%, potassium and sodium chloride (NaCl + KCl) – total 5 wt.% and iron sulphite ($FeSO_3$) – on average 1 wt.%. According to the XRD analysis, the ASRW contained metallic aluminium (Al), iron sulphite ($FeSO_3$), aluminium nitride (AlN), corundum (Al_2O_3), aluminium iron oxide ($FeAlO_3$), magnesium dialuminium ($MgAl_2O_4$), quartz (SiO_2), aluminium chloride ($AlCl_3$) and aluminium hydroxide ($Al(OH)_3$) [1, 16].

Table 1. Chemical composition of ASRW, MK, LSG and Q [wt. %].

Chemical component	ASRW	MK	LSG	Q
Al_2O_3	63.19	25.55	1.03	1.42
SiO_2	7.92	63.66	68.07	96.8
CaO	2.57	3.47	1.39	-
SO_3	0.36	-	-	-
TiO_2	0.53	-	-	-
MgO	4.43	1.97	-	-
Fe_2O_3	4.54	0.51	0.19	0.34
PbO	-		20.02	-
Na_2O	3.84	3.75	8.02	-
K_2O	3.81	0.45	1.17	-
Other	2.60	0.53	0.11	0.49
LOI, 1000°C	6.21	0.21	-	-

LSG was obtained from fluorescent lamp recycling process, which includes lamp classification, glass separation, cleaning from harmful components and glass grinding. The results of grading analysis showed that additionally ground material contains a wide range of particles in dimensions from 8 µm to 30 µm with the average grain size of 26.3 µm. It is comparable with the data obtained from the investigation made by SEM. Ground LSG particles have smooth surface. The specific surface area of LSG ground for 30 min was 1.126 m^2/g.

Quartz sand (Q) was used as an available filler to reduce deformations and dimensional changes of ASC at elevated temperatures. Chemical analysis of Q is given in table 1. The mass ratio between solid raw materials of ASC and Q was from 1:1 to 1:1.67.

Alkali activator was prepared by modification of commercial sodium silicate solution (Ms = 3.22). To obtain alkali activation solution with the required chemical composition (Ms = 1.67), the modification with an addition of sodium hydroxide was performed. Commercially available sodium hydroxide flakes with 97% purity were used.

2.3. Sample preparation and curing conditions

Two series of samples were prepared. Compositions without quartz filler (ASC binders) are presented in Table 2, but with quartz filler (called composites) – in Table 3. Fresh mixtures were cast in 40x40x160 mm prismatic moulds and curing conditions were kept constant at temperature 80°C for 24h. The optimal curing temperature of 80°C was selected for 24 hours to obtain specimens with as few cracks as possible. Components of ASRW – aluminium nitride (AlN) and iron sulphite ($FeSO_3$) – react in the alkali environment and create gases – ammonia and sulphur dioxide, which provide porous structure of material before setting [16]. After 24 hours, samples were demoulded and kept in the ambient environment. To determine dimensional changes at the elevated temperatures, samples were sawn into smaller pieces with dimensions 10x40x40 mm. Density was tested before and after heat treatment at temperatures 800, 900, 1000 and 1100°C.

3. Results and discussion

3.1. Alkali-activated aluminium-silicate binders (ASB)

Due to the chemical reactions and gas evaporation from the mixture before setting, highly porous material was obtained [1]. Selected alkali-activated aluminium-silicate binders (ASB) provided density in the range from 300 to 340 kg/m^3 (table 2). After heat treatment, density of binders did not change significantly except for A10-A1-G1 with bigger amount of glass (LSG) in the composition.

Table 2. ASB compositions and their physical properties before and after heat treatment.

Batch	Composition, mass ratio			Treatment temperature [°C]	Mass changes after heating, [%]	Density, [kg/m^3]	
	MK	ASRW	LSG			before heating	after heating
A10-A1-G0	1	1	0	800	-14.8	340	330
				900	-14.4	340	330
				1000	-13.6	340	320
				1100	-12.8	340	340
A10-A1-G0.5	1	1	0.5	800	-14.5	324	320
				900	-14.4	324	320
				1000	-15.3	323	304
				1100	-14.9	323	311
A10-A1-G1	1	1	1	800	-14.0	300	300
				900	-14.2	310	310
				1000	-13.7	330	340
				1100	-14.0	328	300

Density of A10-A1-G1 decreased from 328 to 300 kg/m^3 after heat treatment at temperature 1100°C (table 2). Bloating of the specimen and volume increase was observed.

DTA and TG tests for the binder specimens with bigger amount of glass (LSG) additive (A10-A1-G1) and samples without glass (LSG) additive (A10-A1-G0) in the composition were performed to explain the dimensional changes during heating process (figure 1 and figure 2). Endoeffect at the temperature of 110°C is explained by lost mechanically associated water and dehydration of N-A-S-H or $Si(OH)_4$ gel, because in the temperature interval 90-130°C up to 80% of water is intensively evaporating from the sodium silicate solution used as activator for production of alkali activated materials. The dehydration of N-A-S-H or $Si(OH)_4$ gel continues till 300-400°C.

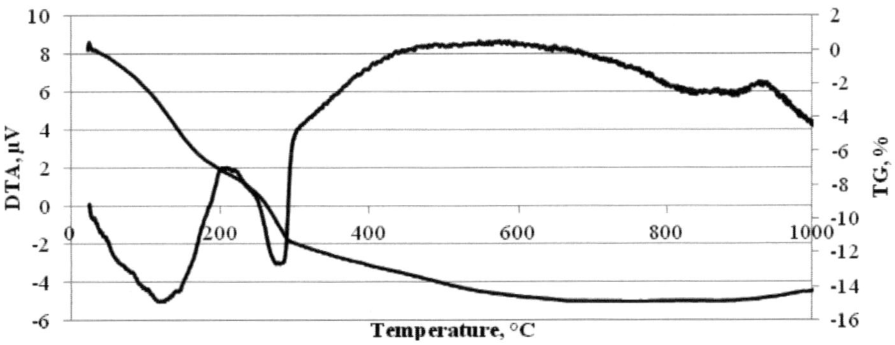

Figure 1. Curves of differential thermal analysis and thermo gravimetric analysis for alkali activated binder (ASB) A10-A1-G0.

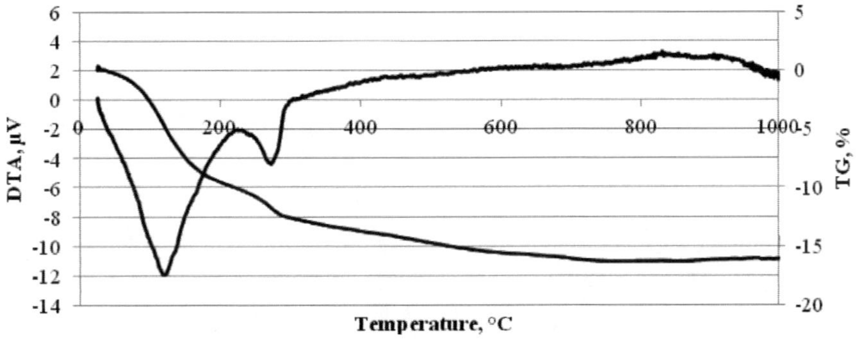

Figure 2. Curves of differential thermal analysis and thermo gravimetric analysis for alkali activated binder (ASB) A10-A1-G1.

There are indications that commercial sodium silicate solution used for preparation of alkali activator releases chemically bonded water in the temperature interval till 600°C - 700°C and the formation of crystals of sodium disilicate is possible. According to XRD, ASRW contains $Al(OH)_3$, as well as aluminium nitride (AlN), which reacts with water vapour and creates additional $Al(OH)_3$ and NH_3 in the temperature interval from 120 to 150°C. Endoeffect observed in the temperature range of 220 to 280°C indicates decomposition of $Al(OH)_3$ (figures 1 and 2). Both exoeffects found in the temperature interval 800-900°C indicate formation of new minerals. It is possible that part of $Al(OH)_3$ and metallic Al reacts with NaOH and creates new mineral $NaAlOH_4$ at the temperature around 800°C (figure 1) and albite ($Na_2O \cdot Al_2O_3 \cdot 6SiO_2$) at the temperature around 900°C. The small endoeffect detected at the temperature 573°C is associated with structural transition of the α-quartz SiO_2 to β-quartz.

According to the data obtained by high temperature microscopy, specimen A10-A1-G0 shows practically no changes before heat treatment up to 800°C (figure 3). Dimensional shrinkage of binders is observed in the temperature interval from 800°C to 1000°C, which compiles around 20% from the total area. In the temperature interval from 1000°C to 1300°C material expands and melts afterwards.

Sample A10-A1-G0.5 has medium LSG amount in the mixture (table 2). Up to temperature 800ºC the sample shows practically no changes. In the temperature range from 800°C to 900°C shrinkage of the material is observed, it constitutes around 10% from the total relative area (figure 4). In the temperature range 900°C - 1080°C material shows no visual changes and shrinkage is constant (10% of the total area). Reaching temperature above 1080°C, the material expands and melts afterwards.

Sample A10-A1-G1 reaches shrinkage around 5% from the total relative area after heat treatment in the interval from 100°C to 750°C. At the elevated temperature 750°C till 800°C the material has lost 10% of the total area, but then from 800°C till 860°C the material expands for about 20%. From 860°C till 950°C the material loses around 15% of the total relative area and then from around 950°C till 1050°C it expands and melts afterwards. It is possible to use this particular material without any volume changes at the working temperature up to 850°C.

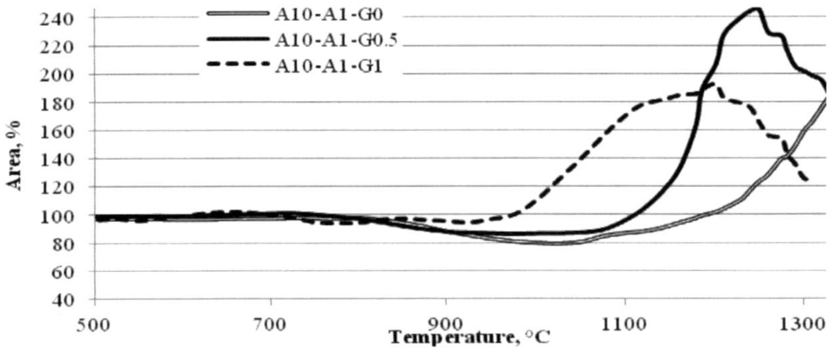

Figure 3. High temperature microscopy curve: relative area change of ASB during heating.

It was concluded that glass (LSG) additive has significant influence on the heat resistance of alkali activated materials. LSG has 8% of Na_2O, sodium oxide is a known flux, if sodium content is increased in the composition, the formation of a melt at a lower temperature is favoured. The binder (A10-A1-G0) has 5% shrinkage of relative area at temperature 840°C, 7.5% –at 868°C and 10% shrinkage –at 887°C, 20% –985°C (figure 3). By increasing the amount of glass (LSG) in the composition, shrinkage of material increases –for specimens A10-A1-G1 shrinkage of 5% is observed already at the temperature of 770ºC, but in case of increasing the temperature up to 1000°C, shrinkage of material does not exceed 5%.

Table 3. Relative area changes of samples at different temperatures.

Composition	Deformation t°, °C	Deformation range, °C	Sphere T, °C	Shrinkage			
				5%	7.5%	10%	20%
A10-A1-G0	1196	not defined	1357	840	868	887	985
A10-A1-G0.5	1169	1169-1349	1272	826	850	878	-
A10-A1-G1	974	974-1286	1174	770	-	-	-
A10-A1-G0-Q3	1186	1186-1199	1199	660	680	1051	1150
A10-A1-G0-Q5	1165	1165-1200	1200	730	1120	1160	1150
A10-A1-G1-Q3	1066	1066-1134	1102	680	720	745	1051
A10-A1-G1-Q5	1165	1165-1234	not defined	1119	1139	1149	1174

By increasing the temperature over 1000ºC, expansion of specimens A10-A1-G1 is observed. Comparing the resulting material to the composition without additive in the material it was concluded that glass (LSG) additive causes shrinkage of the material at lower temperature and therefore reduces material working temperature to 770°C degrees. Mass changes of samples after heat treatment (table 2) correspond to the data obtained by DTA/TG (figures 1 and 2) and HTOM (figures 3 and 5).

3.2. Alkali-activated aluminium-silicate composites (ASC)
Binder compositions (A10-A1-G0, A10-A1-G0.5, A10-A1-G1) with satisfactory dimension changes in the temperature interval from 800 till 1000°C were supplemented with quartz sand filler (Q) and alkali-activated aluminium-silicate composites (ASC) were prepared for the further investigations. Microstructure of hardened ASC looks amorphous and homogenous (figure 4). Additional research was done regarding the investigation of the correlation between curing conditions and microstructure of ASC. However, some micro cracks have been observed in the material structure (figure 4, b). This could be caused by the temperature regime during hardening process of specimens or could have appeared during sample preparation procedure. Micro cracks do not affect strongly the functional properties of ASC due to its high total porosity (82-83%), and low compressive strength (up to 2 MPa).

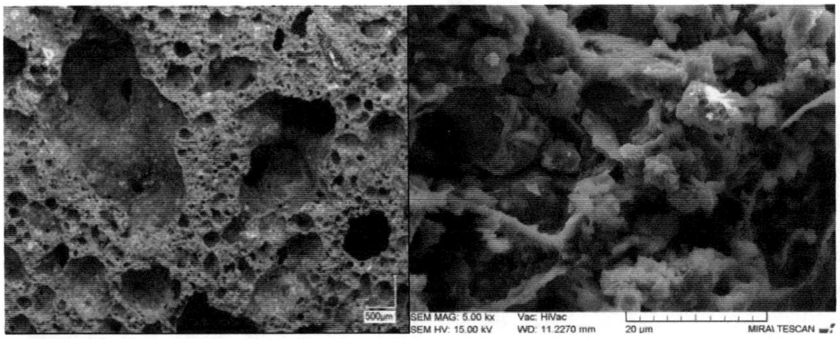

Figure 4. Photographs of ASC A10-A1-G0-Q5 composition's pore structure: a) optical microscopy, b) SEM (magnification 5000x).

The density of alkali-activated aluminium-silicate composites (ASC) with quartz filler is between 510 and 590 kg/m^3 (table 4). Mass loss after heating of all samples containing quartz in the composition was 8-28% depending on the heating temperature.

As the previous studies indicate, quartz sand filler (Q) limits the expansion of the alkali activated material at temperatures above 1000°C, the subsequent high-temperature microscopy studies were also performed for samples A10-A1-G0-Q3, A10-G1-Q3, A10-A1-G1-Q5 and A10-G1-Q5 (figure 5).

Alkali activated binders with different amounts of glass (LSG) additive had been chosen for preparing aluminium-silicate composites (ASC). Quartz sand filler (Q) was added to the mixture in the amount of 3 and 5 weight parts to limit the expansion of materials at the temperature up to 1000°C (table 4). Relative area changes of ASC at different temperatures are given in figure 5. Results show that ASC with increased amount of glass (LSG) and quartz sand filler (Q) in the composition (A10-A1-G1-Q5) have 5% shrinkage at the higher temperature – 1119°C, 7.5% – at 1139°C, 10% – at 1149°C and 20% shrinkage – at 1174°C (figure 5).

Table 4. ASC compositions and their physical properties before and after heat treatment.

Batch	Reactive components (mass ratio)			Filler	Treatment temperature [°C]	Mass changes after heating [%]	Density, [kg/m^3]	
	MK	ASRW	LSG	Q			before heating	after heating
A10-A1-G0-Q3	1	1	0	3	800	-8.66	550	530
					900	-8.63	560	560
					1000	-8.20	540	540
					1100	-8.08	520	530
A10-A1-G0-Q5	1	1	0	5	800	-12.08	550	520
					900	-27.49	580	460
					1000	-27.45	580	440
					1100	-26.58	590	470
A10-A1-G1-Q3	1	1	1	3	800	-18.71	560	540
					900	-26.19	510	530
					1000	-26.59	570	540
					1100	-26.99	540	420
A10-A1-G1-Q5	1	1	1	5	800	-28.37	540	400
					900	-27.87	590	450
					1000	-27.08	590	470
					1100	-28.04	560	480

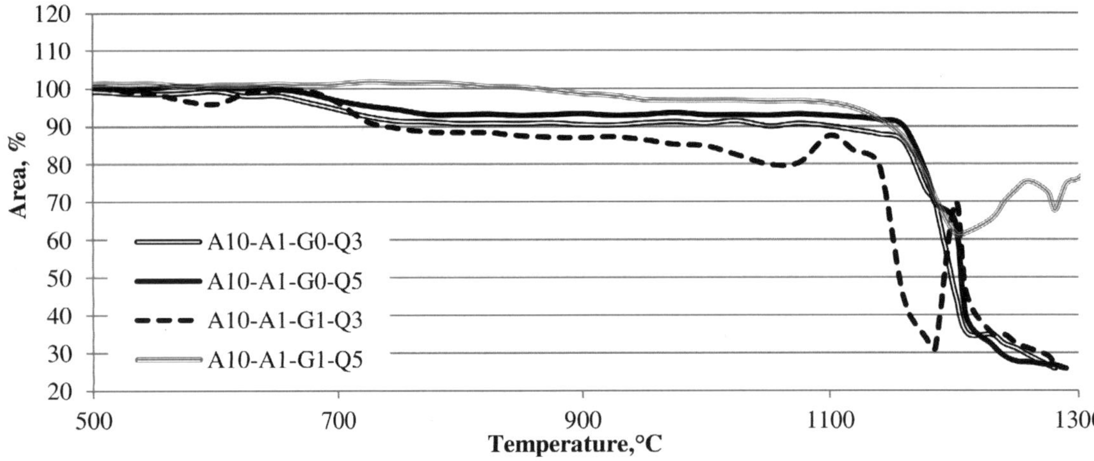

Figure 5. High temperature microscopy curve: relative area change of ASC containing quartz sand filler (Q) during heating.

3. Conclusion

The new type of porous alkali activated aluminium-silicate composite (ASC) was obtained by using alkali activated aluminium-silicate binder (ASB) made of industrial by-products such as aluminium scrap recycling waste and lead-silicate glass waste (LSG) from recycled florescent lamps and natural filler.

It was concluded that glass waste (LSG) additive plays ambiguous role in the composition of alkali activated binders (ASB). On the one hand, it promotes significant shrinkage (5%) and volume changeability already at low temperatures (770°C) and afterwards samples are expanding, when temperature is raised just up to 975°C, but on the other hand, the volume changeability does not increase 5% till expansion temperature. That finding was combined with the results of previous investigation and the quartz sand filler (Q) was selected as an additive for the reduction of expansion. Therefore, alkali activated binders (ASB) with glass (LSG) in the composition were selected as most promising mixture for obtaining heat resistant ASC.

The most important conclusion of this study is that quartz sand filler (Q) slows down the expansion of alkali activated binder (ASB) at temperature higher than 1000°C and significantly prolongs working application of the material. With an addition of quartz sand filler (Q) in the composition (5 weight parts), ASC with density between 540-590 kg/m^3 and shrinkage less than 5% at elevated temperatures up to 1119°C was obtained. Density can be controlled by moulding fresh mixture and alkali content in activator solution. However, due to the low density and specific pore structure development, mechanical properties of the material were limited.

Finally, new material with heat resistance up to 1100°C was obtained. The essential indicator is the ability of quartz sand filler (Q) to slow down the influence of lead-silicate glass waste (LSG) and to prolong working applicability of this material. Therefore, it was concluded that ASC could be used as a new insulation material in the production industry.

4. References

[1] Bajare D, Bumanis G and Korjakins A 2014 New Porous Material Made from Industrial and Municipal Waste for Building Application *Materials Science* **20** 333

[2] Prakash S B, Chang V and Li M 2014 Effect of elevated temperature on strain-hardening engineered cementitious composites *Construction Building Materials* **69** 370-380

[3] Prud'homme E, Michaud P, Joussein E, Clacens J M, Arii-Clacens S, Sobrados I, Peyratout C, Smith A, Sanz J and Rossignol S 2011 Structural characterization of geomaterial foams - Thermal behaviour *J. of Non-Crystalline Solids* **357** 3637

[4] Xu H and Van Deventer J S J 2000 The Geopolymerisation of Alumino-silicate minerals *Int. J. of Mineral Processing* **59** 247

[5] Wu J P, Boccaccini A R, Lee P D and Rawlings R D 2007 Thermal and mechanical properties of a foamed glass-ceramic material produced from silicate wastes *European J. of Glass Science and Technology* **48** 133

[6] Duxson P, Provis J L, Lukey G C and Van Deventer J S J 2007 The role of inorganic polymer technology in the development of 'green concrete' *Cement&Concrete Research* **37** 1590

[7] Lee W K W and Van Deventer J S J 2002 The effect of ionic contaminants on the early-age properties of alkali-activated fly ash-based cements *Cement&Concrete Research* **32** 577

[8] Pacheco-Torgal F, Abdollahnejad Z, Camões A F, Jamshidi M and Ding Y 2012 Durability of alkali-activated binders: A clear advantage over Portland cement or an unproven issue *Construction Building Materials* **30** 400

[9] Elimbi A, Tchakoute H K and Njopwouo D 2011 Effects of calcination temperature of kaolinite clays on the properties of geopolymer cements *Construction Building Materials* **25** 2805

[10] Arellano Aguilar R, Burciaga Díaz O and Escalante García J I 2010 Lightweight concretes of activated metakaolin-fly ash binders with blast furnace slag aggregates *Construction Building Materials* **24** 1166

[11] Temuujin J, Minjigmaa A, Rickard W, Lee M, Williams I and Van Riessen A 2009 Preparation of metakaolin based geopolymer coatings on metal substrates as thermal barriers *Applied Clay Science* **46** 265

[12] Lloyd R R, Provis J L and Van Deventer J S J 2010 Pore solution composition and alkali diffusion in inorganic polymer cement *Cement&Concrete Research* **40** 1386

[13] Zheng L, Wang C, Wang W, Shi Y and Gao X 2011 Immobilization of MSWI fly ash through geopolymerization: effects of water-wash *Waste Management* **31** 311

[14] Abdullah M M A B, Hussin K, Bnhussain M, Ismail K N, Yahya Z and Abdul Razak R 2012 Fly Ash-based Geopolymer Lightweight Concrete Using Foaming Agent *Int. J. of Molecular Sciences* **13** 7186

[15] Zuda L and Cerny R 2009 Measurement of linear thermal expansion coefficient of alkali-activated aluminosilicate composites up to 1000°C *Cement & Concrete Composites* **31** 263

[16] Bajare D, Korjakins A, Kazjanovs J and Rozenstrauha I 2012 Pore structure of lightweight clay aggregate incorporate with non-metallic products coming from aluminium scrap recycling industry *J. of the European Ceramic Society* **32** 141

Acknowledgements
The research work was carried out in the framework of the Latvian Council of Science Project: "Development of sustainable effective lightweight construction materials based on industrial waste and local resources" (No.Z12.0412).

Durability of Alkali Activated Blast Furnace Slag

K Ellis[1,4], NAlharbi[2], P S Matheu[1], B Varela[2] and R Hailstone[3]
[1]Department of Industrial and Systems Engineering, Rochester Institute of Technology, 81 Lomb Memorial Dr, Rochester NY, USA
[2]Department of Mechanical Engineering Rochester Institute of Technology, 76 Lomb Memorial Dr, Rochester NY, USA
[3]Center for Imaging Science, Rochester Institute of Technology, 54 Lomb Memorial Dr, Rochester NY, USA

E-mail: ke4861@rit.edu

Abstract. The alkali activation of blast furnace slag has the potential to reduce the environmental impact of cementitious materials and to be applied in geographic zones where weather is a factor that negatively affects performance of materials based on Ordinary Portland Cement. The scientific literature provides many examples of alkali activated slag with high compressive strengths; however research into the durability and resistance to aggressive environments is still necessary for applications in harsh weather conditions. In this study two design mixes of blast furnace slag with mine tailings were activated with a potassium based solution. The design mixes were characterized by scanning electron microscopy, BET analysis and compressive strength testing. Freeze-thaw testing up to 100 freeze-thaw cycles was performed in 10% road salt solution. Our findings included compressive strength of up to 100 MPa after 28 days of curing and 120 MPa after freeze-thaw testing. The relationship between pore size, compressive strength, and compressive strength after freeze-thaw was explored.

1. Introduction

According the US Geological Survey, 4.18 billion metric tons of cement was produced in 2014 [1]. Due to the sheer size of the concrete market worldwide, the energy and environmental impacts are also large. It has been approximated that 5% of total anthropogenic greenhouse gas emissions stem from cement production [2], [3]. Alkali activated materials are a cementitious material that can be produced from industrial by-products. The chemical composition of the final material is similar to that of Ordinary Portland Cement; however the formation reactions and final chemistry of the two materials differ significantly [4]. This type of material has been identified by the Intergovernmental Panel on Climate Change (IPCC) as a possible alternative to Portland cement that could reduce the environmental impacts of the cement industry [5]. In order for these materials to become a practical

[4] To whom any correspondence should be addressed.

alternative in the cement market further research is necessary to prove durability and resistance of these materials to weathering.

Previous research in the area of alkali-activated materials has shown varied results in the compressive strength of the alkali activated slag (AAS) materials after exposure to freeze-thaw cycles. Good freeze-thaw resistance is considered to be not losing strength or durability after exposure [6]. One study found that the compressive strength increases after freeze-thaw cycling in AAS material [7]. In Portland cement based materials it has been shown that the properties that effect freeze-thaw resistivity are mostly physicomechanical properties like pore structure, pore saturation and entrained air voids. It is thought that this relationship may be similar for AAMs [6]. This research focuses on the characterization of two design mixes by pore structure analysis, and compressive strength measurements under exposure to freeze-thaw conditions in water and salt water. This research will help to shed light on the relationship between the pores present in the materials and the compressive strength results obtained.

2. Experimental methodology

2.1. Sample materials

The samples utilized potassium hydroxide, ground granulated blast furnace slag (GGBFS), mine tailings, and two types of amorphous silica. The potassium hydroxide was sourced from Brainerd Chemical Co. in Tulsa, Oklahoma. The GGBFS came from St. Mary's Cement Company, based in Burlington, ON, Canada. The composition of the GGBFS is shown in table 1.

The mine tailings were provided by NYCO Minerals, based in Willsboro NY, USA. The mine tailings contain the minerals Garnet, Wollastonite and Pyroxene. The NYCO mine tailings were sifted with a 500 micron sieve before use in samples. The amorphous silica was produced by Evonik Degussa Corporation based in Essen, Germany. The water used in the experiments was tap water from the Rochester, NY municipal water supply.

Table 1. Composition of the slag by weight

Oxide	Percentage of slag by weight
CaO	38.3
SiO_2	37
Al_2O_3	8
MgO	10.5
Na_2O	0.025
K_2O	0.043
SO_3	1.7

2.2. Sample preparation

The activating solution was prepared by dissolving amorphous hydrophilic silica (Aerosil 150) in a 45% KOH solution and water to yield a solution with molar ratios of $SiO_2/K_2O = 1$ and $H_2O/K_2O = 50$. The solution was left to mature for 24 hours before used in sample production. In all samples the mass ratio of slag/activating solution was 2. All of the dry ingredients were carefully weighed and mixed together for 1 minute in the Hobart mixer. Then the activating solution was added to the mixture and all of the reactants were mixed in a Hobart mixer until a homogeneous paste was obtained. The paste was cast into ASTM standard 2 in. cement cubes, vibrated for five minutes and left to mature for one hour. After this, the cubes were wrapped in plastic and cured in an oven for 24 hours at 45°C. At the end of the curing time, the cubes were de-molded and placed in saturated lime water until testing. Weight proportions of design mixes studied are shown in table 2.

Table 2. Design mix composition.

Material	Design mix G1	Design mix G5
Slag	2	2
Activating solution	1	1
Mine tailings	2	2
Amorphous Hydrophobic silica (Aerosil R972)	0	0.05

2.3. Characterization of samples

The design mixes were tested at 1.7 and 28 days after curing for compressive strength in a Tinius Olsen Electomatic universal testing machine. Each testing day three cubes of each design mix were tested per ASTM standards.

Samples were stored in saturated lime water for 28 days before the beginning of the freeze-thaw cycling. Samples were placed in a plastic tray. Water was then poured into the tray filling the tray to approximately two cm above the bottom. In the 10% salt water case, the water was mixed with common road salt in a large container and then poured into the tray in solution. The cubes were cycled from -10 °C to 4 °C in 50 cycle intervals using an environmental chamber. The cubes were left for one hour at each reference temperature.

It is known that the pore structure of cementitious materials influences their physical properties and stability[6]. To study this effect, both design mixes were tested after 28 days using two different techniques. In the first technique the porosity of the samples was characterized by the nitrogen adsorption method in a Quantachrome Nova 1000e machine. In the second technique, transmission electron microscopy was used.

In order to evaluate the capacity of the samples to retain strength after being exposed to high temperatures, a thermal shock test was conducted. The cubes were dried in a convection oven at 45 °C for 24 hours. After this, one cube of each design mix was exposed to temperatures of 200 °C, 400 °C, 600 °C, and 800 °C for one hour. After temperature exposure two different cooling techniques were evaluated. One cube was cooled in still air, and another was quenched in water. Once the samples reached room temperature, the compressive strength was measured.

The samples were analysed using a Scanning Electron Microscope (SEM). In this work we have used SPI-MODULE Sputter Coater for both design mixes with gold–palladium under a vacuum of 10 Pa. Once the specimens were ready, their images were taken in a JEOL JSM-6400V SEM. Images were capture at a 7kX magnification using a 20kV bean voltage. Moreover, elemental analysis and mapping were also obtained using energy dispersive X-Ray spectroscopy (EDS) on the specimens. Elemental analysis was conducted at four different points on design mixes G1 and G5 cracks and four other points on the nearby surface. The elemental analysis and mapping were collected from Noran System Six software, which is connected to the EDS X-Ray detector.

Transmission electron microscopy (TEM) was done using a JEOL 2010 electron microscope operating at 200 kV. Specimens for electron diffraction were prepared by grinding the paste to a fine powder with a mortar and pestle, dispersing in isopropyl alcohol, sonicating, and then applying a 10 microliter drop to a standard carbon-film covered copper TEM grid and allowing to air dry. At the microscope, thin sections were used to capture powder diffraction patterns, from which *d* spacings were obtained. The latter were then compared to known spacings of mineral phases that might form under our preparation conditions. Likewise, images of thin sections were obtained for analysis of porosity. ImageJ was used to measure the diameter of the pores.

3. Results and Discussion

3.1. Compressive strength

Compressive strength results over time for design mixes G1 and G5 are shown in figure 1. For each design mix respectively, the strength increases over time. G1 has a higher early compressive strength, but levels out faster over time. Design mix G5 starts out with a lower compressive strength than G1. However, over time G5 gains strength reaching an average of 106 MPa at 28 days.

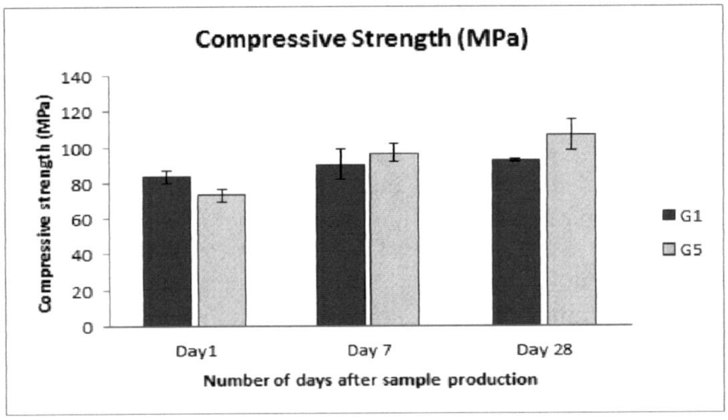

Figure 1. Compressive strength development for design mixes G1 and G5 at days 1, 7 and 28. Error bars are given as one standard deviation.

Figures 2 (in water) and 3 (in salt water) present the compressive strength of both design mixes after thermal cycling in an environmental chamber. In the first experiment (figure 3) design mix G1 lost approximately 15% of the original compressive strength after 100 cycles but design mix G5 showed a 10% increase in strength under the same conditions (final average of 117.64 MPa) after 100 cycles. With the objective of replicating road conditions in New York, the same set of experiments was conducted in a 10% salt water solution. As seen in figure 3, in both cases an unexpected behaviour was observed. Design mix G1 increased 8% of strength (final average of 92.7 MPa) while design Mix G5 increased by 6% (final average of 112.7 MPa) after 100 cycles. This increase in compressive strength under freeze-thaw testing agrees with previous results obtained by Kukko and Mannonen [7].

Figure 4 presents a summary of the residual compressive strength for the samples after thermal shock. As a general trend, the loss in compressive strength is higher for samples quenched in water than those cooled in still air. Samples cooled in still air retained approximately 50% of the original compressive strength up to 600°C but they retained only 30% when quenched in water at the same temperature. The compressive strength retention was insignificant for all cases at 800°C. Thermogravimetric Analysis (TGA) was performed for both design mixes. In both cases the derivative of the weigh curve shows 3 relevant peaks centered at around 80°C, 370°C and 660°C. The first peak is attributed to the evaporation of adsorbed water and dehydroxilation of C-S-H phase, the second is attributed to the decomposition of $Ca(OH)_2$ and the last one is associated with the decomposition of $CaCO_3$.

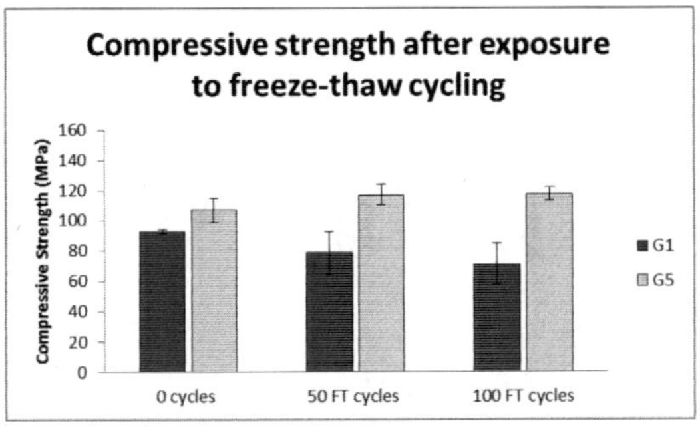

Figure 2. Compressive strength of design mixes G1 and G5 at 50 and 100 thermal cycles in water.

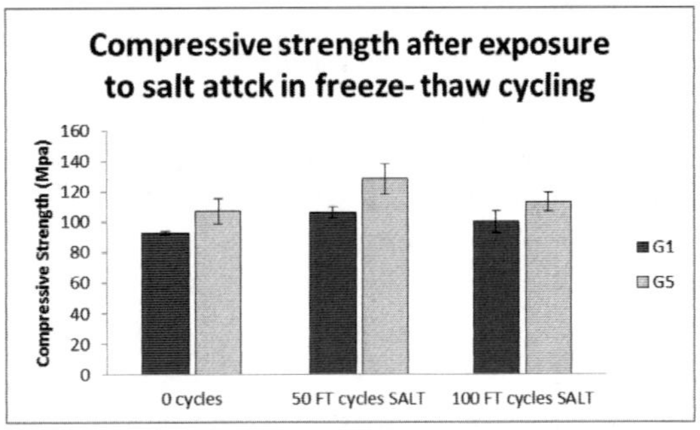

Figure 3. Compressive strength of design mixes G1 and G5 at 50 and 100 thermal cycles in 10% salt water.

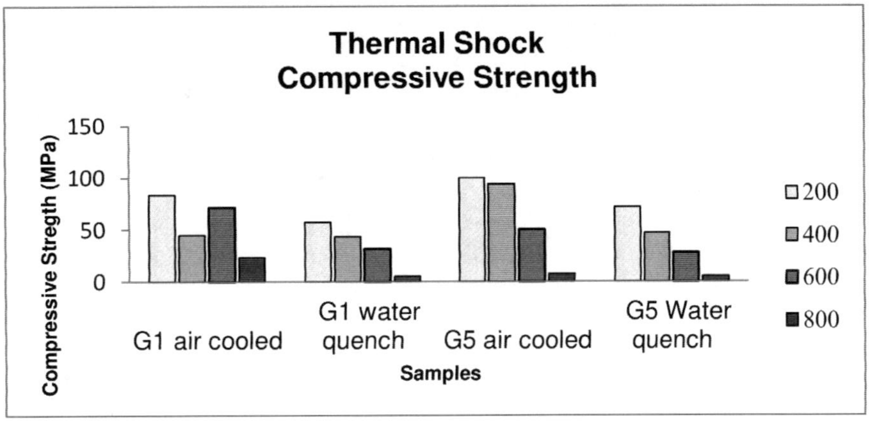

Figure 4. Residual compressive strength for design mixes G1 and G5 after thermal shock at 200°C, 400°C, 600°C and 800°C.

IOP Conf. Series: Materials Science and Engineering **96** (2015) 012004 doi:10.1088/1757-899X/96/1/012004

3.2. Pore size analysis

Table 3 presents the summary of the pore analysis by the Barret-Joyender-Halenda Method for both design mixes. It can be seen that for both mixes the total pore volume is approximately the same as well as the radius of the mesopores (21Å). However there is a significant difference in the total surface area of both design mixes with G5 being the largest. This suggests that design mix G5 may have a larger number of small pores compared to design mix G1.

Table 3. BJH desorption summary.

	Design mix G1	Design mix G5
Surface Area (m^2/g)	31.925	44.837
Pore Volume cc/g	0.052	0.065
Pore Radius Dv(r) (Å)	21.428	21.671

An alternative method for measuring the pore size distribution via Transmission Electron Microscopy (TEM) was evaluated. Images of thin sections were used along with ImageJ to directly measure the pore diameter. The results are shown in figures 8 and 9. Both design mixes show a peaked distribution in the 4 to 6 nm diameter range, with that for G5 being slightly broader.

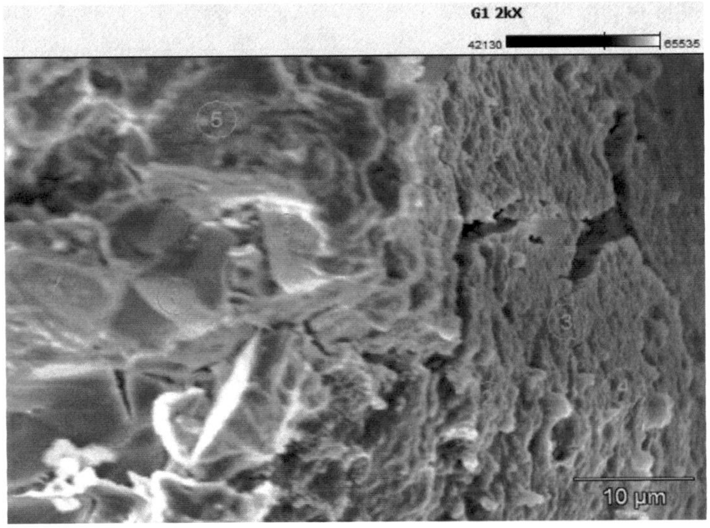

Figure 5. SEM image of G1 showing the interior surface of a crack (right side) and the adjacent surface (left side). The numbers indicate the points where EDS analysis was done (1 through 4 in the crack and 5 through 8 on the surface).

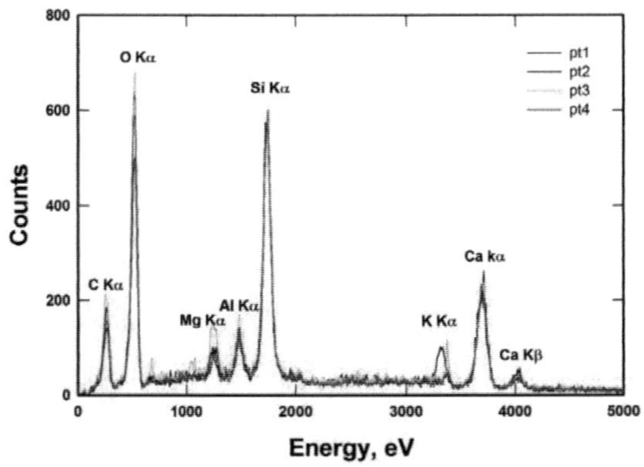

Figure 6. A comparison the EDS spectra for four randomly points in a crack wall of G1. Spectra have been normalized so the Si peak intensity is the same in all four spectra.

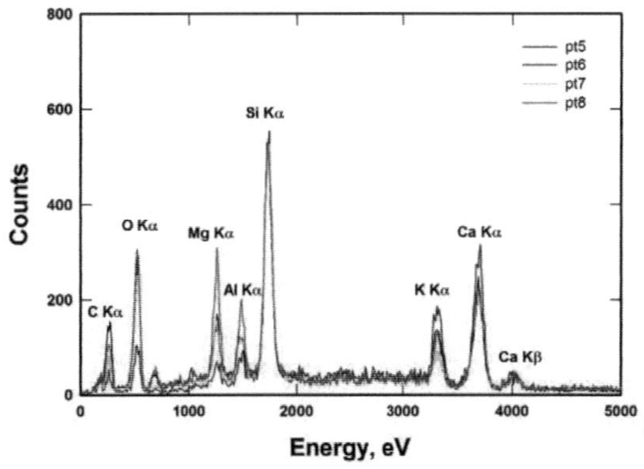

Figure 7. A comparison the EDS spectra for four randomly points on the surface of G1. Spectra have been normalized so the Si peak intensity is the same in all four spectra.

Elemental analysis using EDS was done on both G1 and G5. Figure 5 shows an SEM image of a crack and the adjacent surface area for G1. Spectra were collected from four random points in both the crack and the surface areas. These are compared in figures 6 and 7. The most noticeable difference was that the O peak was much higher for the spectra from the crack (figure 6) than from the nearby surface (figure 7). An identical study on G5 showed smaller differences in the O peak height between the two areas.

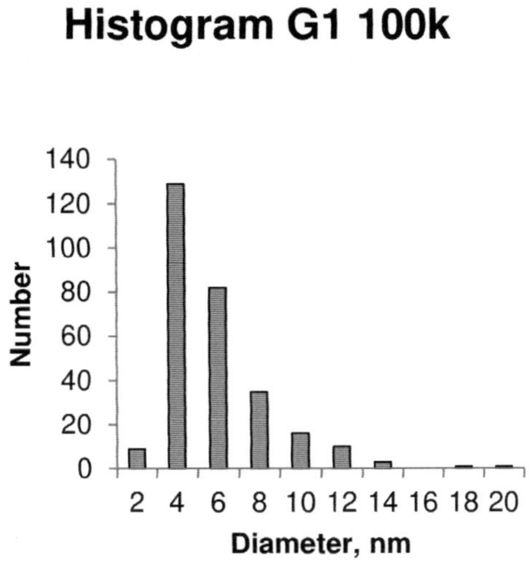

Figure 8. Histograms showing the pore diameter distribution as measured by TEM for G1.

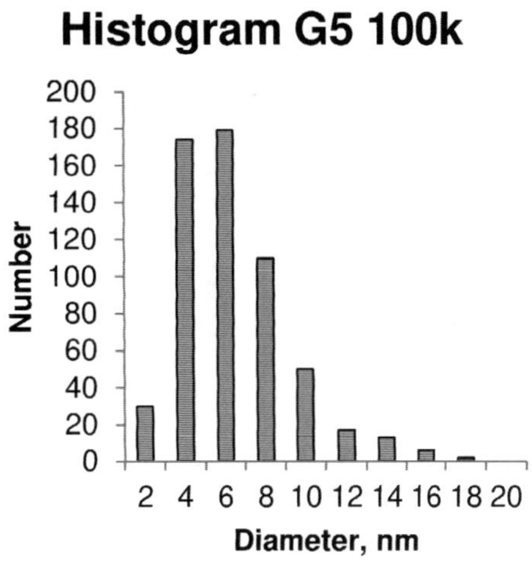

Figure 9. Histograms showing the pore diameter distribution as measured by TEM for G5.

4. Conclusions

A high compressive strength cementitious material was obtained by the alkali activation of blast furnace slag with mine tailings. The average 28 day compressive strength was 92.6 MPa. The addition of amorphous silica in the design mix increased the compressive strength to 106.8 MPa on average, although one sample reached 114 MPa. Both design mixes used 80% by mass of industrial by-products. When exposed to thermal cycling they both increased their compressive strength, reaching an average of 128 MPa after 50 freeze-thaw cycles in a 10% slat water solution. Under thermal shock when the samples are cooled in still air they retained approximately 50% of the original compressive strength up to 600^0C. However, the samples retained only 30% when quenched in water at the same temperature. The compressive strength retention was insignificant for all cases at 800^0C. Results from pore size analysis via Nitrogen adsorption and Transmission Electron Microscopy show that for both mixes the majority of the mesopores are in the range 4 – 6 nm in diameter but there are differences in the mesopore distributions. Moreover, the effect of pore size and distribution in the final compressive strength and durability is inconclusive at this moment.

Overall the preliminary pore analysis via BET support the literature [6] in suggesting that the increase in compressive strength under freeze thaw by the design mix with additional 5% amorphous silica could be caused by the presence of smaller pores. Further research is underway to support this claim.

5. References

[1] Mineral Commodity Summaries: CEMENT, United States Geological Survey, USGS commodity report
[2] Hrvoje Mikulcic M V 2013 CO_2 Emission Reduction in the Cement Industry *Chem. Eng. Trans.* **35**
[3] Bosoaga A, Masek O and Oakey J E 2009 CO_2 Capture Technologies for Cement Industry *Energy Proc.* **1** 133 – 140
[4] Shi C, Roy D and Krivenko P 2006 Alkali-Activated Cements and Concretes (USA: CRC Press)
[5] Climate Change 2007 - Mitigation of Climate Change Cambridge University Press [Available]: http://www.cambridge.org/us/academic/subjects/earth-and-environmental-science/climatology-and-climate-change/climate-change-2007-mitigation-climate-change-working-group-iii-contribution-fourth-assessment-report-ipcc
[6] Provis J L and van Deventer J S J 2014 Alkali Activated Materials vol. 13. (Dordrecht: Springer Netherlands)
[7] Kukko H and Mannonen R 1982 Chemical and mechanical properties of alkali-activated blast furnace slag (F-Concrete) *Nord. Concr. Res.* **1**

IOP Conf. Series: Materials Science and Engineering **96** (2015) 012005 doi:10.1088/1757-899X/96/1/012005

Development of New Cementitious Caterials by Alkaline Activating Industrial by-Products

A Fernández-Jimenez[1,2], I García-Lodeiro[1], and A Palomo[1]
[1]Cement and Recycled Materials Department, Instituto Eduardo Torroja (CSIC), Madrid, 28033, Spain

E-mail: anafj@ietcc.csic.es

Abstract. The alkaline activation of aluminosiliceous industrial by-products such as blast furnace slag and fly ash is widely known to yield binders whose properties make them comparable to or even stronger and more durable than ordinary Portland cement. The present paper discusses activation fundamentals (such as the type and concentration of alkaline activator and curing conditions) as well as the structure of the cementitious gels formed (C-A-S-H, N-A-S-H). The durability and strength of these systems make these materials apt for use in many industrial applications, such as precast concrete elements (masonery blocks, railroad sleepers), protective coatings for materials with low fire ratings and lightweight elements.

1. Introduction

Portland cement-based concrete is today's construction material *par excellence*. It owes this pre-eminence to its mechanical strength, good value for money ratio and performance in general. Nonetheless, Portland cement manufacture raises certain energy and environmental issues, for it calls for temperatures of up to 1400-1500 °C and the extraction of raw materials (with the concomitant destruction of natural quarries), and entails the emission of GHGs such as CO_2 and NO_x. Moreover, concrete is subject to certain durability problems, including the aggregate-alkali reaction and attendant expansion, chloride-induced corrosion in reinforcing steel, retarded ettringite formation, thaumasite formation and so on [1].

One possible way to solve these problems while contributing to better environmental balance (via the re-use of industrial by-products or waste, reducing the cost and potential for pollution inherent in their storage) is to add mineral or industrial by-products to Portland cement clinker. The result is different types of common cements listed in European standard EN 197-1:2000, which specifies the

[2] To whom any correspondence should be addressed.

maximum amount of addition allowed for each cement type.

Another possible, more innovative option is the development of less expensive alternative cements, more respectful of the environment (lower CO_2 emissions, use of industrial by-products, etc.), with performance comparable to or even higher than ordinary Portland cements (OPC). The so-called ***alkaline cements*** [2-5] feature all these properties.

Two key characteristics of alkaline cements are: (i) they don't have clinker contents (free OPC); and (ii) they use alkaline activators. Two major alkali activation models in place can be distinguished by their starting conditions: ***1st model:*** *a*ctivation of ***calcium- and silicon-rich*** materials (Me_2O-MeO-Al_2O_3-SiO_2-H_2O system) as the activation of *blast furnace slag* under relatively moderate alkaline conditions [**2-5**]; ***2nd model***: activation of materials comprising primarily ***aluminium and silicon*** (Me_2O-Al_2O_3-SiO_2-H_2O system) also called geopolymers as the activation of metakaolin or fly ashes [3, 6-8].

The present summary addresses the most relevant aspects of alkaline cements, including composition of the prime materials, reaction mechanisms, characteristics of the reaction products, process variables (such as the type and concentration of alkaline activator and curing conditions), physical-mechanical properties as well as the structure of the cementitious gels formed (C-A-S-H, N-A-S-H). The durability and strength of these systems make these materials apt for use in many industrial applications, such as precast concrete elements (masonry blocks, railroad sleepers), protective coatings for materials with low fire ratings and lightweight elements.

2. Composition of prime materials

The materials most commonly used to manufacture alkaline cements and concretes are blast furnace slag and fly ash. As table 1 shows, the chemical composition of these materials in % of mass.

Table 1. Chemical composition (%) in mass.

	L.O.I.	SiO_2	Al_2O_3	Fe_2O_3	MgO	CaO	SO_3	Na_2O	K_2O	Other
Slag	0.77	37.42	11.47	0.66	10.62	32.71	2.96	1.47	1.19	0.73
Fly ash	2.19	54.42	26.42	7.01	1.79	3.21	0.01	0.59	3.02	1.34

L.O.I. = Loss on ignition.

Vitreous blast furnace slag, waste generated by the steel industry, is formed when the clayey acid gangue from the iron ore and the sulfur ash in the coke combine with calcium and magnesium in the limestone or dolomite used as fluxes. The vitreous blast furnace slag obtained when acid (SiO_2 and Al_2O_3) and basic (CaO and MgO) oxides combine at high smelting temperatures (1600 ºC) and are then abruptly cooled to T< 800 ºC is later ground and stored in silos for shipping. On average, slag has a 90-95 % vitreous content, with network-forming [$(SiO_4)^{4-}$ $(AlO_4)^{5-}$, $(MgO_4)^{6-}$] and modifying (Ca^{2+}, Al^{3+}, Mg^{2+}) cations [2]. Its minority crystalline phases form a solid solution of two melilite family crystals, gehlenite and åkermanite, which crystallize in the tetragonal system.

Fly ash is an industrial by-product generated by coal-fired steam power plants. Before it can be used as a fuel in power plants, coal must be ground to a very fine powder. During combustion, the coal forms coarse ash (called "bottom ash") and other much finer particles that are carried by smoke and collected in precipitators to prevent dispersal into the atmosphere. This is what is known as ***fly ash***. Fly ash particles are characteristically spherical and may be either hollow or filled with other smaller spheres. The ash best suited for alkali activation is type F fly ash ([CaO]< 10%), which has a vitreous content of > 50 % (the vitreous contend can be determined by a chemical attack with HF acid [9-10] or thought a XRD-Rietveltd analysis [11]. Its minority crystalline phases include quartz (5-13 %), mullite (8-14 %) and magnetite (3-10 %), they are basically inert to the alkali- activation process [9-11].

3. Results

A considerable number of variables are involved in alkaline activation, most prominently: the composition of the prime materials, type and concentration of the alkaline activator; liquid/solid ratio; curing temperature and time; and moisture conditions during initial curingThese variables have been amply studied in the literature because they affect the degree of reaction, the mineralogical composition and microstructure of the reaction products and ultimately the physical-mechanical properties of the material. This review focuses specifically on how the type and concentration of the activator, along with temperature and curing time, affect mechanical properties.

3.1. Effect of type and concentration of activator

The solubility of these materials varies depending on the pH of the activating solutions. Solubility is high in acid media, but the instability of the resulting hydrates prevents any cementitious structure from forming. By contrast, at basic pH >11 the material dissolves and forms stable hydrates, with a gradual increase in the binding properties of the material. Therefore, the functions of the activating substance can be summarized as: i) to accelerate material solubility; ii) to stimulate the formation of stable, low solubility compounds; iii) to stimulate the formation of a compact structure.

The most widespread opinion around ***activator concentration*** is that the higher it is, the greater is mechanical strength. Nonetheless, this is not completely accurate: depending in part on the activator used, there may be certain threshold values above which strength does not rise significantly, or may even fall. Moreover, these high doses of alkalis may have adverse effects, increasing efflorescence, brittleness and so on. Nor should the economic factor be overlooked, for higher doses entail higher costs.

In any event, the thresholds apparently vary depending on the material. For materials rich in SiO_2 and Ca_2O, this threshold is in the order range of 3-6 % of Na_2O by mass with respect to the binder [**2-4**]. Much more basic concentrations are needed for materials rich in SiO_2 and Al_2O_3, with Na_2O values in the range of 8-10 % [2-4].

The present discussion focuses on the alkaline activators that the literature regards to be the most effective in terms of mechanical strength and other properties, namely: NaOH, Na_2CO_3 and sodium silicate (waterglass). Some authors [12] have also reported that mixing the activators may have a beneficial effect on strength development.

As figure 1 shows, activator behaviour is very similar in the two systems (Slags and Fly Ash). At any given Na_2O concentration, the highest strength values are consistently obtained when the activator used is a sodium silicate solution (waterglass). In the opinion of a number of authors, waterglass contributes doubly to strength development, both by acting as an alkaline activator and giving rise to the formation of a more compact structure and cementitious gel richer in silica [13]. Variations in the SiO_2/Na_2O ratio are known to significantly modify the degree of polymerization of the dissolved species (monomers, dimers, trimers and so on), the optimal SiO_2/Na_2O ratio in a waterglass solution according to the literature is in the range of 0.75-1.25 [3,13]. While the presence of soluble silica in the system generates no essential change in the reaction mechanism, it does substantially lower the range of the long-term structural order. This plays a very significant role in determining the structure and properties of polymeric gels synthesized using these solutions.

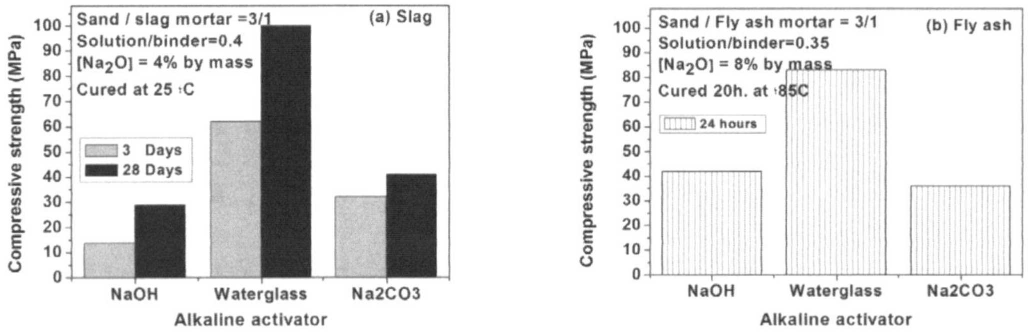

(a) (b)

Figure 1. Variation in compressive strengths in mortars depending on the type of alkaline activator (prismatic specimens 4x4x16 cm): (a) slag mortars; (b) fly ash mortars.

As a rule, sodium carbonate activation yields lower early age strength than when the activator is NaOH or sodium silicate (waterglass). The explanation is associated with the formation of calcium sodium carbonates and/or bicarbonates, which lower the pH and therefore the degree of reaction of the starting material. This effect is attenuated at older ages, however, and some of these mortars eventually exhibit higher strength values than NaOH-activated materials (see figure 1). Briefly, then, the OH^- ions catalyze the dissolution of the vitreous phase of the material. Up to a certain threshold, the presence of soluble silicate ions may have a beneficial effect on mechanical strength, for it occasions the formation of initial gels richer in Si. The presence of carbonate ions may lead to the formation of calcium/sodium carbonates that acidify the system, lowering strength, particularly at early ages.

3.2. Effect of temperature and curing time

Increasing the temperature during initial curing accelerates the reaction rate of these materials: this is known as ***accelerated curing***. *Curing enhances* mechanical strength, particularly early age strength, improves durability, and limits product fluctuations and efflorescence. As in the case of Portland cement, higher temperatures may have both beneficial and adverse effects on mechanical strength development in alkaline cements. Figure 2 shows the effect of thermal pre-curing treatment (time and temperature) in alkali activated slag and fly ash mortars (prismatic specimens 4 x 4 x 16 cm, sand/binder ratio=3/1). Slag mortars were activated with a 4M NaOH solution, solution/ slag ratio = 0.4; the pre-curing process were 20h at two temperatures, 25° and 45 °C, with 95% of r.h. , to be later curing in a chamber (22° ± 1°C at 95% of r.h.). Fly ash mortars were activated with 8M NaOH solution, solution/fly ash ratio = 0.35; the pre-curing process lasted 5h, 12h, 24h and 7 days at 45°C, 65°C and 85°C at 95% r.h. The effect of pre-curing temperature is more relevant in alkali activated fly ash where time and temperature have a considerable impact on the mechanical behaviour. This could be explained by the fact that the solubility of Al and Si increases significantly with temperature but it is not so with calcium compounds. Nonetheless, in alkali activated fly ashes, after a certain threshold value, this rise tapers (see figure 2b). Mechanical strength grows exponentially with temperature [14], considerably reducing the time needed to reach high mechanical strength values.

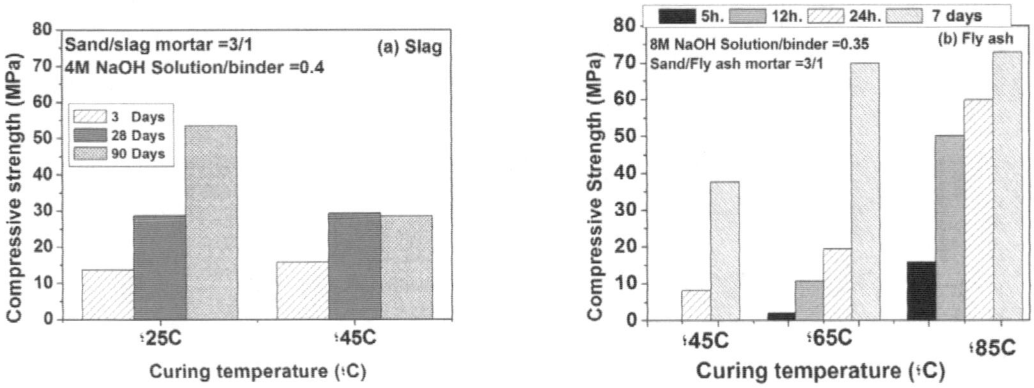

(a) (b)

Figure 2. Effect of time and temperature on the mechanical strength in mortars (specimens 4x4x16 cm) of (a) slag activated with 4M NaOH; (b) fly ash activated with 8M NaOH.

3.3. Reaction products

The majority product forming in Portland cement hydration is a C-S-H gel to which the material primarily owes its mechanical properties. The secondary products include portlandite, ettringite, monosulfoaluminate, etc. In alkaline cements, the process is similar, inasmuch as the main reaction product is a gel with a different composition and structure than the one found in the OPC gel. The secondary products also vary. The type of secondary product depends on the starting materials, type and concentration of the activator, curing conditions and so on. Table 2 compares the most common reaction products formed in these materials [2-5].

Research conducted on fly ash at the Eduardo Torroja Institute has shown that the uptake of small amounts of soluble Si in the system enhances the tendency of the tectosilicate structure to attain high range order in a short period of time. The transformation kinetics of zeolite precursors into crystalline phases (aluminosilicates) are accelerated under these conditions. Nonetheless, when the degree of soluble Si polymerization rises, the gels formed are apparently more amorphous. For both types of materials, then, the optimal SiO_2/Na_2O ratio in a waterglass solution is in the range of 0.75-1.25 [2-5].

Table 2. Reaction products.

	OPC	Alkaline cement		
		Me_2O-MeO-Al_2O_3-SiO_2-H_2O		Me_2O-Al_2O_3-SiO_2-H_2O
Main product	C-S-H	C-A-S-H		N-A-S-H
Secondary product	$Ca(OH)_2$	Hydrotalcite $[Mg_6Al_2CO_3(OH)_{16}\cdot 4H_2O]$		Zeolites: HS, P, Na-CHA, Y, FAU
	AF_m	C_4AH_{13}	$CASH_8$	
	AF_t	C_4AcH_{11}	$C_8A_2cH_{24}$	
	C= CaO	S= SiO_2	A= Al_2O_3	N=Na_2O H= H_2O c=CO_2

The gel forming during the alkali activation of ***blast furnace slag*** is similar to C-S-H gel in which the Al_T is taken up into the structure of the material, replacing a Si tetrahedron in the bridging tetrahedron position. The presence of Al gives rise to longer linear chains as well as to the possibility of occasional inter-chain bonds "*cross-linked*" across Si-O-Si bonds to form plane-like structures (Q^3 nAl units). In other words, in this way C-A-S-H – type gels would form (see figure 3a).

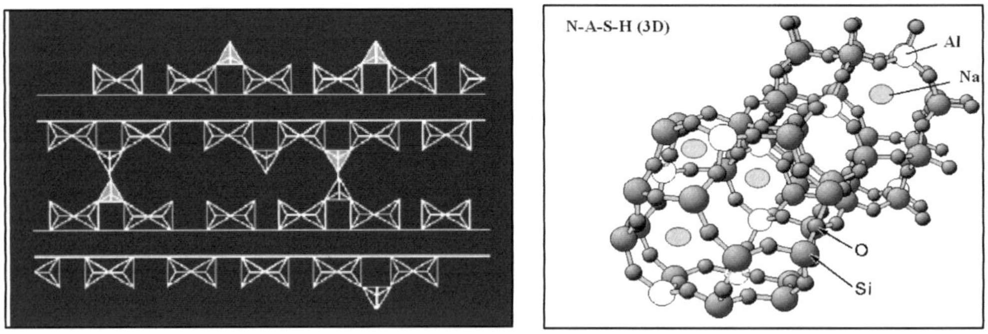

(a) (b)

Figure 3. (a) Structural model for C-A-S-H; (b) structural model for N-A-S-H gel [4-5].

By contrast, the structure of the gels forming ***during the alkali activation of silicoaluminous materials (with low Ca content)*** varies substantially from the foregoing. This gel, a silicoaluminate alkaline inorganic polymer (N-A-S-H) characterized by a three-dimensional structure in which the Si is found in a variety of environments, with predominance of Q^4(3Al) and Q^4(2Al) units, may be identified as a "zeolite precursor". The Si^{4+} and Al^{3+} cations exhibit tetrahedral coordination and are connected by oxygen bonds (see figure 3b). The negative charge on the AlO_4^- group is offset by the presence of alkaline cations (typically Na^+ and K^+). The initial, metastable, Al-rich compound that forms subsequently evolves into phases with a higher Si content [4,5,7].

4. Applications

In this section some applications (made by the authors of this paper) in the precast industry are shown. Figure 4 shows the railway sleepers manufactured with alkali activated fly ash concrete (without OPC) with no need for any significant material change in the normal manufacturing stream used at the factory. The most relevant modification was to increase the curing temperature from 50 to 85-90 ºC. A tunnel was built for this purpose, in which pressurized water vapour was injected (see figure 4). Increasing the curing temperature in these materials does not cause the durability problems observed in Portland cement when cured at T>60 ºC. The sleepers manufactured demonstrated very good behaviour to static and dynamic test [15].

Figure 4. Photographs of industrial-scale manufacture of railway monoblock sleepers using alkali-activated fly ash concrete (with no OPC) [15].

Figure 5 (a) and (b) show alkali-activated fly ash-based inorganic materials usable as a fire-resistant coating for fibreglass or carbon fibre matrices. Despite their lower density, sheets of fibreglass- or carbon fibre-reinforced polymeric resin can develop strength similar to the values characteristic of steel. The main drawback to such materials is their low resistance to high temperatures and their combustibility when directly exposed to flame. The improvements achieved in fibreglass with an alkali-activated fly coating included the following: I) -The primary advantage obtained with the use of this type of coating is non-propagation of the flame; ii)-The resin ignites where the flame is applied but the flame does not propagate upwards. The flame disappears when the resin is consumed.

(a) (b) (c) (d)

Figure 5. Photographs of fire protection test. (a) Fibreglass + epoxy resin, exposure time 8s; (b) coating *Material 1* exposure time 8m; (c) Lightweight materials evenly distributed pores; (d) Physical appearance of the panels manufactured.

Other possibility to manufacture lightweight materials is with alkali-activated fly ash, using a series of air-entraining admixtures to lower the density of this material (see figure 5 c and d). As a rule, they also provide acoustic and thermal insulation. The admixture can react very rapidly with the paste, the resulting material can have densities between 1000 to 500 kg/m^3 and resistances between 10 to 3 MPa latter of cured in a kiln for 20 hours at 85 ºC.

Finally, figure 6 depicts the elements precast (wall blocks and type 6 pavers) with a blend (12% OPC+88% BFS) alkali activated and cured at room temperature. Both 2- and 28-day materials met mechanical strength requirements. The only difference between these elements and the products made with Portland cement worth mentioning was their slightly lower density.

(a) (b)

Figure 6. Production of (a) type 14 wall blocks; (b) type 6 pavers.

Of course, many others works by different authors [2-4, 16] about different applications in the precast industry, ceramics, porous materials and even biotechnologies [17] can be found in literature.

5. Conclusions

Alkali activation of aluminosilicates can be considered a very versatile technology that can be adapted to a variety of industrial applications. Particularly within the construction industry, alkali activation of slag and fly ash (industrial by-products) allows getting a material with the cementing features similar to the Ordinary Portland Cement. Besides, this process offers important economic and environmental benefits when compared with the traditional fabrication of OPC. The possibility to obtain different qualities of materials by modifying not only the binder composition but also the nature of the activator used and the pre-curing treatment deserves special attention.

6. References

[1] Taylor H F W 1990 *Cement Chemistry* (London: Academic Press)

[2] Shi C, Roy D M and Krivenko P V 2006 *Alkali-activated Cements and Concretes* ed. Taylor & Francis (London: U.K.)

[3] Provis J L and Deventer J S J 2009 *Geololymers, Structure, Processing, Properties and Industrial Applications* (Woodhead Publishing Limited)

[4] Pacheco –Torgal F, Labrincha J A, Leonelli C, Palomo A and Chindaprasirt P 2015 *Handbook of Alkali-Activated Cements, Mortras and Concretes* Woodhead Publishing Serires in Civil and Structural Engineering vol.54

[5] Palomo A, Krivenko P, Garcia-Lodeiro I, Kavalerova E, Maltseva O and Fernández-Jiménez 2014 A review on alkaline activation: new analytical perspectives *Materiales de Construcción* **64** e022

[6] Rashad Alaa M 2013 Alkali-activated metakaolin: A short guide for civil Engineer – An overview *Constr. and Build. Mat.* **41** 751 – 765

[7] Fernandez-Jimenez A, Palomo A, Sobrados I and Sanz J 2006 The role played by the reactive alumina content in the alkaline activation of fly ashes *Microp. Mesop, Mater*. **91** 111

[8] Fernández-Jiménez A and Palomo A 2005 Composition and microstructure of alkali activated fly ash binder: Effect of the activator *Cem. Con. Res*. **35** 1984

[9] Fernández-Jiménez A and Palomo A 2003 Characterization of fly ashes. Potencial reactivity as alkaline cements *Fuel* **82** (18) 2259

[10] Ruiz-Santaquiteria C, Fernández-Jiménez A, Skibsted J and Palomo A 2013 Clay reactivity: Production of alkali activated cements *Applied Clay Science* **73** 11 – 16

[11] Fernández-Jiménez A, de la Torre A G, Palomo A, Lopez-Olmo G, Alonso M M and Aranda M A G 2006 Quantitative determination of phases in th e alkali activation of fly ash Part I Potential ash reactivity *Fuel* **85** 625 – 634

[12] Fernandez-Jimenez A and Puertas F 2003 Effect of activator mix on the hydration and strength behaviour of alkali-activated slag cements *Adv. Cem. Res.* **15** (3) 129

[13] Criado M, Fernández-Jiménez A, Palomo A, Sobrados I and Sanz J 2008 Effect of the

SiO_2/Na_2O ratio on the alkali-activation of fly ash. Part II: ^{29}Si MAS-NMR survey *Microp. and Mesop. Materials* **109** 525

[14] Palomo A, Alonso S, Fernández-Jiménez A, Sobrados I and Sanz J 2004 Alkali activated of fly ashes. A NMR study of the reaction products *J. Am. Ceramic. Soc.* **87** 1141

[15] Fernández-Jiménez A and Palomo A 2009 Properties and uses of alkali cements *Revista Ingeniería de Construcción* **24** (3) 213 – 232

[16] Bajare D, Bumanis G, Korjakins A and Sele L 2014 Reuse of Non-Metallic Residues from Aluminium Recycling Industry in Production of Porous Building Materials *Proc. of the First Int. Conf. on Construction Materials and Structures (Johannesburg, Central African Republic, 2014)* pp 136 – 144

[17] Ruģele K, Būmanis Ģ, Bajāre D, Lakevičs V and Rubulis J 2014 Alkaline Activated Material for pH Control in Biotechnologies *Key Engin. Materials* **604** 223 – 226

Acknowledgements

This research was funded by the Spanish Ministry of the Economy and Competitiveness under the projects BIA2013-43293-R and RTC-2014-2351-5. One of the authors worked under a Post-graduate Studies Council grant co-funded by the Spanish National Research Council and the European Social Fund (ESF).

Cements with low Clinker Content

I García-Lodeiro[1,2], A Fernández-Jiménez[1] and A Palomo[1]
[1] Instituto Eduardo Torroja (CSIC), Serrano Galvache 4, 28033, Madrid, Spain

E-mail: iglodeiro@ietcc.csic.es

Abstract. Hybrid alkaline cements are multi-component systems containing a high percentage of mineral additions (fly ash, blast furnace slag), low proportions (<30 %) of Portland clinker and scarce amounts of alkaline activators. The substantially lower amount of clinker needed to manufacture these binders in comparison to ordinary Portland cement is both economically and ecologically beneficial. Their enormous versatility in terms of the raw materials used has made them the object of considerable interest. The present study explored the mechanical strength of binary blends mixes; B1= 20 % clinker (CK) + 80 % fly ash (FA) and B2=20 % clinker + 80 % blast furnace slag (BFS), both hydrated in the presence and absence of an alkaline activator specifically designed for this purpose. The use of the activator enhanced the development of early age strength considerably. All the hydrated matrices were characterised with XRD, SEM/EDX and (^{29}Si and ^{27}Al) NMR. The use of the alkaline activator generated reaction products consisting primarily of a mix of gels ((N,C)-A-S-H and C-A-S-H) whose respective proportions were found to depend upon system composition and initial reactivity.

1. Introduction

Portland cement is the material most widely used in construction today. Its mass production raises substantial energy and environmental issues, however. One of these issues is the manufacturing process, which entails the emission of greenhouse gases such as CO_2 and NO_x. With 0.8 tonnes of CO_2 per tonne of cement manufactured, the industry accounts for 7-8 % of worldwide emissions of this greenhouse gas [1].

One of the approaches to reducing this impact while helping solve other environmental problems (finding a use for industrial by-products or waste that would otherwise be the object of costly and pollution-prone stockpiling) widely accepted today is to include active industrial by-product or mineral additions in Portland cement clinker. Another more innovative option consists of developing alternative, less expensive and where possible less environmentally damaging cements (involving lower CO_2 emissions or the re-use of industrial by-products)whose characteristics or performance are comparable to or even better than ordinary Portland cement (OPC).Such binders are generically

[2] To whom any correspondence should be addressed.

known as "***alkaline cements***" [2-4]. ***Alkaline cements*** are a type of binders formed as the result of dissolving industrial waste or natural aluminosiliceous materials (with amorphous or vitreous structures) in an alkaline medium. When mixed with alkaline activators, these materials set and harden, yielding a material with good binding properties. Based on the nature of the cementitious components, alkaline cements can be grouped under different categories. One of such categories is a new type of binder known as ***hybrid alkaline cements***, formed as a result of alkaline activation of materials with low Portland cement clinker (from 0 to 30 %) and high proportions of mineral additions (slags, fly ash, metakaolin, etc.).

The present study attests to the enormous versatility of low (20 %) clinker, multi-component hybrid cement systems and the good mechanical performance of the respective binders. It also addresses the mineralogical and microstructural characteristics of these hybrid cements when hydrated in the presence of solid alkalis to induce activation in technologically viable times.

2. Experimental

2.1. Materials

The raw materials used included a commercial clinker (CK) supplied by a Spanish cement factory, granulated blast furnace slag (BFS), (from Algeria) and a Slovakian type F fly ash (FA).The chemical composition of the materials given in table 1 was determined by XRF on a PHILIPS PW-1004 X-ray spectrometer. Their mineralogy (XRD-determined) is shown in figure 1.

Table 1. Chemical composition of the materials (XRF, wt% of oxides).

	SiO_2	Al_2O_3	CaO	Fe_2O_3	MgO	Na_2O	SO_3	K_2O	TiO_2	P_2O_5	LoI*
Clinker (CK)	21.83	4.85	61.19	3.20	2.03	0.69	1.11	1.76	0.15	0.48	2.65
Fly ash (FA)	56.51	21.42	3.78	8.57	2.19	0.63	0.99	2.59	0.87	0.15	2.21
Slag (BFS)	36.63	7.25	38.09	0.46	8.51	0.53	2.13	0.62	0.27	0.06	4.88

**LOI: Loss on ignition (1000ºC)*

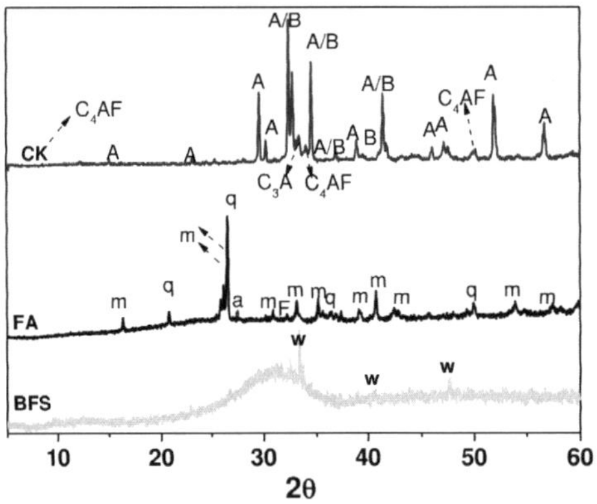

Figure 1. XRD patterns of Raw materials (**CK**: clinker; **FA**: Fly Ash; **BFS**: Blast Furnace Slag) (**Legend**: A: alite (C_3S); B: belite (C_2S); C_3A: tricalcium aluminate; C_4AF: ferrite phase; q: quartz (SiO_2); m: mullite ($3Al_2O_3.2SiO_2$); F: hematite (Fe_2O_3); a; anorthite; w=merwinite ($Ca_3Mg (SiO_4)$)).

2.2. Methodology
Two binders were prepared; **B1**consisting of a blend of 20 % clinker and 80 % fly ash and **B2**consisting of a blend of 20 % clinker and 80 % blast furnace slag. The binders included 4% of solid Na_2CO_3as alkaline activator (see table 2). The pastes obtained by hydrating the binders with water were moulded into six 1x1x6 cm prismatic specimens and cured for 24 hours in a curing chamber (22 °C and 99 % relative humidity). The specimens were subsequently removed from the moulds and stored in the chamber until they reached the test age (2 and 28 days). The w/c ratio for each system was chosen to ensure that all pastes had similar workability. A reserve sample (activator-free) of both blends was kept for use as a reference.

Table 2. Binder composition.

FA	BFS	CK	Activator*	Liquid of hydration	w/c**	Nomenclature
80	--	20	--	water	0.40	B1W
80	--	20	Na_2CO_3	water	0.38	B1A
--	80	20	--	water	0.26	B2W
--	80	20	Na_2CO_3	water	0.26	B2A

*As all the alkali activated blends contained 4 % of Na_2CO_3, the end product contained 96 % of binder and 4 % activator
** water/cement ratio

The specimens were tested to failure on an Ibertest (Autotest–200/10-SW) frame. Hybrid cements were characterised with XRD, SEM/EDX and ^{29}Si and ^{27}Al NMR. The XRD data were acquired with a Bruker AXS D8 Advance diffractometer with a 4-kW high voltage generator and a Cu anode X-ray tube, typically operating at 40 kV and 50 mA. A Jeol5400 electron microscope fitted with an *Oxford*ISIS X-ray analyser was used for the SEM analyses. The NMR spectra were recorded on a Bruker Avance-400 spectrometer (^{27}Al: 104.3 MHz; spinning rate, 10 kHz; 200 acquisitions; reference, $Al(H_2O)_6^{3+}$; ^{29}Si: 79.5 MHz; spinning rate, 10 kHz; 1 000 acquisitions; reference, TMS).

3. Results and Discussion

3.1. Mechanical Strength Behaviour
Figure 2 shows the compressive strength (MPa) findings for the 2- and 28-day binders. The presence of alkaline activator enhances the development of mechanical strength. When these hybrid cements are alkali-activated, up to 80 % of the clinker can be replaced without affecting mechanical performance: both fly ash and slag-containing binders exhibited 28-day compressive strength exceeding 30 MPa.

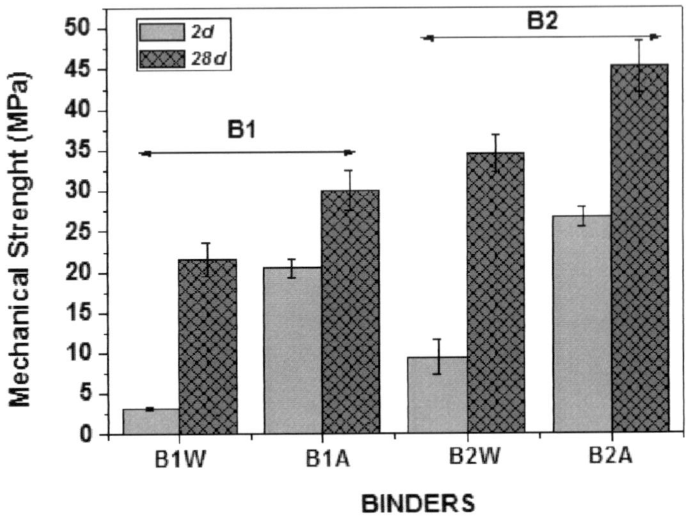

Figure 2. Compressive strength behaviour (MPa) for 2- and 28-day binders (**B1**: 80% FA+20% CK; **B2**: 80% BFS+20% CK; **W**: samples hydrated with no activator; **A**: samples activated with Na_2CO_3).

When the binder consisting of a blend of high ash and low clinker content (*B1*) is hydrated with water in the absence of activators, clinker (20 % of the material) reacts first. Ash, subsequently, reacts with $Ca(OH)_2$ formed during clinker hydration (pozzolanic reaction) ($C_3S+H_2O \rightarrow$C-S-H gel + $Ca(OH)_2$) [5]. Due to low clinker and high ash content, this is a slow process. Hence, the low early strength development.In presence of an alkaline activator, this induces a rise in the aqueous phase pH, hastening fly ash dissolution. Higher pH values intensify reactivity and along with it early strength values (see figure 2) [6].

The process is similar in the system B2 containing slag. Moreover, since slag requires less aggressive activation conditions than ash, mechanical strength values were slightly higher in the slag- than in the ash-bearing hybrid cements. The alkaline medium activates the slag, generating a C-A-S-H gel that contributes to mechanical strength development in the system [7].

3.2. Characterization of different binders

Figures 3(a) and (b) show respectively the XRD patterns for the 2- and 28-day B1 and B2 binders. In system B1(figure 3(a)), all the diffractograms exhibited diffraction lines characteristic of quartz and iron oxides (minority mineralogical phases in the ash), as well as lines associated with calcite precipitation. The signal for alite (clinker) practically disappeared on the 28-day patterns in all the systems. The pastes hydrated with water in the absence of activator (B1W) exhibited the 2-day behaviour expected: lines attributed to portlandite, which declined to near disappearance after 28 days as a result of the pozzolanic reaction.

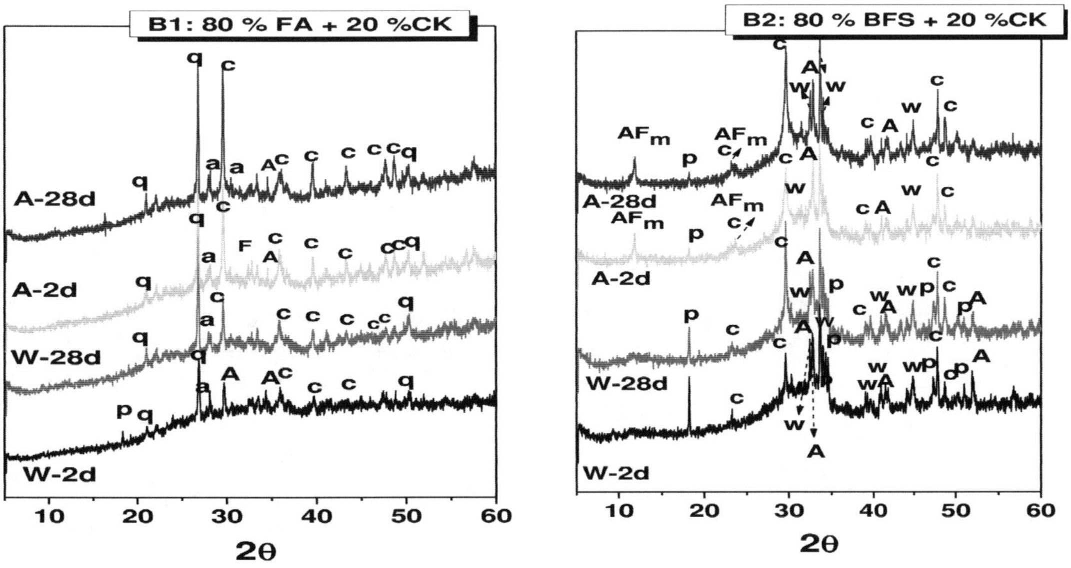

(a) (b)

Figure 3. XRD patterns of **(a) (b) B1 Binders** (80 %FA + 20 % CK) **(c) B2 Binders** (80 % BFS + 20 % CK) (**Legend**: c: Calcite ($CaCO_3$); p:portlandite ($Ca(OH)_2$); AFm: monocarboaluminate (C_4AcH_{11})).

Unreacted clinker was found in the blends containing clinker and slag (B2) (figure 3(b)) with no activator (B2W) and with activator A (B2A). Portlandite was detected in both systems, although in greater amounts in the system hydrated in the absence of activators. In addition, both the 2- and 28-day activated materials exhibited a series of lines attributed to the precipitation of calcium carboaluminate hydrate (AFm), a compound normally observed in alkali-activated slag pastes [8-9].Calcite was likewise identified in both systems. All the XRD patterns also exhibited humps at 2Θ angles of 25-35°, attributed to the formation of a gel regarded as primarily responsible for the mechanical properties of these hybrid cements. Since this technique furnishes little information on gel composition and structure, some of the alkali activated blends (B1A and B2A, at 28 days) were chosen for SEM/EDX and (^{29}Si and ^{27}Al) MAS NMR analysis.

System B1A had a compact matrix in which some of the ash particles had been attacked and others (covered with reaction product) had not (see figure 4(a)).The precipitate observed contained a number of gels not morphologically distinguishable but with different chemical compositions (see EDX microanalysis, figure 4(a)).Essentially two types of gels were present: one rich in Ca, Si and Al (C-A-S-H gel[1], similar to the gel normally identified in ordinary cement hydration) and a more abundant Si- and Al-high gel containing Ca and Na ((N,C)-A-S-H gel[1]), similar to the gel generated in the alkaline activation of fly ash.

[1]C: CaO; S: SiO_2; A: Al_2O_3; N: Na_2O; H: H_2O

System B2A also exhibited a compact matrix (figure 4(b)).Unlike the preceding system, however, here the primary reaction product was a C-A-S-H gel with C/S ratios of around 1 (see EDX microanalysis, figure 4(b)). Gels with lower aluminium contents, known as C-(A)-S-H gels, also appeared sporadically. The micrograph further shows unreacted clinker particles previously observed by XRD (see figure 3(a)).

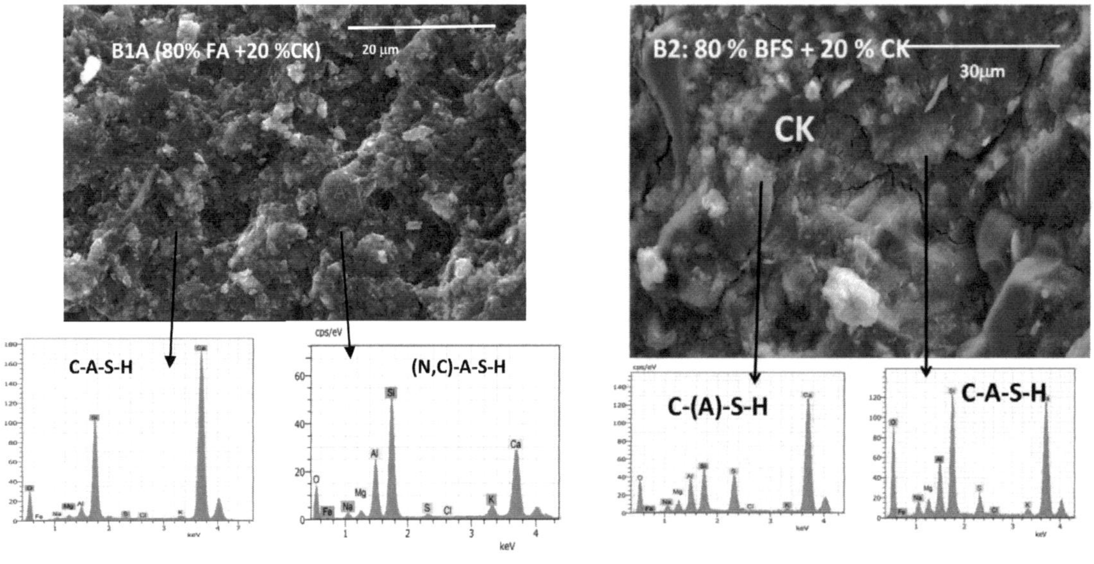

(a) (b)

Figure 4. SEM micrographs at 28 days of **(a) B1A** and **(b) B2A** binders, where we can differentiate the precipitation of C-(A)-S-H, C-A-S-H and (N,C)-A-S-H gels (EDX analyses).

The evidence indicates that the chemical composition of the gels was largely conditioned by the chemical composition of the starting materials. Gels higher in calcium were generated in the systems with high calcium content such as B2 (80 %BFS + 20 %CK), while silicon-rich gels were the major reaction product in the systems high in silicon, such as B1 (80 % FA + 20 %CK).

Figures 5(a) and (b) reproduce the 28-day ^{27}Al MAS-NMR and ^{29}Si MAS-NMR spectra for system B1: raw materials involved in the blend (CK and FA), B1 hydrated with water only (B1W) and alkali activated by Na_2CO_3 (B1A). Figures 5(c) and (d) reproduce the analogous spectra for B2 system (including the spectra associated with CK and in this case BFS).

If we focus our attention on figures 5(a) (^{27}Al spectra), clinker phase shows a narrow signal located at +80 ppm associated with tetrahedral aluminium (Al_T), while anhydrous FA shows a broad and asymmetrical signal located at +56 ppm also attributed to tetrahedral Al (Al_T)[10-12].When the binder was hydrated in the presence or absence of activator, the Al_T signal shifted slightly to +58 ppm and a signal associated with octahedral aluminium (AL_o) appeared between +5 and +10 ppm. As a technique highly sensitive to ^{27}Al nuclei, MAS-NMR is able to identify phases whose small quantity or poor crystallisation makes them nearly invisible to techniques such as XRD. The signal at around +10 ppm might be associated with the formation of calcium aluminate hydrates resulting from the hydration of the C_3A present in the clinker. The shift in and narrowing of the Al_T signal observed in the hydrated materials may be associated with the dissolution of part of the Al present in the fly ash and its inclusion in the reaction products. Its position, at around +58 ppm, would indicate that this aluminium continued to be tetrahedral and surrounded by three or four silicon atoms.

The ^{27}Al spectrum for the slag system (B2) differed from the foregoing (see figure 5(c)).Here the signal attributed to Al_T shifted toward more positive values, around +62 ppm, a position associated with the aluminium present in C-A-S-H gel [7,13].A very intense signal was also observed at +12.7 ppm, associated with the Al_o in the AFm phase precipitate, also identified with XRD (see Figure 1(c)). Interpretation of ^{29}Si MAS-NMR for the system B1 (figure 5(b)) spectra is more complex. The intensity of the signal at -71 ppm (associated with the Q^0 units in the alite + belite present in the clinker) declined in all the hydrated materials, confirming that the calcium silicates in the clinker were reacting. The fly ash spectrum exhibited a wide signal centred at around -109 ppm. This signal contained a series of peaks at around -88, -100, -103 and -110 ppm, associated with the various forms in which the silicate appeared in the fly ash (mullite, vitreous phase, quartz, etc.)[10-12].The intensity of the signals associated with the anhydrous phases declined in the spectra for the hydrated materials, while new (and intense) signals were found in the -80 to -100-ppm range. More specifically, the spectra for all the samples contained a strong signal at around -85/-86 ppm that might be associated with Q^2 units [14].

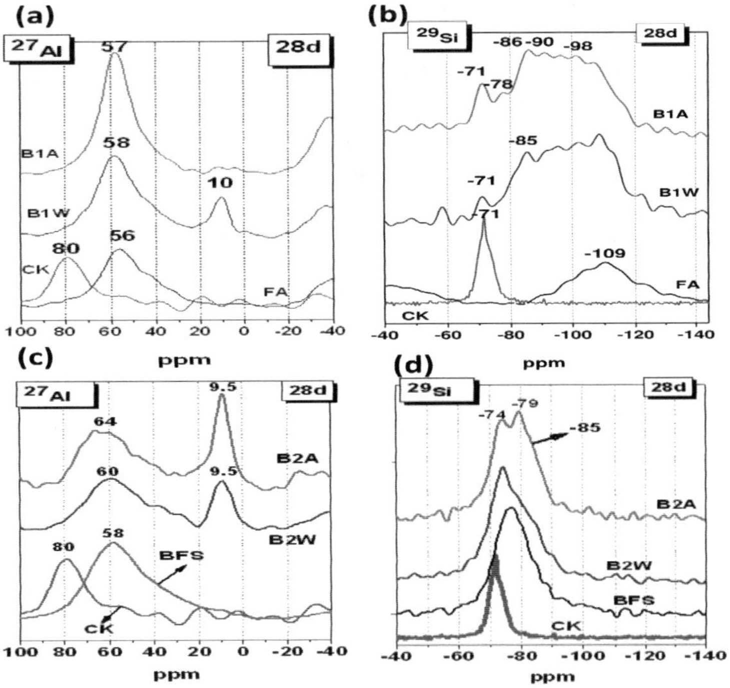

Figure 5. (a)^{27}Al and **(b)**^{29}Si MAS-NMR spectra for **B1 binders** (at 28 days) and **(c)** ^{27}Al and **(d)**^{29}Si MAS-NMR spectra **for B2 binders** (at 28 days).

The component associated with the presence of unreacted fly ash generated an intense signal (see the peak at -110 ppm on figure 5(b)) on the spectrum for the binder hydrated in the absence of a solid alkaline activator (B1W), an indication that under these conditions the ash was scantly reactive. That would explain the low mechanical strength observed in this material (see figure 2). When solid alkaline activator was added (B1A), cleared signals were observed at-79 ppm and -86 ppm (which can be associated with Q^1 and Q^2 units in C -S-H gel), along with other signals at -90 and -98.5 ppm. These spectra proved to be very similar to the ones obtained for similar binders activated with liquid alkalis [15].The increase in the signal at -85 ppm, along with the appearance of new signals, denoted the formation of a more polymerised gel, consisting of Q^3(nAl) or Q^4(nAl) units. The presence of Al in positions at around +58 ppm was suggestive of the presence of Q^4(4Al), Q^4(3Al) and Q^4(2Al) units, associated with N-A-S-H gel precipitation. The possibility ofQ^3(nAl) and Q^4(nAl) signal overlapping cannot be ruled out, however.

The ^{29}Si spectrum for system B2 differed from the foregoing (see figure 5(d)). Here, in addition to the components at-71 and -74 ppm, respectively attributed to the Q^0 units in the clinker and theQ^1 units in the slag [13], signals were observed at -79, -83 and -85 ppm, associated with Q^1,Q^2(1Al) and Q^2 units, respectively, in C-A-S-H gel[14].No signals for more polymerised units were detected, providing further evidence for the formation of C-A-S-H gel in the system.

These findings appear to confirm that in the systems hydrated in the presence of a solid activator (B1A and B2A), the composition of the precipitates depended heavily on the raw materials. Two types of gels formed in the ash and clinker blend; i) a C-S-H gel with aluminium in its structure as a result of its reaction with the clinker (B1A); ii) a N-A-S-H gel with Ca in its structure as a result of the alkaline activation of the ash. The majority product in the slag and clinker blends (B2A) was C-A-S-H gel. The presence of solid activators raised reactivity of ash and slag, furthering the precipitation of the cementitious gels that ensure good mechanical strength of the system.

4. Conclusions

- Viable binders can be derived from cementitious materials with a very low (20 %) clinker content, providing they are hydrated in the presence of solid alkaline activators.
- The alkaline activation procedure proposed here is applicable to different types of raw materials for the production of binders with good cementitious properties.
- The main reaction product generated in the systems depends heavily on the chemical composition of the starting materials. Systems high in calcium (such as 80 % BFS + 20 % CK blends) generate primarily C-A-S-H gel. In systems with higher silica and alumina content (such as 80 % FA + 20 % CK), the main reaction product is a mix of C-A-S-H + (N,C)-A-S-H gels.

5. References

[1] Gartner E 2004 Industrially interesting approaches to "low-CO_2" cements *Cem. Conc. Res.* **34** (9) 1489

[2] Palomo A, Grutzeck M W and Blanco M T 1999 Alkali-activated fly ashes. A cement for the future *Cem.Conc. Res.* **29** 1323

[3] Shi C, Roy D M and Krivenko P V 2006 *Alkali-activated Cements and Concretes* Ed. Taylor & Francis (London: U.K.)

[4] Provis J L and Deventer J S J 2009 *Geopolymers, Structure, Processing, Properties and Industrial Applications* (Woodhead Publishing Limited)

[5] Taylor H F W 1997 *Cement Chemisty* 2nd ed. (London: Thomas Telford)

[6] Garcia - Lodeiro I, Fernández-Jiménez A and Palomo A 2013 Variation in hybrid cements over time. Alkaline activation of fly ash–portland cement blends *Cem.Conc. Res.* **52** 112

[7] Fernández-Jiménez A, Puertas F, Sanz J and Sobrados I 2003 Structure of calcium silicate hydrated formed in alkali-activated slag pastes. Influence of the type of alkaline-activator *J. of American Ceramic Society* **86** (8) 1389

[8] Fernández-Jiménez A, Zibouche F, Boudissa N, García-Lodeiro I, Abadlia M T and Palomo A 2013 Metakaolin-Slag-Clinker Blends. The role of Na^+ or K^+ as Alkaline Activators of Theses Ternary Blends *J. of American Ceramic Society* **96** (6) 1991

[9] Garcia-Lodeiro I, Fernández-Jiménez A and Palomo A 2013 Hydration kinetics in hybrid binders: Early reaction stages *Cem. Conc. Comp.* **39** 82

[10] Palomo A, Alonso S, Fernández-Jiménez A, Sobrados I and Sanz J 2004 Alkali activated of fly ashes. A NMR study of the reaction products *J. of American Ceramic Society* **87** 1141

[11] Fernández-Jiménez A, Palomo A, Sobrados I and Sanz J 2006 The role played by the reactive alumina content in the alkaline activation of fly ashes *Microporous and Mesoporous Materials* **91** 111

[12] Duxon P, Fernández-Jiménez A, Provis J L, Lukey G C, Palomo A and van Deventer J S J 2007 Geopolymer technology: The current state of the art *J. Material Science* **42** 2917

[13] Richardson I G and Groves G W 1997 The structure of calcium silicate hydrate phases present in hardened pastes of white Portland cment/blast furnace salg blends *J. of Materials Science* **32** 4793

[14] Andersen M D Jakobsen H J and Skibsted J 2003 Incorporation of aluminum in the calcium silicate hydrate (C-S-H) of hydrated Portland cements: a high-field ^{27}Al and ^{29}Si MAS NMR study *Inorganic Chemistry* **42** 2280

[15] Palomo A, Fernández-Jiménez A, Kovalchuk G, Ordoñez L M and Naranjo M C 2007 Opc-fly ash cementitious systems: study of gel binders produced during alkaline hydration *J. of Material Science* **42** 2958

Acknowledgements
This research was funded by the Spanish Ministry of the Economy and Competitiveness under the projects BIA2013-43293-R and RTC-2014-2351-5. One of the authors worked under a Post-graduate Studies Council grant co-funded by the Spanish National Research Council and the European Social Fund (ESF).

Prediction of Corrosion Resistance of Concrete Containing Natural Pozzolan from Compressive Strength

A M al-Swaidani[1,2,4], R Ismat[2], M E Diyab[2] and S D Aliyan[3]
[1]Arab International University (AIU), Faculty of Architectural Engineering, Damascus, Syria
[2]Order of the Syrian Engineers, Damascus, Syria
[3]Syrian Arab Organization for Standardization and Metrology (SASMO), Damascus, Syria

E-mail: aydlswaidani@yahoo.fr; a-swaidani@aiu.edu.sy

Abstract. A lot of Reinforced Concrete (RC) structures in Syria have suffered from reinforcement corrosion which shortened significantly their service lives. Probably, one of the most effective approaches to make concrete structures more durable and concrete industry on the whole – more sustainable is to substitute pozzolan for a portion of Portland cement (PC). Syria is relatively rich in natural pozzolan. In the study, in order to predict the corrosion resistance from compressive strength, concrete specimens were produced with seven cement types: one plain Portland cement (control) and six natural pozzolan-based cements with replacement levels ranging from 10 to 35%. The development of the compressive strengths of concrete cube specimens with curing time has been investigated. Chloride penetrability has also been evaluated for all concrete mixes after three curing times of 7, 28 and 90 days. The effect on resistance of concrete against damage caused by corrosion of the embedded reinforcing steel has been investigated using an accelerated corrosion test by impressing a constant anodic potential for 7, 28 and 90 days curing. Test results have been statistically analysed and correlation equations relating compressive strength and corrosion performance have been developed. Significant correlations have been noted between the compressive strength and both rapid chloride penetrability and corrosion initiation times. So, this prediction could be reliable in concrete mix design when using natural pozzolan as cement replacement.

1. Introduction

Pozzolanic materials are being widely used as substitute for Portland cement because of their ecological, economical and performance-related advantageous properties [1-7]. Syria has important volcanic areas. More than 30,000 km^2 of the country is covered by Tertiary and Quaternary-age volcanic rocks [8], among which natural pozzolans constitute a considerable volume with the

[4] To whom any correspondence should be addressed.

estimated reserves of about three-quarters billion tonnes [9]. However, their potential use in making concrete is not well established. The cement produced in the country is almost of CEM I type, although an addition of natural pozzolan up to 5% was frequently used in most local cement plants. Hence, less than 300,000 tonnes of these pozzolans are only exploited annually (the annual production of PC in Syria is about 6 million tonnes) [10].

Compressive strength of concrete is commonly considered its most valuable property, although in many practical cases other characteristics, such as durability and permeability, may in fact be more important [11]. Therefore, designers should not ignore the environmental conditions taking into consideration only the strength as the sole criterion in their structural designs [12].

Chloride-induced corrosion is currently considered to be the most important and most serious deterioration mechanism for RC structures [13-14]. Corrosion of reinforcing steel is an important problem faced by the construction sector in Syria, especially in industrial and marine environments. A lot of RC structures have suffered from this phenomenon, which shortened significantly their service lives.

In order to make the concrete industry much more sustainable, it is imperative to employ some effective approaches in concrete mix design. Probably one of the most effective approaches is to substitute pozzolan for a portion of PC; each 1kg of substitution reduces the emission of CO_2 by about 1 kg, and saves the energy required to produce 1 kg of cement. In addition, these pozzolans could be used to make stronger and more durable concretes [12].

Compressive strength measurements are easy from an experimental point of view. By contrast, evaluation of reinforcement corrosion resistance is time consuming and difficult to deal with. In the study, in order to predict corrosion resistance from concrete compressive strength, compressive strength, rapid chloride penetrability and corrosion initiation times were experimentally investigated. Furthermore, relationships among some investigated properties were analysed in this paper. The study is of particular importance not only for the country in question but also for other countries with similar geology, e.g. Harrat Al-Shaam, a volcanic field covering a total area of some 45,000 km^2, a third of which is located in Syria. The rest covers parts of Jordan and Saudi Arabia.

2. Materials and methods

The natural pozzolan used in the experiments was quarried from Dirat-at-Tulul site, about 70 km southeast of Damascus as shown in figure 1. The mineralogical analysis showed the pozzolan is mainly composed of amorphous glassy ground mass, vesicles, plagioclase and olivine with the following percentages – 20%, 35%, 20% and 25%, respectively (based on the optical estimate). Thin sections of the investigated pozzolan are shown in figure 2. The chemical analysis of pozzolan used in the study is summarized in table 1.

Seven cement samples were prepared in accordance with EN 197-1; one plain PC CEM I (control), three CEM II/A-P samples with three replacement levels of pozzolan (10, 15 and 20%), and three CEM II/B-P samples with three replacement levels (25, 30, 35%), respectively. 5% of gypsum was added to all these cement samples. The clinker was obtained from Adra Cement Plant, Damascus, Syria. Chemical analysis of clinker and gypsum is shown in table 1. All samples were interground by a laboratory grinding mill to a Blaine fineness of 3200±50 cm^2/g. CEM I (the control sample) was designated as CEM I, whereas pozzolan-based cements were designated according to the replacement level. For instance, CEM II/10% and CEM II/35% refer to the cements containing 10% and 35% of pozzolan, respectively.

Table 1. Chemical composition of the used materials.

Chemical composition (by mass, %)	Materials				
	Natural pozzolan[a]	Clinker	Gypsum	Dolomite aggregate	Natural sand
SiO_2	46.52	21.30	0.90	0.42	93.39
Al_2O_3	13.00	4.84	0.07	0.38	0.57
Fe_2O_3	11.40	3.99	0.10	0.10	0.24
CaO	10.10	65.05	32.23	31.40	1.70
CaO_f	-	2.1	-	-	-
MgO	9.11	1.81	0.20	20.46	0.20
SO_3	0.27	0.25	45.29	0.18	1.15
Loss on ignition	2.58	-	21.15	46.48	2.52
Na_2O	2.14	0.60	-	0.06	0.06
K_2O	0.77	0.28	-	0.30	0.05
Cl^-	<0.1	0.05	-	0.021	0.017
Pozzolan activity index [ASTM C 618]	79 (at 7 days) 85 (at 28 days)				

[a.] $SiO_{2(reactive)}$ = 42.22 (measured in accordance with EN 196-2)

Seven concrete mixes were prepared using grading of aggregate mixtures kept constant for all concretes. Aggregates used in the study were crushed dolomite with natural sand added. Chemical composition of the aggregates is illustrated in table 1. Their quantities in $1m^3$ concrete mix based on the oven-dry condition are as follows: 565.5 kg of coarse aggregate, 565.5 kg of medium-size aggregate, 447.5 kg of crushed stone sand and 286.5 kg of natural sand. All concrete mixes were designed to have a water-binder ratio (w/b) of 0.6 and a slump of 150±20 mm. No chemical admixture was added. Concrete cubes (150 mm) and concrete cylinders (100 mm×200 mm) were cast for the determination of compressive strength and evaluation of the penetrability of chloride ions, respectively. The reinforced concrete specimen for the accelerated corrosion tests was 100×200 mm concrete cylinder in which 12 mm diameter steel bar was centrally embedded. The steel bar was embedded into the concrete cylinder so that its end was at least 45 mm from the bottom of the cylinder, and it was coated with epoxy at the exit from the concrete cylinder in order to eliminate crevice corrosion at these locations.

Figure 1. Map of Harrat Al-Shaam and the studied site.

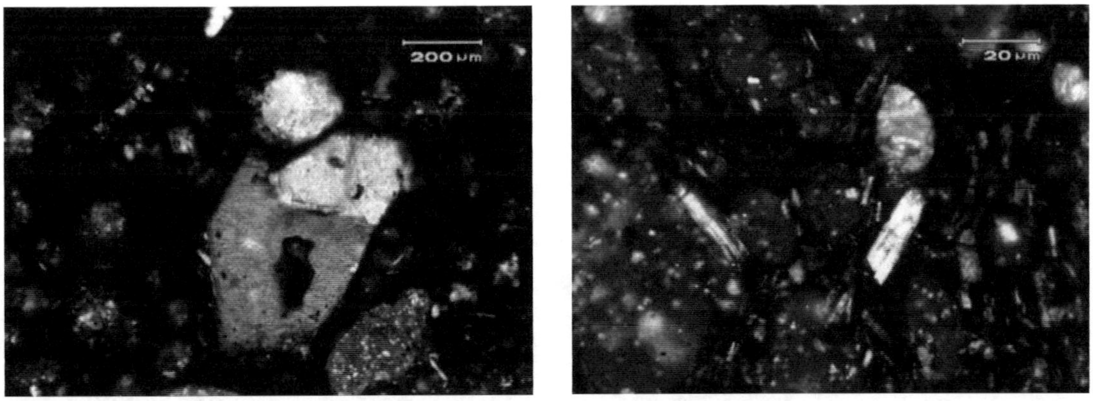

(a) (b)

Figure 2. Thin sections of N pozzolan. a) Microphenocrysts of Olivine in volcanic glass matrix with vesicles, some of which are filled with white minerals; b) Microphenocrysts of elongated plagioclase in volcanic glass matrix with vesicles, some of which are filled with white minerals.

The compressive strength development was determined on 150 mm cubic concrete specimens, in accordance with ISO 4012, 1978, at ages of 7, 28 and 90 days.

The rapid chloride penetrability test was conducted in accordance with ASTM C 1202. Three cylinder specimens of each concrete mix were tested after 7, 28 and 90 days curing.

An accelerated corrosion test was used to compare the corrosion performance of concretes containing cements produced. Similar techniques with little differences were reported by other researchers [15-23]. In the study, RC specimens were immersed in a 15% NaCl solution leveling the half of the concrete cylinder and the steel bar (working electrode) was connected to the positive terminal of a DC power source while the negative terminal was connected to a steel plate (counter electrode) placed near the concrete specimen in the solution. The corrosion process was initiated by impressing a relatively high anodic potential of 12 V to accelerate the corrosion process. Figure 3 shows a schematic representation of the experimental setup for the accelerated corrosion test. The specimen was monitored periodically to see how long it takes for corrosion cracks to appear on the specimen surface. The current readings with time were recorded at 3-4 hours-intervals. Three specimens from each concrete mix were tested after 7, 28 and 90 days curing.

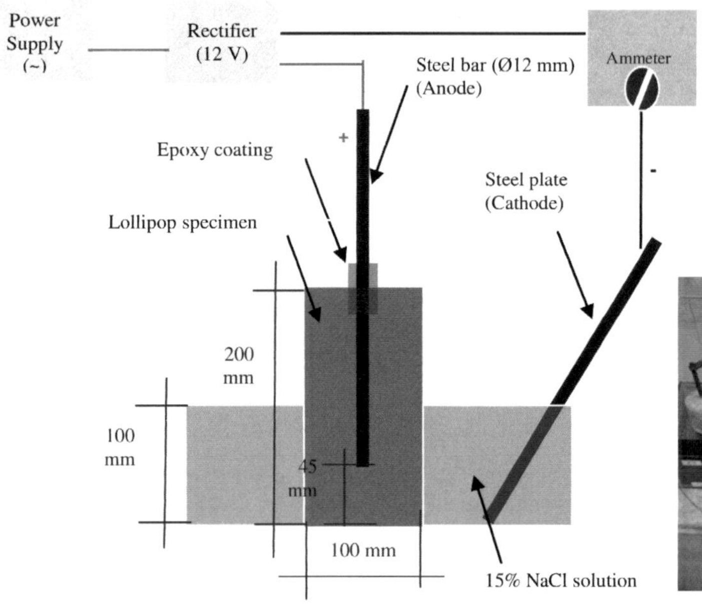

(a) Setup for the accelerated corrosion test.

(b) View of 28 days cured concrete specimens after 4 days of impressing 12 V. potential.

Figure 3. Schematic representation & view of the experimental setup for the accelerated corrosion test.

3. Results and discussion

3.1. Compressive strength of concretes

Results of compressive strength test are plotted in figure 4. As expected, the compressive strength of the concrete increased with curing time with a high rate of strength gains at early ages which gradually decreased at longer ages. Plain cement concrete specimens have higher compressive strengths at any age when compared to pozzolan-based cement concretes. This diminution of strength of pozzolan-based cement concretes was higher at early age and increased with the percentage of pozzolan. The compressive strength at 7 days decreased from 24.9 to 16.0 MPa when CEM I and CEM II/35% of pozzolan were used, respectively. This could be explained by the slowness of the pozzolanic reaction between the glassy phase in pozzolan and the calcium hydroxide released during cement hydration [24]. However, due to the continuation of this reaction and the formation of a secondary CSH, a greater degree of hydration was achieved resulting in strengths after 90 days curing, which were comparable to those of CEM I specimens.

3.2. Rapid chloride penetrability

From the rapid chloride penetrability data, figure 5, it should be noted that the CEM I-based concrete permitted almost 2 or 3 times the coulombs charge, when compared to the concrete containing CEM II/B in spite of the fact that all concretes were made with similar cementitious content and water content. None of concretes has the total charge passed less than 2,000 coulombs after 7 or 28 days curing. This expected result may be due to the high w/b ratio. However, the concretes containing CEM II/B-P showed the best performance among all specimens. According to ASTM C1202, these concretes can be considered as low chloride permeable after 90 days curing.

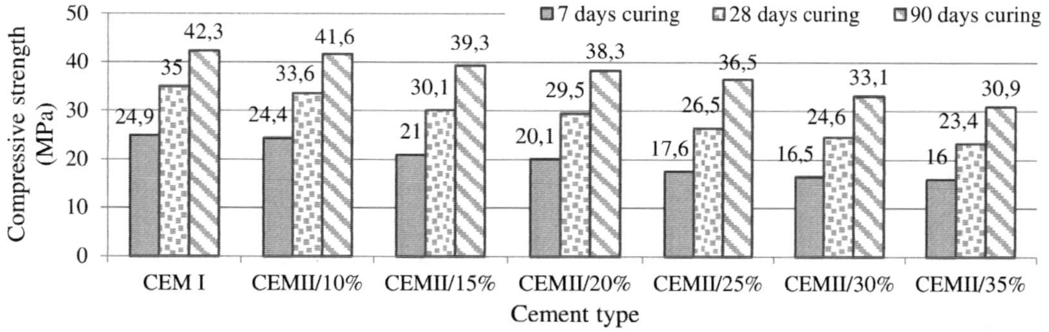

Figure 4. Cubic compressive strength of concrete specimens based on curing age.

The improvement in resistance to chloride penetration may be related to the refined pore structure of these concretes and their reduced electrical conductivity [25]. This, as it was confirmed by many researches [26-28], is due to the secondary pozzolanic reaction which contributes to making the microstructure of concrete denser.

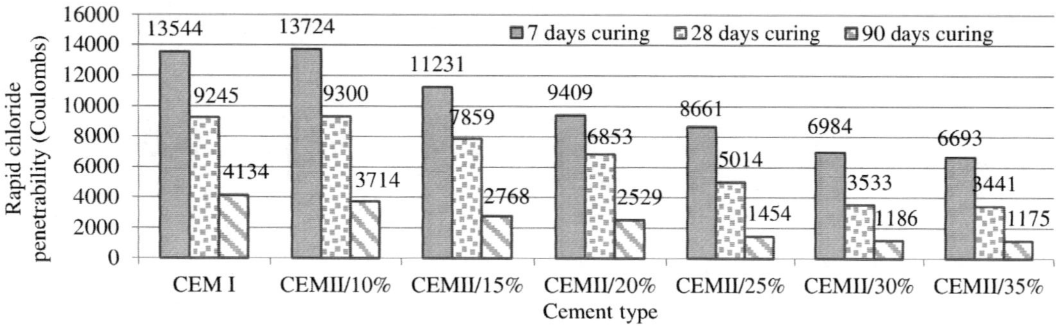

Figure 5. Rapid chloride penetrability of concrete specimens based on curing age.

3.3. Corrosion resistance

The accelerated corrosion behaviour of steel bars embedded in concrete specimens was studied by impressing a constant anodic potential. The current required to maintain the fixed potential was plotted against time and typical curves of corrosion current versus time for the concrete specimens are illustrated in figure 6. As seen from figures 6, current–time curve initially descended till the time period, after which a steady low rate of increase in current was observed, and after a specific time period a rapid increase in current was detected until failure. Almost a similar variation of the corrosion current with time has been also observed by other researchers [15, 19]. The first visual evidence of corrosion was the appearance of brown stains on the surface of the specimens. Cracking was observed shortly thereafter, and it was associated with a sudden rise in the current. Figure 7 presents the average corrosion initiation times required to crack the specimens made with CEM I, CEM II/A-P and CEM II/B-P. Time to initiate corrosion in CEM I concrete specimens was in the range of 52-106 h (2-4.5 days), whereas that in CEM II/B-P was in the range of 83-370 h (3.5-15.5 days), depending on the replacement level and the age at testing. The times of corrosion initiation for the pozzolan-based cement concrete specimens were longer than the plain cement concrete specimens, which indicated that the former provided better protection to steel reinforcement against corrosion. This delay in corrosion time when using blended cements, particularly at longer ages of curing, may be related to the pozzolanic reaction, which contributes to filling the voids and pores in the concrete with an additional CSH gel. This leads to decrease of pore size and to a smaller effective diffusivity for chloride. This can improve the long-term corrosion resistance of reinforced concrete structures and make concrete denser and less permeable [19, 29].

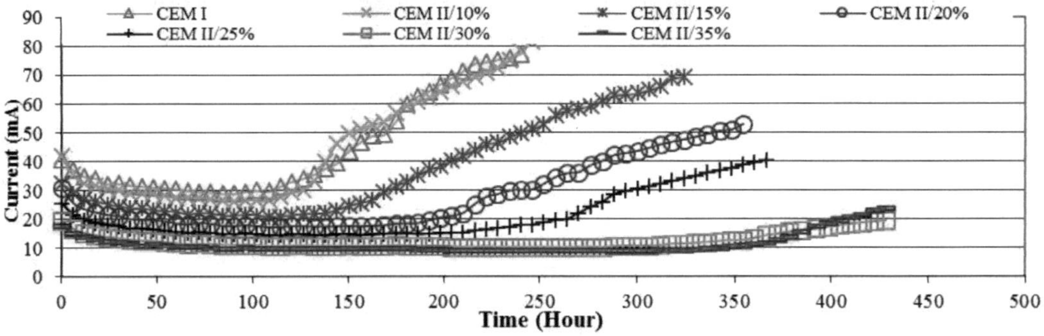

Figure 6. Typical curve of corrosion current versus time of concrete specimens tested after 90 days curing.

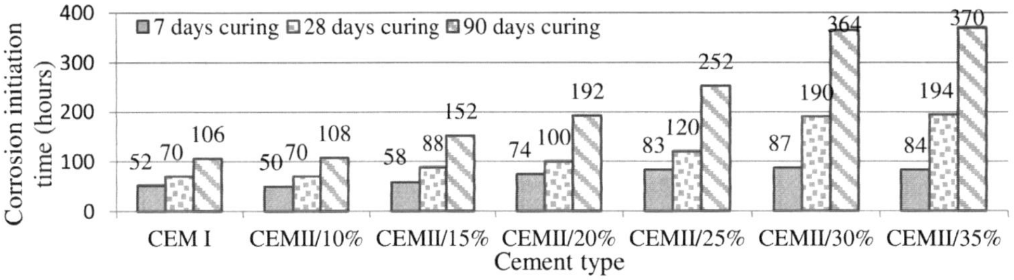

Figure 7. Corrosion initiation times of reinforced concrete specimens based on curing age.

3.4. Corrosion initiation time versus chloride penetrability

It was observed from figure 8 that the corrosion initiation time and chloride penetrability of concrete are closely related. The analysis results showed that there was an excellent correlation between corrosion initiation time and total charge passed through pozzolan-based cement concrete specimens, with a regression coefficient ($R^2 \approx 0.95$). According to Montgomery [30], a regression coefficient, R^2, of more than 0.85 indicates an excellent correlation between the fitted parameters. Increasing the total charge passed through concrete specimens reduces the time to initiate corrosion.

3.5. Correlation between compressive strength and durability-related properties

The rapid chloride penetrability and corrosion initiation times were plotted against the compressive strength in order to develop the relationship between these properties in pozzolan-based cement concretes (CEM II/P). These relationships are illustrated in figures 9-10. The relationships between compressive strength and chloride penetrability for each curing time are illustrated in figure 9. The high correlation coefficients of not less than 0.86 for all curing times indicated that there were significant linear correlations between compressive strength and RCP.

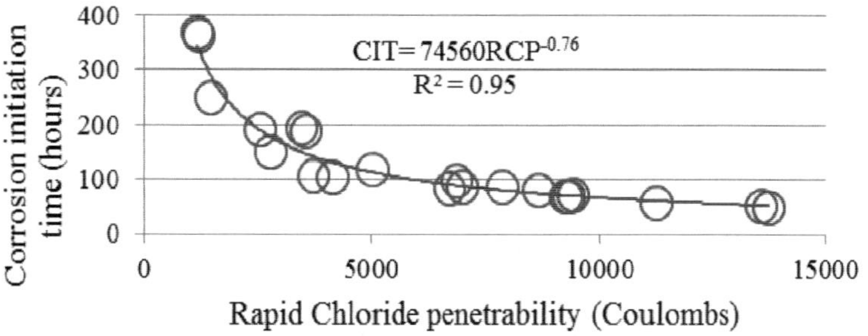

Figure 8. Variation of corrosion initiation time with rapid chloride penetrability.

Based on figure 10, there were good linear correlations between cubic compressive strength and corrosion initiation time for each curing time. It can be seen that after 90 days curing, the corrosion initiation time showed a higher rate of increment indicated by the steep slope of the curve, when compared to both 7 and 28 days curing. It can also be seen that the relationships between compressive strength and each of RPC and CIT are greatly affected by curing time. Contrary to expectations, the increase in strength did not increase the corrosion resistance, as clearly seen in figures 9 and 10. According to the authors, this could be explained as follows: a) firstly, this was in agreement with Neville [11] who indicated that the assumption that 'strong concrete is durable concrete' is not always true, b) secondly, as indicated in the paper, the lower early compressive strength of the blended cement could be attributed to the slow pozzolanic reaction [24], c) thirdly, the low permeability of the blended cement at early ages could mainly be attributed to the filling effect of pozzolan, which can be more finely grained than clinker when they are ground together [31].

The general correlation formulas between each chloride penetrability & corrosion initiation time and compressive strength can be expressed as follows, with regression coefficients (R^2) in brackets:

$$RCP = (-6.83t + 846)f_{cu} + 3.74t^2 - 366t - 4043 \qquad (R^2=0.96) \qquad (1)$$
$$CIT = (-0.26t - 4)f_{cu} + 12.62t + 101 \qquad (R^2=0.97) \qquad (2)$$

where RCP is the rapid chloride penetrability of concrete (in coulombs), CIT is the corrosion initiation time (in hours), f_{cu} is cubic compressive strength of concrete and t is curing age. Correlations between the measured values and the values calculated according to these equations for each of RCP and CIT are illustrated in figure 11 and figure 12, respectively.

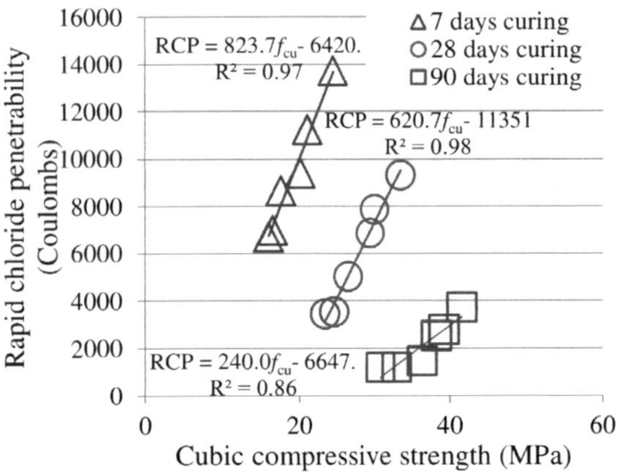

Figure 9. Correlation between RCP and compressive strength.

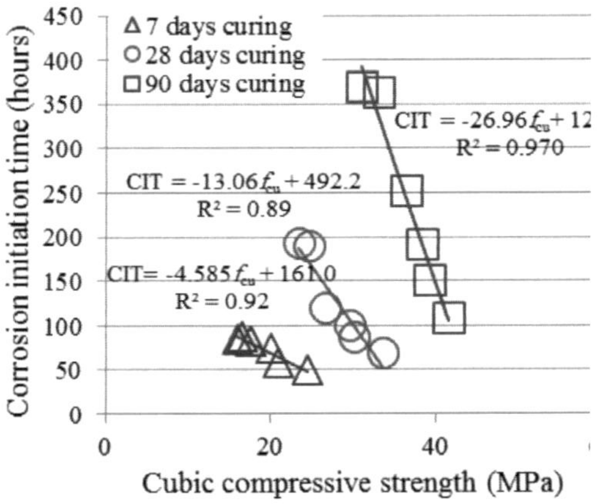

Figure 10. Correlation between CIT and compressive strength.

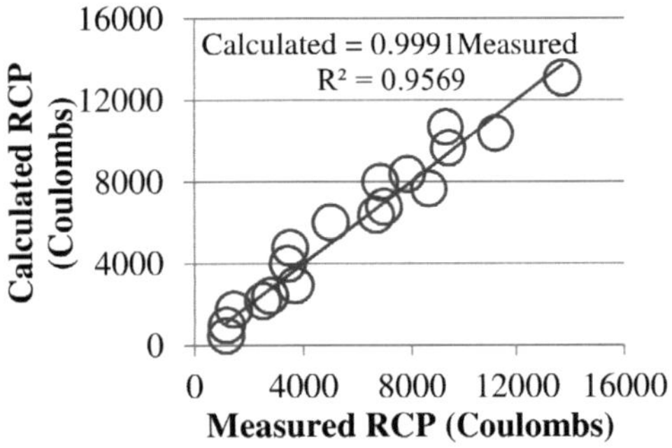

Figure 11. Correlation between measured and calculated RCP values.

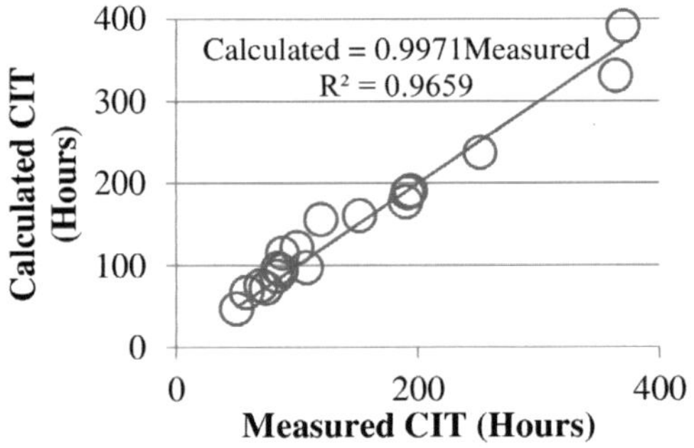

Figure 12. Correlation between measured and calculated CIT values.

4. Conclusion

From the experimental results, the following conclusions could be drawn:

-The cubic compressive strengths of concrete containing pozzolan-based cements were much lower than that of plain cement concrete at all early ages of concrete. However, at 90 days curing, the compressive strengths of pozzolan-based concretes were comparable to those of plain cement concrete.

-The chloride penetrability of pozzolan-based concrete mixes is much lower than that of plain concrete, especially at high replacement levels of pozzolan.

-According to the results of the accelerated corrosion test, concretes produced with pozzolan-based cements decelerated rebar corrosion. Particularly, CEM II/B-P cement types with 25, 30 and 35% pozzolan content were found to delay corrosion significantly. Use of pozzolan at 30% cement replacement level makes the service life of the RC structure under chloride-bearing environments longer twice or thrice when compared to the control.

-A definite correlation is observed between the rapid chloride penetrability test and the accelerated corrosion test, so that one can be estimated from the knowledge of the other.

-Curing has a large influence on both compressive strength and durability properties of pozzolan-based cement concrete. However, the consequences of short-term curing can be more serious for the latter.
-The development of good relationships between compressive strength and corrosion resistance properties of pozzolan-based cement concrete can be of considerable benefit. Compressive strength, which is easier to measure, can be reliability applied to predict some performance-related properties of concretes containing similar pozzolans, including rapid chloride penetrability and corrosion initiation time. However, this prediction can't be used for all pozzolan-based cement concretes.
-Based on the test results, it is suggested that pozzolan can be used up to 30% as a partial substitute for PC in production of blended cements. This addition ratio can reduce the quantity of CO_2 released by Syrian cement plants by about 1.5 million tonnes and the consumed energy by about 30% annually. So, production of green concrete could be promoted.

5. References

[1] Maslehuddin M, Saricimen H and Al-Mana A 1987 *ACI Materials J.* **84** 42–50
[2] Malhotra V M, Carette G G and Sivasundaram V 1994 *Advances in Concrete Technology* (CANMET, Ottawa) pp 925–991
[3] Kouloumbi N, Batis G and Pantazopulou P 1995 *Cem. and Conc. Aggreg.* **17** (1) 18–25
[4] Khan K I and Alhozaimy A M 2010 *Canadian J. of Civil Engineering* **38** 71–78
[5] Moufti M, Sabtan A, El-Mahdy O and Shehata W 2000 *Engin. Geology* **57** 155–162
[6] Walker R and Pavia S 2011 *Materials and Structures* **44** 1139 – 50
[7] Massazza F 2008 *Structure and Performance of Cements* edited by Bensted J and Barnes, Second edition (Taylor & Francis)
[8] *A Guide for Mineral Resources in Syria* 2011 The General Establishment of Geology and Mineral Resources in Syria
[9] *The General Establishment of Geology and Mineral Resources in Syria* Official document nr. (3207/T/9) dated 21.11.2007
[10] The general establishment for cement and building Materials [Available]: www.cemsyria.com
[11] Neville A M 2011 *Properties of Concrete* Fifth edition, Pearson Education
[12] Aitcin P C and Mindess S 2011 *Sustainability of Concrete* (Taylor & Francis)
[13] Rostam S 2006 *Durability of Materials and Structures in Building and Civil Engineering* edited by Yu C W and Bull J W (Taylor & Francis Group)
[14] Shi X, Xie N, Fortune K and Gong J 2012 *Constr. and Build. Mat.* **30** 125–138
[15] Horsakulthai V, Phiuvanna V S and Kaenbud W 2011 *Constr. and Build. Mat.* **25** 54–60
[16] Parande A K, Babu B R, Karthic M A, Kumaar D and Palaniswamy K K N 2008 *Const. and Build. Materials* **22** 127–134
[17] Ha T, Muralidharan S, Bae J, Ha Y, Lee H, Park K and Kim D 2007 *Build. and Environment* **42** 78–85
[18] Saraswathy V and Song H-W 2007 *Constr. and Build. Materials* **21** 1779–84
[19] Guneyisi E et al 2005 *Cement and Concrete Composites* **27** 449–461
[20] Rossignolo J A and Agesini M V C 2004 *Cement and Conc. Composites* **26** (4) 357–380
[21] Shaker F A et al 1997 *Cement and Concrete Research* **27** (5) 7711–20
[22] Khedr S A and Idriss A F 1995 ASCE *J. of Materials in Civil Engineering* **7** (2)
[23] Al-Tayyib A J and Al-Zahrani M M 1990 *ACI Materials J.* **87** (2) 108–113

[24] Mehta P K and Monteiro P J M 2006 *Concrete: Microstructure, properties, and Materials* Third edition (McGraw-Hill)
[25] Talbot C, Pigeon M, Maarchand M and Hornain J *Proc. of the Fifth Int. Conf. on the use of Fly Ash, Silica Fume, Slag, and Natural Pozzolana in Concrete*, edited by Malhotra V M Milwaukee, ACI SP 153, p 125
[26] Rukzon S and Chindaprasirt P 2009 *Int. J. Miner. Metal. Mater.* **16** (2) 242–247
[27] Chindaprasirt P, Chotithanorm P C, Cao H T and Sirivivatnanon V 2007 *Constr. and Build. Materials* **21** 356–361
[28] Gastaldini A L G, Isaia G C, Gomes N S and Sperb J E K 2007 *Cement and Conc. Composites* **21** 356–361
[29] Hossain K M A 2003 *Cement and Conc. Research* **33** 1601–05
[30] Montgomery D C and Peck E A 1982 *Introduction to Linear Regression Analysis* (New York: Wiley)
[31] Yetgin S and Cavdar A 2006 *J. Mater Civ. Eng.* ASCE **18** (6) 813–816

Acknowledgement
The authors gratefully acknowledge the technical and financial support of this research from the management of General Establishment for Cement & Building Materials/Adra Cement Plant. Thanks are also expressed to Chemist Nazeer Adarnaly (GECBM), and Prof. Tamer al-Hajeh, Vice-President of AIU for their appreciated help.

IOP Conf. Series: Materials Science and Engineering **96** (2015) 012008 doi:10.1088/1757-899X/96/1/012008

Study on Properties of Environment-friendly Concrete Containing Large Amount of Industrial by-products

H Fujiwara[1,2], **M Maruoka**[1], **C Sadayama**[1], **M Nemoto**[1], **K Yoshikawa**[1] **and M Yamaji**[1]

[1] Utsunomiya University, 7-1-2, Yoto, Utsunomiya, Tochigi, Japan

E-mail: fhiromi@cc.utsunomiya-u.ac.jp

Abstract. This study aims to reduce CO_2 discharged from the cement and concrete industries by effective use of industrial by-products, such as fly ash, blast furnace slag, and so on. In this paper, the properties of concrete containing large amount of industrial by-products and very small amount of alkaline activator including cement or sludge from ready mixed concrete plant are analyzed. As the result, it was confirmed that concretes containing large amount of industrial by-products can achieve sufficient compressive strength. However, these concretes showed poor frost resistance. It was thought that the reason was coarsening of air void system and this caused their poor frost resistance. Therefore, in order to micronize the air void system and improve frost resistance, the combination of air entraining agent and antifoaming agent was applied. By this method, it was confirmed that the frost resistance of some these concrete improved.

In this study, other properties of these concretes, such as fresh properties and other durability were evaluated and it was confirmed that these concretes show sufficient properties.

1. Introduction

In the recent years, a reduction in the amount of resources required for infrastructure construction has been demanded. Moreover, minimizing the life cycle cost of structures serves to reduce CO_2 emissions. The use of high-strength concrete allows increasing durability of structural components, thus leading to reductions in life cycle costs. However, when large amounts of cement are used in the production of high-strength concrete, the environmental load is high. In cement manufacturing, CO_2, which is a greenhouse gas, is discharged in large quantities. Among the sectors of the Japanese economy, the manufacturing industry in general, and the cement industry in particular, account for 46% and about 4% of CO_2 emissions, respectively. To manufacture 1000 kg of cement, it is supposed that the total amount of CO_2 emissions is 800 kg, including materials, fuel and other sources [1]. For this reason, if the use of Portland cement can be reduced through effective utilization of fly ash and blast furnace

[2] To whom any correspondence should be addressed.

slag, which are industrial by-products, reductions in CO_2 emissions can be realized.

Under these circumstances, some new cement compositions that are more environmentally suitable have been studied recently. One of them is the so-called alkaline cement; it is produced by alkali activating silico-aluminous materials. The research on the mixture of Metakaolin and calcium hydroxide shows that sodium aluminosilicate is the main reaction product. Additionally, the formation of C-S-H gel as a secondary product is also observed [2], [3]. Geopolymer is another secondary product [4]. Some experiments were performed on the blended cements containing 30% Portland cement and 70% fly ash. The powdery material was mixed with two different alkaline solutions for alkaline activation. The mechanical strength developed by this blended cement differed significantly depending on the hydrating solution used. All samples consisted of a mixture of amorphous gels (C-S-H + N-A-S-H gel).

In this research, Portland cement was either not used or the quantity used was greatly reduced, and instead large quantities of industrial by-products such as blast furnace slag and fly ash were used, with the objective of developing high strength concrete with a compressive strength of 60 N/mm^2 after standard curing for 28 days.

To achieve satisfactory performance of such concrete, alkaline activators must be added and their performance is very important.

In this paper, 3 types of calcium hydroxide, 2 types of Portland cement and concrete sludge from ready-mixed concrete plants were used as alkaline activators.

The hardened properties, the fresh properties, and the durability of these concrete were investigated. Low frost resistance was observed in the first test series, and in the second test series, poor frost resistance was improved.

2. Test series I: Fundamental properties of environment-friendly concrete produced using various alkaline activators

2.1. Outline of tests

2.1.1. Materials used.

The materials used are shown in table 1. The main materials were fly ash and blast furnace slag. Anhydrous gypsum was also used as an admixture in order to ensure early strength. Fly ash and blast furnace slag do not harden if alkalinity is not supplied. Therefore, 3 types of calcium hydroxide were used: ultra-fine porous calcium hydroxide (TK) with BET specific surface area of $46m^2/g$, special calcium hydroxide (CHs), and usual calcium hydroxide (CH1) with a purity and cost lower than CHs. The property values of calcium hydroxide are shown in table 2. Concrete sludge (SD) from ready-mixed concrete plants obtained by drying the sludge for 24 hours at 200°C and pulverizing to a fine powder was used as an alkaline activator. In addition, 2 types of cement were used as the most common alkaline activators: ordinary Portland cement (OPC) and a prototype cement (MPC).

IOP Conf. Series: Materials Science and Engineering **96** (2015) 012008 doi:10.1088/1757-899X/96/1/012008

Table 1. Materials.

Materials	Symbol	Type	Density (g/cm^3)
Binder	FA2	Fly ash Type 2 (JIS A 6201)	2.20
	BS4	Blast furnace slag (JIS A 6206, Specific surface area 4000 cm^2/g)	2.90
	AG	Anhydrous gypsum	2.90
Alkaline Activator	TK	Ultra-fine porous calcium hydroxide	2.24
	CHs	Special calcium hydroxide	2.24
	CH1	Type 1 calcium hydroxide	2.24
	SD	Ready-mixed concrete sludge	2.21
	OPC	Ordinary Portland cement (JIS A 5210, similar to CEM I 42.5N)	3.16
	MPC	Prototype cement	3.16
Water	W	Tap water	1.00
Sand	S	Crushed sand	2.63
Gravel	G	Crushed stone	2.64
Chemical Admixture	SP	High-range water reducing agent (Polycarboxylic acid ether)	1.08
	DF	Anti-foaming agent (Polyalkylene glycol conductor)	1.00

The prototype cement had a high quantity of C_3S (alite) compared with OPC, so it was expected to increase the strength. Table 3 shows the chemical compositions and Blaine values.

Table 2. Properties of calcium hydroxide.

Quality	TK	CH_S	CH1
BET specific surface area (cm^2/g)	469000	15000	12000
Average particle size (μm)	6.0	9.4	12.0
CaO (%)	73.4	73.9	Over 70

Table 3. Chemical compositions, Blaine values and mineral compositions of OPC and MPC.

Symbol	Ig-loss (%)	Chemical compositions (%)							Specific surface area (cm^2/g)	Mineral compositions (%)			
		SiO_2	Al_2O_3	Fe_2O_3	CaO	MgO	SO_3	TiO_2		C_3S	C_2S	C_3A	C_4AF
SD	--	--	--	--	--	--	--	--	7540	--	--	--	--
OPC	0.99	20.69	5.52	3.14	63.44	2.03	2.27	0.30	3330	52.97	19.54	9.32	9.56
MPC	1.06	19.24	5.37	3.42	64.82	1.91	2.80	0.28	4780	68.71	3.50	8.44	10.41

Table 4. Mixing conditions.

Slump flow (mm)	Air content (%)	W/(B+A)* (%)	S/(B+A)* (%)	Xv (%)
650±50	≦3.0	20.0	32.0	37.5

*B+A: Binder +Alkaline Activator

Table 5. Powder composition and mix proportions of concrete.

Blend No.	Alkaline activators	W/(B+A)* (%)	Mass ratio (%)				Unit content (kg/m^3)						
			FA2	BS4	AG	A.A**	W	FA2	BS4	AG	A.A**	S	G
1	TK	20	40	40	10	10	168	336	336	84	84	269	990
2	CHs						168	336	336	84	84	269	990
3	CH1						168	336	336	84	84	269	990
4	SD						168	336	336	84	84	269	990
5	OPC						171	342	342	86	86	274	990
6	MPC						171	342	342	86	86	274	990

* B+A: Binder +Alkaline Activator **A.A: Alkaline Activators

2.1.2. Test conditions.

The blending conditions are shown in table 4. The water powder ratio was 20%, the fine aggregate to powder ratio was 32%, and the coarse aggregate absolute volume ratio Xv was 37.5% with the reference to the findings from the previous study [5]. The quantity of high-range water reducing agent was adjusted to obtain a slump flow of 650±50 mm, and the quantity of anti-foaming agent was adjusted to achieve an air content of 3.0% or less. Table 5 shows the powder composition. The mixing proportions of fly ash, blast furnace slag, anhydrous gypsum, and alkaline activator were constant for all mixes.

2.1.3 Method of blending and mixing.

A biaxial forced mixing type of mixer with a nominal capacity of 90 liters was used for mixing the concrete. The method of blending and mixing was to first add the powder and aggregate and carry out dry mixing for 1 minute, then water and water reducing agent were added and mixing was carried out until the target fluidity was obtained.

2.1.4. Test items.

(1) Fresh properties

(a) Slump flow tests

The slump flow tests were carried out in accordance with JIS A 1150. Concrete fluidity was measured in order to evaluate the quantity of SP added to achieve the required slump flow.

(b) Air content tests

Air content tests were carried out in accordance with JIS A 1128

(2) Hardened properties

(a) Compressive strength tests

Compressive strength tests were carried out in accordance with JIS A 1108. The strength was measured at 7 and 28 days after curing in water at 20°C.

(b) Static elastic modulus tests

Static elastic modulus tests were carried out in accordance with JIS A 1149. Measurements were carried out at 28 days after curing in water at 20°C.

(3) Durability

(a) Drying shrinkage tests

Drying shrinkage tests were carried out in accordance with JIS A 1129-3. The test specimens were prismatic test specimens 10×10×40 cm. Measurements were started after curing for 7 days in water at 20°C. The drying conditions were temperature 20°C and 60% relative humidity.

(b) Accelerated carbonation tests
Accelerated carbonation tests were carried out in accordance with JIS A 1153.
(c) Freezing and thawing tests
Freezing and thawing tests were carried out in accordance with JIS A 1148.

2.2. Test results and discussion

2.2.1. Fresh properties.
The fresh property test results are shown in table 6. When TK and SD were used as alkaline activators, the quantity of SP added to obtain the required fluidity was increased in both cases. This is considered to be caused by the following reasons – in the case of TK the particles were extremely fine and the specific surface area was high, so a large quantity of free water and water reducing agent were adsorbed, and in the case of SD its water adsorption is high. In these tests, especially in the case of SD, the dilatancy property was observed. Dilatancy easily occurs in mixtures of powders with uniform particle size and could be a problem in the casting process. The quantity of air was in the range 1.4 to 2.7% as anti-foaming agent was used.

2.2.2. Hardened properties.
The results of the hardened property tests are shown in table 7. In the blends where TK, CHs, CH1, and SD were used as the alkaline activators, it was not possible to achieve 60 N/mm^2 at 28 days or higher. However, when cement was used as an alkaline activator, compressive strengths of about 70 N/mm^2 were obtained. In three cases where calcium hydroxide was used as the activator, the static elastic modulus obtained was slightly smaller.

Table 6. Fresh properties test results.

Blend No.	Alkaline activators	Slump flow (mm)	Air (%)
1	TK	658	1.5
2	CHs	643	1.5
3	CH1	645	1.4
4	SD	690	2.7
5	OPC	710	1.6
6	MPC	678	1.8

Table 7. Hardened properties test results.

Blend No.	Alkaline activator	Compressive strength (N/mm^2)		Static elastic modulus (kN/mm^2)
		7-day	28-day	
1	TK	40	56	35.5
2	CHs	39	53	35.7
3	CH1	34	49	32.0
4	SD	42	52	39.4
5	OPC	55	67	38.3
6	MPC	56	79	37.6

2.2.3. Durability.

(1) Drying shrinkage tests

The results of the drying shrinkage tests are shown in figure 1. In three cases in which calcium hydroxide (TK, CHs, CH1) was used, it was found that the amount of drying shrinkage tended to increase. In particular, it was found that for Blend No. 1, which used TK, the shrinkage was large. Drying shrinkage is caused by the surface tension effect acting between the particles of gel that constitute the cement paste, and the capillary tension effect acting on the meniscus formed by water between the particles. It is considered that by using TK the fine particle component of the powder is increased, so the number of fine voids increases, and hence the amount of shrinkage of Blend No. 1 increased. The amount of shrinkage was also small for Blend No.5 and 6, which used cement as alkaline activator. This is considered to be due to the high compressive strength and high static elastic modulus.

However, the amount of drying shrinkage of all blends was not so large comparing to usual concrete. Therefore, it is thought that it is not such a severe problem.

(2) Accelerated carbonation tests

The variation in carbonation depth is shown in figure 2. In the blend which used SD and OPC as activators, it was found that the progress of carbonation was fast. This is considered to be conditioned by the fact that the calcium hydroxide content was low compared with the other three blends. On the other hand, in the previous research [6], the carbonation depth was about 13 mm at 26 weeks for ordinary concrete with W/C = 55%. Therefore, it is considered that the carbonation resistance is about the same as that for ordinary concrete.

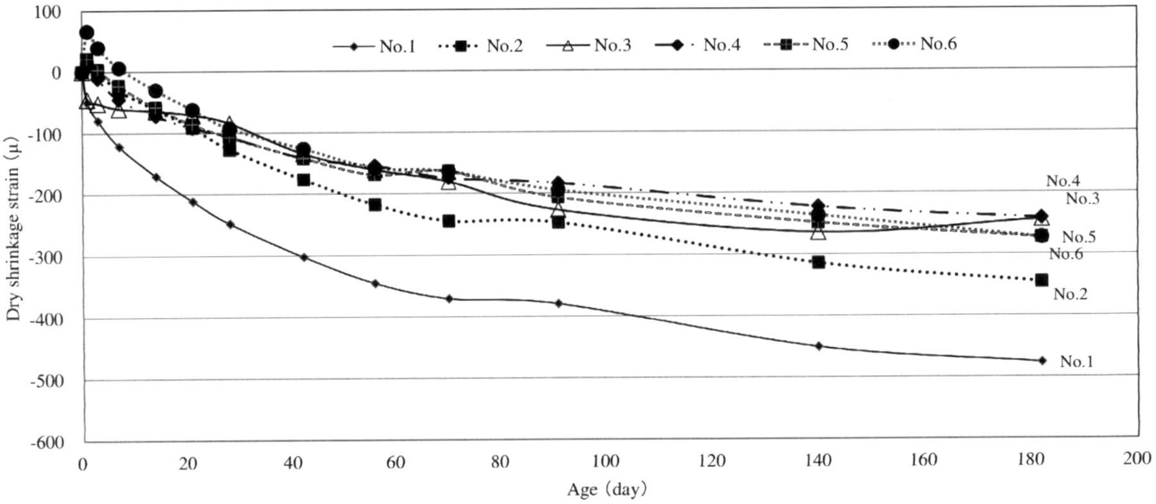

Figure 1. The tests results of drying shrinkage.

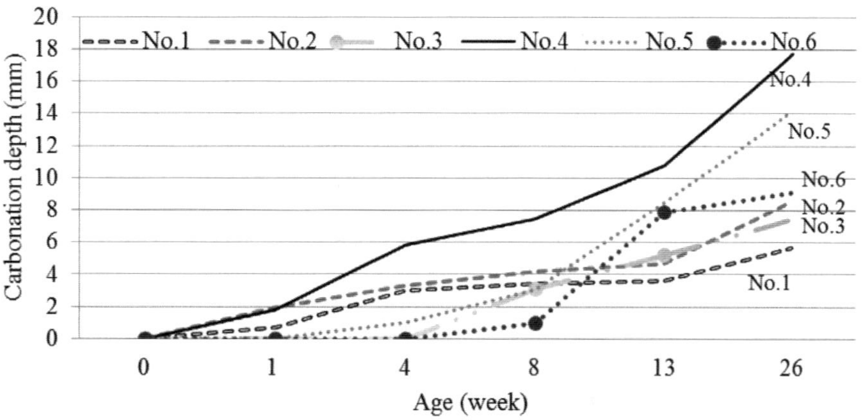

Figure 2. The tests results of carbonation depth.

(3) Freezing and thawing tests

The results of the freezing and thawing test are shown in figure 3. Blends with sufficient frost resistance could not be obtained for all blending conditions. In particular, for three blends that used calcium hydroxide (TK, CHs, CH1) as the alkaline activator the degradation was large after 30 cycles. Normally to increase frost resistance it is effective to entrain 3 to 6% entrained air using an AE agent. However, in these tests an anti-foaming agent was used, with the aim of obtaining high strength concrete with improved resistance to freezing damage. However, for Blend Nos. 5 and 6, which had 28 day strengths of about 70 N/mm^2, the results showed large degradation in frost resistance. Therefore, in the next tests an appropriate amount of air was entrained using AE agent.

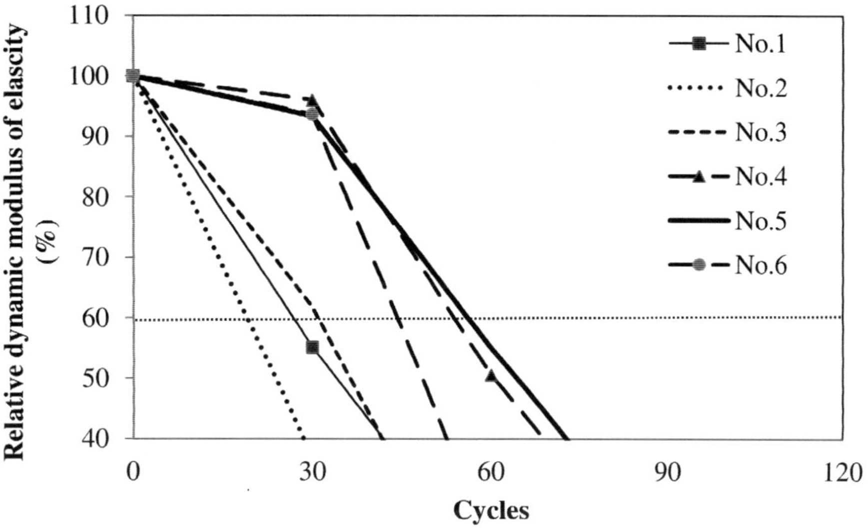

Figure 3. Frost resistance test results.

3. Test Series II: Tests to investigate the improvement in frost resistance in environment-friendly concrete

3.1. Outline of tests
In Test series II, the aim was to improve frost resistance by adding AE agent.

3.1.1. Materials used.
The materials used in this test are shown in table 1. An air entraining agent (AE) was used in addition to the materials used in Series I. And for alkaline activators, CHs, CH1, SD and OPC were used.

3.1.2. Test conditions.
The blending conditions are shown in table 8, and the powder composition is shown in table 9. In order to improve frost resistance, an appropriate quantity of AE agent was added to achieve an air quantity of 3.5±1.0% (All blend) and 4.5±1.0% (OPC blend only). Also, for the same reason the quantity of FA was reduced to 25%, the quantity of BS was increased to 50%, and the quantity of alkaline activator was increased to 15%.

3.1.3. Test items.
The test items of fresh properties were the same as for Test Series I, and compressive strength tests and freezing and thawing test were carried out in the same way as in Test Series I.

Table 8. Mixing conditions.

Slump flow (mm)	Air content (%)	W/(B+A)* (%)	S/(B+A)* (%)	Xv (%)
650±50	3.5±1.0 / 4.5±1.0**	20.0	32.0	37.5

*B+A: Binder +Alkaline Activator, **:OPC blend (Blend No.4-2) only.

Table 9. Powder composition and mix proportions of concrete.

Blend No.	Alkaline activators	W/(B+A)* (%)	Mass ratio (%)				Unit content (kg/m^3)						
			FA2	BS4	AG	A.A**	W	FA2	BS4	AG	A.A**	S	G
1	CHs	20	25	50	10	15	166	208	416	83	125	267	990
2	CH1						166	208	416	83	125	267	990
3	SD						166	208	416	83	125	267	990
4-1	OPC						171	214	428	86	128	274	990
4-2	OPC						168	211	422	84	127	271	990

* B+A: Binder +alkaline activator, **A.A: Alkaline Activators.

Table 10. Fresh property test results and compressive strength.

Blend No.	Alkaline activators	Slump flow (mm)	Air (%)	Compressive strength (N/mm^2)	
				7-day	28-day
1	CHs	495	3.5	43	51
2	CH1	365	3.5	41	54
3	SD	40*	3.4	47	73
4-1	OPC	700	4.0	58	69
4-2	OPC	600	5.2	47	62

*: slump value.

3.2. Test results and discussion

3.2.1. Fresh properties.

Fresh property test results are shown in table 10. The quantity of air was in the range from 3.6 to 5.5%.

3.2.2. Compressive strength.

Compressive strength test results are shown in table 10. For the blend using CHs and CH1, the compressive strength at 28 days was less than 60 N/mm^2, but the same or higher compressive strength was obtained from the other blends. It is considered that a higher compressive strength could be obtained compared with Test Series I by reducing the quantity of fly ash.

3.2.3. Frost resistance.

Frost resistance test results are shown in figure 4. In Test Series II, the quantity of air in the hardened specimens was increased in addition to reducing the quantity of FA. As a result, for the blends that used SD and OPC (Blend No.4-2) as the activator results were obtained that satisfied 300 cycles. For other blend, it was also found that frost resistance was somewhat improved.

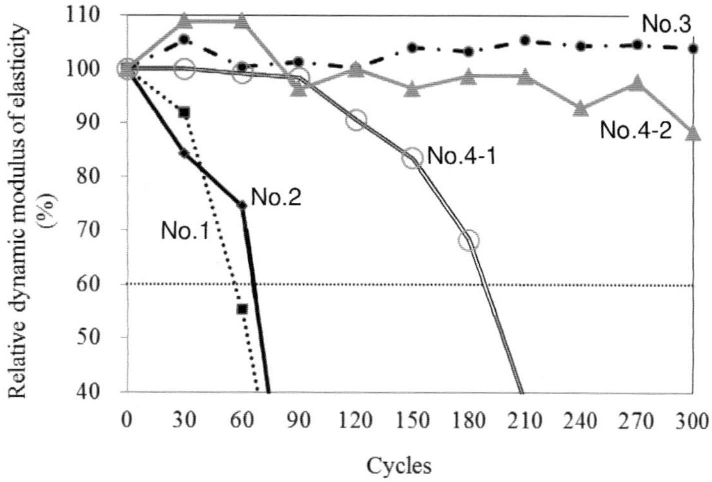

Figure 4. Frost resistance test results.

4. Conclusions

In this research, it was confirmed that concrete containing large amount of industrial by-products can achieve sufficient properties except for frost resistance. However, by combining air-entraining agent and anti-foaming agent adequately, it was confirmed that frost resistance of some these concretes is improved.

1. When TK and SD were used as alkaline activators, the quantity of SP added to obtain the required fluidity was increased in both cases. In the case of SD, the dilatancy property was observed and it could be a problem in the casting process.
2. In the blends where TK, CHs, CH1, and SD were used as alkaline activators, it was not possible to achieve 60 N/mm^2 at 28 days or higher. However, when cement was used as an alkaline activator, compressive strengths of about 70 N/mm^2 were obtained.
3. In three cases using calcium hydroxide, it was found that the amount of drying shrinkage tended to increase. Also, the amount of shrinkage was small for the cases when cement was used as the alkaline activator. The amount of drying shrinkage of all blends was not so large comparing to usual concrete.
4. In the blend which used SD and OPC as activators it was found that the progress of carbonation was fast compared with the other three blends. However, it is considered that the carbonation resistance is about the same as that of ordinary concrete.
5. In Test Series I, sufficient frost resistance could not be obtained for all blending conditions. In Test Series II, the quantity of air was increased in addition to reducing the quantity of FA. As a result, for the blends that used SD and OPC (Blend No.4-2) as the activator results were obtained that satisfied frost resistance requirements.

5. References

[1] Higuchi M 2002 Fundamental study on the environmental agent in environmental impact evaluation of concrete *Proc. of the Japan Concrete Institute* **24** (2) 1532 –36

[2] Alonso S and Palomo A 2001 Alkaline activation of metakaolin and calcium hydroxide mixtures: influence of temperature, activator concentration and solids ratio *Materials Letters* **47** (1-2) 55–62

[3] Palomo A, Fernandez-Jimenez A, Kovalchuk G, Ordonez L M and Naranjo M C 2007 OPC-fly ash cementitious systems: study of gel binders produced during alkaline hydration *J. Materials Science* **42** (9) 2958–66

[4] Shi C and Day R L 1993 Acceleration of strength gain of lime-natural pozzolan cements by thermal activation *Cement and Concrete Research* **23** (4) 824 – 832

[5] Fujiwara H, Maruoka M, Fujimura Y and Hiroshima A 2006 Study on effect of admixture on properties of ultra-high strength concrete *Cem. Science and Concrete Tech.* **60** 441 – 446

[6] Japan Cement Association 2011 Digest report of the technical committee on concrete *Japan Cement Association* pp 79 – 80

Acknowledgment

This work was supported by JSPS KAKENHI Grant Number 23566547.

Evaluation of Strength Characteristics of Laterized Concrete with Corn Cob Ash (CCA) Blended Cement

E E Ikponmwosa[1,2], **M A Salau**[1] **and W B Kaigama**[1]
[1]University of Lagos, Department of Civil and Environmental Engineering, Lagos, Nigeria

E-mail: efe_ewaen@yahoo.com

Abstract: Agricultural wastes are dumped in landfills or left on land in which they constitute nuisance. This study presents the results of investigation of strength characteristics of reinforced laterized concrete beams with cement partially replaced with corn cob (agricultural wastes) ash (CCA). Laterized concrete specimen of 25% laterite and 75% sharp sand were made by blending cement with corn cob ash at 0 to 40% in steps of 10%. A concrete mix ratio of 1:2:4 was used to cast 54 cubes of 150x150x150mm size and 54 beams of dimension 750x150x150mm.

The results show that the consistency and setting time of cement increased as the percentage replacement of cement with CCA increased while the workability and density of concrete decreased as the percentage of CCA increased. There was a decrease in compressive strength when laterite was introduced to the concrete from 25.04 to 22.96N/mm^2 after 28days and a continual reduction in strength when CCA was further added from 10% to 40% at steps of 10%. Generally, the beam specimens exhibited majorly shear failure with visible diagonal cracks extending from support points to the load points. The corresponding central deflection in beams, due to two points loading, increased as the laterite was added to the concrete mix but reduced and almost approaching that of the control as 10% CCA was added. The deflection then increased as the CCA content further increased to 20%, 30% and 40% in the mix. It was also noted that the deflection of all percentage replacement including 40% CCA is less than the standard recommended maximum deflection of the beam. The optimal flexural strength occurred with 10% CCA content.

1. Introduction

The high costs of procuring concrete materials for construction have over the years made the users to compromise quality, resulting in series of collapsed buildings and loss of properties in developing countries. Agricultural by-product pozzolans have been used in the manufacture and application of blended cements [1].

[2] To whom any correspondence should be addressed.

Corn cobs are wastes which are dumped in large quantities during corn season and they impose hazard to the environment in terms of their unpleasant odour and unsightly appearance in open dumps, located in most cities of Nigeria.

Adesanya and Raheem [2] studied the use of CCA blended cement produced in the controlled environment of a factory, to produce concrete specimens. Ettu et al. [3] studied the strength of binary blended cement composites containing corn cob ash. Both studies concluded that the compressive strength of the CCA-blended cement concrete is lower than that of plain concrete (the control) at early curing ages but improves significantly at later ages (after 90 days). Thus, it is important to look for ways of increasing the strength at early ages. Raheem et al. [4] studied the effects of admixtures on the properties of corn cob ash cement concrete. They opined in their results that the incorporation of admixtures in corn cob ash cement concrete generally increases its workability and compressive strength at all ages, irrespective of the type used. Adesanya and Raheem [5] studied the permeability and acid attack of corn cob ash blended cements and concluded that the resistance to chemical attack (HCl and H_2SO4) was improved with the addition of up to 15% CCA. Additionally, Binici, et al. [6] showed that an increase in ash content caused a significant increase in the sodium sulphate resistance of the concrete. This was further buttressed by Binici, et al. [7] that the resistant to sulphate attack was as a result of the more condensed physical structure of CCA compared to that of Portland cement which was shown by microscopic analysis.

Udoeyo et al. [8] investigated some characteristics of concrete containing laterite as partial or full replacement of sand aggregate. They opined in their results that concrete with up to 40% replacement level of sand with laterite could attain design strength of $20N/mm^2$. Hence, they concluded that it is possible to use laterite as replacement for sand in concrete up to 40% replacement level. Oluseyi [9] studied the influence of weather on the performance of laterized concrete. The simulation of weathering condition (wet & dry) was carried out by conditioning laterized concrete cubes to varying temperature and alternate wetting and drying. The results of the investigation showed that generally, the compressive strength of laterized concrete decreases when subjected to alternate wetting and drying. However, he concluded that laterized concrete with a laterite-granite fine ratio of between 40 and 60% conditioned to a temperature range of 75-125°C attained compressive strengths up to $22.52N/mm^2$. This is without the use of Portland cement and there was no addition of any pozzolan.

The current work aims at investigating the strength characteristics of laterized concrete with cement partially replaced by corn cob ash, in view of the fact that laterized concrete and corn cob ash have been established as cost-saving concrete and pozzolan respectively. Also, no known studies to our knowledge, is available on this study.

2. Experimental procedure and methodology

For the constituents of the laterized concrete, the cement was ordinary Portland cement with the properties conforming to the British Standard, BS12 with an average bulk density of about 1297.5 kg/m^3. The coarse aggregate was crushed granite, density of 2460 kg/m^3 with 12mm maximum diameter while the fine aggregate was made up of 75% sharp sand, obtained from an upland river bed and 25% laterite fine, dug from the ground around Ikorodu town, Lagos State, Nigeria. The samples of the aggregate were obtained in accordance with BS1377-1 while the tests on them, reported in table 1, were carried out in accordance with BS 1377-2. The water used was clean, portable and free from impurities. Corn Cob Ash (CCA) was obtained by burning the corn cob in an open drum. Chemical composition and other properties of the corn cob ash (CCA) and cement were determined, to ascertain the pozzolanic or cementitious properties of the materials respectively.

The percentage replacement of cement with corn cob ash was varied in steps of 10% up to 40%. A concrete mix ratio of 1:2:4 (cement/CCA : fine aggregate : coarse aggregate) by weight with water/cement ratio of 0.65 was adopted throughout this work. The normal concrete specimens, that is without laterite and corn cob ash, served as the first control while laterized concrete without CCA was used as the second control for the experiment. The test specimens were rectangular beams of 150 x 150mm in cross-section and 750mm long. Longitudinal reinforcement was provided in the beams. Specimens of each type of concrete mix were prepared in batches of three beams. Three 150mm cubes were also cast with each batch of the beam. The cubes were immersed in water for 7, 14 or 28 days, as appropriate, at the end of the period they were removed and allowed to drain for 2 hours, prior to testing. The beams were cured under damp sand for 3days and kept in the laboratory until the testing age of 45 days, 60 days and 90 days.

The beams were tested on a simply supported span of about 690mm under symmetrical two-point loading in increment of 5 kN using Avery Denison testing machine with maximum capacity of 8500 kN while the cube crushing tests were performed on ELG Budenberg compression machine with maximum capacity of 1560 kN. Dial deflection gauge was placed at mid-span of bottom face of the beam and readings were taken before and after every load increment. During testing, the development of crack, mode of failure, final failure load and central deflection at the ultimate load were noted and recorded for every specimen tested.

3. Results and discussion

3.1. Chemical composition of corn cob ash

Results of chemical composition tests carried out on corn cob ash (CCA) and cement are shown in table 1.

Table 1. Chemical composition of sieved corn cob ash (CCA) and ordinary portland cement (OPC).

Oxides (%)	CaO	SiO_2	Al_2O_3	Fe_2O_3	MgO	Na_2O	K_2O	SO_3	C	LOI
CCA	6.70	44.60	1.42	0.025	0.004	0.050	0.30	0.004	46.73	0.160
OPC	63.58	19.05	4.98	0.64	1.96	0.75	0.43	0.50	0.00	0.018

The results show that the corn cob ash has low CaO content and relatively high SiO_2 content, when compared with cement. Large quantity of carbon was recovered as residue from CCA and thus, CCA can be classified as a pozzlan, according to ASTM C618-12a.

The results obtained indicate that the CCA contained silicon dioxide content of 44.6% which is greater than the recommended minimum requirement of 25% according to BS EN 197-1(2000). The total silicon dioxide, iron oxide, and aluminium oxide (SiO_2+Fe_2O_3+Al_2O_3) content is 46.05%. This value is noted to be slightly below the code (ASTM C618–12a) requirement of 50% minimum for a class C pozzolan. The SO_3 content is 0.004 which is below the maximum value of 4.0% as specified for class C pozzolan; according to ASTM C618-12a. The loss on ignition (0.16) is also below the maximum allowable (6.0) specified. Thus, corn cob ash can be said to be a pozzolan of class C on the basis of its chemical composition. The high carbon content of about 47% could be considerably reduced if the process of combustion (incineration) is improved.

3.2. Physical properties of cement, corn cob ash and aggregates
The physical properties of different concrete materials are summarized in table 2.

Table 2. Summary of physical properties of cement, corn cob ash and aggregates.

Tests	Cement	Corn Cob Ash	Granite	Sharp Sand	Laterite
Moisture Content (%)	0.02	0.25	0.75	6.42	13.48
Water Absorption Capacity (%)	-	-	1.22	16.34	20.11
Specific Gravity	3.08	2.18	2.46	2.56	2.59
Dry Density (kg/m3)	3080	2180	2460	2560	2590
Bulk Density (kg/m3)	1298	923	2478	2724	2939
Coefficient of Uniformity (Cu)	-	-	1.51	3.07	4.55
Coefficient of Curvature (Cc)	-	-	1.07	1.31	1.86
Fineness Modulus (FM)	-	-	2.85	4.09	3.29

The moisture content of granite used was 0.75% while that of sharp sand and laterite were 6.42% and 13.48% respectively. These show that laterite was more saturated than sharp sand by about 7.06%. This moisture content was taken into consideration during the mix design.
The results of sieve analysis and other preliminary tests on the concrete constituent materials, as shown in table 2, indicate that all the fine and coarse aggregates are uniformly graded in accordance with ASTM D2487-11.

3.3. Effect of CCA on workability of concrete
For the consistencies, initial and final setting time tests for all percentage replacements of cement in concrete with CCA were carried out. It is observed that as the percentage replacement of cement with CCA increases, the initial setting time approached the final setting time. The initial setting time increased from 1 hour 33 minutes at control (0% CCA, 0% laterite) to 5 hours 21 minutes at 40% (40% CCA, 25% laterite) replacement level, while the final setting time increased from 2 hour 48 minutes at control (0% CCA, 0% laterite) to 6 hours 23 minutes at 40% (40% CCA, 25% laterite) replacement. However, BS EN 197–1:2000 requirement of 45 minutes minimum initial setting time and maximum of 10 hours final setting time, were met by all the percentage replacement levels. It is concluded that CCA has a retarding effect on concrete; hence, it could be used to delay/retard hydration process in pre-mix concrete requiring transportation to site as well as in arid environment.

The results of slump test carried out on fresh concrete showed that the control mix has the highest average workability value of about 87mm for the 3 batches and could be described as true slump; and the values gradually decreased to 79mm as soon as laterite was added to the mix, but still remained true slump. As corn cob ash was further introduced into the mix at steps of 10% up to 40%, the slump decreased sharply from 79mm to a minimum value of 42mm (for 40% CCA concrete); the type changing from true to shear slump. This could occur as a result of larger surface area, occasioned by the addition of CCA which demanded more water from the mixes; resulting in stiffer concrete with reduced workability

3.4. Effect of CCA on the density and compressive strength of concrete
It can be deduced from table 3 that the normal concrete (0% laterite, 0% CCA) has an average density of about 2445kg/m^3 for all curing ages and there was no significant change in density (from 2445kg/m^3 to 2455kg/m^3) when laterite was added to the mix. When CCA was further introduced to

the mix, density of the cubes decreased to a 2248kg/m^3 (about 8% reduction) for 25% laterite; 40% CCA concrete. This could be due to the fact that CCA has a lower specific gravity than ordinary Portland cement. However, all the density values are within the limit of normal weight concrete (2000-2600kg/m^3), in accordance to BS EN 206-1 (2000).

Table 3 shows that the compressive strength of all the replacements increased with increase in curing age. This is expected just like in normal concrete. Also, there was an insignificant decrease in compressive strength when laterite was introduced into the concrete mix from 19.26 to 17.63N/mm^2, 22.07 to 20.44N/mm^2 and 25.04 to 22.96N/mm^2 for 7 days, 14 days and 28 days respectively. There was a continuous reduction in strength when corn husk ash was further added to the mix from 0-40% at steps of 10% with 40% giving the least strength (8.59N/mm^2, 11.26N/mm^2 and 13.48N/mm^2 for 7 days, 14 days and 28 days respectively) for all the curing ages. The reduction in strength with the introduction of the ash is connected with the low content of CaO and the compound of (SiO_2 + Fe_2O_3 + Al_2O_3).

However, according to ASTM 39, the minimum required compressive strength of concrete to be used for structural purposes is 17N/mm^2. All the percentage replacement levels of cement with CCA, with the exception of 40% (CCA blended concrete) passed the limit. Therefore up to 30% CCA blended cement can be used to make concrete for structural works, depending on the requirements and function of the structure, while 40% CCA blended cement with compressive strength less than 17N/mm^2 can be recommended for non-structural works like blinding, mass concrete filling, etc.

Table 3. Summary of densities and compressive strengths for test samples.

		Percentage replacement of CCA						
	Curing Age (Days)	Normal Concrete	Latetrized Concrete	10% CCA (Laterized)	20% CCA (Laterized)	30% CCA (Laterized)	40% CCA (Laterized)
Density (kg/m^3)	7	2449	2473	2405	2389	2358	2277
	14	2455	2452	2431	2377	2345	2253
	28	2430	2440	2424	2391	2311	2215
	Average	2445	2455	2420	2386	2338	2248
	7	19.26	17.63	14.17	11.85	10.07	8.59
	14	22.07	20.44	18,52	17.33	14.96	11.26
	28	25.04	22.96	21.48	20.15	17.19	13.48

3.5. Strength activity index (SAI)

The strength activity index (SAI) is a measure of the pozzolancity of supplementary cementitious material (SCM) and is measured as the strength relative to the control (laterized concrete), in percent.

Table 4. Strength activity index (SAI) of CCA laterized concrete.

CCA Percentage Replacement of Cement & Sand	Compressive Strength (N/mm2) – SAI (%) Curing Age (Days)		
	7 Days	14 Days	28 Days
0% CCA; 25% Laterite	17.63 – 100	20.44 – 100	22.96 – 100
10% CCA; 25% Laterite	14.17 – 80.15	18.52 – 90.61	21.48 – 93.55
20% CCA; 25% Laterite	11.85 – 62.21	17.33 – 84.78	20.15 – 87.76
30% CCA; 25% Laterite	10.07 – 57.12	14.96 – 73.19	17.19 – 74.87
40% CCA; 25% Laterite	8.59 – 48.72	11.26 – 50.09	13.48 – 58.71

For SCM to be classified as pozzolan, the strength of the blended cement at 7-days and/or 28-days must not be less than 75% of the strength of the control concrete (ASTM C 618-08).

Table 4 shows that at 7 days curing age, lateritic concrete with 10% CCA met the 75% strength of normal laterized concrete with an SAI of about 80.15%. At 14 and 28 days curing ages, apart from lateritic concrete with 10% CCA, the 20% CCA lateritic concrete also met the 75% minimum permissible limit with SAI values of 84.78% and 87.76% at14 days and 28 days respectively. This indicates that laterized concrete and even normal concrete could accommodate CCA as pozzolan. It can be concluded that up to 20% CCA replacement of cement could be used for normal concrete production.

For 30% and 40% CCA replacements of cement in laterized concrete, the SAI for all the ages were below 75%. It was observed that there was no appreciable increase in the SAIs for these replacement levels with increase in age. Thus, 30% and 40% CCA replacement of cement are not recommended for use where strength is a criterion, but can be used for non-structural members like blinding, mass concrete filling, block filling etc.

3.6. Effect of CCA on the flexural strength of beam

Table 5 shows test results of steel reinforced concrete beam specimens under flexural loading with different percentages of CCA.

Table 5. Characteristics in flexure of steel reinforced CCA blended concrete beams.

Composition and Percentage (%) of CCA in Concrete	Beam Characteristics, Curing Age (Days): 45		60		90	
	Failure Load (kN)	Deflection (mm)	Failure Load (kN)	Deflection (mm)	Failure Load (kN)	Deflection (mm)
Normal Concrete (0% laterite, 0% CCA)	220	1.43	240	1.56	250	1.34
Laterize Concrete (25% laterite, 0%CCA)	190	1.73	220	1.94	230	1.59
Laterize Concrete (25%laterite, 10% CCA)	210	1.57	230	1.59	250	1.43
Laterize Concrete (25% laterite, 20% CCA)	180	1.52	210	1.68	220	1.63
Laterize Concrete (25% laterite, 30% CCA)	170	1.55	180	1.58	200	1.52
Laterize Concrete (25% laterite, 40% CCA)	170	1.67	180	1.70	200	1.69

Analysis of test results of the specimens showed that the flexural strength of the beams at 45-, 60- and 90-day curing ages vary with percentage of CCA in the laterized concrete.

For the normal reinforced concrete beams tested (0% CCA; 0% laterite) at 45, 60 and 90 days, table 5 shows that the failure loads are 220, 240 and 250 kN respectively. This indicates a gradual improvement in load bearing performance with curing age. On the addition of 25% Laterite to the concrete mix, the load carrying capacity of the beams decreased to 190, 220 and 230 at 45-, 60- and 90-day curing ages respectively, with corresponding higher values of deflection for identical load levels. The deflection of laterized concrete beam increased generally at all the curing ages.

For reinforced concrete beams having 25% Laterite fines and 10% - 40% CCA at steps of 10%, and cured at 45, 60 and 90 days, results of experiment showed that the failure loads for the beams decreased with increase in percentage of CCA in the mix for all curing ages. Relative to load sustained, the deflection values of the CCA-cement blended samples were quite high at early curing stages but improved considerably in load bearing capacity at an optimum replacement level of 10% addition of CCA as observed for 45, 60 and 90 days curing ages (table 5). The addition of corn cob ash considerably reduced deflection optimally at 10%. The result at 90 days curing age, when the strength became reasonably steady, further confirms the characteristics of a good pozzolan in the material as it helped the concrete to have considerable high strength at that age.

This suggests that the chemistry of interaction of the pozzolanic content of the ash served as an improvement on the load bearing capacity of concrete, irrespective of the addition of laterite. However, there was gradual increase in load bearing capacity, as well as reduced deflection as soon as cement was replaced with CCA at 10%. Addition of 20%, 30% and 40% CCA led to increase in deflection at lower load values for all curing ages. Thus, it can be deduced that 10% gave the optimal as compared to 20-40% replacement levels of cement with CCA.

3.7. Deflection characteristics of the beams.

Figure 1 shows a typical load–deflection curve for beam specimens with and without CCA content at 90-day curing age. This figure bears resemblance with the behaviour of beam specimens having the same composition of materials at 45- and 60-day curing ages.

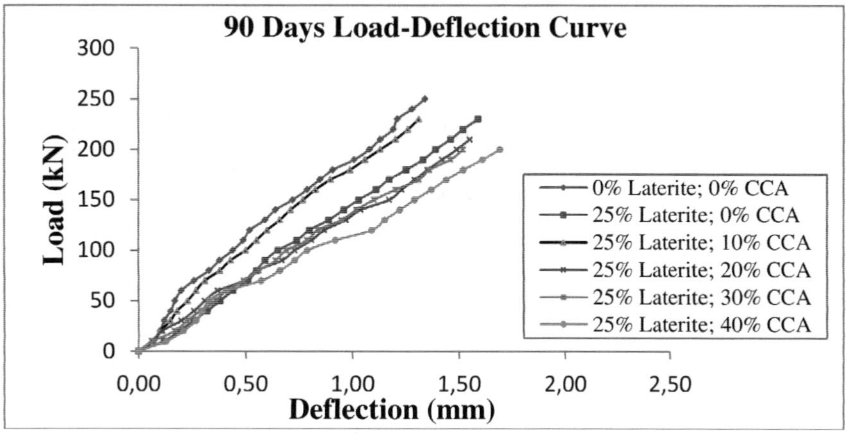

Figure 1. Summary of Load–Deflection Curve for 90 days curing age.

The deflection values were observed at the dial gauge, positioned mid-span of the beam specimens. The observed central deflection exhibited by the beams for different compositions of the materials in the concrete shows similarity in trends.

From Figure 1 above, it is generally observed that deflection increases with increase in CCA content in the concrete mixes irrespective of the curing age. In the same vein, the deflection in laterized concrete without CCA content was observed to be more than the deflection in the laterized concrete mix with 10% CCA content. Generally, for every load applied and at every curing age, plain concrete gave the least deflection before failure. Also, it was observed that for each load case and at predetermined value of loading, deflection reduces with curing age of concrete (which results in increase in strength of concrete), suggesting that the strength of concrete in the beam specimens helps to improve resistance of the beams to flexural loading and deformation.

3.8. Crack pattern and mode of failure

Figure 2. Shear compression failure of a typical beam specimen during flexural test.

Figure 2 shows a typical crack pattern and failure mode of the beams. The specimens had major diagonal shear failure cracks, extending from the supports to the load points in the 2-point load setup.

Largely, the cracks initiated from the tension face of the beams at the supports and propagated rapidly and diagonally toward the load points on the compression zone (compression face) as loading increased. Almost in all cases, the crack patterns were similar. The cracks were observed to widen as applied load increased. Generally, the failure mode of the beams was by shear as shown by the diagonal cracks and to a lower extent, by crushing of concrete on the compression face of the beams. This could be termed 'Shear-Compression Failure' and the cracks referred to as 'Shear-Flexural Cracks'.

4. Conclusion

From the results of the various tests performed, the following conclusions can be drawn:

1. Setting time and consistency of cement increases as percentage replacement of CCA increases. This signifies a retarding property of CCA.
2. Workability of fresh concrete reduces as percentage of percentage of CCA increases. Therefore more water will be required to get workable concrete mixes.
3. The compressive strength of laterized concrete with CCA at all replacement levels increases with increase in curing age which buttresses the fact that compressive strength increases with increase in curing age. Also there was a decrease in compressive strength when laterite was introduced to the concrete mix and a continual reduction in strength when CCA was further added to the mix.
4. The corresponding deflection due to two points loading increases as the laterite is added to the concrete mix but reduces and almost approaching that of the control as 10% CCA is added, before increasing as 20%, 30% and 40% CCA are further added to the concrete mix.
5. Concrete made with 10% CCA has the optimum flexural and compressive strength compared to other replacement levels of CCA and can thus be recommended for use in low cost housing construction.

5. References

[1] Malhotra V M and Mehta P K 2004 *Pozzolanic and Cementitious Materials* (London: Taylor & Francis)

[2] Adesanya D A and Raheem A A 2009 Development of corn cob ash blended cement *Construction and Building Materials* **23** 347–352

[3] Ettu L O, Anya, Arimanwa J I, Anyaogu L and Nwachukwu K C 2013 Strength of binary blended cement composites containing corn cob ash *Int. J. of Engineering Research and Development (IJERD)* **6** (10) 77–82

[4] Raheem A A, Oyebisi S O, Akintayo S O and Oyeniran M I 2010 Effects of admixtures on the properties of corn cob ash cement concrete *Leonardo Electronic J. of Practices and Techn.* **16** 13–20

[5] Adesanya D A and Raheem A A 2010 A study of the permeability and acid attack of corn cob ash blended cements *Construction and Building Materials* **24** (3) 403e409

[6] Binici H, Yucegok F, Aksogan O and Kaplan H 2008 Effect of corncob, wheat straw, and plane leaf ashes as mineral admixtures on concrete durability *J. of Materials in Civil Engineering* **20** (7) 478e484

[7] Binici H, Zengin H, Zengin G, Kaplan H and Yucegok F 2009 Resistance to sodium sulphate attack of plain and blended cement containing corncob ash and ground granulated blast furnace slag *Scientific Research and Essays* **4** (2) 098e106

[8] Udoeyo FF, Iron U and Odim O O 2006 Strength performance of laterized concrete construction and building materials *J. of Engineering Research* **20** (10) 1057–62

[9] Oluseyi L 2007 The Influence of Weather on the Performance of Laterized Concrete *J. of Engineering and Applied science* **2** (1) 129–135

IOP Conf. Series: Materials Science and Engineering **96** (2015) 012010 doi:10.1088/1757-899X/96/1/012010

Cellular Composites with Ambient and Autoclaved Type of Hardening with Application of Nanostructured Binder

V Nelyubova[1,2], N Pavlenko[1] and D Netsvet[1]
[1]Department of Materials Science and Technology, Belgorod State Technological University named after V.G. Shoukhov, Russia, Belgorod, Kostyukov Str., 46

E-mail: vvnelubova@gmail.com

Abstract. The research presents the dimensional and structural characteristics of nonhydrational hardening binders – nanostructured binders. Rational areas of their use in composites for construction purposes are given. The paper presents the results of the development of natural hardening foam concrete and aerated autoclaved concrete for thermal insulating and construction and thermal insulating purposes. Thus nanostructured binder (NB) in the composites was used as a primary binder and a high reactive modifier.

1. Introduction

Due to the accumulated more detailed knowledge of the structure and performance of natural objects at molecular scale, there is a visible tendency in development of basic approaches to producing and application of synthetic materials with nanosized structure.

In spite of the domination of Portland cement as a highly efficient hydraulic binder, it is recognized that its production process is based on the substantial consumption of raw materials and energy and that it asserts enormous pressure on the ecosphere. To reduce the negative impact on the environment, decrease of consumption and more efficient use of Portland cement in concrete are realized. These measures include the use of finely ground cement (FGC), low water demand binders (LWDB), blended cements based on fly ash, granulated blast-furnace slag, microsilica, metakaolin [1–5 and other]. In addition, the development of an alternative to Portland cement "green" eco-cements with fundamentally different mechanisms of hardening was proposed [6–9].

An actual trend, however, is the development of physical-chemical and technological principles to design new nonmetallic materials on the basis of nanotechnologies by oriented structure formation, optimization of physical and chemical principles in manufacture. The principle of structure formation of materials obtained with high concentrated binder system (HCBS) technology underlies the

[2] To whom any correspondence should be addressed.

conception of design of such a technology. In the given system it is nearly 10% of nanosized particles form as a result of mechanochemical activation of raw materials.

The binders studied in this work are produced by wet milling process at temperature 60–80°C at optimal pH level. All these conditions provide the milling process with critical concentration (max dilution) followed by suspension stabilization on the rheological principle, which is mechanical gravitation mixing. At these conditions, polydispersed granular composition as well as a low amount of bonded liquid is achieved. It is a determining factor if we speak about density (porosity), strength and shrinkage of the binder.Binding properties of the nanostructured binders (NB) are demonstrated when the dispersion medium is represented by inorganic acids, chlorides or sols. The latter are formed directly during the binder production process due to the interaction of the phases. In this case the system is usually modified by surface active agents and catalysts dissolved in the dispersion medium. Tweaking the acid-base characteristics of the solid phase of NB allows realizing different binding mechanisms. As a result, NB can be applied in the maximally wide range of building materials [10, 11 and other].

2. Experimental procedure

Particle size distribution in the materials was analyzed using the methods of laser granulometry. The microstructure of the samples was examined by means of scanning and transmission electron microscopy. Analysis of physical and mechanical characteristics of the finished materials was conducted using standard procedures described in the Russian normative literature.

In order to confirm the mechanism of hardening of the binder, chemisorption analyzer «Chemisorb 2750» connected to the gas quadrupole mass spectrometer «QMS-300» was used. The experiment aimed to compare the spectra of water after the temperature-programmed desorption (TPD) from the surface of the examined material.

The samples were preliminary burned in the muffle furnace at 500 °C for 3 hours in order to remove contaminants from the surface of the samples and water. Then, the samples were cleaned in the device «Chemisorb 2750» in TBD mode under the inert atmosphere (He as a carrier gas, flow rate – 20 ml/min) to a temperature of 500 °C at a heating rate of 10 °C/min. The samples were cleaned two times without cooling to the temperature below 100 °C in order to prevent adsorption of residual water from the carrier gas. For comparison, sand-based samples obtained by three different methods were taken: 1 – dry grinding of sand; 2 – sand ground in water and 3 – nanostructured binder (NB).

All studies were conducted using laboratory and analytical base of the Center of High Technologies of Belgorod State Technological University named after V.G. Shoukhov.

3. Features of nanostructured binder

It is known that hardening of binders is based on transition of silanol bonds to silaksan. However, conformation of these processes was impossible up to the present moment. To confirm the obtained data, the samples of binder were examined for formation of bonds on the surface of the comminuted materials using hemisorption analyzer.

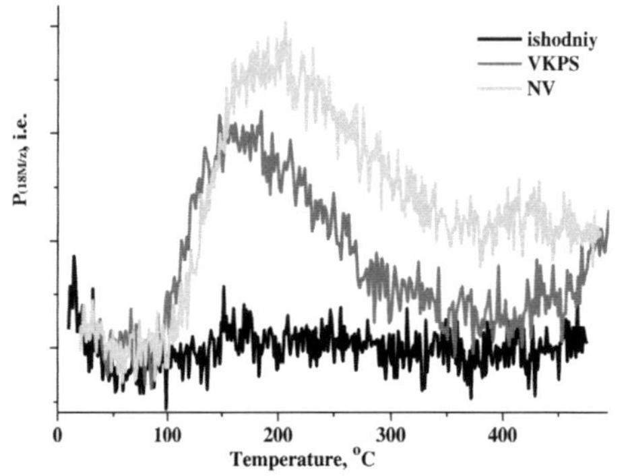

Figure 1. Mass spectra of thermoprogrammed desorption of water obtained from the samples of sand ground by dry (1) and wet method (2), NB (3).

There are 2 forms of desorbed water: lightly bound – with desorption peak 90-360 °C and tightly bound – with peaks in the region of 370 °C. Presence of lightly bound – physically sorbed water – is due to sorption of water by the surface functional groups, tightly bound water is formed by condensation of these functional groups. The intensity of two peaks is stronger for NB, and that indicates a higher content of functional silanol groups on the surface of the sample. Formation of functional groups on the surface of the samples leads to reactivity not only with water but also with other chemical compounds.

Investigation of nano-structure of the NB demonstrated the presence of nano-sized silica components (figures 2, 3). In order to eliminate the contribution of micron- and submicron-sized fractions of the polydisperse binder, the suspension of NB was centrifuged to exclude particles larger than 200 nm.

Centrifugation was carried out for 20 minutes at a speed of 5000 rpm. Next, the resulting centrifugate was analyzed by the dynamic light scattering using a UV-laser analyzer Malvern Zetasizer ZS. The parameters of the colloidal solution were selected from the manufacturer's database for the silica and water as a solvent. The calculation of the distribution of colloidal particles was performed using the manufacturer's software. Figure 2 shows particle size distribution of the centrifuged NB. The main distribution peak corresponds to the particle size of ~ 30 nm (figure 2).

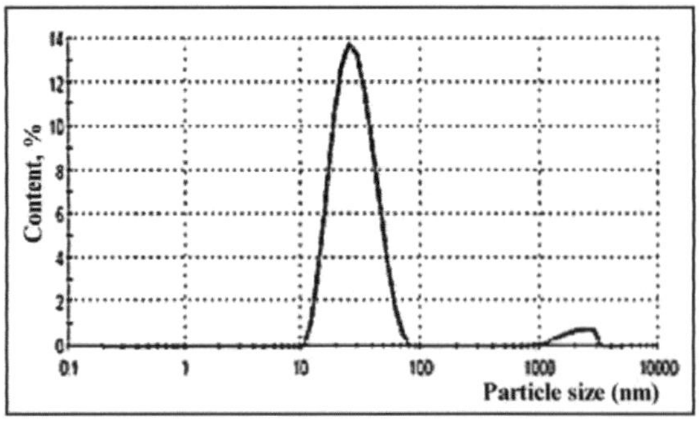

Figure 2. Particle size distribution of NB centrifugate.

The microstructure of NB is characterized by pseudo-uniform distribution of polydisperse particles of a size range from 10 nm to 100 nm of polyhedral morphology within the dispersion medium (figure 3).

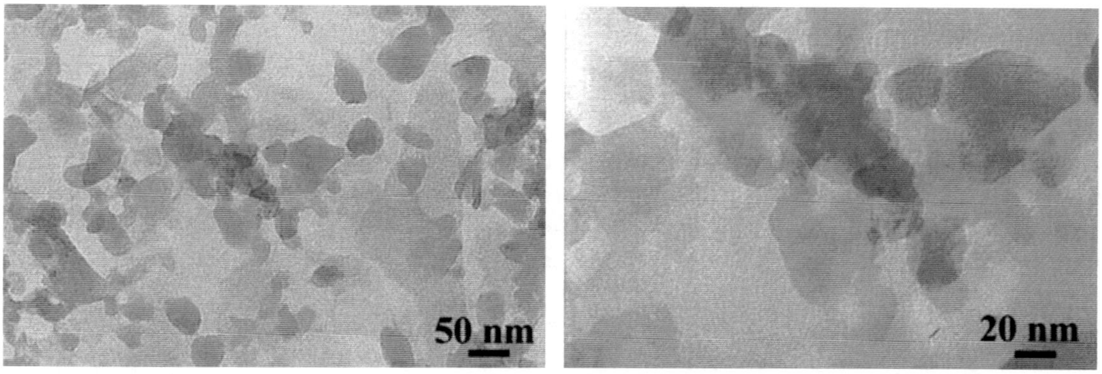

Figure 3. Morphology of nano-dispersed fraction of NB based on silica sand under SEM.

Polydispersity was estimated by an independent method, based on the analysis of the scattering intensity of the mixture of different types of non-interacting particles (the software MIXTURE - IR RAS). The differential particle size distribution curves (normalized to the unity) are shown in figure 4. These binders are characterized by their bimodal distributions with a pronounced peak in the nano-scale region. The NB curve is shifted towards smaller sizes when compared to conventional high concentrated cementing suspension. Maxima of the distributions correspond to the values obtained for the average diameter of spherical nanoparticles.

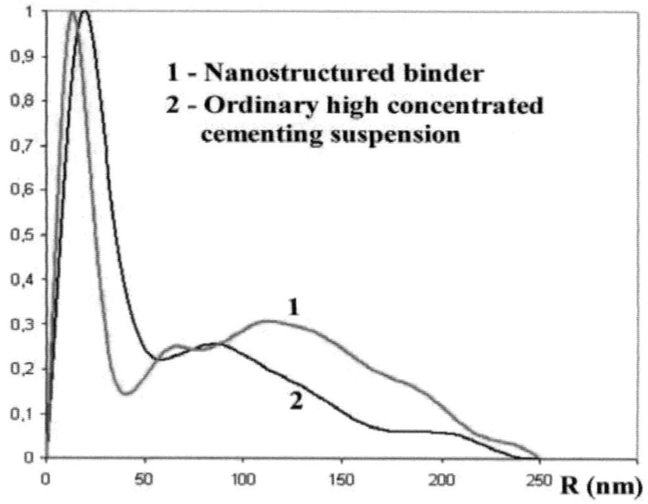

Figure 4. Particle Size Distributions of NB and Reference UDB.

IOP Conf. Series: Materials Science and Engineering **96** (2015) 012010 doi:10.1088/1757-899X/96/1/012010

4. The application of nano-binders in cellular concrete

It became possible to use nanostructured binder as an active silica modifier and the main binder component for production of varying construction materials (figure 5).

The specificity of NB technology allows using a wide range of natural and industrial silica and aluminosilicate rocks as a base raw component. The choice of any given material can be made in view of the regional location of the minerals used, where the binder production is planned. That would allow organization of construction materials production on the NB base in remote regions relative to cement factories.

The main advantages of the nanostructured binder are low cost and high workability. The low cost is conditioned by availability and widespread distribution of raw materials for its manufacturing and, as a result, by reduced transportation costs, no energy costs are incurred for high temperature treatment of raw materials during binder production, it is also characterized by unlimited stock life. High workability occurs due to unique binder properties at low water demand and wide range of exploitation temperatures.

It was established that the application of the nanostructured binder as a modifying agent for autoclaved gas concretes results in improvement of macro- and microstructure of gas concrete before the autoclave treatment and intensification of phase formation under hydrothermal conditions. NB optimizes rheotehnologycal properties of cellular concrete mix, which provides a smooth intense process of gas evolution by increase of the mixture volume, reduction of swelling time, decrease of the thickness of interporous partitions while maintaining the required strength characteristics of the finished products. As a result of research compositions of autoclaved gas concrete for thermal insulating and construction and thermal insulating purposes with density brands D350-D500 and classes for strength B0.75, B2.5 – B5.5 were proposed.

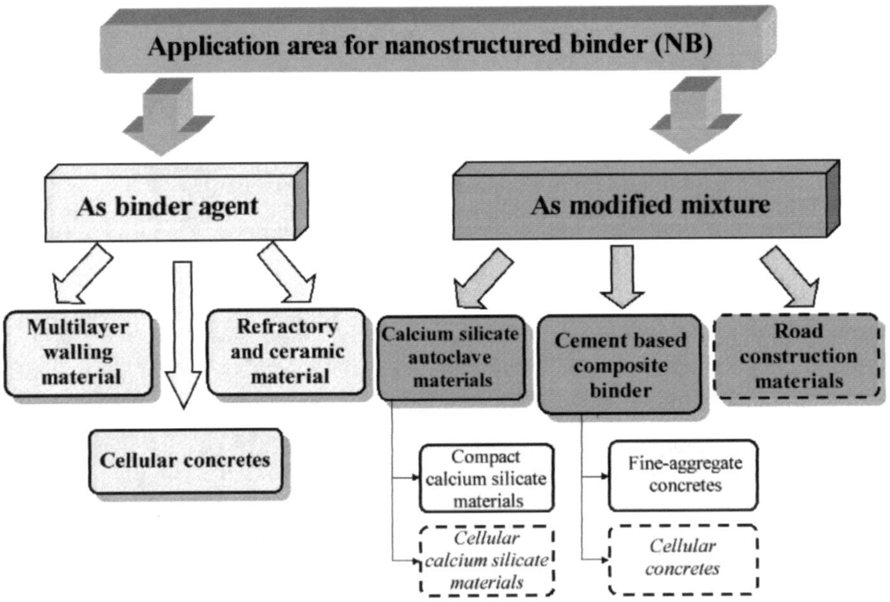

Figure 5. Applications of nanostructured binder.

The specificity of the nanostructured binders allows recommending them as a basic binder for the production of thermal insulating and construction and thermal insulating cellular nonautoclaved concretes obtained by foaming. Foam concrete on the basis of NB has density of 300-900 kg/m^3, compressive strength of 3–12 MPa, coefficient of thermal conductivity of 0.008–0.012 W/(m·K), coefficient of water vapor permeability of 0,23–0,14 mg/(m·h·Pa), moisture sorption of 6–10 mass.%, respectively. The use of NB in the manufacture of foam concrete allows improving technological and economic efficiency of the process due to a significant reduction of manufacturing time of foam concrete products with improved technical performance and thermal characteristics.

Application of NB allows obtaining materials with optimum cellular structure, which is characterized by evenly distributed, polydispersed, locked, deformed in the correct polyhedrons porous, with a glossy surface of the near-porous layer, divided with thin dense interporous partitions-walls. Reduction of porosity of an interporous partition-wall is proved by existence of nanodispersed particles in NB and in moulding systems on its basis. When using NB, the particles of the minimum sizes are located in the gaps between relatively large particles of the matrix system, and that promotes creation of a thin film of the binder on the surface of an air bubble.

5. Conclusion

Taking into consideration the development and rapid growth of monolithic and low cottage construction, cellular concretes are a perspective material for creation of effective walling. It has been revealed that the optimum grain structure, high polydispersion of a solid phase, rheological properties of the NB have a positive effect on formation of a rational cellular structure of heat-insulating concrete and its technical and operational characteristics.

Usage of a nanostructured binder when producing lightweight heat-insulating cellular composites allows carrying out the directed formation of the structure of cellular concrete. Obtaining a rational porous structure of heat-insulating materials is a fundamental task in production of cellular concrete. It influences functional indicators of the product quality, provides decrease in heat conductivity of the final material without considerable reduction of its strength characteristics.

Thus, with regard to the development and application of the new type of nanostructured binder and manufacturing technology in construction materials production, it is possible to reduce energy consumption for production of synthetic composites significantly, i.e. to produce raw mixes with qualitatively new energy requirements, which will provide objective conditions for the introduction of nanomaterials in industrial and civil engineering.

6. References

[1] Kavitha O R, Shanthi V M G, Arulraj P and Sivakumar P 2015 Fresh, micro- and macrolevel studies of metakaolin blended self-compacting concrete *Applied Clay Science* **114**

[2] Bohac M, Palou M, Novotny R, Masilko J et al 2014 Investigation on early hydration of ternary Portland cement-blast-furnace slag–metakaolin blends *Construction and Building Materials J.* **64**

[3] Shebl S S, Seddeq H S, Aglan H A 2011 Effect of micro-silica loading on the mechanical and acoustic properties of cement pastes *Construction and Building Materials J.* **25**

[4] Melo J P, Aguilar A S, Olivares F H 2014 Rheological properties of aerated cement pastes with fly ash, metakaolin and sepiolite additions *Construction and Building Materials J.* **65**

[5] Vance K, Aguayo M, Oey T, Sent G, Neithalath N 2013 Hydration and strength development in ternary portland cement blends containing limestone and fly ash or metakaolin *Cement and Concrete Composite J.* **39**

[6] Yung V N 1951 Fundamentals of technology of binding materials (Moscow: Promstroiizdat) p 540

[7] Sobolev K, Ferrada-Gutiérrez M 2005 *How Nanotechnology Can Change the Concrete World: Part 1*, American Ceramic Society Bulletin 10

[8] Sanchez F, Sobolev K 2010 Nanotechnology in concrete – A review *Construction and Building Materials* **11**

[9] Pivinskii Y E 1990 Ceramic binders and ceramobetons (Moscow: Metallurgiya) p 270

[10] Strokova V V, Cherevatova A V, Pavlenko N V and Nelubova V V 2013 Prospects of Application of Zero-Cement Binders of a Nonhydration Hardening Type *World Applied Sciences J.* **25** (1)

[11] Pavlenko N V, Strokova V V, Gernovskii I V, Cherevatova A V, Nelubova V V, Koshukhova M I and Sobolev K G 2013 The application of nanostructured zero-cement binders in fire-resistant cellular concretes *5th Int. Conf. Nanocon (Czech Republic, Brno, 2013)*

IOP Conf. Series: Materials Science and Engineering **96** (2015) 012011 doi:10.1088/1757-899X/96/1/012011

Pore Distribution and Water Uptake in a Cenosphere–Cement Paste Composite Material

J Baronins[1], J Setina[2], G Sahmenko[3,6], S Lagzdina[4] and A Shishkin[5]
[1]Tallinn University of Technology, Department of Materials Engineering, Ehitajate tee 5, 19086 Tallinn, Estonia
[2]Riga Technical University, Institute of Silicate Materials, Paula Valdena 3, Riga, LV-1048, Latvia
[3]Riga Technical University, Institute of Materials and Structures, Kalku Str. 1, Riga, LV-1658, Latvia
[4]Riga Technical University, Testing Laboratory of Silicate Materials, Azenes Str. 14/24, Riga, LV-1048, Latvia
[5]Riga Technical University, Laboratory of Powder Materials, Azenes Str. 16/20, Riga, LV-1048, Latvia

E-mail: genadijs.sahmenko@rtu.lv

Abstract. Alumina silicate cenospheres (CS) is a significant waste material from power plants that use a coal. Use CS as Portland cement replacement material gives opportunity to control physical and mechanical properties and makes a product lighter and more cost-effective. In the frame of this study, Portland cement paste samples were produced by adding CS in the concentration range from 0 to 40 volume %. Water uptake of hardened samples was checked and pore size distribution by using the mercury porosimetry was determined. In a cold climate where the temperature often falls below 0 °C, it is important to avoid the amount of micrometer sized pores in the final structure and to decrease water absorption capacity of material. In winter conditions, water fills such pores and causes additional stresses to their walls by expansion while freezing. It was found that generally water uptake capacity for cement paste samples decreased up to 20 % by increasing the concentration of CS up to 40 volume %, at the same time, the volume of micrometer sized opened pores increases.

1. Introduction

Coal combustion ashes are one of the key materials in research and practice related to concrete [1]. Cenospheres (CS) are fine spherical particles (microspheres, micro-balloons), formed in coal

---

[6] To whom any correspondence should be addressed.

combustion process in a furnace. Such property of CS like bulk density in the range between 300 to 700 kg/m^3 allows for easy design and building of simple furnaces with separators for different ashes by using the effect of gravity. Therefore, it is not necessary to ensure additional cleaning of CS [2], [3]. An outer diameter of CS may vary from ~1 to 500 μm [4]. Types of CS are classified depending on their chemical composition and raw material. The chemical and mineralogical composition can change the colour of CS from brown to grey or even black, depending on concentration of unburned carbon. According to standard ASTM C618, depending on raw coal, CS is divided into class F (made by burning anthracite and other older coals) and class C (from burning of brown lignite or half-bituminous coals). As required in the standard, content of SiO_2 + Al_2O_3 + Fe_2O_3 for class F (70 %) and class C (50 %) plays the main role in activity of CS, exhibiting a higher pozzolanic activity for class C [5]. Additionally to the type of raw materials, processing temperature, flow rate and other factors in the furnace influence the result regarding the size, type and density of the final spherical ash particles. The models of such particles can be: a) spherical ash particles with an outer diameter, b) a shell type single wall spherical particles (CS), filled with gas, c) a shell type single wall spherical particles filled with smaller particles inside (plerospheres), d) spherical particles with a complex network structure inside [6], [7].

Popular way for utilization of CS is its use in production of lightweight constructions of Portland cement mortar and concrete [8]. Due to the fact that particle sizes of CS can be similar or even smaller than particle sizes of Portland cement and sand, their spherical shape can also reduce the amount and size of the opened pores. Additionally, pozzolonic effect consumes free $Ca(OH)_2$ from hydrated Portland cement by filling pore volume between particle surfaces of CS and Portland cement with insoluble silicate hydrates. All these effects can allow producing a concrete or cementing mortar with less permeability to various gases, vapour and fluids [2], [9]–[11].

As observed by Barbare, Shukla, and Bose in their work [12], the equilibrium moisture content of cenospheres used in their work is about 18 times higher than that of sand, reflecting the porous nature of cenospheres. The flux of water vapour away from both the cenosphere–concrete and the normal concrete shows nonlinear change with moisture content throughout the drying cycle, implying that the pore structure within the concrete strongly influences the drying behaviour. Sorption usually takes place in case of contact of two phases, but sorbents (solid phase) link gases or liquids. In case of dry hardened Portland cement paste immersed into water, chemisorptions occur in addition to physical absorption by forming crystal hydrates. Due to great variety of pores in a Portland cement paste structure, depending on the sizes of pores and related frost resistivity, these are divided into a) small pores (or gel pores) with sizes below 0.04 millimetres and freezing point of water at – 50 °C; b) average size of pores with the diameter between 0.04 to 0.3 mm (micrometer sized or capillary pores) with the water freezing point close to 0 °C and can cause formation of new cracks; c) pores larger than 0.3 mm that are not filled with water fully and do not significantly cause the formation of new cracks [13], [14]. Depending on different experience, scientific medium, deepness of research work and other factors, in different reviews and research works, different pore classifications with similar (but not equal) assessment of pore impact on water absorption capacity and frost resistance of Portland cement pastes, concrete and mortars are considered and proposed [15]–[17].

2. Materials and methods

The Aalborg White CEM I 52.5R cement (Germany) and the mullite containing CS from electric power station "*Troitskaja*" (Russian Federation), supplied by *Biotecha Latvia* (Latvia) were used. The chemical composition and size distribution (determined by using CONTROLS Model 15-D0407 with

Glenammer sieves) are shown in table 1 and table 2. Cement paste samples with water to cement ratio of 0.4 were prepared by mixing fresh cement paste using a laboratory mixer, filled into concrete moulds and vibrated for 10 seconds on the vibration table. Mix compositions are summarised in table 3.

Table 1. Chemical composition of mullite containing cenospheres.

Compound	Mass %
SiO_2	55.8
Al_2O_3	37.9
Fe_2O_3	1.5
CaO	1.9
MgO	0.9
Na_2O	0.7
K_2O	0.3
Loss of ignition – LOI	
20-400 C°	0.5
400-1000 C°	0.1

Table 2. Sieve analysis of used cenospheres CS5 (BS/ISO 3310).

Sieve opening size, mm	Residuals, %
0	0
0.063	0.8
0.075	1.5
0.150	1.9
0.3	95.8
0.6	0

Table 3. Mix design of cement pastes (for production of $1m^3$ cement paste).

Material	Portland cement, kg	Cenospheres CS5, kg	Water, kg
CS5-CP-0	1370	0	545
CS5-CP-5	1296	20	518
CS5-CP-10	1226	41	490
CS5-CP-20	1087	81	435
CS5-CP-30	947	122	379
CS5-CP-40	808	162	323

Cylindrical shape samples with the diameter of 45.5 ± 0.2 mm and height of 41.2 ± 0.3 mm were made for test. After 28 days hardening in water, samples were dried for 24h at 80 ± 5°C by using drying oven *MEMMERT ULE 400* until the constant mass and cooled down to room temperature in desiccators. As observed in this test, application of higher temperature for drying is limited for this material due to formation of new cracks by possible decomposition of hydrated compounds. Densities of dried specimens were recorded before the test (see table 4). Dried samples were fully immersed into a container filled with water. Depth of the water was 45 mm above the top of the samples, while the samples were totally immersed in water. Two plastic wires with the diameter of 2 mm were placed under the samples to ensure maximum contact area with water. Depth of water was controlled and adjusted during every measurement test. Test was executed at 20 ± 5 °C. During the first 60 minutes of the test, the weight of samples was recorded by using scale *Kern EW600-2M* (±0,01g) every 10 minutes. After 60 minutes of test, the weight of samples was recorded every 30 minutes. After 240 minutes, the weight of samples was recorded every 1440 minutes (24 h). Test duration was 12960 min (9 days). Tests were repeated 3 times.

Content of absorbed water W_{Abs} was calculated according to the following equation:

$$\frac{\Delta M}{M_0}(\%) = \frac{M_w - M_0}{M_0} \cdot 100 \quad (1)$$

Where,

M_w – Mass of the sample with water;
M_0 – Mass of the dried sample;
ΔM – Increase in weight of the sample, %.

Pore size distribution and total opened pore volume was determined for 28 days cured samples by using the mercury porosimetry with a porosimeter *Pore Master 33 Quantachrome Instruments* according to ISO 2738.

An important note in this work is that the mercury porosimetry is useful only for determination of a limited range of pore volume (10^{-2} to 950 μm) and there is no reaction between mercury and cement paste samples.

3. Results and discussions

3.1. The influence of concentration of CS on density and water uptake behaviour of cement pastes

Average densities of dried samples are summarised in table 4. Replacing 40% of Portland cement by cenospheres decreases material density up to 23% (from 1.79 to 1.45 T/m^3). As shown in figure 1, all samples filled with CS in the first 10 minutes have uptake more than 50% of total absorbed water in this test. The rate of water uptake for cement paste samples CS5-CP-0 has changed the curve from vertical to horizontal only after 20 minutes from the start moment of the test.

Table 4. Average densities of dried samples before test.

Material	Average densities, T/m^3
CS5-CP-0	1.79
CS5-CP-5	1.71
CS5-CP-20	1.64
CS5-CP-30	1.54
CS5-CP-40	1.45

According to the reviewed literature, a non-linear water uptake curve can be expected in this test, exhibiting a complex path for water trip into a CS containing cement paste pore system. However, in this work the non-linearity has been observed only in first 20 minutes. An observed difference between Portland cement pastes with and without CS could be explained by higher capillary forces into CS containing samples.

After 20 minutes, the remaining unfilled pores and other spaces into structures cement paste samples have uptake water slowly, by increasing the mass by maximum up to 3.5 % in cases of CS5-CP-30 and CS5-CP-40. The total amount of absorbed water decreased with increasing the concentration of CS from 20 to 40 volume %, as compared to Portland cement paste samples without CS, exhibiting up to 3% less water absorption capacity at the same test duration. No significant effect was observed to the sample with 5 volume % of CS, which exhibited a higher water absorption capacity as compared to other samples.

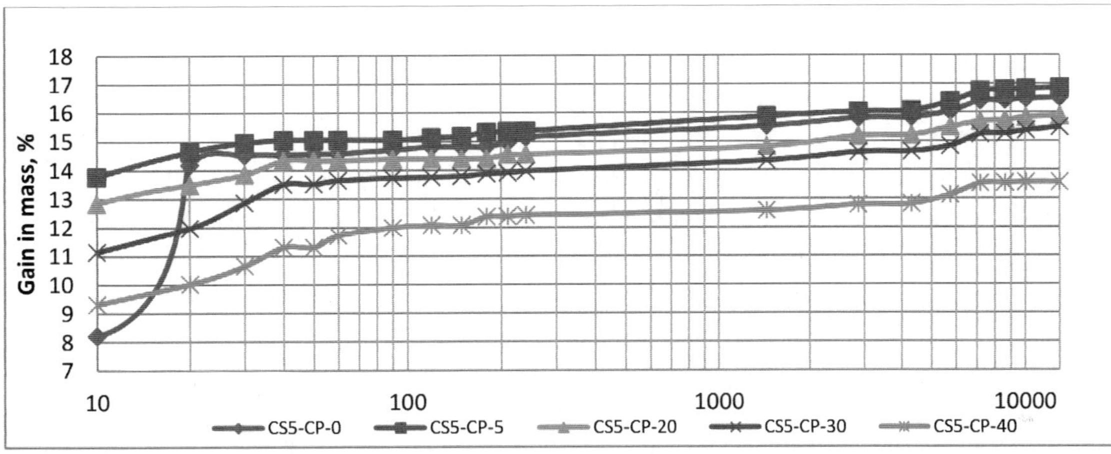

Figure 1. The dynamics of water uptake.

3.2. The influence of concentration of CS on pore size distribution of cement pastes

As shown in table 5, the total opened porosity increased by 7.38 % by increasing the concentration of CS up to 40 volume %. Exception is observed for sample CS5-CP-5, exhibiting similar content of opened porosity as compared to CS5-CP-20. It could be concluded that the cement pastes even with small concentrations of CS significantly reduce the air and vapour discharge ability by using the same preparation method.

As shown in figure 2 (a) and (c), for CS5-CP-0 and CS5-CP-20, similar increase in the total volume of pore with sizes in the range up to $2 \cdot 10^{-1}$ μm (with a highest volume of pores of $5.16 \cdot 10^{-3}$ cm^3/g in this range) was observed. For other samples (see figure 2 (b), (d) and (e)), the pore size range was higher, exhibiting pore diameters up to $1.00 \cdot 10^{-1}$ cm^3/g.

Table 5. The influence of concentration of CS on the total opened porosity determined by mercury porosimetry.

Material	Total opened porosity, %
CS5-CP-0	2.00
CS5-CP-5	5.37
CS5-CP-20	5.11
CS5-CP-30	7.21
CS5-CP-40	9.38

For all samples, a significant increase in content of opened pores was observed in the range of pore sizes from 10 to 500 μm with a clear trend to increase pore volume by increasing the concentration of CS.

In general, an increased volume of micrometer sized pores (in diameter range between 1 to 100 μm) appears for samples with concentration of CS from 20 to 40 volume % (see figure 2 (c), (d) and (e)). As shown in figure 2 (b), a relatively small concentration of CS (5 volume %) decreased the total volume of micrometer sized pores close to 0 cm^3/g.

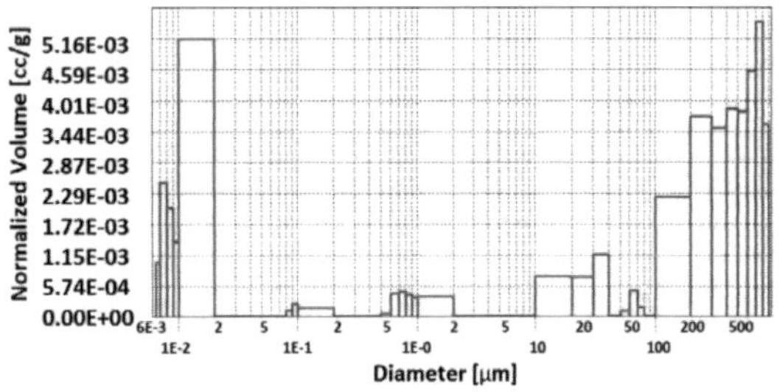

(a) CS5-CP-0

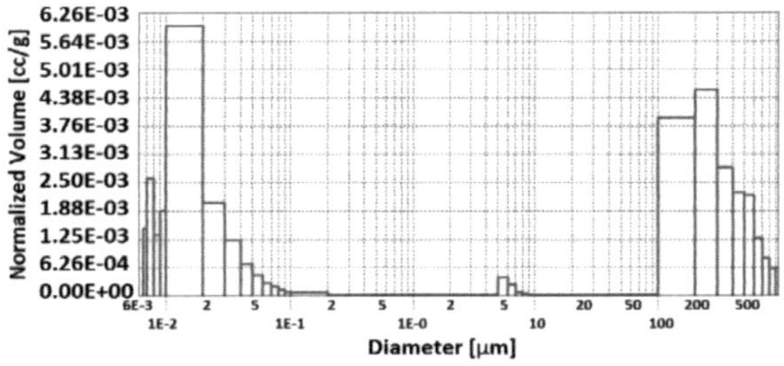

(b) CS5-CP-5

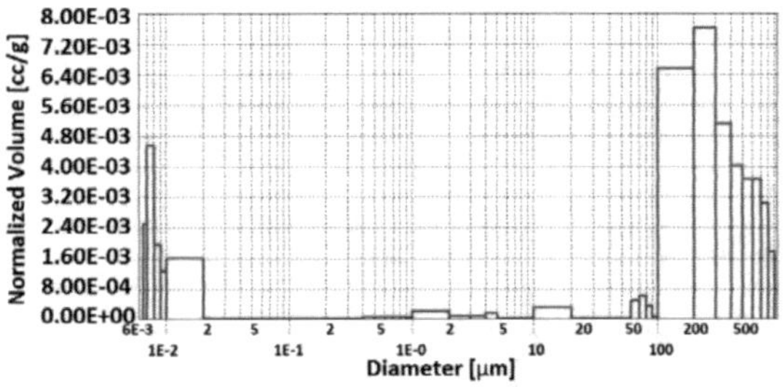

(c) CS5-CP-20

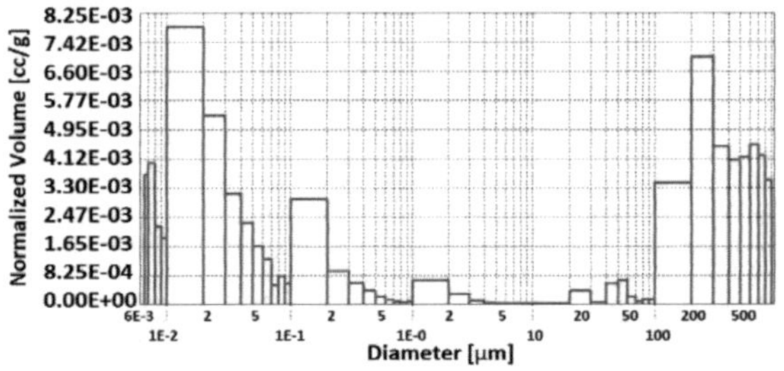

(d) CS5-CP-30

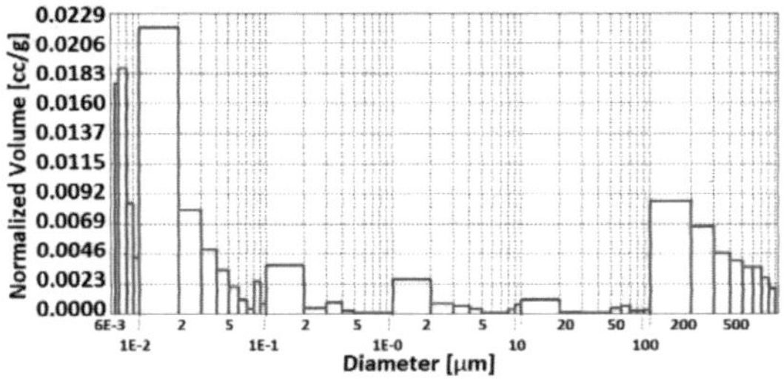

(e) CS5-CP-40

Figure 2. Pore size distribution in the tested samples.

According to the obtained results, it could be concluded that by increasing the concentration of CS in the range from 0 to 40 volume %, the total volume of the opened pores increases. It is expected that by increasing the concentration of CS up to 20 volume %, frost resistance of concrete can be improved due to decreased volume of micrometer sized opened pores. An increase of CS from 30 to 40 volume % can decrease the cement paste durability due to the full filling of micrometer pores with water that could break the pore walls causing the cracks in a cement paste system. Increased porosity may be affected by shrinkage micro cracks in the hydrated cement paste, especially in the contact zone cenosphere – cement paste. A significant factor in mercury porosimetry is relatively high pressure in the work chamber reaching up to 400 MPa. Such pressure can lead to opening of the closed pores and possible collapse of weaker or thinner walls in the pore system.

By comparing the results from pore distribution and water uptake test results, it could be concluded that high relative amount of CS in a Portland cement paste structure provides less water absorption, it is a positive factor for improving frost resistance. On the other hand, high amount of CS in the cold climate can lead to faster formation of new cracks due to increased volume of micrometer sized pores.

4. Conclusions

Generally, replacing Portland cement by cenospheres (CS) (up to 40% by volume) contributes to the reduction of density (up to 23%) and decreasing water absorption capacity up to 3% (by mass) at the same test duration as compared to cement paste sample without CS. Initial water uptake rates for all cement paste samples with CS were higher (have uptake more than 50% of total water in 10 minutes), as compared to Portland cement paste sample without CS (in 20 minutes). Decreased water absorption can have a positive impact on providing durability and frost resistance.

At the same time, by increasing the concentration of CS in the range from 0 to 40 volume %, the total volume of opened micrometer pores determined by mercury porosimetry is higher. An increased volume of "dangerous" (micrometer) pores can cause faster formation of new cracks during the freezing/thawing of wet structures in the cold climate.

Small concentration (5 volume %) of CS did not influence the water uptake capacity and did not increase the volume of "dangerous" (micrometer) pores.

This study has been carried out on a cement paste, which does not fully reflect the behaviour of real concrete, which contains multi-sized aggregate system and surface active admixtures. In the next step of study, durability of real concrete mixtures containing cenospheres would be investigated.

5. References

[1] Bajare D, Bumanis G and Upeniece L 2013 Coal combustion bottom ash as microfiller with pozzolanic properties for traditional concrete *Procedia Eng.* **57** 149–158
[2] A C A Association 2004 *Fly Ash Facts for Highway Engineers* Washington, DC
[3] Mikkanen P 2000 *Fly Ash Particle Formation in Kraft Recovery Boilers Recovery Boilers* (Espoo: VTT Chemical Technology)
[4] Ngu L, Wu H and Zhang D 2007 Characterization of ash cenospheres in fly ash from Australian power stations *Energy & Fuels* **21** (6) 3437–45
[5] Jozic D and Zelic J 2006 The effect of fly ash on cement hydration in aqueous suspensions *Ceramics Silikaty* **50** (2) 98–105
[6] Fenelonov V, Mel'gunov M and Parmon V 2010 The properties of cenospheres and the mechanism of their formation during high-temperature coal combustion at thermal power plants *KONA Powder Part. J.* **28** (28) 189–208
[7] Wesche K ed 1991 *Fly Ash in Concrete: Properties and Performance* Taylor & Francis
[8] Lilkov V, Djabarov N, Bechev G and Petrov O 1999 Properties and hydration products of lightweight and expansive cements Part II *Cem. Concr. Res.* **29** (10) 1641–46
[9] Landman A A 2003 *Literature Review of Fly Ash in Aspects of Solid-State Chemistry of Fy Ash and Ultramarine Pigments* University of Pretoria

[10] Toropovs N, Bajare D, Sahmenko G, Krage L and Korjakins A 2014 The formation of microstructure in high strength concrete containing micro and nanosilica *Key Engineering Materials* **604** 83-86
[11] Baroninsh J, Lagzdina S, Krage L and Shahmenko G 2011 Influence of the dosage of super plasticizer on properties of high performance concrete *IOP Conf. Ser. Mater. Sci. Eng.* **25** 1–6
[12] Barbare N, Shukla A and Bose A 2003 Uptake and loss of water in a cenosphere-concrete composite material *Cem. Concr. Res.* **33** (10) 1681–86
[13] Švinka R and Švinka V 1997 *Silikātu materiālu ķīmija un tehnoloģija* Rīga
[14] *Significance of Tests and Properties of Concrete and Concrete-making Materials, Issue 169, Part 3* ASTM International 1994
[15] Kou S C, Poon C S and Etxeberria M 2014 Residue strength, water absorption and pore size distributions of recycled aggregate concrete after exposure to elevated temperatures *Cem. Concr. Compos.* **53** 73–82
[16] Garbalińska H and Wygocka A 2014 Microstructure modification of cement mortars: Effect on capillarity and frost-resistance *Constr. Build. Mater.* **51** 258–266
[17] Şahmaran M, Özbay E, Yücel H E, Lachemi M and Li V C 2012 Frost resistance and microstructure of engineered cementitious composites: Influence of fly ash and micro poly-vinyl-alcohol fiber *Cem. Concr. Compos.* **34** (2) 156–165

Acknowledgments
The support of the Latvian State Research Programme under grant agreement "INNOVATIVE MATERIALS AND SMART TECHNOLOGIES FOR ENVIRONMENTAL SAFETY, IMATEH" is acknowledged.

IOP Conf. Series: Materials Science and Engineering **96** (2015) 012012 doi:10.1088/1757-899X/96/1/012012

Finite Element Modelling and Analysis of Conventional Pultrusion Processes

P Akishin[1,3], E Barkanov[1] and A Bondarchuk[2]
[1]Riga Technical University, Institute of Materials and Structures, Kalku Str. 1, Riga, LV-1658, Latvia
[2] Southern Federal University, Department of Theoretical and Computational Fluid Dynamics, Rostov-on-Don, Russia

E-mail: pavels.akisins@rtu.lv

Abstract. Pultrusion is one of many composite manufacturing techniques and one of the most efficient methods for producing fiber reinforced polymer composite parts with a constant cross-section. Numerical simulation is helpful for understanding the manufacturing process and developing scientific means for the pultrusion tooling design. Numerical technique based on the finite element method has been developed for the simulation of pultrusion processes. It uses the general purpose finite element software ANSYS Mechanical. It is shown that the developed technique predicts the temperature and cure profiles, which are in good agreement with those published in the open literature.

1. Introduction

Pultrusion is a continuous and efficient process producing composite profiles with a constant cross-section. During pultrusion, fibre reinforcements are saturated with resin in a resin tank and then continuously pulled through a heated die by a puller. Electrical heaters are used for heating of the die in the conventional pultrusion technology. Inside the die, the resin gradually cures and solidifies to form a composite part with the same cross-section profile as in the die. After exit from the die, the finished product is cut to the desired length by saw integrated into the pultrusion tool. Schematic of the process is given in figure 1. An effective set up of pultrusion process is not possible without a numerical analysis of the technological process. A lot of numerical simulations have been done in the last twenty years, most of them focuse on the analysis of the heat transfer and cure [1-5], on the pressure on the inner walls of the die [6,7] and on the problems related to impregnation of reinforcing fibres [8].

[3] To whom any correspondence should be addressed.

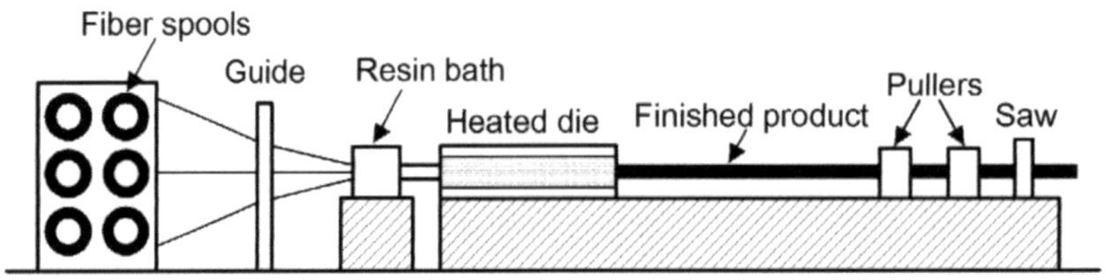

Figure 1. Schematic illustration of the pultrusion process.

2. Theory of the pultrusion process

The object of study in this investigation is part of the pultrusion tool – heated die and composite material in the die.

2.1. Energy equation for the die

Heat transfer in the non-moving die is governed by the following equation in the Cartesian coordinate system:

$$\rho c_p \frac{\partial T}{\partial t} - \frac{\partial}{\partial x}\left(k_x \frac{\partial T}{\partial x}\right) - \frac{\partial}{\partial y}\left(k_y \frac{\partial T}{\partial y}\right) - \frac{\partial}{\partial z}\left(k_z \frac{\partial T}{\partial z}\right) - q_h - q_b = 0 \quad (1)$$

where T is the temperature in Kelvin, ρ and c_p are the density and specific heat of the die material, k_x, k_y, k_z – thermal conductivities in x, y, z directions, q_h – rate of heat generation by heaters, q_b – rate of energy exchange at the boundary.

2.2. Energy equation for the composite

Resin curing is exothermic chemical reaction that gives off heat energy. Therefore it is necessary to introduce the source term in the energy equation for the composite. In pultrusion, the fibres are saturated with resin before entering the die. Therefore it is reasonable to assume that the resin does not flow and is moving with the fibres as a solid composite material. The three-dimensional transient energy equation for the pultruded composite can be written as:

$$\bar{\rho}\bar{c}_p\left(\frac{\partial T}{\partial t} + u\frac{\partial T}{\partial x}\right) - \frac{\partial}{\partial x}\left(\bar{k}_x \frac{\partial T}{\partial x}\right) - \frac{\partial}{\partial y}\left(\bar{k}_y \frac{\partial T}{\partial y}\right) - \frac{\partial}{\partial z}\left(\bar{k}_z \frac{\partial T}{\partial z}\right) - q = 0 \quad (2)$$

where T is the temperature in Kelvin, u is the pull speed, $\bar{\rho}$ and $\bar{c}_p$ are the lumped density and specific heat of the composite material, respectively, $\bar{k}_x$, $\bar{k}_y$, $\bar{k}_z$ are the lumped thermal conductivities in x, y, z directions, and q is the generative term related to internal heat generation due to the exothermic resin reaction.

The lumped thermal properties of the composite are evaluated by the following rules of mixture:

$$\bar{\rho} = (1 - V_r)\rho_f + V_r\rho_r \quad (3)$$

$$\bar{c}_p = \frac{(1-V_r)\rho_f c_{pf} + V_r \rho_r c_{pr}}{\bar{\rho}} \tag{4}$$

$$\bar{k} = \frac{k_f k_r \bar{\rho}}{(1-V_r)\rho_f k_r + V_r \rho_r k_f} \tag{5}$$

where material properties with index *f* are fiber properties, and with index *r* – resin properties.

2.3. Species equation for the resin

The generative term of the composite energy equation related to the resin curing reaction is written as:

$$q = V_r \rho_r H_{tr} R_r \tag{6}$$

where V_r is the resin volume fraction, ρ_r is the resin density, H_{tr} is the total heat of reaction and R_r is the rate of resin reaction:

$$R_r = \left(\frac{\partial \alpha}{\partial t} + u\frac{\partial \alpha}{\partial x}\right) \tag{7}$$

Equation (7) is chemical species equation where α is the degree of cure which is defined as a ratio of the heat amount evolved during the reaction up to present time t, $H(t)$, to the total heat of reaction H_{tr}:

$$\alpha = \frac{H(t)}{H_{tr}} \tag{8}$$

2.4. Curing kinetic model of the resin

Traditionally the resin curing reaction is modelled using the Arrhenius relationship $K(T)$ multiplied by a simple mathematical function $f(\alpha)$ that depends on the resin properties and varies with the resin reaction model:

$$R_r(\alpha, T) = \frac{\partial \alpha}{\partial t} = \frac{1}{H_{tr}}\frac{dH(t)}{dt} = K(T)\cdot f(\alpha) \tag{9}$$

The Arrhenius relationship is written as:

$$K(T) = K_0 \exp\left(-\frac{E}{RT}\right) \tag{10}$$

where T is the temperature in Kelvin, R is the universal gas constant, K_0 and E are, respectively, the frequency factor and the activation energy of the resin.

As a simple mathematical function $f(\alpha)$, 1st-order, *n*-th order, Prout-Tompkins [9] are more frequently used for the formulation of the curing kinetic model:

$$
\begin{aligned}
&f(\alpha) = (1-\alpha) \text{ - first order kinetic model;} \\
&f(\alpha) = (1-\alpha)^n \text{ - } n \text{ - th order kinetic model;} \\
&f(\alpha) = (1-\alpha)^n (1 + K_2 \alpha) \text{ - } n \text{ - th order kinetic model with autocatalysis;} \\
&f(\alpha) = \alpha^m (1-\alpha)^n \text{ - Prout - Tompkins autocatalytic model.}
\end{aligned} \tag{11}
$$

Here K_2, n, m are the kinetic parameters of the resin.

3. Numeric implementation

Numerical technique based on the finite element method has been developed for the simulation of the conventional pultrusion processes. It uses the general purpose finite element software ANSYS Mechanical.

3.1. Numerical procedure based on ANSYS mechanical

The flowchart of the numerical procedure is shown in figure 2. At the beginning of the procedure then $t=0$, α is pre-defined as $\alpha_j^{i=0} = \alpha_0$, where α_0 is equal to 0 or small value depending on the used resin kinetic model, and $T_j^{i=0} = T_0$, where T_0 is room temperature. Here j is the number of FE nodal point (figure 3) and i is the time step index.

The species equation for the resin (7) in numerical procedure is integrated explicitly as:

$$\left[\frac{\partial \alpha}{\partial t}\right]^i = \left[R_r - u\frac{\partial \alpha}{\partial x}\right]^{i-1} \tag{12}$$

Time step is calculated as:

$$\Delta t = \frac{1}{p} \cdot \frac{l}{u} \tag{13}$$

where l is the length of the FE in pull direction, u is the pull speed, p is the number of sub steps chosen by the convergence condition depending on the mesh size and the material properties.

Within each time step, FE software ANSYS is used for the solution of the remaining transient heat conduction problem to obtain the temperature. After obtaining the temperatures, the degree of cure and the effect of the exothermic term are calculated at each nodal point using the nodal control volume (NCV) method realized in the user program. The NCVs are created within the zone defining composites by connecting the centroid of a finite element to the centre points of its surfaces. The boundary of each NCV contains only one FE node. NCV mesh is shown in figure 3. In the NCV the distribution of field variables (T, $\partial\alpha/\partial t$ and α) is assumed constant and unchangeable over a time step Δt.

The rate of the cure for the NCV at any time step i is calculated using (9):

$$\frac{\partial \alpha_j^i}{\partial t} = \left[\frac{\Delta \alpha_j^i}{\Delta t}\right] = K(T_j^i) f(\alpha_j^{i-1}) \tag{14}$$

The degree of cure α for the NCV is obtained continuously as:

$$\alpha_j^i = \alpha_j^{i-1} + \left[\frac{\Delta \alpha_j^i}{\Delta t} \right] \Delta t \tag{15}$$

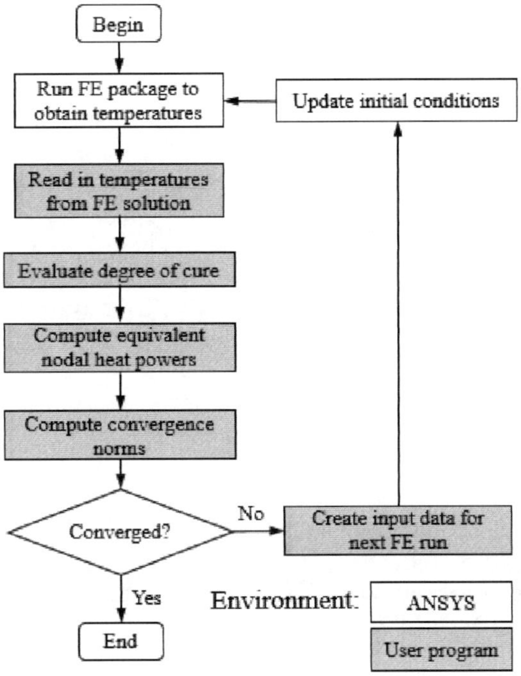

Figure 2. Flowchart of the numerical procedure.

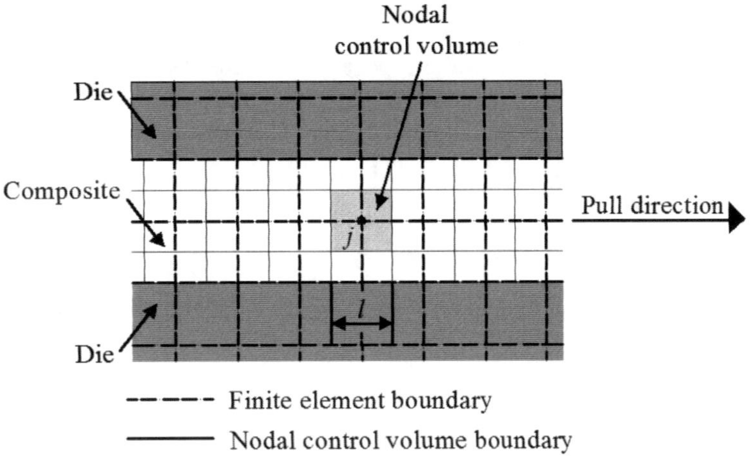

Figure 3. FE/NCV method.

IOP Conf. Series: Materials Science and Engineering **96** (2015) 012012 doi:10.1088/1757-899X/96/1/012012

The exothermic effects of cure reaction are evaluated as the source term for a NCV using (6):

$$q = V_r \rho_r H_{tr} R_r = V_r \rho_r H_{tr} \frac{\partial \alpha_j^i}{\partial t} = V_r \rho_r H_{tr} \left[\frac{\Delta \alpha_j^i}{\Delta t} \right] \tag{16}$$

Movement of resin-saturated composite is simulated as a semi-steady process. The user program shifts the temperature and the curing fields in the composite to distance l in pull direction during creation of initial data for the next FE run.

Iterative procedure shown in figure 2 continues until the difference of temperature and curing fields obtained at the present and previous time steps is less than the set convergence norm.

4. Numerical example

The developed procedure was validated by the temperature and degree of cure fields described in the literature studying the pultrusion of cylindrical rod [1-3], flat plate [3, 4] and I-beam profiles [5]. As an example, the numerical results are given for the pultrusion process of an I-beam.

4.1. Pultrusion of I-beam

This example was chosen to demonstrate an application of the developed numerical procedure for the simulation of the pultrusion process with the temperature control. The temperature control is implemented by turning the heaters on and off.

The scheme of the pultrusion die is presented in figure 4. Dimensions of the pultruded composite profile and die, as well as the locations and dimensions of the heaters and cooling holes are presented in figures 5 - 7. The materials used for production of the I-beam are glass fibers with the fiber volume content of 45% and thermoset vinyl ester resin VE3. Their thermal properties are given in table 1. The rate of resin reaction is described using the Prout-Tompkins curing kinetics model (11) with the parameters presented in table 2.

The pull speed is 30 cm/min. Heating is provided by four electrical heat platens placed on the top and bottom surfaces with the heater's power of 1600 W and two electrical strip heaters placed on both die's sides with the heater's power of 1500 W. The heat platens are insulated except the surfaces contacting with the die. The exterior surfaces of the die, except those under the heaters, and the I-beam are assumed to be exposed to the air and hence are simulated as convective boundaries with a convective heat transfer coefficient of 10 W/m^2·K. The ambient room temperature is 20°C. The die is cooled at the distance of 100 mm from the die entrance by water with the temperature of 20°C to prevent excessive curing at the die entrance.

Table 1. Thermal properties of materials.

	ρ (kg/m^3)	c_p (J/kg·K)	k_z (W/m·K)	$k_x = k_y$ (W/m·K)
Vinyl ester	1100	1640	0.169	0.169
Glass fibers	2560	670	11.4	1.04
Lumped properties (V_r =0.55)	1757	1004	0.4773	0.3748
Steel (die)	7860	486	51	51
Aluminium (heat platens)	2700	896	180	180

Table 2. Kinetics parameters of the resin.

H_{tr} (J/kg)	K_0 (1/s)	E (J/mol)	n	m	α_{max}
398440	186958305	71688	1.2853	0.7147	0.97

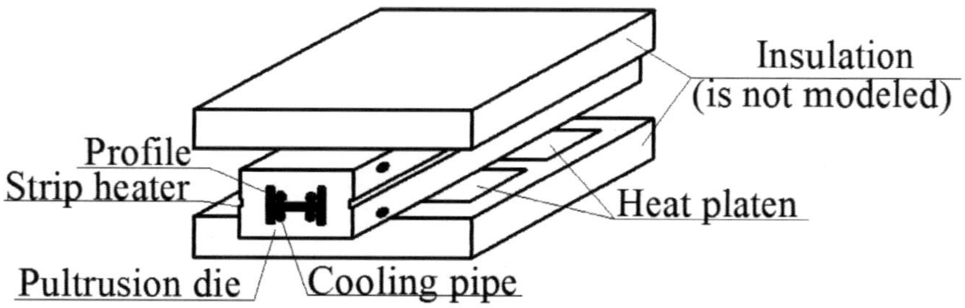

Figure 4. Scheme of the pultrusion die.

Control temperature of the pultrusion process is 123°C. Unfortunately, temperature control points and algorithms of heaters work are not clearly formulated in [5]. The following algorithm was chosen in the simulation: the first heat platen works permanently, the work of the second heat platen and strip heater is controlled – they are switched ON if the maximal temperature on line C within the boundaries of the die is less than the control temperature 123°C and switched OFF otherwise.

The finite element model for the simulation of the I-beam pultrusion was created in ANSYS Mechanical using 3-D thermal solid finite element Solid 70. The element has eight nodes with a single degree of freedom, temperature at each node and the orthotropic material properties. Taking into account the symmetry of the pultrusion tool, only one quarter of it is modelled. The insulation materials surrounding the heat platens are not included into the finite element model, but the surfaces of the platens, except those that contact with the die, are assumed as fully insulated. The strip heaters located on both sides of the die are simulated as surface powers directly applied on the surface of the die. The finite element model is shown in figure 8. An additional length of 285 mm of the I-beam is modelled at the die exit to extend the thermo-chemical analysis to the post-die region. The finite element mesh is regular in pull direction and has 87 elements on the beam. Total number of finite elements is 10660 (1740 for composite, 7480 for die and 1440 for the heaters). The time step defined by (13) is 3 s. During this time the composite travels in the pull direction to the distance equal to the dimension of one element (15 mm). It is necessary to note that iterations (sub-steps) are not performed in the finite element analysis. Initial conditions are applied at time t=0 when all nodal points of the composite have the resin temperature of 20°C and 1^{-10} degree of cure.

Algorithm of heaters work and changes of maximal temperatures on the control lines in time domain are presented in figure 9. It is seen that the pultrusion die has great thermal inertia, because of which maximal temperatures on the control lines vary considerably in the steady manufacturing process. The predicted temperatures and degree of cure distributions at the control lines in general agree with the published results (figures 10 – 13). The largest difference is observed in the temperature predictions on lines A (figure 10) and B (figure 11) located in the die and at the boundary of the die and composite respectively (figure 6). This difference could be explained by different location of temperature control points and algorithm of the heaters work used in [5] and current investigation. It is important to note that there is good agreement between the present and published results for the temperature (figure 12) and the degree of cure (figure 13) distributions in the profile (control line C in figure 6). Higher temperature at the active reaction point is obtained because of a slightly higher rate of the exothermic curing reaction that, in turn, could be explained by different time integration schemes applied in the finite element modelling and different finite element software used for the simulation.

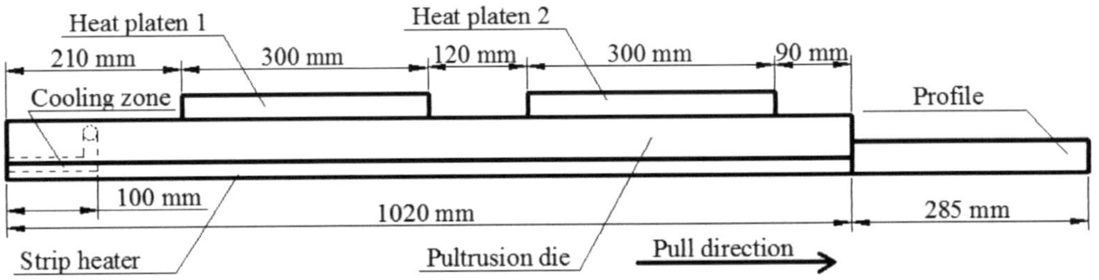

Figure 5. Longitudinal section of the pultrusion die.

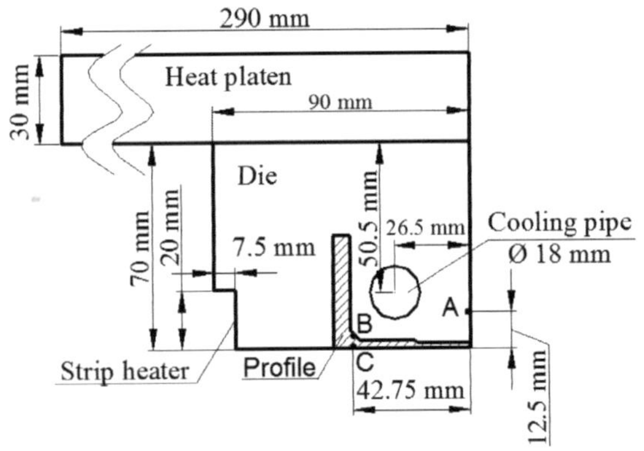

Figure 6. Transverse section of the pultrusion die A-A, B-B, C-C – control lines.

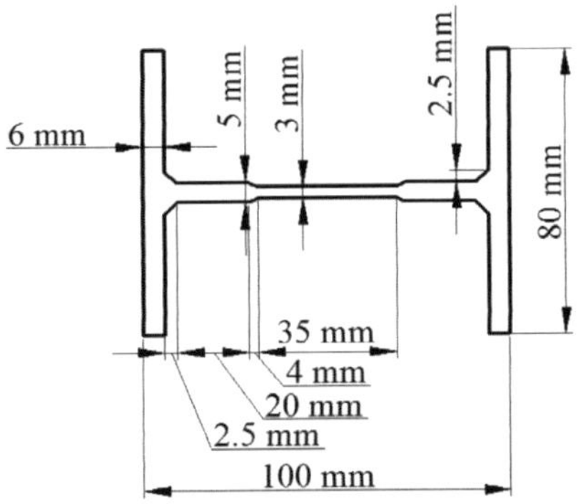

Figure 7. Cross-section of the pultruded profile.

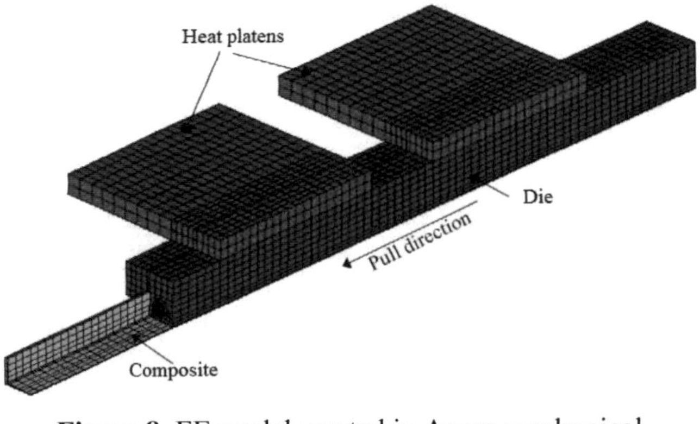

Figure 8. FE model created in Ansys mechanical.

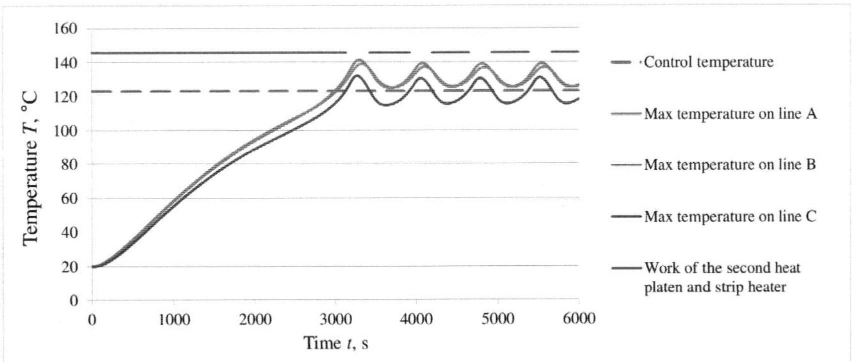

Figure 9. Work of the heaters and maximal temperatures on the control lines A, B and C (within the boundaries of the die).

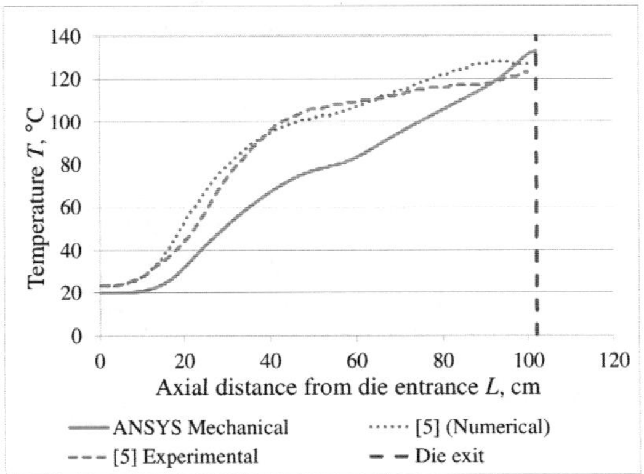

Figure 10. Temperature distribution on line A.

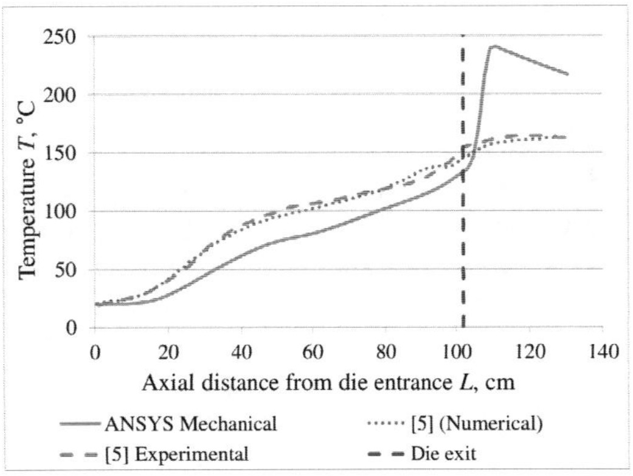

Figure 11. Temperature distribution on line B.

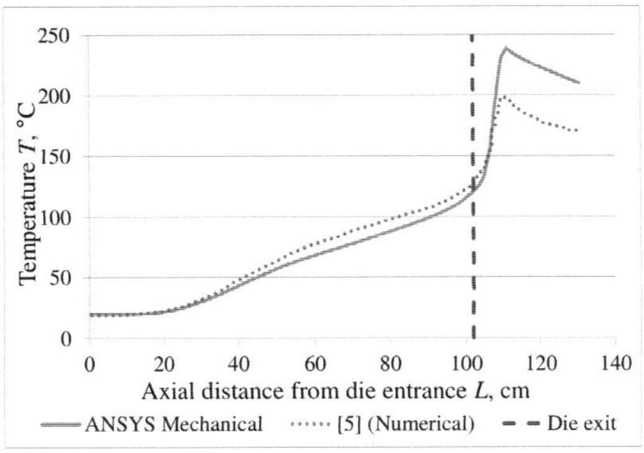

Figure 12. Temperature distribution on line C.

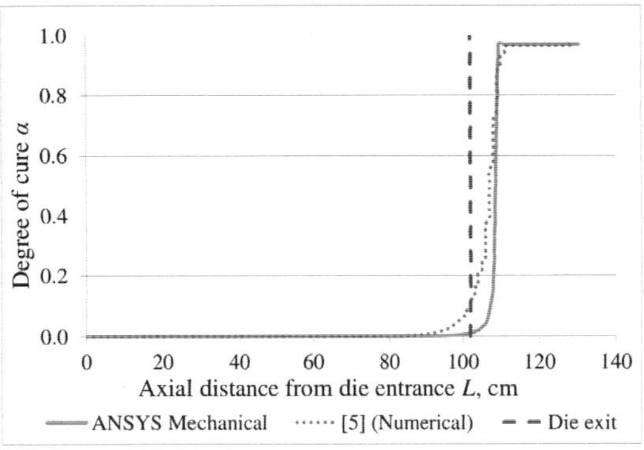

Figure 13. Curing distribution on line C.

5. Conclusions

A numerical procedure based on the general purpose finite element package ANSYS Mechanical has been developed for the thermo-chemical simulation of conventional pultrusion processes. Good agreement between the present finite element results and published numerical-experimental results for both temperature and degree of cure shows that the developed technique can be used successfully for an accurate simulation of the pultrusion processes with different initial data, boundary conditions and new technological requirements.

6. References

[1] Valliappan M, Roux J A, Vaughan J G and Arafat E S 1996 Die and post-die temperature and cure in graphite/epoxy composites *Composites B*. **27** 1–9
[2] Suranto B R, Ye L and Mai Y W 1998 Simulation of temperature and curing profiles in pultruded composite rods *Compos. Sci. Technol.* **58** 191–197
[3] Liu X L, Crouch I G and Lam Y C 2000 Simulation of heat transfer and cure in pultrusion with a general-purpose finite element package *Compos. Sci. Technol.* **60** 857–864
[4] Chachad Y R, Roux J A and Vaughan J G 1995 Three-dimensional characterization of pultruded fiberglass-epoxy composite materials *J. Reinf. Plast. Compos.* **14** 495–512
[5] Liu X-L 2001 Numerical modelling on pultrusion of composite I beam *Composites A*. **32** 663 – 681
[6] Batch G L and Macosko C W 1987 A computer analysis of temperature and pressure distribution in a pultrusion die *Proc. 42nd Annual Conf. and. Expo* (New York: The Composite Institute of the Society of the Plastics Industry) p 12
[7] Sharma D, McCarty T A, Roux J A and Vaughan J G 1998 Investigation of dynamic pressure behavior in a pultrusion die *J. Compos. Mater.* **32** 929–950
[8] Mustafa I, Khomami B and Kardos J L 1999 3-D nonisothermal flow simulation model for injected pultrusion process *AIChE J.* **45** 151–163
[9] Prout E G and Tompkins F C 1944 The thermal decompossion of potassium permanganate *Trans. Faraday Soc.* **40** 488–498

Acknowledgement
The authors gratefully acknowledge the support of European Commission, FRAMEWORK 7 program, contract no. NMP2-SL-2013-609149, project “Development of an Innovative Manufacturing Process for the In-Line Coating of Pultruded Composites (COALINE)”.

IOP Conf. Series: Materials Science and Engineering **96** (2015) 012013 doi:10.1088/1757-899X/96/1/012013

Fibre Reinforced Polymers (FRP) as Reinforcement for Concrete According to German Approvals

R Alex[1]
German Institute of Construction Technology (DIBt), Kolonnenstraße 30B, 10829 Berlin, Germany

E-mail: ral@dibt.de

Abstract. This article demonstrates the possibility of the application of joint principles to develop test programs for national approval or European Technical Assessments of FRP reinforcement for concrete. The limits of different systems are shown, which until now have been approved in Germany.

1. Types of FRP reinforcement for concrete approved in Germany

Approvals in Germany and European Technical Assessments are an efficient means to allow innovative products to enter the market. As standards for the use of such products are present neither in Germany nor in Europe, approvals or European Technical Assessments are the only way to lead to a proof of usability.

In table 1 the types of approvals, which until now exists for FRP reinforcement of concrete in Germany, are listed. Since the materials of on surface bonded and in slits near surface bonded CFRP strips (strips of Carbon Fibre Reinforced Polymers) are the same, it can be considered as the same type of FRP reinforcement. But as it is shown in the next two chapters, the test programs for the two types of approvals with the same type of reinforcement are different.

Table 1. Types of FRP reinforcement approved in Germany.

Subject	Test program	
On the surface bonded strips of FRP (1995)	National test program in Germany exists Draft European Assessment Document (EAD) in work	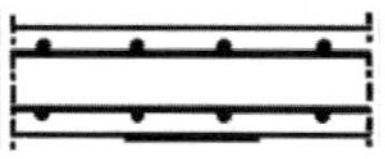
In slits near surface bonded strips of FRP (2002)	National test program in Germany exists Draft European Assessment Document (EAD) in work	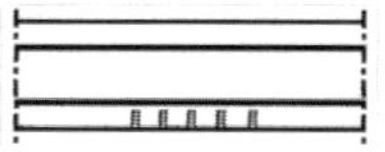
Fibre Reinforced Polymers as reinforcing bars for concrete (Schöck-ComBar 2008)	National test program in Germany exists Draft European Assessment Document (EAD) in work	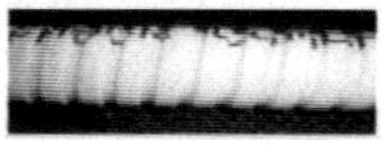
Strengthening with layers of textile concrete (TUDAG 2014)	National test program in Germany exists Draft European Assessment Document (EAD) in work	

[1] To whom any correspondence should be addressed.

Regardless of these differences, we can say that at the moment in Germany three types of reinforcement are permitted to be used:

1. CFRP strips bonded on the surface or in slits of the concrete member;
2. GFRP bars (bars of Glass Fibre Reinforced Polymers);
3. Layers of Textile Reinforced Concrete with reinforcement of Carbon Fibre Textiles.

For all these types of FRP reinforcements there are national testing programs in Germany. European Assessment Documents as the basis for implementation of European Technical Assessments are in preparation.

Table 2. Properties of reinforcement.

Property	Reinforcing Steel	FRP Reinforcement
Tensile strength	known	unknown
Bonding strength	known	unknown
Long term tensile strength in alkaline environment	known in the field of application temperatures (-40 to 100 °C)	Unknown Field of application temperatures has to be defined
Long term bond strength in alkaline environment	known in the field of application temperatures (-40 to 100 °C)	unknown Field of application temperatures has to be defined

Even if confining ourselves to the static structural behavior, it is important to know that the properties of new reinforcements of FRP are determined not only by their short term characteristics as tensile strength and bonding strength to the concrete. Essential are especially the long-term characteristics, which in case of conventional reinforcement often play no role because they are known for a long time for the region of temperature in which they are applied.

2. Strengthening with CFRP strips bonded on the surface of the concrete member

The first approvals in Germany for FRPs with Carbon Fibres were issued in the mid-nineties of the last century. Those approvals exist until now. In most cases FRPs in the form of prefabricated Carbon Fibre Strips are used, which are bonded with epoxy adhesives on the surfaces of the concrete parts for strengthening them (see figure 1).

From the beginning of this approval area, the opinion of the experts was that it is only necessary to verify that the adhesive after aging under stress is better than the concrete to which it is applied. Therefore, tests were conducted on concrete cylinders of 40 to 50 mm diameter and 100 to 150 mm height. The strength class of the concrete was C50/60. Circular specimens of CFRP strip with the same diameter as the concrete cylinder were glued at one end of these cylinders. Afterwards steel dollies with the same diameter as the concrete were glued at both ends. These specimens were loaded with 30% of the surface tensile strength of the concrete for half a year in the climatic chamber with 95% relative humidity and the maximum application temperature (mostly 40 °C). The requirement for these tests was that no failure occurred during the test time and in the following tensile test failure occurred in the concrete. It was practically the proof that the adhesive after aging under load is still better than the concrete. It is not possible to determine a long term tensile strength of the adhesive in this way. But since in the design model (see [1]) only the strength of concrete is used and within the application time of these systems of more than 20 year no case of damage has become known, the testing program for the long-term behaviour has remained.

IOP Conf. Series: Materials Science and Engineering **96** (2015) 012013 doi:10.1088/1757-899X/96/1/012013

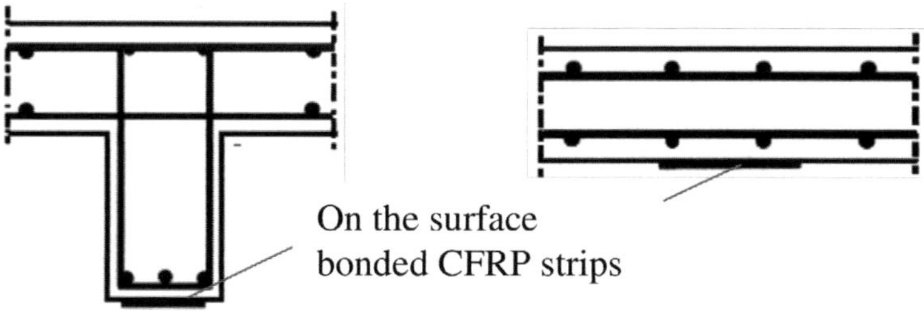

Material of the adhesive:
Epoxy resin consisting of resin (Component A) and hardener (Component B) with glass transition temperatures between 50 and 56 °C.

Material of the CFRP strips:
Matrix: Epoxy resin with glass transition temperatures above 100 °C.
Fibers: Carbon fibers ~68 V% fiber content in CFRP strips.
Short-time strength of CFRP strips: 2000 to 3500 N/mm^2.
E-modulus: 160000 to 200000 N/mm^2.

Figure 1. Materials used for strengthening with CFRP strips bonded on the surface of the concrete member.

Although no cases of damage related to the aging of the adhesive have been recorded, there was a claim with prestressed CFRP strips in Kuala Lumpur which had an influence on the test program. Since it was not a damage case in Germany, DIBt got no detailed information about this case. We only suspected that the reason for failure among others could bealkali sensitivity of some matrix resins of CFRP strips. Therefore, long-term test for CFRP strips in alkaline environment in our national test program has been introduced (see figure 2).

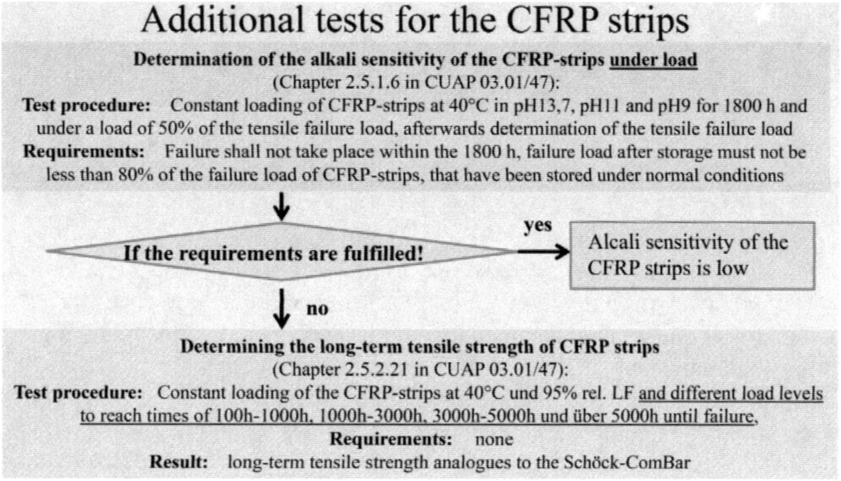

Figure 2. Additional tests to determine the alkali sensitivity or long term tensile strength of CFRP strips in the alkaline environment.

As a result of the investigations, reduction factors related to the short-term strength of the CFRP strips were determined. The values of these factors are based on the results of the tests from 0.85 to 0.9. Since the products of some manufacturers did not fulfil the requirements at pH 13.7, additional information is given in the approval concerning the range of pH values where the reduction factors apply.

3. Reinforcing bar of glass fibre reinforced polymers

A national approval for a reinforcing bar of glass fibre reinforced polymers (GFRP) (see [2]) could be issued in Germany in 2008. For approval of such new objects expert committees have been formed at DIBt, which are composed of experts in the required fields. In the case of this object, scientists in the field of Fibre Reinforced Polymers, structural engineering, reinforced concrete and building materials from three universities of Germany took part in the work of the committee. The discussions on the test program to determine the long term tensile and bond strength of the new reinforcement were particularly difficult. Before issuing the approval (see [2]) of the GFRP reinforcing bar at DIBt, only the structural elements of Fibre Reinforced Polymers carrying loads without contact to concrete were approved. Test programs and evaluation methods for these approvals are based on the recommendations of the construction supervision Association e. V. [BÜV] “Structural plastic elements in the field of construction” (see [3]). The determination of the long-term tensile strength is based on DIN 53768:1990 (see [4]). According to this German standard, an extrapolation of strength test results for long term-tests over 1.5 decades is allowed. If the long-term test, for example, was conducted over 5000 h, the extrapolation to $10^{(\log(5000)+1.5)} \sim 158\,114$ h ~ 18 years can be done according to this standard.

As the strength after 50 to 100 years is of importance in the construction industry, three series of tests were carried out to determine the long-term tensile strength of the ComBAR at three different temperatures (20 °C, 40 °C and 60 °C).

Figure 3. Reinforcing bar of Glass Fibre Reinforced Polymers "Schöck-ComBAR" (short-term tensile strength – 1200 N/mm^2; long-term tensile strength – 580 N/mm^2; E-modulus – 60,000 N/mm^2).

In each series of tests 15 tests were carried out at different stress levels. The longest test duration with the lowest stress level in every test series was over 5000 h. For each test series, the linear regression line of the 5% quantile values of the strength levels in the log (time) - log (stress) diagram was determined in accordance with DIN 53768, Chapter 5. These regression lines have been extended to 10^6 hours (about 114 years). Afterwards, for a given stress level it has been determined which time until failure is obtained from the 5% quantile regression lines at continuous temperatures of 20 °C, 40 °C and 60 °C and whether the Arrhenius relationship for the three values of time until failure is fulfilled:

$$t = 10^{(a/T+b)}$$

where t – the time until failure in [h] of the creep test at a certain stress level; T – the absolute temperature of the creep test in [K] (0 K = -273,15 °C); a, b – constants determined from the creep tests for adjusting the Arrhenius relationship.

Since such Arrhenius relationship could be established, it was agreed between the experts that it is admissible to extrapolate the long-term tensile strength at a maximum continuous temperature of 40°C for an extrapolation of 2.25 decades ($10^{(\log(5000)+2.25)}$ ~ 889 140 h ~ 100 a).

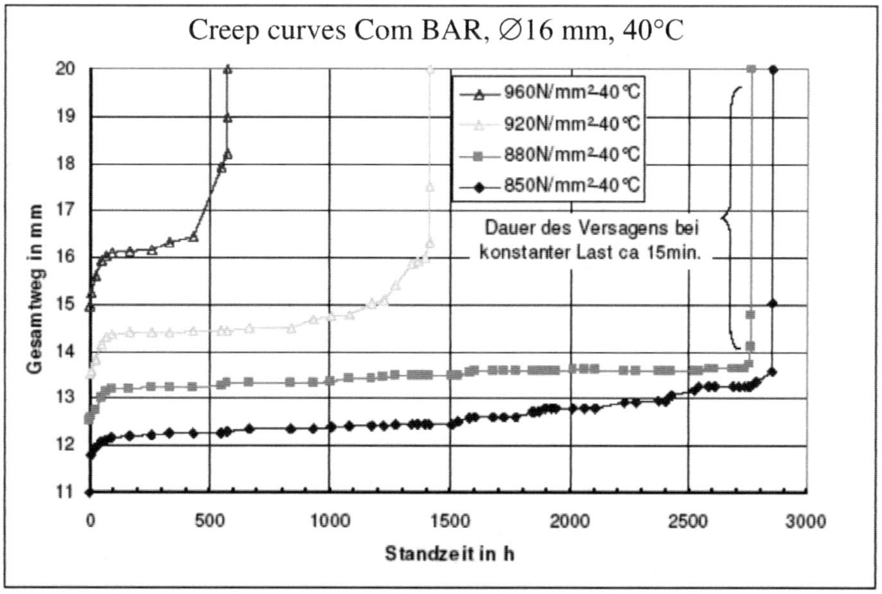

Figure 4. Creep curves typical for the Schöck-ComBAR.

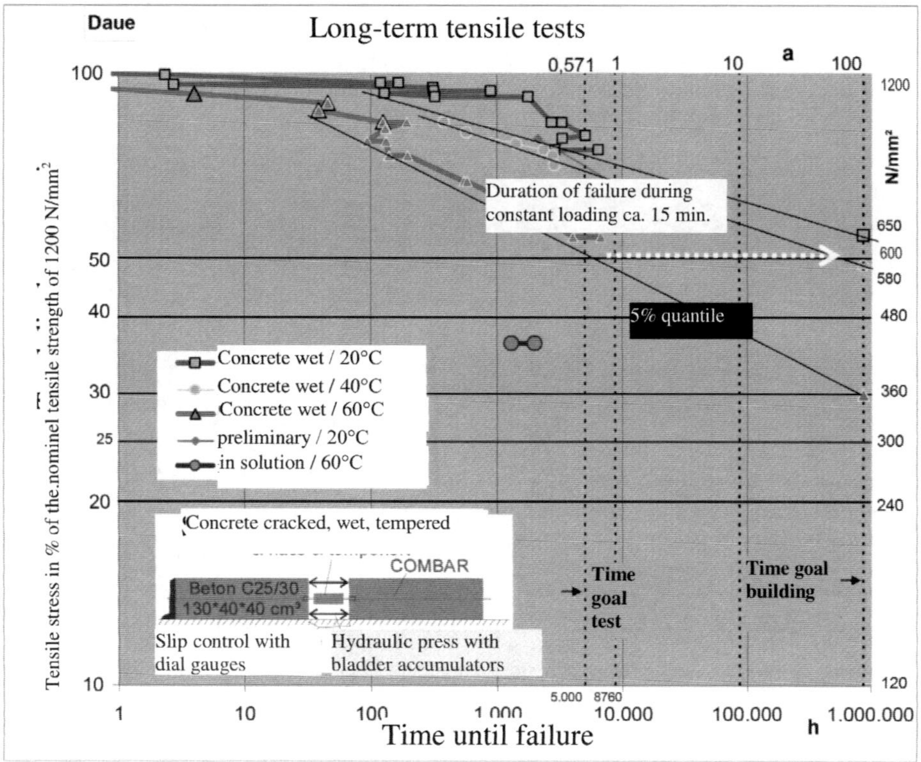

Figure 5. Schematic representation of the measurement points (to improve the overview here connected by lines) and 5%-quantile limit.

The test set used in this case for creep tests to determine the long-term tensile strength is shown in figure 5.

The ComBAR was set in the concrete on both sides with a long anchoring length. One of the two anchor blocks was movable and the other was fixed. Between the two anchor blocks there was a gap. In the middle of this free length the ComBAR was encased by a concrete cylinder. This concrete cylinder was laid in a water bath which could be tempered. After hardening of the anchor blocks and the concrete casing of the ComBAR, the hydraulic presses were operated and the ComBAR was stretched under the respective stress level. At the same time, the water bath was adjusted to the respective temperature. When stretching the ComBAR in the concrete casing cracks are produced (see expanding body test), the tempered water penetrates into the cracks and provides a realistic alkaline environment for the ComBAR. The press forces were regularly checked during the tests and adjusted accordingly, so that over the entire duration of the experiment, a constant stress level in the ComBAR was assured.

As a result of the investigations, a reduction factor related to the short-term strength of the GFRP bar was introduced. The value of this factor is 0.48.

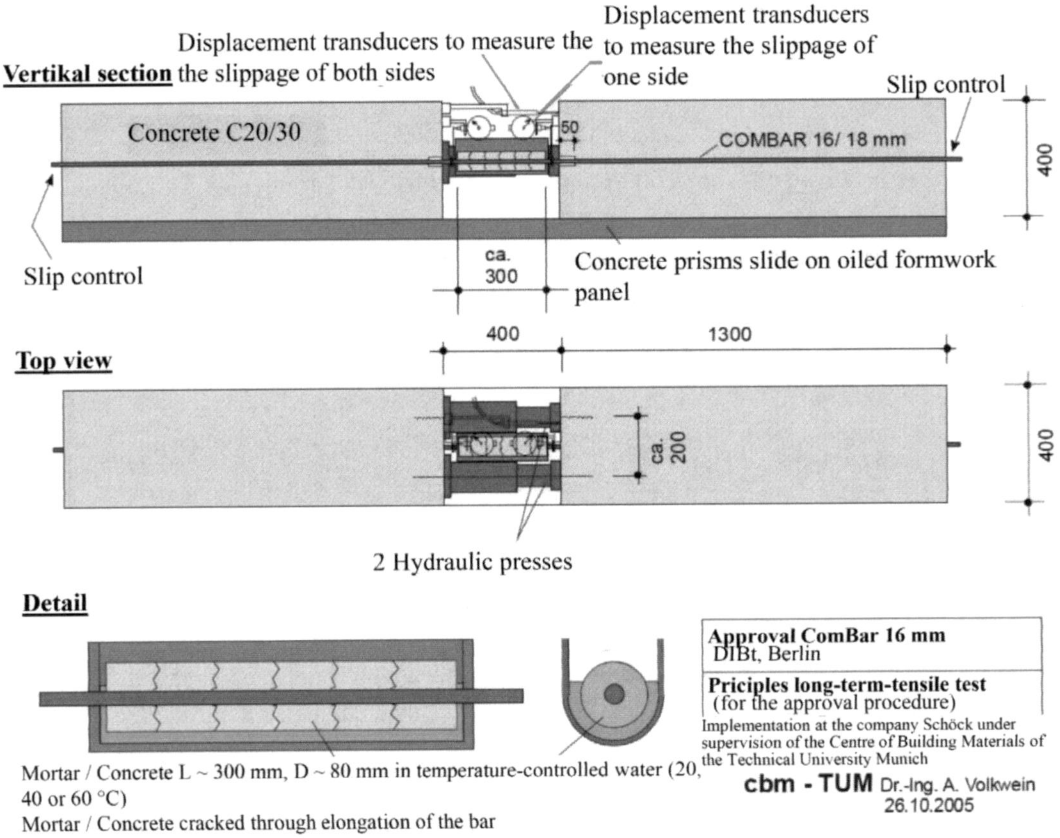

Figure 6. Test set to determine the long-term-tensile strength of the SchöckComBAR.

Furthermore, the bond behaviour had to be examined under continuous load. The procedure and test arrangement (see diagram below) to determine the long-term bond behaviour was similar to the one used to determine the long-term tensile strength (see figure 7 left).

Here, the permanent bond force-slip behavior was again checked at certain continuous load levels (see figure 7 right):

- A2: "Before maximum" of the bond curve (A0) until ca. 60% of maximum load;
- C and D: "After maximum" to approximately 1 mm slip and followed by lowering the load down to 45 kN (level C) resp. 30 kN (level D).

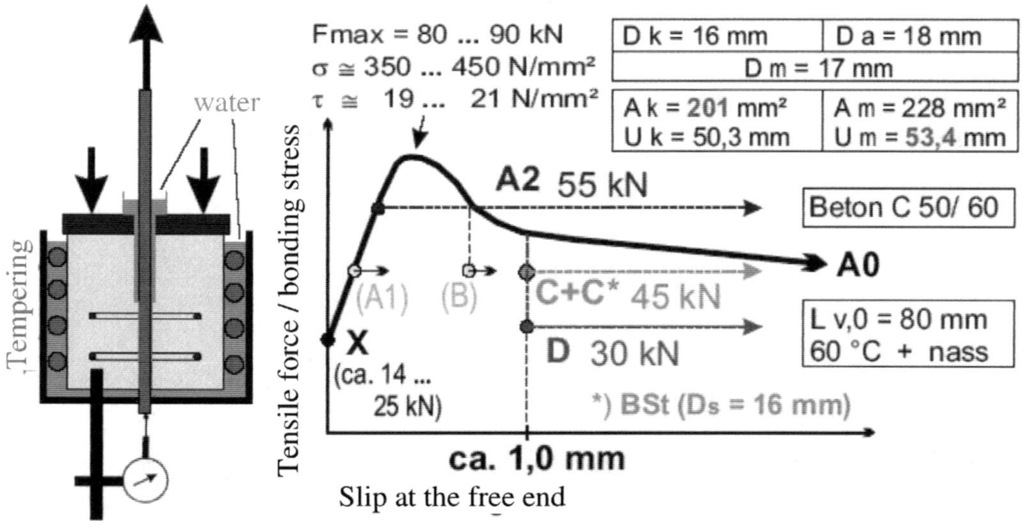

Figure 7. Test set to determine the long-term bond behaviour.

To keep the test effort within the limits, all experiments were carried out at 60 °C. To be on the safe side, the bond strength extrapolated from the results of these tests is used for the application temperature of 40 °C. As a result, the bonding strength corresponding to the load level of 30 kN (level D) was chosen because only at this level the creep curves showed stable behavior similar to the bond creep curves of traditional reinforcing steel. In order to allow such a comparison, three creep bond tests with the same test setup (see figure 7 left) but traditional reinforcing steel were carried out.

4. Strengthening with layers of textile reinforced concrete

The first approval for such a type of reinforcement was issued in July 2014 for the applicant TUDAG (Technical University Dresden Joint-Stock Company) (see [5]). After the sewing, the textile is coated to realise the bond between the single fibres of the multifilament yarns. The approval has been issued for strengthening but this type of reinforcement also may be used for new concrete elements (see figure 8).

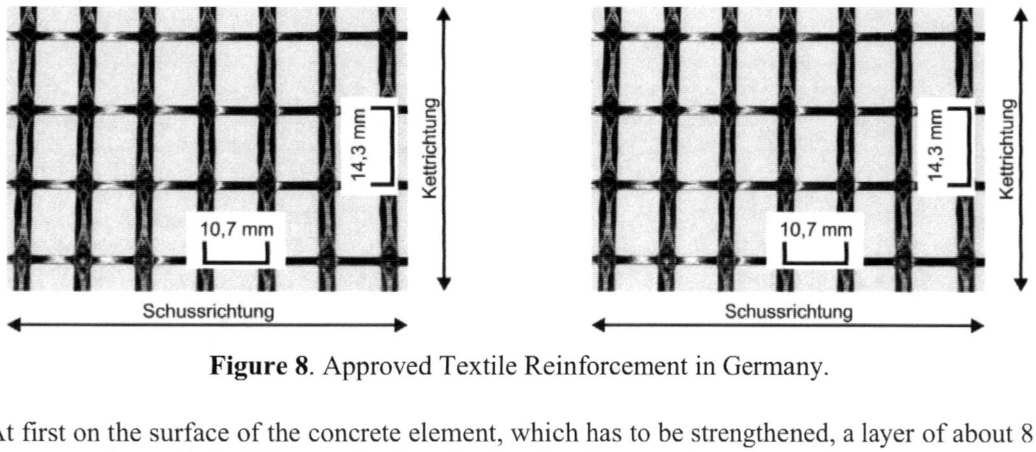

Figure 8. Approved Textile Reinforcement in Germany.

At first on the surface of the concrete element, which has to be strengthened, a layer of about 8 mm fine concrete is applied. This may be done by hand or by spraying with a special spraying method which also is laid down in the approval (see [5] Annex 4). Afterwards the textile is incorporated into

IOP Conf. Series: Materials Science and Engineering **96** (2015) 012013 doi:10.1088/1757-899X/96/1/012013

the freshly applied layer of fine concrete. Then again a layer of fine concrete is applied either to complete the previous fine concrete layer or to incorporate again a Reinforcing Textile in the fresh fine concrete. According to the approval, in this way up to four layers of Textile Reinforcement Concrete may be applied.

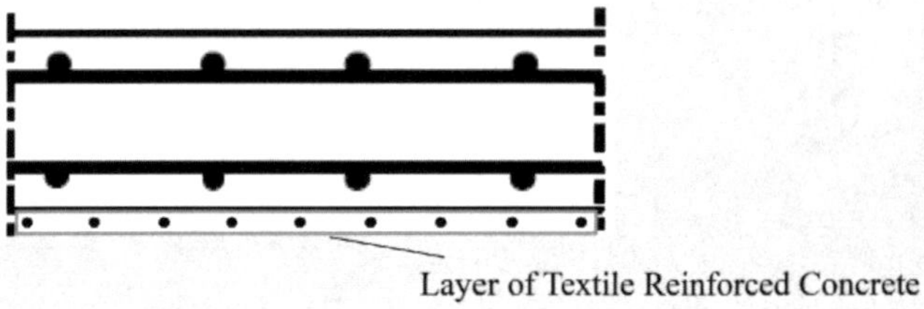

Material fine concrete:
deposited at the DIBt (maximum grain size from 1 to 4 mm) **Material coated Textile:**
Coating: Polymer dispersion with temperature-dependent mechanical behaviour
Fibers: carbon fibers
Fiber content after coating ~35 V%
Short-time strength of coated yarns: 1980 to 2940 N/mm^2
E-modulus of coated yarns: 150.000 to 170.000 N/mm^2

Figure 9. Materials used for strengthening with Textile Reinforced Concrete.

In the test program for this approval it was intended to apply the same procedure for the determination of the long-term tensile strength as it was used for the Schöck-ComBAR. It was found that the behavior of the Textile Reinforced Concrete was very different in comparison to the behavior of the Schöck-ComBAR. In extensive preliminary tests it was detected that at load levels higher than 70% of the characteristic short-term strength in general failure occurs after a few hours, while at loads below 70% of the characteristic short-term strength after 4000 h still no failure occurs. The tested residual strength of these samples was slightly higher than the average value of the reference samples without preloading.

As a result of the investigations, a reduction factor related to the short-term tensile strength of the Textile Reinforced Concrete was introduced. The value of this factor is 0.6. In addition, another factor for the reduction of the bond strength of the textile in the fine concrete was introduced. The value of this factor is 0.21.

5. Conclusions

If new reinforcements for concrete are developed, it is important to define the scope and characteristics of the reinforcement in such a way that it is comparable to the scope and characteristics of the conventional reinforcement. Only in this case the user can determine whether the respective reinforcement is suitable for his purpose or not.

A decisive disadvantage of the reinforcement of Fiber Reinforced Polymers previously approved in Germany is the low maximum application temperature of 40 °C.

A reliable determination of the long-term tensile strength and the long-term bond strength of reinforcement of Fibre Reinforced Polymers is only possible with testing methods, which are equivalent to what has been applied at Schöck ComBar.

The test programs of different types of Fiber Reinforced Polymers as reinforcement for concrete must be coordinated.

The scope of reinforcement of Fiber Reinforced Polymers shall be defined in dependence of the applied testing program and the test results achieved.

The applicants of approvals in Germany but also at European Technical Assessments (ETAs) in the EU should know what options of the test program are available to expand the scope of their approval or ETA.

6. References

[1] DAfStb-Rili VBgB: Richtlinie *Verstärken von Betonbauteilen mit geklebter Bewehrung* Deutscher Ausschuss für Stahlbeton

[2] Z-1.6-238 *Bewehrungsstab Schöck ComBAR aus glasfaserverstärktem Kunststoff* Nenndurchmesser: 16 mm

[3] Tragende Kunststoffteile im Bauwesen [TKB] – BÜV-Empfehlung: Entwurf, Bemessung und Konstruktion–Diskussionsentwurf, Fassung Oktober 2002 (s. Bundesvereinigung der Prüfingenieure für Bautechnik e.V.) Available: http://www.bvpi.de/bvpi-content/ingenieur-box/richtlinien/richtlinien.htm

[4] DIN 53768: 1990: Extrapolationsverfahren für die Bestimmung des Langzeitversagensverhaltens von GFK

[5] Z-31.10-182, Verfahren zur Verstärkung von Stahlbeton mit TUDALIT (Textilbewehrter Beton)

Application of Granulated Blast Furnace Slag in Cement Composites Exposed to Biogenic Acid Attack

M Kovalcikova[1,3], A Estokova[1] and A Luptakova[2]
[1]Technical University of Kosice, Faculty of Civil Engineering, Institute of Environmental Engineering, Kosice, Slovakia
[2]Slovak Academy of Sciences, Institute of Geotechnics, Kosice, Slovak Republic

E-mail: martina.kovalcikova@tuke.sk

Abstract. The deterioration of cement-based materials used for the civil infrastructure has led to the realization that cement-based materials, such as concrete, must be improved in terms of their properties and durability. Leaching of calcium ions increases the porosity of cement-based materials, consequently resulting in a negative effect on durability since it provides an entry for aggressive harmful ions, causing corrosion of concrete. The use supplementary cementing composite materials have been reported to improve the resistance of concrete to deterioration by aggressive chemicals. The paper is focused on the investigation of the influence of biogenic acid attack on the cement composites affected by bacteria Acidithiobacillus thiooxidans. The concrete specimens with 65 wt. % addition of antimicrobial activated granulated blast furnace slag as durability increasing factor as well as without any addition were studied. The experiments proceeded during 150 days under model laboratory conditions. The pH values and chemical composition of leachates were measured after each 30-day cycle. The calcium and silicon contents in leachates were evaluated using X – ray fluorescence method (XRF). Summarizing the results, the 65% wt. addition of antimicrobial activated granulated blast furnace slag was not confirmed to be more resistant.

1. Introduction

Huge amount of concrete and cementitious materials is used in the wastewater systems. Some of the microorganisms in sewer pipes can generate aggressive aqueous solutions which may damage cementitious materials and reduce the material service time [1]. Thus, there is great interest in predicting the corrosion rate and the service life of sewer pipes.

The most prominent biodegradation is biogenic sulfuric acid corrosion on concrete sewer pipes, wastewater collection systems, and treatment plants [2]. The main source of biogenic sulfuric acid is

[3] To whom any correspondence should be addressed.

hydrogen sulfide, produced by sulfate reducing bacteria (SRB, e.g. *Desulfovibrio* sp.). The latter are active under anaerobic conditions and reduce oxidised sulfur compounds to H_2S. The transformation of H_2S into sulfuric acid occurs after the sorption of H_2S from the sewer atmosphere into the concrete, or the biofilm on the surface of the pipelines above the water line, under aerobic conditions. Once the H_2S has reached the atmosphere, it may react with oxygen to elemental sulfur, which is deposited on the slime-layer, coating the walls. Sulfur is a substrate for many thiobacilli, such as *Acidithiobacillus thiooxidans*, *Acidithiobacillus neapolitanus*, *Acidithiobacillus intermedius* [3, 4], which metabolise it into sulfuric acid. The sulfuric acid produced is aggressive and may attack the inner surface of the concrete pipe and other parts of the treatment and transportation facilities. Ensuing, corrosion products like gypsum ($CaSO_4 \cdot 2H_2O$) and ettringite ($CaO \cdot Al_2O_3 \cdot 3CaSO_4 \cdot 32H_2O$) can be formed [5]. These expansive products can lead to increased internal pressure resulting in small cracks. Furthermore, the corroded materials can be removed by the flow of sewage, which also results in the acceleration of corrosion [6].

Ground granulated blast furnace slag is a by-product of the iron-making process and because of its high calcium silicate content it has excellent cementitious content and has been used in the construction industry for years as a replacement for ordinary Portland cement [7]. The replacement of Portland cement by blast furnace slag is an advantageous practice to produce construction materials of improved long-term mechanical properties and durability at relatively lower costs [8]. Use of ground granulated blast furnace slag as a cementitious material blended with Portland cement is based on its activation with alkalis (mainly $Ca(OH)_2$) released from hydration of the Portland cement [9, 10]. When cement reacts with water, it hydrates and produces calcium silicate hydrate (CSH), the main component to the cement strength, and calcium hydroxide ($Ca(OH)_2$). When ground granulated blast furnace slag is added to the mixture, it also reacts with water and produces CSH from its available supply of calcium oxide and silica. A pozzolanic reaction also takes place which uses the excess SiO_2 from the slag source, $Ca(OH)_2$ produced by the hydration of the Portland cement, and water to produce more of the desirable CSH making slag, a beneficial mineral admixture to the durability of concrete [11].

In the experiments discussed in this paper, the influence of the addition of granulated blast furnace slag on the resistance of the concrete to biogenic sulfuric acid was investigated.

2. Materials and methods

2.1. Concrete samples

The effects of sulphur-oxidizing bacteria *Acidithiobacillus thiooxidans* on concrete samples with addition of ground granulated blast furnace slag were investigated during 150 days under model conditions.

The concrete samples used in the experiment were prepared using cement CEM I 42.5 N as one of the common cements. The composition of concrete mixtures has been chosen to be similar to the concrete composition used in sewerage systems. Mixtures A and B were designed to meet the requirement of at least C25/30 concrete strength class in accordance with EN 206-1. The concrete samples have been prepared considering exposure class XA1 - slightly aggressive chemical environment. This corresponds to conditions when the concrete is subjected to an aggressive chemical action that takes place in the natural soil and in wastewater with the speed of the water, low enough to be accepted for hydrostatic factor of impact. Mix proportion of two different concrete mixtures is shown in table 1.

Table 1. Mix proportion per 1 m^3 of concrete mixtures.

Component		CEM I 42.5 [kg]	Addition of slag [kg]	Water [L]	Aggregate fraction [kg]		
					0/4 mm	4/8 mm	8/16 mm
Mixture	A	360	-	162	825	235	740
	B	126	234	162	825	235	740

The standardized concrete specimens with dimensions of 100 x 100 x 400 mm were prepared in accordance with EN 206-1, cured for 28 days in water environment and afterwards cut into small prisms with dimensions of 50 x 50 x 10 mm. The test specimens were slightly brushed in order to remove polluting particles, sterilized in 70% ethanol for 24 hours and dried at 80^0C to constant weight before use in the cultivation experiments.

2.2. Microorganisms
In the experiment, the bacterial culture of sulphur-oxidising bacteria – *Acidithiobacillus thiooxidans*, isolated from the acid mine drainage (the shaft Pech, the locality Smolnik, Eastern Slovakia) was used. The selective nutrient medium Waksman and Joffe (pH 4.0) was used for the preparation of the active bacterial culture as well as for the isolation and cultivation of *Acidithiobacillus thiooxidans* in the presence of the concrete samples [12]. Its composition is described in table 2.

Table 2. The selective nutrient medium by Waksman and Joffe.

Chemical compound	Amount
$CaCl_2.6H_2O$	0.25 g/L
$(NH_4)_2SO_4$	0.2 g/L
K_2HPO_4	3.0 g/L
$MgSO_4.7H_2O$	0.5 g/L
$FeSO_4.7H_2O$	Trace amount
S°	10.0 g/L
Distilled water	Up to 1000 mL

The prepared concrete samples were exposed to the medium of activated bacteria. The volume ratio of solid concrete sample to the liquid phase was set to 1:10. Experiments carried out in covered glass jars in an aerobic atmosphere at laboratory temperature using liquid selective culture medium inoculated at 7 day intervals with bacteria *Acidithiobacillus thiooxidans*. The 20 % vol. inoculum of bacteria *Acidithiobacillus thiooxidans* relative to the total of the liquid phase was used. pH value of liquid phase was kept at an optimal level of 4.0. After each 30 day-immersion period, the change in pH as well as the dissolved concentration of calcium ions was measured in leachates.

2.3. Analytical methods
The chemical composition of leachates was analysed during the experiments by X-ray fluorescence analysis (XRF). SPECTRO iQ II (Ametek, Germany) with SDD silicon drift detector with resolution of 145 eV at 10 000 pulses was used for the analysis. The primary beam was polarized by Bragg crystal and Highly Ordered Pyrolytic Graphite - HOPG target. The samples were measured during 300 and 180 s at voltage of 25 kV and 50 kV at current of 0.5 and 1.0 mA under helium atmosphere by

using the standardized method of fundamental parameters for leachates. pH changes were measured by pH meter FG2- FiveGo (Mettler-Toledo, Switzerland).

3. Results and discussion

Leaching of calcium and silicon ions from the concrete matrix due to the action of an aggressive environment on concrete samples causes degradation of concrete structures [13]. For this reason, leached out amounts of calcium and silicon ions during the experiment were studied.

3.1. Ca leaching due to biological corrosion

Changes in the concentration of calcium ions in leachates of samples A and B are illustrated in table 3.

Table 3. Concentrations of dissolved Ca ions.

Days of exposition	Ca concentration (mg/L)	
	Sample A	Sample B
30	684.5	705.6
60	832	840.5
90	691.8	755.4
120	719.6	799.8
150	793.3	838.4

The concrete sample A affected by bacteria *Acidithiobacillus thiooxidans* was found to have better leaching performance of calcium ions over a whole period of the experiment. After the experiment, the final concentrations of dissolved Ca ions were determined of 793.3 mg/L and 838.4 mg/L for sample A and B, respectively.

Comparison of quantities of calcium ions leached out due to biological corrosion from samples made of two different mixtures during 150-day experiments corresponding to 1 g of concrete sample is shown in figure 1.

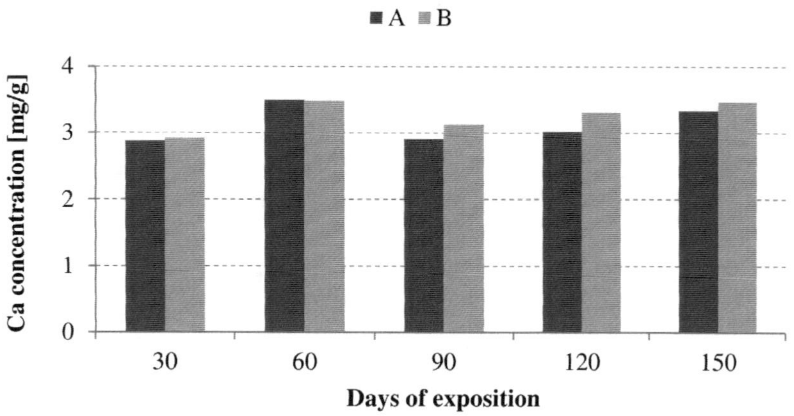

Figure 1. Dissolved calcium ions corresponding to 1 g of concrete samples.

The most intensive leaching of calcium ions (3.497 mg/g) during the 150 days of exposition was observed for concrete sample A after 60 days of exposition as it can be seen in figure 1. The Ca ions leaching was observed to be the lowest (2.877 mg/g) for sample A after 30 days of exposition (figure 1). The concentration of dissolved Ca ions in leachates was observed to be higher for sample B compared to the sample A during the experiment except the measurement after the 60-day exposition.

3.2. Si leaching due to biological corrosion

Quantities of dissolved silicon ions during the experiment are illustrated in table 4.

Table 4. Concentrations of dissolved Si ions.

Days of exposition	Si concentration (mg/L)	
	Sample A	Sample B
30	876.9	784
60	1386	1184
90	484.9	488.2
120	669.3	819.3
150	1042	1361

Table 4 illustrates the changes in the concentrations of Si ions in the leachates of samples A and B during the experiment. The concentration of silicon ions in liquid phase for sample A was lower till 60 days of the experiment and subsequently the concentration increased when compared to sample B.

Leaching performance of silicon in sulfate environment for concrete samples A and B corresponding to 1 g of concrete samples is illustrated in figure 2.

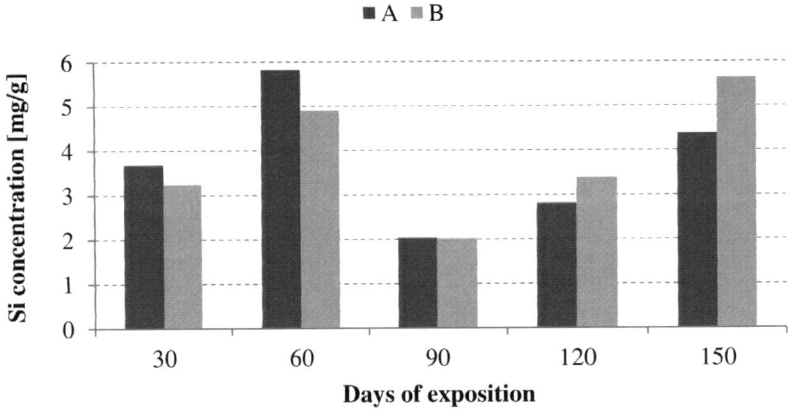

Figure 2. Dissolved silicon ions corresponding to 1 g of concrete samples.

Based on the leached-out masses of silicon after the experiment, sample B was found to be less durable when compared to sample A. Achieved concentrations of dissolved silicon ions were in the range of 2.038 - 5.826 mg/g and 2.021 – 5.634 mg/g for sample A and B, respectively. The concrete sample B with slag addition was found to have better leaching performance of silicon ions than sample

A during 90 days of experiment. The lower durability of sample B with slag addition was detected after 90 days of experiment. As reported in our previous study [14], granulated blast furnace slag based cement composites in terms of silicon ion leaching are more durable in short term experiment. The final concentration of Si ions in leachate of sample A affected by bacteria *Acidithiobacillus thiooxidans* (5.634 mg/g) was observed to be 1.29 times lower than for sample B.

However, to definitely confirm better performance of the analysed samples in the sulphate environment in terms of the dissolved calcium and silicon ions, the calculation of the percentage of leached out ions related to the total ion content and calculations of the leaching rates is needed and is in progress.

4. Conclusion

This paper presents the partial results of the study of the deterioration of concrete samples caused by microbiological attack. Attention was paid to the leachability of the most important components of cement matrix – silicon and calcium compounds.

Many factors, such as elemental chemical composition, replacement rate of cement with slag and slag particle size induce a great variability of the efficiency of the admixture. The 65% wt. addition of antimicrobial activated granulated blast furnace slag was not confirmed to be more resistant, so further research is, indeed, necessary.

5. References

[1] Gutiérrez-Padilla M, Bielefeldt A, Ovtchinnikov S, Hernandez M and Silverstein J 2010 Biogenic sulfuric acid attack on different types of commercially produced concrete sewer pipes *Cement and Concrete Research* **40** (2) 293–301

[2] Yuan H, Dangla P, Chatellier P and Chaussadent T 2015 Degradation modeling of concrete submitted to biogenic acid attack *Cement and Concrete Research* **70** 29–38

[3] Truper H G 1984 *Microorganisms and the Sulphur Cycle* Muller K A and Krebs B (eds) Sulphur, its Significance for Chemistry, for the Geo-, Bio-, and Cosmosphere and Technology, Studies in inorganic chemistry (Amsterdam: Elsevier Science Publishers) p 351–365

[4] Kelly D P 1982 Biochemistry of the chemolithotrophic oxidation of inorganic sulfur *Philosophical Transactions of the Royal Society London Series B* **298** 499–528

[5] Bock E and Sand W 1986 Applied electron microscopy on the biogenic destruction of concrete and blocks-use of the transmission electron microscope for identification of mineral acid producing bacteria. Bayles J, Gouda G R and Nisperos A (eds.) *Proc. of 8th Int. Conf. on Cement Microscopy, International Cement Microscopy Association (Duncanville, Texas)*

[6] Mori T, Nonaka T, Tazaki K, Koga M, Hikosaka Y and Noda S 1992 Interactions of nutrients, moisture and pH on microbial corrosion of concrete sewer pipes *Water Research* **26** (1) 29–37

[7] Topçu İ B 2013 High-volume ground granulated blast furnace slag (GGBFS) concrete *Eco-Efficient Concrete* 218–240

[8] Escalante-Garcia J I, Castro-Borges P, Gorokhovsky A and Rodriguez–Varela F J 2014 Portland cement-blast furnace slag mortars activated using waterglass: Effect of temperature and alkali concentration *Construction and Building Materials* **66** 323–328

[9] Gjorv O E 1989 Alkali activation of a Norwegian granulated blast furnace slag American Concrete Institute, Detroit *Proc. 3th Int. Conf. Trondheim, Norway*, 2, SP114-73 p 1501–

1517

[10] Hakkinen T 1993 Influence of slag content on the microstmcture, permeability and mechanical properties of concrete *Cement and Concrete Research* **23** 407–421

[11] Wild S, Kinuthia J M, Robinson R B and Humphreys I 1996 Effects of ground granulated blast furnace slag (ggbs) on the strength and swelling properties of lime-stabilized kaolinite in the presence of sulphates *Clay Minerals* **31** 423–433

[12] Karavajko G I, Rossi G, Agate A D, Groudev S N and Avakyan Z A 1988 Biogeotechnology of metals, Centre of projects GKNT, Moscow

[13] Sand W, Dumas T and Marcdargent S 1994 Accelerated biogenic sulfuric-acid corrosion test for evaluating the performance of calcium-aluminate based concrete in sewage applications *Microbiologically Influences Corrosion Testing*, ASTM, Philadelphia, Pennsylvania, USA 234–249

[14] Ondrejka Harbulakova V, Estokova A, Stevulova N, Terpakova E and Foraiova Katarina 2013 Deterioration of concrete composites placed in several pipelines evaluated by calcium and silicon ions changes *SGEM 2013: 13th Int. Multidisciplinary Scientific Geoconference: Ecology, Economics, Education and Legislation (Bulgaria, Albena, 16–22 June 2013)* vol 1 p 267–274

Acknowledgments

The authors are grateful to the Slovak VEGA Grant Agency for financial support of the project VEGA 2/0145/15.

IOP Conf. Series: Materials Science and Engineering **96** (2015) 012015 doi:10.1088/1757-899X/96/1/012015

The Investigation of Properties of Insulating Refractory Concrete with Portland Cement Binder

A Kudžma[1,2], V Antonovič[1], R Stonys[1] and J Škamat[1]
[1]Vilnius Gediminas Technical University, Scientific Institute of Thermal Insulation, Linkmenų Str.28, Vilnius, LT-08217, Lithuania

E-mail: andrius.kudzma@vgtu.lt

Abstract. The present work contains the results of experimental study on properties of insulating refractory concrete created on the basis of Portland cement (PC) binder and modified with microsilica (MS). The experimental compositions were made using Portland cement, lightweight aggregates (expanded clay and vermiculite) and microsilica additives. It was established that MS additives enable significant improvement of mechanical properties and thermal shock resistance of PC-based insulating concrete with values comparable to insulating refractory concrete based on calcium aluminate cement.

1. Introduction

In heating equipment, such as water boilers, boiler utilizers, etc., the operating temperature usually does not exceed 600–1000 °C. Therefore, the lining of such units is made of insulating refractory concretes with calcium aluminate cement (CAC) with Al_2O_3 content not exceeding 40 percent. One of the possible ways to reduce the cost of the above-mentioned refractory concrete is the replacement of the alumina cement with Portland cement (PC), which is approximately by four times cheaper than the alumina cement.

Refractory concretes with PC and dispersive additives have been widely used in various countries from 1950 to 1980. Concretes were made with fireclay, expanded clay, vermiculite, perlite and other fillers. However, main disadvantages of these concretes are particularly high (70%) decrease of compressive strength after firing the concrete at 600–800 °C, relatively low operating temperature – 1100 °C [1], risk of explosion during first drying and heating [2]. Moreover, thermal shock resistance of these concretes is very low, it often reaches only 5–12 cycles of heating and cooling in water. For these reasons, the use of concrete fell considerably, and in most countries it is practically no longer used.

[2] To whom any correspondence should be addressed.

When determining thermal resistance of hardened PC paste, the effect of $Ca(OH)_2$ formed during its setting on the fired cement paste and concrete failure should be taken into account [3, 4]. $Ca(OH)_2$ crystal hydrates with a size of 10^{-6} m lose water when heated and are replaced by CaO crystals with the characteristic size of 10^{-9} m [5]. Since CaO surface is very big, it relatively quickly re-hydrates. Under humid conditions, this process is even faster. It is stated in [6] that during the second hydration, the volume of $Ca(OH)_2$ increases by 44%. This causes full decomposition of cement stone.

In order to use PC in refractory concretes, the binder composites consisting of PC and various dispersive additives are used (average particle size is close to the average size of cement particles ~15 μm). At high temperature the dispersive additives react with CaO. The latter reacts with silica, alumina and iron oxides most easily.

The dispersive additives and combinations of heat-resistant binder are used with PC not only for CaO binding, but also for the improvement of physical and mechanical properties of composites and increase of fire resistance. Various additives were used for that purpose: quartz, refractory clay, dispersive fireclay, metallurgical slag, used ground refractory concrete, chromite, etc. [7].

Interesting results were obtained analyzing the PC mixtures with aluminium hydroxide additives ($Al(OH)_3$) [8], which showed that the thermal characteristics of such binder can match the properties of heat-resistant compositions with CAC. The paper reveals that during the hardening of the composition $Al(OH)_3$ with an average particle size of 4.5 μm reacts with cement components resulting in the formation of $Ca_3Al_2(SiO_4)(OH)_8$ crystalline hydrate. At temperature of 1200 °C $Al(OH)_3$ initiates the formation of calcium- and aliumo-silicates as well. Most importantly, in the composition with $Al(OH)_3$ during the curing and drying at 105 °C less calcium hydroxide than in the composition with pure PC is formed. After firing at 500 °C, free calcium oxide was not found in the substance.

PC compositions with SiO_2 and *Aloxil* additives (silica-aluminium compositions with different amounts of Al_2O_3, with particle size of ~3 μm) were analyzed in the paper [9]. It was found that fire resistance of the composition with *Aloxil* 80 (Al_2O_3 content 78%) has increased only slightly (~50 °C), and fire resistance of the composition containing the additive SiO_2 is much lower than that of Portland cement. It should be stated that all the components used improve the quality of refractory concrete matrix. However, when commenting on these works [8, 9], it should be noted that the dispersive components with aluminium oxide are very expensive compared with the price of PC, therefore, their use in large amounts is not economically justified.

As it is known, the SiO_2 micro-dust is an effective pozzolan which binds $Ca(OH)_2$ well. Moreover, this additive increases the strength characteristics of PC materials (mortars, concretes) [10, 11]. Most research on the effect of temperature on the properties of compositions of PC and SiO_2 micro-dust have been carried out to determine the impact of fire on simple building materials. However, the results obtained are also important when dealing with problems related to the application of Portland cement in heat resistant materials. Previous research by the authors [12] as well as the analysis of others publications showed that PC with an effective additive such as microsilica can be used for creation of various compositions of refractory concretes. The insulating refractory castable concrete with expanded clay and vermiculite filler and binder of Portland cement has been analyzed in this paper. The research results on concrete with fire resistant calcium aluminate cement are provided for comparison.

2. Experimental set-up

The following materials were used in the present work:

- Portland cement (PC) CEM I 42.5 R produced by "Akmenės cementas" (Lithuania): main mineral phase – C_3S (~59%); minor phases – C_2S (~14%), C_3A (~9%), C_4AF (~11%), where C denotes – CaO, A – Al_2O_3, F – Fe_2O_3, S – SiO_2. Specific surface by Blaine method – 3200 cm^2/g; chemical composition is given in Table 1.
- Calcium aluminate cement (CAC) "Istra 40" produced by "Calucem" (Croatia): main mineral phase – CA; minor phases – C_4AF, C_2AS and $C_{12}A_7$. Specific surface by Blaine method – 3000 cm^2/g; chemical composition is given in Table 1.
- Vermiculite (1-3 mm) produced by "Vermipu OY" (Finland): bulk density – 95–100 kg/m^3.
- Expanded clay (0-4 mm) produced by "Palemono keramika" (Lithuania): ~820 kg/m^3.
- Dispersive expanded clay produced by "Palemono keramika" (Lithuania): ~930 kg/m^3.
- Microsilica (MS) RW-Fuller produced by "RW Silicium GmbH" (SiO_2 – 96.1%): 410 kg/m^3.
- Deflocculant (NT) sodium tripolyphosphate $Na_5P_3O_{10}$.

Table 1. Chemical composition of cements, wt. %.

	CaO	SiO_2	Al_2O_3	Fe_2O_3	MgO	Na_2O	K_2O	SO_3	L.O.I.
PC	54.90	20.50	5.41	4.41	3.62	0.12	0.89	0.98	2.73
AC	37–40	≤ 6	38–42	13–17	< 1.5	-	-	< 0.4	-

The temperature of exothermic reaction (EXO) during the hydration and hardening of concretes was determined according to the "Alcoa Industrial Chemicals" method [13]. The samples were prepared by mechanical mixing (2 min) of dry compositions (Table 2) with water and were cast in the moulds (100×100×100 mm) under light vibration (10 sec). The cast sample (EXO specimen) weighted about 1.5 kg. The filled mould was put into the insulating box and thermocouple was stuck into a freshly mixed composition. The temperature variation inside the mix was recorded along with the time of measurement with the help of a data logger connected with thermocouple and PC.

The cubes of the size 70×70×70 mm were formed for all types of concrete. After 3 days of normal curing, the samples were dried at a temperature of 110±5 °C for 48 h. Next, they were kept for 5 h under the investigated temperatures of 800 °C and 1000 °C in an electronic controller furnace and cooled. The shrinkage of samples was measured after heating at each temperature (105 °C, 800 °C and 1000 °C).

Thermal shock resistance of the investigated concretes has been evaluated by heating-cooling test, where a concrete sample (70×70×70 mm) is heated for 60 min at 800 °C and then cooled for 20 min in air current. Such cycles are repeated until the samples break or lose 20% of their mass and if it does not break, residual strength value is measured.

3. Results and discussion

The experimental concrete compositions have been created on the basis of conventional composition (B0) of lightweight refractory concrete containing 40% of alumina cement, 50% of lightweight coarse aggregates (vermiculite and expanded clay) and 10% of dispersive filler (see Table 2). In experimental compositions (B1, B2 and B3), Portland cement has completely replaced alumina cement and dispersive expanded clay has been partially replaced with 1.5% and 2.5% of MS (B2 and B3 respectively). Deflocculant NT was used to decrease water consumption in MS-containing compositions (B2, B3).

Table 2. Concretes compositions, wt. %.

	B0	B1	B2	B3
MS	0	0	1.5	2.5
Expanded clay	40	40	40	40
Dispersive expanded clay	10	10	8.5	7.5
Vermiculite	10	10	10	10
CAC	40	0	0	0
PC	0	40	40	40
NT	0	0	0.35	0.35
Water *	32	33	31	31

* - over 100 % of dry components

The kinetics of binder hydration process is mainly determined by the mineralogical phases present in the cement. PC minerals, according to their reactivity, may be ordered as $C_3A > C_3S > C_4AF > C_2S$ [14]. For CAC minerals, this order is $C_{12}A_7 > CA > CA_2 > CA_6$ [15]. Thus, the presence of very reactive C_3A causes early beginning of hydration in B1 composition: PC-based concrete hydration processes occurred after ~60–80 minutes (figure 1). In CAC-based concrete, less reactive CA phase dominates and hydration starts later – after ~150 minutes with significant increasing of the heat release rate after ~4 hours. Exoterm profiles presented in figure 1 show that the hydration process for both CAC-based (B0) and PC-based concrete without MS additive (B1) develops during the first 6–7 hours with the main exopeaks occurring after ~360 minutes for B0 and after ~460 minutes for B1. Hydration of B0 is accompanied by more intensive heat release with maximum process temperature reaching 70 °C after ~6 hours, while for B1 much less main exothermic effect was obtained with the temperature reaching 37–40 °C. Very similar curves were obtained for PC-based compositions (B2 and B3) containing MS and deflocculated with NT. The combination of MS additive with NT deflocculant in denoted proportions (Table 2) has led to the decrease in the main exothermic effect temperature and to significant hydration process delay: main exothermic effect duration has increased about two times and the exothermic peak has appeared after ~16–17 hours.

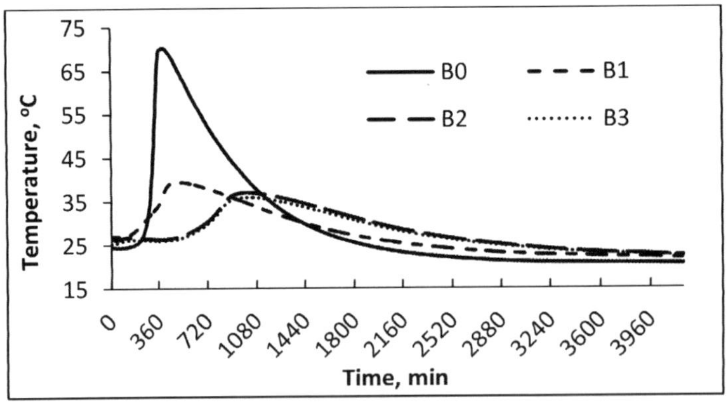

Figure 1. Development of EXO reaction in concretes.

The initial density of concretes after normal curing at 20 °C ranged from 1630 to 1680 kg/m^3 (see figure 2) with the lower value corresponding to PC-based composition without MS additive (B1) and with the upper value obtained on experimental compositions containing MS (B2 and B3). The reference sample (B0) showed medium value of 1660 kg/m^3. After drying (110 °C) accompanied by the evaporation of weakly bound water, the samples lost ~14–17 % of their density and the most significant decrease in density was obtained on B1 composition having the biggest water consumption. After firing (800 and 1000 °C), where destruction of hydraulic bonds and evaporation of chemically combined hydraulic phases take place, concretes gain their final density. Thus, after firing at 800 °C as well as at 1000 °C, the density of B0 and MS-unmodified B1 is 1270 kg/m^3, while for B2 and B3 containing MS additive final density increased by 4% and has a value of 1320 kg/m^3. Thus, PC-based cement with MS ultrafine filler tends to form slightly denser structure compared with CAC-based reference and PC-based cement without MS, what may be attributed to the MS ability to provide better structure compaction due to its morphology.

Heating of the samples is followed by dehydration, when, firstly, free water is removed and then dissociation of hydrates takes place along with some possible polymorphic transitions and formation of the new phases. It all causes local volume changes – expansion and/or shrinkage, determining the total deformation of the sample. Here, the gradual shrinkage has been determined on CAC-based concrete with the increase of heating temperature (figure 3, B0). The deformation after drying does not differ significantly for all the samples and is in the range between -0.15 (for B1) and -0.24 (B2). More noticeable shrinkage was determined on PC-based samples after firing at 800 °C with the increased values for the samples modified with MS. However, maximum deformation did not exceed ~1.1% and further heating (at 1000 °C) did not increase the shrinkage of PC-based samples: on the contrary, the deformation slightly reduced, most likely due to expansion of new formed phases.

Developing of structural discontinuity during heating caused by dehydration and differences in thermal deformation between the aggregates and the binder lead to the loss in strength with increasing temperature. For CAC concrete (B0), gradual wane in cold crushing strength (CCS) from 15 MPa to 9 MPa is established when the sample is exposed to temperature between 20 and 1000 °C (see figure 4). For PC-based samples (B1, B2 and B3), the nature of the curve is different and increasing strength is obtained after drying. It has been shown already in [16] that "*when temperature is lower than 350 °C (for ordinary Portland cement concrete)..., the strength of the concrete is increased due to accelerated hydration and strengthening effect caused by drying*". Further increasing temperature lead to the noticeable loss in CCS, largely caused by different expansion of the aggregates and the binder along with possible rehydration. The residual CCS for B1 composition after firing at 800 °C was 9.3 MPa and dramatic drop to 4.5 MPa has been established after firing at 1000 °C; such values yield significantly to CAC concrete.

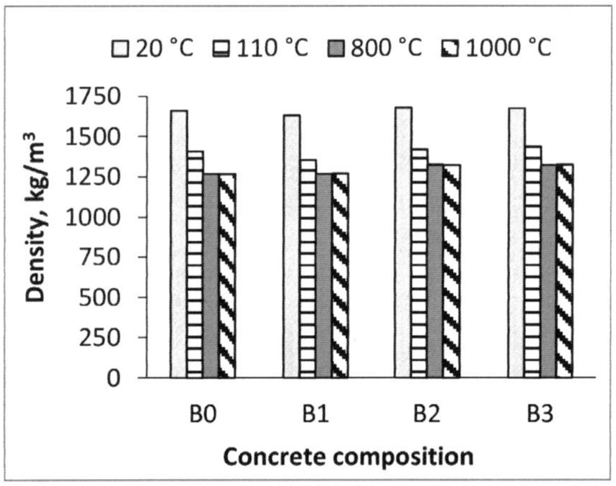

Figure 2. Density changes in concretes after curing (20 °C), drying (110 °C) and firing (800 °C, 1000 °C).

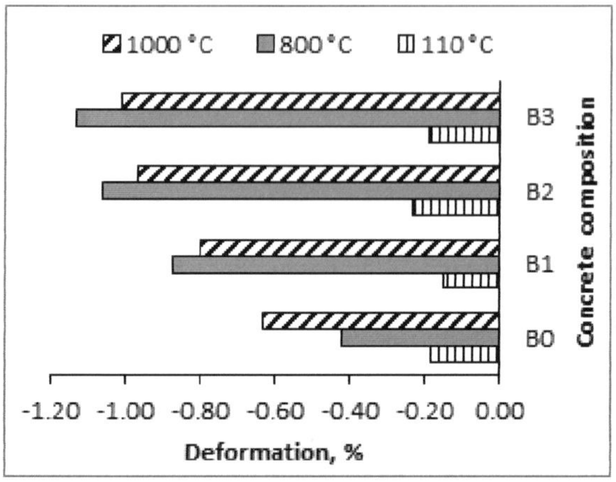

Figure 3. Concrete shrinkage after drying (110 °C) and firing (800 °C, 1000 °C).

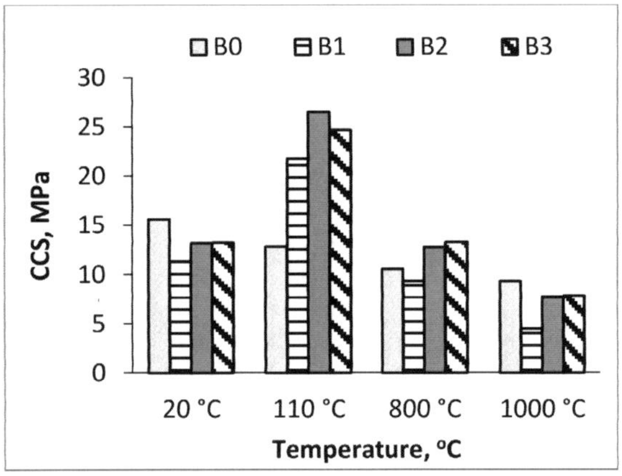

Figure 4. CCS of concrete after curing (20 °C), drying (110 °C) and firing (800 °C, 1000 °C).

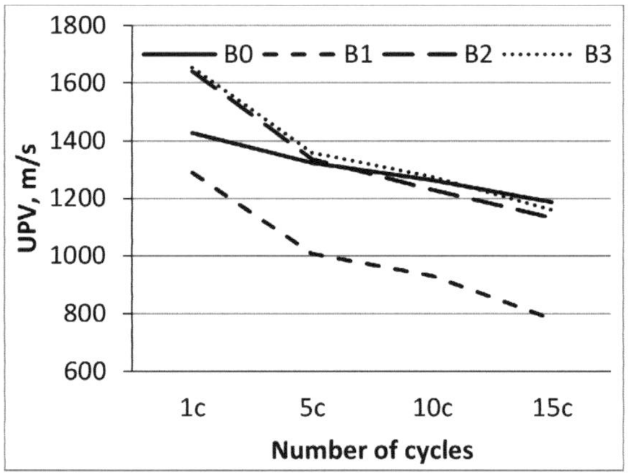

Figure 5. Ultrasonic pulse velocity variation after heating-cooling cycles.

Compressive strength of PC concrete has been notably improved with MS additives. Thus, the following increase in CCS values has been obtained on B2 and B3 compositions compared with B1: 17% (for both) – after normal curing; 22% and 15% (respectively) – after drying; 38% and 43% – after firing at 800 °C; 71% and 73% – after firing at 1000 °C. The obtained results show that positive impact of MS additive on compressive strength of PC concrete is expressed more significantly when concrete is exposed to high temperatures (800 and 1000 °C). It must be also noted that better CCS values here correspond to the higher MS content in the composition. Moreover, CAC concrete yields to MS-containing PC concrete in strength after firing at 800 °C: 22% (B2) and 27% (B3) better CCS values have been obtained on PC concrete with MS additives compared with CAC composition.

Heating-cooling procedure and measuring of ultrasonic pulse velocity (UPV) after 1, 5, 10 and 15 heating-cooling cycles have been employed to evaluate thermal shock resistance of experimental compositions. The results of UPV measurements are presented in figure 5. Ultrasonic wave flows faster in materials with better compaction; the structural damages disturb ultrasonic wave propagation and decrease its velocity. Here, sloping UPV curves illustrate clearly decreasing UPV related to developing of microstructure discontinuity occurring during heating of all the investigated samples. The arrangement of UPV values after the first stage of thermal shock test is in a good correlation with CCS values obtained after firing at 800 °C, being B0 between B1 (lower value) and B2, B3 (upper values). The UPV in B1 decreases critically with further thermal cycles. The loss of the UPV in MS-modified compositions (B2 and B3) has the same nature: first they showed more rapid decrease in UPV compared with B0 and after 5 cycles their results became equal to UPV, exhibited by B0. Finally, B1 showed much less UP velocities while very similar UPV values have been obtained on B0, B2 and B3.

Table 3. The residual CCS of concrete after 15 heating-cooling cycles, %.

Concrete composition	Residual CCS, %
B0	66
B1	66
B2	78
B3	73

Moreover, after 15 cycles, the CCS test showed less loss in compressive strength for MS-modified PC concrete compared with CAC-based composition: residual CCS of 78% and 73% (from CCS after firing at 800 °C) has been obtained for B2 and B3 respectively (see Table 3). Thus, it all testifies good suitability of MS additives in improving of thermal resistance of PC concretes.

4. Summary

1. Compressive strength and thermal shock resistance of the concrete with Portland cement have been significantly improved by addition of microsilica.
2. The properties of the created PC-based insulating refractory concretes containing expanded clay aggregates and modified with microsilica additives have been obtained similar to those of insulating refractory concrete with calcium aluminate cement.

5. References

[1] 1983 *The guide to construction of thermal units from refractory concrete* (Moscow: Strojizdat)
[2] Seifert H and Tomsu F 2001 *On the past and the future of refractory castables. Silika* **5–6** 178–183
[3] Nekrasov K D 1957 *Refractory Concrete* (Moscow: Promstrojizdat)
[4] Harada T, Takeda J, Yamane S and Faramura F 1972 Strength, elasticity and thermal properties of concrete subjected to elevated temperatures *Int. Seminar on concrete for nuclear reactors* (vol 34 Am. Constr. Inst.) pp 377 – 406
[5] Vydra V, Vodak F, Kapičkova O and Hoškova Š 2001 *Cement and Concrete Research* **31** 1023–26
[6] Hung Y, Wong Y, Poon C and Anson M 2001 *Cement and Concrete Research* **31** 1065 – 73
[7] Goberis S and Antonovič V 2007 *Refractory Chamotte Castables* (Vilnius: Technika)
[8] Ševčik V and Škvara J 2002 *Ceramics – Silicaty* **46** 104 – 109
[9] Haubler K and Seifert H 2001 Mikrofiller for Portland cement bond *XIV Int. Conf. on refractory*

castables (*Prague, 23-24 October 2001*) pp 123 – 131
[10] Feldman R and Cheng H 1985 *Cement and Concrete Research* **15** 943 – 952
[11] Rao G 2001 *Cement and Concrete Research* **31** 1141 – 46
[12] Antonovič V and Goberis S 2003 *Materials Science = Medžiagotyra* **9** 379 – 382
[13] 1999 *Calcium Aluminate Cements. Cemet Test Methods. Exothermic Reaction (EXO)* (Alcoa Industrial Chemicals)
[14] Kishi T and Maekawa K 1995 Multi – Component Model for Hydration Heating of Portland Cement *Translation from Proc. of JSCE* (*Japan, November 1995*) **29** pp 97 – 109
[15] Parr Ch 2005 *J. Tech. Assoc. Refract.* **25** 78 – 88
[16] Shi C, Roy D and Krivenko P 2006 *Alkali-Activated Cements and Concretes* (CRC Press) pp 213 – 216

Mechanical and Acoustical Properties of Bushings Made of Low-Alloyed Materials and Used in Brake Systems of Transport Vehicles

V Mironov[1,3], **P Stankevich**[1], **A Tatarinov**[1], **V Zemchenkov**[1] **and I Boiko**[2]
[1]Riga Technical University, Laboratory of Powder Materials, Kalku Str. 1, Riga, LV-1658, Latvia
[2]Riga Technical University, Faculty of Transport and Mechanical Engineering, Institute of Mechanical Engineering, Ezermalas Str. 6k, Riga, LV-1006, Latvia

E-mail: viktors.mironovs@gmail.com

Abstract. The increasing freight volumes and the speed of transport, as well as increasing demands with regard to safety have made the search and development of new anti-friction materials a topical task. The main requirements to the bearing assemblies include reliability, durability, simplicity and ease of maintenance. The paper considers the features of the manufacture of braking sliding sleeves/ bushings for transport vehicles made from sintered low-alloy powder materials based on Fe-C-Ni with Ni and Mo contents of 0.3%. The paper presents data on the microstructure of the sintered samples of the products and research of some physical and mechanical properties. Surface quality and friction coefficient were evaluated by surface profiling and tribological tests, correspondingly. Particular attention is paid to the use of ultrasonic testing for quality control of the products.

1. Introduction

Sliding sleeves and bushes made of sintered powder materials are widely used in the manufacturing of machinery, railway and various vehicles. Braking assemblies with the braking leverage transmission have specific operating conditions. It presents a system of levers that transmit forces arising from the pressure of the compressed air on the piston of the brake cylinder to the friction units of the brake system. The connection units of the brake system experience great dynamic and vibration loads. In addition, increased friction increases in the connection units during braking and brake release. In order to reduce friction, bearing bushes with heightened anti-friction properties are used. The use of metal powder

---

[3] To whom any correspondence should be addressed.

bushings(MPB) allows extension of the service life of bearing units several times at the same time reducing operational, maintenance and repair costs [1,2].

Given the importance of braking system assemblies, it is necessary to improve the quality assurance of the components. By means of non-destructive testing, in particular, ultrasonic testing (UT), it is possible to predict the mechanical properties of the braking elements, to assess the presence of pores and defects. It was shown in [3] that the main feature of sintered products distinguishing them from powder metals is the presence of free surface of the pores as a major factor in the development of phase transformations. The free pore surface manifests itself, on the one hand, as the stress relaxation space, and, on the other hand, as a place where various defects emerge. In this case, it is advisable to use UT methods.

One of the main advantages of MPB is the presence of pores on their surface that facilitates the formation of a stable oil film. As a result of the pre-dip of MPB in hot oil, the friction surface is provided by the lubricating film for a long time. MPB are effective substitutes of bearing products made of bronze and brass. The optimal physical and mechanical properties were established to be the following: porosity in the range of 15 – 20%, oil absorption – not less than 1.8%, the compression strength –180 MPa. To improve the strength properties, alloy powder materials containing nickel and manganese in the range from 1.5 to 2.5% are commonly used [4]. This provides an increase in strength and hardness. However, it should be emphasized that these additives significantly increase the cost of metal powder products [5].

The purpose of this paper was to examine the possibility of manufacturing of MPB for leverage brake system in transport using low-alloy powder compositions [6] with the contents of Ni and Mo up to 0.3% and to assess the impact of residual porosity on the mechanical and tribological properties of MPB. It has been tasked to evaluate the use of UT to control the degree of porosity and the presence of defects.

2. Experimental studies

2.1. Mechanical properties and microstructure

Powder mixtures containing the alloying elements Ni, Cu, and Mo were used to produce the samples (table 1). Mixtures AHC100.29 + Cu + Ni + C produced by Hoganas AB (mixture 1) and the same mixture with an increased Cu content (mixture 3) [7] were assumed as basic materials. MPB samples were made also from the low-alloy experimental mixture with Ni and Mo content less than 0.3% and reduced phosphorus content (mixture 2).

Table 1. Specification of materials for manufacturing of MPB samples.

Mixture	Components [mass fraction by weight, %]								Powder properties		
	Fe,%	Ni,%	Cu, %	Mo, %	S, %	P, %	C, %	Kenno-lube	Fluidity [s/50 g]	Bulk density [g/cm^3]	Compactibility [g/cm^3]
Mixture 1	94	2.0	2.0	0.5	0.2	0.1	1.2	1.0	38	3.18	7.05
Mixture 2	96.5	0.22	2.27	0.28	0.04	0.01	0.68	0.7	40	3.15	6.85
Mixture 3	94	2.5	2.0	0.5	0.2	0.1	0.7	1.0	40	3.05	6.90

Pressing of MPB samples specified in table 2 was carried out in a rigid mold using a hydraulic press at a pressure of 600 MPa. Sintering was performed in a protective atmosphere during 40 minutes at a temperature of 1120^{0}C.

Table 2. Main geometrical parameters and some physical and mechanical properties of MPB.

Outer diameter [mm]	Wall thickness [mm]	Height [mm]	Porosity [%]	Hardness [HB]
40	4	30	15	120
80	6	80	18	90

The results of the study of physical and mechanical properties of the MPB samples made of mixtures specified in table 1 are shown in figure 1. The compression strength of the samples was determined according to ISO2739: 2012 [8].

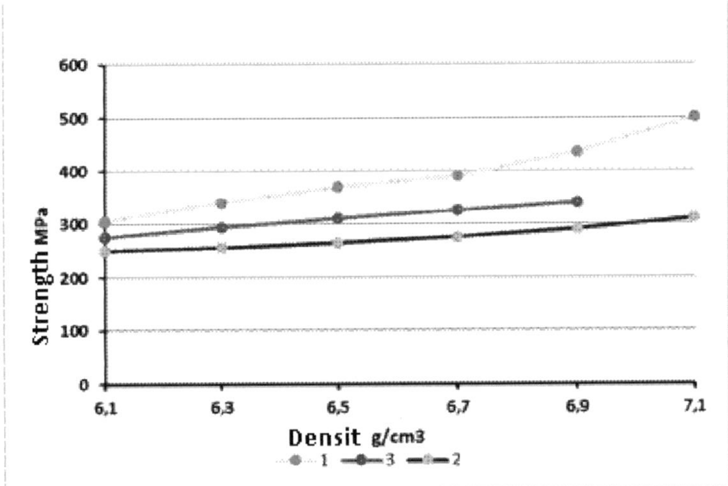

Figure 1. Compression strength of MPB samples vs. density.

The sintered material of low-alloy powder (mixture 2) has a strength somewhat lower than the control material (mixture 1). However, given the significantly lower content of costly components in mixture 2, it can be recommended for the MPB production.

The microstructure of a sample from mixture 2 reflects the uniformly distributed porosity (figure 2).

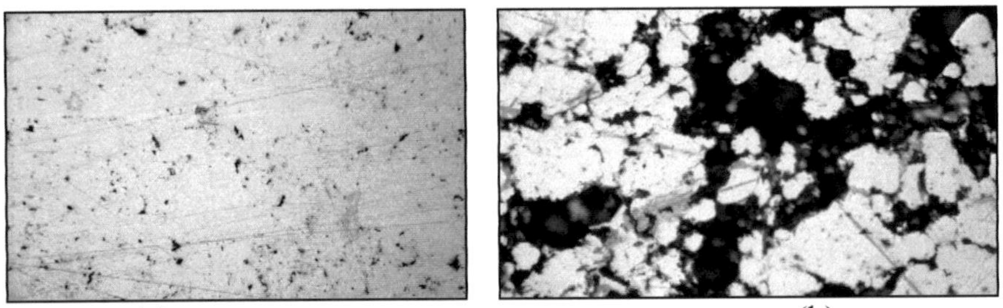

(a) (b)

Figure 2. Microstructure of MPB of low-alloy powder (Mixture 2): A -x50;B – x500.

Studies have shown that when mixtures 1 and 3 are used for MPB production, the problems with obtaining homogeneous samples by their chemical composition and structure appear, since the mixtures are produced by mixing the starting powders. Mixture 2 is made by spraying a steel melt with the predetermined chemical composition followed by the addition of solid lubricant. Therefore, to obtain MPB for brake systems designed to operate in the conditions of impact loading and complex stress, it is advisable to use a low alloy powder composition, such as mixture 2. However, the final conclusion can be drawn based on the tribological properties.

2.2. Surface quality and tribological properties

Surface quality of the low-alloyed MPB was investigated by using "Taylor Hobson Ltd" 3D measurement system. 2D and 3D measurements with stylus instrument were achieved by basic "stepping" method (figure 3, figure 4). Main roughness and surface texture parameters according to ISO 4287 "Surface texture: Profile method - Terms, definitions and surface texture parameters" and EN ISO 25178-2 "Surface texture: Areal - Part 2: Terms, definitions and surface texture parameters" respectively are given in table 3.

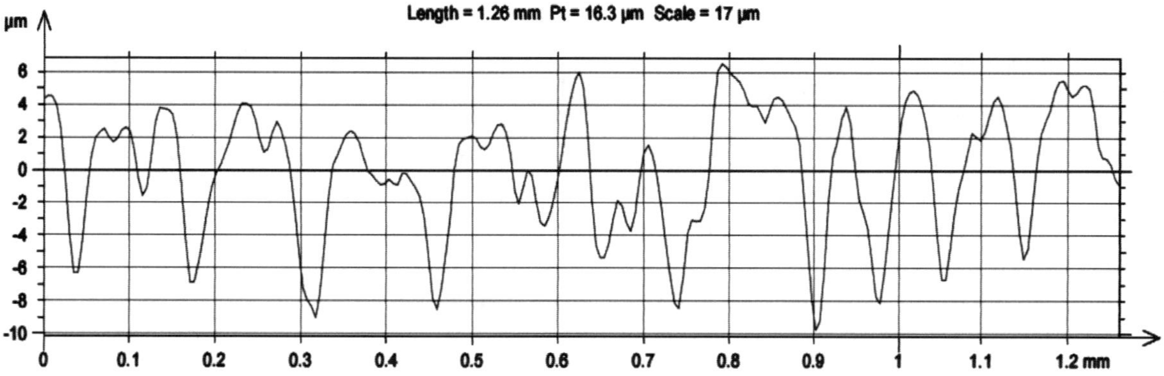

Figure 3. Surface profilogram of a low-alloyed powder part.

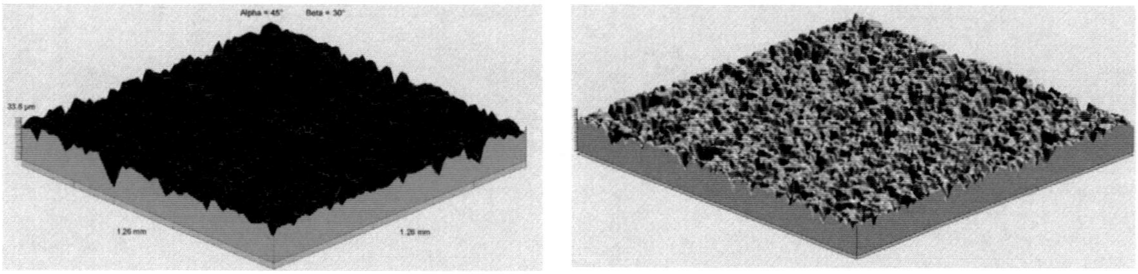

Figure 4. View of 3D surface image and photo simulation of the low-alloyed powder part.

Table 3. Roughness and surface texture parameters of the low-alloyed powder parts.

2D amplitude roughness parameters	Ra, [µm]	2.37
	Rz, [µm]	11.2
3D amplitude texture parameters	Sa, [µm]	2.65
	Ssk	-1.56
3D spatial parameter	Str	0.766
3D functional parameter	Vmc, [mm^3/mm^2]	0.0187
3D functional parameter	Sfd	2.56

The results of the measurement confirm that powder part of the new low-alloyed powder material has a relatively high surface quality. It should be mentioned that the surface is sufficiently close to isotropic since the value of parameter Str (texture aspect ratio of the surface, measures the isotropy of the surface) is close to 1 (maximal value of Str). The highest value of functional parameter Vmc (core material volume of the scale-limited surface, indicates a measure of the material forming the surface between various heights) proves a good bearing ability of the surface due to great core material volume in contact during normal exploitation time (after wear-in). The negative value of amplitude parameter Ssk (skewness of the scale-limited surface, represents the degree of symmetry of the surface heights about the mean plane) confirms good lubricant retention ability.

Tribological study was done using tribometer of “CSM Instruments” (Switzerland). The friction coefficient was evaluated using “ball-on-disk” testing without lubrication, when the tested example was used as a disc, but a high-strength steel ball with the contact radius 3 mm was used as a counter-body. The main parameters of testing were: force – 3 N; measurement period– 20 s, linear velocity – 0.05 m/s.

The friction coefficient as a function of time, laps (number of rotation cycles) and length of friction path for compacted and sintered powder parts are shown in figure 5. The stable curve of friction coefficient with a relatively small fluctuation along the length of friction is observed. The visual examination after testing in order to reveal any traces of wear or plastic deformation of the surface revealed that the standard friction process occurred during testing. The average value of friction coefficient for low-alloyed MPB was $\mu = 0.22$.

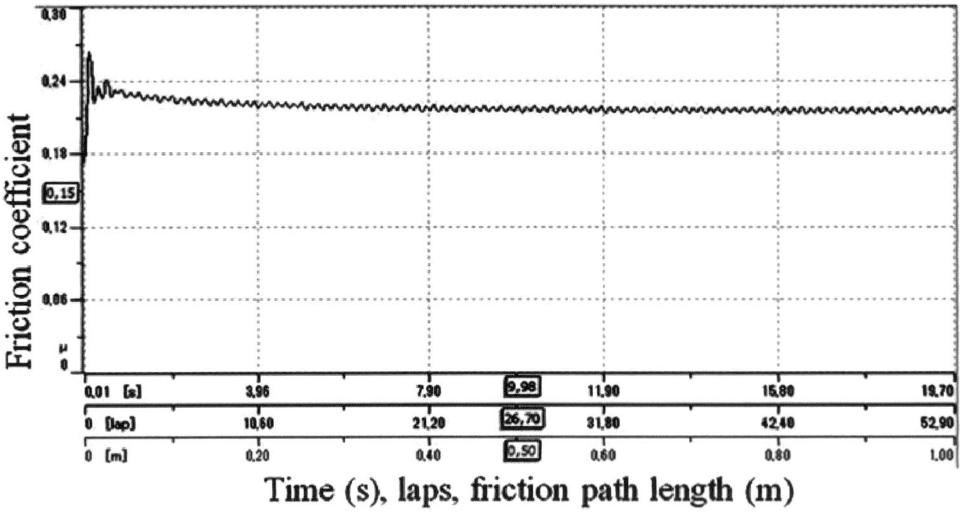

Figure 5. The friction coefficient as a function of time, laps and length of friction path for sintered powder part.

2.3. Ultrasonic testing of MPB

To assess the degree of homogeneity of MPB and the possibility of determining the possible cracks by means of ultrasound, the TOF (time-of-flight) method was used. Traditionally, the method demonstrates its efficacy in the detection of cracks in welds and metal products [8]. The method is based on sounding the zone between the transmitter and the receiver, and the analysis of signals generated by direct propagation of ultrasound, bottom reflections and diffraction reflections from the cracks if they appear. The purpose of our measurements was simultaneous estimation of heterogeneity of material properties of the PMB products and locations of cracks. For comparative examination, PMB samples with longitudinal cracks by height and PMB without visible defects were chosen.

Measurements were carried out in the surface transmission mode at a fixed acoustic base between the transmitter and the receiver of 20 mm (figure 6). The base was oriented perpendicular to the longitudinal axis of MPB to be perpendicular to the direction of the crack. To perform the measurements on the cylindrical surface, transducers with point contact were used requiring a small amount of liquid lubricant. Ultrasound excitation signals were sinusoidal pulses with duration of 2 periods at a carrier frequency of 1 MHz. The samples were measured in 4 sectors around the circumference and in 3 zones by height.

Figure 6. Illustration of ultrasonic measurements: on the left – a sample with a crack; on the right – the application of the sensors.

The results of ultrasonic measurements on the MPB surface showed that the samples were heterogeneous. The ultrasound velocity related to the longitudinal wave varied in different zones of the samples from 3.75 to 5.00 km/s, showing this heterogeneity related to the composition and porosity distribution and caused by the technological process. The distribution of ultrasound velocity of in the samples (figure 7) indicates a greater variability of properties by height than in the zones of perimeter. There was a sharp increase of ultrasound velocity in one of the end zones caused by the non-uniform powder compaction during manufacturing.

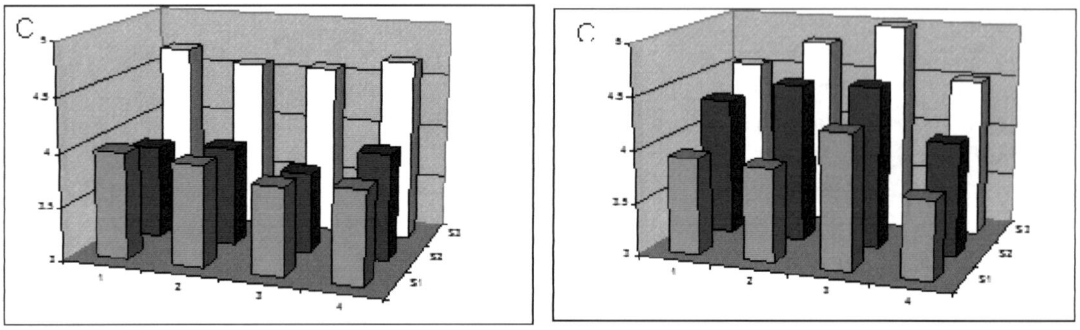

Figure 7. Distribution of ultrasound velocity C (km/s) in the perimeter (zones 1-4 by axis X) and by height (zones S1-S3 by axis Y) in two MPB samples.

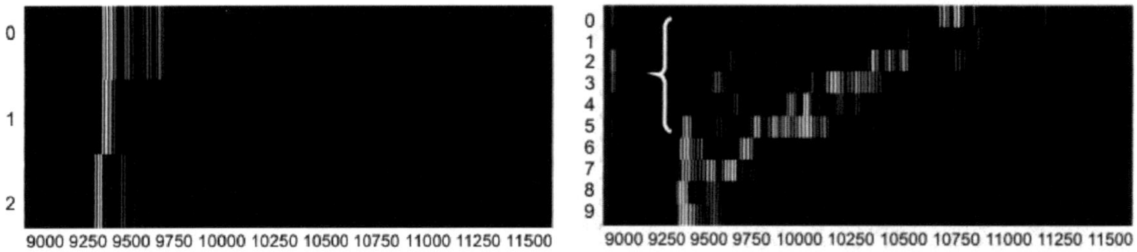

Figure 8. Plots of ultrasonic signals by samples' height (see description in the text): in the left–the sample without defects, in the right–the sample with a longitudinal crack. Brace marks the height of the crack.

In the area of the crack, ultrasonic signals were recorded in 10 zones by the sample's height, in order to observe the changes of ultrasonic signals related to the development of cracks. Figure 8 shows the two-dimensional plotting of plurality of ultrasonic signals obtained by height, where the abscissa axis is time of ultrasound propagation, the axis of ordinates is the zone of the sample's height, and the brightness corresponds to the signal amplitude. The two-dimensional plot of the ultrasonic signals shows a clear detection of the cracks by the lack of the direct propagation signal and the relative gain of delayed echoes from the opposite end of the sample. Thus, a simple analysis of signal plots can indicate the appearance of cracks very illustratively.

3. Conclusions

1. The use low-alloyed iron-based powders can improve strength characteristics of products without increasing their cost.
2. The achieved 2D surface roughness parameters, 3D texture parameters and the friction coefficient (average value 0.22) confirmed the appropriateness of the new low-alloy metal powder material on the iron base for producing bushing for braking systems.
3. TOF method of ultrasonic testing is able to detect cracks in the PMB, and also is promising to control the spatial deviations of the material properties in the PMB products.

4. References

[1] Ermakov S S and Vyaznikov N F 1990 Powdered steels and products *Mashinostroyenie*

[2] Dorofeev Y G, Geydarov Y G and Mamedov A T 1987 Peculiarities of obtaining of low-alloyed powder steel for constructional and anti-friction purpose *Poroshkovaya Metallurgiya* **9** 42 – 47

[3] Bernstein M L, Felgina S B, Kostyrko L N, Napara-Volgina S G and Radomyselskiy I D 1976 Investigation of properties of low-alloyed sintered steels *Sintered Construction Materials* (Kiev) pp 93 – 98

[4] Causton R J and Brian J W 1991 Performance charasteristics of a new sinter-hardening low-alloy steel *Industrial Heating* **58** 38 – 41

[5] MPIF Standard 35, Materials Standards for PM Structural Parts, 2007 Edition, MPIF

[6] Brian W J, Lindsley B, Howard G.R and Narasimhan K S 2009 Lean Hybrid Low-Alloy PM Molybdenum Steels *Hoeganaes Corp. USA* [Available]: www.gkn.com/hoeganaes/

[7] Hoganas Iron and Steel Powders for Sintered Components 2002 *Hoganas AB* p 394

[8] Manjula K, Vijayarekha K, Venkatraman B and Karthik D 2012 Ultrasonic Time of Flight Diffraction Technique for Weld Defects: A Review *Research J. of Applied Sciences, Engineering and Technology* **4** (24) 5525 – 33

Acknowledgements

This work has been partially supported by the European Social Fund within the project "Development of multifunctional nanocoatings for aviation and space techniques constructive parts protection" No.2013/0013/1DP/1.1.1.2.0/13/APIA/VIAA/027. In this work, RTU students Armands Leitans and Janis Lungevics were involved.

Performance of Lightweight Concrete based on Granulated Foamglass

M Popov[1,4], L Zakrevskaya[1], V Vaganov[2], S Hempel[3] and V Mechtcherine[3]
[1]Vladimir State University, Department of Construction Production, Vladimir, Russia
[2]Vladimir State University, Department of Physics and Applied Mathematics, Vladimir, Russia
[3]Institute of Construction Materials, TU Dresden, Dresden, Germany

E-mail: popov.maxim.1989@gmail.com

Abstract. The paper presents an investigation of lightweight concretes properties, based on granulated foamglass (GFG-LWC) aggregates. The application of granulated foamglass (GFG) in concrete might significantly reduce the volume of waste glass and enhance the recycling industry in order to improve environmental performance. The conducted experiments showed high strength and thermal properties for GFG-LWC. However, the use of GFG in concrete is associated with the risk of harmful alkali-silica reactions (ASR). Thus, one of the main aims was to study ASR manifestation in GFG-LWC. It was found that the lightweight concrete based on porous aggregates, and ordinary concrete, have different a mechanism of ASR. In GFG-LWC, microstructural changes, partial destruction of granules, and accumulation of silica hydro-gel in pores were observed. According to the existing methods of analysis of ASR manifestation in concrete, sample expansion was measured, however, this method was found to be not appropriate to indicate ASR in concrete with porous aggregates. Microstructural analysis and testing of the concrete strength are needed to evaluate the damage degree due to ASR. Low-alkali cement and various pozzolanic additives as preventive measures against ASR were chosen. The final composition of the GFG-LWC provides very good characteristics with respect to compressive strength, thermal conductivity and durability. On the whole, the potential for GFG-LWC has been identified.

1. Introduction

Due to the tightening of the requirements for energy efficiency in Russia, the construction industry needs materials which provide not only the necessary load-bearing capacity of structures, but also have low thermal conductivity. The lightweight concrete with porous aggregates might be used for such purposes [1].

[4] To whom any correspondence should be addressed.

Currently, the technology of waste glass processing into a highly porous granulated inorganic thermal insulation material – granulated foamglass (GFG) – has been actively developing in Russia. GFG can be used as an aggregate for lightweight concretes. It has a uniform distribution of cells, enclosed pores and a pronouncedly rough surface. Lightweight concrete based on GFG is characterized by high physical-mechanical and thermal characteristics [2–4]. Also, the application of GFG in concrete might significantly reduce the volume of waste glass and enhance the recycling industry in order to improve its environmental performance.

Due to the high content of amorphous silica in GFG, there is a necessity for a more thorough investigation of possible harmful alkali-silica reactions (ASR) in lightweight concretes properties, based on granulated foamglass (GFG-LWC). The problem of ASR in concretes has been investigated for many years. Usually, this problem occurs most frequently in the concrete containing heavy and fine reactive aggregates [5, 6]. ASR in lightweight concrete with GFG were also investigated by some scientists [7–9]. Some research results in these studies contradict each other. Currently, there is an absence of a clear understanding regarding the safety of GFG performance in cement composites, causing the need for studying ASR in GFG-LWC more comprehensively. Thus, the main aim was to study ASR manifestation in GFG-LWC.

2. Materials and methods

The following materials were used in the investigation at hand: GFG "Neoporm" from company "STES-Vladimir" (Russia), cement CEM 42.5 and CEM 42.5 NA (low alkali) from company Schwenk (Germany), fly ash from company Powerment (Germany) and microsilica from company Elkem (Norway). The chemical compositions of the materials are given in table 1.

Table 1. The chemical compositions of raw materials, the content of oxides by weight, %.

	CaO	SiO_2	Al_2O_3	SO_3	Fe_2O_3	MgO	Na_2O_{eq}	Others
Cement CEM I 42.5	63.38	18.66	6.31	3.31	3.03	3.01	0.87–1.01	1.29–1.43
Cement CEM I 42.5 NA	63.49	19.41	6.25	2.81	3.27	3.07	0.47–0.62	1.08–1.22
GFG	9.75	69.87	1.72	0.09	0.35	4.12	13.49	0.61
Microsilica	0.45	97.33	0.18	0.15	0.06	0.12	0.37	1.34
Flyash	4.91	51.63	26.69	0.83	7.43	1.04	4.08	3.39

The compressive strength of GFG-LWC was investigated according to the Russian Government Standard (RGS) 12730. The sample size used was 100x100x100 mm. Thermal conductivity of GFG-LWC was tested according RGS 7076. In this method a stationary heat flow through the concrete sample (directed perpendicular to the front face of the sample)was created and the density of the heat flow as well as the temperature of the opposite face of the sample were measured. The sample size used was 100x100x20 mm. The potential reactivity of GFG to ASR was determined according to RGS 8269, which methods are similar to those of ASTM 1260, i.e. the chemical analysis of GFG and the expansion tests on concrete prisms. Conditions of the experiments are described in the section "Results and Discussion". The size of the prepared specimens for the expansion tests was 160x40x40 mm. Scanning electron microscopy (SEM), energy dispersive X-ray spectroscopy (EDS), X-ray diffraction (XRD), X-ray fluorescence spectroscopy (XRF), and inductively coupled plasma atomic emission spectroscopy (ICP-AES) were used for characterizing the morphology, crystallinity and chemical composition of concrete constituents.

3. Results and discussion

Foamglass is a silicate material manufactured by thermoplastic methods. The results of the conducted XRF and XRD analysis show the content of amorphous silica in GFG to be about 70%. According to the RGS 8269, the granules were soaked in 1 M NaOH solution at a temperature of 80 °C for 24 hours. Concentration of silicon ions in the solution after soaking amounted to more than the allowed limit of 50 mmol/l [10]. This means that the GFG likely had a capability of interacting with cement alkalis. Therefore, further experiments on the concrete expansion were required.

Optimum compositions of GFG-LWC were developed. To do this, features and properties of GFG were taken into account, such as physic-mechanical characteristics of GFG depending on its fractional composition, features of cellular structure and the roughness of granules surfaces. GFG-LWC compositions with density of 400 to 800 kg/m^3, compressive strength of 2.2 to 6.5 MPa and thermal conductivity of 0.09–0.15 W/(m·K) were obtained. Table 2 shows the final GFG-LWC compositions and their main characteristics.

The density of the samples for concrete expansion tests was 700±50 kg/m^3. The experimental conditions of the expansion tests on concrete prisms were as follows: A) in the solution of 1M NaOH at 80 °C for 14 days; B) in a climatic chamber at 40 °C and 100% relative humidity for 12 months. The aggregate is considered as reactive if the extension of concrete samples in tests A and B exceeds the permissible limits 0.1% and 0.04%, respectively.

The experimental results show that the relative expansion of concrete prisms amounted to 0.055% and 0.031% for experiments A and B respectively, thus, the limit values were not exceeded (figure1). According to the conditions of the test, this result indicates the suitability of GFG for use in cement composites.

Table 2. Properties of developed GFG-LWC.

Composition			Density,	Compressivestren	Thermalconductivity,
GFG, kg/m^3	Cement, kg/m^3	W/C	kg / m^3	gth, MPa	W / mK
180	120	0.83	400	2.4	0.09
170	200	0.65	500	3.6	0.11
160	280	0.57	600	4.4	0.13
140	360	0.55	700	4.9	0.14
130	440	0.52	800	5.5	0.15

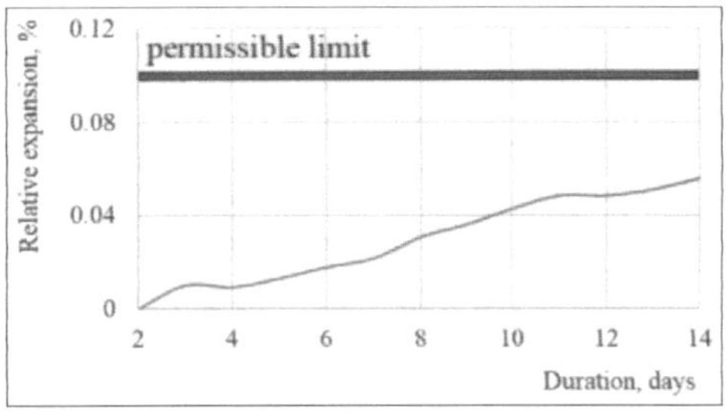

(a)

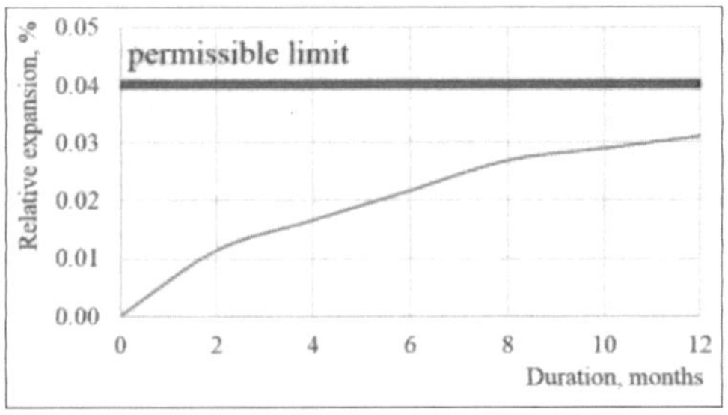

(b)

Figure1. The results of expansion test on concrete prisms.

The main part of the amorphous silicon oxide was transformed into a calcium silicate hydrate during the interaction with alkalis of cement. Shrinkage cracks could be seen in the drying samples after the experiment. In turn, the aggregate structure had a similar fracture pattern after passing test B, but the degree of destruction and the amount of silica hydrogel was greatly reduced (figure 2 c, d). Thus, results show that the expansion of samples is not an appropriate factor for evaluating ASR in concrete with reactive porous aggregates. Microstructural analysis and testing of the concrete strength is needed to evaluate the degree of damage due to ASR.

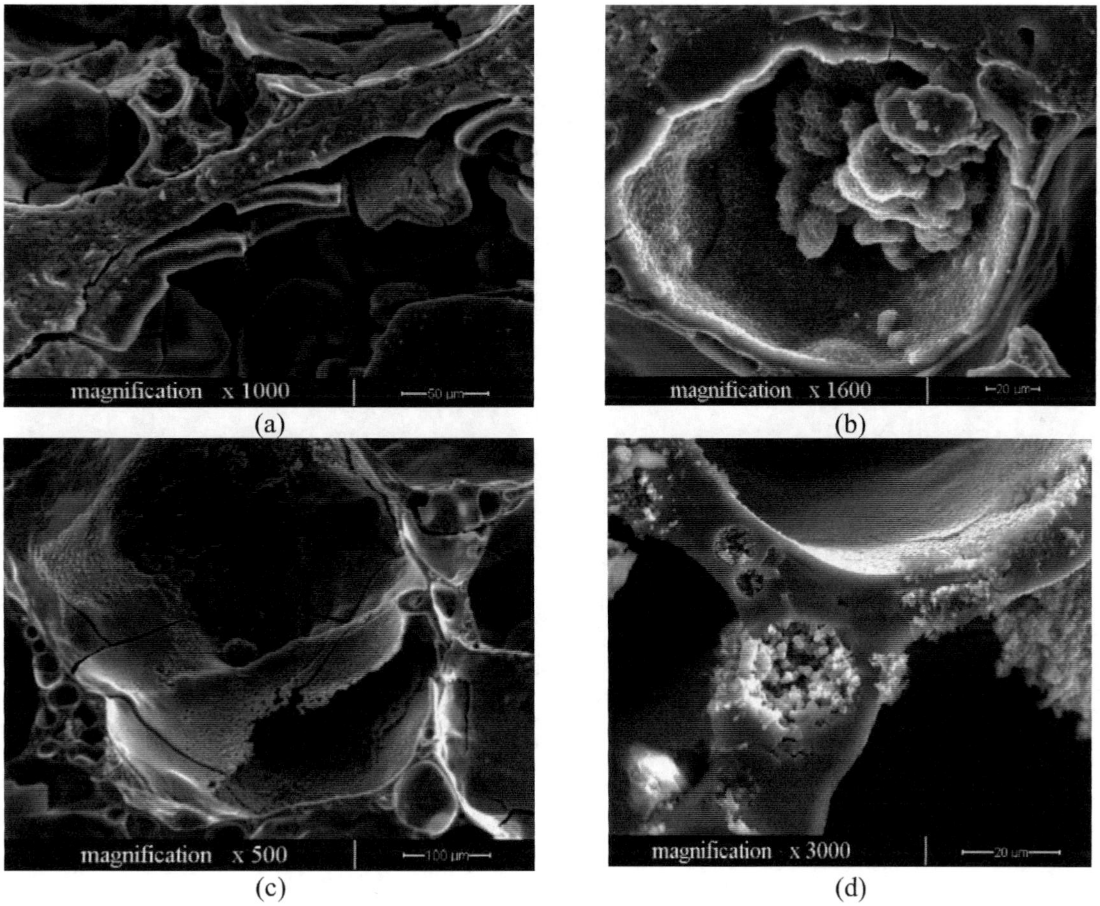

Figure 2. The microstructure of GFG-LWC after: a), b)storing in the 1M NaOH solution at 80 °C for 14 days; c, d) storing in a climatic chamber at 40 °C and 100% relative humidity for 12 months.

The mechanism of reactive porous aggregate interaction with alkalis of concrete has been proposed on the basis of the obtained results. This mechanism differs from the ASR mechanism in the ordinary concrete, which is described in detail in the literature, see, for example [6]. In ordinary concrete ASR occurs in the interaction zone of aggregate and cement stone, see figure 3 a. Silicic acid salts, covering the surface of the aggregate with a semipermeable coating containing high content of calcium, which absorbs water, expands in volume and subsequently creates internal stresses in concrete. It leads to cracking in concrete. In contrast to this, in the concrete with porous aggregates, ASR is a microstructural transformation of silica aggregate in the surface layer to the calcium silicate hydrate. Silicic acid salts may accumulate inside the pores of the aggregate without the formation of reaction products at the interaction zone of aggregate and cement stone (figure 3 b). Thus, ASR in the GFG-LWC does not lead to the occurrence of internal osmotic pressure. It contributes only to its partial destruction inside porous aggregate particles.

Thermal conductivity of GFG-LWC did not change due to ARS after the experiment (measured accuracy is 10%). Damage of aggregate grains due to ASR leads to reduction in strength characteristics of the concrete, while the formation of silica hydrogel in the pores of GFG causes an increase in the concrete density. Depending on the concrete composition, the compressive strength decreased by 20 to 30% and the density increased by 2 to 8% in comparison with the reference samples.

Additionally, the effects of some preventive measures against ASR were investigated. Low alkali cement as well as addition of fly ash and microsilica were considered in this respect. The effectiveness of each measure was evaluated in terms of the difference in relative extension of prisms, reduction in strength and increase in density of samples after passing test B, while the modified material was compared to the control composition of GFG-LWC. The results shown in table 3 demonstrate that the addition of microsilica was the most effective measure.

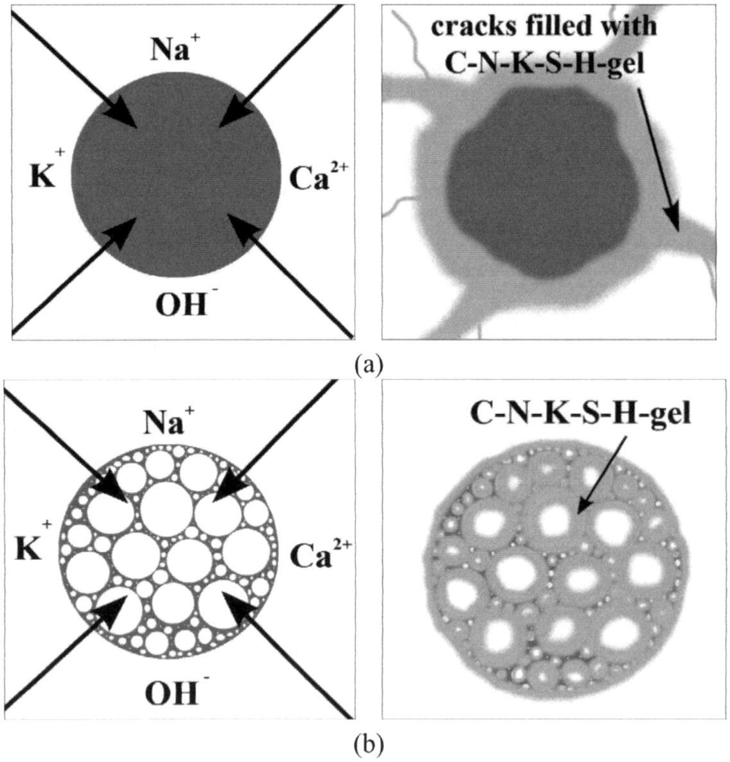

Figure 3. Scheme of ASR in concrete when using: a) dense aggregates; b) porous aggregates.

Table 3. Effectiveness of measures against ASR circumstances (in percent in comparison with control composition of GFG-LWC).

Measure against ASR	Decrease in expansions of concrete prisms	Reduction in compressive strength	Increase in density
Lowalkalicement	38.1	10.4	1.8
Fly-Ash	32.5	8.6	0.5
Microsilica	65.1	32.9	4.9

The obtained data were in agreement with microstructural investigations of the samples after passing the tests. According to the microstructural study, concrete modification by microsilica improved the uniformity of theinteraction zone between aggregate and cement stone, reduced the number of shrinkage cracks of the pore walls and decreased the volume of silica hydrogel formed in granule pores.

The final composition of the GFG-LWC, i.e. that modified with microsilica, exhibits a high compressive strength as well as low thermal conductivity and good durability with respect to ASR. The obtained GFG-LWC is suitable for use in energy efficient buildings.

4. Conclusions

The main finding of the investigation at hand can be summarized as follows:

1. GFG has the capability of interacting with cement alkalis.
2. The GFG-LWC samples showed relatively low expansion of concrete prisms as a result of ASR. However, the expansion is not the determining factor for microstructural damage due to ASR when reacting aggregates are porous. In such cases, microstructural analysis and testing of the concrete strength are needed to evaluate the degree of concrete deterioration due to ASR.
3. The interaction mechanism of reactive porous aggregates with alkalis of concrete has been proposed on the basis of the obtained results. ASR does not lead to build-up of internal pressurein the GFG-LWC.
4. The effectiveness of low-alkali cement and use of various pozzolanic additives was tested as preventive measures against ASR. Microsilica revealed the best preventive capacity.
5. The compositions of GFG-LWC with the density of 400 to 800 kg/m^3, compressive strength of 2.2 to 6.5 MPa, heat conductivity of 0.09 to 0.15 W/(m·K)were developed for use in construction industry.

5. References

[1] Bazhenov Yu 2002 *Technology of Concrete* (Moscow: ASV Publ) p 500

[2] Yu Q, Spiesz P and Brouwers H 2013 Development of cement-based lightweight composites – Part 1: Mix design methodology and hardened properties *Cem. and Conc. Compos.* **44** 17–29

[3] Al-Sibahy A and Edwards R 2012 Mechanical and thermal properties of novel lightweight concrete mixtures containing recycled glass and metakaolin *Constr. and Buil. Materials* **31** 157–167

[4] Zhang M and Gjirv O 1990 Characteristics of lightweight aggregates for high-strength concrete *ACI Mater J.* **88** (2) 150

[5] Moskvin V and Royak F 1962 Corrosion of concrete under the action of cement alkali on aggregate silica *Gosstroyizdat* p 164

[6] Brykov A 2009 ASR and concrete corrosions *Cement and its Applications* **5** 31–37
[7] Limbachiya M, Meddah M and Fotiadou S 2012 Performance of granulated foam glass concrete *Constr. and Buil. Mat.* **28** (1) 759–768
[8] Mladenovič A, Šuput J, Ducman V and Škapin A 2004 Alkali-silica reactivity of some frequently used lightweight aggregates *Cem. and Conc. Research* **34** 1809–16
[9] Bumanis G, Bajare D, Locs J and Korjakins A 2013 Alkali-silica reactivity of foam glass granules in structure of lightweight concrete *Constr. and Buil. Mat.* **47** 274–281
[10] GOST 8269-0-1997 Mountainous rock road-metal and gravel, industrial waste products for construction works – methods of physical and mechanical tests

Investigation of Mechanical Properties of Steel Fibre-Reinforced Concrete

A Ryabchikov[1,4], V Tamme[2] and M Laurson[3]
[1]Estonian University of Life Sciences, Department of Rural Building, Tartu, Estonia
[2]Estonian University of Life Sciences, Department of Forest Industry, Tartu, Estonia
[3]Savekate OÜ, Tartu, Estonia

E-mail: alexander.ryabchikov@emu.ee

Abstract. Steel fibre-reinforced concrete (SFRC) is widely used in the structural elements of buildings: industrial floors, slabs, walls, foundation, etc. When a load is applied to a fibre-reinforced composite consisting of a low-modulus matrix reinforced with high-strength, high-modulus fibres, the plastic flow of the matrix under stress transfers the load to the fibre; this results in high-strength, high-modulus material which determines the stiffness and stress of the composite. In this study the equivalent flexural strength, equivalent flexural ratio $R_{e,3}$ and the compressing strength of SFRC are investigated. Notched test specimens with five different dosages of steel fibres (20, 25, 30, 35, 40 kg/m^3) were prepared using industrial concrete. Determination of flexural tension strength was carried out according to the EU norm EVS-EN 14651:2005+A1:2007. The equivalent flexural strength and subsequent equivalent flexural ratio $R_{e,3}$ of SFRC with a dosage of 20, 25, 30, 35 kg/m^3 similar to their average values and with a dosage of 40 kg/m^3 were 31% higher than their average values. The compressive strength of the steel fibre-reinforced concrete was slightly higher compared to plain concrete, except specimens with the dosage of 40 kg/m^3 where the increase was 30%.

1. Introduction

Steel fibre-reinforced concrete (SFRC) is currently used in a wide range of applications, including bridges, industrial floors, walls, tunnels and structural slabs [1,2,3]. The FRC is a concrete made primarily of hydraulic cements, aggregates and discrete reinforcing fibres. A variety of tests have been performed to determine the actual characteristics and advantages of fibrous materials [4,5,6]. Fibres suited to reinforcing composites are made of steel, glass and organic polymers [7]. Different fibres differently affect the properties of concrete. Addition of synthetic fibres prevents cracking caused by plastic shrinkage and plastic setting, and provides tensile strength in the initial phase of hardening [7].

[4] To whom any correspondence should be addressed.

Synthetic fibres do not affect the load-bearing capacity of construction elements. Addition of steel fibres provides concrete with higher flexural tensile strength, higher cracking resistance, higher impact resistance and higher resistance to volume shrinkage [7]. Steel fibres are effectively used for reinforcement of edge and corner regions, where shear stresses are higher. In the case of using steel fibres as primary reinforcement, cracking of concrete caused by plastic setting under reinforcement rods is eliminated.

The SFRC has advantages over traditionally reinforced concrete in civil engineering. Steel fibres are added to concrete mix and become an integral part of fresh concrete. Thus SFRC can be then pumped like plain concrete and there is no need for preparing the reinforcement frame. Reinforcement by steel fibres can improve resistance to shrinkage cracking and durability of reinforced concrete structures. Addition of steel fibres sufficiently improves the post-cracking process of concrete [1,2,8,9]. High wearing of the concrete pumping equipment can be mentioned as a disadvantage of SFRC.

The design and analysis of structures containing SFRC depend on its mechanical parameters. The objectives of the present study were: 1) to determine the flexural behaviour of industrial C25/30 class of concrete with five different steel fibre dosages: 20, 25, 30, 35, 40 kg/m3; 2) to determine the effect of steel fibre dosage on the ultimate compressive strength.

Fracture energy is an important characteristic in describing the properties of concrete and in determining the design criteria for concrete structures with SFRC. Fracture energy is defined as the area under the load-deflection curve per unit fractured surface area [4].

2. Experimental work

2.1. Test specimens

Test specimens with five dosages of crimped steel fibres SAVEX 1/50 (20, 25, 30, 35, 40 kg/m^3) were manufactured by the concrete production company Rudus AS, they represented industrial class C25/30 concrete with water/cement ratio of 0.5. The mix details of concrete are presented in table 1. Twelve specimens were prepared with each dosage of fibres.

Table 1. Mix details.

Material	Quantity (kg/m^3)
Cement	258
Sand	843
Gravel 4-12	362
Gravel 4-16	288
Water	142

Specimens for bending tests were prisms conforming to EN 12390-1 [10] with a nominal size: section width 150 mm and depth and length 600 mm. The mould was filled up to approximately 90 % of the height of the test specimen and the mix was compacted by external vibration. After compacting the mould was topped up and levelled off. All specimens were demoulded after about 48 h and stored in water at room temperature up to 28 days, and were then removed and stored until testing. All specimens were notched before testing. Test specimens were rotated by 90° around their longitudinal axis and were then sawn through at the mid-span of the specimen. The width of the notch was 5 mm and the depth 25 mm (figure 1b).

The test specimens for the compressing tests with dimensions 150×150×150 mm were sawed out after bending tests from the parts of beams. Four specimens were prepared with each dosage of fibres. Specimens for determination of the compressing strength of the plane concrete were cast at the same time as were casted beams for bending tests.

2.2. Experimental procedure

Experiments were carried out in accordance with EN 14651:2005+A1:2007 [11]. An Instron 3369 testing machine (figure 1a) with a capacity of 50 kN was used for the bending tests. The specimens were tested at deflection control rates so that deflection increased at a constant rate of 0.08 mm/min. When deflection equalled to 0.13 mm, the machine was operated so that deflection increased at a constant rate of 0.21 mm/min. The test was terminated at a deflection value of 5 mm. Deflection was measured at the mid-span by means of a video extensometer to 0.001 mm.

The yoke arrangement was designed and manufactured for holding the assisting equipment of the video extensometer at the mid-span of the specimen. A schematic diagram illustrating the test set-up is presented in figure 1. Holes were introduced using a masonry drill at the mid-height of the beams at the supports to install screws to hold the yoke arrangement. The yoke arrangement held assisting equipment so that deflections were measured directly from the specimen. Mid-span deflection readings depending on the load were recorded at the rate of 5 Hz.

Compressing tests were carried out in accordance with EVS-EN 12390-3:2002 [12].

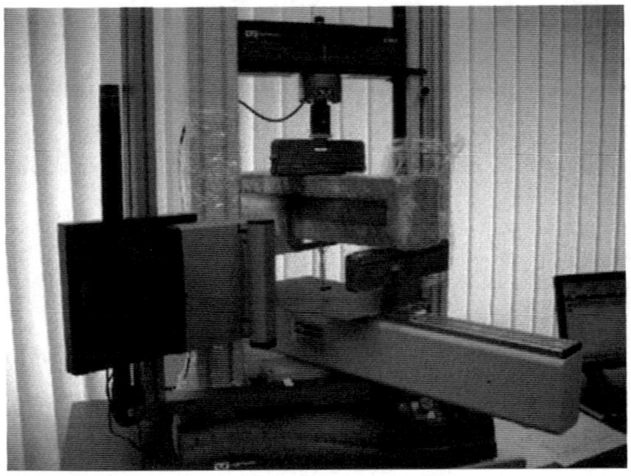

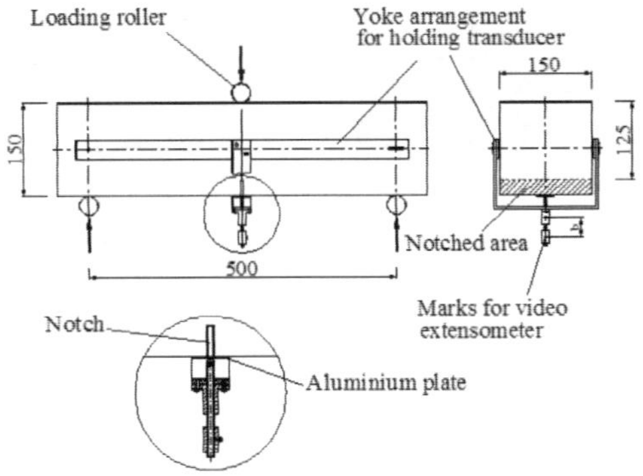

(a) (b)

Figure 1. (a) Testing machine Instron 3369 and (b) test set-up for a notched beam specimen.

3. Results and discussion

Five experiment series S20, S25, S30, S35, S40 were carried out with a steel fibres content of 20, 25, 30, 35, 40 kg/m^3, respectively. A typical bending diagram representing the dependence of central deflection on the load is shown in figure 2. Strain is obtained according to basis b (figure 1b) of video extensometer.

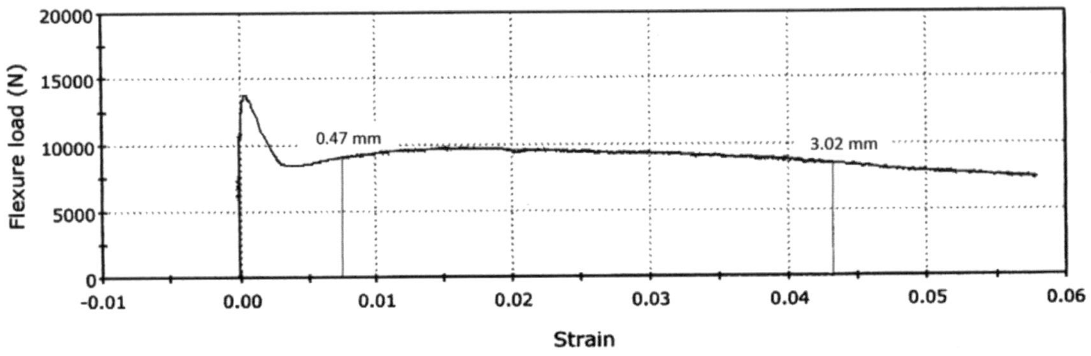

Figure 2. Experimental diagram representing the dependence of central deflection on load.

One of the advantages of using steel fibres is improvement of the post-cracking behaviour of concrete. The appearance of the first crack during flexural loading is clearly the evidence of the ultimate stress of plain concrete. After the first crack, the strength of the specimen diminishes. During this process steel fibres bridge the crack preventing its propagation. A specimen with a bridged crack can successfully carry on external load which in some cases can even be higher than the ultimate load of plain concrete.

In accordance with the standard EVS-EN 14889-1:2006 [13], the minimum dosage of fibres in concrete must be such that a residual flexural strength ($f_{0.47}$) of 1.5 MPa is achieved at 0.47 mm of central deflection and a residual flexural strength ($f_{3.02}$) of 1.0 MPa at 3.02 mm of central deflection. F_L is the load at the limit of proportionality, which corresponds to the formation of the first crack. The results of experiments and the calculations of five experiment series are presented in table 2.

Table 2. Data of specimens and results of experiments.

Series No	Dimensions of section (mm)		Residual flexural strength (MPa)			Equivalent flexural strength	
	Height	Width	f_L	$f_{0.47}$	$f_{3.02}$	$f_{e,3}$ (MPa)	$R_{e,3}$ (%)
S20	126	155	4.23±0.13	2.20±0.34	1.68±0.26	2.24±0.29	53±7
S25	125	156	4.30±0.13	2.20±0.44	1.78±0.35	2.18±0.37	51±8
S30	126	155	4.39±0.08	2.18±0.30	1.74±0.27	2.14±0.29	49±7
S35	125	154	4.47±0.13	2.35±0.16	1.94±0.16	2.30±0.15	51±2
S40	125	154	4.42±0.14	2.90±0.49	2.48±0.43	2.91±0.47	65±9

The equivalent flexural strength of SFRC was determined according to the code of the Japanese Society of Civil Engineers – JSCE-SF4 [14]. The equivalent flexural strength $f_{e,3}$ is determined from the area below the load-deflection curve until measured deflection becomes 1/150 of the specimen's span. This value of stress can be used directly for designing SFRC elements. Further toughness of SFRC can be estimated by the equivalent flexural ratio $R_{e,3} = f_{e,3}/f_L$. The higher is the value of $R_{e,3}$, the higher is load capacity and ductility.

Compressing tests were carried out on concrete samples at the age of eight months and the results are presented in table 3.

Table 3. Results of compressing tests.

Series No	Average dimensions of section (mm)		Ultimate load (kN)	Average compressing strength (MPa)
S20	151.2	153.1	924	39.9
S25	151.4	153.5	990	42.6
S30	150.8	153.6	841	36.3
S35	150.4	155.0	886	38.0
S40	150.7	150.8	1069	47.0

Four specimens from each series were tested under the compressive load. Compared to compressive strength f_c = 36.0 MPa of plain concrete, there is a slight increase in the compressive strength of SFRC specimens. Highest compressive strength is observed for the specimens of series S40.

It has been shown [15] that failure in compression occurs in a localized zone as a shear or cleavage fracture. The failure mode of concrete specimens can be considered as resulting from local tensile mechanisms, or from a combination of tensile and shear mechanisms. Mainly three-dimensionally distributed steel fibres can markedly reinforce both shear failure (oblique cracks) and splitting failure (vertical cracks). At early age, in comparison with plain concrete, steel fibres can significantly increase the duration of the peak load. In the long term, for hardened concrete, steel fibres have no influence on the duration at peak load. Damage of concrete is initiated because of distributed micro-cracks caused by the shrinkage of cement paste around aggregates [4].

4. Conclusions

The equivalent flexural strength, equivalent flexural ratio $R_{e,3}$ and compressive strength of SFRC with steel fibres Savex 1/50 were determined in this study. The equivalent flexural strength and subsequently equivalent flexural ratio $R_{e,3}$ of SFRC with a dosages of 20, 25, 30, 35 kg/m^3 were similar and with dosage of 40 kg/m^3 was 31% higher compared to their average values. In all cases, the obtained stress values can be used directly for designing of SFRC elements, as results meet the requirements of code [13]. The compressive strength of the steel fibre-reinforced concrete was slightly higher compared to plain concrete, except S40 where increase was 30%.

5. References

[1] Banthia N, MacDonald C and Tatnall P 1999 *Structural Applications of Fiber Reinforced Concrete. ACI International, SP-182* (Mich: Farmington Hills) p 257
[2] Altun F, Haktanir T and Ari K 2007 Effects of steel fiber addition on mechanical properties of concrete and RC beams *Construction and Building Materials* **21** 654–661
[3] Chiaia B, Fantilli A P and Vallini P 2009 Combining fiber-reinforced concrete with traditional reinforcement in tunnel linings *Engineering Structures* **31** 1600–06
[4] Köksal F, Sahin Yu, Gencel O and Yigit I 2013 Fracture energy-based optimisation of steel fibre reinforced concretes *Engin. Fracture Mechanics* **107** 29–37
[5] Kim J, Kim D J and Zi G 2013 Improvement of the biaxial flexure test method for concrete *Cement & Concrete Composites* **37** 154–160
[6] Pajak M and Ponikiewski T 2013 Flexural behavior of self-compacting concrete reinforced with different types of steel fibers *Construction and Building Mat.* **47** 397–408
[7] Andrzej M B 2008 Fibre reinforced cement-based (FRC) composites after over 40 years of

development in building and civil engineering *Composite Structures* **86** 3–9

[8] Khaloo A R and Afshari M 2005 Flexural behavior of small steel fibre reinforced concrete slabs *Cement & Concrete Composites* **27** 141–149

[9] Altun F, Haktanir T and Ari K 2007 Effects of steel fiber addition on mechanical properties of concrete and RC beams *Construction and Building Mat.* **21** 654–661

[10] EVS-EN 12390-1:2012 Testing hardened concrete. Part 1: Shape dimensions and other requirments for specimens and moulds *Estonian centre for standardisation* (Tallinn)

[11] EVS-EN 14651:2005+A1:2007 Test method for metallic fibre concrete. Measuring the flexural tensile strength (limit of proportionality (LOP), residual) *Estonian centre for standardisation* (Tallinn)

[12] EVS-EN 12390-3:2009 Testing hardened concrete - Part 3: Compressive strenght of test specimens *Estonian centre for standardisation* (Tallinn)

[13] EVS-EN 14889-1:2006 Fibres for concrete - Part 1: Steel fibres - Definitions, specifications and conformity *Estonian centre for standardisation* (Tallinn)

[14] JSCE-SF4 1984 Methods of tests for flexural strength and flexural toughness of fibre reinforced concrete *Japan Society of Civil Engineers*

[15] Ding Y and Kusterle W 2000 Compressive stress-strain relationship of steel fibre-reinforced concrete at early age *Cement & Concrete Composites* **30** 1573–79

Influence of Plasticizer Amount on Rheological and Hydration Properties of CEM II Type Portland Cements

J Šeputytė-Jucikė[1,3], I Pundienė[1], A Kičaitė[2] and J Pranckevičienė[1]
[1]Vilnius Gediminas Technical University, Scientific Institute of Thermal Insulation, Linkmenu Str. 28, LT-08217, Vilnius, Lithuania
[2]Vilnius Gediminas Technical University, Sauletekio Str. 11, LT-10223, Vilnius, Lithuania

E-mail: jurga.seputyte@vgtu.lt

Abstract. The article analyzes the effect of plasticizer (based on polycarboxilates) amount (0.3 – 1.2% wt. of cement) on the rheological and hydration properties of two Portland cements pastes: CEM II/A-S 42.5N and CEM II/A-LL 42.5N. Increase of plasticizer amount reduces viscosity of CEM II/A-LL 42.5N cement paste from 3 to 12 times, where viscosity of CEM II/A-S 42.5N cement paste reduces from 5 to 20 times. The optimum plasticizer dose (0.3%) in case of CEM II/A-S 42.5N and (1.2%) in case of CEM II/A-LL 42.5N was established. Calorimetry studies have shown that plasticizer reduces the wetting heat release rate in CEM II/A-LL 42.5N cement twice and in CEM II/A-S 42.5N cement – by 25%. Plasticizer prolongs the maximum heat release rate time by 16 h in CEM II/A-LL 42.5N samples and reduces heat release rate by 19%. In CEM II/A-S 42.5N cement samples plasticizer prolongs maximum heat release rate time by 14.5 h and increases heat release rate by 15%. The goal of this study is to analyze the effect of the dosage of the most widely used plasticizer on solubility characteristics, rheological and hydration properties of two cements CEM II/A-S 42.5N and CEM II/A-LL 42.5N to establish the optimum dose of plasticizer in cements pastes.

1. Introduction

Concrete has been known as a versatile and reliable construction material for many years. Despite this its production and final product's properties constantly change due to the newly acquired knowledge, development of products and technologies and environmental issues. Seeking to make the production of cements less costly and to address environmental issues, various additives are used, including limestone and slag. It was estimated that production of cements causes (5 – 7)% [3, 6] of the total quantity of CO_2 emitted to the environment. Therefore recently the number of researches of various

[3] To whom any correspondence should be addressed.

IOP Conf. Series: Materials Science and Engineering **96** (2015) 012019 doi:10.1088/1757-899X/96/1/012019

additives which may partly substitute Portland cements has increased. Blended cement is a well-known component of concrete. CEM II blended Portland cements, in comparison with ordinary CEM I Portland cements have ecologic advantages, since the use of admixtures reduces fuel costs and CO_2 emissions into the atmosphere during Portland cement production and somewhat solve waste utilization problems [1, 2, 3, 4]. That'- is why it is expected that the future world production of CEM II blended Portland cement will be continuously increasing. In general, fillers are incorporated in order to complete the granulometric distribution of cement, to enhance its granular packing factor and to block up capillary pores [4]. Whereas limestone and slag fillers used in composite cements are very fine and they impose an impact on rheological properties of grout, the use of plasticizers is a necessity. Plasticizers are widely used in concrete and have become common components of concrete. Various plasticizers have been widely offered to the users, therefore choosing an option and amount of plasticizer in blended cements is essential [5, 6, 7].

According to their chemical composition plasticizers may be classified into the following main groups: sulphonated naphthalene formaldehydes, sulphonated melamine formaldehydes, modified lignosulphonates, polycarboxylates, polycarboxylate ethers [8, 9]. Researchers have performed several cement and superplasticizer compatibility researches and have analyzed rheology of grouts [10, 11]. Flowability of fresh mixtures of concrete and mortars may be influenced by various factors, related to chemical and mineralogical composition of Portland cements, as well as their fineness, type and quantity of superplasticizers, as well as ambient temperature [12]. Pozzolanic and cementitious materials, when mixed with Portland clinker and water, produce C–S–H similar to that generated from the hydration of calcium silicate of clinker [13]. This reaction can be significantly influenced by the type and amount of plasticizer. Moreover, plasticizers may affect (impede or promote) hydration of the silicate and alumina phases of clinker grains. And it can lead to a lower strength at early ages. Superplasticizers can interact with cement particles physically and chemically. They can also react with hydration products of the cement. It is reported that some of them can lower the initial hydration rate by reacting with C_3A preferentially, while others delay the hydration reaction of C_3S component [14]. Among the plasticisers lignosulphonate base works like retarders. It prolongs the hydration period of cements.

It is reported that the lignosulphonate base plasticisers greatly modify the initial rheological properties and the final compressive strength of cement base materials [15]. Chandra et al. [16], have suggested that the hydration retardation effects of superplasticisers in cement/cement admixtures is due to the interaction of calcium ions [or $Ca(OH)_2$ particles] formed during C_3S hydration with sulphonate ions in superplasticisers, or due to superplasticiser micellization of water available for the hydration of the silicates. Results from a study of lignosulphonic acid used, - as a plasticiser [17] shows that a plasticiser has dual role; one mainly of a kinetic inhibitor (poison) in cement hydration mechanism and the other of a dispersant. It was found that the decrease of rheological parameters depends on fine fraction contents in cement. It was determined that lignosulphonates may cause retarded strength development and reduced flexural strength of slag cement concrete [18]. Addition of lignosulphonates and superplasticizer lowers rheological parameters of cement pastes containing blast furnace slag [19]. It is proved, that the slag–cement formulation demonstrates a lower viscous property under the influence of the plasticizer additive, than ordinary Portland cement [19]. Limestone presence in the binding system determines the acceleration of cement initial hydration, especially of tricalcium silicate [2]. As one can see, a considerable number of researches were performed, however, there still remain some uncertainties during selection of a suitable plasticizer for cements of type CEM II. That' - is why choosing an option of plasticizer for limestone cement and slag cement are essential. The goal

IOP Conf. Series: Materials Science and Engineering **96** (2015) 012019 doi:10.1088/1757-899X/96/1/012019

of this study is to analyze the effect of the dosage of the mostly used plasticizer on solubility characteristics, rheological and hydration properties of two cements CEM II/A-S 42.5N and CEM II/A-LL 42.5N to establish the optimum dose of plasticizer in cements pastes.

2. Materials and research methods

Two-types of Portland cements, namely CEM II A-LL 42.5 N (L) and CEM II/A-S 42.5 N (B) – type cement, meeting LST EN 197-1: 2011 requirements, manufactured by Akmenės Cementas (Lithuania) were tested in the research. The main properties of both cements are provided in Tables 1 and 2 below. Fineness (by Blaine) of L cement is 4080 cm^2/g, of B cement – 3860 cm^2/g.

Table 1. Mineral compositions of cements in %.

Cement type	C_2S	C_3S	C_3A	C_4AF	MgO	SiO_2	$CaCO_3$	$CaSO_4\cdot 2H_2O$	K_2SO_4	$Na_2O_{ekv.}$
B	10.84	60.25	5.85	10.70	2.05	0.51	2.75	2.94	3.34	0.77
L	7.44	55.30	6.25	10.55	3.15	0.32	11.15	2.19	2.91	0.74

Polycarboxylate ester based plasticizer (PCE) was used in research. Electric conduction (EC) and pH measurements in PCE solution were tested by MPC 227 device, (pH electrode InLab 410, measuring accuracy 0.01 S/cm, EC electrode InLab 730, measuring range 0 S/cm – 1000 S/cm). PH and EC characteristics of PCE are 4.22 and 2.41 mS/cm. In the study different concentration solutions were used, where PCE amount changes from 0.3 to 1.2% in the distilled water. The PCE amount range was selected according to the manufacturer recommendation. Increasing of the amount of PCE leads to reductione of pH from 3.9 to 3.6 and increas of EC from 82 to 250 mS/cm.

Cement paste was mixed in a planetary mixer according to EN 196-1. The mixing procedure was as follows: cement (1kg) and 3/4 of required water were mixed for 2 min. at low speed, after that the remaining water and plasticizer were added and cement paste was mixed further for 1 min. at high speed.

The influence of PCE on dynamic viscosity of cement pastes was tested using the SV-10 vibro-viscosimeter. With SV-10 vibro-viscosimeter it is possible to define the dynamic viscosity of pastes up to 12,000 mPa·s with 0.01 mPa·s accuracy in a very small amount of paste (35 ml). At the same time, the device records the temperature changes in cement paste with 0.1 °C accuracy with the purpose to monitor the PCE effect on the cement paste temperature. W/C ratio in cement pastes in all cases was constant 0.28. Cement pastes with PCE contents dosed from 0 to 1.2% (depending on cement amount) were prepared for the tests. The dynamic viscosity of the prepared pastes was measured instantly and then after 5, 10, 15, and 20 min. In order to evaluate PCE impact on rheological and hydration characteristics of on L and B cements, 10 cements pastes compositions have been made (table 2). Calorimetric investigations of samples L-0, L-0.6 and B-0, B-0.6 were performed, using Toni CAL III equipment. Pure L-0 and B-0 cements compositions of proportions 0.28 parts of distilled water and 1part of cements were studied. For compositions with PCE additive preparation was as follows: 0.28 parts of distilled water, 0.06 parts of PCE and 1 part of cements. The heat evolution curves of all pointed compositions were registered up to 48 h.

Table 2. Cement paste compositions by weight (in %).

Batch	L-0	L-0.3	L-0.6	L-0.9	L-1.2	B-0	B-0.3	B-0.6	B-0.9	B-1.2
Cement L	100	99.7	99.4	99.1	98.8					
Cement B						100	99.7	99.4	99.1	98.8
PCE, g		0.3	0.6	0.9	1.2		0.3	0.6	0.9	1.2
W/C ratio	0.28	0.28	0.28	0.28	0.28	0.28	0.28	0.28	0.28	0.28

3. Test results

Cement paste viscosity measurements is very informative, as it is possible to monitor the kinetics of cement pastes variation at selected time interval until the viscosity reaches the limit value of instrument readings. Fluctuations of dynamic viscosity illustrate the binding process of cement pastes (deterioration of rheological properties and loss of workability) during the hydration process and formation of new compounds. The results of PCE impact on L and B cement pastes' viscosity are presented in figure 1. Monitoring shows, - that viscosity of pure L cement samples paste increased within the testing period (from 610 Pa·s to 1600 Pa·s (figure 1). Effect of PCE additives on viscosity of L cement paste (figure 1a) shows, - that increase of PCE additive amount in the paste (from 0.3 to 1.2%) immediately reduces viscosity from 270 to 43 Pa·s. The last measurement shows, - that viscosity in the pastes increases to 830 and 430 Pa·s. It is observed that the higher amount of PCE additive is there in the paste, the lower the viscosity. It has been noticed that the difference in viscosity using 0.6, 0.9 and 1.2% of PCE additives in the paste is small, thus we can conclude that amount of 0.6% of PCE additive is optimal and it is enough for maintain constant level of viscosity.

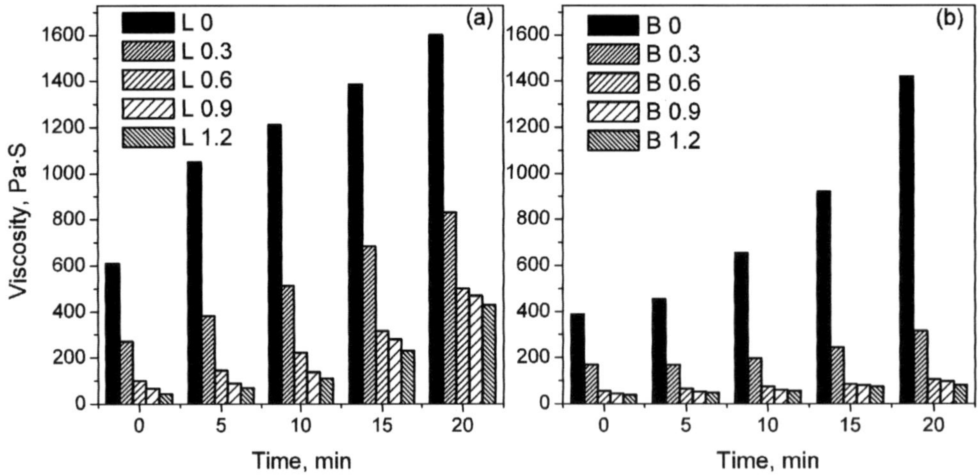

Figure 1. Changes of viscosity in L (a) and B (b) cement paste with PCE additive.

Meanwhile, viscosity of B cement (B-0 sample) is lower right after mixing, 390 Pa·s, than in L-0 sample. Noticeable growth of viscosity up to 1420 Pa·s has been observed after 10 – 15 min. The observed differences in viscosity in L and B cements may be conditioned by their mineral compositions. As it is described in [19], the influence of limestone on rheological properties of the paste depends not only on mineral composition of clinker (amount of C_3A) but also on the amount of

MgO. According to the data presented in table 1, the amount of MgO in L cement is higher than in B cement. Growth of PCE additive in the paste of B cement (from 0.3 to 1.2%) reduces viscosity from 170 Pa·s to 30 Pa·s (figure 1b) and during last measuring viscosity increases up to 315 Pa·s and 80 Pa·s respectively. It is obvious that PCE additive decreases viscosity of B cement more significantly than L cement paste. We can assume that less than 0.6% of PCE additive is enough for decrease of viscosity.

During the viscosity research, measurements of changes in temperature of pastes were performed. Seeking for better presentation of on-going processes, changes in temperature were recorded immediately after the mixing procedure, when the temperature of all samples was the same, i.e. 20 °C, then it was recorded in 1 min and 5, 10, 15 and 20 min. When the 1-st minute of the research passed (figure 2a), the highest temperature was achieved in sample L-0.3 – 21.7 °C, meanwhile temperature of other samples was as follows: L-0.6 – 21.5 °C, L-1.2 – 21.35 °C, and L-0 – up to 21 °C. It is obvious that the minimum quantity of PCE causes higher temperature, lower quantity of PCE causes lower temperature, and the temperature of the samples with no additives seems to be the lowest. It means that higher quantity of PCE reduces wetting heat of sample L. During 19 min reduction of temperature of the samples was observed. After 20 min. the temperature of sample L-0.3 was 20.3 °C, L-0.6 – 20.2 °C, L-1.2 – 20 °C, and L-0 – 19.55 °C.

Different tendencies have been noticed in case of sample B. When 1st min. of the test passed (figure 2b), the highest temperature was reached in the sample B-1.2 with the highest amount of PCE 25.37 °C. It is noticed that the temperature of the sample B-0 was higher than samples L-0 (by almost 2 °C), and it was equal to – 22.98 °C. If the quantity of PCE is consistently increased, the temperature of the samples rises. It means that clinker minerals of cement B (in total their quantity in cement B exceeds by 8.28% the quantity in cement L) generates more heat than cement L during dissolution and additive PCE interacts with dissolving minerals of the cement and causes growth of the temperature of the sample. Because of cement minerals interaction with PCE additive, temperature of the sample rises even more and the viscosity drops. During the last 19 min. temperature of the samples drops, however, in 20 min. the recorded temperature of sample B-0 is 20.16 °C, and of B-1.2 – 22.67 °C. It shows that temperature of the samples from cement B drops faster than in case of samples of cement L.

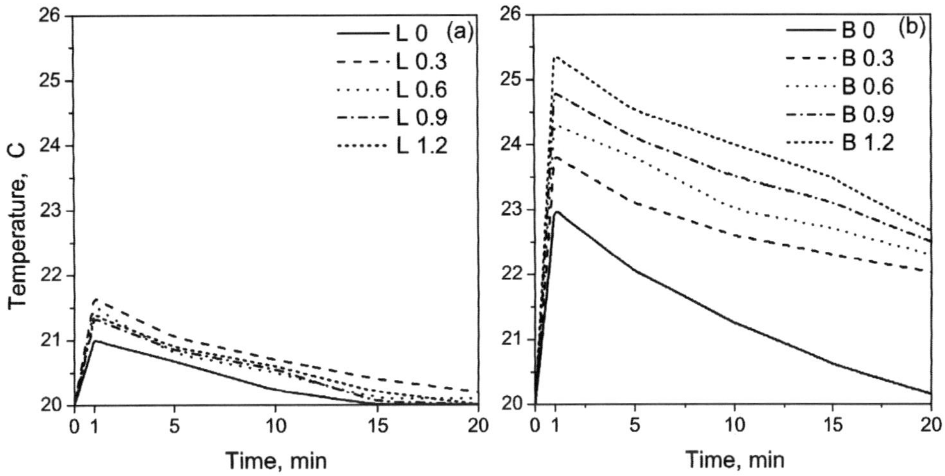

Figure 2. Changes of viscosity in L (a) and B (b) cement paste with PCE additive.

Some correlation between viscosity and temperature changes can be seen. One of the factors which influence interaction of cement and PCE is mineral composition of the cement. If the cement contains more cement clinker minerals, wetting heat during dissolution and correspondingly temperature of the sample rises. The heat generated during cement hydration activates the impact of PCE. Cement L generates less heat during dissolution process, - than cement B, therefore viscosity of the sample is higher than of cement B. Additive PCE a little raises temperature of L cement samples, and this may be the reason why reduction in viscosity in this samples is slower than in case of B cement samples, where additive PCE significantly influences increase of temperature and reduction of viscosity.

For the purpose to further clarify the effect of PCE additive on both cements pastes and to explain the conducted viscosity tests, measurements of the EC were performed within 2 h. It has been explained that EC depends on the amount of ions in solution and well describes the course of the hydration process during cement mineral dissolution [16, 17]. Scientists singled out three main periods of EC changes. During the first period fast increase of EC is observed, which is interpreted as solution of cement minerals and ion transit to the solution. During the second period EC stabilization is observed. The third period consists in hydrate generation and reduction of EC.

EC values in L-0 sample during the first period (~20 min) grow from 17.3 to 19.0 mS/cm (figure 3a). In the course of the second period (next 20 min) EC does not change noticeably and during the third period, EC reduction is visible. Increase of PCE additive (figure 5), in L cement paste from 0.3 to 1.2%, - slightly decreases the primary EC values (from 16.4 to 15.6 mS/cm). Less amount of PCE additive in the paste defines a shorter first period and quicker EC growth at the end of the period. Larger amount of PCE defines a longer second period. At the end of the second period the difference among all pastes EC is slight (0.5-0.7 mS/cm) and during the third period a larger amount of PCE defines lower EC values in the paste.

Cements mineral solution in the B-0 cement paste slightly differs from mineral solution of L cement. EC in the B-0 cement paste at the start of the first period (figure 3b) is by almost 2.3 mS/cm less than in the L-0 paste and its value is just 15.0 mS/cm. Perhaps this can be explained by lower amount of C_3A in B cement (table 1). During the second period, EC values quickly grow and at the end of the third period EC rise to 19.0 mS/cm, like in L-0 paste. Same as in case of L cement, EC changing in B cement paste is minimal during the second period and during the third period EC diminishes. Same as in the case with L cement paste, increase of PCE additive amount in B cement paste a little decreases EC values (from 14.6 to 13.8 mS/cm). The shorter is the first period and the superior is the rate of EC increase at the end of the period, the lower amount of PCE additive is there in the paste. The larger amount of PCE additive extends the second period; EC also decreases during the third period. We can conclude that additive PCE retards transition of cements minerals ions to the solution. It can be stated that despite the type of cement, PCE additive, characterized by its moderate EC and acidic pH, decrease the values of EC in the paste, retards transition of ions to the solution, and decreases viscosity of the paste. Efficacy of PCE additive depending on the type of cement is different: for B cement 0.3 – 0.6% is enough and for L cement larger amount – 0.9 – 1.2% is the most efficient.

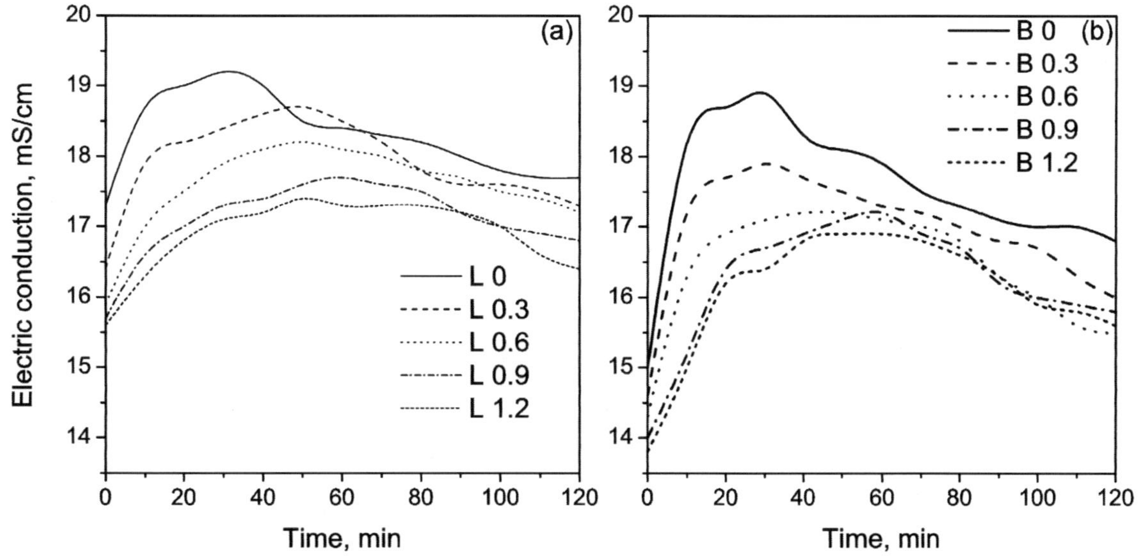

Figure 3. Changes of EC in L (a) and B (b) cement paste with PCE additive.

Calorimetric test well describes the differences between L and B cements and PCE additive influence on cement hydration process. Investigation shows (figure 4a, 4b), - that wetting heat release rate in both cements samples is different: in B-0 sample 12.4 J/g and in L-0 sample 8.6 J/g. May be it is due large quantity of gypsum in B cement and its reaction with C_3A.

The highest heat release rate (10 J/gh) in L-0 sample was reached after 12.5 h and in B-0 sample (8.6 J/gh) – after 11.5 h of hydration. Noticed differences may be conditioned by different fineness and mineralogical composition of both cements, containing different amount of C_3A, C_2S and C_3S. It is known that most of plasticizers retard the hydration of the Portland cements [33, 35, 21, 34]. Comparing with rezults of L-0 sample, PCE additive (figure 4a) reduces a wetting heat release rate from 8.6 J/g to 4.4 J/g. These results verify that wetting heat has a direct influence on the development of temperature, decline of viscosity and EC values changes in cement pastes (figures 2, 3). PCE additive increases time required to achieve maximum heat release rate in L cement samples from 12.5 to 17.5 h. In L-0.6 sample heat release rate increases to 8.1 J/gh. It is stated that PCE additive decelerates hydration process more in L than in B cement. Possibly, addition of limestone may reduce the potency of PCE additive [11].

Investigations show that in comparison with B-0 sample (figure 4 b) PCE additive in B cement reduces wetting heat release rate from 12.4 J/g to 9.8 J/g. The maximum heat release rate in the B-0.6 sample is reached after 14.5 h, in B-0 sample – after 11.5 h.

The highest heat release rate in sample B-0.6 is 9.8 J/gh. Probably bigger total amount of C_3A in B cement determines quicker action of PCE additive on the hydration process in B cement than in L cement, because in the study significance of C_3A amount on cement hydration in presence of plasticizer has been pointed out [22].

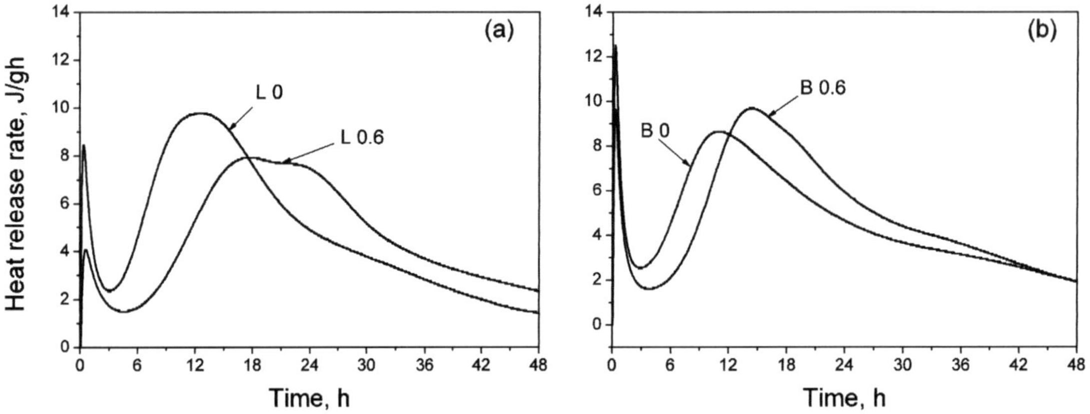

Figure 4. Heat release rate in L (a) and B (b) cement paste without and with PCE additive.

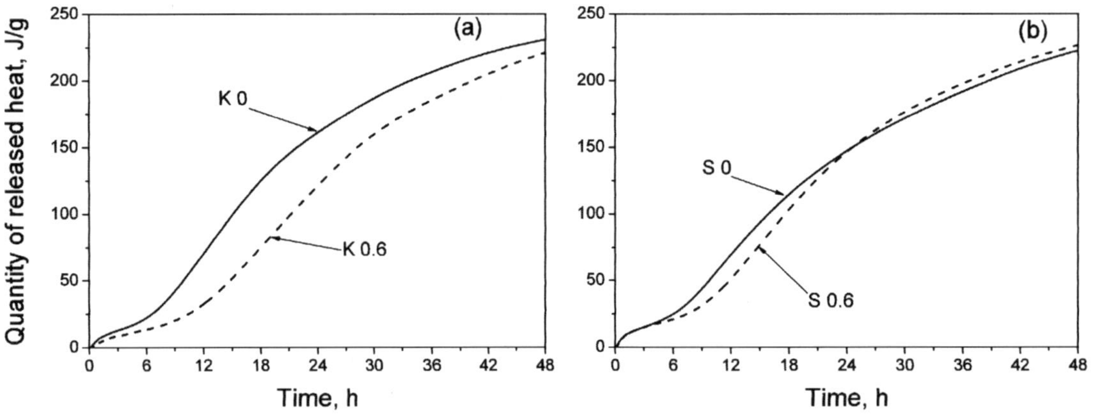

Figure 5. Amount of released heat in L (a) and B (b) cement paste without and with PCE additive.

Total heat value curves confirmed by data of figure 5a indicate that PCE additive in L cement impedes heat release by almost 6 h in comparison with L-0 sample. PCE additive influences faster heat releases in B cement, than in L cement samples. After one day total heat value in L-0 sample reaches 175 J/g, in L-0.6 sample reduces by one fifth – 133 J/g. After two days of measurements the total heat value in sample L-0 was 240 J/g, while in L-0.6 sample 225 J/g. PCE additive impedes heat release in B cement by almost 2 h (figure 5 b). Heat release starts increasing in a B-0.6 sample after this time is over. After one day we can see that the value of released heat in both samples (B-0, B-0.6) becomes equal and reaches 150 J/g. After two days of measurements total heat value in sample B-0 reaches 225 J/g and little differs from B-0.6 sample. We can conclude that L cement is more sensitive to pH and EC values of plasticizer, than B cement. This fact attests that these two cements need attentive choosing of plasticizers and it is confirmed by researches [24, 25, 26, 27]. It is possible to generalize that the observed temperature, viscosity and EC changes tendencies in pastes basically reflect in calorimetric tests.

4. Conclusions

It was found that mineral composition of the cement is one of the important factors affecting the cement and plasticizer interaction. If cement contains more cement clinker minerals, wetting heat and temperature in the sample paste rise. In case of cement CEM II/A-S 42.5N, - containing more cement clinker minerals, wetting heat and temperature of the paste are higher than in CEM II/A-LL 42.5N cement. Cement CEM II/A-LL 42.5N generates less heat than cement CEM II/A-S 42.5N, therefore viscosity of the sample is higher than that of cement CEM II/A-S 42.5N. It can be assumed that higher wetting heat in the cement paste more markedly activates PCE additive and as the consequence this leads to viscosity drop.

It has been established that increase of PCE additive (from 0.3 to 1.2%) reduces viscosity of the CEM II/A-LL 42.5N cement paste after 20 min of mixing from 1600 to 430 Pa·s and EC – from 17.7 to16.4 mS/cm, where viscosity of CEM II/A-S 42.5N cement paste reduces from 1420 to 80 Pa·s and EC – from 16.8 to 15.6 mS/cm. During hydration PCE additive decreases the wetting heat release rate in CEM II/A-LL 42.5N cement from 8.6 to 4.4 J/g and in CEM II/A-S 42.5N cement from 12.4 to 8.1 J/g.

Additive PCE prolongs the time, required for reaching the maximum wetting heat release rate in CEM II/A-LL 42.5N cement samples from 12.5 to 17.5 h, in CEM II/A-S 42.5N cement samples – from 11.5 to 14.5 h. It is determined that in comparison with pure CEM II/A-LL 42.5N sample the heat release rate in CEM II/A-LL 42.5N sample with PCE additive is reduced by 19%, and the total heat value – by 6% in the samples after 48 h of hydration. In CEM II/A-S 42.5N cement samples PCE additive increases heat release rate by 15% and 4% total heat value in samples after 48 h of hydration. It is assessed that efficacy of PCE additive depending on the type of cement is different: for CEM II/A-S 42.5N cement 0.3-0.6% is enough and for CEM II/A-LL 42.5N cement larger amount – 0.9-1.2% - is the most efficient.

5. References

[1] Justs J, Wyrzykowski M, Winnefeld F, Bajare D and Lura P 2014 Influence of superabsorbent polymers on hydration of cement pastes with low water-to-binder ratio *J. Therm. Anal. Calorim.* **115** 425–432
[2] USGS 2013 Mineral commodity summaries *United States Geological Survey* 39
[3] Damtoft J S, Lukasik J, Herfort D, Sorrentino D and Gartner E M 2008 Sustainable development and climate change initiatives *Cem. Concr. Res.* **38** 115–127
[4] Valenza J J and Sher GW 2006 Mechanism for salt Scaling *J. Am. Ceram. Soc.* **89** 1161–1179
[5] Menendez G, Bonavetti V and Irassar E F 2003 Strength development of ternary blended cement with limestone filler and blast-furnace slag *Cem. Concr. Compos.* **25** 61–67
[6] Scrivener K L and Kirkpatrick R J 2008 Innovation in use and research on cementitious material *Cem. Concr. Res.* **38** 128–136
[7] Tsivilis S, Chaniotakis E, Kakali G and Batis G 2002 An analysis of properties of Portland limestone cements and concrete *Cem. Concr. Compos.* **24** 371–378
[8] Zhang X and Han J 2000 The effect of ultra–fine admixture on the rheological property of cement paste *Cem. Concr. Res.* **30** 827–830
[9] Usherov-Marshak O 2005 *Himichesekie i mineral'nye dobavki v beton* 2005 Har'kov, Kolorit 280
[10] Nawy E G 2001 *Fundamentals of High-Performance Concrete* second edition Wiley J and Sons p 441

[11] Hanna E, Luke K, Perration D and Aitcin P C 1989 Rheological behaviour of portland cement in the presence of a superplasticizer *Proc. of the third Int. Conf. on Superplasticizer and Other Chemical Admixture in Concrete (Ottawa, Malhotra) VM 171–188*

[12] Hallal A, Kadri E H, Ezziane K, Kadri A and Khelafi H 2010 Combined effect of mineral admixtures with superplasticizers on the fluidity of the blended cement paste *Constr. Build. Mater.* **24** 1418–23

[13] Golcaszewski J and Szwabowski J 2004 Influence of Superplasticizers on Rheological Behaviour of fresh Cement Mortars *Cem. Concr. Res.* **34** 235–248

[14] Sakai K, Watanabe H, Suzuki M and Hamazaki K 1992 Properties of granulated blast-furnace slag cement concrete Fly ash, silica fume, slag, and natural pozzolans in concrete *Fourth Int. Conf. (Turkey, Istanbul, 3–8 May 1992)* pp 1367–83

[15] Agarwal K, Masood I and Malhotra K 2000 Compatibility of superplasticizers with different cements *Constr. Build. Mater.* **14** 253–259

[16] LSB 2011 *Composition, Specifications and Conformity Criteria for Common Cements* Lithuanian Standard Board, LST EN 197-1: 2011m Cement - Part 1 p 38

[17] Pundienė I, Stonys R, Goberis S and Antonovič V 2007 Properties of suspension and pastes of different types of microsilica with various deflocculants *Mater. Sci-Medzg.* **13** 317–321

[18] Pundienė I, Goberis S, Antonovič V and Stonys R 2007 Study of the possible application of low quality microsilica in medium cement heat-resistant concrete with deflocculant Castament FS 20 *Refract. Ind. Ceram.* **48** 298–302

[19] Mejlhede J, Freiesleben H, Coats A M and Glasser F P 1999 Chloride ingress in cement paste and mortar *Cem. Concr. Res.* **29** 1497–1504

[20] Voglis N, Kakali G, Chaniotakis E and Tsivilis S 2005 Portland-limestone cements. Their properties and hydration compared to those of other composite cements *Cem. Concr. Compos.* **27** 191–196

[21] Aydin S, Aytaç A H and Ramyar K 2009 Effects of fineness of cement on polynaphthalene sulfonate based superplasticizer-cement interaction *Constr. Build. Mater.* **23** 2402–08

[22] Caszewski J and Szwabowski J 2004 Influence of superplasticizers on rheological behaviour of fresh cement mortars *Cem. Concr. Res.* **34** 235–48

[23] Duval R and Kadri E H 1998 Influence of silica fume on the workability and the compressive strength of high-performance concretes *Cem. Concr. Res.* **28** 533–547

[24] Binici H, Temiz H and Kose M M 2007 The effect of fineness on the properties of the blended cements incorporating ground granulated blast furnace slag and ground basaltic pumice *Cem. Concr. Res.* **27** 1122–28

[25] Bonavetti V, Donza H, Menendez G, Cabrera O and Irassar E F 2003 Limestone filler cement in low W/C concrete: A rational use of energy *Cem. Concr. Res.* **33** 865–871

[26] Burgos-Montes O, Palacios M, Rivilla P and Puertas F 2012 Compatibility between superplasticizer admixtures and cements with mineral additions *Constr. Build. Mater.* **31** 300–309

[27] Sugiyjama T, Rittchichanauy W and Tsuji Y 2008 Experimental and numerical modelling of chloride penetration and calcium dissolution in saturated concrete *Cem. Concr. Res.* **38** 49–67

IOP Conf. Series: Materials Science and Engineering **96** (2015) 012020 doi:10.1088/1757-899X/96/1/012020

Modification in Cay Concrete Properties During Fluid Flow Permeability Measurement

F Solomon[1,2] **and S O Ekolu**[1]
[1]University of Johannesburg, Department of Civil Engineering Science, Auckland Park 2006, Johannesburg, South Africa

E-mail: Fitsummoa@gmail.com

Abstract. In this paper, two methods consisting of triaxial water permeability and water penetration were used to evaluate the changes occurring in the pores of clay concretes during the tests. Triaxial permeability is generally used for concrete with higher permeability while concretes with very low permeability are suited for the penetration method.
Clay concrete specimens of 0 to 40% clay content were used in the study. The concrete mixes had water-to-cement ratios (w/c) of 0.70, 0.75, 0.80, 0.85, and the cementitious content 380 and 450 kg/m^3. Results show that concrete gains moisture during wetting at a much faster rate than loses it during subsequent drying. This could be explained by the contribution of suction pressure created upon drying. When water penetration pressure is applied, more water is driven into pore space that could be responsible for changing the network of the voids. Pore structure during drying may certainly be different in size and shape than its form during wetting, leading to a consequent effect on the permeability of the clay concretes. The modification could be one reason why the moisture gain percentage in clay concretes was higher than in normal concretes.

1. Introduction

Natural soil binder in clay-bonded stabilized gravel produces a clay concrete. Partial replacement of cement with clay leads to systems that possess properties intermediate between those of clay concrete and a Portland cement concrete. Similarly, partial replacement of the soil binder by asphalt leads to water proofed granular soil stabilization, and complete replacement by bituminous and filler leads to bitumenious concrete. In like manner, there exist different types of concretes which include clay concrete, lime concrete, resin concrete, gypsum plasters concrete, and others. Partial or complete substitution of clay binder with other cementing agents can be also done in sand-clay and clay-mortar systems. The use of such cementing materials is limited by availability, cost, and susceptibility to local

[2] To whom any correspondence should be addressed.

climatic conditions, mixing, placing, and densification with the available implements at the site of the construction.

In concrete, both the physical structure of concrete and the state of water in pores influence these processes. Hydrodynamics of porous materials, which considers a porous body as a continuum, may be used to obtain equations which define these transport processes [1]; however, empirical laws, such as Darcy's equation, are often applied [2-5] widely for concrete. The theoretical descriptions of the transport processes generally form the basis to measure the transport properties of concrete. This study provides details of test methods which can be used to measure various transport characteristics of clay concretes.

2. Background

Both the characteristics of concrete and the environmental conditions significantly influence permeation properties. Some of the parameters, which might influence permeation tests, will be mentioned while discussing test methods. As the specific influence of these factors on each of the test techniques is beyond the scope of this paper, only their general effects are included in this study. The following factors should be given adequate attention while planning a permeability test program:

- Driving force and duration of test.
- Ambient conditions at the time of test (temperature and humidity).
- Moisture content of the test specimen.

One way or the other, the effect of these factors has to be taken into account in any permeation test. Permeability coefficients are expressed in m/s, so that water viscosity which is a function of the pore size is not a factor. The temperature of the environment surrounding the permeability set up contributes to pressure stability, so in this work, testing temperature was maintained at 23°C and relative humidity at 52%. Finally, de-aired water was used for the experiment as well as for applying confining pressure. The use of air-free water was intended to avoid the effect of air that could affect flow and stability of pressure.

2.1. Test pressure and duration of test

In steady-state permeability tests, it is generally accepted that a constant rate of flow is established. However, the flow need not be constant for the entire duration of the test and there may not be a linear relationship between the applied pressure and the flow being measured. The following changes may occur during permeability testing [6]:

- Impurities in the permeating water may cause silting and chemical action with the materials in concrete;
- Silting may occur due to particles carried from one part of the concrete and deposited in the pores lower down;
- Swelling of the cement may take place;
- Further hydration of cement is likely;
- Calcium hydroxide is washed to the bottom surface and carbonated by the atmosphere, thus forming an impermeable layer. Figure 1 shows a set up of permeability used in this work. In figure 1(c), it can be seen that calcium hydroxide has leached from the concrete sample, during water penetration test.

2.2. Moisture content of concrete

Humidity is significant in permeability measurements; however, it is difficult to account for its effect without the knowledge of the moisture content of concrete. Therefore, the effect of moisture content is generally given importance in relation to the humidity effect on permeation tests on concrete [3,7]. Moisture content reduces the flow path in the case of gas flow tests. It also influences other permeation mechanisms such as sorptivity, diffusivity, etc. The effect of moisture has been reported in various published works. Whereas the influence of moisture in laboratory studies is usually eliminated by drying the samples, the following methods can be resorted to in field applications.

- Precondition the surface by force drying;
- Independent measurement of moisture content prior to testing.

The effectiveness of these techniques needs to be investigated thoroughly before either of them can be used as a standard procedure. At present, only isolated data exist.

2.3. Steady and non-steady state conditions flow

For measuring permeability, the most commonly used method is a permeability cell. Permeability cells of various specifications and dimensions are used to admit fluid under pressure to one side of the specimen and measure the flow either at the inlet or at the outlet [8,9].

Permeability tests provide a means for measuring the true permeability. The basic requirement is that a specimen, usually a core, should be sealed on its curved face so that between its two opposite parallel faces the flow of a liquid or a gas can be promoted by an applied pressure. Under steady-state conditions, the coefficient of permeability is calculated from the knowledge of sample geometry and fluid characteristics, and the measurement of flow rate and applied pressure. However, if a steady-state condition cannot be established due to either low permeability of the test specimen or limitations of the test conditions, a non-steady state flow can be used to obtain a permeability index.

2.3.1. Steady state flow test.

The test determines the rate of flow of water at steady-state conditions for the given test pressure and sample geometry. Using these data, the coefficient of water permeability or the intrinsic permeability is calculated.

2.3.2. Test procedure.

Permeability test essentially consists of saturating a test specimen and measuring the rate of flow of water through it due to a pressure gradient as in triaxial cell in figure 1 (a). Whereas for water penetration, the most commonly used test set up is shown in figure 1 (b) and (c) consisting of a permeameter cell to hold the sample, a set of inlet controls to admit water at the specified test pressure head while measuring the inflow and a set of outlet controls to allow the discharge from test specimen to be monitored along with the outlet pressure.

(a) Triaxial cell (b) Permeameter cell (c) Leached calcium hydroxide

Figure 1. Typical permeability triaxial and permeameter cells.

2.4. Design of the permeability cell

A suitable permeability cell should withstand test pressure without any deformation. As high driving pressures are commonly used with low permeability materials, special attention must be given to the seal on the side of the specimen. Where confining pressures are used to seal the sides of the specimen, the ratio of the confining pressure to the driving pressure must be suitably chosen. This ratio is known to have an effect on the rate of flow at low confining pressures [10].

2.4.1. Test specimen.

Permeability of concrete test requires fully saturated specimens. If unsaturated specimens are used, the time taken to establish the steady state will be longer than normal and, hence, the duration of the test will increase. This is not advisable for concrete containing partially hydrated cement particles. The size of the test specimen may pose another problem. If thin specimens are used in order to improve the degree of saturation, it may adversely affect the reliability of the permeability test. The height of the sample may be taken as three times the maximum aggregate size in order to reduce the scatter of results [10]. The diameter of test specimens also depends on the maximum aggregate size, and here a minimum size of 50 mm is preferable [10].

2.4.2. Driving pressure.

Although high test pressures may accelerate the test and establish the steady_-state rapidly they may result in the modification of the pore structure [10]. Leaching, associated with high pressures used in the water permeability test, may result in an increase of permeability.

3. Details of the experiment

This experimental study forms part of a much wider research program, undertaken in the laboratory, on cementitious materials. Clay concrete specimen were cast and cored. Mixes of water-to-cement ratio (w/c): 0.70, 0.75, 0.80, 0.85, and the cementitious content 380 and 450 kg/m^3 were prepared, as shown in Table 1. The cement used in the concrete was CEMI 42.5N. The clay soils were obtained from Springs/Brakpan (RD) and from Soweto (S2M). Two clay types, RD and S2M were used, respectively classified as reddish sandy silty clay and deep red sandy silty clay.

Table 1. Clay- cement concrete mixtures.

Mix	W:C	Clay (%)	Density (kg/m^3)	Cement	Clay	Water	Building Sand	River sand	Stone(19mm)
CM1	0.70	0	2235	350	0	245	380	380	880
RD1	0.70	10	2235	315	35	245	380	380	880
S2M2	0.70	20	2235	280	70	245	380	380	880
S2M4	0.70	40	2235	210	140	245	380	380	880
CM2	0.75	0	2253	350	0	263	380	380	880
S2M7	0.75	20	2253	280	70	263	380	380	880
CM3	0.80	0	2144	280	0	224	380	380	880
S2M14	0.80	40	2144	168	112	224	380	380	880
CM4	0.85	0	2158	280	0	238	380	380	880
S2M18	0.85	30	2158	196	84	238	380	380	880

During permeability tests, the outflow volume and duration time for the tests were recorded. Mass changes before and after the tests were measured by using a laboratory balance of accuracy to the nearest 0.01 g. Specimen were 75 mm thick and 100 mm in diameter. The samples were oven-dried at 50°C, until a weight change of less than 0.1% over 24 h was observed [8]. The tests were set up for two methods consisting of triaxial water permeability and water penetration.

3.1. Testing

Tests to choose appropriate triaxial driving pressures were conducted in accordance with CRD-C 163-92 [11]. A typical non-Darcian flow is shown by an experimental program done at the beginning of the permeability test on triaxial cell as shown in Table 2 and figure 2 following more than 35 days of recordings measured on triaxial cell at driving pressures (P_d) ranging from 50 to 600 kPa, and their confining pressure (P_c). The control mix 1(CM1) was used to conduct experimental set up tests for driving pressures.

Table 2. Set up tests for selection of driving pressures.

Time of steady (hr)	P_c(kPa)	P_d(kPa)	Q(ml/s)	K*E-10 (m/s)
20-42	60	50	0.33	6.89
48-53	110	100	0.14	1.48
25-30	210	200	0.056	0.29
75-95	410	400	0.035	0.092
194-380	610	600	0.083	0.014

3.2. Verification of Darcy's law

In the literature, both equations (1) and (2) have been referred to as Darcy's law when in reality they are not. The general case of seepage in two dimensions of x-z planes is given by Darcy's law which can be written in generalized form of equation (1) to (2).

$$V = k\frac{dp}{dx} \tag{1}$$

where V is velocity of flow, k is Darcy's permeability coefficient, p is pressure. Darcy's law simply states that for vertical flow through a filter bed of uniform cross-section A, is

$$V_z = -k\frac{dH}{dz} \tag{2}$$

where $H = z + \frac{p}{\rho g}$, H is head applied to specimen, such that downstream pressure is zero, z is the elevation of the position above the chosen datum (elevation head), ρ is density of liquid and g is gravitational constant.

This does not necessarily relate to the usual form of Darcy's law, which usually states that certain invariant relations exist between flow, Q (discharge through specimen) and the properties of the specimen (gross area A, length L), the head loss across the specimen ΔH, and a parameter k – a constant for any particular concrete of a particular curing history. In the following equation (3) and plots in figure 2 are shown in relation to Darcian and non-Darcian fluid flow.

$$\frac{Q}{A} = k\frac{\Delta H}{L} \tag{3}$$

where $\frac{Q}{A}$ *is the flux, discharge per unit area.*

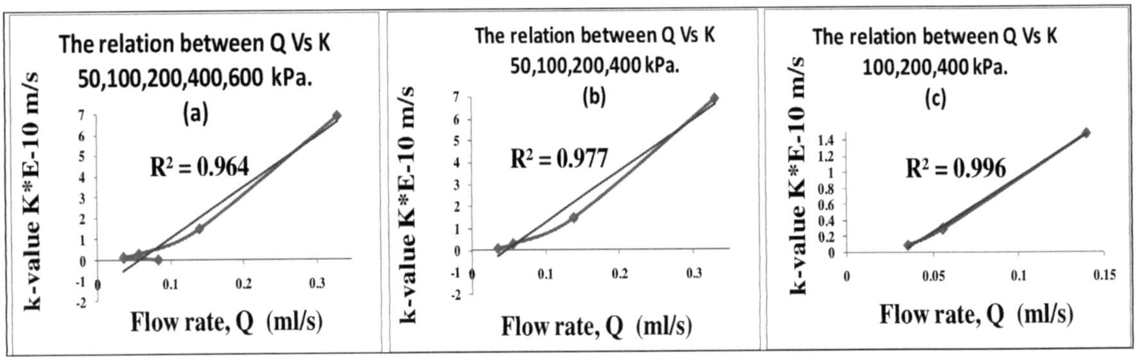

Figure 2. A typical non-Darcian (in cases of a and b) and Darcian flow (c) are shown at the beginning of the experimental permeability test program.

3.3. Moisture content determination

The mass of each specimen (saturated surface-dry condition) was recorded before and after the water permeability test to reveal any change that might have occurred during the test i.e. moisture gain (ΔM) = $M_{after} - M_{before}$ and after oven-drying loss (ΔM) = $M_{after} - M_{before}$. The moisture content and weight at steady-state flow of the specimen was determined after triaxial water permeability test, then each specimen was oven-dried at 50°C, and weighed.

4. Experimental data

4.1. Constant head water permeability on triaxial cell

4.1.1. Water permeability at lower pressure and higher head on triaxial cell.
Water permeability by flow experiment was conducted using a triaxial cell at 200 kPa driving pressure. Figure 3 indicates that at lower w/c ratios the trend in duration of test time could take longer than at higher mix ratios. The values are acceptable and relevant since the cross-sectional area of pore space available for the conduction of water is at its maximum and hence requires less time to reach steady-state condition. In this case, the first pores to empty are the largest and most interconnected and, consequently, the most conductive to water. Interestingly, at higher w/c i.e. above 0.80, the normal concretes and clay concretes show similar fluid flow behaviour when tested at 200 kPa.

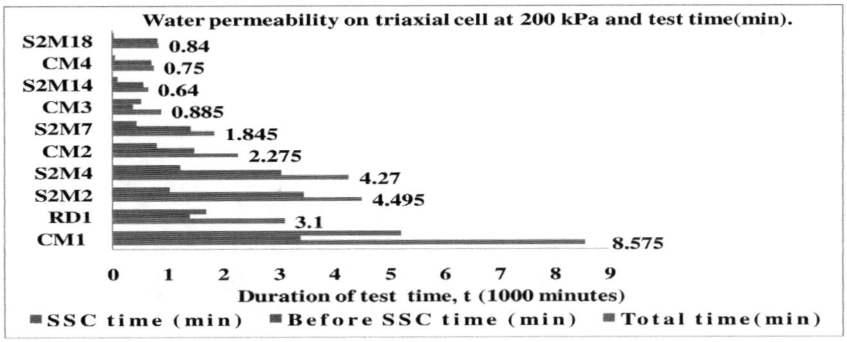

Figure 3. Water permeability and test duration at 200 kPa on triaxial cell.

In contrast to the higher pressure at 400 kPa, the S2M14 at w/c = 0.80 behaved differently as shown in figure 4. The observation is commonly attributed to the fact that hydraulic conductivity is directly related to the volume fraction of the pore space available for water flow, which is directly influenced by moisture content or degree of saturation. In cases of water penetration, since the samples were not saturated, Darcy's law does not apply. Instead, modified Valenta equation is employed for unsteady state condition. It should be noted that the coefficient of permeability obtained by the penetration method is dependent on the duration of test [9].

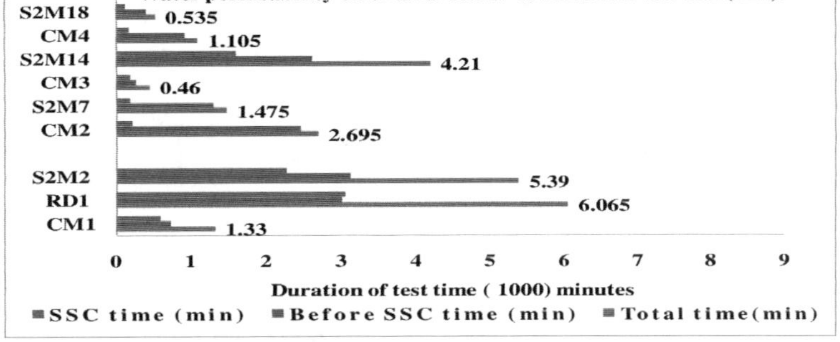

Figure 4. Water permeability at 400 kPa on triaxial cell.

4.2. Moisture content gain and loss during permeability testing
The moisture data from permeability test are shown in table 3 and figure 5. Mass gain and loss were recorded before and after the test. In each case, the data consist of moisture gained after triaxial wetting followed by oven drying and water penetration rewetting. In triaxial wetting condition, the concrete matrix is completely saturated and the matric suction is zero. The saturated moisture content is at the maximum gain and the hydraulic conductivity is at a steady-state value. The saturated hydraulic conductivity is at maximum for the system since the cross-sectional area of pore space available for the conduction is water-filled. After sample oven-drying (following completion of triaxial test and before water penetration test), the concrete matrix sustains a finite amount of suction prior to desaturation. A further drying leads to increase in suction and release of moisture in the pores.

Table 3. Moisture gain or loss percentage.

Moisture gain and loss (%)	CM1	S2M2	RD1	S2M4	CM2	S2M7	CM3	S2M14	CM4	S2M18
Wetting in triaxial test	0.30	0.13	0.38	0.15	0.35	0.33	0.43	0.52	0.37	0.48
Oven drying loss 50°C	7.80	8.30	8.80	9.20	8.70	9.30	8.50	9.30	8.70	9.50
Rewetting penetration test	8.10	8.50	8.90	9.90	9.0	9.60	8.60	9.60	8.80	10.50
Net mass	0	0.20	0.10	0.70	0.30	0.30	0.10	0.30	0.10	1.0

The negative values shown in figure 5 under increasing suction show a significant decrease in moisture content i.e. hydraulic conductivity is effectively reduced to zero at 100% desaturated state. It is notable that re-wetting during water penetration test gave a higher moisture content than that obtained during triaxial wetting. These observations indicate a significant increase in pore space of the samples as a result of the tests. This pore volume increase would likely occur as a result of applied pressures during permeability testing and suction during oven-drying. In both cases, significant damage appears to occur in the pores structure of the concretes. It is also notable that clay concretes exhibit the most significant changes in pore volume leading to the observed higher moisture content under water penetration test, compared to that of the corresponding normal concrete.

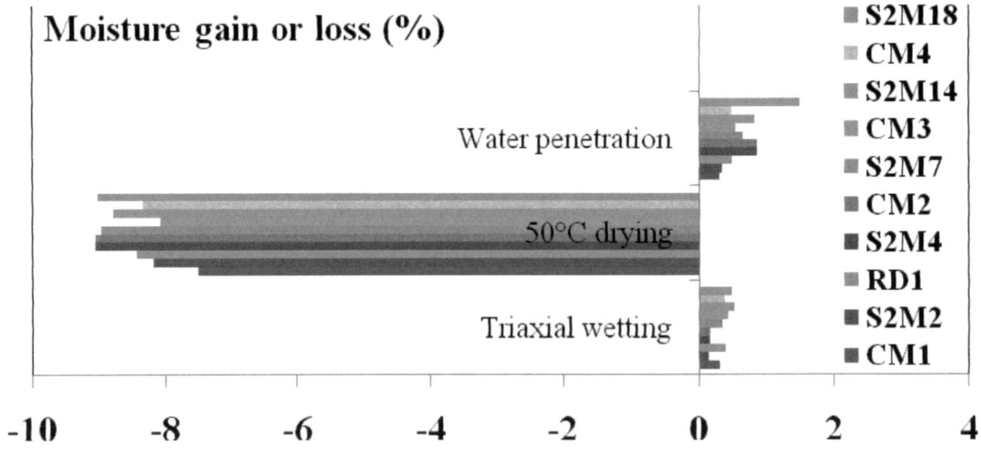

Figure 5. Moisture gain or loss condition.

6. Conclusions

The analysis illustrates the changes in extent of moisture content of specimens after permeability tests were done, as well as oven-drying influence on the degree of saturation. One cycle consisted of wetting during water permeability, oven-drying at 50°C, and water penetration. Results show that the concrete mixes gain moisture during wetting at a much faster rate than loses it during subsequent drying. This could be explained by the contribution of suction pressure created upon drying. When water penetration pressure is applied, more water is driven into pore space and it could be responsible for changing the network of the voids. The moisture gain and loss can be explained in terms of conservation of mass so that the net amount should be zero. It is clear that the pore sizes and shapes at the stage of oven-drying are different from those that are observed and filled during penetration test. The modification could be the main reason why the moisture gain percentage in clay concretes was higher than in normal concretes.

7. References

[1] Bear J et al 1968 *Physical Properties of Water Percolating and Seepage* Arid Zone Research XXIX, UNESCO, place de Fontenoy, Paris p 465

[2] Bamforth P B 1987 The relationship between permeability coeffiients for concrete obtained using liquid and gas *Magazine of Concrete Research* **39** (138) 3–11

[3] Dhir R K, Hewlett P C and Chan Y N 1989 Near surface characteristics of concrete: Intrinsic permeability *Magazine of Concrete Research* **41** (147) 87–97

[4] Kelham S 1988 A water absorption test for concrete *Magazine of Concrete Research* **40** (143) 106–110

[5] Nyame B K 1985 Permeability o normal and lightweight mortars *Magazine of Concrete Research* **37** (130) 44–48

[6] Ramachandran V S and Beaudoin J J 2001 *Handbook of Analytical Techniques in Concrete Science and Technology, Principles, Techniques and Applications* (Ottawa, Ontario, Canada) p 718 – 719

[7] Abbas A, Carcasses M and Olivier J P 1999 Gas permeability of concrete in relation to its degree of saturation *Materials and Structures* **32** (215) p 3–8

[8] Basheer P A M 1993 A brief review of methods for measuring the permeation properties of concrete in-situ *Build Structures Institute Civil Engineering* **99** (1) 74–83

[9] Concrete society 1987 *Permeability Testing of Site Concrete-a Review of Methods and Experience* Report No. 31. The concrete society

[10] Hooton D and Wakeley L D 1989 Influence of test conditions on water permeability of concrete in a triaxial cell *Materials Research Society Symposium Proc.* ed. Roberts L R and Skanly J P **137** 157–164

[11] CRD-C 163-92 1992 Standard test method for water permeability of concrete using triaxial cell *US Federal Standards*

IOP Conf. Series: Materials Science and Engineering **96** (2015) 012021 doi:10.1088/1757-899X/96/1/012021

Permeability of Clay Concretes

F Solomon[1,2] **and S O Ekolu**[1]
[1]University of Johannesburg, Department of Civil Engineering Science, Auckland Park 2006, Johannesburg, South Africa

E-mail: Fitsummoa@gmail.com

Abstract. This paper presents an investigation on the effect of clay addition on water permeability and air permeability of concretes. Clay concrete mixes consisted of 0 to 40% clay content incorporated as cement replacement. Flow methods using triaxial cells and air permeameters were used for measuring the injected water and air flows under pressure. It was found that the higher the clay content in the mixture, the greater the permeability. At higher water-cement ratios (w/c), the paste matrix is less dense and easily allows water to ingress into concrete. But at high clay contents of 30 to 40% clay, the variation in permeability was significantly diminished among different concrete mixtures. It was confirmed that air permeability results were higher than the corresponding water permeability values when all permeability coefficients were converted to intrinsic permeability values.

1. Introduction

Concrete is inherently a porous material. This arises from the use of water in excess of that required for the purpose of hydration in order to make the mix sufficiently workable. It also results from the difficulty of removing all the entrapped air voids during compaction. Permeability of concrete generally refers to the rate at which fluid medium can penetrate concrete. It plays an important role in the long-term durability of concrete [1]. In spite of its importance, permeability is not clearly specified in the criteria for designing concrete or mortar even if its characterization is increasingly recommended by experts working in the field of materials. In literature review, among the numerous permeability measurement possibilities, there is no agreement on a single test and the medium (liquid or gas) to be used. According to Darcy's law and assumptions, permeability is an intrinsic property of a material, since it does not have to depend on the injected fluid. Measuring material permeabilities of the cement matrix present some difficulties and phenomena, which are not completely understood at the present time. For example, it is not known why the intrinsic permeability of certain types of concrete is sometimes a thousand times lower when measuring with water rather than gas, whereas the

[2] To whom any correspondence should be addressed.

values for rocks are virtually identical [2]. Many researchers have attributed this behaviour to the self-healing phenomena due to the hydration of previously unreacted cement. However, specimens with no remaining non-hydrated cement also exhibit self-healing [3]. More commonly, the difference between gas and water permeability values is explained by the theory of gas slippage or the Klinkenberg effect [4]. The presence of this effect results in an increase in the apparent medium permeability, when the mean free path of the gas molecules is sufficiently large compared to the pore diameter, and thus depends on the mean gas pressure and the porous network. Another factor, frequently mentioned in the literature [5,6], is the moisture content of the material, which can obviously affect the gas permeability. One advantage of clay concretes is that actual water permeability tests can be conducted, whereas conventional concrete of low water - cement ratio are too impermeable for practical measurement of water permeability.

2. Experimental

This experimental study forms part of a much wider research program, undertaken in the laboratory, on cementitious materials. Mixes of water-to-cement ratio (w/c): 0.70, 0.75, 0.80, 0.85, and the cementitious content 380 and 450 kg/m^3 were prepared, as shown in table 1. The cement used in the concrete was CEM I 42.5N. The clay soils were obtained from Springs/Brakpan (RD) and from Soweto (S2M).Two clay types, RD and S2M were used, respectively classified as reddish sandy silty clay and deep red sandy silty clay.

Table 1. Clay-cement concrete mixtures used.

Mix	W:C	Clay (%)	Density (kg/m^3)	Cement	Clay	Water	Building Sand	River sand	Stone (19mm)
CM1	0.70	0	2235	350	0	245	380	380	880
RD1	0.70	10	2235	315	35	245	380	380	880
S2M2	0.70	20	2235	280	70	245	380	380	880
S2M4	0.70	40	2235	210	140	245	380	380	880
CM2	0.75	0	2253	350	0	263	380	380	880
S2M7	0.75	20	2253	280	70	263	380	380	880
CM3	0.80	0	2144	280	0	224	380	380	880
S2M14	0.80	40	2144	168	112	224	380	380	880
CM4	0.85	0	2158	280	0	238	380	380	880
S2M18	0.85	30	2158	196	84	238	380	380	880

The air and water permeability tests were applied to cored concrete specimen of 100 mm diameter and 50 or 75 mm length. The samples were oven-dried at 50°C, until a weight change of less than 0.1% over 24 h was observed. Conventionally, drying at 105°C has been the method of choice [6]. However, it is well recognized that microstructure of materials may be altered at elevated temperature, for that reason 50°C was used [7]. The tests were set up for different methods using triaxial cell for permeability of water and air, as well as falling head test for air permeability.

2.1. Measuring techniques

2.1.1 Non-steady and steady state water permeability on triaxial cell.
After water curing for 28 days, the 150 mm cube samples were cut into cores of 100 mm diameter and 75 mm length of specimen. Tests were carried out in accordance with CRD-C 163-92 [8]. Each specimen was polished to the required finish, encased in a rubber membrane, and set up in the 100 mm diameter triaxial chamber. The chamber was then filled with water. A pressure of 210 kPa was applied to the chamber (cell pressure) and simultaneously a pressure of 200 kPa was applied to the pore fluid in the specimen (back pressure) by applying pressure to the drain lines connected to perforated caps located at the top and bottom of the specimen. Pressure of 10 kPa differentials was necessary to prevent the flow of water between the membrane and the specimen. The reading for the constant pressure head at 200 kPa was recorded until the steady state condition was reached. Then following the same procedure as above, the pressure head was increased to 410 kPa and 400 kPa respectively for cell pressure and back pressure. The readings were then taken until the steady-state condition was reached. The calculation was carried out considering the geometry of the sample (A, area perpendicular to the flow and L, length of specimen) and material properties. The measured properties, subject to the back pressure at the base of the specimen (P_b, kPa) and hydraulic gradient (i) were change in volume (q, m^3), elapsed time (t, s), rate of flow (Q, m^3/s), coefficient of permeability (k, m/s).

The hydraulic gradient (i), Q (m^3/s) and k_w (m/s) computed from triaxial test form [8], are as follows:

$$i = \frac{\Delta \; pressure}{length \; of \; specimen} \tag{1}$$

$$Q = \frac{\Delta q}{\Delta t} \tag{2}$$

$$k_w = \frac{q}{t}\frac{1}{Ai} \tag{3}$$

If the test specimen is fully saturated with water and there are no chemical or physical interactions between concrete and water during the test, then the permeability obtained is the intrinsic permeability of the concrete. Therefore, this value can be used to determine the coefficient of permeability for any other fluid by using the relationship:

$$K = \frac{k\,\eta}{\rho g} \tag{4}$$

where K is the intrinsic permeability (m^2), k is the coefficient of permeability (m/s), η is the viscosity of water (10^{-3} N.s/m^2), ρ is the density of water (1000 kg/m^3), and g is the acceleration due to gravity (9.81 m/s^2). For water at a temperature of 20°C, the relationship between coefficient of water permeability k_w (m/s), and the intrinsic permeability K (m^2) is:

$$K = \frac{k_w}{9.75 \times 10^6} \tag{5}$$

2.1.2. Steady-state air permeability on triaxial cell.

After conditioning the samples for drying to acceptable limit, tests were carried out on the same samples on triaxial cell. The only change from water permeability test set up was that the cell pressure (water) was at 1000 kPa and back pressure (air) at 200 kPa. In this case, only back pressure subjected to 200 kPa was considered due to the limitation of cell pressure gauges to 1200 kPa at full operation. The triaxial cell used to measure the air permeability coefficient was developed and reported in ASTM STP publication [9]. Air was applied through the specimen from the base to the top; perforate caps were used at each end of the specimen. The air passing through the specimen was collected in a graduated burette having an air-water interface. In order to maintain the air at atmospheric pressure, the two legs of the u-tube were adjusted continuously to keep the air-water interfaces at the datum level (centre of the specimen).

The air outflow readings were recorded and used to calculate the rate of air flow through the sample until steady-state condition. The following formula was applied to calculate either intrinsic permeability or coefficient of permeability.

The intrinsic permeability of concrete, using air is usually defined by the equation eq. (6)

$$K = \frac{2.Q.L.\eta.P_a}{A(P_b^2 - P_a^2)} \tag{6}$$

where Q is the flow rate (m^3/s), η is the viscosity of the air (Ns/m^2), A is the cross-sectional area of the specimen (m^2), and L is the length of the specimen (m). P_a is the atmospheric pressure of 1 bar [10,11], and P_b is back pressure. For air at 20°C, with a value of $\eta = 1.81 \text{ X } 10^{-5}$ Ns/m^2, the above equation can be written as:

$$K = \frac{3.62 \times 10^{-10}\, Q.L}{A(P_b^2 - 1)} \tag{7}$$

$$k_{air} = \frac{\rho.g.K}{\eta_d.L} \tag{8}$$

where η_d is the dynamic viscosity (Ns/m^2) for air at 20°C , with a value of $\eta_d = 1.81 \text{ x } 10^{-5}$ Ns/m^2, and density of air (ρ) = 1.205 kg/m^3.

Therefore, for air at a temperature of 20°C, the relationship between coefficient of air permeability k_{air} (m/s), and the intrinsic permeability K (m^2) is:

$$k_{air} = 6.5 \text{ x } 10^5 \text{ K} \tag{9}$$

2.1.3. Falling head air permeability test.

At the appropriate time after casting of mixtures, sufficient samples were oven dried at 50°C. Samples of different sizes were tested, i.e. 100 mm diameter x 75 mm specimen length and 100 mm diameter x 50 mm specimen length. Falling head test using air medium was carried out with Blight apparatus [12,13]. Note that 1 mm mercury is equal to 0.133 kPa. The coefficient of permeability (m/s) is calculated using the equations below:

$$k = \frac{W.V.g}{RA}\left(\frac{L}{\theta.t}\right)\log e^{(h_o/h)} \quad \text{when mercury height manometer is used [12]} \tag{10}$$

$$k = \frac{W.V.g}{R.A}\left(\frac{L}{\theta.t}\right)\ln\left(\frac{P_o}{P}\right) \quad \text{when pressure gauge is used [13]} \tag{11}$$

where k is the coefficient of permeability (m/s), W is molecular mass of air (28.97g/mol), V is volume of air under pressure in permeameter (m^3), g is acceleration due to gravity (9.81m/s^2), R is Universal Gas Constant (8.313Nm/Kmol), A is superficial cross-sectional area of sample (m^2), L is specimen length, θ is absolute temperature (K), t is time (s) for pressure to decrease from h_o to h or P_o to P, h_0 and h are mercury manometer height at the beginning and at the end of the test (mmHg) respectively.

3. Experimental results

Table 2 presents result of the water permeability and air permeability tests on clay concrete. The results allow (i) comparison of the influence of different test methods and fluid types, (ii) clay types and their contents, (iii) mixtures. The influence of the fluid types on measurement values indicated that the falling head method for all control mixes gave consistent permeability results when the fluid air was used rather than water. On the other hand, water triaxial test for clay concretes showed an increasing trend for water permeability results. As expected, permeability of the mixtures increases with increase in water-cement ratio, regardless of the test method employed. In all cases, the incorporation of clay into the mixtures led to the corresponding increase in the permeability of the concrete, for a given water cement ratio. This was evident in both clay types RD and S2M. For example, mixes S2M2 and S2M4 containing 20% and 40% clay contents had triaxial permeability values measured at 200 kPa, of 0.44xE-16 m^2 and 0.71xE-16 m^2 respectively. To obtain an overall view of the influence of these various factors, the results given in table 2 have also been plotted in figure 1 which shows clearly that for 0.70 w/c, clay content increase led to rise in permeability. However, when high S2M clay content of 30-40% was used in the mixes, the permeability seems not to be significantly influenced by w/c as seen in figure 1 for mixes S2M18, S2M14, and S2M4. The steady-state triaxial permeability gives lower permeability values despite the high applied pressure of 400 kPa over the non-steady state. These observations are related to the saturated condition of the samples under steady-state condition in which all pores would be water-filled, compared to the partial saturation under non-steady state condition that would have more open pore spaces. Also, in the non-steady state, triaxial permeability increased more rapidly as the w/c of mixes increased. On the contrary, the corresponding air permeability on triaxial cell increased at a much lower rate.

It is also interesting to note that the intrinsic air permeability was found to be higher than intrinsic water permeability. This is particularly evident for both test methods done on air triaxial permeability at 200 kPa as well as falling head air permeability. In accordance with literature [4], it would be expected that air permeability methods would give higher values than water permeability tests. In this work, all the pressures used, i.e., 100 to 400 kPa, are generally low pressures so such effects as gas slippage are evident when the gas pressure is low or the concrete has low permeability. Similarly, the properties of the air or fluid could be modified differently at high pressures. To measure the values of gas permeability which are representative of the true intrinsic value, tests need to be carried out at high pressure (greater than 1 MPa). Thus the intrinsic permeability coefficients obtained using a liquid are a better representation of the pore structure of concrete. These and other effects need further investigations with respect to the results determined in this work.

Table 2. Intrinsic permeability of clay concretes measured using different test methods.

	Non-steady state triaxial permeability at 200 kPa	Steady-state triaxial permeability at 400 kPa	Air permeability, on triaxial at 200 kPa	Falling head air permeability at 100 kPa	Falling head air permeability at 100 kPa
Test ID	Wtx cell200/75	Wtx cell400/75	Atx cell 200/75	Air fall head 100/75	Air fall head 100/50
Mixes /Permeability	K_wxE-16 (m^2)	K_wxE-16 (m^2)	K_{air}xE-16 (m^2)	K_{air} xE-16 (m^2)	K_{air} xE-16 (m^2)
CM1	0.03	0.01	1.38	2.42	1.75
RD1	0.50	0.03	-	9.63	1.05
S2M2	0.44	0.38	-	2.42	1.91
S2M4	0.71	0.36	4.79	4.30	19.20
CM2	0.24	0.04	1.80	4.77	3.12
S2M7	0.77	0.30	-	10.10	1.82
CM3	2.27	0.73	3.12	3.37	2.38
S2M14	1.58	1.14	7.89	17.10	7.11
CM4	1.31	0.58	3.75	5.46	2.84
S2M18	1.15	0.78	-	11.70	60.30

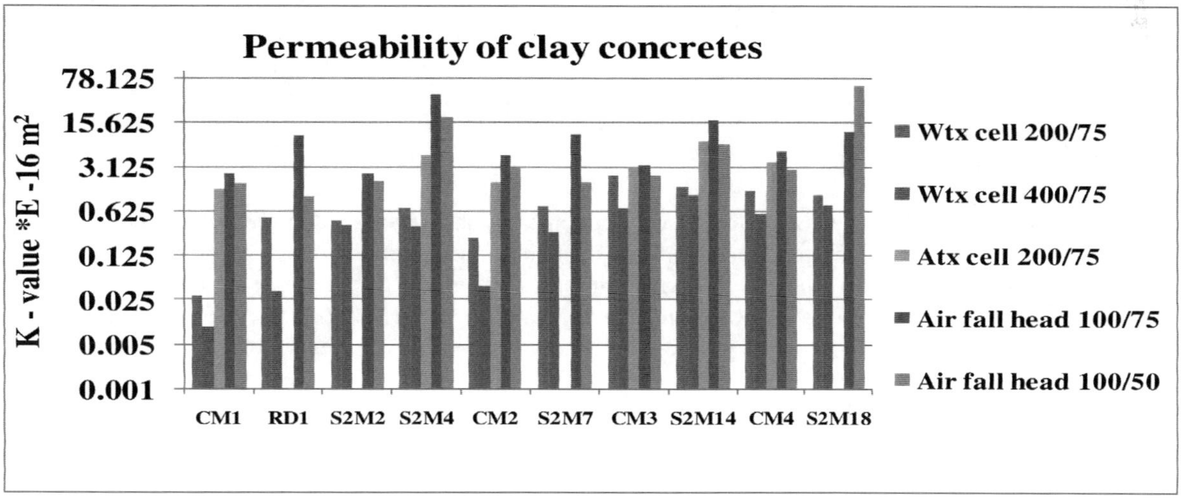

Figure 1. Water and air permeability of normal and clay concretes.

4. Discussion of results

The permeability of concrete is not a simple function of its porosity, but depends also on the size, distribution, shape, tortuosity, and continuity of the pores. In addition, the cement particle size distribution affects the concrete permeability [14]. In the case of clay, the factors that affect the pore system of concrete are (a) the filler effect and (b) migration of fine elements. The two factors influence the total volume and size distribution of pores and finally affect the concrete permeability. The results show that air permeability is closely related to water permeability at higher w/c ratio, i.e., at 0.80 and 0.85 (w/c). From figure 1, results also show that increase in clay content from 20% to 40% gave no corresponding rise in water permeability. On the other hand, at low w/c 0.70 results seem to be affected by the size and nature of pores. In any case, further investigation is needed. Considering all the above measurements, it is clear that, in the case of high w/c ratio (above 0.60), the presence of clay does affect both air and water permeabilities.

In the previous work reported in [15], it was shown that clay concretes have benefits concerning mechanical and physical properties as well as the workability characteristics and the strength development of concrete for low cost applications. It was reported that clay-cement concrete mixtures with a maximum of w/c = 0.80 and 20 to 30% clay replacement would fulfil the strength requirements for low-cost, low strength applications including housing, roads and dams. The mixes used in the current work had compressive strength greater than 6 MPa.

The measurement of permeability of concrete is not only a function of the porosity or pore network of the concrete but also the properties of the fluid. In the case of water permeability measurements, if any unhydrated cement is present, the water will cause hydration to commence, the rate at which this occurs being determined in part by the pressure applied. Unhydrated cement is surrounded by hydration products which have to be penetrated if the water is to reach the unhydrated core. Pressure aids this process. In addition, the flow of water through the concrete can result in silting the pores by loose particles in the capillary system.

The influence of these processes is to reduce the porosity of the concrete as the test proceeds, hence reducing the measured value of permeability. Furthermore, the specimen itself will have variable permeability over the test area.

5. Conclusions

Three permeability tests were conducted on clay concrete samples using water and air. Proportions of 10 to 40% clay were incorporated as cement replacements in concrete mixes. Two different types of clays were used in the study. The results show that clay concretes generally increase the permeability of the mixtures. The higher the clay content, the higher the permeability. At higher w/c ratios, the paste matrix is less dense and easily allows water to ingress into concrete. However, at high clay contents of 30 to 40% clay, the variation in permeability was significantly diminished among different concrete mixtures.

The steady-state triaxial permeability gives lower permeability values despite the high applied pressure of 400 kPa over the non-steady state. These observations are related to the saturated condition of the samples. Also, the driving pressure increase during permeability test could dislodge hydrated particles, causing the blocking of pores and resulting in lower permeability.

6. References

[1] Jackson N 1978 *Civil Engineering Materials* Third edition (London: MACK Press Ltd)

[2] Skoczylas F and Henry J P 1995 A study of the intrinsic permeability of granite to gas *Int. J. of Rock Mechanics and Mining Sciences & Geomechanics* Abstracts **32** (2) 171–179

[3] Hearn N, Hearn, Detwiler R J and Sframeli C 1994 Water permeability and microstructure of three oldconcretes *Cement and Concrete Research* **24** (4) 640–663

[4] Klinkenberg L J 1941 The *Permeability of Porous Media to Liquids and Gases, Drilling and Production Practice* (New York: American Petroleum Institute) pp 200–213

[5] Abbas A, Carcasses M and Olivier J P 1999 Gas permeability of concrete in relation to its degree of saturation *Materials and Structures* **32** (215) 3–8

[6] Dhir R K, Hewlett P C and Chan Y N 1989 Near surface characteristics of concrete: intrinsic permeability *Mag. Concr. Res.* **41** (147) 87–97

[7] Ballim Y A 1994 *Curing and the Durability of Concrete,* Dept of Civil Engineering, PhD Thesis, University of the Witwatersrand, Johannesburg, South Africa

[8] CRD-C 163-92 1992 Standard test method for water permeability of concrete using triaxial cell, US Federal Standards

[9] Matyas E L 1967 Air and water permeability of compacted Soils, permeability and capillarity, ASTM STP 417 *Am. Soc. Testing Mats.* p 168

[10] Cabrera J G and Lynsdale C J 1988 A new gas permeability for measuring the permeability of mortar and concrete *Magazine of Concrete Research* **40** (144) 177–182

[11] Gowripalan N et al 1989 Effect of Curing on Durability, Concrete International, ACI, Phladephia p 310–311

[12] Blight G E 1977 A falling head permeameter for testing asphalt. Trans. of the S.A. Inst. of Civil Engineers p 123–126

[13] Ballim Y A 1991 Low cost, falling head permeameter for measuring concrete gas permeability. Concrete/Beton *J. of the Concrete Society of South Africa* **61** 13–18

[14] Neville A M 1996 Properties of concrete, fourth and final edition (England: Addison Wesley Longman)

[15] Solomon F and Ekolu S 2012 Strength behaviour of clay-cement concrete and quality implications for low - construction materials *Proc. of the Third Int. Conf. on Concrete Repair, Rehabilitation and Retrofitting (Cape town, South Africa, 3-5 September 2012)*

Damping Properties of Sandwich Truss Core Structures by Strain Energy Method

M Wesolowski[1,4], S Rucevskis[2], R Janeliukstis[2] and M Polanski[3]

[1]Department of Structural Mechanics, Koszalin University of Technology, Koszalin, Poland

[2]Institute of Materials and Structures, Riga Technical University, Riga, Latvia

[3]Department of Advanced Materials and Technologies, Military University of Technology, Warsaw, Poland

E-mail: miroslaw.wesolowski@tu.koszalin.pl

Abstract. Sandwich panel structures with stiff face sheets and cellular cores are widely used to support dynamic loads. Combining face sheets made of carbon fibre reinforced plastics (CFRPs) with an aluminium pyramidal truss improves the damping performance of the structure due to viscoelastic character of CRFP composites. To predict the damping characteristics of the pyramidal truss core sandwich panel the strain energy method is adopted. The procedure for evaluating the damping of the sandwich panel was performed using commercial finite element software NASTRAN and MATLAB. Non-contact vibration tests were performed on the real sandwich panels in order to extract the modal characteristics and compare them with the numerical predictions.

1. Introduction

Sandwich structures with a pyramidal truss core are widely used in the form of plates, beams, or cylindrical shells. Their principal advantage is the high specific stiffness and strength ratio with respect to mass. The strength and stiffness characteristics of the sandwich structures have been already studied exclusively [1, 2]. The exceptional stiffness properties of the sandwich structures are due to high bending stiffness of the face sheets and high shear stiffness of the core. Little work, however, was done on the dynamic response of truss core sandwich structures which includes both resonant and damping characteristics [3, 4]. The present work aims to predict the modal loss factors of the sandwich panels made of two parent materials: carbon fibre reinforced plastics (CFRPs) and an aluminium alloy. To predict the damping characteristics the strain energy method was used [5, 6, 7] and combined with the finite element solutions. To validate the numerical predictions the non-contact vibration tests were performed on the real sandwich panels in order to extract their dynamic characteristics.

[4] To whom any correspondence should be addressed.

2. Materials and samples

Damping properties of a pyramidal truss core sandwich panel was investigated in the present study. Three sandwich panels were manufactured for the experimental purposes. The panels had length l, width a, and hight h (Table 1). A global Cartesian co-ordinate system (x, y, z) was located in the corner of the panel, with z axis upward and normal to the face layer (Figure 1). Each sandwich panel was composed of two parent components: (1) a laminated carbon fibre reinforced plastic (CFRP) was used for the face sheets; (2) the aluminium alloy PA6 was used for the pyramidal truss core. The face sheets were cut out of a long unidirectional CRFP tape, which were proven to be good candidates for structural stiffness improvement [8]. Additionally, the CFRP composites improves the damping performance of the whole structure due to viscoelastic character of the matrix. The face sheets were composed of 8 layers having material properties as listed in Table 2 and the total thickness t. For each layer the lamina co-ordinate system $(1, 2, 3)$ was defined with the direction 1 along the lamina fibres, 2 transverse to this direction, and 3 - through the thickness direction. A lamination angle (Φ) of a single layer was defined between the x-axis and the 1-axis. The pyramidal truss core was assembled from separate pieces which were cut out of an aluminium plate by means of water-jet cutting. The pieces were bonded together with a thermoset epoxy adhesive to form a continuous pyramidal truss core. The material properties of PA6 alloy are given in Table 2. The parent components were bonded together to form a sandwich panel with the same thermoset epoxy adhesive as used for the core manufacture. The epoxy adhesive was applied locally on the face sheets where the contact with the core existed.

3. Damping model

The damping prediction of truss core sandwich panel is evaluated using the modal strain energy method [5, 7]. According to the method, the specific damping capacity (SDC) ψ_n is defined as the ratio of the total dissipated energy ΔU_n to the maximum strain energy U_n stored in a structure during a stress cycle at n^{th} mode of vibration:

$$\psi_n = \frac{\Delta U_n}{U_n} = 4\pi\zeta_n = 2\pi\eta_n \tag{1}$$

where η is the modal loss factor and ζ is the damping ratio.

3.1. Strain energy

It is proposed to express the total modal damping of the sandwich panel as the sum of the damping fractions form the individual parent components. The energy stored in the structure is invariant of the coordinate system, therefore the strain energy of each parent component is considered in a different coordinate system. The total strain energy stored in a sandwich panel may be written as (the mode

Table 1. Sandwich panels used in the experimental investigation.

Sample	l [mm]	a [mm]	h [mm]	t [mm]	Φ [deg]
Sandwich-1	380.3	50.0	27.8	1.4	0
Sandwich-2	380.1	50.0	27.8	1.4	0
Sandwich-3	380.2	50.0	27.8	1.4	0

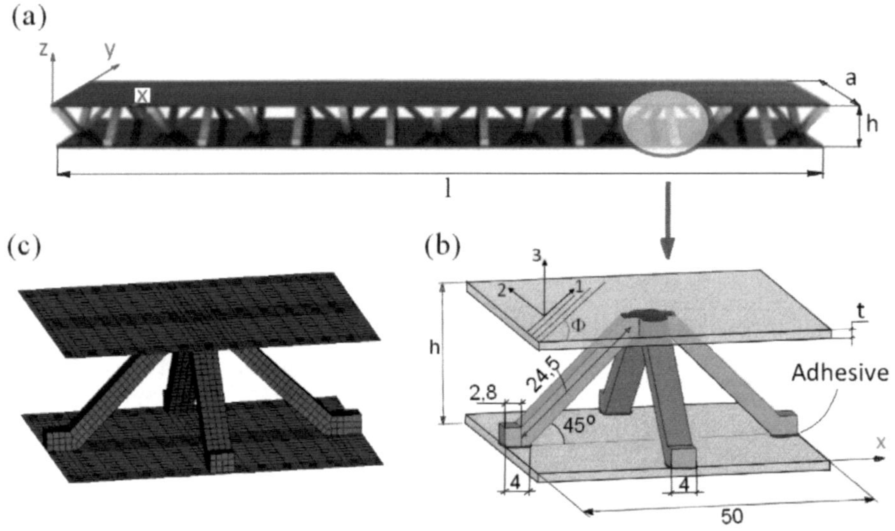

Figure 1. Sandwich panel with pyramidal truss core: (a) Manufactured sample; (b) Unit cell of the truss core sandwich panel; (c) Finite element model of the unit cell - dimensions in [mm].

number has been dropped for a notation convenience):

$$ {}^{s}U = {}^{l}U + {}^{c}U \tag{2} $$

where ${}^{l}U$ and ${}^{c}U$ are the strain energy stored in the laminated face sheets and the aluminium core, respectively. The strain energy stored in a single finite element e of the thin laminated face sheets is the sum of the energies stored in the given lamina directions:

$$ {}^{l}U^{e} = {}^{l}U_{11}^{e} + {}^{l}U_{22}^{e} + {}^{l}U_{12}^{e} \tag{3} $$

For a finite element composed of i layers, the element's energies are:

$$ {}^{l}U_{11}^{e} = \sum_{k=1}^{i} {}^{l}U_{11,k}^{e}; \quad {}^{l}U_{22}^{e} = \sum_{k=1}^{i} {}^{l}U_{22,k}^{e}; \quad {}^{l}U_{12}^{e} = \sum_{k=1}^{i} {}^{l}U_{12,k}^{e} \tag{4} $$

where ${}^{l}U_{ij,k}^{e}$ $(ij = 11, 22, 12)$ is the strain energy stored in the kth layer of element e, and is related to the strain and stress components as:

$$ {}^{l}U_{ij,k}^{e} = \int_{V_k} \{\varepsilon_{ij}\}_k^T \{\sigma_{ij}\}_k dV_k \tag{5} $$

where V_k is the layer volume. Thus the total strain energy stored in the laminated face sheets made of M elements is:

$$ {}^{l}U = {}^{l}U_{11} + {}^{l}U_{22} + {}^{l}U_{12} \tag{6} $$

where:

$$ {}^{l}U_{11} = \sum_{e=1}^{M} {}^{l}U_{11}^{e}; \quad {}^{l}U_{22} = \sum_{e=1}^{M} {}^{l}U_{22}^{e}; \quad {}^{l}U_{12} = \sum_{e=1}^{M} {}^{l}U_{12}^{e} \tag{7} $$

The aluminium truss core is mostly subject to the transverse shear stress [4]. Therefore the strain energy stored in the finite element p of the core, in the global coordinate system, is given as:

$$ {}^{c}U^{p} = {}^{c}U^{p}_{xz} + {}^{c}U^{p}_{yz} \tag{8} $$

And the total transverse shear strain energy stored in the core made of J elements is:

$$ {}^{c}U = {}^{c}U_{xz} + {}^{c}U_{yz} \tag{9} $$

where

$$ {}^{c}U_{xz} = \sum_{p=1}^{J} {}^{c}U^{p}_{xz}; \quad {}^{c}U_{yz} = \sum_{p=1}^{J} {}^{c}U^{p}_{yz} \tag{10} $$

The total strain energy of the sandwich panel is given as:

$$ {}^{s}U = {}^{l}U_{11} + {}^{l}U_{22} + {}^{l}U_{12} + {}^{c}U_{xz} + {}^{c}U_{yz} \tag{11} $$

3.2. Dissipated energy

The energy dissipated by k^{th} layer of the element e is given as:

$$ \Delta^{l}U^{e}_{k} = \psi_{11,k}{}^{l}U^{e}_{11,k} + \psi_{22,k}{}^{l}U^{e}_{22,k} + \psi_{12,k}{}^{l}U^{e}_{12,k} \tag{12} $$

where $\psi_{ij,k}$ $(ij = 11, 22, 12)$ are the specific damping capacities of the material used for layer k. The energy dissipated by element e is given as:

$$ \Delta^{l}U^{e} = \sum_{k=1}^{i} \Delta^{l}U^{e}_{k} \tag{13} $$

And the total dissipated energy by the laminated face sheets is:

$$ \Delta^{l}U = \sum_{e=1}^{M} \Delta^{l}U^{e} \tag{14} $$

Accordingly, the energy dissipated by element p of the aluminium truss core is given as:

$$ \Delta^{c}U^{p} = \psi_{xz}{}^{c}U^{p}_{xz} + \psi_{yz}{}^{c}U^{p}_{yz} \tag{15} $$

For the isotropic material $\psi_{xy} = \psi_{xz} = \psi_{yz} = \psi_{G}$, and therefore Eq. 15 simplifies to:

$$ \Delta^{c}U^{p} = \psi_{G}\left({}^{c}U^{p}_{xz} + {}^{c}U^{p}_{yz}\right) \tag{16} $$

The total dissipated energy by the aluminium truss core is:

$$ \Delta^{c}U = \sum_{p=1}^{J} \Delta^{c}U^{p} \tag{17} $$

Thus, the total dissipated energy by the sandwich panel is given as:

$$ \Delta^{s}U = \Delta^{c}U + \Delta^{l}U \tag{18} $$

The specific damping capacity of the sandwich panel which deforms at the nth mode of vibration, is as follows:

$$ {}^{s}\psi_{n} = \frac{\Delta^{s}U_{n}}{{}^{s}U_{n}} \tag{19} $$

4. Numerical model

The finite element (FE) method was used to develop the numerical model of the sandwich panel. The FE model of the laminated face layers based on the First Order Shear Deformation Theory (FSDT). Shell layered linear elastic elements CQUAD4 were used to model the laminated face sheets. 3D CHEXA linear elastic elements were used to develop the FE model of the aluminium truss core. The face layers and truss core elements were connected together by node fit at the locations where the contact spots between the elements existed (Figure 1). The epoxy adhesive was not modelled in the present study. The eigenvalue problem for undamped free vibrations of the sandwich panel was represented by $\left(\mathbf{K}-\omega_n^2\mathbf{M}\right)\Psi_n=0$, where **K** and **M** are the global stiffness and mass matrices of sandwich panel, respectively; Ψ_n are the eigenvectors (mode shapes) corresponding to eigenvalues $\omega_n = 2\pi f_n$, with f_n being eigenfrequencies. The commercial finite element code NASTRAN was used to solve the eigenvalue problem. The Real Eigenvalue Analysis using Lanczos eigensolver was applied to determine eigenvalues and corresponding eigenvectors of the undamped vibrations of the FFFF sandwich panel (all edges free).

5. Experimental procedure

In the present investigation a non-contact method for vibration sensing is applied using a scanning laser vibrometer POLYTEC PSV-400-B (Figure 2a). The PSV-400-B system consists of: the scanning head PSV-I-400 LR; the controller OFV-5000 (with internal signal generator), the junction box PSV-E-400, PC with data acquisition board and PSV Software 8.8. Additionally the power amplifier Bruel&Kjaer type 2732 is employed and a piezoelectric disc actuator (PZT) or loudspeaker is used to drive a structure into vibrations. The required data from the vibration test are frequency response functions (FRFs) which serve for the estimation of modal loss factors (Figure 2b). The comprehensive description of the operating procedure of the PSV-400-B system and its proper assemblage has been given elsewhere [9, 10, 11]. The estimation of the loss factor is based on the real part of the measured mobility FRF. The sandwich panels' loss factor (${}^s\eta_n$) for the mode with resonance frequency (f_n) are calculated as:

$$
{}^s\eta_n = \frac{1-\left(\frac{f_n^b}{f_n^a}\right)^2}{1+\left(\frac{f_n^b}{f_n^a}\right)^2} \tag{20}
$$

where n stands for a mode number, f_n^a and f_n^b are frequencies estimated from the real part of the FRF (Figure 4). Prior to the damping prediction of the sandwich panel, the constitutive stiffness and damping data of the parent components were estimated. To do so, the resonant method [12] was used to estimate the stiffness properties. Laser vibrometer was used for vibration sensing and loudspeaker for the excitation purposes. The dynamic results from the resonant method were then combined with

Table 2. Properties of the constituent materials.

Material	E_1[GPa]	E_2[GPa]	G_{12}[GPa]	ν_{12}	ν	ρ[kg/m^3]	ψ_{11}[%]	ψ_{22}[%]	ψ_{12}[%]	ψ_{xy}[%]
CFRP	142.2	10.5	7.8	0.38	-	1540.0	0.43	5.95	6.62	-
PA6	66.2	-	-	-	0.25	2700.0	-	-	-	0.31

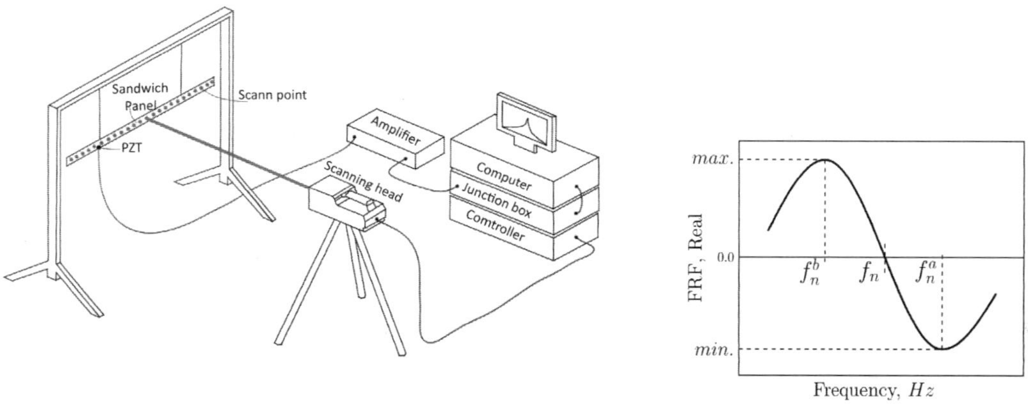

(a) **(b)**

Figure 2. Experimental procedure: (a) Experiment set-up for vibration testing; (b) Data acquisition for loss factor estimation.

energy method as proposed in [3] to find the specific damping capacities of the parent components. Seven beams made of CFRP with unidirectional alignment of fibres as $[0]_8$ and 4 aluminum beams were used to estimate the constitutive properties (Table 2).

6. Damping prediction

The finite element solution on the FE model of the sandwich panel is performed at first. The selected eigenfrequencies and eigenvectors are then extracted - first 3 bending modes. The strain and stress components corresponding to selected eigenvectors are stored in a text file. To do so, each shell element has been decomposed to separate layers and ε_{ij} and σ_{ij} of the layers were extracted. In addition, the element dimensions were preserved. Since the layer thickness was known, the volume of the layer was calculated and multiplied by the strain and stress components giving the strain energy of the layer. Summing up all the layers of all the elements gave the strain energy stored in the face layers of the panel. The procedure was repeated for the core, with the exception that no element decomposition was required and only shear strain and stress components were stored. The total strain energy of the panel is the sum of the energies stored in the face sheets and truss core. Then, the specific damping capacities of each layer and core material are used to calculate the dissipated energy components. The specific damping capacity of the sandwich panel is then calculated according to Eq.19 and the modal loss factor according to Eq.1. The postcalculations of the finite element data were performed using MATLAB software.

7. Results and conclusions

Three sandwich panels were subjected to vibration tests in order to extract the real part of the FRFs. The excitation was provided by the PZT disc. The necessary frequency information was then stored and the loss factors of the first three bending modes (Figure 3) were calculated according to Eq.20. The mean value of the loss factor was calculated and compared with the corresponding values predicted numerically (Table 3). A good coincidence of the results can be read from Table 3, although the predicted eigenfrequencies are overestimated in most cases and all the loss factors are underestimated. The major source of the results discrepancy is the lack of the epoxy adhesive included in the model. The energy dissipated by the adhesive makes the experimental loss factors higher than the predicted ones. Additionally, the lack of adhesive makes the model stiffer and increases the predicted eigenfrequencies. From the experimental point of view, the major drawback arises from the lack of vacuum chamber during vibration tests. In such case the loss factors increase due to air - structure interaction. Finally, the proper estimation of both stiffness and damping properties of the constituents encloses the sources of results discrepancy. Summing up, the numerical model improvement for the adhesive material and the tests done in a vacuum chamber should bring better results agreement.

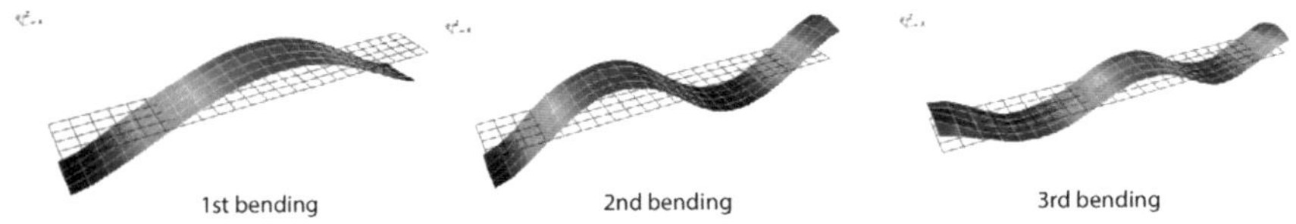

Figure 3. Experimental mode shapes of the truss core sandwich panel - Sandwich-2.

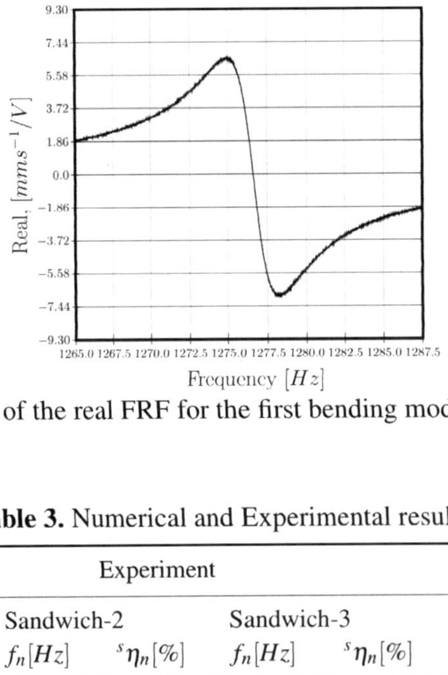

Figure 4. Zoomed fraction of the real FRF for the first bending mode of the Sandwich-1 panel.

Table 3. Numerical and Experimental results.

	Experiment							Numerical prediction	
	Sandwich-1		Sandwich-2		Sandwich-3		Mean	FEM model	
Mode	$f_n[Hz]$	$^s\eta_n[\%]$	$f_n[Hz]$	$^s\eta_n[\%]$	$f_n[Hz]$	$^s\eta_n[\%]$	$^s\eta_n[\%]$	$f_n[Hz]$	$^s\eta_n[\%]$
1	1296.89	0.32	1277.55	0.36	1274.50	0.47	0.38	1299.50	0.35
2	2172.52	0.70	2132.80	0.69	2117.35	0.75	0.71	2166.30	0.60
3	2927.66	0.88	2861.88	0.87	2838.75	0.77	0.84	2764.30	0.75

References

[1] Deshpande V. S. and Fleck N. A. 2001 Collapse of truss core sandwich beams in 3-point bending *Int. J. Solids. Struct.* **38** 6275

[2] Tochukwu G. and Deshpande V. S. and Wadley H. N. G. 2013 Mechanical response of carbon fiber composite sandwich panels with pyramidal truss cores *Composites Part A* **47** 31

[3] Yang J. and Xiong J. and Ma L. and Wang B. and Zhang G. and Wu L. 2013 Vibration and damping characteristics of hybrid carbon fiber composite pyramidal truss sandwich panels with viscoelastic layers *Compos. Struct.* **106** 570

[4] Yang J. and Xiong J. and Ma L. and Zhang G. and Wang X. and Wu L. 2014 Study on vibration damping of composite sandwich cylindrical shell with pyramidal truss-like cores *Compos. Struct.* **117** 362

[5] Adams R. D. and Bacon D. G. C. 1973 Measurement of the flexural damping capacity and dynamic Young's modulus of metals and reinforced plastics *J. Phys. D: Appl. Phys.* **6** 27

[6] Maheri M. R. and Adams R. D. and Hugon J. 2008 Vibration damping in sandwich panels *J. Mater. Sci.* **43** 6604

[7] Berthelot J.-M. and Sefrani Y. 2004 Damping analysis of unidirectional glass and Kevlar fibre composites *Compos. Sci. Technol.* **64** 1261

[8] Glodkowska W. and Ruchwa M. 2010 Static Analysis of Reinforced Concrete Beams Strengthened with CFRP Composites *Archives of Civil Engineering* **2** 111

[9] Kovalovs A. and Wesolowski M. and Barkanov E. and Gluhihs S. 2009 Application of macro-fibre composite (MFC) as a piezoelectric actuator *J. Vibroeng.* **11** 105

[10] Rucevskis S. and Wesolowski M. and Chate A. 2009 Damage detection in laminated composite beam by using vibration data *J. Vibroeng.* **11** 363

[11] Barkanov E. and Wesolowski M. and Hufenbach W. and Dannemann M. 2015 An effectiveness improvement of the inverse technique based on vibration tests *Compos. Struct.* **146** 152

[12] Wesolowski M. and Barkanov E. and Rucevskis S. and Chate A. and La Delfa G. 2009 Characterisation of elastic properties of laminated composites by non-destructive techniques *Proc. Int. Conf. on Composite Materials (Edinburgh)* **F5:12**

Acknowledgments

The research leading to these results has received the funding from the Faculty of Civil Engineering Environmental and Geodetic Sciences, Koszlain University of Technology (grant no. 504-01-03), and from the Latvia state research programme under grant agreement "Innovative Materials and Smart Technologies for Environmental Safety, IMATEH".

IOP Conf. Series: Materials Science and Engineering **96** (2015) 012023 doi:10.1088/1757-899X/96/1/012023

Application of Hemp Hurds in the Preparation of Biocomposites

J Cigasova[1,2], N Stevulova[1], I Schwarzova[1], A Sicakova[1] and J Junak[1]
[1]Technical University of Kosice, Faculty of Civil Engineering, Institute of Environmental Engineering, Vysokoskolska 4, 042 00 Kosice, Slovakia

E-mail: julia.cigasova@tuke.sk

Abstract. Hemp is a controversial bio-product with promising performance as a sustainable building material. The fact that hemp is an organic, natural product makes it highly relevant in the present reality of global pollution and struggle for coping with planetary warming. The construction sector is among the leading industries when it comes to energy consumption, release of CO_2; it is responsible for great amounts of waste and pollution. The research and implementation of sustainable building materials is a crucial necessity in the modern times. Hemp (Cannabis sativa) is an agricultural crop that can be used as a building material in combination with conventional or alternative binders. Hemp composites have many advantages as a building material, but it is not load-bearing and must be used in combination with a load-bearing wooden frame. Despite this disadvantage, hemp composite materials offer several of appropriate properties, namely: low density, good thermal insulation, antiseptic and breathability. This paper studies the possibility of preparing the lightweight composites based on hemp hurds (treated and/or untreated) as a filler and alternative MgO-cement as a binder. Properties of hemp composites are characterized by mechanical and physical methods.

1. Introduction

Agricultural biomass, also referred to as lignocellulosic material, is produced in billions of tons around the world every year. There are various types of agricultural biomass across the world that can potentially be used as a raw material in different applications, such as oil palm trunks, bagasse, coconut coir, bamboo, kenaf and hemp. Mostly, this biomass is found in the form of residual stalks from crops, leaves, roots, seeds, seed shells, etc. Types of biomass can be divided into main groups depending on the part of the plant which they are extracted from, i.e. bast (stem), leaf, fruit (seed) and straw as it is shown in figure 1. The composition of these organic fibres varies from one plant species to another. In addition, the polymer constituent composition in a single plant varies among species and

[2] To whom any correspondence should be addressed.

even in different parts of the same plant. It depends on the plant age, development growth, environment and other condition [1]. The properties of biomass vary considerably depending on the fibre diameter, structure, degree of polymerization, crystal structure and source and on the growing conditions. For decades, biomass raw materials have been historically used for ancient tools, food source, construction materials and textiles and as a source of energy. However, recently there has been a dramatic increase in the use of plant fibres for the development of environmentally friendly materials, especially as a reinforcing agent in polymeric composite materials in substitution of synthetic fibres like glass fibres [2]. This situation is largely spurred by environmental awareness, ecological consideration and technological advances.

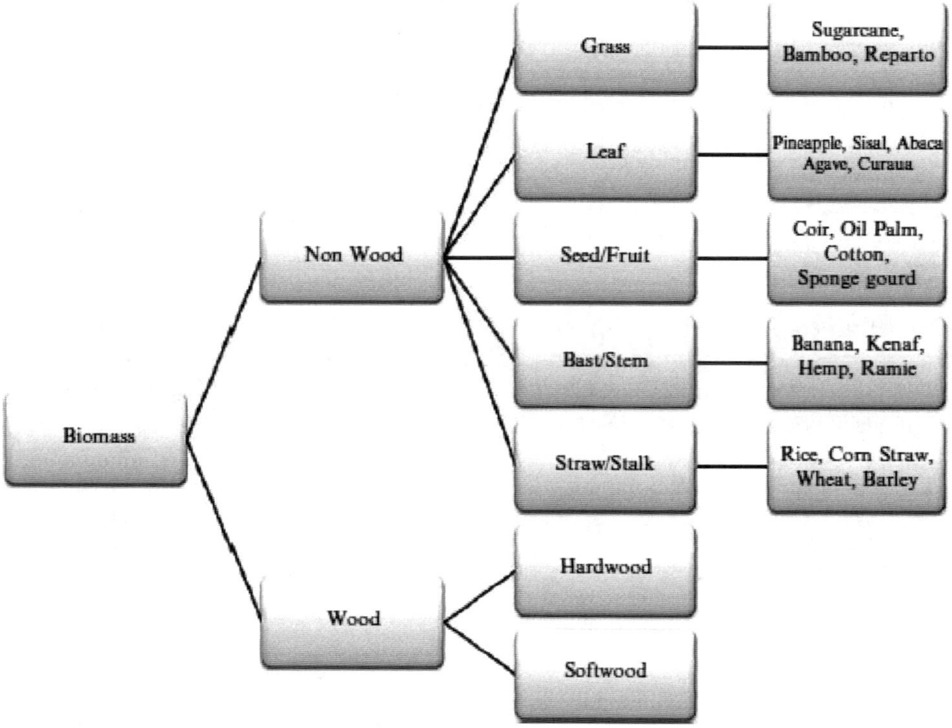

Figure 1. Classification of agricultural biomass [2].

Compared to a glass fibre, biomass offers many advantages due to its unique characteristics such as low cost, low energy consumption, zero CO_2 emission, low abrasive properties, low density, biodegradability, non-toxicity and continuous availability [3]. However, biomass fibres also have certain drawbacks, especially when considering their application in composites. They have high moisture absorption and poor compatibility with polymer and/or inorganic matrix, which is responsible for poor mechanical and thermal properties. Modification or treatment (chemically, physically or mechanically) of the fibre is needed to enhance the performance of biomass in different multiple applications [4]. In the past few decades, the development of new materials that involve natural resources as the raw material, especially as a composite material, has accelerated. Nowadays, a large number of interesting applications is emerging for these materials due to recent progress in technologies, biomass material development, genetic engineering, and composite science technology that offer significant opportunities for an exploration and development of improved materials from renewable resources usable in various applications such as biocomposites, pulp and paper, construction, automotive, medical, packaging, aerospace, pharmaceutical and biomass energy production [5].

This article deals with utilization of hemp hurds (as waste material from bast fibre production) in preparation of lightweight composites. Hemp hurds were chemically or physically treated to remove impurities from the surface of the fibres as well as amorphous substances. The effect of treatment of fibres and the influence of hardening time on mechanical and physical properties of the prepared biocomposites were monitored.

2. Experiment

2.1. Materials

2.1.1. Hemp hurds.

Hemp is an annual plant which provides two materials used in civil engineering: hemp hurds (granular form of hemp descending from the inner woody core) and hemp fibres (fibrous form of hemp descending from the bark-like bast fibres).

The original material studied in this paper is hemp hurds coming from the Netherlands Company Hempflax. The hemp hurds used had wide particle size distribution (8-0.063 mm). The mean particle size was calculated to be 1.94 mm. Bulk density of hemp material was 117.5 kg•m^{-3}. Average moisture content of the used hemp hurds was determined by weighing the sample before and after drying in an oven at 70°C until constant moisture content of 10.78% was reached.

The second sample used in the experiment was hemp hurds chemically modified in an alkaline environment. Dried original hemp hurds were soaked in 1.6 M NaOH solution during 48 h and then neutralized by 1% vol. acetic acid. Hemp hurds were then washed with water until the pH value was 7. The physical modification of dried hemp hurds was made by their thermal treatment in hot water (100 °C) for 1 h. Subsequently, hemp hurds slices were dried and their surface was obviously distraught and shredded.

2.1.2. Binder.

Alternative binder MgO-cement was used for the preparation of the specimens. Binder consisted of milled magnesium oxide (MgO), silica sand (SiO_2) and sodium hydrogen carbonate ($NaHCO_3$) [6].

2.2. Preparation of specimens

Experimental mixtures were prepared according to the recipe published in [7]. Mixture recipe is presented in figure 2. The mixtures were homogenized in the classical concrete mixer and the standard steel cube forms with dimensions 100mm x 100mm x 100mm were used for the preparation of the bodies. Two days later the composites were taken out of the forms and cured under laboratory conditions (18ºC, 65 % of relative humidity) during 28, 60 and 90 days. The number of tested specimens for each set was 3.

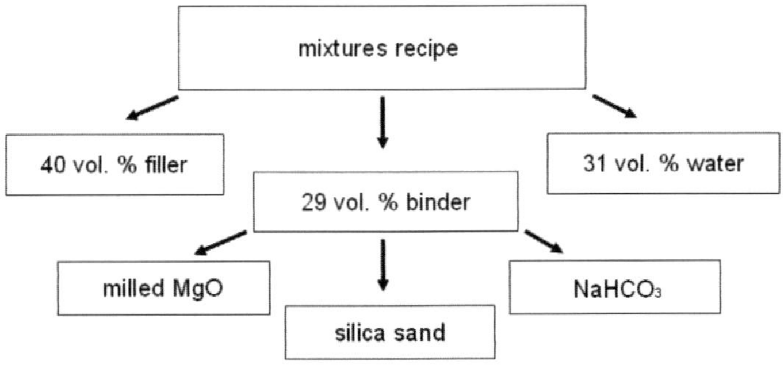

Figure 2. Scheme of mixture recipe.

2.3. Testing methods

For all hardened composites, the following properties were measured: density, thermal conductivity coefficient, compressive strength and water absorbability. Density was determined in accordance with standard STN EN 12390-7 [8]. The thermal conductivity coefficient of samples as the main parameter of heat transfer was measured by the commercial device ISOMET 104 (Applied Precision Ltd, Slovakia). The measurement is based on the analysis of the temperature response of the studied material to heat flow impulses. The heat flow is induced by electrical heating using a resistor heater having direct thermal contact with the surface of the sample. The water absorption of hardened composites was determined after their short-term storage in water during 1 hour in accordance with standard STN EN 12087/A1 (727056) [9]. After immersion period, the specimens were taken out from the water and were reweighed. Water content of the composite samples was determined from the weight differences between wet and dry test samples.

Compressive strength of all composites was determined using the instrument ADR 2000 (ELE International, England) according to standard STN EN 12390-3 [10].

3. Results and Discussion

In this paper, the impact of chemical and physical modification of the hemp hurds slices and hardening time on the selected physico-mechanical properties (density, compressive strength, thermal conductivity, water absorption) of biocomposites with alternative binder MgO-cement was studied. The results of testing the parameters of hardened composites (after 28, 60, 90 days) are presented. As it can be seen in figure 3, the values of compressive strength of hemp composites depend on the time of hardening as well as on the surface modification of hemp hurds slices. Hemp composites based on untreated original hemp hurds have higher values of compressive strength (1.84 – 4.9 MPa) in comparison with the composites based on physically treated hurds (1.12 – 2.74 MPa). The same behaviour of composites has been observed in the case of hardened specimens with chemically treated hemp hurds, strength values were lower than values of physically treated filler (1.08 – 1.80 MPa). One of the reasons leading to reduced values of strength parameter of the hardened composites based on modified filler is the high alkalinity of MgO-cement and poor adhesion of biomass aggregate with binder particles as a result of release of extractable substances inhibiting the process of hydration, too [11, 12]. The setting impairment caused by extracts and/or degradation by products could be explained by various phenomena like adsorption on hydrated or unhydrated MgO-cement particles, ionic complexation or the formation of a thin layered barrier around binder grains by precipitation [13].

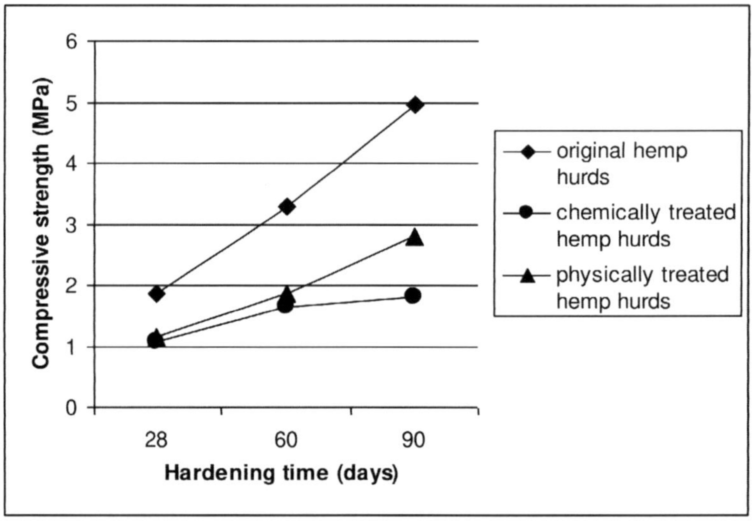

Figure 3. Dependence of compressive strength of the biocomposites on the time of hardening.

The density values of biocomposites ranged from 910 to 1200 kg•m^{-3}, which puts them in the category of lightweight composites. The linear dependency of the compressive strength on bulk density of hemp hurds composites was confirmed [14].

The hemp composite samples based on modified hemp hurds have lower values of thermal conductivity coefficient. The values in the range 0.071-0.085 W•m^{-1}•K^{-1} were recorded for physically treated samples. Composites with chemically modified hemp hurds reached the values of thermal conductivity coefficient in the range of 0.058 – 0.072 W•m^{-1}•K^{-1} which were lower than those of reference composites based on original hemp slices (0.08-0.098 W•m^{-1}•K^{-1}). All measured conductivity values of the prepared biocomposite samples are in the range acceptable for thermal insulation materials.

The water contents in the prepared composite samples after their short-term immersion in water, shown in figure 4, decreased along with the increasing time of hardening. Water absorbability values of composites are influenced by treatment methods of hemp hurds. The highest values of water content (13.9 – 26 wt. %) are demonstrated by biocomposites based on unmodified hemp hurds for each hardening time. The results confirmed that the absorbability decreases over time; that may indicate the fact that hydration products of the binder gradually fill the vacant interior spaces in the composite. The reduction of water absorption during aging could be explained by additional hydration during cement matrix hardening and creation of a denser structure inside the paste.

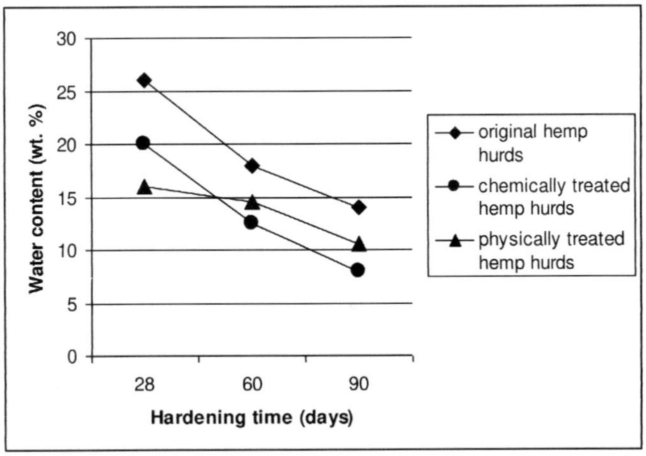

Figure 4. Dependence of water content in biocomposites on the time of hardening.

As it can be clearly seen from figure 4, the specimens with modified hemp hurds have observably lower water absorption values than the composites based on unmodified fibres. It means that the chemical and/or physical modification of hemp hurds reduced water absorption of the lightweight composites. According to literature data [15], physical treatment changes structural and surface properties of the fibres but it does not extensively change their chemical composition. On the other hand, chemical modification led to significant changes in defibrillation of bundles, chemical and phase composition and cellulose crystallinity [16].

4. Conclusions

This paper has analysed the application of hemp hurds as organic filler in unmodified and modified (physically or chemically) form in building materials. Thermal treatment (cooking in hot water) and alkali activation (leaching in NaOH solution) were used as the methods for removal of organic and inorganic loosely bound contaminants from the surfaces of hemp hurds. The influence of the used method on hemp hurds slices on the mechanical (compressive strength) and physical properties (density, thermal conductivity, water absorbability) of the hardened composites was observed. The following conclusions can be made:

The compressive strength of biocomposites prepared with hemp hurds increased with increasing time of hardening. However, physical and chemical treatment of hemp hurds has a negative impact on compressive strength of composites in comparison to composites based on untreated organic filler.

The values of thermal conductivity of composites prepared with unmodified and modified hemp hurds predetermine this material to its use as a thermal insulation. The samples based on chemically modified hemp hurds had the lowest values of thermal conductivity coefficient.

Hot water treatment and alkali activation of hemp hurds affect the water absorption properties of the hardened composites. Lower water contents in composites were observed in the case of specimens containing hemp hurds with the treated surface.

Further investigations are needed to understand and explain the influence of natural fibres as filler in composites structure by using physical method.

5. References

[1] Kumar R, Choudhary V, Mishra S and Varma I 2008 Banana fibre-reinforced biodegradable soy protein composites *Front Chem China* **3** (3) 243 – 250

[2] Alwani M S, Khalil H A, Asniza M, Suhaily S S, Amiranajwa A N and Jawaid M 2014 Agricultural Biomass Raw Materials: The Current State and Future Potentialities *Biomass and Bioenergy* **1** (5) 77 – 100

[3] Guimarães J, Frollini E, Da Silva C, Wypych F and Satyanarayana K 2009 Characterization of banana, sugarcane bagasse and sponge gourd fibres of Brazil *Ind. Crops Prod.* **30** (3) 407–415

[4] Pandey J K, Ahn S, Lee C S, Mohanty A K and Misra M 2010 Recent advances in the application of natural fiber based composites *Macromol. Mater. and Engin.* **295** (11) 975 – 989

[5] Lau K T, Ho M P, Au-Yeung C T and Cheung H Y 2010 Biocomposites: their multifunctionality *Int. J. Smart Nano Mater.* **1** (1) 13 – 27

[6] Kidalova L, Terpakova E, Priganc S and Stevulova N 2011 Possibilities of using hemp shives in light-weight composites *Conf. CEST 2011: 12th Int. Conf. on Environ. Science and Techn.* pp 547 – 552

[7] Bydzovsky J 2009 Utilization of rapidly renewable raw materials in building materials *Proc. of the 12th Int. Scientific Conf., Brno Technical University, Brno Czech Republic* pp 43 – 46

[8] STN EN 12390-7 2011 Testing hardened concrete, Part 7: Density of hardened concrete

[9] STN EN 12087/A1 2007 Thermal insulating products for building applications, Determination of long term water absorption by immersion

[10] STN EN 12390-3 2010 Testing hardened concrete, Part 3: Compressive strength of test specimens

[11] Diquélou Y, Gourlay E, Arnaud L and Kurek B 2015 Impact of hemp shiv on cement setting and hardening: Influence of the extracted components from the aggregates and study of the interfaces with the inorganic matrix *Cement & Concrete Comp.* **55** 112 – 121

[12] Vaickelionis G and Vaickelioniene R 2006 Cement hydration in the presence of wood extractives and pozzolan mineral additives *Ceram – Silik* **50** (2) 115 – 122

[13] Jolicoeur C and Simard M A 1998 Chemical admixture – cement interactions: phenomenology and physico –chemical concepts *Cem. Concr. Compos.* **20** (2–3) 87–101

[14] Cigasova J, Stevulova N, Terpakova E, Sicakova A and Junak J 2012 Possibility of Using Chemically Modified Hemp Shives in the Preparation of Lightweight Composites *Proc. of the 1st Int. Conf. for PhD Students in Civil Engineering: New Researcher Generation with Challenges in Civil Engineering Cluj-Napoca (Romania, 4-7 November 2012)* pp 414 – 420

[15] Renouard S, Hano Ch, Doussot J, Blondeau J P and Lainé E 2014 Characterization of ultrasonic impact on coir, flax and hemp fibers *Materials Letters* **129** 137 – 141

[16] Stevulova N, Cigasova J, Estokova A, Terpakova E, Geffert A, Kacik F, Singovszka E and Holub M 2014 Properties Characterization of Chemically Modified Hemp Hurds *Materials* **7** (12) 8131 – 50

Acknowledgments
The authors are grateful to the Slovak Grant Agency for financial support of the project VEGA1/0277/15.

IOP Conf. Series: Materials Science and Engineering **96** (2015) 012024 doi:10.1088/1757-899X/96/1/012024

Intrinsic Hydrophobicity of Rammed Earth

M Holub[1,4], C Stone[2], M Balintova[1] and R Grul[3]
[1]Technical University of Kosice, Institute of Environmental Engineering, Faculty of Civil Engineering, Vysokoskolska 4, Kosice, 042 00 Slovakia
[2]Technical University of Kosice, Faculty of Civil Engineering, Building Construction Laboratory, Vysokoskolska 4, Kosice, 042 00 Slovakia
[3]Technical University of Kosice, External relation and marketing office, Faculty of Civil Engineering, Vysokoskolska 4, Kosice, 042 00 Slovakia

E-mail: marian.holub@tuke.sk

Abstract. Rammed earth is well known for its vapour diffusion properties, its ability to regulate humidity within the built environment. Rammed earth is also an aesthetically iconic material such as marble or granite and therefore is preferably left exposed. However exposed rammed earth is often coated with silane/siloxane water repellents or the structure is modified architecturally (large roof overhangs) to accommodate for the hydrophilic nature of the material. This paper sets out to find out optimal hydrophobicity for rammed earth based on natural composite fibres and surface coating without adversely affecting the vapour diffusivity of the material. The material is not required to be waterproof, but should resist at least driving rain. In order to evaluate different approaches to increase hydrophobicity of rammed earth surface, peat fibres and four types of repellents were used.

1. Introduction

Rammed earth, also known as taipa [1] (Portuguese), tapial (Spanish), and pisé (de terre) (French), is a technique for building walls using natural raw materials such as earth, chalk, lime or gravel.

It is an old building method that is coming to the fore in the recent years as people seek more sustainable natural building materials and economic building methods.

Rammed earth walls are simple to construct, non combustible, strong, durable and exhibit good thermal properties.

By 2000 BCE, rammed earth building techniques were commonly used for walls and foundations in China [2] (some can be observed even now, figure 1), but the evidence of early using can be seen on Neolithic archaeological sites dating back to 5000 BCE. Rammed earth buildings are found all over

---

[4] To whom any correspondence should be addressed.

the world, in a range of environments that include temperate and wet regions of northern Europe [3], semiarid deserts, mountain areas and the tropics. The availability of useful soil and a building design appropriate for local climatic conditions are the factors that favour its use.

Rammed earth constructions also exhibit negative properties resulting from the material used. Rammed earth is well known for its vapour diffusion properties, its ability to regulate humidity, which is a positive factor for the internal building environment, but only to a certain extent. Due to the permeable and hydrophilic nature of rammed earth building materials, water can easily penetrate through the construction. For this reason water penetration becomes an important factor which affects the durability of rammed earth based construction. Modern approach focused to improve the durability of rammed earth structures includes usage of small amount of cement as a stabiliser. Taking into consideration the fact that cement is also a hydrophilic material, it cannot solve the problems associated with water penetration that damages thermal insulation properties of rammed earth buildings.

Figure 1. The ruins of a Han Dynasty (202 BCE – 220 CE) Chinese watchtower made of rammed earth at Dunhuang, Gansu province.

At present, there exist two approaches that can improve the durability against natural weathering. In the first stage of building the rammed earth construction it is possible to use *internal admixtures*, mainly silicon water-repellent admixtures that effectively minimise water movement within the capillaries. Another benefit is that vapour permeability and surface appearance of the construction are not affected.

In the case when the construction is built, a post-*surface treatment* can be applied. Silicone water-repellents are useful sealers for rammed earth constructions. This approach is based on the penetration of the sealer into the capillary system and reaction with the substrate via strong siloxane bonding providing a long-term protection against natural weathering [4]. Silicone is a highly reactive material which forms a polysiloxane molecular lining within the capillary walls of the masonry substrate. Commonly used silicone sealers include siliconate, alkyalkoxysilane and siloxane (see figure 2) [4].

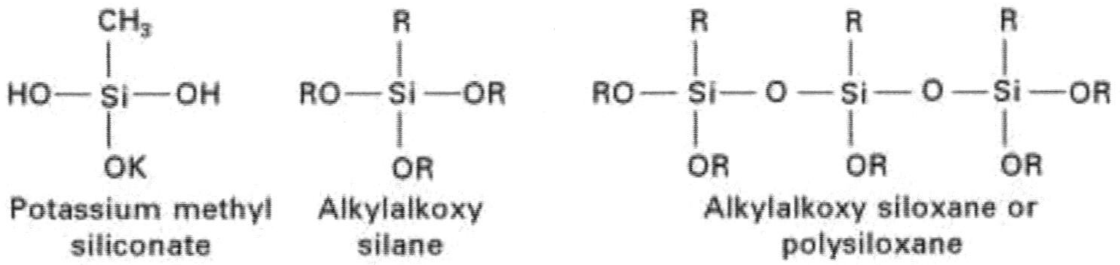

Figure 2. Silicone water-repellent sealers for masonry substrate.

The presented article is focused on the second approach mentioned. Four types of repellent were used to study the hydrophobic/hydrophilic properties of rammed earth. One sample with hydrophobic peat fibres (as internal admixture) was also tested. Samples were studied by goniometric surface hydrophobicity measurements of water drop on the rammed earth surface.

1.1. Theoretical background
Regarding the decision on surface hydrophobicity measurements used in this article, brief foundations have to be presented. There are three types of forces acting upon a liquid drop on a surface (see figure 3).

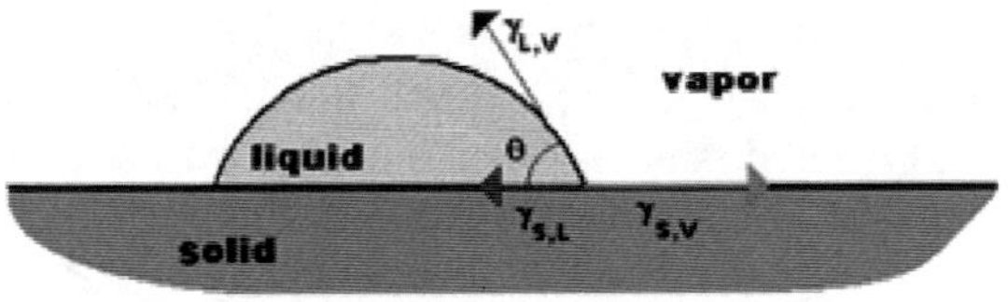

Figure 3. Forces acting on a liquid droplet on a solid.

Following simple goniometric rules, Young [5] derived that the contact angle θ_Y is given by relation (1):

$$\cos \Theta_Y = \frac{\gamma_{SV} - \gamma_{SL}}{\gamma_{LV}} \quad (1)$$

where γ_{IJ} denotes surface tension (energy per unit surface) of the interface IJ and where *S*, *L* and *V* designate solid, liquid and vapor phases.

Different approaches can be used for measuring contact angles of non-porous solids, a goniometric approach and a tensiometric approach, with both having their advantages and their drawbacks. Another approach is used when measuring the angles of porous substrates, involving the use of a tensiometer and the Washburn method [6].

The results presented in this contribution are achieved by applying a goniometric method, which is based on the analysis of the shape of the liquid drop. The contact angle can be found directly by measuring the angle formed between the solid and the tangent of the drop of an image made of the drop.

IOP Conf. Series: Materials Science and Engineering **96** (2015) 012024 doi:10.1088/1757-899X/96/1/012024

The large advantage of goniometry comes from its relative simplicity. It can be used for almost any solid, as long as it has a relatively flat portion or a regular curvature and can be fitted on the stage of the instrument. The main disadvantage of this approach is the subjectivity of the researcher in assigning the tangent line. This problem was eliminated by computer analysis of the droplet shape, which will be described later.

2. Materials and methods

2.1. Samples preparation

For the application of surface treatment, a reference samples consisting of components listed in table 1 were prepared.

Table 1. Percentage composition of basic components of rammed earth samples.

Component	Content (%)
Dredged sand (fraction 0/2 mm)	50
Lime gravel (fraction 0/4 mm)	20
Clay	15
Silt	15

Variable speed mixer was used to ensure homogeneity and mixing of the above listed components with water and Portland cement (type III, 32.5 R) in the ratio 82:10:8.

The mixture was placed in a steel cylinder of 101.3 mm diameter and height of 80 mm and subsequently compressed with powered tamper. Once removed from the molds, samples were dried at laboratory temperature for two weeks before application of the surface treatment.

Another type of sample (P1) was also prepared adding of 5% hydrophobic peat fibers to the mixture and processed under the same conditions.

After hardening of the rammed earth material, the surface of the samples was treated with different types of sealers and dried once again to ensure the homogeneity of the surface. For better clarity the samples are marked according to different types of sealers (see table 2).

Table 2. Different types of repellent sealers used in surface treatment.

Label	Repellent type
Ref.	Reference sample without repellent
S703	Sikagard – 703 W (silane/siloxane based repellent)
G1	Geolak – transparent varnish (mixed with water in ratio 2:1)
G2	Geolak – transparent varnish (not mixed with water)
A1	Aqua lackspray
S1	GrandX Silicon oil spray

2.2. Measurements set-up and method

Measuring contact angles with a high level of precision usually requires high-tech contact angle goniometers that can perform a great number of automated measurements (N=50-100) per drop, thus reducing the error on each returned average value [7].

In this article, simplified experimental apparatus according to [8] was used. The simple experimental apparatus provides a convenient alternative to commercial goniometers because the method allows measurements with sufficient precision to be obtained while being accessible in terms of cost and ease of construction. A schematic view of the set-up is given in figure 4.

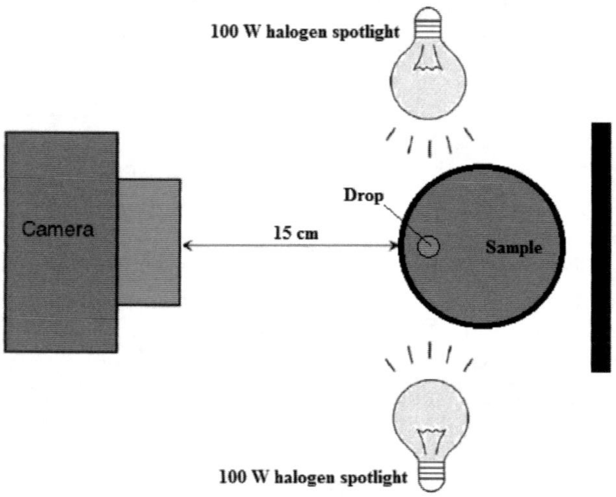

Figure 4. Sketch of experimental set-up used to measure contact angles.

The optical part includes 21.1 megapixels full-frame CMOS digital camera (Canon EOS 5D Mark II) with EF 24 – 105 mm f/4.0L IS USM wide-to-telephoto zoom lens. A highly absorptive background located behind the sample and a couple of halogen spotlights (100 W) positioned on the sides of the sample were used to make the drop shape sharp, which is necessary for measurement precision as well as for image processing.

Every measurement was performed duplicate applying 50 μL deionized water by laboratory pipette on the sample surface to create an equal droplet. After applying of the water and stabilization of the drop (10s), the picture was made. The pictures were processed and exported twice (original and optically zoomed, see figures 5 and 6).

The profiles of the drops were automatically fitted using ImageJ software using the contact angle plug-in.

Figure 5. Original picture of the drop on the silane/siloxane treated rammed earth surface.

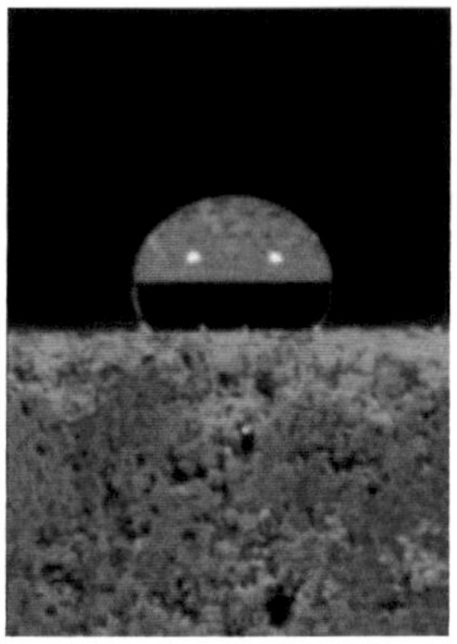

Figure 6. Optically zoomed drop on the silane/siloxane treated surface.

3. Results

On a hydrophobic surface, water forms distinct droplets. As the hydrophobicity increases, the contact angle of the droplets with the surface increases. Surfaces with contact angles greater than 90° are designated as hydrophobic. The theoretical maximum contact angle for water on a smooth surface is 120° – 130°. Micro-textured or micro-patterned surfaces with hydrophobic asperities can exhibit apparent contact angle exceeding 150° and are associated with super hydrophobicity and the "lotus effect".

The fitted contact angles using ImageJ software are presented in table 3 and also depicted in figures 7 – 12. From the results it is evident that the reference sample represents ordinary surface with "typical wetting". During the measurement procedure (wetting the surface), different shapes of drop on different parts of the sample were observed. This fact was caused by heterogeneity of the surface (different smoothness, presence of the cracks). It can be said that the working procedure has an impact on the hydrophilicity/hydrophobicity of the sample. This negative factor should be eliminated by the treatment of the surface, where whole surface is impregnated by equal layer of the repellent. The same contact angle was also calculated for G2 sample and thus was proved that it is necessary to mix the Geolak with the water. After the recommended mixing of Geolak with water and surface treatment, sample (G1) exhibited hydrophobic properties (see figure 10). Similar properties were achieved applying Aqua lackspray (contact angle of 93°).

Table 3. Calculated contact angles.

Label	Calculated contact angle and deviation (°)	
Ref.	61.3°	(+/- 2.5°)
S703	123.8°	(+/- 1.0°)
G1	93.2°	(+/- 1.5°)
G2	61.8°	(+/- 1.7°)
A1	91.7°	(+/- 0.9°)
S1	132.5°	(+/- 0.5°)
P1	0°	-

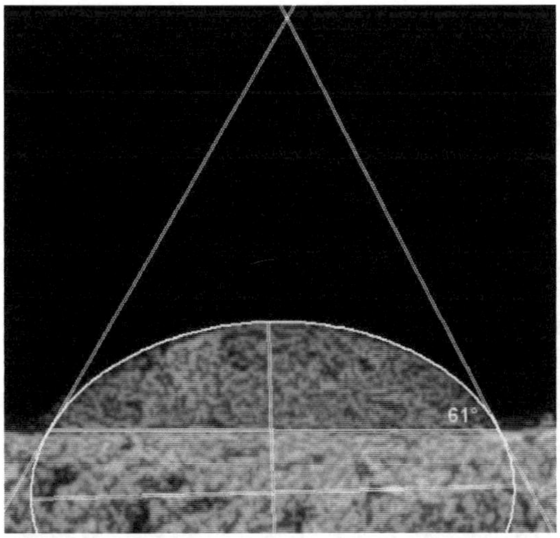

Figure 7. The shape of the drop and calculated contact angle (Ref. sample).

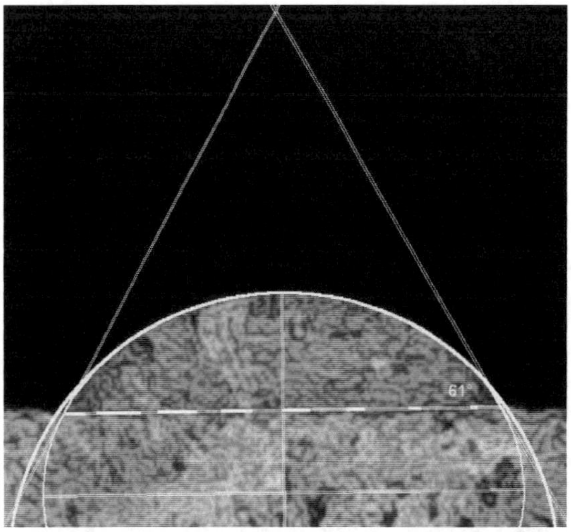

Figure 8. The shape of the drop and calculated contact angle (G2 sample).

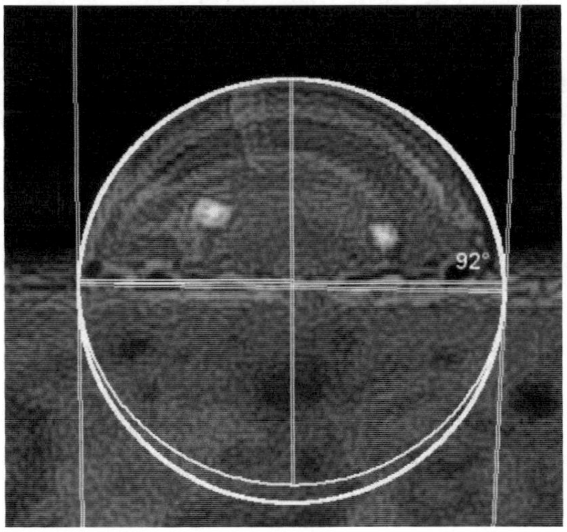

Figure 9. The shape of the drop and calculated contact angle (A1 sample).

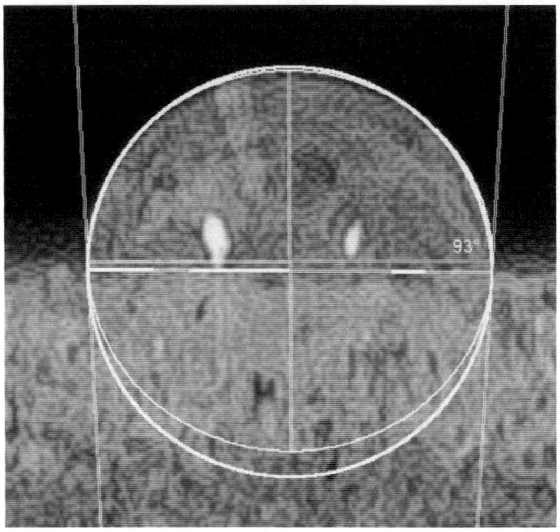

Figure 10. The shape of the drop and calculated contact angle (G1 sample).

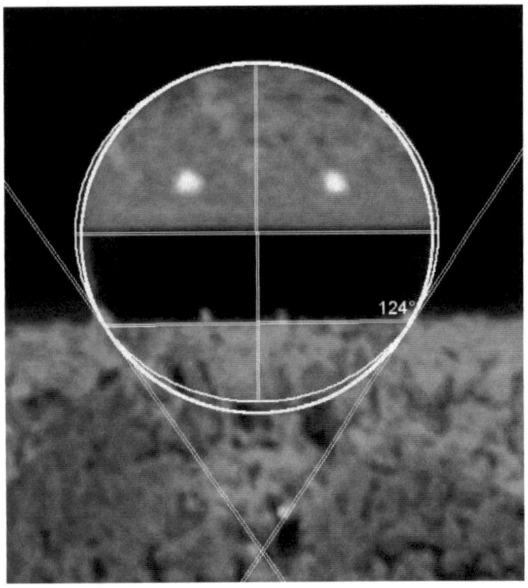

Figure 11. The shape of the drop and calculated contact angle (S703 sample).

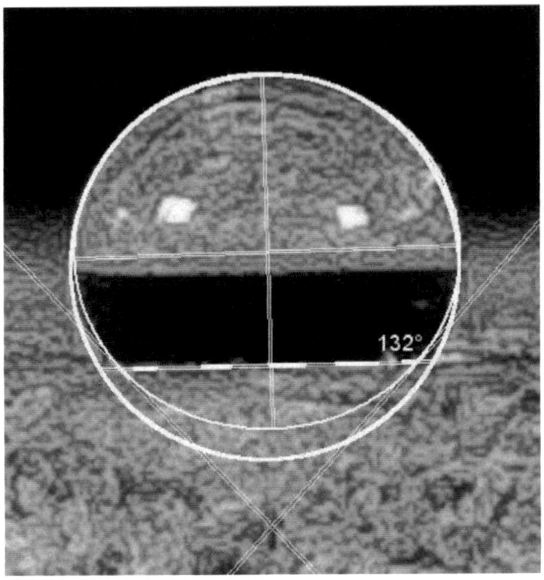

Figure 12. The shape of the drop and calculated contact angle (S1 sample).

Higher contact angles (up to 130°) were achieved by commercial silane/siloxane and silicone repellents (figures 11 and 12) designed for building construction materials. The cost of these repellents is a negative aspect of this approach. Usage of hydrophobic peat fibres proved to be an inappropriate way to increase the hydrophobicity of rammed earth, it was not possible to measure the contact angles due to high hydrophilic nature of the sample.

4. Conclusions

We used a simple, efficient, and inexpensive method for contact angle measurements of the rammed earth samples that proved to be adequate. The experimental results, summarized in table 3, were obtained with ImageJ software (with contact angle plug-in), which returned precise and stable values of the contact angles.

From the results it is evident that pure rammed earth represents ordinary surface with "typical wetting", which can be improved by different types of repellents. Application of commercial silane/siloxane and silicone repellents can sufficiently protect the rammed earth constructions for a long period. In this case, we can observe 100 % increases in contact angle values, from 61.3° of original sample to 123.8° and 132.5° after silicone and silane/siloxane treatment, respectively. Treatment by Geolak (G1 sample) was also successful and surface exhibited hydrophobic properties (contact angle of 92°). Similar properties were achieved applying Aqua lackspray (contact angle of 93°). Hydrophobic peat fibres proved to be an inappropriate internal admixture to increase the hydrophobicity of rammed earth.

5. References

[1] Fleming J, Honour H and Pevsner N 1991 *The Penguin Dictionary of Architecture* (London: England Penguin Books)

[2] Fu X, Guo D, Liu X, Pan G, Qiau Y and Sun D 2002 *Chinese Architecture* (ed. Steinhardt, Nancy Shatzman) The Culture and Civilization of China (New Haven: Yale University Press)

[3] Keable R 2012 *Rammed Earth Lecture Theatre, CAT* (London: Rammed Earth Consulting)

[4] Hall M R, Lindsay R and Krayenhoff M 2012 *Modern Earth Buildings: Materials, Engineering, Construction and Applications* (Cambridge: Woodhead Publishing Limited)

[5] Young T 1805 An Essay on the Cohesion of Fluids *Philosophical Transactions of the Royal Society of London* **95** 65 – 67

[6] Lai S C S 2003 *Mimicking Nature: Physical Basis and Aartificial Synthesis of the Lotus – Effect* (Leiden: Universiteit Leiden)

[7] Barankin M D, Gonzalez E, Habib S B, Gao L, Guschl P C and Hicks R F 2009 *Langmuir* **25** 2495 – 2500

[8] Buvailo A, Xing Y, Keuleyan S, Prakash V, Eftekhari – Bafrooei A and Borguet E 2012 Contact Angle Measurements Using a Simplified Experimental Setup *J. of Chem. Educ.* **87** 1403 – 07

Acknowledgements

This publication is the result of the implementation of the Project "University Science Park TECHNICOM for Innovation Applications Supported by Knowledge Technology, ITMS: 26220220182", supported by the Research & Development Operational Programme funded by the ERDF and funded by Grant No. 1/0563/15 of the Slovak Grant Agency for Science.

Possibilities of Using Cellulose Fibres in Building Materials

V Hospodarova[1,2], N Stevulova[1] and A Sicakova[1]
[1]Technical University of Kosice, Faculty of Civil Engineering, Institute of Environmental Engineering, Kosice, Slovakia

E-mail: viola.hospodarova@tuke.sk

Abstract. Nowadays, utilization of wastes from agriculture, paper production and building construction is becoming increasingly important due to environmental concerns. Material recycling is a growing trend in the development of building materials; some waste materials can be used in construction as secondary raw materials. The demand for natural non-renewable raw materials is increasing rapidly, therefore, wastes as resources for secondary raw materials can be a good substitute in the production processes. In this way, the shortage of natural raw materials can be supplemented. Construction industry uses secondary raw materials very effectively thereby substituting virgin materials. One of the interesting secondary raw materials is waste coming from natural plant fibres. In this paper, characterization of cellulose fibres from wood pulp, waste paper and their use in cement composites are considered. Technically important parameters of hardened composites are determined and tested (density, water absorbability and compressive strength).

1. Introduction

In the recent years, much more attention has been paid to sustainable, green and environmentally friendly materials for various applications. In this regard, the application of renewable and biodegradable biomass fibres reinforced composite materials has been developed to get subsequent generation of sustainable and green materials in this field [1].

Producing more sustainable and environmentally friendly materials has gained the attention of researchers at the international level. It includes a search for alternatives to petroleum-based materials. The development of biomaterials holds great promise to mitigate many of the sustainability problems, offering the potential of renewability, biodegradation, and a path away from harmful additives. The research reported here involves the development of efficient techniques for the production of natural fillers as reinforcing agents for composite applications. Chemical composition, morphological structure, physical and thermal properties of the fibres can be evaluated to estimate their ability to be used as fillers in biocomposite applications [2].

[2] To whom any correspondence should be addressed.

The cellulosic fibres are produced in large quantities, different morphologies and sizes, derived from renewable resources, such as agro-industrial and waste material, and available at reduced cost as compared to synthetic fibres (i.e., polypropylene and polyvinyl alcohol fibres).The increase in the agro-industrial activities generates a large quantity of waste in this sector. The use of lightweight cellulosic pulp from agro-industrial by-products as reinforcement in low cost building materials is an interesting strategy for managing these by-products, from the point of view of their mechanical properties, environmental and social-economic impacts [3].

The use of wood fibres and recycled paper fibres in cement composites has many advantages. Wood fibre reinforced composites are easy to cast to mold into a desired shape, they are resistant to fire, as well as resistant to harmful effects of sunlight, rain, and insects. These composites also have low thermal conductivity, a great degree of processing flexibility, in addition, they might help eliminate environmental pollution by recycling wood fibre. Some of the advantages of wood fibres in cement fibre composites are availability, high tensile strength, relatively high modulus of elasticity, and the well-developed technology to easily extract the fibres from wood for use in the composites. The main disadvantage is their vulnerability to chemical decomposition of certain wood chemical constituents in the alkaline cement environment [4].

The primary advantages of using lignocellulosic fibres as additives in cement are low density, low cost, nonabrasive nature, high filling levels possible, low energy consumption, and wide variety of fibres available throughout the world. With an increasing worldwide shortage of wood resources, there has been a strong trend to produce composite products using recycled paper, non-wood plant materials and agricultural residues. Among the possible alternatives, the development of pulp and paper industries and bio-composites using recycled paper is currently at the centre of attention [5].

The objective of this study was to determine physical and mechanical properties of fibre-cement composites and compare those properties with the properties of composites made without fibres. The density, water absorbability and compressive strength properties of hardened cement-based composites were analyzed.

2. Materials and methods

2.1. Materials

In this experiment, a renewable raw material based on cellulosic fibres was used. Cellulose fibres - Greencel (figure 1) were obtained from bleached wood pulp (white sample GW-500) and unbleached recycled waste paper (grey sample G-700T). The Greencel fibres came from company Bukoza Invest Ltd (Hencovce, Slovakia). Two types of samples were tested. Physical and chemical properties of cellulose fibres are shown in table 1.

Figure 1. White wood pulp and grey waste paper cellulose fibres.

Table 1. Properties of cellulose fibres Greencel.

Cellulose fibres	Cellulose content [%]	Bulk density [kg/m^3]	Max. length [μm]	Dry matter [%]	Ash [%]	pH	Colour
GW-500	99.5	60-80	500	93	0.5	6±1	white
G-700T	80	40-70	600	93	20	7±1	grey

Portland cement CEM I 42.5 R (Holcim Slovakia Ltd) was employed the binding agent. Natural silica sand of the fraction 0–0.6 mm was used as filler into the mixtures. Water for the preparation of cement mixtures was used in accordance with standard STN EN 1008 [6].

2.2. Preparation of composites

The cement composites consisting of cement, sand, water and cellulose fibres were prepared according to the recipe given in Table 2. The cellulose fibres substituted 0.2%, 1.0% and 5.0% weight of filler (sand). Fibres from wood pulp (GW-500) were used in the first set of experiments. A partial sand replacement by fibres from waste paper (G-700T) was performed in the second experimental set. Reference sample (RF) was prepared without addition of cellulosic fibres. Water-cement ratio (w/c) was 0.75 for each sample.

Table 2. Mixture proportion of fibre-cement composites.

Mixture samples		CEM I 42.5 R (wt.%)	Sand (wt.%)	Water (wt.%)	Cellulosic fibres (wt.%)	
					GW-500	G-700T
Alternative 1	WP1	21.05	62.96	15.79	0.2	-
	WP2	21.05	62.16	15.79	1.0	-
	WP3	21.05	58.16	15.79	5.0	-
Alternative 2	RP1	21.05	62.96	15.79	-	0.2
	RP2	21.05	62.16	15.79	-	1.0
	RP3	21.05	58.16	15.79	-	5.0
Reference mixture RF		21.05	63.16	15.79	-	-

Preparation of fibre reinforced cement composites was carried out in two steps. At first, cellulose pulp was dispersed in water by mechanical stirring (approximately 50 wt.% of water). Subsequently, cement, sand and remaining amount of water were added, and mixing continued to allow uniform fibre dispersion in the mixture. After mixing, the cement pastes were immediately cast into the standard steel block forms with dimensions 40 mm x 40 mm x 160 mm, which were used for preparation of specimens. The bodies were cured for 2 days in the indoor climate at approximately +18°C and then they were removed from the moulds. After that time, the specimens were held under PVC foil for 26 days. On completion of the cure, specimens were tested at 28 days after production. The cement composites were weighed and their density, water absorbability and compressive strength were tested. Three specimens were prepared for each physical and mechanical test.

2.3. Testing methods

The density of 28 days hardened composites was determined in accordance with standard STN EN 12390-7 [7]. Water absorption test was carried out by immersing the composites in water bath (PE container) at laboratory temperature (20°C). The specimens were reweighed and afterwards were dried in the oven at 80°C up to constant weight for the following measurement of water absorption. Water absorbability was determined in agreement with standard STN 73 1316 [8]. Compressive strength was the mechanical property determined after 28 days curing and it was performed using the instrument ADR 2000 (ELE International, England) according to standard STN EN 12390-3 [9].

The mean values of density, water absorbability and compressive strength of each specimen were calculated as the average of the three measured values.

3. Results and discussion

3.1. Density of cellulosic composites

As shown in Table 3, density values of composites based on two types of cellulosic fibres (wood pulp composites WP and recycled waste paper composites RP) were lower in comparison to reference sample (RF). Density of composite samples in the first and second experimental sets was not influenced by fibre amount in specimens. Differences in density of composites with various portions of cellulose fibres were too small (0.2; 0.4 and 0.6 %) when compared with each other.

Table 3. Physical and mechanical parameters of fibre-cement composites.

Sample	Density [kg/m^3]	Compressive strength [MPa]
WP1	1940±3	13.84±0.89
WP2	1945±3	17.79±0.62
WP3	1945±5	18.72±0.81
RP1	1947±4	18.51±0.83
RP2	1935±5	16.28±0.65
RP3	1955±5	21.85±0.72
RF	2096±7	26.44±0.59

3.2. Water absorbability of cellulosic composites

The influence of wood pulp and recycled paper fibres on water absorbability of cement based composites is shown in figure 2. It is evident that the water absorbability of fibre-cement composites increased with higher amount of cellulose fibres in the mixture. Dependence of water absorbability on the quantity of fibres up to 5.0 % in composites has a linear character. Values of water absorbability of composite specimens both alternatives were measured in the range of 12.10–12.70 ± 0.15 %. All the wood pulp based composite specimens seem to have slightly higher water absorbability than the specimens with waste paper fibres. The sample with recycled waste paper fibres RP1 with fibre content 0.2 % achieved the lowest water absorbability. In comparison with the reference sample, it was higher by 2.11 %. Composite WP3 had the highest water absorbability (12.70 %). This value was higher by 7.17 % compared to the RF sample.

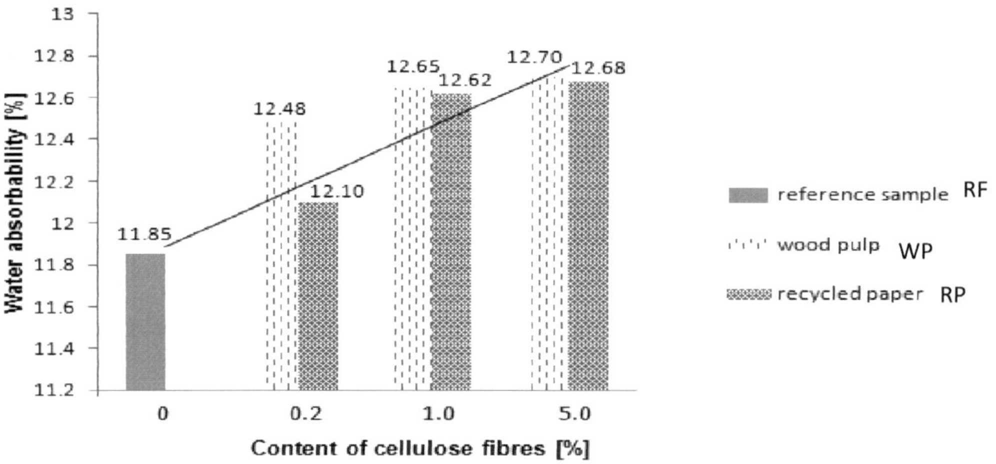

Figure 2. Influence of cellulosic fibres content on water absorbability.

3.3. Compressive strength of cellulosic composites

The variation of the compressive strength values of composites with different amount of cellulosic fibres is evident (Table 3). These values reached 52-83 % of strength parameter value for the reference composite. The results show that compressive strength increases with an increasing fibre content in the given range up to 5.0 wt.%. This confirmed a known fact that wood fibre content is the major factor affecting the properties of composites.

Composites of the second set show higher values of strength parameter (16.28-21.85 MPa) when compared to compressive strength values of the first set (13.84-18.72 MPa). This fact can be explained by different nature of surface fibres. Similar results were obtained in paper [10], where the impact of unbleached and bleached wood pulp on cement composite properties was studied. In the case of cellulosic fibres from wood pulp, no active centres are present on the surface of fibres due to bleaching process and therefore interaction between surface fibres and cement particles is weaker.

Figure 3 illustrates linear relationship between the compressive strength and density of composites based on two types of cellulosic fibres.

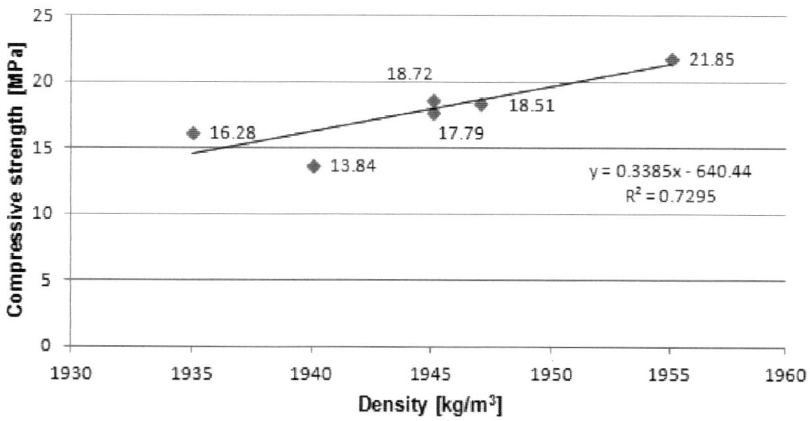

Figure 3. Linear relationship between the compressive strength and density of cellulosic cement-based composites.

4. Conclusion

In this paper, physical and mechanical properties of fibre-cement composites were investigated. Wood pulp and waste paper fibres were used in various proportions (0.2; 1.0; 5.0 %). Cellulosic fibres are a more effective lightweight raw material. Addition of cellulosic fibres led to the reduction of density of cement composites, slightly increased water absorbability and compressive strength depending on the amount of cellulose fibres as well as the type of cellulose fibres in cement composites. Wood and waste paper cellulose fibres have approximately the same effect on the properties of fibre cement composites, but there are slight differences in the density of composites 0.2; 0.4 and 0.6 % in comparison to each other. Waste paper fibres provide lower water absorbability than wood fibres but the differences between values of water absorbability are very small (3.04; 0.23; 0.15%). The values of compressive strength of composites with waste paper fibres show better results in comparison with wood pulp composites as well (25.23; 8.48; 14.32 %). Properties of fibre cement composites depend on the nature of cellulose fibres and the processing methods. In accordance with the previous research, the results show that physical and mechanical properties of composites are determined by the properties of cellulosic fibres. Further research on cellulose fibres cement composites is needed to better understand the properties and behaviour of fibre-cement composites and to enable their wide implementation in sustainable building materials such as non-load bearing building products.

5. References

[1] Khalil H A, Davoudpour Y, Islam M N, Mustapha A, Sudesh K, Dungani R and Jawaid M 2014 Production and modification of nanofibrillated cellulose using various mechanical processes: a review *Carbohyd. Polym.* **99** 649 – 665

[2] Johar N, Ahmad I and Dufresne A 2012 Extraction, preparation and characterization of cellulose fibres and nanocrystals from rice husk *Ind. Crop. Prod.* **37** 93 – 99

[3] Mármol G, Santos S F, Savastano H, Borrachero M V, Monzó J and Payá J 2013 Mechanical and physical performance of low alkalinity cementitious composites reinforced with recycled cellulosic fibres pulp from cement kraft bags *Ind. Crop. Prod.* **49** 422 – 427

[4] Pehanich J L, Blankenhorn P R and Silsbee M R 2004 Wood fiber surface treatment level

effects on selected mechanical properties of wood fiber – cement composites *Cement Concrete Res* **34** 59 – 65
[5] Ashori A, Tabarsa T and Valizadeh I 2011 Fiber reinforced cement boards made from recycled newsprint paper *Mat. Sci. Eng.* **528** 7801 – 04
[6] STN EN 1008: 2003 *Mixing water concrete* Specification for sampling, testing and assessing the suitability of water, including water recovered from processes in the concrete industry, as mixing water for concrete
[7] STN EN 12390-7: 2011 *Testing hardened concrete* Part 7: Density of hardened concrete
[8] STN 73 1316: 1989 *Determination of moisture content, absorptivity and capillarity of concrete*
[9] STN EN 12390-3: 2010 *Testing hardened concrete* Part 3: Compressive strength of test specimens
[10] Kidalova L, Stevulova N and Geffert T 2014 Possibility of using wood pulp in the preparation of cement composites *Selected Scientific Papers – J. of Civ. Engin.* **9** 51 – 58

Acknowledgments
The authors are grateful to the Slovak VEGA Grant Agency for financial support of the project VEGA 1/0277/15.

IOP Conf. Series: Materials Science and Engineering **96** (2015) 012026 doi:10.1088/1757-899X/96/1/012026

Sapropel as a Binder: Properties and Application Possibilities for Composite Materials

V Obuka[1,3], M Šinka[2], M Kļaviņš[1], K Stankeviča[1] and A Korjakins[2]
[1] Department of Environmental Science, University of Latvia, Raina Blvd. 19, LV 1586 Riga, Latvia
[2] Riga Technical University, Faculty of Civil Engineering, Institute of Materials and Structures, Kalku Str. 1, Riga, LV-1658, Latvia

E-mail: vaira_obuka@outlook.com

Abstract. Recent development trends largely look for possibilities of a wider use of natural materials and local resources. In this perspective, the use of organic rich lake sediment – sapropel – as a binding material in line with other environmentally friendly filling materials can be considered as a challenge. Sapropel itself is a valuable resource with multiple areas of application, for example, medicine, veterinary, agriculture, livestock farming, balneology, cosmetic applications, construction, and its application options have been widely studied in the 20th century in the Baltic countries, Ukraine and Russia. Birch wood fibre and sanding dust, hemp shives, 'Aerosil' are used as a filler and three types of sapropel are used as a binder in making composites. After material preparation and curing, physical and mechanical properties - density, thermal conductivity, compressive and flexural strength, were determined and compared to the data in the literature, and the opportunities to use them in the ecological construction were considered. The obtained results give insight into possibilities to use sapropel as a raw material, which can be considered as prospective material for construction materials and design products.

1. Introduction

Sapropel is a partially renewable geological resource [1], it is a fine-grained organic-rich sediment or sedimentary rock, it refers to lacustrine environment inland waters [2]. Organic rich lake sediment – sapropel – is a valuable resource of natural origin. In Latvia, its estimated reserves amount to 700-800 million m^3, 1.5 billion m^3 underlie the peat layer, 2 billion m^3 in total [1]. Sapropel, like, for example, peat has extensive opportunities and can be used in many fields, which vary depending on the composition of sapropel and its properties, and the availability of resources. Sapropel can be used in

[3] To whom any correspondence should be addressed.

various economic fields such as agriculture, medicine, veterinary medicine, construction, livestock farming, balneology, cosmetic applications. Sapropel has adhesive properties with high abilities to glue as well as shape holding ability; it can be used as a binder for manufacturing of environmentally friendly, natural nanostructured materials. Development of composites made from sapropel as a binder and birch wood sanding dust and fibre, hemp shives as a filler contributes to the rational use of natural resources, recultivation of the lakes and production of renewable energy, obtaining materials with the possibilities to manufacture ecological construction products.

Sapropel is the bottom deposits of fresh water bodies containing organic matter up to more than 15% [3]. Sapropel is formed from residues of the plankton and benthos rests – phytoplankton and zooplankton [4]. These elements multiply in large quantities, especially in standing or poorly through running shallow overgrown lakes [5].

One of the most important properties of moist sapropel is a colloidal suspended phase structure, which determines the ability of sapropel organic colloidal particles to absorb large quantities of water, so it has a high moisture capacity, which amounts to 70-97% [6] and low filtration rate [7]. Relative humidity of sapropel is associated with organic matter and its value increases with the content of organic matter [8].

In this research, significant characteristics of organic rich lake sediments are adhesive properties and water repellence [9], [10], [11]. Adhesive capacity of sapropel is determined by the presence of animal and vegetable residues. Green algae shells consist mostly of cellulose, which has weak decomposing properties. Organic sapropel proportion of matter is formed by green algae and is rich with cellulose but is poor in minerals, it consists of ash content, low humic substance level content, formed mainly by peat meristem [12]. It should be noted that the adhesive properties give higher organic nitrogen amounts of sapropel, also including free amino acids [4].

Composition of sapropel can be affected by molecular structure of humic substances and their quantity, respectively, if the content of humic substances increases then ramification of peripheral parts of molecules also increases. This contributes to the emergence of strong links between the molecules of the material at the time of creation. Molecules of humic substances remain flexible, malleable composite material comprises particles capable of providing material durability and high strength [4]. Therefore, it is rational to use sapropel with the above-mentioned properties as an adhesive for various ecological construction materials – plaster/finishing materials and thermal insulation materials, which could replace the traditionally used materials – stone wool and expanded polystyrene. Adhesive properties of sapropel are important in the production of building materials, i. e., if cold techniques compaction and techniques with hot glue presses at elevated temperature and pressure are applied [13].

The aim of this study is to explore different options to produce sapropel-wood fibre, sapropel-hemp shives, and sapropel-wood sanding dust composites and to determine the optimal composition ratio of certain raw material components and to characterize the resulting material properties.

2. Materials and methods

2.1. Sapropel sampling

In this research, organic rich freshwater sediments (further – sapropel) were used. Sapropel sediments were sampled from three lakes in Latvia – Padelis, Pilcenes and Pilvelu, located in Rezekne District, Latgale Region.

Table 1. Characteristics of the sapropel samples.

Lake	Sample Nr.	Moisture,%	Organic matter,%	Carbonates,%
Padelis	Sample 1	85.97	15.27	35.57
Pilvelu	Sample 2	94.99	84.51	1.26
Veveru	Sample 3	97.66	86.25	1.18

Sapropel samples differ from one another in terms of moisture (%), organic matter (%) and carbonates (%). Characteristics of the sapropel samples are shown in table 1. For example, Lake Padelis sapropel sample contains 35.57% carbonates, moisture is 85.97%, but colour – pale gray-pink and its density is 1.24 g/cm^3. Lake Pilvelis sapropel sample is dark greenish brown with homogeneous and jelly-like structure and density of 1.10 g/cm^3. Lake Veveru sapropel sample moisture level is high – 97.66%, it has low density – 1.08 g/cm^3 and organic matter reaches 86.25%.

2.2. Loss on ignition

Loss on ignition (LOI) method was applied in order to estimate moisture content, content of carbonate matter and organic matter of sediments [14]. Moisture content of sapropel was determined after drying at 105°C, following organic matter estimation at 550°C for 4 h. The content of mineral substances was determined after heating at 900°C for 2 h.

2.3. Fillers for composite materials

Birch wood sanding dust (also known as wood dust) and fibre (also known as wood fibers), and hemp shives were selected as fillers for production of composite materials. Birch wood sanding dust and fibre is industrial by-product from JSC "Latvijas finieris" – a plywood manufacturing company. Birch wood fibre is up to 15 mm long and up to 0.1 mm thick in diameter. In addition, hemp shives ("Bialobrezeskie") were taken from "Zalers" Ltd. Hemp shive slices were maximal 5.5 cm long and up to 0.6 cm thick in diameter. In producing of composites, an additional thickening additive filler was used – colloidal silica product "Aerosil". It is a filler that creates a smooth mixture, often in combination with other fillers.

2.4. Sample preparation and curing

For the developed composite materials raw sapropel was used as a binder (adhesive). Mixing of sapropel-filler mass was done manually until homogeneous and smooth mixture was reached at the stage where filler was fully covered with sapropel. Sapropel was mechanically treated by mixing together with electrical hand mixer until smooth and homogeneous material was formed. Metal mould with dimension of 30×30 cm and with adjustable height was used for composite material production. The mixture of raw materials was laid in by layers in mold for more dense composite material structure, higher mechanical strength and for minimizing final product shrinkage. Sapropel-filler samples were cured at the temperature of 80 – 105°C for 36 – 72 hours.

2.5. Thermal conductivity test

Before the thermal conductivity test, the density of the samples was calculated by weighting the samples and measuring the dimensions. Thermal conductivity was measured using LaserComp FOX 600 heat-flow measurer. Test settings were 0°C upper and 20°C lower plate. Automatic determination of sample thickness was chosen for this study.

2.6. Compressive and flexural strength tests

For testing compressive and flexural strength, the samples were specially prepared (sawed in necessary dimensions). The dimensions were 27-60×27-55×27-55 mm cubic forms for compressive tests (parallel and crosswise to the tamping direction) and 27-60×27-55×120 mm pieces for flexural strength tests. Mechanical tests were performed on ZWICK Z100 universal testing machine. For compressive strength, a stress at 10% deformation was recorded, for compressive crosswise and flexural strength – until failure. For sapropel-hemp shives, a layer of gypsum was spread over the interfaces to ensure even pressure application.

3. Results and discussion

3.1. Thermal conductivity test results

Paying particular attention to shrinkage cracks, thermal properties of the composite materials (wood sanding dust and 'Aerosil' with sample 2) were tested by heating the material in the oven. Shrinkage cracks impacted the quality of material for further tests, but in this case gypsum was used to fill the cracks and to do a thermal conductivity test. The measurement results also indicate a higher thermal conductivity of the material, that is, 0.080 W/m*K. This value can be explained by the fact that all the cracks were not sufficiently filled, gypsum also influenced the result, as thermal conductivity of gypsum is around 0.18 W/m*K. Thermal conductivity test of sapropel – filler composite was carried out by changing the types of sapropel and fillers. The results obtained are shown in table 2.

Table 2. Thermal conductivity of the studied materials.

Material: binder-filler	Density, kg/m^3	Thermal conductivity, W/m*K
Sample 1 and sample 3 - hemp shives	191	0.063
Sample 1 and sample 2 - hemp shives	200	0.059
Sample 3 - wood fibre	153	0.055
Sample 3 - wood fibre	202	0.060
Sample 3 - wood sanding dust	214	0.061
Sample 3 – wood sanding dust - 'Aerosil'	376	0.080

According to the results obtained, the composite made from sapropel sample 3-wood fibre (density 153 kg/m^3) has the best results of thermal conductivity. From visual aspect, it is a material different from another sample 3 sapropel-wood fibre (figure 2) with density 202 kg/m^3, because it has a denser structure and it has comparatively better resistance with respect to deformation. Results indicate that these composites have similar characteristics, thereby they have similar possibilities of use and potential. Thermal conductivity of air-dried composites is relatively low because of the organic origin of raw materials, detailed cellural mixed structure and homogeneous fibres with interconnected and open pores.

Figure 1. Sample 1 and sample 3 – hemp shives.

Figure 2. Sample 3 – wood fibre.

While sapropel-hemp shives composites (figure 1 and 4) are characterized by a heterogeneous structure due to different size of particles, having cavities and uneven composition with weaker inclusions, deforming more quickly. However, composite (density 376 kg/m^3) made from wood sanding dust and 'Aerosil' with sample 3 sapropel binder, and composite made from sample 3 sapropel and wood sanding dust (figure 3) with density 214 kg/m^3, formed shrinkage cracks during drying indicating the inferiority of the technology. Composite structure is made of densely grouped wood sanding dust particles with a sapropel binder.

In the study of sapropel-sawdust and peat-sawdust composite materials [15], the results obtained in thermal conductivity measurements were 0.067 W/m*K and 0.060 W/m*K. The study considered the freezing cycles and moisture of the tested materials. It shows that sapropel-sawdust thermal conductivity coefficient becomes lower after freezing cycles. If the composite material is resistant to freezing cycles, this material is applicable for Latvian conditions [15]. When compared with the results obtained, where composite's humidity is from air - dry state and a moisture saturated material (12%), thermal conductivity coefficient for sapropel-sawdust composite material is 0.050 – 0.060 W/m*K and for peat-sawdust composite material is 0.055 – 0.064 W/m*K, respectively, which is very similar to the results of this research. In the study of sapropel-straw panels where composition is similar to the composites created in this research and the samples from the study of sapropel and peat sawdust composite material [15], the result obtained is 0.055 W/m*K and it is stressed that with varying fillers and sapropel ratio, it is possible to produce a more effective thermal conductivity material [16].

Figure 3. Sample 3 – wood sanding dust.

Figure 4. Sample 1 and sample 2 – hemp shives.

In the study by Binici and colleagues on the insulation materials made from the sunflower stems, textiles and agricultural by-products as a filler, epoxy resin was used as the binder for better fiber strength and binding efficiency. An average thermal conductivity coefficient of 0.1642 W/m * K was obtained. However, the thermal conductivity coefficient of the sample with less epoxy resin added and made only from sunflower stems and sunflower stem fibers reached 0.0728 W/m*K [17]. In research of bamboo fiber and polyester composite Mounika et al. conclude that thermal conductivity coefficient decreases with the increase in the proportion of fiber. Thermal conductivity coefficient ranged from 0.185 to 0.196 W/m*K [18]. Literature on materials based on natural fibres from renewable raw material resources (flax, hemp, wood, bamboo, sheep's wool) shows that they possess very good sound and thermal insulation properties. This is due to material's low density and natural character of input fibres ("airy", lightweight material) [19], [20].

In the study about lime-hemp concrete, a variety of binders were compared, including metakaolin, obtained by burning kaolin clay (40% by mass) at 800°C and dolomitic lime (60% by mass), produced by "Saulkalne" Ltd. 8 different types of hemp shives were used as filler. The obtained results shows lime-hemp concrete material has a density of 312-337 kg/m^3, thermal conductivity of the material is more diverse than its density – from 0.0718 to 0.0778 W/m*K [21].

In the research about the use of agricultural waste in sustainable construction materials, Madurwar et al. compared the measurements of insulation materials of different origin. The results ranged from 0.046 to 0.056 W/m*K for rice husks to 0.118 to 0.240 W/m*K for oil palm leaves. The authors concluded that there are significant similarities between the corn's cob and the extruded polystyrene (XPS) material in terms of microstructure and chemical composition. Materials made from rice husk and coconut coir demonstrated the best results [22]. Previously discussed results of thermal conductivity showed similar values comparing with the results obtained in the current study.

3.2. Mechanical resistance and stability test results

The results obtained from the compressive deformation test (figure 5) show that composites where sample 3 sapropel was used as a binder with a filler of birchwood sanding dust and 'Aerosil' have the best result among the analysed samples. Compressive deformation results vary from 0.724 MPa sample 3 sapropel-wood sanding dust composite to 0.674 MPa for sample 3 sapropel-wood sanding dust and 'Aerosil' as a filler, while compressive strength crosswise shows respectively 0.669 MPa and 0.760 MPa.

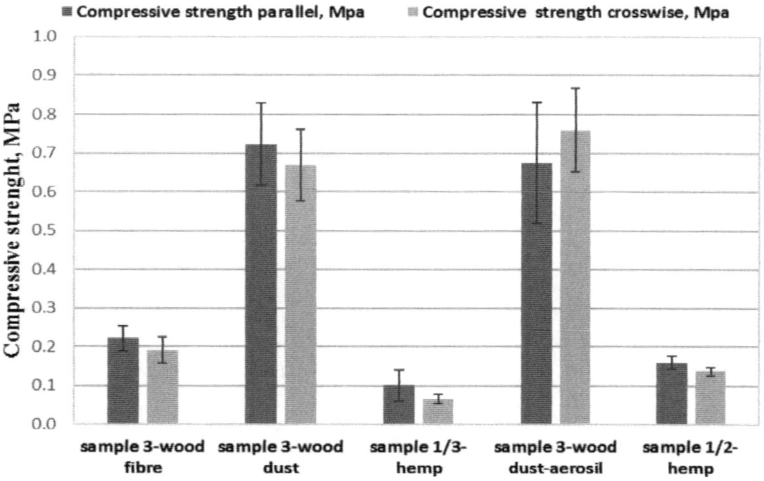

Figure 5. Compressive strength parallel and crosswise of different composites.

Composites made from hemp shives and wood fibre as a filler in compressive deformation obtain the rate of 0.221MPa for sample 3 sapropel-wood fibre and hemp shives materials 0.101 (Sample 1 and 3 – hemp shives) and 0.159 MPa (Sample 1 and sample 2 – hemp shives), while compressive strength crosswise shows respectively 0.191 MPa, 0.066 MPa and 0.138 MPa. The aforementioned materials have relatively lower rates due to lower intensity of filler and binder bonding. The mixing of the binder with the filler is more complex, because for the binder it is more difficult to enter the filler structure as its particles are larger. This was observed with wood fibre composite and sample 3 as a binder. However, filler (birch wood sanding dust and 'Aerosil'), which contains a mass of dust particles, mixes with the binder evenly, entering filler's structure. The structure of the material is smoother and will increase the mechanical resistance of the composite. Compressive strength results are influenced by the mode of sample preparation as they are sawn before, disrupting the structure of the material. The use of various fillers and binders affects physical and mechanical properties of composite materials, so changing the type of fillers and binders can induce change in physical and mechanical properties.

Comparing to the compressive strength results, the flexural strength results (figure 6) show that the material strength is relatively lower. It can be seen considering the results that the samples which have been made from hemp shives show lower values, the reason could be granulometric composition, as there are many large shives that create voids and uneven composition with weaker inclusions resulting in faster deformation of the samples [21]. In the process of making wood fibre-sapropel composite, the mixing of the binder with the filler is more complex, because for the binder it is more difficult to enter the filler structure, forming tangles and air gaps. Therefore, there are voids which can reduce mechanical strength of the obtained material.

Composites made from hemp shives and wood sanding dust as a filler in flexural deformation strength obtain a value of 0.069 MPa for sample 3 sapropel – wood fibre and 0.164 MPa (sample 3 sapropel – wood sanding dust) and 0.203 MPa (sample 3 sapropel – wood sanding dust – 'Aerosil'), while compressive strength crosswise shows 0.066 MPa for sample 3 – wood fibre 0.182 MPa for sample 3 sapropel – wood sanding dust – 'Aerosil', respectively. These results are relatively lower due to the lower intensity of the filler and binder bonding. The mixing of the binder with the filler is more complex, because for the binder it is more difficult to enter the filler structure as its particles are larger, longer (with each of different size). For all measurements, the values are higher using mix of sample 1 and sample 2 sapropel, as the reason could be higher sapropel adhesive capacity and lower amount of moisture.

In the study of sapropel and peat sawdust composite material [15], similar composites were created and results show that the type of filler and binder, as well as the preparation technology, change composite compressive and flexural strength. Accordingly to the indicated results, average compressive resistance is 0.03 MPa, but compressive resistance of peat-sawdust composite (activated peat binder) is 0.3 MPa.

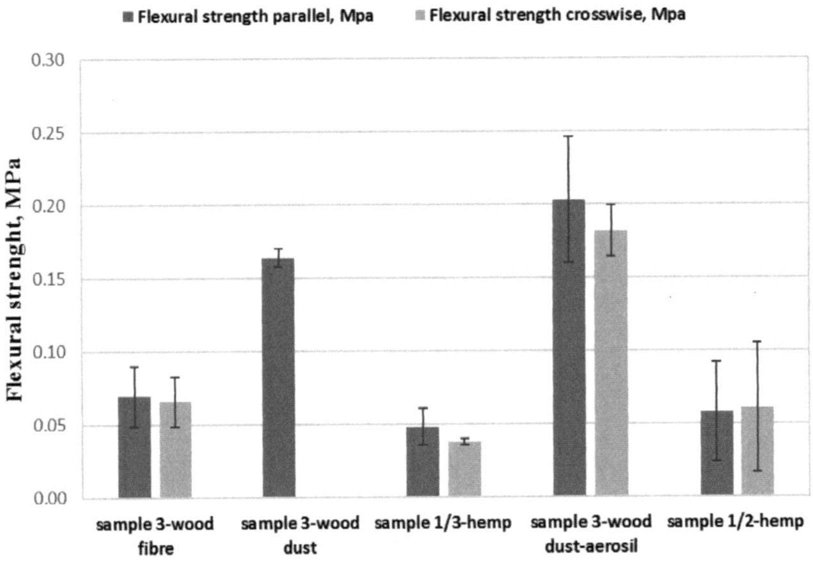

Figure 6. Flexural strengths parallel and crosswise of different composites.

Comparing the results of this work with the research into sapropel–concrete, the results of the current study show lower rates of wood fibre and hemp shives materials. During preparation 1 % NaOH solution was added. The obtained results of the studies showed that compressive strength of absolutely dry composite materials is 0.55 MPa, while that of air-dried – 0.56 MPa [11]. In the study about lime-hemp concrete, a variety of binders including metakaolin and dolomitic lime were compared. The obtained results show that compressive strength of different lime-hemp concrete materials ranges from 0.140 to 0.337 MPa, flexural strenght – from 0.021 to 0.059 MPa [21]. In the study on the composite material, which was created from agricultural by-products, it was concluded that the highest strength indicators were shown by durian peel and coconut-fiber composites, which range from 2.9 to 36 MPa [22]. In turn, the results of the study on the composite of sunflower stems and epoxy resin show compressive strength tests scores of 0.283 to 0.312 MPa. Flexural deformation strength results are from 0.06 to 0.09 MPa. High porosity is the reason of low mechanical strenght of the derived materials.

4. Conclusions

Using natural materials and local resources, such as sapropel, as well as industrial by-products such as birch wood sanding dust and fibre, and hemp shives it is possible to develop environmentally friendly composite materials for the construction industry, adjusting for the need for utilization. Granulometric composition of the particles, surface area and other characteristics of the material used as a filler have an effect on the binding with sapropel. Composite materials, which are made of birch wood sanding dust, 'Aerosil' and sample 2 and sample 3 sapropel, are characterized by relatively high mechanical strength, shape holding ability and easily amenable texture imprint. The composites sample 3 – wood sanding dust, sample 3 – wood sanding dust and 'Aerosil' during heating formed shrinkage cracks, thereby showing technological inferiority, as well as the need for evenly temperature raise and reduction of moisture in sapropel samples.

5. References

[1] Segliņš V and Brangulis A 2002 *Latvijas Zemes Dzīļu Resursi* (Riga, Valsts ģeoloģijas dienests)
[2] Emeis K C 2009 Sapropel *Encycl. of Paleoclim. and Anc. Environm* (Springer, Dordrecht)
[3] Kurzo B, Hajdukiewicz K and Krasnoberskaya O 2004 Relationships of sapropel formation in lake–mire complexes of Belarus *Limnolog. Review* **4** 125–132
[4] Kurzo B V 2005 *Regularity of Development and Problems in Using Sapropel* (Minsk)
[5] Lācis A 2003 Sapropelis Latvijā *LU* 61 *Scientific Conf. (Riga, Latvia)*
[6] Vimba B 1956 *Sapropeļa termiskā šķīdināšana un iegūto produktu ķīmiskais raksturojums* doctoral theses (Riga, Latvia)
[7] Liužinas R, Jankevičius K and Šalkauskas, M 2005 Improvement of lake sapropel quality: a new method *Geogrfijos Metraštis* **38** (2) 44–55
[8] Stankevica K, Rutina L, Burlakovs J and Klavins M 2014 Multi proxy chemical properties of freshwater sapropel *Geophysical Research Abstracts* **16** EGU2014-7885-1
[9] Shtan S 2005 *Ozernye sapropeli i ih kompleksnoe osvoenie* (Moskva: Moskovskogo Gosudarstvennogo gornogo universiteta)
[10] Brakšs N, Alksne A, Āboliņš J and Kalniņš A 1960 Sapropeļa un kūdras humīnskābes kā saistviela koksnes atlikumu izmantošanā *Latvijas PSR Zinātņu Akadēmijas Vēstis* **10** (159) 101–108
[11] Gružāns A 1960 Sapropeļbetons *LLA raksti* **9** 547–561
[12] Stankevica K and Klavins M 2013 Sapropelis un tā izmantošanas iespējas *Sc. J. of RTU: Mat. Sc. and Appl.Chem.* **29** 109–126
[13] Brakšs N and Miļins N 1960 Pētījumi par sapropeļa kompleksas izmantošanas iespējām ar hidrolīzes un saustvaices metodi *LLA raksti* **9** 471–479
[14] Heiri O, Lotter A and Lemcke G 2001 Loss on ignition as a method for estimating organic and carbonate content in sediments: reproducibility and comparability of results *J. of Paleolim.* **25** 101–110
[15] Obuka V, Korjakins A, Brencis R, Preikss I, Purmalis O, Stankevica K and Kļavins M 2013 Sapropl peat, sapropel shavings insulation plates and those properties *Sc. Journal of RTU: Mat. Sc. and Appl.Chem.* **29** 127–136
[16] Kasperiūnaite D and Navickas J 2006 Sapropel derivatives in new light *Radiation Interaction with Material & Its Uses in Technologies* Lithuanian p 2–32

[17] Binici H, Eken E, Dolaz M, Aksogan O and Kara M 2014 An environmentally friendly thermal insulation material from sunflower stalk, textile waste and stubble fibres *Constr. and Building Mat.* **51** 24–33

[18] Mounika M, Ramaniah K, Ratna Prasad V A, Mohana Rao K and Hema Chandra Reddy H 2012 Thermal conductivity characterization of bamboo fiber reinforced polyester composite. *J. Mater. Environ. Sci.* **3** (6) 1109–16

[19] Zach J, Hroudová J, Brožovský Z K, Gailius A 2013 Development of thermal insulating materials on natural base for thermal insulation systems. *Procedia Engin.* **57** 1288–94

[20] Kymalainen R, Sjoberg M 2008 Flax and hemp fibres as raw materials for thermal insulations *Buil. and Envir.* **43** 1261–69

[21] Sinka M, Radina L, Sahmenko G, Korjakins A and Bajare D 2015 Enhancement of lime-hemp concrete properties using different manufacturing technologies *First Int. Conf.on Bio-based Building Materials (In press)*

[22] Madurwar M V, Ralegaonkar R H and Mandavgane S A 2013 Application of agro-waste for sustainable construction materials: A review. *Constr. and Build. Mater.* **38** 872–878

Acknowledgments
The research leading to these results has received the funding from Latvia State Research Programme under grant agreement "INNOVATIVE MATERIALS AND SMART TECHNOLOGIES FOR ENVIRONMENTAL SAFETY, IMATEH".

Thermal Insulation from Hardwood Residues

I Sable[1,4], U Grinfelds[1], L Vikele[1], L Rozenberga[1], M Zeps[2] and S Luguza[3]
[1] Latvian State Institute of Wood Chemistry, 27 Dzerbenes Str., Riga, LV 1006, Latvia
[2] Latvian State Forest Research Institute "Silava", 111 Rigas Str., Salaspils, LV 2121, Latvia
[3] Latvian University of Agriculture, Forest Faculty, 11 Akademijas Str., Jelgava, LV 3001, Latvia

E-mail: inese.sable@inbox.lv

Abstract. Adequate heat is one of the prerequisites for human wellbeing; therefore, building insulation is required in places where the outside temperature is not suitable for living. The climate change, with its rising temperatures and longer dry periods, promotes enlargement of the regions with conditions more convenient for hardwood species than for softwood species. Birch (*Betula pendula*) is the most common hardwood species in Latvia. The aim of this work was to obtain birch fibres from wood residues of plywood production and to form low-density thermal insulation boards. Board formation and production was done in the presence of water; natural binder, fire retardant and fungicide were added in different concentrations. Board properties such as density, transportability or resistance to particulate loss, thermal conductivity and reaction to fire were investigated. This study included thermal insulation boards with the density of 102-120 kg/m^3; a strong correlation between density and the binder amount was found. Transportability also improved with the addition of a binder, and 0.1-0.5% of the binder was the most appropriate amount for this purpose. The measured thermal conductivity was in the range of 0.040-0.043 W/(m·K). Fire resistance increased with adding the fire retardant. We concluded that birch fibres are applicable for thermal insulation board production, and it is possible to diversify board properties, changing the amount of different additives.

1. Introduction

Adequate heat is one of the prerequisites for human survival; therefore, building insulation is required in places where the outside temperature is not suitable for living. For thousands of years, people have used natural materials for that purpose – moss, clay, straw, wool, herbs, etc. In the last century, however, a lot of synthetic materials have been developed in this application area, but considering the

[4] To whom any correspondence should be addressed.

global tendency to reduce CO_2 emissions, humanity comes again to green-thinking [1]. In addition, natural materials in construction create a pleasant microclimate for living. A modern thermal insulation material should have air permeability together with good thermal insulation properties and easy handling. Thermal insulation boards from soft wood fibres are part of the global market nowadays [2]. The climate change, with its rising temperatures and longer dry periods, promoted enlargement of the regions with conditions more convenient for hardwood species than for softwood species [3]. Manufacturers can change the focus on hardwood products. It is a known fact that bonding of internal and external hydroxyl (OH) groups of cellulose provides holding fibres together. Evaporation of water creates capillary force, which reduces the distance between the fibres, thereby hydrogen bonds and van der Waals attraction hold fibres together after drying [4]. Nevertheless, only 2% of all OH groups are involved in interfibre bonding [4]; so, additives are preferable. Insulation fibre boards and panels from softwood fibres are available on the market, but hardwoods with an average fibre length of 1-2 mm [5] are considered inappropriate for this product. Birch (*Betula pendula*) is the most common hardwood species in Latvia – its stock was 160 million m^3 in 2014, being 24% of the total wood stock of our country [6]; so, it can be a sufficient source of fibres.

The aim of this work was to obtain birch fibres from wood residues of plywood production and to form low-density thermal insulation boards. The impact of additives on density, transportability and thermal conductivity was also investigated.

2. Materials and methods

Residues of plywood production were the raw material for birch fibres, which were obtained in the thermo mechanical pulping (TMP) process at Institut für Holztechnologie (Dresden, Germany) using refiner 12-1 CPH (Andritz, Austria). The fibre material was studied and boards were made at the Latvian State Institute of Wood Chemistry, Laboratory of Cellulose.

Scanning electron microscopy was performed using SEM Vega Tescan 5136MM (Tescan Brno, Chech Republic). The fractional composition of the fibre material was determined, using mechanical sieving in a shaker Retsch A200 (Retsch GmbH, Germany) with several mesh sizes. The moisture content of all materials and substances was determined with Kern MLB 50-3N (Kern & Sohn Gmbh, Germany).

A solution of binder was made, dissolving a specific amount of potato starch (Aloja-Starkelsen, Latvia) in cold water, obtaining 2 wt.% starch solution and then mixing with boiling water till gelatinization. Sodium tetraborate decahydrate $Na_2B_4O_7{\cdot}10H_2O$ (FR1) and boric acid H_3BO_3 (FR2) (both granulated, Etimadan, Turkey) were used as a fire retardant (FR) and a fungicide, dissolved in cold water before adding to fibres. Amount of additives in boards was calculated regarding the weight of absolutely dry fibres; 0.1 – 2% in case of starch and 0 – 20% in case of FR.

Fibre board formation occurred as follows: fibres and water were mixed together with a weight ratio of 0.03:1, and starch gel, fire retardant and/or fungicide solutions were added to the mixture. Pulp was transferred to the board forming equipment, made of plywood with a hydro isolation layer inside, which let water drain by gravity, and then extruded to a 4-4.5 cm board thickness. A wet board was dried for 24 h at 60°C, then cooled, weighted and measured, and its density was calculated (see figure 1).

Figure 1. Thermal insulation board from birch fibres.

The test for imitation of transportation was conducted: 140 x 140 mm samples were cut from fibre boards. To provide the material for particulate loss only from the side faces, the upper and lower faces were coated. Weighted samples were put in a sieve (diameter 200 mm, height 25 mm, mesh size 1 mm) and placed in a shaker Retsch A200 (Retsch GmbH, Germany). The 3-D throwing motion mode was used with the amplitude 70 and vibration height 1.5 mm for 120 min. The weighting procedure and calculating of mass reduction were performed after shaking.

Thermal conductivity was determined according to ISO 8301 in a heat flow meter Linseis HFM 200/300/600 (Linseis GmbH, Germany).

For the reaction to the fire test ~40g of board sample was cut, fixed on non-combustible wire and placed in open fire (gas blow torch Rofire Piezo 1900, Rothenberger industrial, Germany) for 30 seconds simultaneously recording the weight loss dynamics and final weight loss was calculated.

3. Results and discussion

The fibre mass after the TMP process consisted of fibres, particles and fines (small size particles from damaged fibres) of different size (figure 2). The fraction with the size of 0.5-2 mm was 44.5% of the obtained pulp; there were fibres and fibre fragments in this fraction. The fraction with the size of 2-3.25 mm constituted 16%, and fibres, fibre bundles and vessel fragments were found there. There were 0.5% of particles bigger than 3.25 mm and 29.5% of fines in the pulp.

Thermal insulation fibre-boards with the density of 102-120 kg/m^3 were made from TMP birch fibres. Starch was added in different amounts to improve the adhesion of fibres. There was a correlation between the amount of the added starch and the density of the board (figure 3). More added starch caused higher density; however, these changes were caused directly by the increase of the total weight of the substance. The role of starch in strengthening the inter-fibre bonds [4] also should be taken into account – this phenomenon can cause an increase of board density.

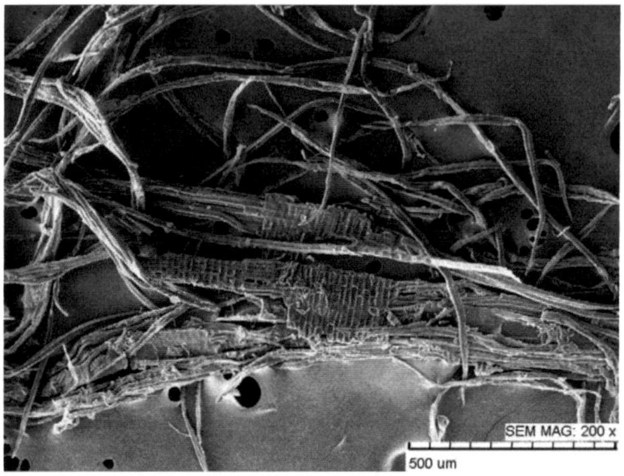

Figure 2. TMP fibres of birch (SEM).

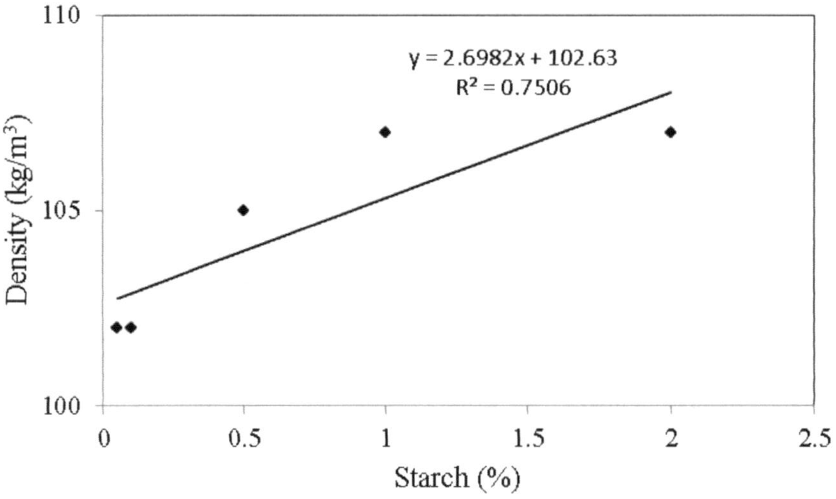

Figure 3. Density of a thermal insulation fibre-board, made from birch TMP fibres, depending on the amount of the binder.

The effect of starch as a binder was shown in a transport imitation or shaking test, where a decrease of mass reduction was observed (figure 4). A rapid decrease in mass reduction was observed in the case, when the amount of starch was increased from 0.1 to 0.5 %, but continuing adding of starch did not change the result significantly. This proved the effect of starch as a binder on strengthening the inter-fibre bonds.

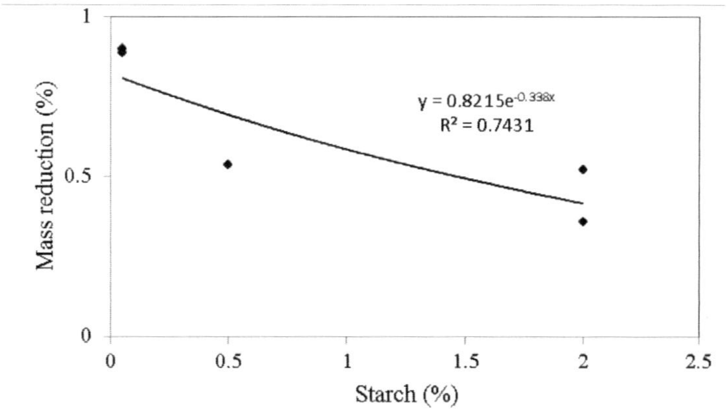

Figure 4. Decrease in mass reduction of a thermal insulation birch wood fibre-board during the "shaking" test (transport imitation). Influence of the concentration of the binder – starch.

The measured thermal conductivity of the obtained boards was within the interval 0.040-0.043 W/(m·K) and slightly varied depending on the additives (figure 5), but the differences were not significant. These results are in accordance with [7], where the thermal conductivity of 0.040 W/(m·K) for all natural fibre-boards was stated. As wood fibre-boards that are on the market nowadays have similar thermal conductivity, we can presume that our samples are appropriate for using in thermal insulation for residential houses [2].

The response of fibre-board samples to fire flame is shown in figure 6. A control sample was made only from birch fibres; no adhesives or fire retardants were added. Differences in the starting weight should not be taken into account; all samples were identically prepared and weight differences were caused by adding starch and a fire retardant.

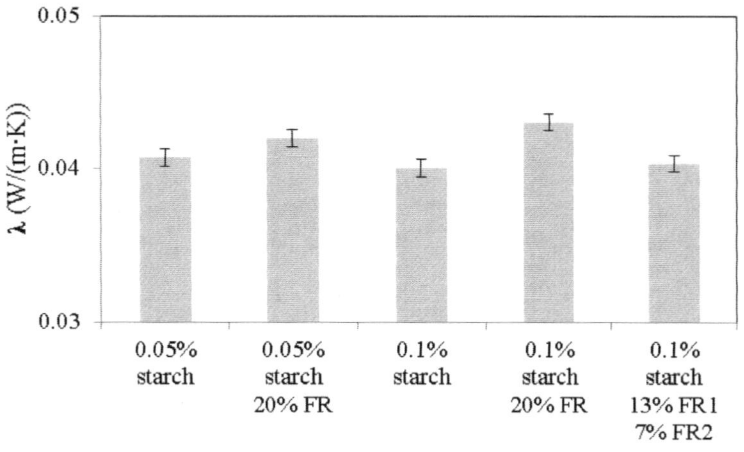

Figure 5. Thermal conductivity of birch fibre-board, depending on additives (FR – fire retardant).

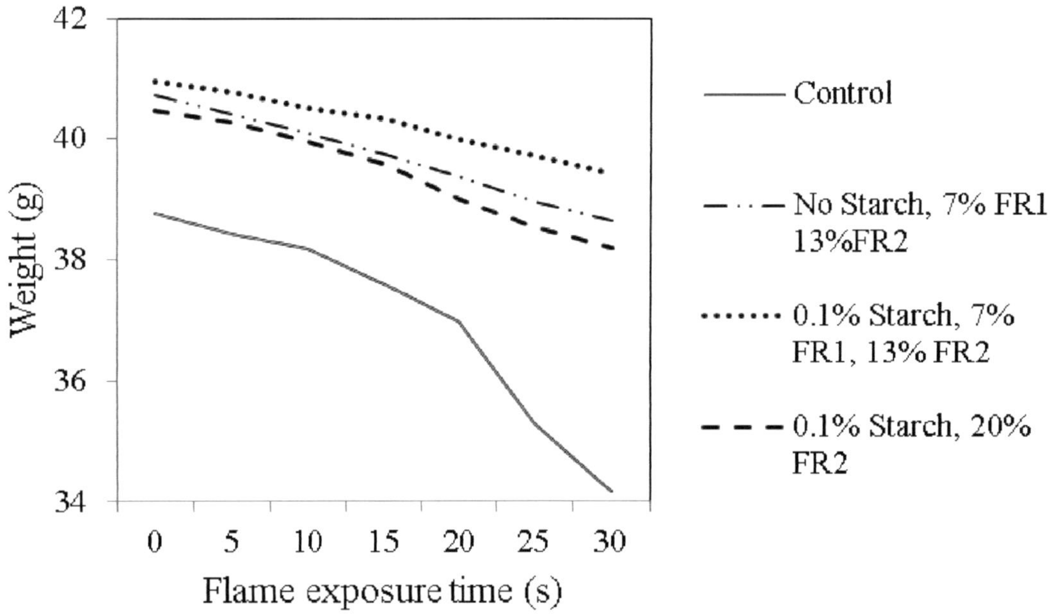

Figure 6. Response to fire of a birch thermal-insulation board with and without additives (FR – fire retardant).

The weight of the sample decreased with the time of flame exposure. Samples with a fire retardant lost their weight evenly, about 4-6% of the initial number after 30 s of flame exposure, whereas the control sample lost 12% of weight and had a sharp break point at 20 s. These results proved the efficiency of the tested fire retardants to reduce the combustibility of thermal insulation fibre-boards from birch TMP fibres. Preliminary comparing tests were done with thermal insulation materials, available on the market and results showed equal response to fire.

4. Conclusions

It is possible to obtain fibre-board from birch TMP fibres for thermal insulation, using starch as a binder. Transportability improved with the addition of starch and 0.1-0.5% was the most appropriate amount for this purpose. Thermal conductivity of boards was in the range 0.040-0.043 W/(m·K). Fire resistance depended on the composition of the fire retardant. Birch fibres from the thermo-mechanical process are found to be applicable for thermal insulation board production. It is possible to diversify board properties, changing the amount of different additives. It is a good opportunity to obtain a product of high added value from wood residues instead of burning them to produce only energy.

5. References

[1] Gellert R 2010 Natural fibre and fibre composite materials for insulation in buildings *Mater. for Energy Effic. and Thermal Comfort in Build.* ed Hall M (Woodhead Publishing) pp 22–256

[2] Asdrubali F, D'Alessandro F and Schiavoni S 2015 A review of unconventional sustainable building insulation materials *Sustainable Materials and Technologies* In Press (Online)

[3] De Borst K and Bader T K 2014 *J. Theor. Biol.* **345** 78–91

[4] Hubbe M A 2006 *Bioresorces* **1** (2) 281–318

[5] Dai D and Fan M 2014 Wood fibres as reinforcements in natural fibre composites: structure, properties, processing and applications *Natural Fibre Compos.* ed Hodzic A and Shanks R (Woodhead Publishing) pp 3–65
[6] Forest Statistic 2014 *The State Forest Service Report* [Available]: http://www.vmd.gov.lv/valsts-meza-dienests/statiskas-lapas/publikacijas-un-statistika/meza-statistikas-cd?nid=1049#jump
[7] Kaemmerlen A, Asllanaj F, Sallée H, Baillis D and Jeandel G 2010 *Int. J. Therm. Sci.* **49** (11) 2169–76

Acknowledgments
This work was supported by the Latvian Council of Science project “Adaptive capacity of forest trees and possibilities to improve it” (No 454/2012).

Surface Treated Natural Fibres as Filler in Biocomposites

I Schwarzova[1,2], **N Stevulova**[1], **E Singovszka**[1] **and E Terpakova**[1]
[1]Technical University of Kosice, Faculty of Civil Engineering, Institute of Environmental Engineering, Vysokoskolska 4, 042 00, Kosice, Slovakia

E-mail: ivana.schwarzova@tuke.sk

Abstract. Biocomposites based on natural fibres as organic filler have been studied for several years because traditional building materials such as concrete are increasingly being replaced by advanced composite materials. Natural fibres are a potential replacement of glass fibres in composite materials. Inherent advantages such as low density, biodegradability and comparable specific mechanical properties make natural fibres an attractive option. However, limitations such as poor thermal stability, moisture absorption and poor compatibility with matrix are challenges that need to be resolved. The primary objective of this research was to study the effect of surface treatment on properties of hemp hurds like a natural lignocellulosic material and composites made thereof. Industrial hemp fibre is the one of the most suitable fibres for use in composite materials because of its good specific properties, as well as it being biologically degradable and CO_2 neutral. Improving interfacial bonding between fibres and matrix is an important factor in using hemp fibres as reinforcement in composites. In order to improve interfacial bonding, modifications can be made to the hemp fibres to remove non-cellulosic compounds, separate hemp fibres from their bundles, and modify the fibre surface. This paper contains the comparison of FTIR spectra caused by combination of physical and chemical treatment of hemp material with unmodified sample. Modification of hemp hurds was carried out by NaOH solution and by ultrasonic treatment (deionized water and NaOH solution were used as the cleaning mediums).

1. Introduction

Principles of sustainable construction of the buildings bring new requirements to develop sustainable materials. The use of construction and building materials made from renewable resources is generally regarded as an indispensable option so the construction industry can become more sustainable. The use of wood species in the production of cement composites is particularly interesting for the construction industry. This represents a large opportunity in the field of vegetable fibres that is cost-effective and environmentally friendly. Therefore, promoting the use of cementitious building

[2] To whom any correspondence should be addressed.

materials reinforced with vegetable fibres could be a way to achieve sustainable and more eco-efficient construction [1].

In this respect, there is a large group of materials of plant origin, which are rapidly renewable raw materials that can be used as a suitable reinforcement component to lightweight composites [2]. Traditional building materials are increasingly being replaced by advanced composite materials in accordance with sustainable development requirements. Fiber reinforced polymers and fibre reinforced cement represent this group of new composites with advantageous properties [3].

Lignocellulosic biomass has become a promising alternative source of materials for industrial applications. Cellulose, hemicelluloses and lignin are the major components of biomass. Cellulose and hemicelluloses (holocellulose) are generally referred to as polysaccharides. Waxes, tannins and phenols belong to extractives [4].

Natural vegetable fibres such as hemp, jute, sisal, bamboo, coir, kenaf and others have potential to be used as organic reinforcement and/or filler material in composites. Composites based on natural and fast renewable resources, especially cellulosic materials, are increasing in importance due to their numerous advantageous properties for application in sustainable building constructions [3].

Composite materials based on natural cellulosic fibres with inorganic binder represent a group of lightweight materials providing a healthy living in buildings [5].

Growing interest in using hemp hurds as waste material from bast hemp fibres production into lightweight composites is recorded in the recent years. Interesting properties such as thermal, mechanical, acoustic and antiseptic including low density, biodegradability, low costs and ecological suitability of this plant raw material predetermine its use in a function of organic filler into composite [6].

The technical hemp (Cannabis Sativa L.) is the source of two types of fibres; bast fibres and woody fibres (hurds). The properties of hemp fibres depend on their chemical composition. The bast fibres contain larger amounts of cellulose compared to the hemp hurds. On the contrary, contents of hemicelluloses and lignin as amorphous substances are higher in hurds [7].

Cellulose in hemp hurds is identified as the main structural component of the fibre, which is present mainly in the crystalline phase. Hemicelluloses and lignin present mostly in the amorphous phase, which plays an important role in controlling fibre properties. One of the key problems of successful application of plant fibres is their heterogeneity and hydrophility resulting in high moisture sorption sensitivity of biomaterial. Hydroxyl groups in the structure of cellulose, hemicelluloses and lignin are responsible for the hydrophility of the plant material [8]. To decrease hydrophility and modify cellulosic composition of hemp fibres, chemical and/or physical [9-10] treatment of the natural material surface are applied.

In this paper, the results concerning Fourier transforms infrared spectroscopy (FTIR) study of hemp hurds after physical, chemical and physico-chemical treatment in comparison to the reference sample are given.

2. Materials

In the experiments, the technical hemp hurds coming from the Dutch company Hempflax was used as organic filler into biocomposites. The used material was polydispersive with wide particle length distribution (8-0.063 mm) and its density was 117.5 kg.m^{-3}. These hemp hurds contain more hurds material than bast fibres. The content of polysaccharide component (holocellulose) is 77.3 %. The amounts of cellulose and hemicelluloses like holocellulose components are 42.6 and 34.7 %, respectively. Other components present in hemp are lignin (23.7%), compounds soluble in toluene and ethanol (2.8 %) and ash (1.8 %). The structure of hemp hurd slices was examined by optical microscopy (NIKON SMZ 1500, Japan). The photomicrograph (figure 1) of hemp hurds shows compact and unfibrilled fibre structure. The presence of dirt and waxes on the material surface is observed.

Figure 1. Micrograph of hemp hurds slice (40x magnification).

For surface treatment process of hemp hurds, deionized water and sodium hydroxide (Chemapol, Slovakia, p. a.) solution were used as ultrasonic cleaning medium.

3. Methods

Prior to treatment and in order to ensure constant humidity content, fibres were dried at 80°C for 24 h in the drying oven.

3.1. Fibre surface treatment

Most of modification processes of dried hemp hurds were realized by ultrasonic treatment process for 1h. To compare the impact of physical treatment by ultrasound, the same experiments were carried out under static conditions without ultrasonic action. Deionized water, 0.1 M and 0.2 M NaOH solution, was used as a cleaning medium in the experiments. In all cases of surface treatment, the ratio s:l was 1:10. An ultrasonic bath TESON 10 (Tesla, Slovakia, 220 V, 50 Hz, 650W of power output) was used for ultrasonic cleaning process of hemp hurds. All hemp hurds samples were dried at 80°C after their treatment.

3.2. Fourier transform infrared spectroscopy

Fourier transform infrared spectroscopy (FTIR) measurements were carried out on Bruker Alpha Platinum spectrometer with Attenuated Total Reflectance (ATR) technique (BRUKER OPTICS, Germany). Total 24 scans were performed on each sample in the range of 400-4000 cm^{-1}. FTIR spectroscopy is capable of detecting structural changes in biomaterial. It provides information about the presence or absence of specific functional groups or formation of new functional groups and can give an even deeper insight into the fibre structure. FTIR allows identification of the main components of cellulose.

4. Results and discussion

FTIR spectroscopy provides information about the structure of the main components in hemp hurds. The FTIR spectra for chemically treated hemp hurds samples compared to the reference sample are shown in figure 2. As it can be seen, significant infrared spectral differences are observed in case of

IOP Conf. Series: Materials Science and Engineering **96** (2015) 012028 doi:10.1088/1757-899X/96/1/012028

hemp hurd surface modification with 0.2 M solution of NaOH. Therefore, this sample was selected for the further FTIR spectra comparison. In figure 3, the spectra of the physical, chemical and physico-chemically treated hemp hurds are compared with the original sample.

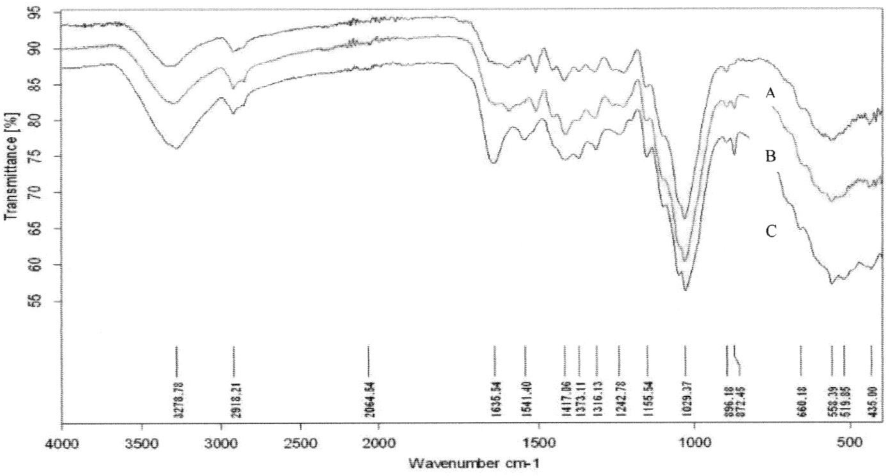

Figure 2. FTIR spectra of hemp hurds: reference (A); chemically modified with 0.1 M NaOH solution (B); 0.2 M NaOH solution (C).

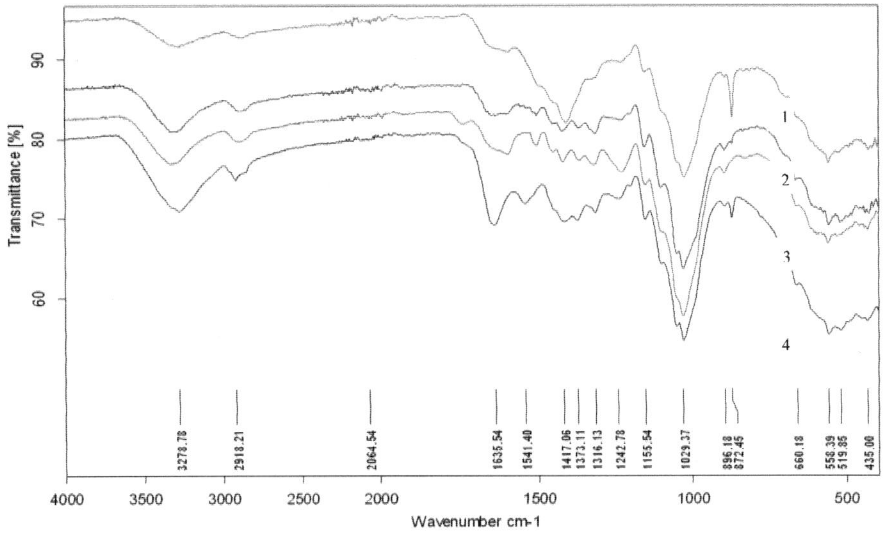

Figure 3. FTIR spectra of hemp hurds: reference (1); ultrasonic modified with deionized water (2); chemically modified with 0.2 M NaOH solution (3); ultrasonic modified with 0.2 M NaOH solution (4).

Main infrared spectral differences are observed in figures 2 and 3, which allow identifying the structural changes in lignocellulosic material after chemical, physical and physico-chemical treatment.

The changes in FTIR spectra due to all treatments of hemp hurds are shown in two regions between 3600 – 2800 cm^{-1} and 1800 to 900 cm^{-1}. Peak positions corresponding to vibrations of function groups present in the studied hemp material are consistent with the literature data published for vegetable fibres [11]. The range of wave number of 3570-2900 cm^{-1} is characteristic for stretching vibration of O-H and C-H bonds in polysaccharides. Based on [12], a broad band in the spectra range of 3490 - 3170 cm^{-1} represents the complex vibrations of hydroxyl stretching of inter- and intra-molecular hydrogen bonds. For the hemp hurds sample treated in NaOH solution, the peak intensity located near 3300 cm^{-1} is less sharp (figure 2). Its intensity decreases in the case of 0.2M NaOH treatment of hemp hurds due to mercerization. The peak characteristic for waxes and oils present at 2851 cm^{-1} is present in all samples but after physico-chemical treatment is obviously less marked.

Many absorption bands corresponding to vibration of various functional groups present in hemp components are observed in the region of 1800 to 900 cm^{-1}. This range was employed to characterize the structure of hemicelluloses, lignin, but mainly of cellulose. Only the most visible differences in the spectra of the treated samples compared to the reference hemp hurds are presented and discussed.
One of such differences is the modification of the absorption bands in the range of 1700-1750 cm^{-1}. Two signals are attributed to the stretching vibration of an unconjugated functional group of C=O or COOH, where peak 1733 cm^{-1} corresponds to hemicelluloses and 1742 cm^{-1} – to pectins and waxes [13]. These peaks have partially disappeared after treatment with NaOH in accordance with [14]. Indeed, NaOH treatment is known to remove hemicelluloses [15]. The peak at 1030 cm^{-1} belongs to C-C, C-OH, C-H ring and side group vibrations in hemicelluloses and pectin.

According to [16], lignins in hemp have most powerful absorption bands at the frequency range 1605 –1639 cm^{-1} (in our spectra 1594 – 1635 cm^{-1}). Based on the observation of the sharp peaks located at 1594 cm^{-1} and 1507 cm^{-1} (C-C stretching from aromatic ring) in FTIR spectra, ultrasonic treatment and also NaOH treatment led to partial removal of lignin. According to literature data [15], lignin cannot be totally removed by the alkaline process. As it is evident from figure 3, the peaks typical for lignin were clearly observed at 1319 cm^{-1} in all samples.

Typical bands assigned to cellulose were observed at 896 cm^{-1} and in the region of 1630 – 1160 cm^{-1}. The absorption band at 896 cm^{-1} is assigned as C-O-C stretching vibration of glycosidic bonds in polysaccharides. The absorption bands assigned to cellulose at 1424 and 1373 cm^{-1} come from $-CH_2-$ and CH bending vibrations. The peaks at 1337 and 896 cm^{-1} represent O-H bending vibrations in cellulose. The band about 1320 cm^{-1} corresponds to $-CH_2-$ wagging vibration, which distinguished between amorphous and crystallized cellulose [17].

5. Conclusion

The present paper is devoted to FTIR investigation of hemp hurds changes caused by chemical or physical treatment and by combination of these treatments. As it has been shown, the non-cellulosic components such as hemicelluloses, lignin and waxes after all treatment processes of hemp hurds were partially removed in dependence on the treatment procedure and operating conditions.
The best result of degradation of these non-cellulosic materials was reached by physico-chemical treatment (ultrasonic alkaline treatment in 0.2 M NaOH solution).

Whereas this physico-chemical treatment is in the initial stage of research, it is necessary to carry out additional analysis (changes in chemical composition, thermal analysis, X-ray diffraction measurement) and also to study the impact of these hemp hurd treatments on the properties of biocomposites prepared with organic filler and inorganic matrix.

6. References

[1] Glasser F, Marchand J and Samson E 2008 Durability of concrete: Degradation phenomena involving detrimental chemical reactions *Cem. Concr. Res.* **38** 226–246

[2] De Bruijn P B, Jeppsson K H, Sandin K and Nilsson C H 2009 Mechanical properties of limehemp concrete containing shives and fibres *Biosystems Engineering* **103** 474–479

[3] Azwa Z N, Yousif B F, Manalo A C and Karunasena W 2013 A review on the degradability of polymeric composites based on natural fibres *Materials and Design* **47** 424–442

[4] Xu F, Yu J, Tesso T, Dowell F and Wang D 2013 Qualitative and quantitative analysis of lignocellulosic biomass using infrared techniques: a mini-review *Applied Energy* **104** 801-09

[5] Preikss I, Skujans J, Adamovics A and Iljins U 2013 Evaluation of hemp (Cannabis Sativa L.) quality parameters for building materials from foam gypsum products *Chemical Engin. Transactions* **32** 1639–44

[6] Stevulova N et.al 2014 Properties characterization of chemically modified hemp hurds *Materials* **7** 8131–50

[7] Gumuskaya E, Usta M and Balaban M 2007 Carbohydrate components and crystalline structure of organosolv hemp (Cannabis sativa L.) bast fibers pulp *Bioresource Technology* **98** 491–97

[8] Sreekala M S, Kumaran M G, Joseph S and Jacob M 2000 Oil palm fiber reinforced phenol formaldehyde composites: Influence of fiber surface modifications on the mechanical performance *Appl. Compos. Mater.* **7** 295–329

[9] Cordeiro N, Ornelas M, Ashori A, Sheshmani S and Norouzi H 2012 Investigation on the surface properties of chemically modified natural fibers using inverse gas chromatography *Carbohydrate Polymers* **87** 2367–75

[10] Renouard S, Hano Ch, Doussot J, Blondeau J P and Lainé E 2014 Characterization of ultrasonic impact on coir, flax and hemp fibers *Mater. Lett.* **129** 137–141

[11] Fan M, Dai D and Huang B 2012 Fourier transform infrared spectroscopy for natural fibers *Fourier Transform - Materials Analysis* InTech 45–68

[12] Carilo F, Colom X, Suňol J J and Saurina J 2004 Structural FTIR analysis and the thermal characterization of lyocell and viscose-type fibers *Eur. Polym. J.* **40** 2229–34

[13] Biagiotti J, Puglia D, Torre L and Kenny J M 2004 A systematic investigation on the influence of the chemical treatment of natural fibers on the properties of their polymer matrix composites *Polymer Composites* **25** 470–479

[14] Liu Y 2013 Recent progression Fourier transform infrared (FTIR) spectroscopy study of compositional, structure and physical attributes of developmental cotton fibers *Materials* **6** 299–313

[15] Kostic M, Pejic B and Scundric P 2008 Quality of chemically modified hemp fibers *Bioresour Technol.* **99** 94–99

[16] Wang B, Sain M, Oksman K 2007 Study of structural morphology of hemp fiber from the micro to the nanoscale Applied Composite Materials **14** 89–103

[17] Fackler K, Stevanic J S, Ters T, Hinterstoisser B, Schwanninger M and Salmen L 2011 FT-IR imaging spectroscopy to loclise and characterise simultaneous and selective white-rot decay within sprude wood cell *Holzforschung* **65** 411–420

Acknowledgements

The authors are grateful to the Slovak Grant Agency for financial support of the project VEGA 1/0277/15.

Hemp Thermal Insulation Concrete with Alternative Binders, Analysis of their Thermal and Mechanical Properties

M Sinka[1,2], G Sahmenko[1], A Korjakins[1], L Radina[1] and D Bajare[1]
[1]Riga Technical University, Faculty of Civil Engineering, Institute of Materials and Structures, Kalku Str. 1, Riga, LV-1658, Latvia

E-mail: maris.sinka@rtu.lv

Abstract. One of the main challenges that construction industry faces today is how to address the demands for more sustainable, environmentally friendly and carbon neutral construction materials and building upkeep processes. One of the answers to these demands is lime-hemp concrete (LHC) building materials – carbon negative materials that have sufficient thermal insulation capabilities to be used as thermal insulation materials for new as well as for existing buildings. But one problem needs to be overcome before these materials can be used on a large scale – current manufacturing technology allows these materials to be used only as self-bearing thermal insulation material with large labour intensity in the manufacturing process. In order to lower the labour intensity and allow the material to be used in wider applications, a LHC block and board production is necessary, which in turn calls for the binders different from the classically used ones, as they show insufficient mechanical strength for this new use. The particular study focuses on alternative binders produced using gypsum-cement compositions ensuring they are usable in outdoor applications together with hemp shives. Physical, mechanical, thermal and water absorption properties of hemp concrete with various binders are addressed in the current study.

1. Introduction

The construction industry has to satisfy the growing demand for sustainable and environmentally friendly construction materials. One of such materials is hemp-lime concrete (LHC) [1] – thermal insulation material with good thermal insulation properties (0.07 to 0.14W/m*K) [2][3], negative CO_2 emissions [4][5][6], excellent hydric properties[7][4], great acoustic capabilities [8], and good fire resistance [9].

At present, LHC material is mostly produced in-situ as a frame filling material for new low-rise buildings [10]. In order to decrease labour intensity in the manufacturing process, it is necessary to

---

[2] To whom any correspondence should be addressed.

produce LHC blocks and boards. This way requires the binders different from the classically used ones, as they show insufficient mechanical strength for this new use.

Traditionally, pure lime or lime based binders with hydraulic additives [11] have been used for LHC materials. These materials show insufficient hardening of the core of the constructional element because of insufficient carbonation. This is an unacceptable property for LHC blocks and boards as they need to be sawn in right sizes on site.

One of the possible ways to improve mechanical stability of hemp composite is to use other types of binders which provide more homogenous material structure. In the current study, alternative binders based on gypsum composite are tested, to assess their applicability together with hemp shives; the mechanical, thermal and water absorption properties of hemp concrete materials are addressed.

2. Materials

2.1. Hemp Shives

Hemp shives are a by-product of hemp-fibre manufacturing process, they are the wooden inner core of the hemp stalk. They are separated from the outer long fibres, which are the primary product of hemp fibre production that is mainly used in the production of paper, textiles and biogredable plastics [12]. Hemp shives are regarded as waste or by-product [13], it is used as bedding in horse stables or as thermal insulation building material.

Hemp shives have largely different physical properties, which are influenced by hemp breed, growing and harvesting conditions, processing technology, etc. In the current study, only one type of hemp shives is used focusing on different binders. The hemp shives used are from "Zalers" Ltd., flax and hemp processing company in Kraslava County. Granulometric composition, density, thermal conductivity and moisture content can be seen in table 1.

Table 1. Properties of shives.

Fibre	>20mm	10-20mm	0.63-10mm	Dust	Density, kg/m^3	Thermal cond. W/m*K	Moisture, %
3.3%	3.2%	8.0%	84.2%	1.2%	72.26	51.87	6.77%

2.2. Binders

Typically, lime is used as a binder in LHC materials. It can be pure hydrated lime, natural hydraulic lime or formulated lime – hydrated lime with added pozzolans to enhance hydraulic properties of the binders. Commonly used pozzolans are either natural (volcanic ashes, pumices) or artificial (silica fume, metakaolin). Natural hydraulic lime or formulated lime shows significantly higher strength than hydrated lime and in combination with hemp shives, hydrated lime lacks the mechanical rigidity required for it to be used as a binder for LHC building blocks [14].

In the frame of this work, hemp composites with different mineral binders were produced and compared. Standard (reference) compositions were prepared using the following binders:

- NHL 2 is feebly hydraulic natural hydraulic lime (St. Astier);
- DL60 is dolomitic lime, produced locally by "Saulkalne" Ltd in combination with pozzolanic admixture – metakaolin – obtained by burning of kaolin clay at 800^0 C [14];
- GCP is a gypsum-cement-pozzolanic composition, which was elaborated and was widely used in 1970's and 1980's. Such compositions combine rapid hardening of the gypsum binder and water

resistance of a cement binder. In this case, gypsum with addition of cement CEM II 42.5N from "Cemex" Ltd. and metakaolin is used, as well as with "Tradical" Ltd. HB commercial lime binder.

Table 2. Compositions of mixtures.

Type	Hemp shives	DL 60	NHL2	Gypsum	Cement	Metakaolin	Tradical HB	Water	Shives:binder ratio
DL	1	0.83	-	0.08	-	0.27	-	1.8	0.85
NHL	1	-	2	-	-	-	-	1.86	0.50
NHL-W	1	-	2.5	-	-	-	-	2	0.40
GCP	1	-	-	1.2	0.4	0.4	-	2	0.50
GCP-W	1		-	1.38	0.46	0.46	-	2.08	0.40
GC-TR	1	-	-	1	0.32	0.32	0.58	2.08	0.45

3. Methods

3.1. Preparation, mixing and moulding

Before the mixing process, moulds are prepared. They are made of moisture-resistant plywood, dimensions 350x350 mm, before the tests the moulds are greased with formwork oil to minimize friction when the samples are demoulded.

All ingredients are measured in weight and weighed using electric scales. Mixing is done using laboratory drum mixer, first the shives are mixed with the half of all the necessary water, then the binder and the rest of the water are added. Total mixing time is 3 minutes for each mixture.

Compositions of mixtures can be seen in table 2. Traditionally when using natural hydraulic lime, the shives: binder ratio of around 0.4 – 0.5 is used, so in the current study ratios of 0.4 and 0.5 were chosen. The quantity of the water added was set to achieve wet mixture that sticks together when pressed.

Dolomitic lime have also been used as a binder for LHC [15], but has shown weakness of inner strength of material, as LHC blocks with dolomitic lime have rigid outer shell but when sawn in pieces show an unbounded inner core, which is attributed to chemical reaction between organic hemp shives and alkaline lime binder. So in the present study, an admixture of metakaolin and gypsum is added to enhance the early strength of the material.

Cement has also been used as a binder for hemp concrete materials and has shown promising results regarding mechanical strength, but the weakness of the cement is its environmental impact, which is a lot more significant than that of a lime binder. This impact could be minimized if same strength material could be produced with higher shives: binder ratio, which would mean a lower amount of binder used.

Gypsum is another type of the mineral binder. As it is soluble in water [16], it has limited use compared to the hemp concrete binder, as it is used only as indoor plaster [17] with hemp fibre additions for enhancement of flexural strength [19], or as foam gypsum with hemp fibre reinforcement [18]. In order to use gypsum as a binder for hemp concrete it needs to be improved by addition of cement [16] and pozzolans [17]. These additions decrease the initial strength of gypsum, but in the long term improve its strength and durability in outdoor applications. This gypsum-cement-pozzolan (GCP) mixture, on the other hand, has much higher initial strength than lime or cement binders, and lower alkalinity of the binder decreases the amount of organic hardening retardants emitted from shives.

Two different shives: binder ratios are used for GCP mixtures – 0.4 and 0.5 and one mixture with ratio 0.45 where some of the binder is replaced with commercially available Tradical HB binder designed for use with hemp shives. It consists of hydrated lime with pozzolanic admixtures.

During the mixing it was observed that some of the mixtures form lumps during the process. This formation of has also been observed by other authors [20], and in the current study it was mostly detected during the mixing of DL sample – the lumps can be seen in figure 1. This can be explained by the higher shives: binder ratio.

Figure 1. Compositions with different binders – (from the left) - GC-TR, GCP, DL, GCP-W.

After mixing, the mixture was put in moulds, filling in layers slightly tamping each layer by hand. In the end, a plate with around 10 kg weight was put on top. Demoulding was done on the next day leaving the samples horizontally, after they gained enough strength they were raised up vertically. Samples were kept in laboratory conditions (20±2 ^{0}C and 40±10 %RH) and weighted until they didn't show any change in weight, which was on average about a month after demoulding.

Figure 2. Water absorption test – GCP-W (left) and DL (right) samples.

3.2. Testing of the samples

When the weight of the samples stabilized, their thermal conductivity was measured. This was done using LaserComp's FOX600 heat flow meter, according to the guidelines of LVS EN 12667. After these tests, the samples were left in laboratory conditions for two years – from April 2013 till April 2015, to allow the binders to fully set and for lime to carbonate.

After this period, the samples were tested for compressive strength and water absorption. First of all, the samples were sawn into four 170x170 mm pieces. This process already allowed making some conclusions, as inner cores of DL, NHL and NHL-W samples were soft and only partially bounded, the NHL sample was too damaged during the process and was not used for further tests. Gypsum-based mixtures showed similar rigidity throughout the sample.

After cutting and preparing the test samples, water absorption test according to LVS EN 1609 standard was carried out. It required samples to be weighed, immersed in water 10 mm deep for 24 hours, then dried for 10 minutes by placing them on a sieve at 45^0 angle and weighed again. The result expressed as water absorbed in kg on immersed area can be seen in table 3.

Table 3. Results of the tests.

Mix name	Shives:binder ratio	Density, kg/m^3	Water absorption, kg/m^2	Compr. strength, kPa	Compr. strength (W), kPa	Thermal conductivity, W/m*K	Compr. strength ratio
DL	0.85	290	16.085	0.081	0.030	0.069	2.667
NHL	0.50	320	-	-	-	0.084	-
NHL-W	0.40	355	20.079	0.067	0.047	0.093	1.437
GCP	0.50	310	14.800	0.126	0.059	0.088	2.137
GCP-W	0.40	325	10.615	0.200	0.082	0.097	2.422
GC-TR	0.45	300	10.874	0.146	0.079	0.082	1.870

After the water absorption test, compressive strength was tested for all samples, the maximum strength was registered as 10% deformations according to LVS EN 826 guidelines. The tests were done using ZWICK Z100 universal testing machine, with applied pressure at 10mm/min, force-deformation diagram was recorded. The tests were done for both water saturated (Compr. strength (W)) and dry samples (Compr. strength), the results are summarized in table 3.

4. Results and discussion

4.1. Water absorption

Differences in water absorption can be seen in both figure 2 and figure 3. From table 3 it can be seen that density doesn't correlate with water absorption as it could be expected, because the sample with highest density (NHL-W) has the higher water absorption, but the lowest one (DL) doesn't have the lowest absorption.

Water absorption somewhat correlates with compressive strength, this could be explained by the differences of the binder nature. As more hydraulic binders (containing more cement) showed similar rigidity throughout the material, the lime binders were softer in the inner core and more porous, which allowed larger water uptake. This water uptake of lime-based binders should be taken in account when considering the protection of LHC construction (mortar, overhang, etc.).

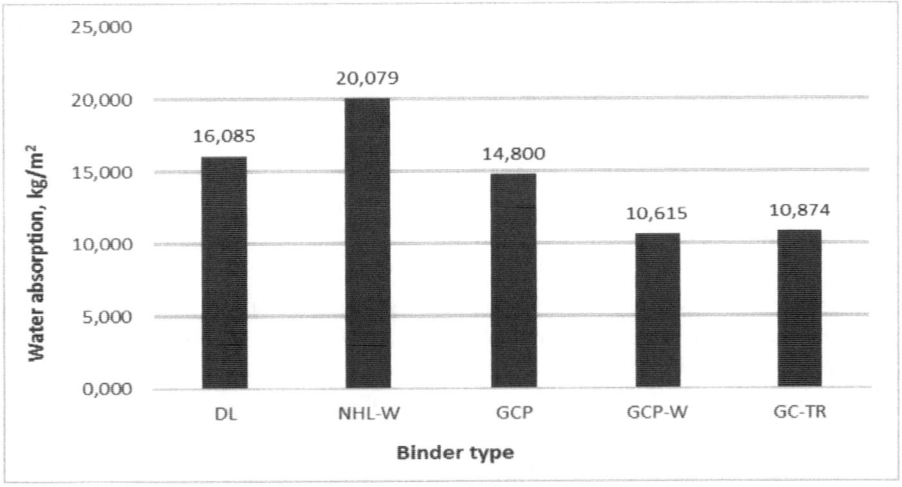

Figure 3. Water absorption dependence on the binder type.

4.2. Thermal conductivity

As can be seen from figure 4, thermal conductivity is in direct correlation with density. This has also been shown by other authors [21]. The obtained results of 0.069 W/m*K at 292.32 kg/m^3 are slightly lower than those of other authors – 0.074 W/m*K at 269 kg/m^3 [21] and 0.086 W/m*K at 250 kg/m^3 [22].

This can be explained by looking at figure 4 – all the lime-based materials are below the trend line and all gypsum-based are above. This can be explained by the differences in the inner core of the materials, which was unhardened and porous in the lime material, but rigid and monolithic – in the gypsum-based materials. This unhardened and porous structure ensured that there were fewer chemical and physical bounds between the particles, which in turn lead to the production of materials with lower thermal conductivity.

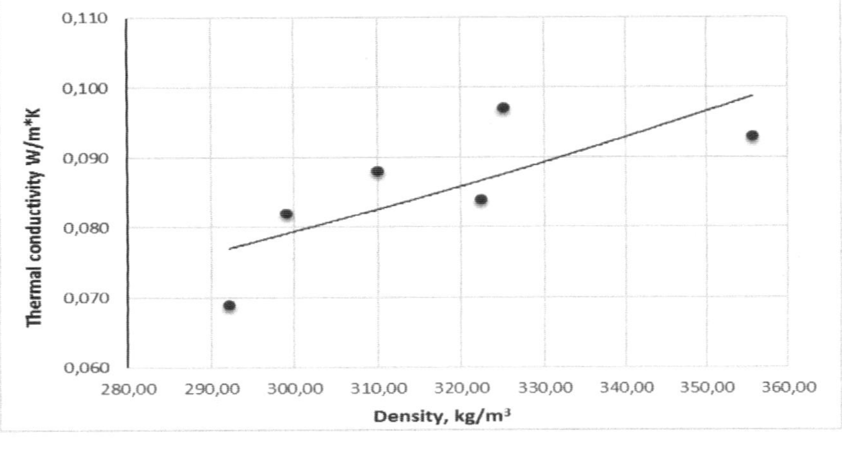

Figure 4. Thermal conductivity and density of compositions.

4.3. Compressive strength

After sawing of the samples, first conclusions about the strength of the samples were made. First, lime-based samples display poor inner strength, as in previous tests [14], but in this study the samples were cured for 2 years in laboratory conditions, but had not displayed any significant improvement. It means that low carbonisation is not the reason for this low strength as previously thought, as the samples did not carbonate for two years.

Insufficient water amount as the reason of low material strength has been put forward as one of the theories [23], but this theory should be dismissed. As it can be seen from table 2, the amount of spare water for NHL and GCP samples was similar, but the GCP samples show a rigid inner core. The main reason for the unhardened lime binder is that chemical compositions – pectins and other organic compounds are emitted in the reaction between organic hemp shives and alkaline lime binder environment [24] [25]. And these polysaccharides serve as retarders of lime and cement binding reactions, in the case of NHL binders not only temporarily slowing the process but permanently stopping it [26].

But this organic chemical reaction influences the gypsum-based binder far less, because, firstly, gypsum doesn't create alkaline environment, it is pH neutral, so the organic compounds are created in lower amounts. Secondly, gypsum has fast initial setting times, which allows it to harden before the organic compounds are even created, as lime samples have shown that these retardation processes happen during several first days of setting of the LHC [14].

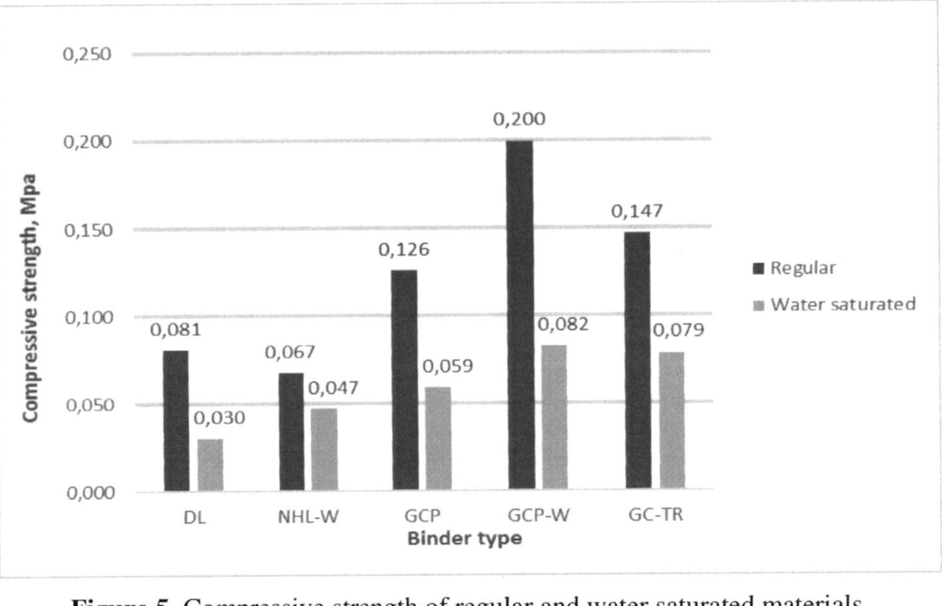

Figure 5. Compressive strength of regular and water saturated materials.

Results of the compressive strength testing can be seen in figure 5 and table 2. The top achieved results (0.2 MPa at 325kg/m^3) is similar to other tests [27][28][14], but as the density was achieved with extra binder, but not pre-compression of the material, the results from previous tests suggest that similar results could be achieved with lower amount of the GCP binder and pre-compression of the material.

It can also be seen from the results that addition of 14% gypsum to DL binder didn't have a significant influence on the strength, tests with different ratios is necessary to make final conclusions, but for now it seems that gypsum content was insufficient to neutralize the negative effect of emergence of organic compounds.

From figure 5 it can also be concluded that addition of cement and pozzolanic mixtures to the gypsum binder had positive effect on its resistance to water solubility, as the compressive strength ratio of regular: water saturated samples was higher than that of NHL-W, but lower than DL.

5. Conclusions

The following conclusions can be summarised within the study:

- Weakness of the inner core of LHC material that has been attributed to insufficient carbonization and low water content is most likely the fault of organic polysaccharide compounds that are emitted in the reaction between hemp shives and alkaline binders.
- As gypsum-based binders create less alkaline environment and have high early strength, organic sugar binding retarders do not affect these binders significantly.
- Gypsum binders can be enhanced with addition of cement and pozzolanic mixtures to be useable as binders for outdoor conditions.
- As thermal conductivity showed correlation with density as observed previously, it can be suggested that pre-compression of the material could lead to lesser amount of binder used at similar densities which could improve the materials' environmental impact.
- Water absorption is largely based on the type of binder used – the more hydraulic the binder, the less water is absorbed.

6. References

[1] Tronet P, Lecompte T, Picandet V and Baley C 2014 Study of lime hemp composite precasting by compaction of fresh mix – An instrumented die to measure friction and stress state *Powder Tech.* **258** 285-96

[2] Evrard A, Flory-Celini C, Claeys-Bruno M and De Herde A 2014 Influence of liquid absorption coefficient on hygrothermal behaviour of an existing brick wall with Lime-Hemp plaster *Build. and Envir.* **79** 90-100

[3] Chabannes M, Benezet J, Clerc L and Garcia-Diaz E 2014 Use of raw rice husk as natural aggregate in a lightweight insulating concrete: An innovative application *Constr. and Build. Mat.* **70** 428-38

[4] Faure P, Peter U, Lesueur D and Coussot P 2012 Water transfers within Hemp Lime Concrete followed by NMR *Cem. and Concr. Res.* **42** 1468-74

[5] Pretot S, Collet F and Garnier C 2014 Life cycle assessment of a hemp concrete wall: Impact of thickness and coating *Build. and Envir.* **72** 223-31

[6] Rahim M, Douzane O, Tran Le AD, Promis G, Laidoudi B, Crigny A, Dupre B and Langlet T 2015 Characterization of flax lime and hemp lime concretes: Hygricproperties and moisture buffer capacity *Energ. and Build.* **88** 91-99

[7] Collet F and Pretot S 2014 Experimental highlight of hygrothermal phenomena in hemp concrete wall *Build. and Envir.* **82** 459-66

[8] Gle P, Gourdon E and Arnaud L 2012 Modelling of the acoustical properties of hemp particles *Constr. and Build. Mat.* **37** 801-11

[9] Benfratello S, Capitano C, Peri G, Rizzo G, Scaccianoce G and Sorrentino G 2013 Thermal and structural properties of a hemp–lime biocomposites *Constr. and Build. Mat.* **47** 745-54
[10] Collet F, Chamoin J, Pretot S and Lanos C 2013 Comparison of the hygric behaviour of three hemp concretes *Energ. and Build.* **62** 294-303
[11] Arnaud L and Gourlay E 2012 Experimental study of parameters influencing mechanical properties of hemp concretes *Constr. and Build. Mat.* **28** 50-56
[12] Guessasma S, Bassir D and Hedjazi L 2015 Influence of interphase properties on the effective behaviour of a starch-hemp composite *Matr. and Design* **65** 1053-63
[13] Colinart T, Glouannec P and Chauvelon P 2012 Influence of the setting process and the formulation on the drying of hemp concrete *Constr. and Build. Mat.* **30** 372-80
[14] Sinka M, Sahmenko G and Korjakins A 2014 Mechanical properties of pre-compressed hemp-lime concrete *J. of Sustain. Arch. and Civil Engin.* **3** 92-99
[15] Sinka M and Sahmenko G 2013 Sustainable thermal insulation biocomposites from locally available hemp and lime *Environment. Tehcnology Resources: Proc. of the 9th Int. Sc. and Pr. Conf.* **9** 73-77
[16] Camarini G and De Milito J 2011 Gypsum hemihydrate–cement blends to improve renderings durability *Constr. and Build. Mat.* **25** 4121 – 25
[17] Colak A 2002 The long-term durability performance of gypsum–Portland cement–natural pozzolan blends *Cem. and Concr. Res.* **32** 109-115
[18] Brencis R, Skujans J, Iljins U and Preikss I 2011 Research of hemp fibrous reinforcement effect to bending strength and sound absorption of foam gypsum *3rd Int. Conf. CIVIL ENG.`11 Pr.* **1** 32-37
[19] Dalmay P, Smith A, Chotard T, Sahay-Turner P, Gloaguen V and Krausz P 2009 Properties of cellulosic fibre reinforced plaster: influence of hemp or flax fibres on the properties of set gypsum *J. of Matr. Sc.* **45** 793-803
[20] Bruijn P and Johansson P 2014 Moisture transport properties of lime-hemp concrete determined over the complete moisture range *Biosys. and Eng.* **122** 31 – 41
[21] Balciunas G, Vejelis S, Vaitkus S and Kairyte A Physical Properties and Structure of Composite Made by Using Hemp Hurds and Different Binding Materials *Proc. Eng.* **57** 159-66
[22] Collet F and Pretot S 2014 Thermal conductivity of hemp concretes: Variation with formulation, density and water content *Constr. and Build. Mat.* **65** 612-619
[23] Arizzi A, Cultrone G, Brümmer M and Viles H 2015 *Constr. and Build. Mat.* **75** 375-82
[24] Diquelou Y, Gourlay E, Arnaud L, Kurek B 2015 *Cem. and Concr. Comp.* **55** 112-121
[25] Sellami A, Merzoud M and Amziane S 2013 *Constr. and Build. Mat.* **47** 1117-24
[26] Sedan D, Pagnoux C, Smith A and Chotard A 2008 *J.. of the Eur. Cer. Soc.* **28** 183-92
[27] Walker R and Pavía S 2014 Moisture transfer and thermal properties of hemp–lime concretes *Constr. and Build. Mat.* **64** 270-276
[28] Elfordy S, Lucas F, Tancret F, Scudeller Y and Goudet L 2008 *Constr. and Build. Mat.* **22** 2166-23

Acknowledgments
The research leading to these results has received the funding from Latvia state research programme under grant agreement "INNOVATIVE MATERIALS AND SMART TECHNOLOGIES FOR ENVIRONMENTAL SAFETY, IMATEH".

Feasibility of Integrated Insulation in Rammed Earth

C Stone [1,3]**, M Balintova**[2] **and M Holub**[2]
[1]Technical University of Kosice, Faculty of Civil Engineering, Building Construction Laboratory, Vysokoskolska 4, Kosice, 042 00 Slovakia
[2] Technical University of Kosice, Faculty of Civil Engineering, Institute of Environmental Engineering, Vysokoskolska 4, Kosice, 042 00 Slovakia

E-mails: clayton.stone@tuke.sk

Abstract. Building Codes in Europe stipulate strict thermal performance criteria which any traditional rammed earth recipe cannot meet. This does not infer that the material itself is inferior; it has many other face saving attributes such as low embodied energy, high workability, sound insulation, fire resistance, aesthetics, high diffusivity and thermal accumulation properties. Integrated insulation is experimented with, to try achieve a 0.22 [W/(m^2.K)] overall coefficient of heat transfer for walls required by 2015 Slovak standards, without using external insulation or using technologically complex interstitial insulation. This has the added aesthetic benefit of leaving the earth wall exposed to the external environment. Results evaluate the feasibility of this traditional approach.

1. Introduction

With the advent of thermal criteria for building codes, thermal comfort as a criterion is superseded by efficiency measured in kWh/m^2 p.a. which is a product of a buildings shape, size, airtightness and thermal resistance of the envelope and its parts. Contemporary stabilized rammed earth (SRE) is an evolutionary product of traditional rammed earth (RE) methods and materials, often incorporating steel reinforcing and rigid insulation, enhancing the structural and energy performance of the walls while satisfying building code parameters. Interstitial thermal insulation is known to be effective for many types of building envelopes made of sandwich or composite structures. The implementation of interstitial insulation in stabilized rammed earth (SRE) has received significant attention in Research and development as stringent thermal requirements for building codes continue to increase. Thermal conductivity tests, conducted by M.A. Hall, on composite SRE walls with extruded polystyrene (XPS) insulation demonstrated that the combination of a high mass wall with the low conductivity of foam

[3] To whom any correspondence should be addressed.

IOP Conf. Series: Materials Science and Engineering **96** (2015) 012030 doi:10.1088/1757-899X/96/1/012030

insulation resulted in a wall that had a lower thermal conductivity than a solid earth wall or an earthen wall with insulation located at only the internal or external face, while improving the mass performance of the wall as a whole [1, 2]. However, due to the complexity and technological limitations of implementing (SRE) interstitial sandwiches in situ, the authors were inspired to analyse if it is holistically feasible to integrate bulk insulation in rammed earth.

2. Sample production

All tested soil components were obtained locally and oven-dried to a constant mass at a temperature of 60 °C as opposed to 110 °C to prevent potential changes in the material structure. The Portland cement (type III) content was 10 % by weight. The clay and silt (30 % content) was pulverized into a coarse powder, passing through an 8 mm sieve, and then mixed with sand and gravel using a variable speed mixer to ensure uniformity between batches.

Table. 1 Determination of OMC for 8 [%] cement content.

CS0108 Sample: 8%

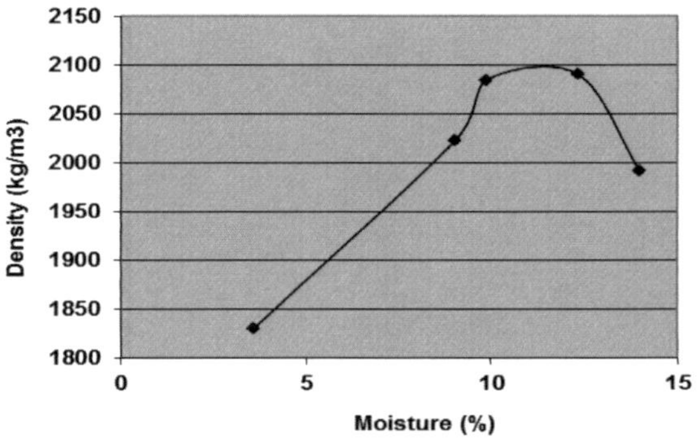

	1	2	3	4	5
Volume of material in cylinder [m^3]	0.000947	0.000947	0.000947	0.000947	0.000947
Mass of cylinder with material [kg]	6.46	6.754	6.834	6.889	6.815
Empty mass of cylinder [kg]	4.665	4.665	4.665	4.665	4.665
Mass of moist material [kg]	1.795	2.089	2.169	2.224	2.15
Bulk density of material [kg/m^3]	1895.459	2205.913	2290.39	2348.469	2270.337
Mass of receptacle + moist sample [g]	0.0620	0.0797	0.0570	0.0420	0.0736
Mass of receptacle + dry sample [g]	0.0612	0.0765	0.0555	0.0401	0.0694
Mass of receptacle [g]	0.0388	0.0411	0.0403	0.0247	0.0394
Mass of dry material [g]	0.2240	0.0354	0.0152	0.0154	0.0300
Mass of moisture in the material [g]	0.0080	0.0320	0.0150	0.019	0.0420
Moisture content [%]	3.571429	9.039548	9.86842	12.33766	14.000
Dry bulk density of material [kg/m^3]	1830.099	2023.04	2084.67	2090.545	1991.515

Eleven kilograms of soil were used to make one test specimen that measured 32.5 x 32.5 x 5.5 cm. A controlled amount of water was added to raise the soil to optimum moisture content (OMC) which is the moisture content at which a material reaches its maximum dry density (MDD) for a given compactive effort. This is said to influence the strength and durability of the material. Table 1 shows the MDD and OMC of the reference sample. Columns one through five represent the five test points on the line curve and represent increased moisture content for the same recipe [3]. The material for all test samples was compacted using a pneumatic tamper with a 55 mm head.

Component soils were blended in ratios and then graded to obtain their particle-size distribution parameters plotted logarithmically with respect to percentage (by dry mass) of the total specimen on a linear scale, see figure 1. STN EN 1015-1/A1 (72 2441) and STN EN 933-2 (72 1186) were used to classify the granular materials and sedimentation into different physical dimensions using a number of meshes and sieves [4; 5; 6].

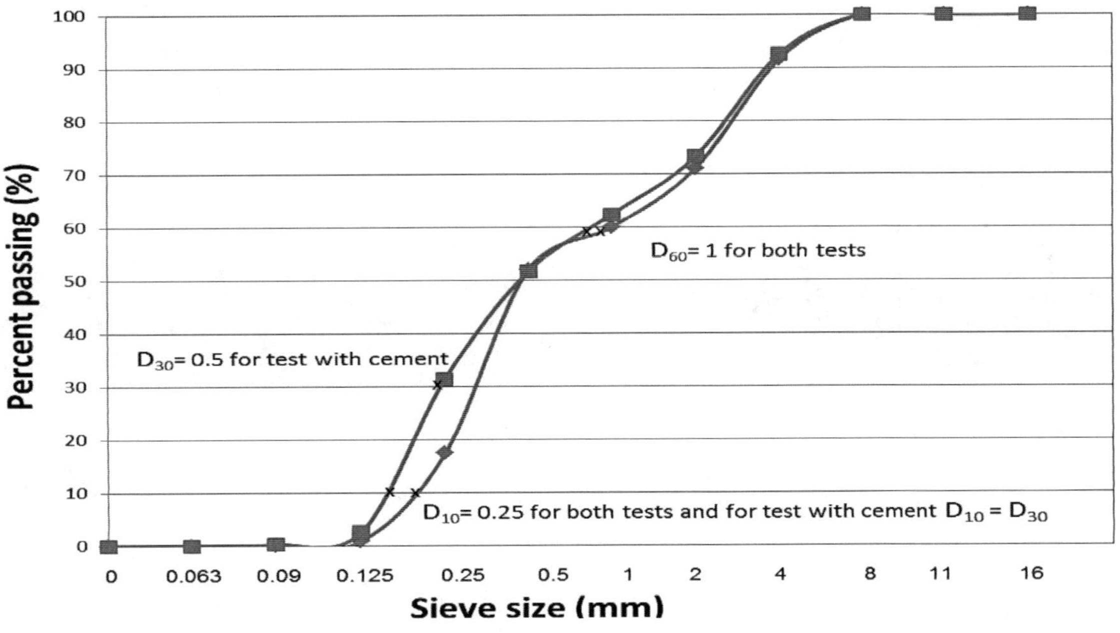

Figure 1. Cumulative particle-size distribution curve (Red line: cement included; blue line: cement omitted).

2.1. Selection of thermal insulation

Insulation had to fulfill several criteria. These included that the material be ecological; natural based; inherently hydrophobic, but not a vapour barrier; be relatively incompressible and withstand the impact of ramming. Foam glass, clay pellets and cork were amongst those considered, but hydrophobic expanded perlite seemed the most promising. Batches were defined by the percentage of bulk volume of thermal insulation that was put in the form. Reference batches had no insulation, while each subsequent batch had an increase of 20 % bulk insulation while the maximum feasible case had a 100 % volume of bulk insulation relative to the volume of the form. Figure 2 illustrates a freshly removed sample left and a dried sample right ready for testing. Figure 3 overleaf shows the preparation of the test specimens, equipment used and formwork for the test specimen.

Figure 2. Left: A freshly removed sample, Right: A sample after curing until reaching a constant mass.

(a) (b) (c)

Figure 3. (a) soil and expanded perlite mixture; (b) mixing and compacting equipment; (c) braced formwork with test specimen.

3. Determination of the coefficient of thermal conductivity

Thermal conductivity was measured using a calibrated Stirolab heat flow meter with compliant Lm 305 software. Accuracy was set to 99 % with only a 1 % permissible deviation. Each test lasted 3600 seconds and every specimen was tested twice. The first test stabilized the high mass material while the second test produced the final measurement results.

IOP Conf. Series: Materials Science and Engineering **96** (2015) 012030 doi:10.1088/1757-899X/96/1/012030

3.1. Test Procedure

The Stirolab lambda meter works by placing a test sample between a warm and cold plate. The standard setting is 0 ºC for the bottom plate and 20ºC for the top plate producing a 20 K temperature difference. The temperatures of the plates are kept constant while the amount of energy required to maintain constant temperatures depends on the conductivity of the material. The conductivity of the material is therefore directly proportional to the amount of energy required to maintain a constant temperature between the plates. Thus thermal conductivity is calculated as:

$$\lambda = \frac{j.d}{\Delta T} \qquad \text{(W/m.K)} \qquad (1)$$

Where

j	- is the heat flow	(W/m^2)
d	- is the thickness of the test specimen	(m)
ΔT	- is the temperature difference	(K)

Figure 4 shows the simultaneous thermal stabilization curve of the reference specimen for the warm and cold plates.

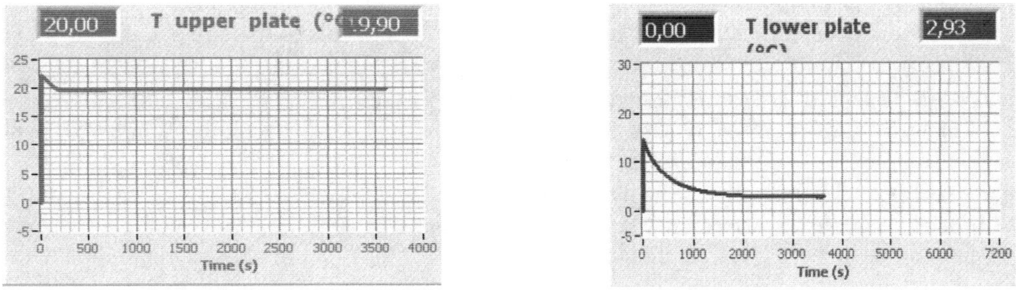

Figure 4. Heat flow meter approaching steady-state: Left, warm plate; Right, cold plate.

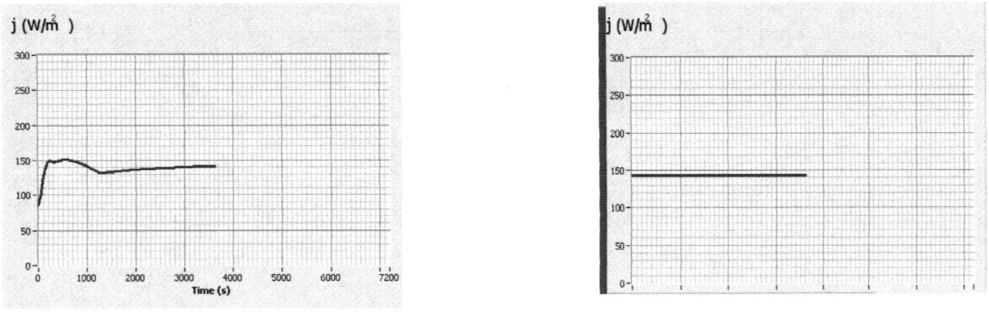

(a) (b)

Figure 5. (a) stabilizing the heat flow; (b) energy required to maintain steady-state measurement.

Figure 5 below depicts the heat flow needed by the warm and cold plates to maintain a constant temperature.

4. Results and discussion

Table 2 was drawn to define the dry mass, density, quantity of insulation expressed as a percentage of total volume and statistical thermal conductivity of each test specimen. These are statistical values averaged over multiple samples. The volume of each sample was approximately 0.0058 m^3.

Table 2. Determination of OMC for 8 [%] cement content.

Percentage bulk insulation [%]	Weight [kg]	Density of samples [kg/m^3]	Coeff. Thermal conductivity [W/(m.K)]
0	12.304	2037.4	0.510
20	12.099	2022.4	0.499
40	11.224	1823.5	0.462
60	10.694	1656.7	0.413
80	9.187	1530.5	0.412
100	8.784	1500.8	0.410

Conditions of the feasibility study were that the walls should have a coefficient of thermal conductivity that is low enough to meet thermal transmittance targets without sacrificing load bearing ability. Therefore, multiple Schmidt hammer tests were performed as an indicator of strength for each corresponding test sample. The Schmidt hammer was used because the shape of the specimens did not conform to compression tests via crushing. The tests were performed in the direction of the ramming and showed that each 20 % increase in perlite dramatically reduced the strength. The reference sample had a characteristic strength of almost 4.94 MPa, the 20 % had strength of 2.45 MPa and all subsequent samples had strengths lower than the minimum required for identification. Ignoring this fact and focusing only thermal conductivity and Thermal Admittance related values [7], the results were still disappointing. Figure 6 shows the heat flows due to unit swing in the internal environmental temperature that would occur over a 24 h period.

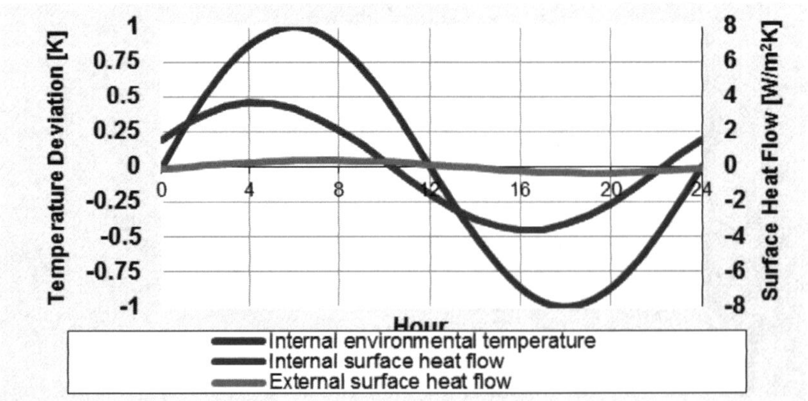

Figure 6. Heat flow of the 100% bulk volume specimen for a 1 K temperature variation over a 24 hour period.

5. Conclusions

The paper experimented with the feasibility of integrating expanded perlite into SRE to reasonably attain a 0.22 [(W/(m^2.K)] overall coefficient of heat transfer for walls required by 2015 Slovak standards while maintaining minimum strength requirements. The experiment failed on two accounts. The first failure was that rammed gains its strength from its density. Intermolecular forces weaken if the perlite is compressed. Rammed earth technology is impact based and most bulk forms of insulation are either too brittle or compressible to make this a viable alternative. Secondly, although a dramatic increase in thermal insulation produced a rapid reduction in strength, it only produced a marginal improvement in thermal resistance. This can be explained by using the 40 % volume and 100 % volume specimens as examples. The 40 % perlite per volume specimen produced the greatest improvement in thermal performance as it is assumed there was enough rammed covering the perlite to prevent significant compression of individual perlite particles. However, the 100% volume of perlite mixed with rammed earth was easily compressed producing only a marginal improvement over the previous layer. A wall thickness in excess of 1.7 meters would be required to attain a thermal transmittance of 0.22 [W/(m^2.K)] making the use of perlite in this way unfeasible. In addition, the authors doubt whether any other form of integrated particle insulation could be used to attain 0.22 [W/(m^2.K)] overall coefficient of heat transfer, but does not rule out that low density cob with large thicknesses might achieve these values.

6. References

[1] Hall M A 2008 Assessing the moisture-content-dependent parameters of stabilized earth materials using the cyclic-response admittance method *Energ. Build.* **40** 2044–51

[2] Allinson D and Hall M 2007 Investigating the optimisation of stabilised rammed earth materials for passive air conditioning in buildings *Proc. for the Int. Symposium of Earthen Structures (Bangalore)*

[3] NZS 4298: 1998 Materials and workmanship for earth buildings – incorporating amendment no.1. Standards New Zealand (Wellington, New Zealand,)

[4] Bagona M, Stone C and Simicek J 2014 Dynamic thermal properties of uninsulated rammed earth building envelopes *Advanced Materials Research: ABCM 2013* **855** 85–88

[5] STN EN 1015-1/A1 (72 2441): *Methods for Testing Mortar for Masonry,* Part 1: Determination of particle size distribution (by sieve analysis), 2007, (Slovak)

[6] STN EN 933-2 (72 1186): *Tests for Geometrical Properties of Aggregates*. Part 2: Determination of particle size distribution. Test sieves, nominal size of apertures, 1999, (Slovak)

[7] Stone C and Katunský D 2015 Dynamic thermal properties of uninsulated rammed earth envelopes *Pollack Periodica* **10** (1) 103–112

Acknowledgements

This publication is the result of the Project implementation: University Science Park TECHNICOM for Innovation Applications Supported by Knowledge Technology, ITMS: 26220220182, supported by the Research & Development Operational Programme funded by the ERDF and funded by Grant No. 1/0563/15 of the Slovak Grant Agency for Science.

Influence of Building Envelope Thermal Mass on Heating Design Temperature

B Gaujena[1] **A Borodinecs**[2,3], **J Zemitis**[2] **and A Prozuments**[2]
[1]Riga Technical University, Institute of Building Production, Department of Civil Engineering, Kalku Str. 1, Riga, LV-1658, Latvia
[2]Riga Technical University, Institute of Heat, Gas and Water Technology, Department of Heat and Gas technology, Kalku Str. 1, Riga, LV-1658, Latvia

E-mail: anatolijs.borodinecs@rtu.lv

Abstract. The stability of indoor air parameters is a very important factor, essential for such institutions as museums, schools and hospitals. Nowadays the use of renewable energy for space heating became one of the top priorities in modern building design. The active and passive solar energy as well as heat pumps are widely used nowadays. However, such technologies have a limitation in cold climates and often are not able to cover maximal heating loads.
This paper is devoted to analysis of influence of building envelope's properties and outdoor air parameters on indoor air thermodynamic parameters stability in winter time. It presents analysis of thermal mass impact on building energy performance and indoor air parameter stability in cold climate. The results show that the thermal mass of building envelope is able to cover extreme winter temperatures as well as in case of emergency heat supply break.

1. Introduction

Since the early 1970ies, the U-value of building envelope has reached reasonable minimums. The latest publications show the potential for the wide implementation of different types of vacuum heat insulation materials and systems with very low heat conductivity coefficient. But in most cases the main benefits of such systems are not significant reduction in U-value but rather minimization of thickness of external envelope. According to passive house concept, the U-value is 0.1 W/(m^2·K) for walls, roofs and 0.8 W/(m^2·K) for windows [9]. In addition, solar heat gains for passive heating are widely implemented in the energy efficient building concept using the well-known principles of solar geometry [11]. So traditionally it is assumed that energy efficient buildings are supposed to have maximally high thermal and solar resistance of building envelope. At the same time, very low heat loses during winter time can lead to overheating in spring/autumn period due to high internal heat

[3] To whom any correspondence should be addressed.

gains, increasing thermal discomfort. Incorrect windows size and placement in addition to incorrect use of shading approach can cause space overheating during hot and sunny days in summer time [8]. In practice, high thermal resistance is only needed in coldest winter days in the countries with cold climate or on hot summer days. The author [6] already presented principal solutions for changeable thermal properties of building envelope for accumulated heat discharge for passive cooling strategies and heat loses minimization during winter days.

In the last decade, the use of thermal mass for reduction of cooling/heating loads and improvements of thermal comfort has become a widely discussed research topic in different climatic zones. The research [11] shows that the timber-framed houses have higher daily amplitudes of humidity and temperature than the heavyweight. Also [4] shows the positive influence of thermal mass on energy savings and stability of indoor air temperature during the winter time. In turn, Kosny (2001) research estimates that energy savings for the whole building made from high thermal mass envelope is up to 8% in heating dominated region. Study [18] shows possible savings for heating and cooling by using masonry with smaller surface absorptivity. At the same time savings also depend upon utility incentives, building construction, climate, and the type of air conditioning system [2].

The thermal mass became a topical issue in the 1930ies [16]. Later the research on thermal mass was continued by [15, 5] in the 1950ies. All these researches considered thermal mass as a possibility for energy accumulation in winter time in cold climatic zones and creation of comfort indoor parameter in summer time in hot climatic zones. Before the 1960ies, thermal mass was topical in terms of energy accumulation due to the absence of centralized heating systems. The apartments were heated manually by inhabitants using local wood boilers and it was important that during the day between operation of boiler indoor air temperature did not drop below dew-point. The thermal mass for hot climatic zones was considered as a measure for creation of comfort indoor air parameters mainly due to lack of air-conditioning systems.

Nowadays, when energy prices went up and sustainable construction is a priority, all above-mentioned benefits of thermal mass became also topical in modern construction. But it should be noted that modern HVAC systems have advanced automatic as well as manual control possibilities in order to adjust indoor air parameters according to inhabitant needs and outdoor air parameters. The inappropriate combination of high thermal mass and modern HVAC automation systems can lead to such negative effect as:

- ✓ building overheating under rapid outdoor air temperature fluctuation;
- ✓ increase of energy consumption for building heating/cooling after the night/weekend indoor temperature reduction;
- ✓ building overheating in hot summer days when accumulated heat has not been discharged during the night.

It can be concluded that nowadays the choice of optimal thermal properties of building envelope for indoor air thermodynamic parameters stability became a great challenge for architects and HVAC designers. In order to use the benefits of the thermal mass, special control of heating system should be introduced on the basis of weather analysis and forecast.

Recent research [19] has proved that currently used winter design temperatures are overestimated. The above-mentioned research provides recommendation for winter design outdoor air temperature in Kaunas. According to the research study, the existing winter design temperatures exceed research recommendation by 33%-40%.

IOP Conf. Series: Materials Science and Engineering **96** (2015) 012031 doi:10.1088/1757-899X/96/1/012031

2. Methods

Fifty buildings were selected for the experiment; they were divided into two groups: lightweight construction buildings and massive construction buildings. For each building, measurements were conducted during the winter period from 2009 to the winter months of 2013. The heat consumption of the buildings was assessed during the work. For the purposes of the study, the data were obtained from heat supply organizations regarding the heat consumed during the heating season, while building owners and other public institutions provided information regarding the number of inhabitants and technical information regarding the building enclosures. In analyzing the data obtained in each building group, measurements of the building enclosures and measurements of indoor ambient temperature and air humidity were conducted for each building by installing two measuring instruments: Testo 635-2 at the building enclosures and Testo 175-H2 in the middle of the room.
The total area of the described lightweight construction buildings is 235.8 m^2, the total area of the massive construction buildings – 8684.3 m^2.

To analyze the measurement data in real conditions, the work process involved the use of simulation software Computational Fluid Dynamics (CFD) Autodesk Simulation CFD 2013, which is based on a process that combines the real conditions with mathematical calculations. The software allows assessing activity by using modeling. The simulation software uses extensive calculations to ensure that the final result is as close as possible to the real conditions. The analysed room model is presented in figure 1.

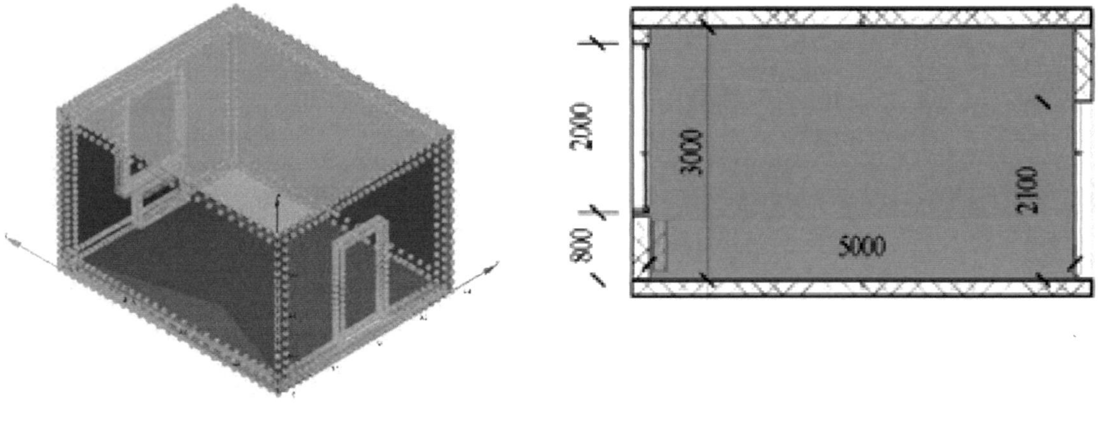

Figure 1. Analysed room model.

3. Interconnection between building elements

As a rule, the building envelope is a passive element and the HVAC system is an active element regarding the possibility of regulation in order to ensure optimal indoor air parameters. In case a building has high thermal mass, the building envelope can work as an active element and insure stability of air thermodynamic properties. In order to use positive effect of thermal mass, the choice of HVAC working regimes and capacity should be based on outdoor air conditions and properties of the building envelope. The paper [17] also present approach that adopts building envelope, air change rate and heating/cooling loads according to the requirement regarding indoor air parameters.

Taking into consideration the above-mentioned studies, figure 2 demonstrates interconnection between building elements, indoor air temperature (IAT) and outdoor air conditions.

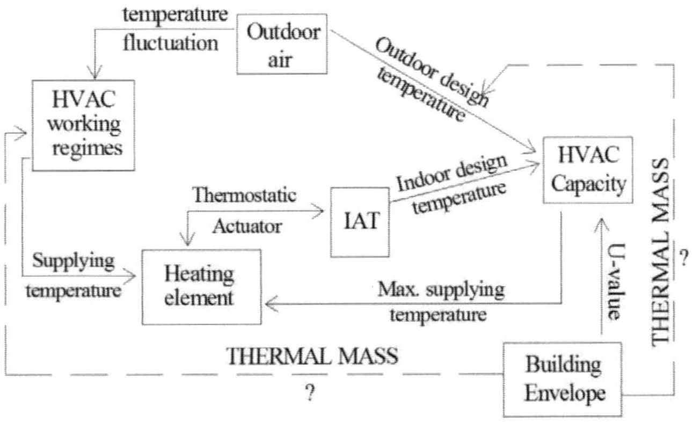

Figure 2. Interconnection between building elements.

4. Calculation of heating source capacity

For calculation of heating boiler capacity, constant temperature is usually used. The correctly chosen outdoor temperature allows improving energy performance of heating system and ensuring stability of indoor air thermodynamic parameters. At the moment, there is no clear methodology how to choose outdoor air temperature for calculation of heating boiler capacity.

Before 2000 for evaluation of heating system capacity the following range of outdoor temperatures was taken into account [1]:

- ✓ average temperature of the coldest 5 days;
- ✓ average temperature of the coldest 3 days;
- ✓ average temperature of all coldest days;
- ✓ absolutely minimal temperature.

The approach that had existed before the 1990ies was developed for external envelope with heat transfer coefficient 0.7 – 1.4 W/(m^2·K). In addition, the heating systems were without any regulation possibilities. So there was no any outdoor temperature sensors, circulation pumps, etc. The group of 5-9 multi-storey apartment buildings was connected directly to one centralized heat substation. In such case the impact of building envelope thermal mass was crucial as regards the indoor air parameters stability.

The choice of outdoor temperature depended on heat inertia of external building envelope before the 1990ies. In general, heat inertia is calculated using the following equation (1):

$$D = \sum R_i \cdot \sqrt{\frac{2\pi\lambda\gamma c}{z}} \tag{1}$$

Where,
z – time period of heat flow fluctuation; R – thermal resistance, (m^2·K)/W. S thermal capacity, W/(m^2·K).

At the same time, the study by (Olsen 2008) presents the following equation (2):

$$S=c\cdot q\cdot t \tag{2}$$

IOP Conf. Series: Materials Science and Engineering **96** (2015) 012031 doi:10.1088/1757-899X/96/1/012031

Where,

t – thickness of material, m.

The proposed [14] equation does not take into account the time period of heat flow fluctuation and material thermal conductivity. Table 1 shows the outdoor design temperature for Riga City according to Latvian building code LBN003-01 "Building Climatology" [13]. The average temperature of heating period is 0 °C.

Table 1. Outdoor design temperature for Riga city and heating boiler capacity.

	Outdoor design temperature °C according to existing practice in Latvia	Heating boiler heat capacity %
Average temperature of the coldest 5 days (probability 0.92)*	– 20.7	100
Average temperature of the coldest 5 days (probability 0.98)	– 24.6	119
Absolutely minimal temperature (once in 10 years)	– 31	150
Absolutely minimal temperature (once in 50 years)	– 34.8	168
Heating season's average temperature	0	49

* now used for calculation of heating systems' maximal capacity

One of the main tasks of a designer is to choose correct outdoor air design temperature. Currently only average temperature of the coldest 5 days with probability 0.92 (-20.7°C) is used in Latvia in order to calculate the capacity of heating boiler and heating elements. This approach does not take into account the heat capacity of building external and internal elements that can lead to indoor temperature reduction below set point under extreme outdoor temperatures. For example, under Latvia climatic condition the outdoor air parameters below design condition can reach consequent 17 hours long time period. Figure 3 presents the outdoor temperature fluctuations for February 14 – 16, 2010 in Riga.

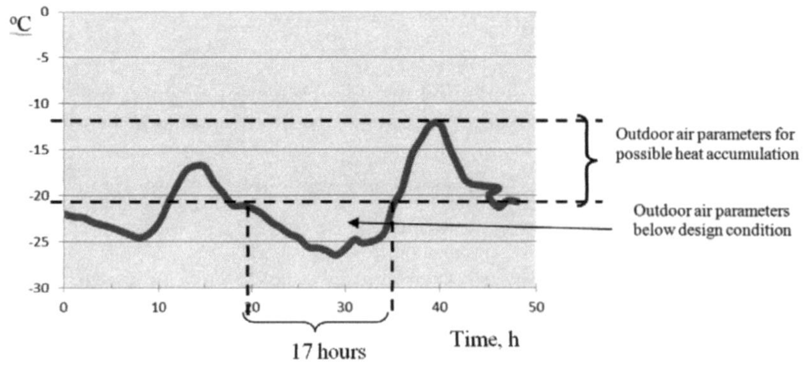

Figure 3. Outdoor temperature fluctuations in February 2012 in Riga.

IOP Conf. Series: Materials Science and Engineering **96** (2015) 012031 doi:10.1088/1757-899X/96/1/012031

In order to find the number of hours below design condition, it is necessary to use hourly spread of outdoor air parameters. The example of average temperatures and relative humidity during the 10-year monitoring period for the capital city of Latvia Riga is given in figure 4. This table shows monthly iteration of outside air parameters in three time zones of the day: from 0.00 till 8.00, from 8.00 till 16.00 and from 16.00 till 24.00.

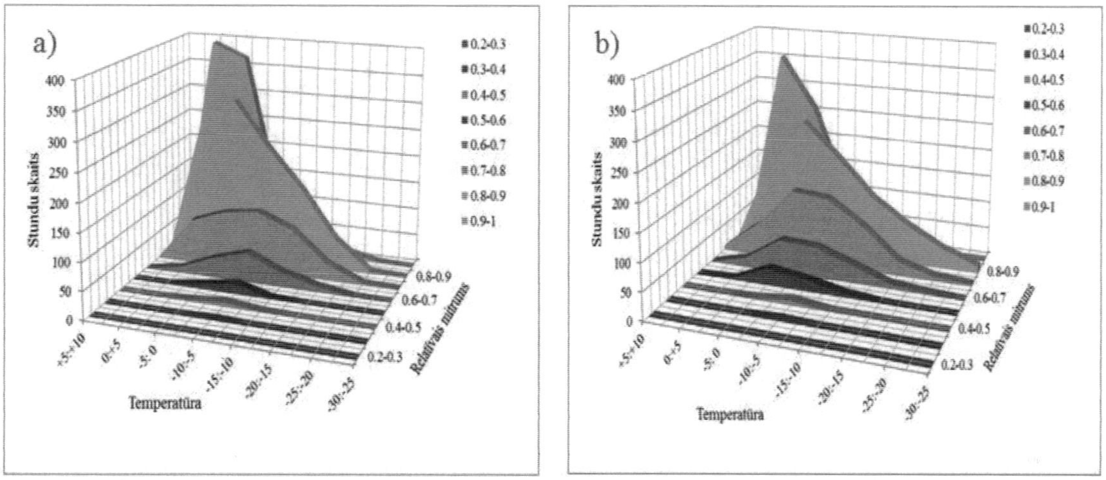

Figure 4. Number of hours in November (a) and December (b) at the respective temperature and relative humidity for three daily time zones in Riga, Latvia (during 10-year monitoring period).

In case of Latvia, the outdoor air temperature was below heating system design temperature for 305 hours during the 10-year weather monitoring period. That means 30 hours per average year. It is the task for a building owner to define correctly the requirements for stability of indoor air parameters and the designer consequently should select the best ratio between thermal mass and heating system maximal capacity. At the same time it should be mentioned that thermal mass has an important effect on stability of indoor air temperature in case of heat supply break [10].

5. Influence of thermal mass on indoor air parameters stability in winter time

The enclosure of the lightweight construction house has been renovated with a heat insulation layer. The construction of the load-bearing walls consists of 180 mm thick wood beams with ecowool cladding and a 50 mm stone wool heat insulation layer on a framework of 5 cm thick planks on the outside. According to the calculation, the thermal resistance of the wall R = 5.089 (m2K)/W.

Practical measurements have shown that outdoor ambient temperature ranged from -5.2°C to +2.7°C, and the difference between the wall internal surface temperature and the indoor ambient temperature was 0.68 or 3.7%. Indoor relative humidity fluctuated from 35.1% to 42.08%, which is an acceptable range for human comfort. Indoor ambient temperature during the study fluctuated by as much as 5.8 °C, which is a relatively large range, and the fluctuations are frequent, which is dictated by the outdoor ambient temperature and may cause discomfort to the people in the room.

The software Autodesk Simulation CFD 2013 was used to simulate a model corresponding to the characteristics of the building with the experimentally obtained data and in accordance with the previously described selection of the room model (figure 5).

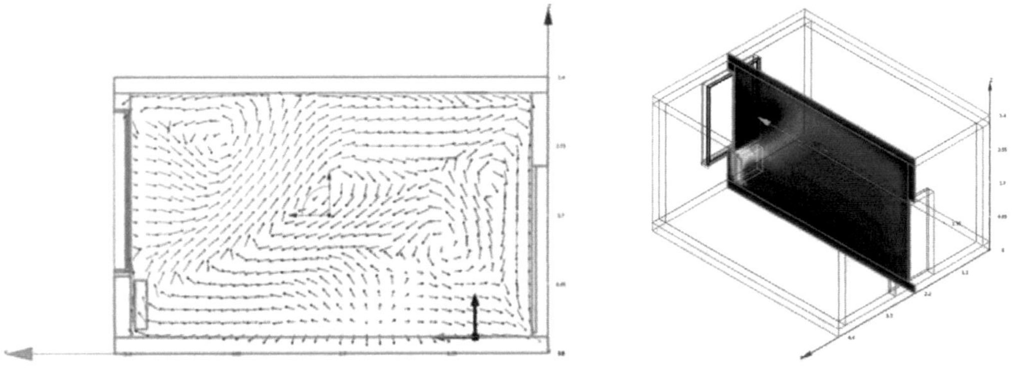

Figure 5. Air flow direction and air temperature distribution in the room.

Air flow in the building is shown with the help of vectors and corresponds to actual indoor air flow exchange.

The enclosure of the massive construction house is cast reinforced concrete with a heat insulation layer. The load-bearing wall of the building consists of 300 mm thick cast reinforced concrete with a 100 mm thick insulation layer of stone wool. The interior lining is a single layer of plaster; on the outside, there is painted plaster on a grid. Three people inhabit the room where the measuring instruments have been placed.

During the practical measurements, outdoor ambient temperature ranged from -10.1 °C to +5.9 °C, and the difference between the wall internal surface temperature and the indoor ambient temperature was 1 or 5.1%. Indoor relative humidity ranged from 31.4% to 51.7%, which is an acceptable range for human comfort.

Indoor ambient temperature during the study fluctuated by 2.9 °C, which is a small range, and the fluctuation period was gradually rising or falling depending on the surface temperature; consequently, a person staying in the room feels comfortable.

As the chart shows, heat flow is variable in the middle of the room, which evidences constant movement of the warm air following the air flow vector lines. However, temperature in the middle of the room in accordance with the model has been modelled precisely and corresponds to the experimental measurement: -19.4°C (figure 6).

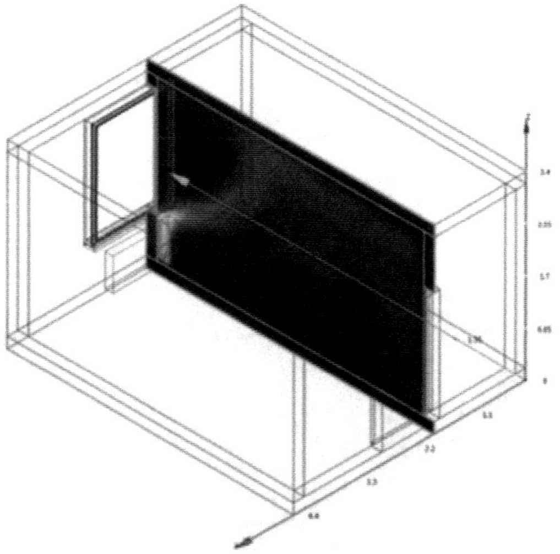

Figure 6. Indoor temperature distribution in 3D format.

6. Optimization of heating system, taking into consideration the solidity of buildings

The thermal stability of a room is the ability to reduce indoor ambient temperature fluctuations upon fluctuations in the heat flow of the heating system. The smaller the fluctuation range of ambient temperature in the room at the same conditions, the more stable it is. Thermal stability is most linked with the solidity of the building enclosure; if the wall is more massive, it maintains heat better. The lower the inside surface temperature changes at the same ambient temperature range, the more stable it is, and vice versa. As the acceptable range of temperature fluctuations is ± 3 °C, using previously experimentally proven data, it can be assumed that this value for premises with lighter enclosures fluctuates within a much larger range than for massive enclosures. Heat transfer rate fluctuations of the heating device are assessed based on the variation coefficient m, which is determined in accordance with the formula [Tabunikov] (3):

$$m = \frac{Q_{max} - Q_{min}}{2Q_z} \tag{3}$$

where: Q – the maximum heat transfer rate of the heating system, Wh; Q – the minimum heat transfer rate.

The value m depends on the heating system and its operation. The coefficient m has a great significance in determining the indoor temperature fluctuation range, i.e. in assessment of the indoor thermal stability.

It was previously believed that if district heating is used, the value of coefficient m is 0.1, but in finite element computational dynamics software, it was proven that the coefficient varies both at different temperatures and depending on whether the external wall is lightweight or massive. The obtained results are presented in table 2.

In order to determine a more accurate value of the variation coefficient, like before, a model was simulated with the effect of external weather conditions.

Table 2. Calculation values of coefficient m at various outdoor ambient temperature values.

t °C / Construction	0°C – (10°C)	(-10°C) – (-15°C)	(-15°C) – (-20.7°C)
Light weight construction	0.06	0.20	0.30
Massive wall construction	0.10	0.23	0.36

7. Conclusion

The indoor temperature fluctuation range for massive constructions is ±4°C, for lightweight constructions: ± 6 °C. The fluctuation range of massive constructions also exceeds the permissible range of ± 3 °C, but only by one degree, which might not be felt in practice, while for lightweight constructions the range is exceeded twofold, which causes discomfort in the premises.

The mathematical model of buildings developed as part of the thesis proved that it is unnecessary for the heating systems of massive construction buildings to use the lowest temperature of 5 days, because when the temperature temporarily falls below -15 °C, the indoor temperature does not decline sharply; the same cannot be said about lightweight construction buildings – their indoor temperature, in accordance with experimental data, fluctuates along with the outdoor temperature.

A detailed model can be used for the processing of experimental data – as proven in this work, by using a simplified, approximated model, it is possible to obtain real values of weather conditions and indoor parameters.

The experimental measurements are dependent on the outdoor air parameters, which are variable and differ each year. Because of that, the obtained results may differ in other time periods, and therefore the final conclusions and results provide only general information, which may slightly vary from year to year.

Thermal stability is an important property of building constructions; if it is impossible in an emergency to quickly restore heat supply, the massive construction walls are able to maintain heat in the building for 8 h or 33% longer that a lightweight construction building.

8. References

[1] Akmens P and Kreslins A 1995 *Buildings Heating and Ventilation* (Riga: Zvaigzne ABC) p 68

[2] Braun J, Montgomery W and Chaturvedi N 2001 Evaluating the performance of building thermal mass control strategies *HVAC&Research* **7** (4) p 403–428

[3] Eicker U, Fux V, Bauer U, Mei L and Infield D 2008 Facades and summer performance of buildings *Energy and Buildings* **40** p 600–611

[4] Ferrari S 2007 Building envelope and heat capacity: re-discovering the thermal mass for winter energy saving *Proc. 2nd PALENC Conf. and 28th AIVC Conf. on Building Low Energy Cooling and Advanced Ventilation Technologies in the 21st Century (Greece, Crete island, September 2007)* p 346–351

[5] Fokin F K 2006 *Building Physics of Building Envelope* (Moscow: ABOK-PRESS) p 256 (in Russian, original publication in 1937, second edition in1953, reprinted in 2006)

[6] Givoni B 1994 Passive and low energy cooling of building, Van Nostr and Reinhold p 258–262

[7] Gratia E and Herde A 2004 Natural ventilation in a double skin facade *Energy and Buildings* **36** pp 137–146

[8] Hens H 2010 Zero- and plus energy buildings, do not forget economics and grid stability *Proc. of the 1st Central European Symposium on Building Physics (Poland, Cracow, 13–15 September 2010)* p Appendix A1 – A6

[9] Heinonen J and Vuolle M 2003 *The Performance Simulations of Double-Skin Facades in Office Building with Hybrid Ventilation Systems* Helsinki University of Technology, Department of Mechanical Engineering, Laboratory of Heating, Ventilating and Air-Conditioning, Annual Report pp19–358

[10] Karbauskaite J, Stankevicius V, Burlingis A and Morkvenas R 2008 The assessment of freezing risk in apartment buildings after heat supply break *Proc. The Nordic Symposium on Building Physics (Denmark, Copenhagen, 16-18 June 2008)* pp 1341–1347

[11] Korpi M, Kalamees T, Vinha J and Kurnitski J 2008 The influence of exterior walls on the level and stability of indoor humidity and temperature in detached houses *Proc. The Nordic Symposium on Building Physics (Denmark, Copenhagen, 16-18 June 2008, 2008)* p 1413–1420

[12] Kosny J, Petrie T, Gawin D, Childs P, Desjarlais A and Christian J 2001 *Energy Benefits of Application of Massive Walls in Residential Buildings*, ASHRAE: Buildings VIII Walls III – Principles p 1–9

[13] Latvian Building Code LBN 003-01 *Building Climatology* Riga 2001

[14] Olsen L 2008 Heat capacity in relation to the Danish building egulaiton *Proc. The Nordic Symposium on Building Physics (Denmark, Copenhagen, 16-18 June 2008)* p 1349–1355

[15] Sokolov V S 1953 Unsteady heat exchange in construction (Moscow: Profizdat) p 333

[16] Vlasov O 1927 Flat temperature waves *Newsletter of Thermal-Technical Institute* **3** (26)

[17] Zeng R, Wang X, Di H, Jiang F and Zhang Y 2011 New concepts and approach for developing energy efficient buildings: Ideal specific heat for building internal thermal mass *Energy and Buildings* **43** pp 1081–90

[18] Sami A, Al-Sanea M F and Zedan S N 2013 Al-Hussain. Effect of masonry material and surface absorptivity on critical thermal mass in insulated building walls *Applied Energy* **102** pp 1063–70

[19] Bruzgevičius P, Burlingis A, Stankevicius V, Pupeikis D, Norvaisiene R and Banionis K 2013 Investigations into thermal capacitance of the building envelope *J. of Sustainable Architecture an Civil Engineering* (Kaunas: Technologija, Kaunas University of Technology) **1** (2) pp 40–49

Energy Efficiency and Sustainability of Different Building Structures in Latvian Climate

A Jakovičs[1,2], S Gendelis[1] and L Bandeniece[1]
[1]University of Latvia, Department of Physics and Mathematics, Zellu Str. 8, Riga, LV-1002, Latvia

E-mail: andris.jakovics@lu.lv

Abstract. Five experimental test buildings have been built in Riga, Latvia. They are identical except external walls for which different mainly regional building materials are used. Calculated *U*-values of the other walls, floor and ceiling are the same for each test building. Initial moisture influences the relative humidity of indoor air, which can be higher in the initial time period; as a result, heat transmittances are also very different and cause different heating/cooling energy consumption. Overheating risk in summer exists for test buildings with the smallest thermal inertia. Both summer and heating seasons have been analysed and differences between five test houses have been discussed in details.

1. Introduction

This article presents selected results of the test building monitoring project. The choice of building material for five different walls of experimental constructions was determined by the objective to find the best possible application of high-quality insulation materials produced from the local resources. The expected outdoor and indoor air parameters and energy consumption monitoring data allow analyzing the buildings in which five different construction materials have been used. Similar studies were performed on a variety of building design solutions and material effect on energy consumption and indoor climate also in other countries, e.g. in Finland, Spain [1, 2].

Five test buildings have been built with different wall materials, identical in terms of design, geographic location and engineering solutions (figure 1). For all types of exterior walls with ventilated facade, the calculated *U*-value is 0.16 $W/(m^2K)$. Constant air exchange rate of 0.45 h^{-1} is provided by the air-air heat pump, which is used also for heating and cooling. More detailed information about project is summarized in [3]. The basic materials used for the walls are as follows (see figure 2):

[2] To whom any correspondence should be addressed.

Figure 1. Experimental buildings.

- perforated ceramic blocks (440 mm) with stone wool insulation outside (type CER);
- aerated concrete blocks (375 mm) with stone wool layer outside (type AER);
- plywood panels with stone wool filling (200 mm) and fibrolite (70 mm) inside (type PLY);
- perforated ceramic blocks (500 mm) filled with insulating granules (type EXP);
- laminated beams (200 mm) with stone wool insulation layer and wood panelling (type LOG).

To provide the same set temperature and air exchange, every building is equipped with an air-air heat pump with outdoor air intake option. All test stands are equipped with the same set of sensors. 40 different sensors include temperature and humidity (T/H) sensors, air velocity flow sensors, solar radiation sensors, energy meters, heat flow sensors. The locations of the main air temperature and humidity sensors are shown in figure 3 as black dots. The data logger collects all sensor data including data from the electric energy meter. To collect meteorological data, a weather station is installed on the top of a test stand with the separate data logger. Details of the developed measuring system can be found in [3].

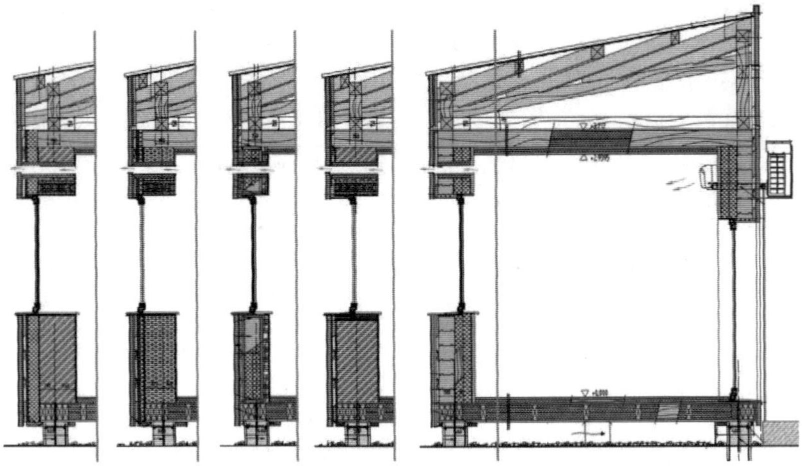

Figure 2. Cross-section of the buildings.

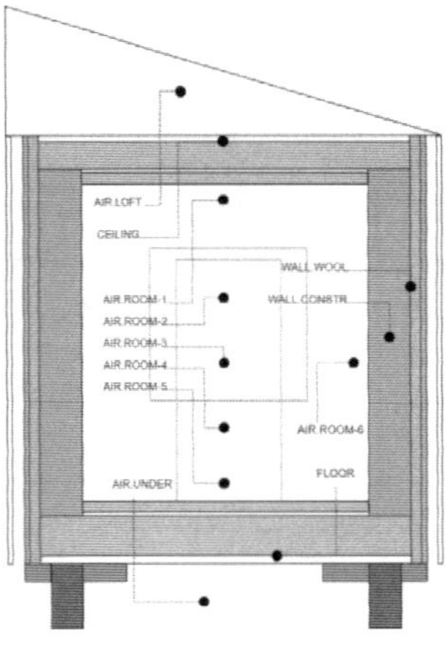

Figure 3. The location of T/H sensors.

2. Air exchange rate measurements

The main aim of the project is to determine and analyse energy consumption for all test buildings, therefore it is very important to evaluate all the heat losses. One of them is convection heat loss through ventilation opening and construction joints, which can be characterized by the air exchange rate in the room. Tracer gas method and special measuring system including multipoint sampler/doser and photo acoustic gas monitor are used for this purpose.

Table 1. Measured air exchange rate *n*.

Test building	*n* (1/h)
LOG	0.45±0.03
LOG[*]	0.03±0.01
EXP	0.48±0.02
AER	0.50±0.03
CER	0.43±0.04
PLY	0.44±0.01

[a] with sealed ventilation opening.

Experimental studies of actual air change were made in all test buildings at least for 24 hours for every building. The obtained results (table 1) show that the actual air exchange rate with switched on ventilation system in all test buildings is within the range of 0.43…0.50 (1/h). An additional measurement was carried out with the switched off ventilation system and sealed ventilation opening; this study shows that air change in this case is very close to zero (see [5]). The general finding of this experiment is that test buildings are very air-tight and the actual air exchange rates *n* with the switched on ventilation system are very close, which means that more than 90% of actual air exchange is a result of mechanical ventilation system operation.

3. *U*-value measurements

The measurements of heat transmittance U (W/m2K) are made for constructions in all test buildings just after manufacturing of the test buildings and during operation period (figure 4). Measurements are carried out using long-term monitoring of heat flux density and temperature difference. The variation in the determined *U*-values is mainly caused by different humidity conditions of a building structure (see next chapter).

The resulting value range is a result of several measurement cycles (up to 5 for each construction). Comparing experimental results with the calculated (designed), it is seen that walls in PLY building, as well floors and ceilings in all buildings are very close. The measured *U*-value for walls in LOG building is lower than the calculated value; this can be explained by low moisture content in timber constructions just after manufacturing, which is slightly increasing. Very high values of heat transmittance obtained for EXP building can be explained by mistakes in the manufacturing process of ceramic blocks, resulting in the highest volume of ceramics and the highest thermal conductivity.

Based on the experimental result obtained, it is estimated that the *U*-values for all the masonry constructions (especially for the aerated concrete) in the first year of operation are very closely linked to the moister content in the structures. In the next years, heat transmittance was decreasing, this means also reduction in conduction heat losses from the buildings. Slight changes in heat transmittance due to variable moisture content in autumn and spring periods have also been observed.

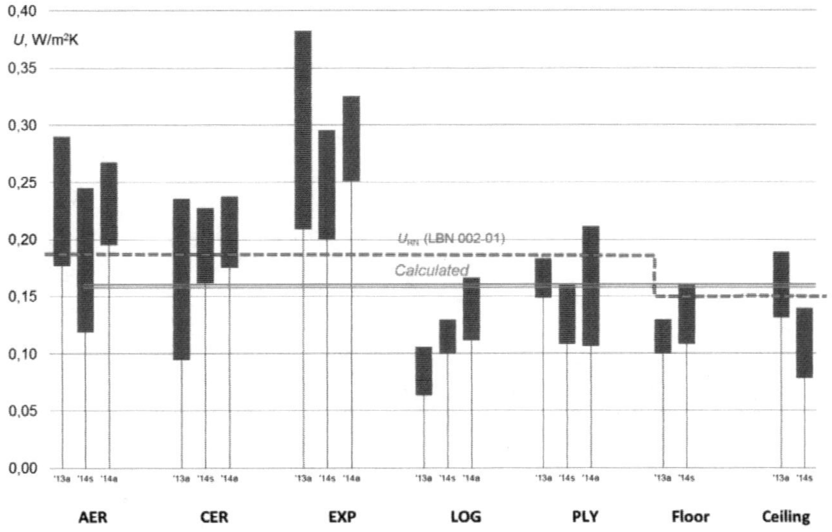

Figure 4. Measured *U*-values for different constructions (*a* – autumn, *s* – spring periods).

4. Air humidity measurements

Temperature and relative air humidity (RH) in different places of the test buildings (in the air as well as in the building structures) are controlled by several sensors (figure 3). Characteristic values of air humidity in all buildings (the data from sensor located in the middle of a room) for all monitored period are displayed in figure 5. The highest relative air humidity values (by 10-15%) are observed for AER type building built from aerated concrete after one year operation under set ventilation conditions. This difference decreases in time and after two heating seasons it becomes very close to other buildings due to relatively slow drying of the construction.

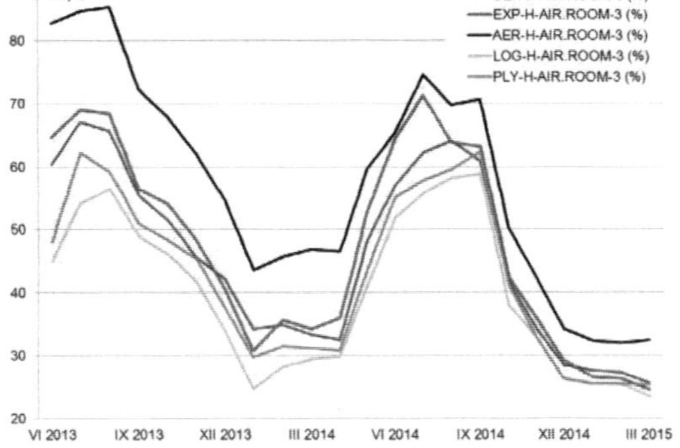

Figure 5. Measured relative air humidity RH (%) in the middle of the room.

As the test buildings are well ventilated, the measured air humidity depends mainly on moisture content in the building structures. However, humidity measurements in the structures or near their surfaces are also very important, e.g. the data from humidity sensors located under the window sill and between mineral wool layer and the main material (figures 6, 7) allows to better understand the drying processes and moisture transport in building constructions [6]. As seen from the graphs, the relative air humidity in AER and EXP buildings is very high and reaches even 100% in autumns under the window sill. The measured relative humidity near mineral wool layer for AER, LOG and PLY buildings (figure 7) is very high throughout the year (more than 65%); seasonal changes are clearly seen for LOG and PLY buildings.

Moisture has a significant negative influence on building structures, not only in terms of increasing of thermal transmittance, but also condensation and mould growing. More information about mathematical modelling and analysis of the condensation risk and mould growth in test buildings is summarised in [6, 7].

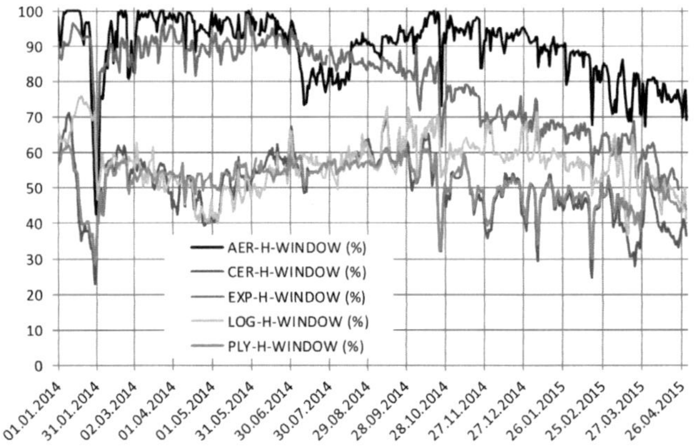

Figure 6. Measured RH (%) under the window sill.

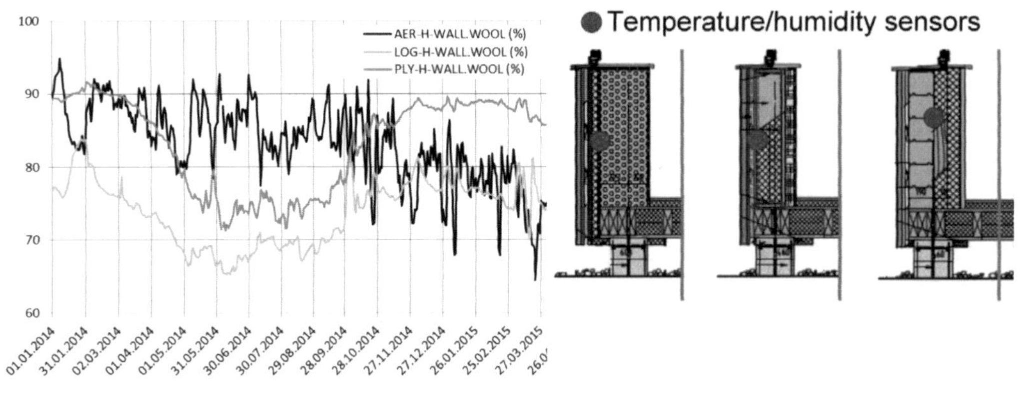

Figure 7. Measured RH (%) near mineral wool layer.

5. Energy consumption measurements

After two years of project running, huge amount of data from all types of sensors has been collected. It is possible to get the most interesting and representative results by analysing the heating and cooling energy consumption for different test buildings.

At the beginning of heating season, couple of months after the test buildings were built, energy consumption in buildings AER and EXP was higher than in another three buildings with practically the same consumption. Graphs in figure 8 show the increase of difference in heating energy consumption for all test buildings during one heating month. The main reason is the increased conduction heat losses through walls in AER building due to higher moisture content. Increasing of U-value for walls also in EXP building has an effect on heating energy consumption, which is 25% higher.

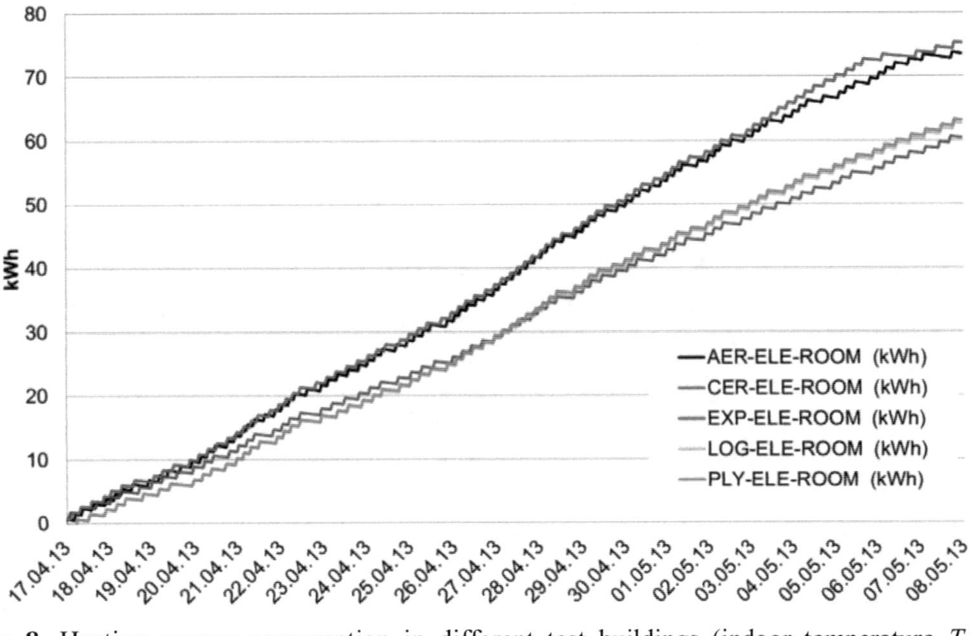

Figure 8. Heating energy consumption in different test buildings (indoor temperature T_{in}=19°C, n=0.45 h^{-1}).

At the same time, cooling energy consumption in hot summer days shows than the buildings which consumed more heating energy (AER and EXP) are very energy efficient. It can be explained by two factors:

- highest thermal transmittance for those buildings (see figure 4), which means more intense heat transmittance from the room;
- drying of relatively humid structures (see figure 6), which requires additional energy for evaporation.

6. Overheating analysis

The role of heat capacity is very important to reduce peaks of temperature fluctuation, especially in summer days when solar radiation influences the indoor temperature very significantly. The effect of rapid increase of indoor temperature during direct solar radiation is also called "overheating"; an example of this process is shown in figure 9, where temperature in the test buildings is not controlled, only ventilation system is on. As it is seen, the maximum temperature inside LOG building is 5°C higher than in AER building.

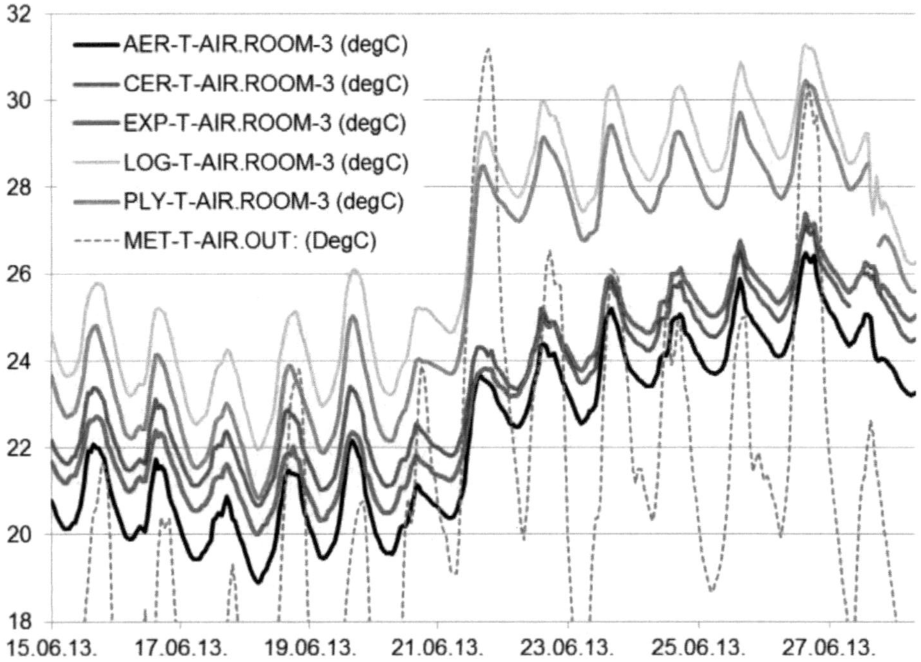

Figure 9. Indoor temperature (°C) demonstrated overheating of light-weight test buildings (LOG and PLY) without cooling.

7. Conclusion

Long-term monitoring of various physical parameters in 5 different low energy test buildings in the Latvian climate shows that:

- The calculated and measured heat transmittance for building structures may vary mainly due to different moisture content, which means also increased heating energy consumption.
- After two years of operation, wet constructions dry out and room's air humidity decreases; it means decreasing in heating energy consumption for the next heating seasons.
- Thermal mass of a building structure is a very important factor, which affects increasing of indoor temperature in the buildings without cooling systems in summer time.

More information about actual research results including thermal comfort analysis, estimation of actual energy efficiency for different heating systems (heat pumps) and indoor air quality may be found in publication [8-11] or at the project's web site *www.eem.lv* (in Latvian).

IOP Conf. Series: Materials Science and Engineering **96** (2015) 012032 doi:10.1088/1757-899X/96/1/012032

8. References

[1] Vinha J 2007 Hygrothermal performance of timber-framed external walls in finnish climatic conditions: A method for determining the sufficient water vapour resistance of the interior lining of a wall [*PhD Thesis*] University of Tampere, Tampere

[2] Cabeza L, Castell A, Medrano M, Martorell I, Pérez G and Fernández I 2010 Experimental study on the performance of insulation materials in Mediterranean construction *Energy and Buildings* **42** (5) 630–636

[3] Dimdiņa I, Jakovičs A, Gendelis S and Kļaviņš J 2013 Testing of energy-efficient building envelope materials in natural conditions *Proc. of World Sustainable Energy Days (Wels, Austria)*

[4] Greitans M, Grunde U, Jakovics A and Gendelis S 2013 Web-based real-time data acquisition system as tool for energy efficiency monitoring *Proc. of 21st Telecommunications Forum (TELFOR) (Belgrade, Serbia)*

[5] Gendelis S, Jakovičs A, Nitijevskis A and Ratnieks J 2013 Comparison of different air tightness and air exchange rate measurements in very small test buildings *Proc. of 34th AIVC Conference: Energy Conservation Technologies for Mitigation and Adaptation in the Built Environment (Athens, Greece)*

[6] Ozolins A, Jakovics A, Ratnieks J and Gendelis S 2013 Numerical modelling of humidity and temperature conditions in buildings with different boundary structures *Proc. of the 11th REHVA World Congress & 8th Int. Conf. on IAQVEC - CLIMA 2013: Energy Efficient, Smart and Healthy Buildings (Prague, Czech Republic)*

[7] Ozolinsh A and Jakovich A 2013 Risk of condensate formation and mould growth in buildings under Latvian climate conditions *Latvian J. of Physics and Technical Sciences* **5** 44–53

[8] Gendelis S, Jakovics A, Dzenis T and Bandeniece L 2015 analysis of thermal comfort conditions and actual energy efficiency for different heating systems in test buildings *Proc. of REHVA Annual Conf.: Advanced HVAC and Natural Gas Technologies (Latvia, Riga)*

[9] Kazjonovs J, Sipkevics A, Jakovics A, Dancigs A, Bajare D and Dancigs L 2014 Performance analysis of air-to-water heat pump in Latvian climate conditions *Environmental and Climate Technologies* **14** (1) 18–22

[10] Apine I, Orola L and Jakovics A 2015 Effect of building envelope materials on indoor air quality in low energy test houses *IJESD* **6** (12) 952–957

[11] Ozolins A and Jakovics A 2014 Thermal comfort in summer in low energy buildings *Proc. of the Proc. of the 10th Nordic Symposium on Building Physics (Lund, Sweden)*

Acknowledgments

This research is done with financial support of ESF, project realized by the University of Latvia, No. 2013/0027/1DP/1.1.1.2.0/13/APIA/VIAA/007.

Moisture and Thermal Conductivity of Lightweight Block Walls

R Joosep[1]
Estonian University of Life Sciences, Institute of Forestry and Rural Engineering, Kreutzwaldi 1, 51014, Tartu, Estonia

E-mail: ragnar@disainprojekt.ee

Abstract. This article examines thermal properties of lightweight block walls and their changes over the course of time. Three different types of lightweight blocks and two types of heat insulation are used in construction. Aeroc aerated concrete blocks are in use, as well as compacted LECA (Lightweight Expanded Clay Aggregate) Fibo blocks made from burned clay and Silbet blocks produced from oil shale ash. Expanded Thermisol EPS60F polystyrene plates and glass wool Isover OL-P plates are used for thermal insulation. The actual and computational values of thermal conductivity and the water draining properties of walls over time are compared in this article. Water draining from glass wool walls is relatively fast. Water-draining can take over a year in polystyrene insulated walls. All four wall constructions can be used as external walls, but care must be taken regarding the moisture content of the blocks during construction (the construction should be handled with care to minimise the moisture in the blocks), especially in polystyrene board-insulated walls.

1. Introduction

Based on Estonian weather conditions (colder period of five days in Tallinn -26.5^{0}C, in Tartu -31.9^{0}C), the recommended maximum thermal transmittance [2] of the external walls of a small house is u = 0.12...0.22 W/m^2 ^{0}K [1]. This value will be lowered at least one time in five years period. In order to meet this standard, it is necessary to insulate walls made of small blocks. The required insulation thickness is calculated based on the maximum allowable thermal transmittance, the heat conduction of stone wall material and thermal conductivity of surface layers. Thermal transmittance calculations are done on the basis of a material's thermal conductivity provided by the manufacturer (λ_{10}, which is the thermal conductivity at +10^{0}C in a laboratory). It is recommended that calculations are in accordance with the building material real thermal conductivity (λ_n). These figures are in accordance with a plate of glass wool (Isover OL-P) λ_{10} = 0.037 and λ_n = 0.041 W/m ^{0}K [9].

---

[1] To whom any correspondence should be addressed.

There are often difficult climatic conditions (rain, cold, etc.) while doing masonry, insulation and finishing work. The aim of this study has been to determine how low temperature and humidity during construction will impact the wall thermal condition throughout the term of the building use. It is generally known that moisture in the wall material reduces directly thermal transmittance. It is necessary to identify possible water condensation and moisture dry-out conditions in walls. The aim of this study has been also to determine whether such walls may be used only with well moisture leading glass wool (moisture resistance factor $\mu = 1.0$) or also in tight polystyrene ($\mu = 60$).

In order to study the technical heat characteristics of a wall built in early winter, it is necessary to measure the wall's real thermal conductivity and relative air humidity, as well as the temperature of a cross-section of different layers of the wall. A wide variety of methods are available for exploring the technical characteristics of a wall. De Gracia et al (2011) [3] built different thermal boundary wall designs and traditional designs of wall cubes (2.4 x 2.4 x 2.4 m) to investigate such properties. They compared the thermal performance of the walls in different cubes and after reaching a stable temperature, they calculated the thermal transmittance of the walls (u figures).

Skujans et al (2007) [5, 6] studied thermal conductivity by measuring the heat of multilayer porous gypsum plate walls with a heat flow measurement plate and the temperature in different layers with thermocouples. Air temperature was also measured, as was the temperature in different layers of the wall. Thermal conductivity was calculated according to the measurements obtained from the wall. The wall was tested using this method in a laboratory and outdoors. The differences of the results were within error limits.

Four different test walls were built in window openings of the laboratory in order to investigate the technical condition of the walls. The laboratory room made one side of the wall and the other side was exposed to the external environment. By simultaneously measuring the wall's thermal physical characteristics (temperature and moisture on the surface of the wall, and in various layers, heat flow through the wall), comparable thermal conductivity data were obtained for four different wall constructions.

A new short time method makes it possible to determine the water vapour diffusion coefficient versus relative humidity function during a single experiment. The basic idea of the method consists in exposing a material specimen to different climate conditions concerning relative humidity and monitoring the moisture level in the specimen, whereas the experiment is carried out under isothermal conditions. The basic difference between the proposed method and all other methods for determination of water vapor diffusion coefficient is that it provides the relative moisture level within the analyzed specimen. This greatly simplifies the data evaluation procedure because the methods of inverse analysis known in heat transfer and liquid moisture transfer problems can be used with only slight modifications. It must be taken into account that this method was tested on one material only [17].

In conclusion, the thermal diffusion is of no importance for building physical applications, leaving vapor pressure as the sole significant transport potential for the diffusion of water vapor in porous materials [18] [19].

Also concrete masonry units (CMU) are in use to build thermal efficient walls. CMUs are available in a variety of configurations. Some are simple and consist of only a single material, while others have interlocking paths of structural and insulating materials. Within this experiment, simple two-core hollow block CMUs (common in the U.S.) and more advanced multicore and interlocked CMUs (common in Europe) were used. As a result, the use of lightweight concretes has improved the thermal performance of the walls more than using complex insulations structures [15].

IOP Conf. Series: Materials Science and Engineering **96** (2015) 012033 doi:10.1088/1757-899X/96/1/012033

Thermal conductivity and elastic module are fundamental properties of the porous building materials. For isotropic materials the relative thermal conductivity or the relative elastic modulus can theoretically be calculated from the other via the cross-property relation as soon as one of them has been measured [16].

2. Materials and methods

2.1. Used Materials

To solve the assigned task (water drying out and U-value change in the usage conditions of outside walls built in autumn and winter) different outer walls were built in place of laboratory windows. Small blocks most widely used in Estonia – Aeroc, Fibo, Silbet – were used to build the walls, whose main technical indicators are presented in table 1. Density in table 1 was taken from the manufacturer data.

Table 1. General properties of lightweight blocks [11, 12, 13].

Block / Property	Aeroc Classic	Fibo F-3	Silbet
Density [kg/m^3]	425	650	500
Thermal conductivity λ_{10} [W/mK]	0.10	0.19	0.12
Thermal conductivity λ_n [W/mK]	0.134	0.24	0.16
Diffusion constant μ	5–10	5–10	6–10

Two types of insulation materials were used to insulate walls. The general properties of these materials are presented in table 2.

Table 2. General properties of thermal insulation materials [9, 10].

Insulation mat. / Property	Thermisol EPS60F	Glass wool ISOVER OL-P
Density [kg/m^3]	65	77
Diffusion constant	60	1
Thermal conductivity λ_{10} [W/mK]	0.0375	0.037
Thermal conductivity in building λ_n [W/mK]	0.039	0.041

Sakret PM Super Diffusion constant is 15/35 and Thermal conductivity λ_{10} is 0.83 W/mK. Different outer walls were built from these materials in place of laboratory windows. The constructions of these walls are in table 4. The thermal conductivity of the walls was calculated with two thermal conductivity numbers, λ_{10} and λ_n.

Thermal conductivity is heat-energy transfer from the hotter body (or body part) to the colder body (body part) as a result of particle interactions. To put it simply – the thermal conductivity of the wall shows how fast the wall cools down at various internal and external temperatures. The lower the number, the better insulation qualities it has and the less we must heat the rooms. The calculated thermal transmittance of the walls are calculated in accordance with EVS-EN ISO 6946:2008 Building components and building elements – Thermal resistance and thermal transmittance – Calculation method [8].

Table 4. Structure of tested walls.

Material	Wall A1	Wall A2	Wall F1	Wall S1
Plaster Sakret [mm]	5+5	5+5	5+5	5+5
Insulation [100 mm]	Isover OL-P	Thermisol EPS60F	Isover OL-P	Isover OL-P
Block 250 mm	Aeroc classic	Aeroc classic	Fibo F3	Silbet
Plaster Sakret [mm]	5	5	5	5
Total thickness of the wall [mm]	365	365	365	365
Thermal transmittance U, by λ_{10} [W/m^{2}K]	0.186	0.187	0.238	0.201
Thermal transmittance U, by λ_n [W/m^{2}K]	0.223	0.217	0.273	0.239

Thermal transmittance (U-value) in test walls was calculated based on the inside and outside temperature and heat flow through the wall.

$$U = \frac{q}{\Delta T}, \text{ W/(m}^2\text{K)} \tag{4}$$

where:

q heat flow rate, W/(m^{2}K);

ΔT the difference of temperature between the indoor and outdoor environments.

2.2. Test methods

The walls were erected at the end of December 2007 according to figure 1 and measurements began in early January 2008.

The walls were erected in the north-east direction and shielded from direct sunlight. Aeroc blocks were transported directly from the factory. Other test blocks were taken from shop, where they were kept in external weather conditions.

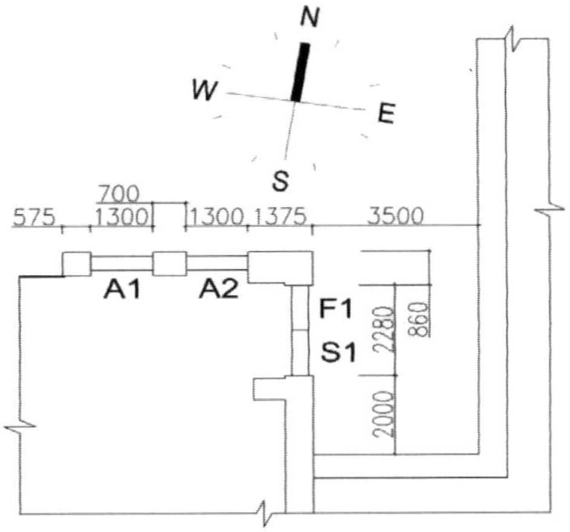

Figure 1. Alignment of walls in window openings of the laboratory.

To measure the walls' thermal physical characteristics, sensors were placed in the walls during the construction. Sensors (temperature and moisture on the surface of the wall, and in various layers, heat flow through the wall) and other equipment were placed according to figure 2.

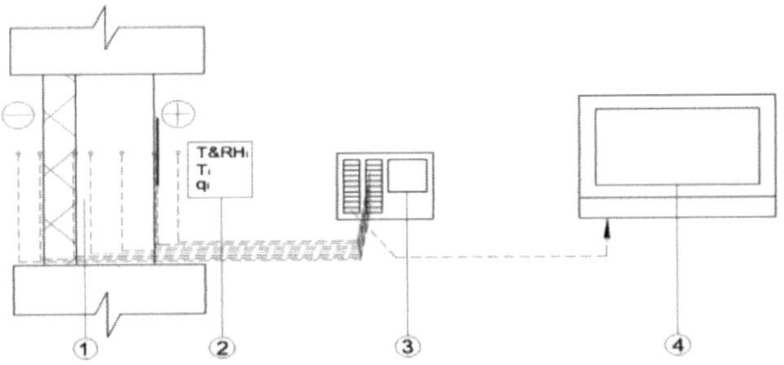

Figure 2. The measurement devices to determine the construction element's heat transfer coefficient: 1) construction element, 2) sensors, 3) data logger, 4) computer with software.

Temperature sensor, relative humidity sensor, heat flow plate and data logger with software were used. The placement of sensors for wall types A1, F1 and S1 are shown in figure 3a and for wall A2 – in fgure 3b.

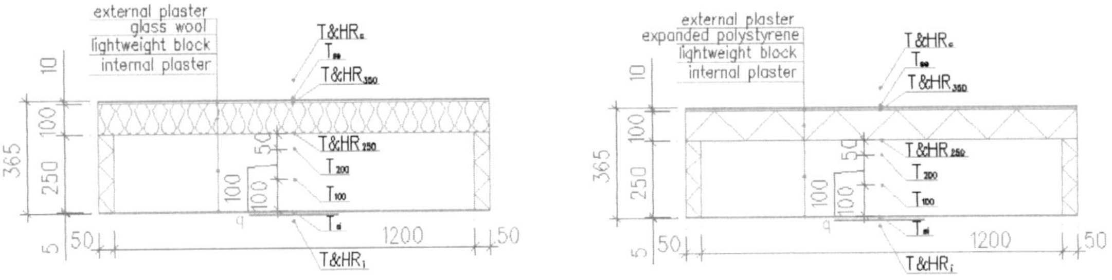

a) b)

Figure 3. Sensors placement schemes for insulated lightweight walls.

a) – wall A1, wall F1 and wall S1 (glass wool); b) – wall A2 (polystyrene), where
$T\&RH_c$ – temperature and relative humidity of outside air;
T_{se} – temperature of external wall surface;
$T\&RH_{350}$ – relative humidity and temperature between external plaster and insulation;
$T\&RH_{250}$ – relative humidity and temperature between insulation and block;
T_{200} – temperature of inside block, 200 mm from inside surface;
T_{100} temperature of inside block, 100 mm from inside surface;
T_{si} – temperature of wall inside surface;
$T\&RH_i$ – temperature and relative humidity of inside air
q – heat flow

The test objective was to study the temperature, the relative air humidity in the wall and in the ambient environment, and to measure the heat flow through the wall. Based on the measurements, the wall's thermal conductivity was calculated, as was its change during wall moisture regime change. In addition, the moisture dry-out dynamics was calculated for the autumn–winter period for walls with glass wool insulation (walls A1, F1, S1) or polystyrene (wall A2).

3. Results and discussion

3.1. Temperature and moisture in wall

Data was recorded in the data logger at 60-minute intervals with the sensors located as shown in figure 3. Measurements were performed for 500 days from 1 January 2008.

Monthly average temperatures were relatively high and the average monthly temperature between thermal insulation layer and masonry did not fall below $+10^0$C. The relative humidity content between the insulation layer and masonry in Aeroc was 100% at the beginning of the test, but in Silbet and Fibo walls 70 to 80%. The difference is explained by the fact that Aeroc blocks still had production moisture inside, but Silbet and Fibo blocks were stored in the laboratory in a 20 degrees for several weeks.

Wall A1

The relative humidity between insulation layer and blocks was 99% more than ~ 32 days after construction.

Outer plaster had more than 99% relative humidity ~ 148 days after construction. During this period, the temperature was below 0 degrees on 26 days (-8°C min.).

Wall A2

Relative humidity between insulation layer and blocks was 99% more than ~ 376 days after construction (the lowest temperature of +11 degrees).

Outer plaster had more than 99% relative humidity ~ 66 days after construction. During this period, the temperature was below 0 degrees on 11 days (-8°C min.).

Wall F1

Amount of relative humidity in the outer plaster was more than 99% after ~ 74 days of construction. During this period, the temperature was below 0 on 7 days (-6°C min.).

Wall S1

The relative humidity between insulation layer and blocks was 80% during the construction of the wall. Due to room heating in wintertime, the humidity started to decrease. The humidity level in the summer was sufficient and started to decrease again during the next heating period.

As the thickness of the insulation in the walls was sufficient, the 0-point of the walls was inside the insulation layer, and therefore minus degrees temperature could not get in the blocks. In the polystyrene board wall (wall 2) the moisture content remained at more than 99% between the insulation and blocks throughout the period of the experiment. However, since there was no extreme frost during the test period, and the thickness of the insulation was sufficient, the temperature between the blocks and insulation did not drop below 0 degrees.

This would suggest that in case of insufficient thickness of insulation, the blocks would have had to endure several freeze cycles.

In our climatic conditions it is necessary to add extra insulation layer outside the block wall. Even when we use multicore or multilayer insulated blocks. To have U value under 0.12 W/m2K means that R-value must be over 8.3 m2K/W.

3.2. Calculation of wall condensation risks

Moisture moves in various ways in the wall. A wall "breathing" cannot be estimated with the diffusion constant of a material [14]. The walls' ability to allow movement of moisture is not important, but the location between each layer of the wall is. The outer wall of a room does not have to easily let out moisture, because it is known that 99% of the moisture must be removed by ventilation.

The outer walls of buildings must be airtight and therefore there cannot be ventilation through the walls [1]. It is important to build walls that prevent water condensation. A common occurrence in calculating condensation is the Glaser method, which is described in EVS-EN ISO 13788:2012[7].

It is necessary to find the diffusive moisture barriers of the layers of the walls:

$$S_d = \mu \cdot s, \text{[m]} \tag{5}$$

where:

μ diffusive barrier constant of material

s thickness of layer [m].

It should be Insured that the wall layers of vapour barriers reduce from the inside out. To compose graphs, partial moisture pressures are necessary in the external and internal environments:

$$P = P_S \cdot \varphi, \text{[Pa]} \tag{6}$$

where:

P_S saturated vapour pressure [Pa]

φ relative air humidity [%]

The abovementioned values will be taken during the condensation period:
Outdoor climate -10^{0}C, 80% relative humidity; indoor +20^{0}C, 50% relative humidity. After measuring the saturation vapour pressure value for each layer found by temperature, we can create the necessary charts and find the amount of condensation and dry-out water that occurs.

It is important in the construction of walls that condensation between wall layers is avoided. It depends on the material's diffusion constant, the thickness of the layer, heat conductivity, temperatures, humidity, and moisture.

All wall types have one condensation level and it is located under the exterior (cold) layer of plaster. The water amount emerging during the condensation period (W_T) and the amounts of water drying out in the drying period (W_V) are shown in table 5 [7].

Table 5. Amount of condensation and dry-out water in walls.

Wall	$W_{T,}$ [kg/m^2]	$W_{V,}$ [kg/m^2]
A1	0.67	2.33
A2	0.20	2.11
F1	0.45	2.23
S1	0.31	2.16

According to standard EVS-EN ISO 13788:2012 [7], the following should be observed:

- the condensate during the condensation period must dry out during the drying period;
- wall construction must not be damaged due to condensate;
- condensate amount must not exceed 2.4 kg/m^2;

- in case of capillary non-absorbent construction materials, condensation is restricted to 0.5 kg/m^2.

In these types of walls all requirements are met and all of them can be used in real constructions.

4. Summary

This study showed that moisture dry-out from wall depends on the thermal insulation type. Under EPS boards placed on relatively wet lightweight blocks, relative humidity remained at 100% for nearly a year, while the walls that were covered with mineral wool dried up faster, and this moisture level persisted for up to two months.

In calculation of a wall's thermal conductivity, a minimum of 5% moisture in lightweight blocks should be considered, and preferably an even greater year-round moisture level.

The diffusion of all observed wall types is similar. The tests showed that there was condensate under the external plaster in all types of walls, but it all dries out in summer. EPS board insulation walls also had condensate under the insulation.

All four wall structures can in principle be used for external walls, but care must be taken in the moisture content of the blocks during construction, especially in polystyrene board-insulated walls. These walls made of blocks should be dried for some time before being covered with insulation and plaster.

5. References

[1] Estonian Minister of Economic Affairs and Infrastructure 03.06.2015 regulation no 55 The building energy performance requirements, §12

[2] Freitas de V P and Barriera E 2012 Heat, air and moisture transfer terminology - parameters and concepts *CIB Int. Council for Research and Innovation in Buildings and Constructions. CIB-W040 Heat and Moisture Transfer in Buildings CIB Publication 369* p 12–44

[3] Gracia de A, Castell A, Medrano M and Cabeza L F 2011 Dynamic thermal performance of alveolar brick construction system *Energy Conversion and Management*

[4] Declaration of performance No 103CPR2013-07-01 Sakret PM Super G-M5 [Available]: www.sakret.ee

[5] Skujans J, Vulans A, Iljins U and Aboltins A 2006 Measurements of heat transfer of multi-layered wall construction with foam gypsum *Applied Thermal Engineering* **27** (7) 1219–24

[6] Skujans J, Iljins U, Ziemelis I, Gross U, Ositis N, Brencis R and Veinbergs A 2010 Experimental research of foam gypsum acoustic absorption and heat flow *Chemical Engin. Transactions* **19** 79–84

[7] EVS-EN ISO 13788:2012 Hydrothermal performance of building components and building elements – Internal surface temperature to avoid critical surface humidity and interstitial condensation – Calculation methods

[8] EVS-EN ISO 6946:2008 Building components and building elements – Thermal resistance and thermal transmittance – Calculation method

[9] ISOVER OL-P EC Certificate of Conformity 0615-CPD-222984G [Available]: www.isover.fi

[10] Sertifikaatin nro 129/00, 30.03.2000 [Available]: www.thermisol.fi

[11] Certificate of Compliance, 1403-CPD-0027 [Available]: www.aeroc.ee

[12] Declaration of performance Fibo3 (DoP-EE-40) [Available]: www.weber.ee

[13] Suvi-Anne S and Ahokas R 2003 *Lämmoöneristys ohjeet* RT RakMK-21217

[14] Iila R 2008 *Kas soojustatud välissein peaks ka hingama?* Keskkonnatehnika 2/08 pp 52–53

IOP Conf. Series: Materials Science and Engineering **96** (2015) 012033 doi:10.1088/1757-899X/96/1/012033

[15] Urban B, Engelmann P, Kossecka E and Kosny J 2011 Arranging insulation for better thermal resistance in concrete and masonry wall systems *9th Nordic Symposium on Building Physics*
[16] Matiašovský P and Bagel L Correlation between thermal conductivity and elastic modulus of porous building materials – power law functions of porosity, *Institute of Construction and Architecture, Slovak Academy of Sciences, Slovakia 9th Nordic Symposium on Building Physics*
[17] Pavlík Z, Žumár J, Pavlíková M, Jerman M and Černý R A transient method for determination of water vapor diffusion coefficient of building materials as function of relative humidity *Czech Technical University in Prague Czech Republic 9th Nordic Symposium on Building Physics*
[18] Janssen H 2011 Thermal diffusion of water vapor in porous materials: true or false? *Department of Civil Engineering, Technical University of Denmark 9th Nordic Symposium on Building Physics*
[19] Janssen H 2011 Thermal diffusion of water vapor in porous materials:fact or fiction? *Int. J. of Heat and Mass Transfer* **54** 1548–62

IOP Conf. Series: Materials Science and Engineering **96** (2015) 012034 doi:10.1088/1757-899X/96/1/012034

Prediction of Moisture Distribution in Closed Ribbed Panel for Roof

A Kukule[1,2], **K Rocens**[1]
[1]Riga Technical University, Department of Structural Engineering, Kalku Str. 1, Riga, LV-1658, Latvia

E-mail: aiva.kukule@rtu.lv

Abstract. Nowadays one of the possibilities to improve energy efficiency is the use of building elements with low air permeability, for example, sandwich panels with steel sheeting. However, these panels have one important disadvantage – a relatively small load-bearing capacity. This can be prevented by reinforcing the panel with antiseptized birch plywood ribs. For wood-based materials prediction of hygrothermal performance is important to avoid rot. Currently the methodology of ISO 13788:2012 is widely used assuming that moisture flux passes through the building envelope of any material. This assumption is not completely accurate with regard to a closed structure where no penetration of ambient humidity is possible. Therefore, in order to predict the distribution of moisture in such structure with the surfaces exposed to different temperatures and to assess the hazards of rot for plywood ribs, a methodology for closed building envelope is presented. To provide insight into expected results according to both methodologies, estimation for individual case with constant environmental conditions is given. According to the methodology for the closed building envelope no free water will occur. Therefore, it is believable that also no rot will be observed. This is contrary to the assessment according to the methodology of ISO 13788:2012, which predicts condensation.

1. Introduction

Nowadays, the when increase of energy resource costs is observed, more pronounced attention is paid to building envelope energy efficiency. To increase energy efficiency, elements with low air permeability are used more and more often. The second trend is the usage of prefabricated structural elements, thus reducing the labor-intensity at the building site. All these requirements are met by sandwich panels. However, these panels used as part of the combined roof structure may have insufficient load-bearing capacity required for covering the span resulting in additional purlins,

[2] To whom any correspondence should be addressed.

increased costs and labor intensity. One possibility to prevent that is reinforcing the panel by placing antiseptized birch plywood ribs into the insulation layer of the panel, which is a cost-effective solution. [1]

Rot is an important disadvantage of plywood and other wood-based materials; it can be caused by the appearance of free water. [2] Therefore, the analysis of moisture movement in the building envelope is required. Nowadays, the methodology of ISO 13788:2012 is widely used for this purpose. In accordance with this methodology, it is assumed that moisture flux is possible through the envelope of any material. This assumption does not reflect the actual situation in a closed structure such as a panel with steel sheeting. According to the methodology of this standard and also to the information presented in [3], the difference of water vapour pressure on both sides of the building envelope is the sole significant moisture transport potential for diffusion of water vapour. Whereas in accordance with [4] at least two moisture transport potentials should be taken into account where one of them is a temperature gradient since it can induce moisture flux directed contrary to the flux under water vapour pressure or moisture content gradients.

Therefore, the model for determination of moisture distribution in the closed building envelope has been developed, in which no moisture penetrates from the surrounding environment. Moisture flux under temperature gradient is also taken into account.

2. Prediction of condensation according to the methodology of ISO 13788:2012

According to the methodology of ISO 13788:2012, no moisture distribution across the depth of the panel is determined, only the probability of condensation is evaluated. The calculations are based on comparison of the partial water vapour pressure and the water vapour saturation pressure. At the interfaces of different materials where the former exceeds the latter, condensation occurs.

A constant one-dimensional moisture flux from the environment with higher partial vapour pressure to the environment with lower partial vapour pressure occurs due vapour diffusion:

$$J = \delta_0 \left(\frac{p_i - p_c}{s_{d,T} - s_{d,c}} - \frac{p_c - p_e}{s_{d,c}} \right) \tag{1}$$

where J – moisture flux (kg/(m^2·s)); $\delta_0 = 2 \cdot 10^{-10}$ (kg/(m·s·Pa) – water vapour permeability of air with respect to partial vapour pressure; $p_i = p_e + \Delta p$ – internal vapour pressure (Pa); $p_e = \varphi_e \cdot p_{sat}$ – external vapour pressure (Pa); Δp – internal vapour pressure excess depending on the internal humidity class (Pa); φ_e – relative external air humidity (-); p_{sat} – water vapour saturation pressure (Pa); $p_c = p_{sat}$ – partial water vapour pressure at condensation interface (Pa); $s_{d,T}$, $s_{d,c}$ – respectively total water vapour diffusion-equivalent air layer thickness of the whole construction and of the layers from the external face to the condensation interface (m); $s_d = \mu \Delta x$, where μ – water vapour resistance factor (-); Δx – material layer thickness (m).

The water vapour saturation pressure p_{sat} at temperature T_C (°C) can be given by:

$$p_{sat} = 610{,}5 e^{\frac{17{,}269 T_C}{237{,}3 + T_C}} \text{ if } T_C \geq 0\,°C \tag{2}$$

$$p_{sat} = 610{,}5 e^{\frac{21{,}875 T_C}{265{,}5 + T_C}} \text{ if } T_C < 0\,°C \text{ [8]} \tag{3}$$

IOP Conf. Series: Materials Science and Engineering **96** (2015) 012034 doi:10.1088/1757-899X/96/1/012034

Accumulated moisture content per area at the condensation interface M_c (kg/m^2) after time Δt (s) can be determined by the following equation:

$$M_c = J \cdot \Delta t \quad (4)$$

3. Methodology for determination of moisture distribution in the closed building envelope

In the developed methodology for prediction of moisture distribution in the closed building envelope such as sandwich panel reinforced by plywood ribs the following assumptions are made:

- no moisture from environment air can penetrate inside the panel (the total moisture content is constant);
- built-in moisture flux under temperature and moisture gradients is one-dimensional;
- initially moisture is evenly distributed over the entire volume of the plywood rib.

Since the concentration of the built-in moisture due the moisture movement across the depth of the plywood rib is time-varying, moisture distribution after time Δt is calculated in several steps. At the start of calculations, the rib is divided into n layers. Within each layer wood characteristics are determined depending on mean moisture M_n (%) content and temperature $T_{K,n}$ (K) in the respective layer. Within one step it is assumed that characteristics of wood in each layer are constant.

Moisture distribution in plywood rib can be predicted by using the following algorithm:

1) Determine characteristics of wood in each layer by equations given below:

- density of wood ρ_n (kg/m^3) if moisture content of wood $M_n \leq 30\%$:

$$\rho_n = 0{,}957\,\rho_{12}\frac{100 + M_n}{100 + 0{,}6M_n} \quad (5)$$

where ρ_{12} – wood density at moisture content $M = 12\%$ (kg/m^3) [5]

- specific gravity of wood G_n (–):

$$G_n = \frac{\rho_n}{\rho_w\left(1 + \frac{M_n}{100}\right)} \text{ [6]} \quad (6)$$

- porosity of wood $v_{a,n}$ (–):

$$v_{a,n} = 1 - G_n(0{,}667 + 0{,}01M_n) \text{ [7]} \quad (7)$$

- thermal conductivity of wood λ_n (W/(m·K)) if $M_n < 40\%$:

$$\lambda_n = 418\left[G_n(5{,}18 + 0{,}096M_n) + 0{,}57v_{a,n}\right]\cdot 10^{-4} \text{ [6]} \quad (8)$$

- thermal resistance of layer R_n (m2·K)/W:

$$R_n = \frac{\Delta x}{\lambda_n} \quad (9)$$

where Δx – layer thickness (m). [8]

2) Determine mean temperature $T_{K,n}$ (K) in each layer:

$$T_{K,n} = 273.15 + \left[T_e + \frac{\sum_{n=1}^{n} R_n - 0.5R_n}{R_T} (T_i - T_e) \right] \quad (10)$$

where T_e – temperature of external air (°C); T_i – temperature of internal air (°C); ΣR_n – total thermal resistance of layers from the external face to the analyzed layer ((m^2·K)/W); R_n – resistance of the analyzed layer ((m^2·K)/W); R_T – total thermal resistance ((m^2·K)/W). [8]

3) Determine bound water diffusion coefficient transverse to the wood fiber direction $D_{T,n}$ (m^2/s) in each layer:

$$D_{T,n} = \frac{0{,}7 \cdot 10^{-5} \cdot \exp\left[-\left(\frac{9200 - 70M_n}{RT_{K,n}} \right) \right]}{(1 - v_{a,n})(1 - \sqrt{v_{a,n}})} \quad (11)$$

where – $R = 2$ cal/mol – universal gas constant.

4) Determine the change of moisture content in each layer. Before that the moisture flux rate shall be known. One dimensional non-isothermal moisture flux in wood (in the hygroscopic range) may be written in the following form:

$$J = -\frac{G\rho_w}{100} \left[D_T \left(\frac{M}{RT_K + 70M} \right) \left(\frac{E_b}{T_K} \right) \frac{\Delta T_K}{\Delta x} + D_T \frac{\Delta M}{\Delta x} \right] \quad (12)$$

where J – moisture flux (kg/(m^2·s)); G – specific gravity of wood (–); ρ_w = 1000 kg/m^3 – normal density of water; D_T – bound water diffusion coefficient transverse to the fiber direction (m^2/s); M – moisture content of wood (%); R = 2 cal/mol – universal gas constant; T_K – temperature (K); $E_b = 9200 - 70M$ – activation energy for bound water diffusion in the transverse direction (cal/mol); x – distance in the direction of moisture flux (m). [9]

If the moisture flux rate is known, the concentration changes Δc (kg/m^3) in time may be characterized by the Fick's second law:

$$\frac{\Delta c}{\Delta t} = -\frac{\Delta J}{\Delta x} \quad (13)$$

where Δt – time (s); ΔJ – difference between incoming and outgoing moisture flux rate in layer (kg/(m^2·s)); Δx – layer thickness (m).

Equation (14) can be derived from equations (12) and (13) to determine the change of moisture concentration in each layer Δc_n (kg/m^3) after time Δt (s):

$$\Delta c_n = \frac{\Delta t}{\Delta x} \cdot \Delta J = \frac{\Delta t}{\Delta x}\left(J_{n-1,n} - J_{n,n+1}\right) =$$
$$= -\frac{G_n \rho_w \Delta t}{100(\Delta x)^2}\left\{\left[D_{T,n-1}\left(\frac{M_{n-1}}{RT_{K,n-1} + 70M_{n-1}}\right)\left(\frac{E_{b,n-1}}{T_{K,n-1}}\right)\left(\frac{T_{K,n-1} - T_{K,n}}{\Delta x}\right) + D_{T,n-1}\left(\frac{M_{n-1} - M_n}{\Delta x}\right)\right]\right.$$
$$\left. - \left[D_{T,n}\left(\frac{M_n}{RT_{K,n} + 70M_n}\right)\left(\frac{E_{b,n}}{T_{K,n}}\right)\left(\frac{T_{K,n} - T_{K,n+1}}{\Delta x}\right) + D_{T,n}\left(\frac{M_n - M_{n+1}}{\Delta x}\right)\right]\right\} \quad (14)$$

where indexes "n", "n-1" un "n+1" – sequence number of the layer.
5) Determine the change of moisture content ΔM_n (%) in each layer:

$$\Delta M_n = \frac{\Delta c_n \cdot 100\%}{G_n \rho_w} \quad (15)$$

6) Determine total moisture content M_n (%) in each layer:

$$M_{n,i} = M_{n,i-1} + \Delta M_n \quad (16)$$

where index "i" – the number of calculation step.
7) Repeat this algorithm in the next step by taking into account the changes of moisture content in each layer.

During calculations it is necessary to control that no naturally impossible situations occur when between two layers with increasing moisture content there is one with decreasing moisture content or vice versa. In such case it would be necessary to repeat the calculation by reduced step duration.

The method described above can be used for determination of moisture distribution if no water freezing can be observed, i.e. at temperatures below zero moisture content does not exceed relative amount of non-freezing moisture in the wood M^* (%). It can be characterized by the minimum temperature in layer $T_{C,n}$ (°C):

$$M^* = 12 + 19.5\exp\left(0.055T_{C,n}\right) \quad (17)$$

This equation is valid for wood in the temperature interval from –50 °C up to 0 °C. The relative amount of non-freezing water at temperature –50 °C is equal to 13.2% and reaches its maximum value 31.5% at 0 °C. [10]

4. Characteristics of the research object

To compare the expected results obtained by both methods (ISO 13788:2012 and the method for closed building envelope), the sandwich panel with polystyrene insulation, steel sheeting and birch plywood ribs is analysed, see figure 1.

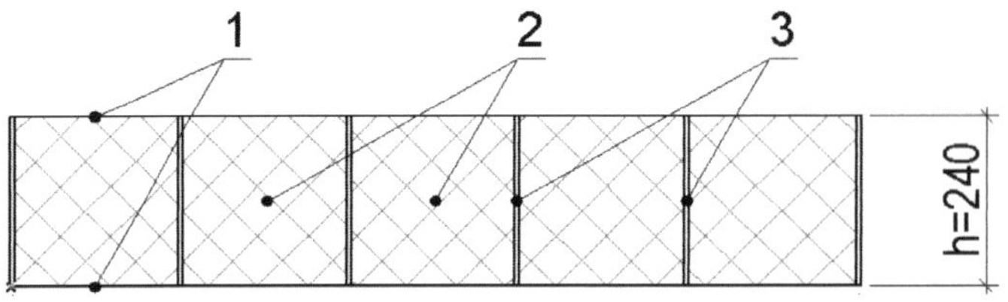

Figure 1. Cross section of ribbed sandwich panel: 1 – steel sheeting; 2 – polystyrene insulation; 3 – birch plywood ribs.

The calculation is carried out according to the methodologies given in sections 2 and 3. Both methodologies are applied for moisture analysis of 1 month (31 days = 2592000 seconds) long period. The assumed external temperature T_e = -7,6 °C is January mean temperature in Aluksne which leads to greatest amount of moisture on the condensate interface if compared with other months. [11] The interior temperature is assumed T_i = + 18 °C.

In case of calculation according to the methodology of ISO 13788:2012, the following input data are assumed: external air humidity $\varphi_e = 0.87$; internal vapour pressure excess $\Delta p = 270$ Pa for internal humidity class I and $\Delta p = 1080$ Pa for internal humidity class IV [8]; density of birch plywood $\rho = 700$ kg/m^3; thermal conductivity $\lambda = 0.174$ W/(m·K). Water vapour resistance factor $\mu = 40$ is taken as for solid wood rather than for plywood because the directions of adhesive layers are parallel to the flux of moisture. For steel sheeting of thickness 0.5 mm this factor is assumed $\mu = 10^6$. [12] During calculations it is assumed that cross section of the panel is divided into layers whose boundaries coincide with the boundaries of the material. Initially it is assumed that partial water vapour pressure increases linearly across the depth of the panel. Since the partial water vapour pressure p_n cannot exceed the saturated water vapour pressure p_{sat}, the straight line p_e–p_i is redrawn as a broken line p_e–p_c–p_i, see figure 2. The condensate will occur at point p_c.

In case of panel analysis as for closed building envelope, plywood ribs made of birchwood with built-in moisture M = 12% and density ρ_{12} = 640 kg/m^3 are assumed. [5] Initially the calculations are done by dividing plywood rib into 10 layers. After repeating of calculations by dividing the panel into 20 layers it was concluded that the results differed by no more than 1.1% and no finer division of plywood rib is necessary. Therefore the thickness of each layer is assumed Δx = 0.012 m. The calculation is carried out in 31 steps. Duration of each step is one day (86,400 seconds). To evaluate the influence of great temperature gradient on moisture distribution, additional case is analyzed when exterior temperature is $T_e = -41$ °C. This temperature conforms to the absolute minimum of air temperature in Daugavpils that is possible once per 50 years and is the lowest temperature in Latvia. [11] These latter conditions may not be used for characterizing average conditions per month but only for demonstrating boundary values and distribution of moisture content in case of very low temperature.

5. Results

5.1. Results obtained according to the methodology of ISO 13788:2012

Results obtained according to the methodology of ISO 13788:2012 are shown in figure 2. The condensate interface is between insulation layer and exterior steel sheeting. After 31 days in case of moisture class I 0.19 g/m^2 of condensate and 1.12 g/m^2 in case of moisture class IV will be stored at condensate interface. In case of moisture class I condensate will occur in the layers of thickness 2.4

cm at the cold side of the panel and 12 cm in case of moisture class IV. As the condensation is predicted, appearance of rot is also not excluded.

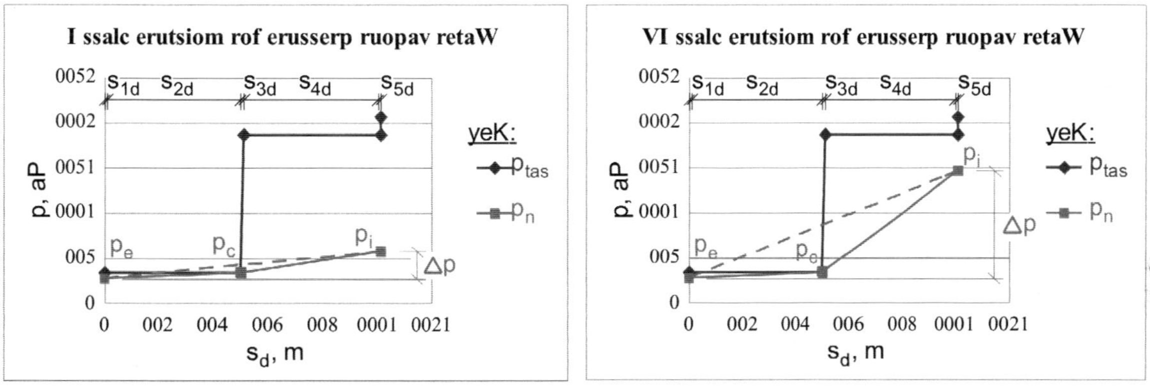

Figure 2. Water vapour pressure in birch plywood rib along tickness of the panel for moisture classes I and IV: $s_{d1} = s_{d5} = 0$ m, $s_{d2} = s_{d4} = 500$ m, $s_{d3} = 9.6$ m – water vapour diffusion-equivalent air layer thickness of external and internal surfaces, of steel sheeting and of plywood rib, respectively; p_{sat} – water vapour saturation pressure; p_e – external vapour pressure; p_i – internal vapour pressure; p_c – partial water vapour pressure at condensation interface; Δp – internal vapour pressure excess depending on the internal humidity class.

5.2. Results obtained according to the methodology for the closed building envelope

Two cases are analyzed – with external temperature T_e = -7.6 °C and T_e = -41.0 °C. According to equation (17) in both cases bound water will not freeze throughout all volume of the plywood rib and its flux will not be held up. During calculations it was found that moisture flux due the temperature gradient was 2 to 10^5 times greater than moisture flux caused by the moisture gradient at the end and at the start of the analyzed time period, respectively.

Moisture distribution in plywood rib is shown in figure 3. Numbering of the layers starts form the cold side of the panel. If T_e = -7.6 °C, maximum moisture content after 31 days will be 12.05% at layer No. 1. If T_e = -41.0 °C, maximum moisture content after the same period will be 12.03% at layer No. 17. Moisture flux due to temperature gradient increases with increasing temperature and vice versa. As a result, decreased wetting of the panel's cold side and increased drying of the panel's warm side can be observed if the case with exterior temperature T_e = -41.0 °C is compared to the case with exterior temperature T_e = -7.6 °C. In both cases the change of moisture content in the middle layers is relatively small if compared with outer layers of the plywood rib. This can be explained by the fact that the middle layers receive and return the moisture to adjacent layers while the outer layers only receive or only return the moisture.

As the moisture content at moisture concentration areas is much less than moisture content at fiber saturation point (M = 25 – 35% [2]), no free water will be observed in plywood ribs. Moreover, it can be believed that moisture concentration will be much less because in natural conditions continuous changes of external conditions will induce the change of moisture flux direction. Therefore, it is reasonable to assume that no rot will occur.

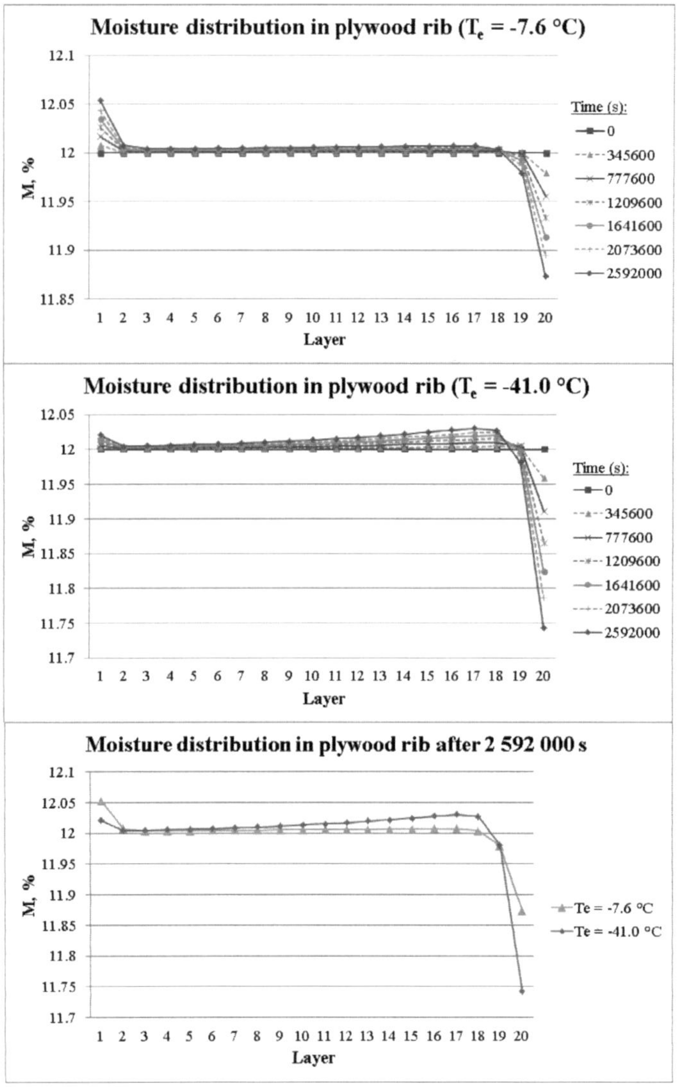

Distribution of moisture content in plywood rib M, %

No.	Te = -7.6 °C	Te = -41.0 °C
1	**12.05***	12.02
2	12.01	12.00
3	12.00	12.00
4	12.00	12.01
5	12.00	12.01
6	12.00	12.01
7	12.00	12.01
8	12.00	12.01
9	12.01	12.01
10	12.01	12.01
11	12.01	12.02
12	12.01	12.02
13	12.01	12.02
14	12.01	12.02
15	12.01	12.02
16	12.01	12.03
17	12.01	**12.03***
18	12.00	12.03
19	11.98	11.98
20	11.87	11.74

Note.
* – maximum moisture content in plywood rib after 31 days or 2,592,000 seconds.

Figure 3. Moisture distribution in plywood rib devided into 20 layers along the depth of the panel.

6. Conclusions

Methodology for prediction of moisture distribution in the closed building envelope is presented. An insight into prospective results obtained according to this methodology is given by analysing a sandwich panel with steel sheeting, polystyrene insulation and birch plywood ribs at constant internal and external conditions. Comparison with the results according to ISO 13788:2012 methodology is also made.

During the calculations according to the methodology for the closed building envelope it was found that in case of sandwich panel temperature gradient may be the main moisture transport potential. For the analyzed cases, moisture flux under temperature gradient was even 10^5 times greater than moisture flux due moisture gradient. Results show that at moisture concentration areas moisture content of a birch plywood rib changes from 12.00% to 12.05%. This is more than two times lower if compared with moisture content at the fiber saturation point. Therefore, no free water will be observed in plywood ribs. Consequently, theoretically it can be assumed that no rot will occur in this case.

Opposite conclusions can be drawn from the results obtained in accordance with the methodology of ISO 13788:2012. According to these results, condensation may occur in the panel. It may create favorable conditions for rot.

7. References

[1] Šliseris J 2013 *Netradicionālu koksnes kompozītu konstrukciju elementu un to aprēķinu metodikas* (Rīga: Rīgas Tehniskā Universitāte) p 165
[2] Adan O C G and Samson R A 2011 *Fundamentals of Mold Growth in Indoor Environments and Strategies for Healthy Living* (Netherland: Wageningen Academic Publishers) p 523
[3] Janssen H 2011 Thermal diffusion of water vapour in porous materials: Fact or fiction? *Int. J. Heat Mass Transfer* **54** 1548–62
[4] Nilsson L O 2013 *Moisture Transport Under a Temperature Gradient–Some old and new Studies* (Sweden: University of Lund) p 14
[5] Ugolev B N 2007 *Drevesinovedehie i Lesnoe Tavarovedenie* (Mosskva: Izdatelstvo Mosskovskogo gosudarstvennogo universiteta lesa) p 351
[6] Siau J F 1984 *Transport Process in Wood* (New York: Springer-Verlag) p 255
[7] Avramidis S, Hatzikiriakos S G and Siau J F 1994 An irreversible thermodynamics model for unsteady-state nonisothermal moisture diffusion in wood *J. Wood Science and Technology* **28** 349–58
[8] EN ISO 13788:2012 Hygrothermal performance of building components and building elements –Internal surface temperature to avoid critical surface humidity and interstitial condensation– Calculation methods (ISO 13788:2012)
[9] Siau J F 1983 A proposed theory for nonisothermal unsteady-state transport for moisture in wood *J. Wood Science and Technology* **17** 75–77
[10] Chudinov B S 1984 *Voda v Drevesini* (Sibir': Nauka) p 272
[11] Latvijas būvnormatīvs LBN 003-15 *Būvklimatoloģija*
[12] Latvijas būvnormatīvs LBN 002-15 *Ēku norobežojošo konstrukciju siltumtehnika*

Acknowledgement. The research leading to these results has received the funding from the Latvian State Research Programme under grant agreement "Innovative materials and smart technologies for environmental safety, IMATEH".

Evaluation of Time of Scour at Guide Banks in Plain Rivers

O Lauva[1], G Klive[1] and B Gjunsburgs[1,2]
[1]Riga Technical University, Water Engineering and Technology Department, Kalku Str. 1, Riga, LV-1658, Latvia

E-mail: gjunsburgs@bf.rtu.lv

Abstract. Analysis of the literature shows that there are no methods or formulas to calculate the equilibrium time of scour near elliptical guide banks. In the formulas used in calculating the equilibrium time of scour at piers or abutments, different parameters are not taken into consideration. The differential equation of the bed sediment movement in clear-water was used and the method for computing the equilibrium time of scour near elliptical guide banks was elaborated. New hydraulic threshold criterion is proposed for the calculation of equilibrium time of scour. Analysis of the proposed method shows the influence of the following parameters on the equilibrium time of scour: contraction rate of the flow, Froude number, bed layering, sediment movement parameters, local flow modification, relative local and critical velocities ratio, and relative depth. Computer modeling results were compared with the equilibrium times of scour which were calculated by the presented method and they were in good agreement.

1. Introduction

Different authors in order to predict the depth of the bridge foundations (calculate how deep the foundation will be in the river bed) use formulas where the equilibrium time of scour is one of the main parameters. Incorrect prediction of the depth of scour and, consequently, the level of the foundation for abutments, piers, guide banks or spur dikes may lead to severe damages of the bridge structures and cause considerable economic and financial losses.

The equilibrium time of scour at bridge piers, abutments and spur dikes was studied, among others, by Melville & Chiew [1], Ballio & Orsi [2], Lauchlan et al. [3], Coleman et al. [4], Gjunsburgs & Neilands [5], Dey & Barbhuiya [6], Grimaldi et al. [7], Cardoso & Fael [8], Gjunsburgs et al. [9], Ghani et al. [10], Mohammadpour et al. [11], Abou-Seida et al. [12], Gjunsburgs et al. [13].

Scour evaluation at clear-water conditions never ceases completely, so threshold criteria have to be found when the scour development in time has reduced to a negligible value. Various threshold criteria proposed in the literature usually assume that the equilibrium has been reached when the depth of

[2] To whom any correspondence should be addressed.

scour evaluation is less than 5% of the pier diameter within a period of 24 hours [1] or less than 5% of the flow depth or abutments length [4], or again less than 5% of the 1/3 of the pier diameter [7]. All these criteria for the equilibrium time of scour reference only on the geometrical size of the bridge structures.

Analysis of the literature shows that today there are no methods or formulas to predict the equilibrium time of scour near elliptical guide banks at clear-water conditions, while in the available formulas for calculating the equilibrium time of scour at piers and abutments some important parameters of the flow and river bed are not taken into consideration.

The aim of this paper is to find the equilibrium time of scour near elliptical guide banks in clear-water conditions by using the differential equation of the bed sediment movement in clear-water. Solution of that equation, which describes scour development in time, allows easily finding the equilibrium time.

New hydraulic criterion is proposed to determine the equilibrium time of scour. It has been found that equilibrium time of scour depends on the following parameters: contraction rate of the flow, Froude number, bed layering, grain size diameter, local flow velocity near structure, ratio of local velocity to critical one, and are changing with the relative depth of scour.

The equilibrium time of scour in tests is calculated by using grain size diameter d_{50}, which is found from the uniform sand granulometric curve. The test results of scour evaluation in time with duration of 7 hours were prolonged till the equilibrium stage of scour. Computer modeling of the equilibrium depth and time of scour by early proposed method by Gjunsburgs et al. [13] was done and compared with the equilibrium time of scour at the elliptical guide bank calculated by the presented method.

2. Experimental setup

The tests were carried out in a flume 3.5 m wide and 21 m long. The tests were carried out under open-flow conditions, while studying the flow distribution between the channel and the floodplain.
The fixed bed tests were performed for different flow contractions and Froude numbers in order to investigate the local velocity and the water level changes in the vicinity of the guide bank and along it [14].

The aim of the sand bed tests was to study the scour process, the changes in the local velocity, the effect of different hydraulic parameters, flow contraction rate, grain size, stratification of the model bed and scour development in time.

The openings of the bridge model were 50, 80, 120, and 200 cm. Flow contraction rate Q/Q_b (where Q is flow discharge and Q_b is discharge in the bridge opening under open-flow conditions) varied respectively from 1.56 to 5.69. The depth of water on the floodplain was 7 cm and 13 cm, while the Froude numbers varied from 0.078 to 0.124. The tests were carried out under clear-water conditions. The sand was placed 1 m up and down of the contraction of the flume. The mean grain size was 0.24 mm and 0.67 mm.

The condition that $Fr_R = Fr_f$ was fulfilled; where Fr_R and Fr_f are the Froude numbers for the plain river and for the flume, respectively. The tests in the flume lasted for 7 hours, the length scale was 50 and the time scale was 7.

The tests were carried out with one floodplain model and one side contraction of the flow. The dimension of the upper part of an elliptical guide bank, particularly the length, was calculated according to the Latishenkov [15] method and was found to be dependent on the flow contraction rate and the main channel width. The length of the lower part of the guide bank was assumed to be half of the upper part.

3. Method

The differential equation of equilibrium for the bed sediment movement in clear-water conditions has the form:

$$\frac{dW}{dt} = Q_s \tag{1}$$

where: W = volume of the scour hole at the elliptical guide bank, which, according to the test results, is equal to $1/5\pi m^2 h_s^3$; t = time; and Q_s = sediment discharge out of the scour hole.

The volume and shape of the scour hole are independent of the contraction rate of the flow [16]. The left-hand part of Eq. (1) can be written as:

$$\frac{dW}{dt} = \frac{3}{5}\pi m^2 h_s^2 \frac{dh_s}{dt} = ah_s^2 \frac{dh_s}{dt} \tag{2}$$

where: h_s = scour depth; m = steepness of the scour hole; a=$3/5\pi m^2$.

Sediment discharge was determined by the Levi [17] formula:

$$Q_s = AB \cdot V_{l\,el}^4 \tag{3}$$

where: $B = mh_s$ describes the width of the scour hole; V_{lel} = local velocity at the elliptical guide bank with a plain bed; and A = a parameter in the Levi [17] formula.

The discharge across the width of a scour hole before and after the scour is determined as follows [15]:

$$Q_f = Q_{sc} \tag{4}$$

where: Q_f = discharge across the width of the scour hole with a plain bed; Q_{sc} = discharge across the scour hole with scour depth h_s.

Now we have:

$$mh_s h_f V_{l\,el} = \left(mh_s h_f + \frac{mh_s}{2} h_s \right) \cdot V_{lt\,el} \tag{5}$$

where: mh_s = width of the scour hole; h_f = water depth in the floodplain; h_s = scour depth; and V_{ltel} = local flow velocity at the elliptical guide bank with a plain bed at scour depth h_s.

From Eq. (5), the local velocity for any depth of scour is:

$$V_{lt_{el}} = V_{l\,el} \left(1 + \frac{h_s}{2h_f} \right)^{-1} \tag{6}$$

The critical velocity at the plain bed V_0 can be determined by the Studenicnikov [18] formula V_0=$3.6d_i^{0.25}h_f^{0.25}$, where: d_i = grain size of the bed materials.

The critical velocity V_{0t} for any depth of scour h_s and for the flow bended by the bridge crossing embankment is:

$$V_{0t} = \beta \cdot 3.6 \cdot d_i^{0.25} \cdot h_f^{0.25} \left(1 + \frac{h_s}{2h_f}\right)^{0.25} \tag{7}$$

where: β = coefficient of critical velocity reduction near the elliptical guide bank, because of flow circulation, determined using the Rozovskyi [19] approach.

At a plain river bed the formula for $A = A_1$ is presented as (Eq. 3)

$$A = \frac{5.62}{\gamma}\left(1 - \frac{\beta V_0}{V_{lel}}\right)\frac{1}{d_i^{0.25} \cdot h_f^{0.25}}, \tag{8}$$

where: γ = specific weight of the sediments.

Parameter A depends on the scour, local velocity V_{lel}, critical velocity V_0 and grain size of the bed material which changes during the floods:

$$A_i = \frac{5.62}{\gamma}\left[1 - \frac{\beta V_0}{V_{lel}}\left(1 + \frac{h_s}{2h_f}\right)^{1.25}\right] \cdot d_i^{-0.25} \cdot h_f^{-0.25}\left(1 + \frac{h_s}{2h_f}\right)^{-0.25} \tag{9}$$

where: $\frac{\beta V_{0t}}{V_{lt\,el}} = \frac{\beta V_0}{V_{lel}}\left(1 + \frac{h_s}{2h_f}\right)^{1.25}$.

Then we replace V_{lel} in Eq. (3) with the local velocity at any depth of scour V_{ltel} from Eq. (6). The parameter A in Eq. (3) is replaced with the parameter A_i from Eq. (9). The sediment discharge upon the development of the scour is:

$$Q_s = A_i \cdot mh_s \cdot V_{lt\,el}^4 = b \cdot h_s \left(1 + \frac{h_s}{2h_f}\right)^{-4} \tag{10}$$

where: $b = A_i m V_{lel}^{\;4}$.

The hydraulic characteristics, such as contraction rate of the flow, the recalculated critical velocity βV_0 and local velocity V_{lel}, grain size in different bed layers, sediment discharge, and the depth, width and volume of the scour hole, varied during the floods.

Taking into account Eq. (2) and Eq. (10), the differential Eq. (1) can be written in the form:

$$ah_s^2 \frac{dh_s}{dt} = b \cdot h_s \left(1 + \frac{h_s}{2h_f}\right)^{-4} \tag{11}$$

After separating the variables and integrating Eq. (11), we have:

$$t = D_i \int_{x_1}^{x_2} h_s \left(1 + \frac{h_s}{2h_f}\right)^4 dh_s \tag{12}$$

$$D_i = \frac{a}{b} = \frac{3}{5} \frac{\pi \cdot m \cdot}{A_i \cdot V_{l\ el}^4} \tag{13}$$

where: $x_1 = 1 + h_{s1} / 2h_f$ and $x_2 = 1 + h_{s2} / 2h_f$ are relative depths of scour.

After integration with new variables $x = 1 + h_s / 2h_f$, $h_s = 2h_f(x-1)$ and $dh_s = 2h_f dx$ we obtain:

$$t = 4D_i h_f^2 (N_i - N_{i-1}) \tag{14}$$

where: $N_i = 1/6x_i^6 - 1/5x_i^5$, $N_{i-1} = 1/6x_{i-1}^6 - 1/5x_{i-1}^5$, and $x = 1 + h_s/2h_f$ are the relative depths of scour. Using Eqs. (9), (13) and (14), which contain the equilibrium depth of scour, it is therefore possible to find the equilibrium time of scour near elliptical guide banks:

$$t_{equil} = 4D_{equil.} .h_f^2 (N_{equil.} - N_1) \tag{15}$$

The sequence to calculate the equilibrium time of scour is the following:
The equilibrium depth of scour at elliptical guide banks is found [14]:

$$h_{equil} = 2h_f \left[\left(\frac{V_{lel}}{\beta V_0}\right)^{0.8} - 1\right] \cdot k_\alpha \cdot k_m \tag{16}$$

where: V_0=3.6$d_i^{0.25} h_f^{0.25}$ = critical velocity at the plain bed; k_α = a coefficient depending on the flow crossing angle; and k_m = a coefficient depending on the side-wall slope of guide banks.
Using value h_{equil}, hydraulic threshold criterion, it is possible to find values A, D, N_{equil} and finally t_{equil}. When local velocity V_{lt} becomes equal to the recalculated critical velocity βV_{0t}, then $A_{equil} = 0$, $D_{equil} = \infty$ and $t_{equil} = \infty$. Criterion to evaluate the threshold is needed to appoint to calculate the equilibrium time of scour.

4. Analysis of the method

To analyze the method, Equations (14) and (15) are transformed to a form that shows clearly that they contain dimensionless parameters and characteristics of the flow and riverbed:

$$N_i = \frac{t_i}{4D_i h_f^2} + N_{i-1} \tag{17}$$

Eq. (17) is transformed in to a more detailed form, expressing the parameters for value D_i:

$$N_i = \frac{2t_i A \varphi^4 g^2}{\pi m} \frac{\Delta h^2}{h_f^2} + N_{i-1} \tag{18}$$

From Eq. (18) value t_i is expressed and it reads as follows:

$$t_i = \frac{(N_i - N_{i-1})(\pi mh_f^2)}{2A\varphi^4 g^2 \Delta h^2} \tag{19}$$

In the general form Eq. (18) can be written as:

$$N_i = \frac{2A_1\varphi^4 g^2}{\pi m}\frac{h^2}{h_f^2}\left(\frac{d}{h_f}\right)^{-0.25}\left\{\frac{P_k}{2}\left[\left(\frac{Q}{Q_b}\right)^2 - 1\right] + \frac{1}{2}P_{kb}\left(\frac{Fr}{i}\right)^{-0.5}\left[\left(\frac{Q}{Q_b}\right)^2 + 1\right] + P_{kb}\right\}^2 \cdot t_i + N_{i-1} \tag{20}$$

From Eq. (20) value t_i is expressed and it reads as follows:

$$t_i = (N_i - N_{i-1})(\pi mh_f^2)\left(\frac{d}{h_f}\right)^{0.25} \cdot \left(\left\{\frac{P_k}{2}\left[\left(\frac{Q}{Q_b}\right)^2 - 1\right] + \frac{1}{2}P_{kb}\left(\frac{Fr}{i}\right)^{-0.5}\left[\left(\frac{Q}{Q_b}\right)^2 + 1\right] + P_{kb}\right\}^2 \cdot 2A_1\varphi^4 g^2 h^2\right)^{-1} \tag{21}$$

where: Q/Q_b – flow contraction rate; $P_k = V_k^2/gh$ – kinetic parameter of the flow in contraction in open-flow conditions; V_k – flow velocity in contraction (under the bridge); $P_{kb} = V^2/gh_f$ – kinetic parameter of the open flow in natural conditions; V – approach flow velocity; Fr/i – ratio of the Froude number to the river slope; h/h_f – relative depth of the flow; h – average depth of the flow in the contracted section; and h_f – water depth in the floodplain.

In the general form, the equilibrium time of scour is a function of the following parameters:

$$t_i = f\left(\frac{Q}{Q_b}; P_k; P_{kb}; \frac{Fr}{i}; \frac{d}{h_f}; \frac{\beta V_0}{V_{l\,el}}; \frac{h}{h_f}; \frac{h_s}{h_f}; N_{i-1}\right) \tag{22}$$

where: Q/Q_b – flow contraction rate; P_k – kinetic parameter of the flow in contraction in open-flow conditions; P_{kb} – kinetic parameter of the open flow; Fr/i – ratio of the Froude number to the river slope; d/h_f – dimensionless grain size; $\beta V_0/V_{lel}$ – ratio of the recalculated critical velocity at which the sediment movement starts to the local velocity; β – coefficient of critical velocity reduction near the elliptical guide bank, because of flow circulation; h/h_f – relative flow depth; h_s/h_f – relative scour depth; N_{i-1} – calculated parameter, which is taking into account depth of scour in previous time step.

5. Results

At the head of the elliptical guide bank, we observe a concentration of streamlines, a sharp drop in water level, and a local increase in velocity. Locally modified flow near the guide banks forms the scour hole. With the scour depth increase, the local velocity is reducing and the critical one is increasing.

The ratio of the recalculated critical velocity to the local one at the head of the elliptical guide bank is accepted as the threshold criterion in the equilibrium time of scour calculation. According to the computer modeling results, the scour stops when the local velocity V_{ltel} becomes equal to the recalculated critical velocity βV_{0t} or the ratio of those velocities becomes equal to 1, and the equilibrium is equal to infinity. The threshold criterion for equilibrium time of scour calculation was checked and accepted equal to:

$$\frac{\beta V_{0t}}{V_{lt\,el}} = \frac{\beta V_0}{V_{lel}} \left(1 + \frac{h_{equil.}}{2h_f}\right)^{1.25} = 0.985222 \tag{23}$$

Analysis of the method presented and the test results confirmed the influence of the contraction rate of the flow, Froude number, bed grain size diameter, relative local and critical velocities ratio, as well as relative depth of scour on the equilibrium time of scour.

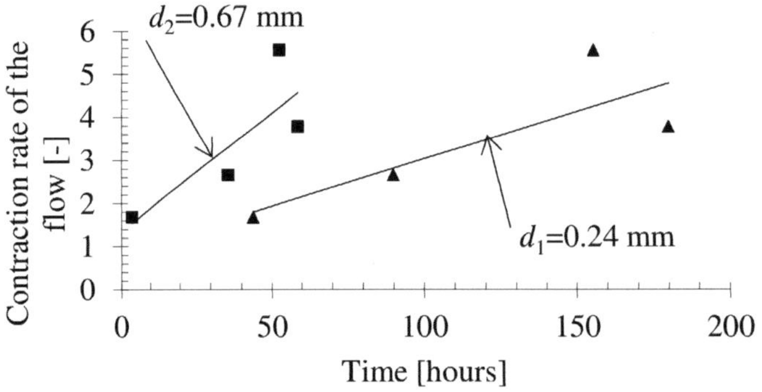

Figure 1. Contraction rate of the flow influence on the equilibrium time of scour. Tests EL3, EL6, EL9, & EL12 with two different sand grain sizes d_1=0.24 mm and d_2=0.67 mm.

Flow contraction creates a series of events that also has an impact on the equilibrium time of scour, thereby if the flow contraction rate Q/Q_b increases, this leads to an increased equilibrium time of scour, consequently, the greater the contraction rate of the flow Q/Q_b value is, the greater the equilibrium time of scour value becomes. Since fine (d_1=0.24 mm) sand particles are more easily scoured away, it takes longer time to achieve the equilibrium stage than in the case with coarse (d_2=0.67 mm) sand.

The ratio of the recalculated critical velocity to the local one $\beta V_0/V_{ltel}$ depends on the contraction rate of the flow Q/Q_b, therefore with an increase of the contraction rate of the flow the equilibrium time of scour is increasing (figure 1).

With an increase in the Froude number *Fr*, there is also an increase in the equilibrium time of scour, the greater the Froude number *Fr* value becomes, the further the scouring process continues in the scour hole, resulting also in an increased equilibrium time of scour. With finer sand the equilibrium time is greater, the scouring process takes longer to achieve the equilibrium stage; with coarser sand, on the other hand, the scouring process ends more quickly, resulting in a lesser equilibrium time of scour value. The greater the grain size diameter of the river bed is, the shorter is the equilibrium time of scour.

Flow velocity is one of the fundamental scouring agents; when the local flow velocity V_{lel} at the elliptical guide bank exceeds the critical value of sediments, the scouring process begins. Since coarser sand particles are heavier, it is more difficult to scour them away, so with an increase of the relative velocity of the flow $V_{ltel}/\beta V_{0t}$ the increase in the equilibrium time of scour is medium, however for the finer sand the increase in the equilibrium time is accelerating with the increase of the relative velocity of the flow $V_{ltel}/\beta V_{0t}$. So the greater the local flow velocity V_{ltel} is and, at the same time, the smaller the recalculated critical flow velocity βV_{0t} for the finer sand is, the greater the equilibrium time of scour will become.

When the scouring process continues in the scour hole, it takes longer time to achieve the time to equilibrium scour. In case of coarser sand (d_2=0.67 mm), the depth of scour is achieved faster, thus resulting in lesser relative depth of scour and at the same time lesser time to equilibrium scour, however, with finer sand (d_1=0.24 mm), on the contrary, it takes more time to achieve the equilibrium scour depth, since also the scour depth increases, consequently increasing the relative depth of scour $h_{equil.}/h_f$. When the relative depth of scour $h_{equil.}/h_f$ becomes greater, the equilibrium time of scour increases as well (figure 2).

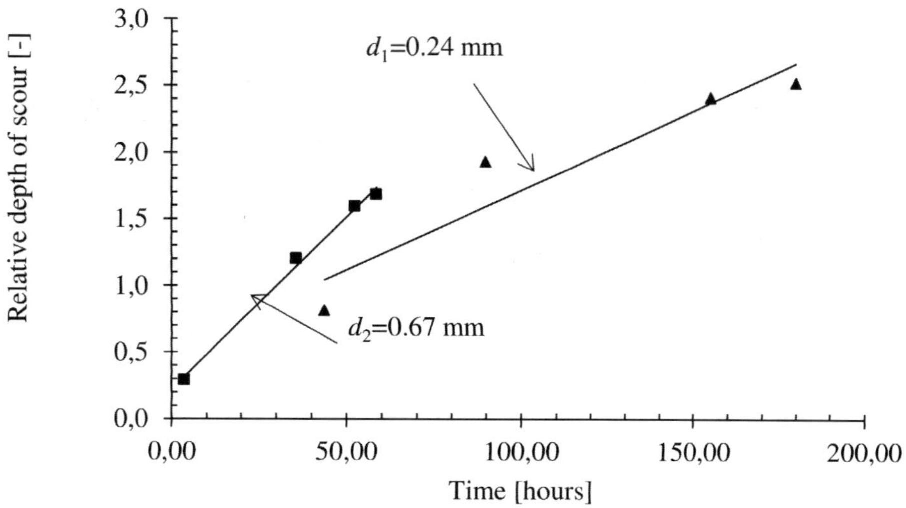

Figure 2. Equilibrium time versus the relative equilibrium depth of scour with sand particle sizes d_1=0.24 mm and d_2=0.67 mm.

Using the threshold criteria (Eq. 23), the equilibrium depth of scour h_{equil} (Eq. 16), A, D, N_{equil} and finally the equilibrium time t_{equil} (Eq. 15) is calculated.

Table 1. Comparison of equilibrium time of scour calculated by computer modeling and by the proposed method for d_{50}=0.24 mm.

TEST	Q/Q_b	D	N_i-$N_{i\text{-}1}$	$t_{comp.}$ [hours]	$t_{form.}$ [hours]	t_c/t_f	□V_{0t}/V_{lt}	Fr
EL1	5.27	104.71	1.89	96.0	93.33	1.03	0.985	0.078
EL4	3.66	166.54	1.37	92.1	107.42	0.86	0.985	0.078
EL7	2.60	450.47	0.20	45.0	42.96	1.05	0.985	0.078
EL10	1.56	957.28	0.04	18.0	18.89	0.95	0.985	0.078
EL2	5.69	52.28	5.46	132.0	134.25	0.98	0.985	0.103
EL5	3.87	47.49	4.09	100.8	91.36	1.10	0.985	0.103
EL8	2.69	130.76	1.55	90.0	95.57	0.94	0.985	0.103
EL11	1.66	619.50	0.10	30.5	29.96	1.02	0.985	0.103
EL3	5.55	40.54	8.14	153.6	155.18	0.99	0.985	0.124
EL6	3.78	39.32	9.72	151.2	179.83	0.84	0.985	0.124
EL9	2.65	55.08	3.47	84.0	89.87	0.93	0.985	0.124
EL12	1.67	467.54	0.20	45.0	43.76	1.03	0.985	0.124

Comparison of some results of the equilibrium time of scour calculated by the method presented and found by the computer modeling for d_{50}=0.24 mm are presented in table 1.
Computer modeling of the scour evaluation in time near elliptical guide banks at clear-water conditions was used [14] to prolong the test results to the equilibrium depth and time of scour.
Comparison of the equilibrium times of scour calculated by the computer modeling and by the Eq. (15) has been made; the results are in good agreement (table 1).

6. Conclusions

The flow pattern at the head of the elliptical guide bank was modified, a concentration of streamlines, a sharp drop in water level, local increase in the velocity, circulation and scour hole were observed. Locally modified flow near the head of the guide bank forms a scour hole.
The equilibrium depth of scour development under steady flow conditions can be reached in equilibrium time.

Analysis of the literature shows that there are no methods or formulas to calculate the equilibrium time of scour near elliptical guide banks.
The differential equation of the bed sediment movement in clear-water was used and the method for computing the equilibrium time of scour near elliptical guide banks was elaborated. The test results in the flume with duration of 7 hours were prolonged till the equilibrium stage by scour evaluation in time calculations [16].

It has been confirmed by the method elaborated that the equilibrium time of scour depends on the contraction rate of the flow, Froude number, grain size diameter, local flow velocity near the elliptical guide bank, and the ratio of local velocity to critical one, and is changing with different relative depth of scour. Dependence of the equilibrium time of scour on some parameters is presented in figures (1-2).

With the scour depth increase, the local velocity is reducing and the critical one is increasing. According to Gjunsburgs et al. [14], the scour evaluation stops when the local velocity V_{ltel} becomes equal to the recalculated critical velocity βV_{0t} and the ratio of those velocities becomes equal to 1. In that case the parameters become equal to $A_{equil} = 0$, $D_{equil} = \infty$ and the equilibrium time goes to infinity $t_{equil} = \infty$.

The equilibrium time of scour in tests is calculated by using grain size diameter d_{50}, which is found from the uniform sand granulometric curve. The equilibrium time for other grain size diameters, for example, d_{10}, d_{16}, d_{84}, d_{90} of the same uniform sand will be very different.

As scour evaluation at clear-water conditions never ceases completely, the threshold criterion is needed to be accepted, when the scour development in time has reduced to a negligible value. The new criterion as $\beta V_{0t}/V_{lt\ el}$ = 0.985222 is checked and accepted for the equilibrium time of scour calculations.

The threshold criteria and the values h_{equil}, A, D, N and finally the equilibrium time of scour t_{equil} can be calculated with Eq. (15).

Computer modeling results were compared with the equilibrium time of scour for grain size d_{50} calculated by the method presented (Eq. 15) and they were in good agreement (table 1).

7. References

[1] Melville B W and Chiew Y M 1999 Time scale for local scour at bridge piers *J. of Hydraulic Engin.* **125** (1) 59-65

[2] Ballio F and Orsi E 2001 Time evolution of scour around bridge abutments *Water Engin. Resources* **2** (4) 243 – 259

[3] Lauchlan C S, Coleman S E and Melville B W 2001 Temporal scour development at bridge abutments *Proc. of the XXIX Congress of the Int. Association of Hydraulics Research (Beijing, 2001)* pp 738 – 745

[4] Coleman S E, Lauchlan C S and Melville B W 2003 Clear water scour development at bridge abutments *J. of Hydraulic Research* **42** (5) 521 – 531

[5] Gjunsburgs B and Neilands R 2005 Local velocity at bridge abutments on plain rivers *Proc. River Flow 2004, in Greco (Napoly, Italy 2004)* ed. Carravetta & Della Morte pp 443-448

[6] Dey S and Barbhuiya A K 2005 Time variation of scour at abutments *J. of Hydraulic Engin.* **131** 11-23

[7] Grimaldi C, Gaudio R, Cardoso A H and Calomino F 2006 Local scouring at bridge piers and abutments: time evolution and equilibrium. *Proc. River Flow 2006 (Lisbon, Portugal 2006)* ed. Ferreira, Alves, Leal & Cordoso pp 1657 – 64

[8] Cardoso A H and Fael C M S 2010 Time to equilibrium scour at vertical wall bridge abutments *Proc. of the ICE-Water Management* **163** (10) 509 – 513

[9] Gjunsburgs B, Jaudzems G and Govsha E 2010 Scour at elliptical guide banks under stratified bed conditions: equilibrium stage. *Proc. of Int. Scientific Conf. of People, Buildings and Environment (Krtiny 2010)* pp 24-30

[10] Ghani A A, Azamathullah H M and Mohammadpour R 2011 Estimating time to equilibrium scour at long abutment by using genetic programming *3rd Int. Conf. on Managing Rivers in the 21st Century: Sustainable Solutions for Global Crisis of Flooding, Pollution and Water Scarcity, Rivers (Penang, Malaysia, 6 – 9 December 2011)* pp 369 – 374

[11] Mohammadpour R, Ghani A A and Azamathullah H M 2011 Estimating time to equilibrium scour at long abutment by using genetic programming *3rd Int. Conf. on Managing Rivers in the 21st Century: Sustainable Solutions for Global Crisis of Flooding, Pollution and Water Scarcity, Rivers (Penang, Malaysia, 6 – 9 December 2011)* pp 369 – 374

[12] Abou-Seida M M, Elsaeed G H, Mostafa T M and Elzahry E F 2012 Local scour at bridge abutments in cohesive soil *J. of Hydraulic Research* **50** 171 – 180
[13] Gjunsburgs B, Lauva O and Neilands R 2014 Equilibrium time of scour near structures in plain rivers *The 9th Int. Conf. Environmental Engin. (Vilnius, Lithuania 22–23 May)*
[14] Gjunsburgs B, Govsha E and Neilands R 2006 Scour development at elliptical guide banks during multiple floods *Harmonizing the demands of art and nature in hydraulics, 32nd Congress of IAHR (CD) (Italy, Venice 1 - 6 July 2006)* pp 1-10
[15] Latishenkov A M 1960 *Questions of artificially contracted flow* (Moscow: Gostroizdat)
[16] Gjunsburgs B, Govsha E and Neilands R 2006 Local Scour at the Elliptical Guide Banks *ICSE-3, 3rd Int. Conf. on Scour and Erosion (Netherland, Amsterdam 1 – 3 November 2006*) pp 120-128
[17] Levi I I 1969 *Dynamics of River Flow* (Moscow: Gidrometeoizdat)
[18] Studenicnikov B 1964 *Scouring capacity of the flow and methods of channel calculation* (Moscow: Stroiizdat)
[19] Rozovskij J L 1957 *Water movement at the bends of the open flow* (Kiev: Academy of Science)

Modelling Flood Losses to Buildings: Relationship between Room Dimensions and Depth of Flooding

M Tuscher[1,2] **and T Hanák**[1]
[1]Brno University of Technology, Faculty of Civil Engineering, Veveří 95, 602 00 Brno, Czech Republic

Email: tuscher.m@fce.vutbr.cz

Abstract. This paper discusses the evaluation of flood losses to buildings using loss curves. The objective of the research was to ascertain whether there is a relationship between the depth of flooding and the amount of unit loss set with regard to the impact of room dimensions. The research was conducted on a sample of 154 model situations for two different room shapes. The results confirm that the depth of flooding, considered in the context of the impact of room dimensions, has an influence over the accuracy of setting the amount of loss, and that the significance of this influence increases with increasing depth. These results can help achieve a more accurate evaluation of flood losses using loss curves or indicators when settling, for example, insurance claims filed in consequence of large-scale flooding.

1. Introduction

Quick and accurate evaluation of flood losses to buildings using loss indicators or loss curves makes it possible to process a large number of insurance claims in a relatively short period of time. It represents an effective method of evaluating the amount of loss to a specific building that allows for a quick payment of the insurance claim and at the same time decreases the costs for the insurance companies. However, the accuracy of this method is questionable, as loss indicators and curves cannot fully replace evaluation using an itemised budget. The accuracy of the outcomes of this method therefore requires special attention.

This paper examines certain factors that have an impact on the accuracy of calculation of flood losses to buildings. More specifically, the objective of this paper is to examine the relationship between the depth of flooding and the dimensions of rooms of various shapes and sizes. By doing so, the paper presents partial outcomes of a long-term research focused on this topic.

---

[2] To whom any correspondence should be addressed.

The paper is structured in the following way: first, research background is introduced, it is followed by research methodology. Afterwards, the obtained results are presented and discussed. The conclusion of the paper summarizes the main results of the research and mentions research limitations as well as future research directions.

2. Research background

The topic of the relationship between the geometry of a building and construction costs has already been examined in the current body of knowledge. For example, Cunningham states that building geometry has a major impact on costs, as it is influenced by the size, shape and complexity of the object [1]. A building project should therefore be drafted so as to be cost-effective from the point of view of the investor, using for example indexes evaluating shape-effectiveness of a building [2]. In this respect, the results of Belniak *et al.* show that a 0.1 change in the value of the VOLM index (Volume – block compactness index evaluating building shape) corresponds to a 6% change in constructions costs [3].

It can therefore be assumed that building geometry will also have an impact on the amount of future losses, whether caused by a natural disaster or another factor; this paper examines only flood losses. It is necessary to mention that apart from depth of flooding, losses caused by floods are also influenced by a number of other variables, such as the duration of flood, bear load sub-grade [4], velocity [5], material and structural characteristics of the building and the stability of structures in case of multiple floods [6].

As it was proved during the preceding stage of the research [7], *room dimensions have an impact on the amount of unit loss in relation both to the size of the room and to its shape*; moreover, *the combination of these two factors may in extreme cases have a fundamental impact on the amount of unit loss*. In this respect, it seems desirable to examine the extent to which these relationships have any impact during the modelling of flood losses for different depths of flooding.

3. Research methodology

The preceding stage of research defined building categories [8] in accordance with the use of the buildings (A – houses, B – housing units, C – common and basement areas of apartment buildings; due to structural and material characteristics of the solution, this paper examines only categories A and B), room size groups (altogether 14 groups) [7] and a side ratio (*SR*) indicator [7] defined by the following relationship:

$$SR = \frac{l}{w} \tag{1}$$

with variables l and w representing room dimensions (l = length, w = width). This paper compares the outcomes for 2 *SR* values: $SR = 1$ (square-shaped room), $SR = 0.1$ (rectangle-shaped room). The core parameters of the examined sample are listed in table 1.

Table 1. Parameters of the examined set.

Room size group	Area [m^2]	Side dimensions l, w with side ratio 1.0 [m]	Side dimensions l, w with side ratio 0.1 [m]	
		$l = w$	l	w
RG1	4	2.000	0.632	6.325
RG2	6	2.449	0.775	7.746
RG3	8	2.828	0.894	8.944
RG4	10	3.162	1.000	10.000
RG5	12	3.464	1.095	10.954
RG6	14	3.742	1.183	11.832
RG7	16	4.000	1.265	12.649
RG8	18	4.243	1.342	13.416
RG9	20	4.472	1.414	14.142
RG10	22	4.690	1.483	14.832
RG11	24	4.899	1.549	15.492
RG12	26	5.099	1.612	16.125
RG13	28	5.292	1.673	16.733
RG14	30	5.477	1.732	17.321

For $SR = 1$, it holds that $l = w$; for $SR = 0.1$, it holds that $l/w = 0.1$. The following structural and material specifications were set for the examined rooms: vertical structures consist of a masonry wall with plaster and paint, horizontal structures consist of a concrete panel and laminate composite flooring. These specifications are identical to the ones used in the previous research [7].

For the purposes of modelling damage to structures, the parameters of the flood also had to be specified. At this stage of research, the depths of flooding used in the models ranged between 0.00 m and 2.50 m. Other flood factors (duration of flooding, flow velocity, etc.) were not taken into account at this stage. The depth of flooding was recorded for each 250 mm increase in depth; overall, the analysis assessed 154 scenarios for 11 different depths of flooding for a specific SR value. The scale of damage to structures can be specified in this context. In the case of floors, the structures in question are the tread layers and baseboards; in the case of walls, they are plasters and paints. Damage to horizontal structures remains constant for various depths of flooding, while damage to vertical structures increases in proportion to the depth of flooding. This stage of research disregards damage to door wings and frames. For the purposes of loss modelling, depth of flooding of 0 m entailed only the damage to horizontal structures, with no damage to vertical structures (disregarding any possible damage to plaster by capillary action).

Afterwards, itemised budgets corresponding to the estimated level of damage were developed for the individual scenarios to set the costs needed for the repair of the damaged structures. For the flooding of masonry walls (vertical areas), the budget includes costs of work related to the removal of plaster, high-pressure cleaning and disinfection of the masonry, manual plastering using lime stucco plaster and final surface treatment using two layers of antimicrobial paint. Flooding of floors (horizontal areas) was modelled specifically for laminate composite flooring. The work included in the budget consists of mounting and removing laminate floating flooring, baseboards and bottom counterbalance layer, just as in [7]. Calculation of costs related to transport of rubble and material on the construction site was added to the work done on both vertical and horizontal structure. Afterwards, unit loss for 1 m^2 of room floor area was calculated from the resulting amount of total loss.

The price of work and materials was calculated using the KROS plus budgeting programme with 17.00 price database [9]. The materials used are priced at the standard level. All prices for work and material are listed without VAT (exchange rate CZK 1 = EUR 0.0365 as of 6 May 2015). In the next step, loss curves for the individual room size groups were created in Excel programme for *SR* = 1 and *SR* = 0.1.

4. Results and discussion

Processing 154 scenarios for the defined *SR* (altogether 308 model situations) resulted in a large data file with information about unit losses per 1 m^2 of floor area. Thanks to a sufficient amount of data, it was possible to represent the data graphically in the form of lines. Altogether, there are 14 lines with each line representing one room size group. Figure 1 shows the amount of unit loss per 1 m^2 in relation to changing depth of flooding for *SR = 1* (square room). Figure 2 shows the same for *SR = 0.1*.

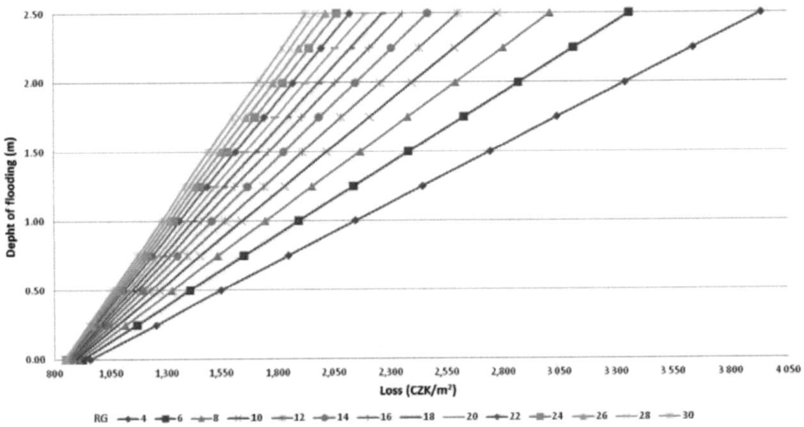

Figure 1. Loss curves for *SR* = 1.

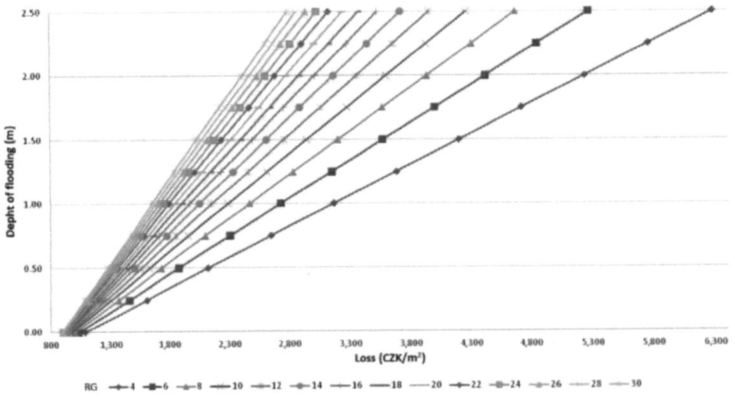

Figure 2. Loss curves for *SR* = 0.1.

Both charts clearly show that the loss per 1 m^2 of floor area increases linearly with the increasing depth of flooding of the room. In general, it holds that the larger the area of the room, the lower the unit loss and vice versa. This dependence arises out of the changing proportion of the amount of loss resulting from damage to horizontal and vertical structures in the total loss. The closer *RS* is to 1, the lower the unit loss and vice versa – the larger the difference between the length and the width of the room, the higher the value of the unit loss. This dependence is valid regardless of the size of the room. The reason is a longer room circumference related to the lower side ratio value; the primary consequence of this is an increase in the areal extent of damage to vertical structures (plaster, paint). These connections were proved at the previous stage of research and are confirmed at this stage [7].

It is interesting to note that an increase in the depth of flooding corresponds with a more profound difference between unit losses, for example, when comparing different *RG* or *RS*. This is clearly presented in table 2.

Table 2. Difference in unit loss per m^2: RS/RG/depth of flooding.

Room size group	Room area [m^2]	Side ratio	Depth [m]	Unit loss [CZK/m^2]	Comparison base	Increase in unit loss value
RG1	4	1.0	0.0	954.879	Base 1	100
RG1	4	1.0	2.5	3938.731	Base 2	100
RG1	4	0.1	0.0	1084.532	Base 1	114
RG1	4	0.1	2.5	6274.205	Base 2	159
RG14	30	1.0	0.0	843.536	Base 3	100
RG14	30	1.0	2.5	1933.085	Base 4	100
RG14	30	0.1	0.0	890.879	Base 3	106
RG14	30	0.1	2.5	2785.880	Base 4	144

Table 2 shows that the significance of the relative difference of unit loss per m^2 for different *RS* (1 m and 0.1 m) within one room size group increases with increasing depth of flooding. Specifically, the amount of unit loss for *RG1* and depth of flooding of 0 m is 14% higher for *RS* = 0.1 m than for *RS* = 1 m. At the same time, the amount of unit loss for the depth of flooding of 2.5 m is 59% higher for *RS* = 0.1 m than for *RS* = 1 m. This dependence can also be seen in the case of values related to *RG14* as well as for situations when *RS* is constant and *RG* is changing.

It can therefore be concluded that the impact of room dimensions on the amount of unit loss increases in significance with increasing depth of flooding. This is caused by the changing ratio of damage to vertical and horizontal structures.

Future analyses should take into account some of the other factors that have an impact on the amount of flood losses to buildings. Flow velocity might be a suitable choice in this respect, as reported by [10]. Similarly, the authors in [11] confirmed on the case of a 2-storey masonry building with basement that high water flow velocity increases the water damage ratio.

5. Conclusion

This paper focused on the modelling of flood losses to buildings using loss curves. Specifically, its objective was to ascertain whether there is a relationship between the depth of flooding and the amount of unit loss set with regard to the impact of room dimensions. The results proved that the impact of room dimensions on the ability to evaluate the loss with the required accuracy becomes more significant with increasing depth of flooding. In other words, the higher the depth of flooding, the greater the inaccuracy in the setting of the unit loss, unless room dimensions are taken into account during the evaluation process.

There are three main limitations to this study that need to me mentioned. First, the research does not take into account other parameters that may influence the amount of unit loss (such as flow velocity or duration of flooding). Second, only the floor and vertical structures of the model rooms were taken into account; doors and windows were not included in the model. Third, the relationships described in this paper are valid for individual (separate) rooms, but not for whole buildings. Therefore, future research should be directed not only to accounting for doors and windows, but also to modelling various material specifications so as to assess the significance of the examined factors on the level of the whole building.

More detailed future research into the use of loss curves and indicators could contribute to a more accurate choice of parameters for loss assessment, which is not only used for settling insurance claims but could also be helpful for the prediction of potential future losses and adjustment of loss rates.

6. References

[1] Cunningham T 2013 *Factors Affecting the Cost of Building Work - An Overview* Dublin, Institute of Technology

[2] Belniak S, Leśniak A, Plebankiewicz E and Zima K 2013 The influence of the building shape on the costs of its construction *J. of Financial Manag. of Property and Constr.* **18** (1) 90–102

[3] Zima K and Plebankiewicz E 2012 Analysis of the building shape erected in Krakow and its impact on construction costs *Organization, Technology and Management in Construction - an Int. J.* **4** (1) 411–419

[4] Korytárová J and Hromádka V 2010 Assessment of the flood damages on the real estate property in the Czech Republic area *Agricultural Economics* **56** (7) 317–324

[5] Kreibich H, Piroth K, Seifert I, Maiwald H, Kunert U, Schwarz J, Merz B and Thieken A H 2009 Is flow velocity a significant parameter in flood damage modelling? *Natural Hazards and Earth System Sciences* **9** (5) 1679–92

[6] Gjunsburgs B, Govša J and Neilands R 2008 Assessment of flood damage risk for engineering structures in river floodplains *Global Changes and Water Resources: Confronting the Expanding and Diversifying Pressures: 13th World Water Congress pp* 1-10

[7] Tuscher M and Hanák T 2015 Evaluation of Flood Losses to Buildings: Effect of Room Dimensions *Periodica Polytechnica Social and Management Sciences* Online published: 02-06-2015

[8] Tuscher M and Hanák T 2014 Problematika oceňování povodňových škod z pohledu pojišťovny s využitím škodního ukazatele *Juniorstav 2014* Brno pp 1–5

[9] ÚRS Praha. Kalkulační a rozpočtovací program KROS plus 2014 (Calculation and Budgeting Software KROS plus 2014)

[10] Pistrika A K and Jonkman S N 2010 Damage to residential buildings due to flooding of New Orleans after hurricane Katrina *Natural Hazards* **54** (2) 413–434

[11] Custer R and Nishijima K 2015 Flood vulnerability assessment of residential buildings by explicit damage process modelling *Natural Hazards* **78** (1) 461–496

Acknowledgments

The project presented in this paper is supported by research grant FAST-J-15-2706 entitled "Refining the Calculation Methodology Used to Determine Damage to Buildings Caused by Floods".

IOP Conf. Series: Materials Science and Engineering **96** (2015) 012037 doi:10.1088/1757-899X/96/1/012037

Development of the One Centimeter Accuracy Geoid Model of Latvia for GNSS Measurements

J Balodis[1,4], G Silabriedis[1], D Haritonova[1,2], M Kaļinka[1,2], I Janpaule[1], K Morozova[1], I Jumāre[1], I Mitrofanovs[1,2], J Zvirgzds[1,3], J Kaminskis[1,2] and I Liepiņš[1,3]

[1] Institute of Geodesy and Geoinformatics, University of Latvia, Riga, Latvia

[2] Riga Technical University, Faculty of Civil Engineering, Department of Geomatics, Kalku Str. 1, Riga, LV-1658, Latvia

[3] Latvian Geospatial Information Agency, Riga, Latvia

E-mail: janis.balodis@lu.lv

Abstract. There is an urgent necessity for a highly accurate and reliable geoid model to enable prompt determination of normal height with the use of GNSS coordinate determination due to the high precision requirements in geodesy, building and high precision road construction development. Additionally, the Latvian height system is in the process of transition from BAS-77 (Baltic Height System) to EVRS2007 system. The accuracy of the geoid model must approach the precision of about ~1 cm looking forward to the Baltic Rail and other big projects. The use of all the available and verified data sources is planned, including the use of enlarged set of GNSS/levelling data, gravimetric measurement data and, additionally, the vertical deflection measurements over the territory of Latvia. The work is going ahead stepwise. Just the issue of GNSS reference network stability is discussed. In order to achieve the ~1 cm precision geoid, it is required to have a homogeneous high precision GNSS network as a basis for ellipsoidal height determination for GNSS/levelling points. Both the LatPos and EUPOS® – Riga network have been examined in this article.

1. Introduction

A highly accurate and reliable geoid model is needed to enable a prompt determination of normal height with the use of GNSS coordinate determination due to the high precision requirements in geodesy, building and high precision road construction development and other industries [1, 2]. Several Latvian geoid models have been developed [3-6], however, maximum precision and reliability should still be achieved in order to strive towards the compliance with the requirements of the

[4] To whom any correspondence should be addressed.

European Vertical Reference System (EVRS). Currently, the Latvian height system is in the process of transition from BAS-77 (Baltic Height System) to EVRS2007 system. The accuracy of the geoid model must be of about ~1cm. Various software [7, 8] and verified data sources are planned to be used including the enlarged set of GNSS/levelling data, gravimetric measurement data and, additionally, the vertical deflection measurements over the territory of Latvia [9]. However, in order to achieve the ~1 cm precision geoid, it is required to have a homogeneous high precision GNSS network as a basis for ellipsoidal height determination for GNSS/levelling points.

Several DGNSS measuring campaigns have been carried out with an aim to determine the ETRS89 coordinates of the Latvian geodetic reference network. Since 1992 the Nordic countries and Nordic Geodetic Commission (NKG) helped all three Baltic Countries. Latvian geodesists really appreciated this aid at the stage when the country still had neither equipment, nor software or experience.

However, the situation changed in 2006 when Latvia succeeded in developing two RTK reference station networks - LatPos and EUPOS® – Riga. The LatPos RTK reference station coordinates were determined by the State Land Service (old) and by the Latvian Geospatial Information Agency (LGIA new) by using four "0 order" reference stations of the fundamental LKS92 network which coordinate values were obtained by NKG. The EUPOS® – Riga RTK network coordinates were determined by using the LatPos network as a reference in order to be included in National LKS92 reference network.

Both network service providers are members of the EUPOS® initiative, which was launched in 2002 by the Central and Eastern Europe National Mapping and Cadastre Agencies under the guidance of German geodesists. The ECC is hosted by the FÖMI Satellite Geodetic Observatory, Hungary, with an aim to perform the quality analysis of National RTK networks of EUPOS® countries. The Latvian data for ECC has been computed at the Institute of Geodesy and Geoinformatics (University of Latvia). The IGS (International GNSS Service) data reduction methods have been applied and the IGS and EPN (European Permanent Network) reference data have been used. The obtained results have been regularly sent to ECC for quality control and the same results have been used to analyse the time series of LatPos and EUPOS® – Riga RTK reference station coordinates. All together the reference station coordinates have been calculated for the time span of more than 8 years. The IGS/EPN reference station coordinates from the surrounding countries have been used for GNSS observation reduction. The Bernese software v. 5.0 and recently 5.2 has been applied. Actually, the EPN network densification is performed – it has been concluded by EUREF (European Reference Network) officials at the recent EUPOS® meeting in Warsaw (2015).

In spite of the fact that LKS92 is accepted as an ETRS89 National realization, there is a difference in LKS92 and ETRS89 coordinate values. As a consequence, different sets of coordinates have been recognized for LatPos and EUPOS® – Riga RTK networks – LKS92 computed by LGIA and ETRS89 computed by the Institute of Geodesy and Geoinformatics (GGI).

2. Past of LKS92

The first GPS measuring campaign was carried out in 1992. There were no any GPS receivers in Latvia at this time and Latvian surveyors were facing great difficulties in connection with the coming land reform. The Latvian coordinate information in the Soviet times was top secret and it was governed by the Soviet army officials. Just local reference systems for towns and cities were used by civil engineering under the strong control of GUGK (the main governance office for geodesy and cartography). However, no reference network information was available for citizens and services throughout the country.

IOP Conf. Series: Materials Science and Engineering **96** (2015) 012037 doi:10.1088/1757-899X/96/1/012037

Table 1. Comparison of LatPos old and LatPos new coordinates.

#	Dome	LatPos-old (LGIA)			LatPos-new (LGIA)			Differences (old-new)		
		Northing	Easting	Up	Northing	Easting	Up	N	E	Up
1	AIZK	273457.120	576759.481	117.521						
2	ALUK	368767.725	684444.147	238.264	368767.729	684444.151	238.275	-0.004	-0.004	-0.011
3	ANNI	311454.995	500600.707	80.485						
4	BALV	336401.495	697627.256	145.656	336401.489	697627.262	145.667	0.006	-0.006	-0.011
5	BAUS	251271.224	511487.303	69.364	251271.229	511487.312	69.395	-0.005	-0.009	-0.031
6	DAGD	222580.559	719741.434	223.093	222580.586	719741.425	223.110	-0.027	0.009	-0.017
7	DAU1	194384.031	657579.771	130.222	194384.059	657579.760	130.229	-0.028	0.011	-0.007
8	DAUG	194449.908	657323.506	132.876						
9	DOB1	277175.287	456046.659	92.322	277175.283	456046.668	92.303	0.004	-0.009	0.019
10	DOBE	276183.153	455883.751	85.063						
11	GULB	339572.577	666649.007	160.447						
12	IRBE	381137.180	371472.549	45.872	381137.166	371472.548	45.823	0.014	0.001	0.049
13	JEK1	263838.390	614761.957	116.091	263838.408	614761.955	116.010	-0.018	0.002	0.081
14	JEKA	263562.752	615239.602	117.159	263562.799	615239.552	117.195	-0.047	0.050	-0.036
15	KREI	321251.274	505410.480	41.467	321251.278	505410.478	41.467	-0.004	0.002	0.000
16	KUL1	316987.952	375188.230	62.265						
17	KULD	315769.973	375951.712	70.826						
18	LIEP	266927.776	315796.822	51.327						
19	LIMB	374452.065	542746.163	115.389	374452.065	542746.164	115.385	0.000	-0.001	0.004
20	LIPJ	266705.121	316436.124	45.291	266705.144	316436.125	45.364	-0.023	-0.001	-0.073
21	LODE	333120.229	599645.419	228.509	333120.232	599645.410	228.505	-0.003	0.009	0.004
22	LUNI	311894.457	507088.292	49.989	311894.457	507088.292	49.989	0.000	0.000	0.000
23	LVRD	291847.661	550573.361	86.497	291847.660	550573.366	86.518	0.001	-0.005	-0.021
24	MADO	303239.921	635457.432	174.181	303239.934	635457.430	174.194	-0.013	0.002	-0.013
25	MASK	306730.036	511685.055	52.617						
26	MAZS	413259.202	562588.955	80.385	413259.190	562588.958	80.373	0.012	-0.003	0.012
27	OJAR	309093.596	504631.306	41.961	309093.593	504631.314	41.962	0.003	-0.008	-0.001
28	PLSM	362909.360	631206.583	141.134	362909.378	631206.579	141.146	-0.018	0.004	-0.012
29	PREI	241689.618	668648.827	159.127	241689.638	668648.821	159.163	-0.020	0.006	-0.036
30	REZ1	269831.781	705617.372	187.278	269831.799	705617.363	187.312	-0.018	0.009	-0.034
31	REZE	268589.721	706252.072	170.508						
31	RIGA	311667.766	503574.998	34.730	311667.766	503574.998	34.730	0.000	0.000	0.000
33	SALP	302645.984	524265.958	73.416	302645.984	524265.958	73.416	0.000	0.000	0.000
34	SIGU	334306.011	554012.160	133.899	334306.016	554012.156	133.906	-0.005	0.004	-0.007
35	SLD1	280522.727	407976.999	143.668	280522.715	407977.007	143.692	0.012	-0.008	-0.024
36	SLDS	282873.495	406505.079	144.780						
37	TALS	345708.180	414728.448	115.060	345708.168	414728.454	115.073	0.012	-0.006	-0.013
38	TKMS				313969.186	449024.820	94.007			
39	VAIV	312895.928	479790.517	59.792	312895.928	479790.517	59.792	0.000	0.000	0.000
40	VAL1	377920.047	584171.780	76.003	377920.046	584171.776	76.013	0.001	0.004	-0.010
41	VALM	377914.495	584166.371	75.648						
42	VANG	315918.823	514011.967	67.209	315918.828	514011.972	67.209	-0.005	-0.005	0.000
43	VENT	363589.677	352632.012	39.359			Average	-0.006	0.002	-0.006
							Min	-0.047	-0.009	-0.073
							Max	0.014	0.050	0.081

The EUREF 92 GPS campaign was initiated by the General Directors of Nordic Geodetic and Mapping Authorities at a meeting in Tallinn, May 1992, as a contribution to the establishment of a new geodetic reference network [10, 11]. The campaign was organised by Nordic Geodetic Commission and the results were computed by NKG people. Four reference points in Latvia were positioned in ETRS89 system with EUREF accuracy estimates of class C. The GPS network in Latvia was densified by the State Land Service and by the Latvian Geospatial Information Agency [3]. The NKG GPS campaigns were performed two more times where the above mentioned 4 points in Latvia were included and Latvian own receivers were used by Latvian surveyors in those last campaigns. The first one was carried out in 2003 and the results were published [12]. No results have been published on the campaign of 2008.

LatPos RTK network was developed in 2006. The LatPos station coordinates were determined by LGIA staff using the "0 order" reference network as the basis ("LatPos old" in this article). The new values of coordinates were introduced by LGIA starting from 3 August 2011 ("LatPos new" in this article) [13]. Between the two sets of coordinates the Northing component differs in interval [-0.047, 0.014] meters, Easting component [-0.009, 0.050] and Up component [-0.073, 0.081] correspondingly (table 1).

At the Institute of Geodesy and Geoinformatics the daily values of coordinates of both the LatPos and *EUPOS® – Riga* stations were computed for the time span from 2007 till 2014. The coordinates of IGS and EPN station RIGA has been calculated as well. The SINEX weekly solutions have been regularly sent to ECC for quality control. The Bernese 5.0 software and the IGS/EPN reference network have been used.

3. GNSS measurement reduction for the one cm geoid model

The GNSS/levelling point network densification was carried out by GGI at the end of 2014 and the beginning of 2015 to develop one cm geoid model. For this purpose, the LatPos and EUPOS® – Riga station ETRS89 coordinates epoch 1989.0 were calculated at the situation of 2015.0. The IGS08 coordinates were taken for IGS/EPN reference stations including their coordinate velocities.

Eurasian plate velocities are taken into account by reducing the results to epoch 1989.0, but intraplate velocities for Latvian stations are known only for the period 2007.0-2015.0. Therefore, the intraplate situation for Latvian stations is fixed to the date 2015.0. The comparison of the ETRS89 results with a National LKS92 coordinates are shown in table 2 and in figure 1.

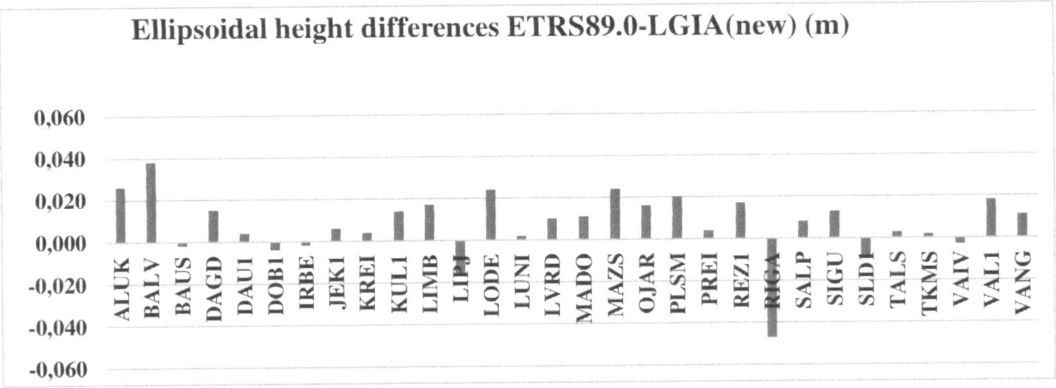

Figure 1. Differences in ellipsoidal height values.

IOP Conf. Series: Materials Science and Engineering **96** (2015) 012037 doi:10.1088/1757-899X/96/1/012037

In 2003, the European Commission accepted the ETRS89 as an obligatory spatial reference system in the EU. Consequently, the official geodetic terrestrial reference system for National networks is the ETRS89 and its actual frame.

Table 2. Comparison of ETRS89 and LatPos new coordinates.

#	DOME	ETRS89 at 2015.0 (R08)			LGIA (new)			Diff. ETRS89-LGIA		
		N	E	h (ell)	N	E	h (ell)	N	E	h (ell)
1	ALUK	368767.796	684444.107	238.301	368767.729	684444.151	238.275	0.067	-0.044	0.026
2	BALV	336401.549	697627.213	145.705	336401.489	697627.262	145.667	0.060	-0.049	0.038
3	BAUS	251271.296	511487.259	69.393	251271.229	511487.312	69.395	0.067	-0.053	-0.002
4	DAGD	222580.646	719741.376	223.125	222580.586	719741.425	223.110	0.060	-0.049	0.015
5	DAU1	194384.123	657579.706	130.233	194384.059	657579.760	130.229	0.064	-0.054	0.004
6	DOB1	277175.354	456046.614	92.299	277175.283	456046.668	92.303	0.071	-0.054	-0.004
7	IRBE	381137.243	371472.497	45.821	381137.166	371472.548	45.823	0.077	-0.051	-0.002
8	JEK1	263838.471	614761.903	116.016	263838.408	614761.955	116.010	0.063	-0.052	0.006
9	KREI	321251.347	505410.437	41.471	321251.278	505410.478	41.467	0.069	-0.041	0.004
10	KUL1	316988.017	375188.188	62.279	316987.952	375188.230	62.265	0.065	-0.042	0.014
11	LIMB	374452.136	542746.114	115.402	374452.065	542746.164	115.385	0.071	-0.050	0.017
12	LIPJ	266705.220	316436.067	45.347	266705.144	316436.125	45.364	0.076	-0.058	-0.017
13	LODE	333120.307	599645.359	228.529	333120.232	599645.410	228.505	0.075	-0.051	0.024
14	LUNI	311894.527	507088.247	49.991	311894.457	507088.292	49.989	0.070	-0.045	0.002
15	LVRD	291847.727	550573.316	86.528	291847.660	550573.366	86.518	0.067	-0.050	0.010
16	MADO	303239.998	635457.379	174.205	303239.934	635457.430	174.194	0.064	-0.051	0.011
17	MAZS	413259.258	562588.907	80.397	413259.190	562588.958	80.373	0.068	-0.051	0.024
18	OJAR	309093.661	504631.259	41.978	309093.593	504631.314	41.962	0.068	-0.055	0.016
19	PLSM	362909.443	631206.532	141.166	362909.378	631206.579	141.146	0.065	-0.047	0.020
20	PREI	241689.701	668648.776	159.167	241689.638	668648.821	159.163	0.063	-0.045	0.004
21	REZ1	269831.860	705617.312	187.329	269831.799	705617.363	187.312	0.061	-0.051	0.017
22	RIGA	311667.780	503575.003	34.683	311667.766	503574.998	34.730	0.014	0.005	-0.047
23	SALP	302646.053	524265.910	73.424	302645.984	524265.958	73.416	0.069	-0.048	0.008
24	SIGU	334306.082	554012.105	133.919	334306.016	554012.156	133.906	0.066	-0.051	0.013
25	SLD1	280522.789	407976.951	143.683	280522.715	407977.007	143.692	0.074	-0.056	-0.009
26	TALS	345708.241	414728.400	115.076	345708.168	414728.454	115.073	0.073	-0.054	0.003
27	TKMS	313969.255	449024.767	94.009	313969.186	449024.820	94.007	0.069	-0.053	0.002
28	VAIV	312895.994	479790.470	59.789	312895.928	479790.517	59.792	0.066	-0.047	-0.003
29	VAL1	377920.113	584171.726	76.031	377920.046	584171.776	76.013	0.067	-0.050	0.018
30	VANG	315918.896	514011.930	67.220	315918.828	514011.972	67.209	0.068	-0.042	0.011
43	VENT	363589.677	352632.012	39.359			Average	0.066	-0.048	0.007
							Min	0.014	-0.058	-0.047
							Max	0.077	0.005	0.038

4. Intraplate movements

Comparing the yearly average values of ETRS89 coordinates it is possible to conclude that the station movement velocities are variable from year to year. Unfortunately, some stations were displaced in other sites in 2011. Just 7 stations in the countryside and 5 stations in Riga City were not displaced. The site movements are depicted in figures 2-4 for countryside and figures 5-7 for Riga City.

All the stations during 8 years are moved North by ~ 2.5 cm by evaluating the Northing intraplate movement in figure 2. In figure 3 it is seen that behavior of PREI and BALV stations in the movement of East component differs from others by 3-5 mm. Starting from year 2010 (figure 4), the PREI station has down lifted by about 1.6 cm. BAUS, LUNI, KREI and VANG stations from 2007 till 2010 have uplifted by 8 mm (figures 4 and 7), but after 2010 till 2014 moved down to the initial situation of 2007.

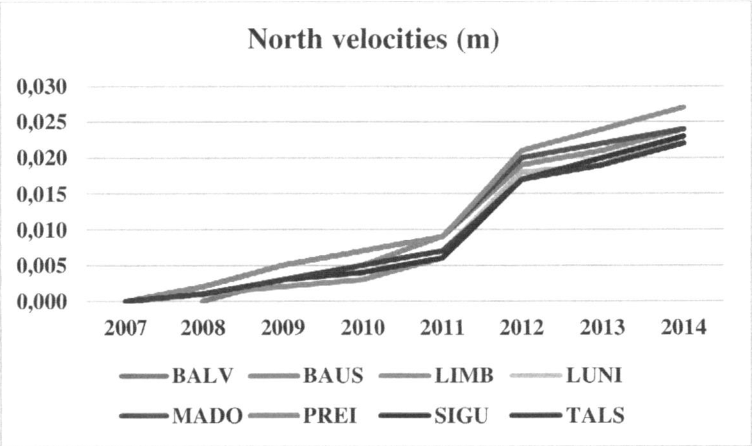

Figure 2. Site movements North.

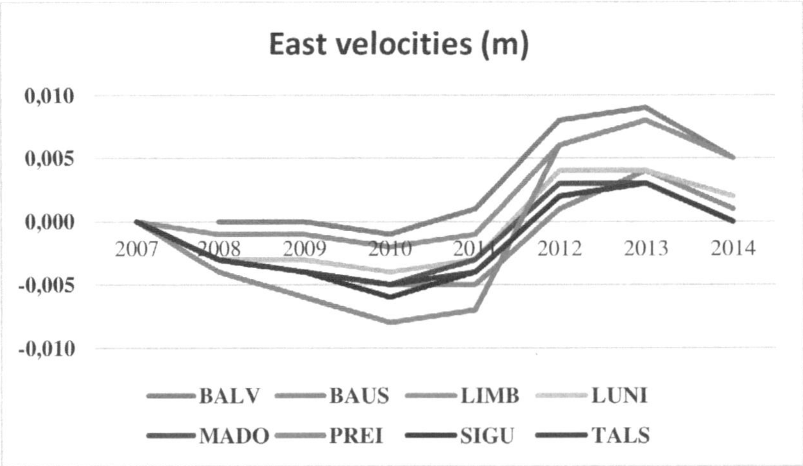

Figure 3. Site movements East.

The Riga City situation on the right bank of the Daugava River is quite a different from the situation on the left bank. The coordinates of IGS/EPN station RIGA changed sharply on Day 107 of Year 2011 when the receiver's antenna parameters were changed. ANNI and MASK stations were operational just till May 2011. It is known from geological maps that OJAR site is very close to the tectonic fault.

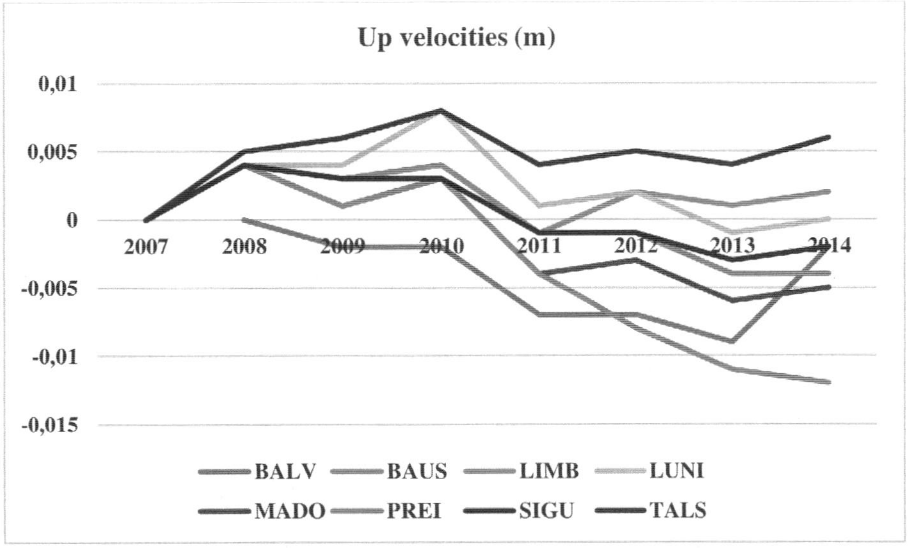

Figure 4. Site movements Up.

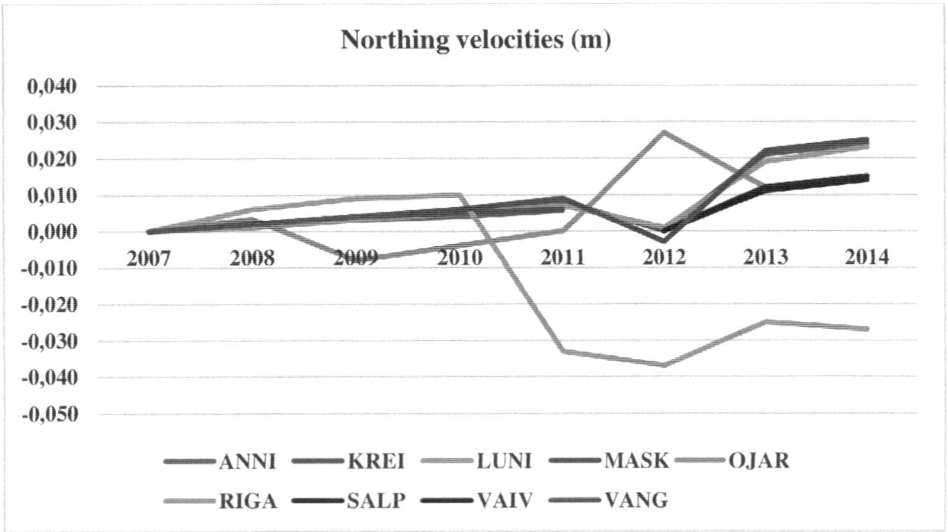

Figure 5. Site movements North in Riga City.

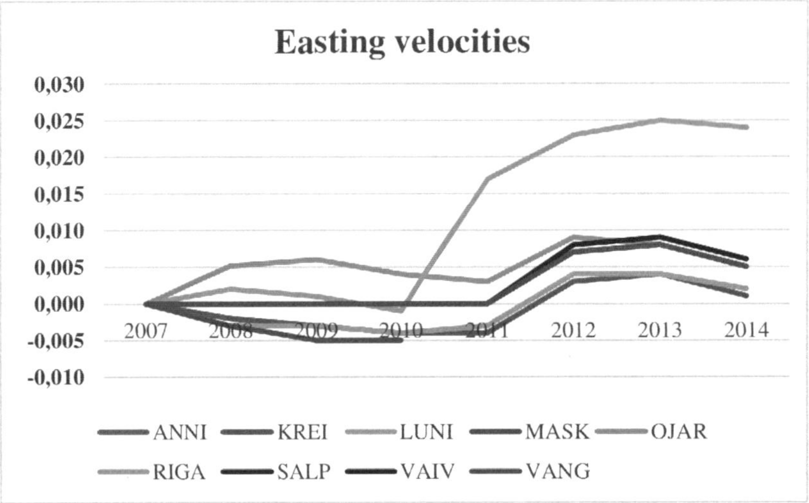

Figure 6. Site movements East in Riga City.

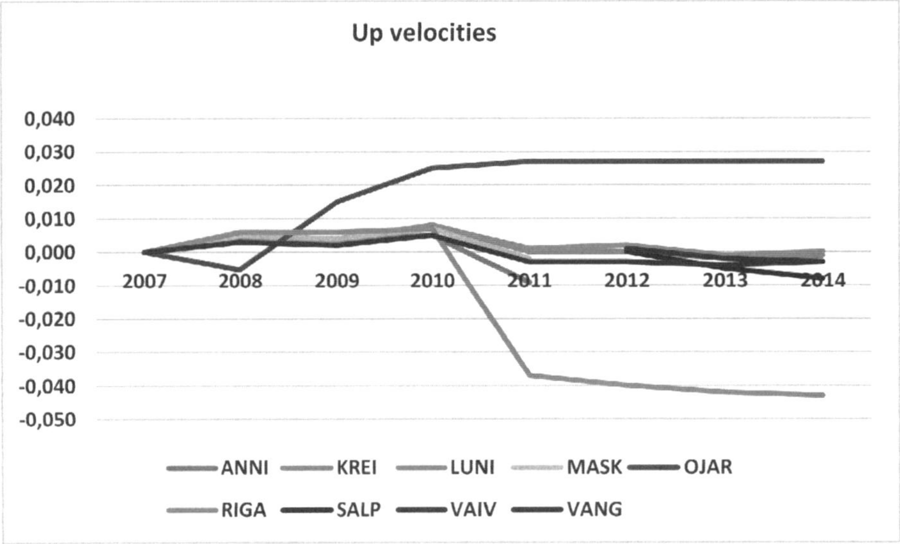

Figure 7. Site movements Up in Riga City.

5. Conclusion

The development of one cm geoid model is still underway and not all the development details are discussed in this article. The GNSS/levelling point network densification is under development. In table 2 the ETRS89 coordinates are depicted, which will be used as a basis for the geoid model. The transformation to another epoch will occur only within the framework of the knowledge on the intraplate movements.

6. References

[1] Ellmann A 2012 Considerations on the further improvements of regional geoid modeling over the Baltic countries *Geodezija ir Kartografija* **36** (1) 5–15

[2] Hirt C 2013 RTM Gravity forward-modelling using topography/bathymetry data to improve high-degree global geopotential models in the coastal zone *Marine Geodesy* **36** (2) 1–20

[3] Kaminskis J 2010 *Geoid Model for Surveying in Latvia FIG Congress 2010 (Sydney, Australia, 11-16 April 2010*) [Available]: http://www.fig.net/pub/fig2010/papers/fs01c%5Cfs01c_kaminskis_4066.pdf

[4] Jäger R, Kaminskis J, Strauhmanis J and Younis G 2012 Determination of quasi-geoid as height component of the geodetic infrastructure for gnss positioning services in the Baltic States *Latvian J. of Physics and Technical Sciences* **3** 5–15

[5] Janpaule I 2013 Methods for the determination of high precision Latvian geoid model *Summary of the Doctoral Thesis* (Riga: RTU) p 24

[6] Aleksejenko I, Sakne J, Kalinka M, Reiniks M, Kalantaite A, Krikstaponis B, Parseliunas E K, Petroskevicius P and Viskontas P 2012 The united geodetic vertical network of Latvia and Lithuania *Geodesy and Cartography* **38** (1) 9–19

[7] Jäger R 2007 DFHRS (Digital FEM Height Reference Surface) – A rigorous approach for the integrated adjustment and fitting of height reference surfaces *Forschung Aktuell* (Karlsruhe: Hochschule Karlsruhe Technik und Wirtschaft) pp 78–82

[8] Sjöberg L E and Hunegnaw A 2000 Some modifications of Stokes' formula that account for truncation and potential coefficient errors *J. of Geodesy* **74** (2) 232–238

[9] Abele M, Balodis J, Janpaule I, Lasmane I, Rubans A and Zariņš A 2012 Digital Zenith Camera for Vertical Deflection Determination *Geodesy and Cartography* **38** (4) pp 123–129

[10] Madsen F and Madsen B 1992 A new geodetic network in the Baltic countries *National Survey and Cadastre* Denmark pp 1–12

[11] Madsen F and Madsen B 1993 A new GPS-network in the Baltic Countries *Report on the Symp. of the IAG Subcommission for Europe (EUREF) (Hungary, Budapest*) pp 17–19

[12] Jivall L, Kaminskis J and Paršeliūnas E 2006 3. Improvement and extension of ETRS89 in Latvia and Lithuania based on the NKG 2003 GPS campaign *Geodesy and Cartography* **33** (1) pp 13–20

[13] Latvijas Ģeotelpiskās informācijas aģentūra [Available]: http://www.lgia.gov.lv

Acknowledgment

ERAF activity 2.1.1.1. "Support science and research". Project "One centimeter accuracy Latvian geoid model for GNSS measurements" r.2014/0039/2DP/2.1.1.1.0/14/APIA/VIAA/012.

European Vertical Reference System Influence in Latvia

A Celms[1,2], I Reķe[1] and A Ratkevičs[1]
[1]Latvia University of Agriculture, Liela Str. 2, Jelgava, LV-3001, Latvia

E-mail: armands.celms@llu.lv

Abstract. Till 1 December, 2014, in Latvia the heights were determined in Baltic Normal Height System 1977. The national height system is determined by the Cabinet of Ministers and internal laws. Now for the change of the national height system to reconcile it with the European Vertical Reference System, amendments to the laws and regulations have been developed, but so far only the amendment to the Geospatial Information Law is in force, the amendment to the regulation of the Cabinet of Ministers is still not approved. This amendment declares the Latvia Height System based on the European Vertical Reference System in Latvia as the national height system.
For height transformation, there is a transformation formula for each European country. After calculations it is seen that height difference between Baltic Normal Height System 1977 and European Vertical Reference System depends on point location in the territory (coordinates). This unequal height difference between both height systems will cause unequal height values on border connection points between Baltic countries. The aim of the research is to evaluate the European Vertical Reference System in Latvia. To reach the aim the following tasks are set: 1) to evaluate the components of transformation formulas; 2) using the transformation formulas to calculate height differences between Baltic Normal Height System 1977 and the European Vertical Reference System realization EVRF2007 for the territory of Latvia and also between Baltic Normal Height System 1977 and the Latvia Normal Height System; 3) to get height differences in the European Vertical Reference System on the borderlines of Latvia – Estonia and Latvia – Lithuania.

1. Introduction

Levelling network is a forming element of the national height system. Levelling network ensures realization of various functions in the national economy. Using the Class I levelling results for any kind of transformations, it is necessary to know the accuracy of the established levelling network (Celms et al., 2013). Since 1 December 2014 Latvia uses the realization of the European Vertical Reference System in Latvia – Latvia Normal Height System as the national height system. Lithuania and Estonia still use Baltic Normal Height System 1977 as the national height system. In Latvia this is

[2] To whom any correspondence should be addressed.

IOP Conf. Series: Materials Science and Engineering **96** (2015) 012038 doi:10.1088/1757-899X/96/1/012038

determined by the Cabinet of Ministers. The use of the European Vertical Reference System in every European Union member state is determined under Directive 2007/2/EC of the European Parliament and of the Council of 14 March 2007 establishing an Infrastructure for Spatial Information in the European Community (INSPIRE) and European Commission Regulation (EU) No 1089/2010 of 23 November 2010 implementing Directive 2007/2/EC of the European Parliament and of the Council as regards interoperability of spatial data sets and services. Annex II, article 1.3.3. „Compound Coordinate Reference Systems" determines that the European Vertical Reference System shall be used to express gravity-related heights within unit's geographical scope.

2. Methodology of research and materials

To facilitate transformation from Baltic Normal Height System 1977 to the European Vertical Reference System, Federal Agency for Cartography and Geodesy *(BKG – Bundesamt für Kartographie und Geodesie),* Germany, and the Reference Frame Sub-Commission for Europe (EUREF) have developed a transformation formula for realization of EVRF2007 in the territory of Latvia:

$$H_{(II)} = H_{(I)} + a_1 + a_2 \cdot M_o \cdot (LAT - LAT_o) + a_3 \cdot N_o \cdot (LON - LON_o) \cdot \cos(LAT)$$

$H_{(I)}$: height in the source system [m];

$H_{(II)}$: height in the target system [m];

M_o: radius of curvature in the meridian of GRS80 [m] in P_o;

N_o: radius of curvature perpendicular to the meridian of GRS80 [m] in P_o;

LAT: latitude in ETRS89 [radian];

LON: longitude in ETRS89 [radian];

$P_o(LAT_o,LON_o)$: Reference point of the transformation;

a_1...vertical translation [m];

a_2...slope in the direction of the meridian [radian];

a_3...slope in the direction perpendicular to the meridian [radian] (Description of national…).

In the European Coordinate Reference Systems Database CRS-EU the values of transformation formula for the territory of Latvia are available:

Mo = 63840416.7 m; No = 6393195.1 m; LAT_o = 56°58' = 0.99426 radian; LON_o = 24°53'= 0.43430 radian; a_1 = 0.15374 m; a_2 = 0.01558 sec and a_3 = 0.01174 sec.

Amendment to the regulations by the Cabinet of Ministers determines the Latvia Normal Height System LAS-2000,5, the transformation formula for the Latvia Normal Height System and its parameters. The transformation formula and some of its parameters are the same as the transformation formula and parameters for EVRF2007 realization in the territory of Latvia. For unknown reasons, for the territory of Latvia slope in the direction of the meridian $a_2 = 7.99066182789555\ E^{-0008}$ m and slope in the direction perpendicular to the meridian $a_3 = -9.48289473646151\ E^{-0008}$ m are different from all parameters for EVRF2007. The amendment defines these two parameters in meters, but probably it must be a mistake, because parameters a_2 and a_3 can be determined in radians or seconds. For the calculations of height difference, the authors of research adopted the values of both parameters in radians.

3. Results and discussion

After applying the transformation formula, height difference between Baltic Normal Height System 1977 and the European Vertical Reference System is not a constant value in the entire territory of Latvia, it differs from 135 mm in the south-east part of the country to 169 mm in the north-west part of the country (see figure 1) and depends on point location in the territory (coordinates).

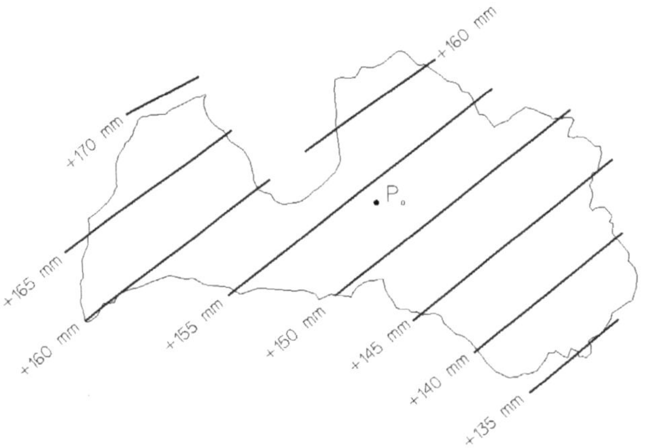

Figure 1. Height difference between Baltic Normal Height System 1977 and the European Vertical Reference System realization EVRF2007 for the territory of Latvia.

After calculations using the parameters for Latvia Normal Height System, the height differences between Baltic Normal Height System 1977 and Latvia Height System are from 125 mm in the south-east part of the country to 173 mm in the north-west part of the country (see figure 2). The differences constitute 48 mm, which is higher than primarily determined by EVRF2007.

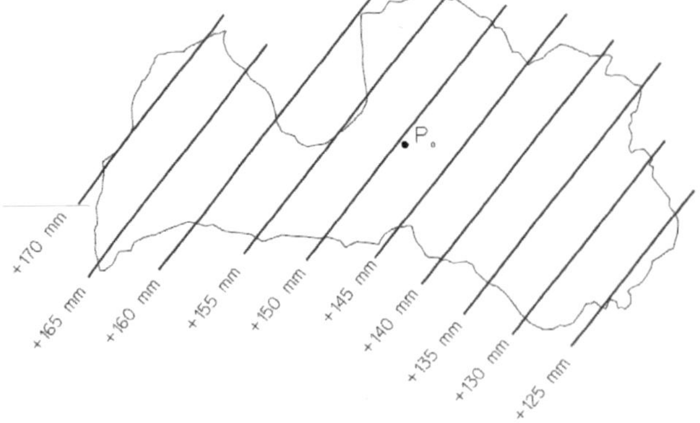

Figure 2. Height difference between Baltic Normal Height System 1977 and Latvia Normal Height System LAS-2000,5.

In the European Coordinate Reference Database it is also possible to calculate the height difference between Baltic Normal Height System 1977 and the European Vertical Reference System realization EVRF2007 for the territory of Lithuania and Estonia. Figure 3 shows the results of the calculation.

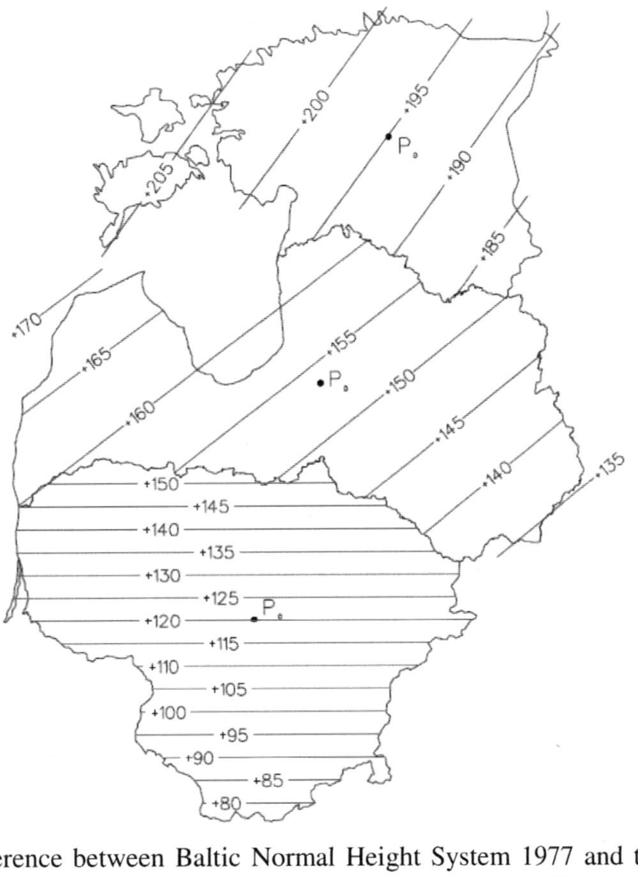

Figure 3. Height difference between Baltic Normal Height System 1977 and the European Vertical Reference System realization EVRF2007 for the territory of Lithuania, Latvia, Estonia and locations of transformation points P_0.

As seen in figure 3, in Lithuania the difference between height systems changes in the south – north direction and will be 80 mm in the south part till 150 mm in the north part of the country. In Estonia the difference between height systems changes in the same direction as in Latvia – from 185 mm in the south-east part till 207 mm in the north-west part of the country.

As it is seen in figure 3, there is one serious problem – the European Vertical Reference System will cause unequal height values at border connection points (see figure 4) – on the state border the height difference of the same point in the height system of Latvia and Estonia or Latvia and Lithuania can reach 33 mm (see table 1).

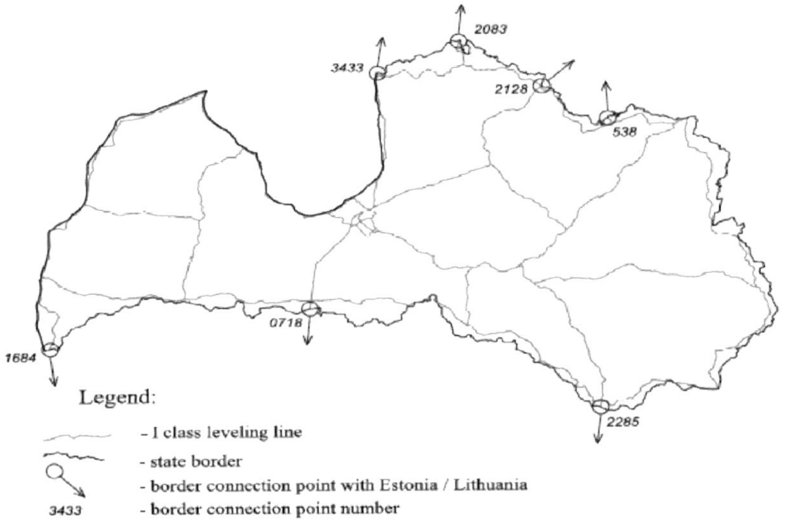

Figure 4. 1st class leveling network border connection points with Estonia and Lithuania.

Table 1. Height difference of 1st class leveling network border connection points.

Border	Point No.	Point height (BAS-77), m	Point height (EVRF2007) in Latvia, m	Point height (EVRF2007) in Estonia/Lithuania, m	Difference between heights of EVRF2007 in Latvia and Estonia/Lithuania, mm
LV-EE	3433	3.21	3.373	3.407	-33.9
	2083	72.37	72.532	72.564	-32.0
	2128	50.93	51.087	51.120	-33.3
	538	78.64	78.793	78.826	-33.4
LV-LT	1684	11.01	11.170	11.150	19.6
	718	96.4	96.560	96.560	-0.5
	2285	138.9	139.037	139.030	7.4

Table 1 shows the height difference of 1st class leveling network border connection points. On Latvia – Estonia border, height difference in the European Vertical Reference system realization EVRF2007 in both countries will be from 32.0 to 33.9 milimetres, but on Latvia – Lithuania border connection point heights in Latvia and Lithuania in the European Vertical Reference System realization EVRF2007 will be from -0.5 mm to 19.6 mm. So the same point on the border will have different height value in each country's realization of the European Vertical Reference System EVRF2007. There will be no more height connections between the Baltic countries.

For the practical testing, 2 geodetical points were mesaured with the GNSS instruments at the same time– Latvia 1st class leveling network point *1727* and Lithuania 1st class leveling network point *10284*. Point *1727* is situated at about 5 km from Latvia – Lithuania border line and point *10284* – at 12 km (see figure 5). The measurements were made in the static mode for 3 hours.

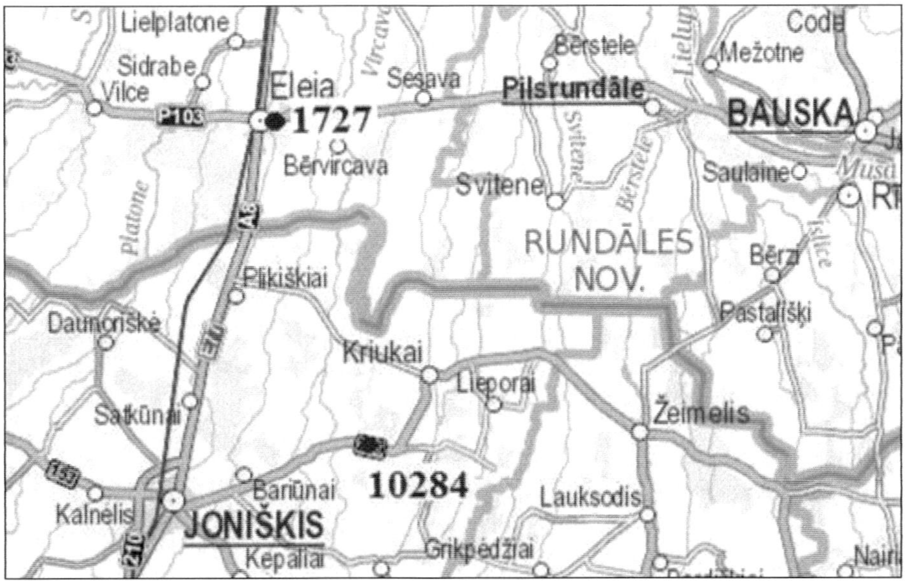

Figure 5. Location of geodetic points *1727* and *10284*.

After data adjustment, point heights in Baltic Normal Height System 1977 were obtained (see table 2).

Table 2. Point *1727* and *10284* heights.

Point	H (BAS-77), m		Height difference (BAS-77) between 1727 and 10284, m		H (EVRF2007), m	Height difference (EVRF2007) between 1727 and 10284, m
	measured with GNSS	from database	using GNSS measurements	using data from database		
1727	32.498	32.380	14.872	15.027	32.53	15.03
10284	47.370	47.407			47.56	

The height difference between both points in Baltic Normal Height System 1977 using Global Navigation Satellite System (GNSS) is 14.872 m but calculating with heights from Latvian and Lithuanian state geodetic network database the height difference is 15.027 m – 15.5 cm bigger. The reasons for 15.5 cm difference could be the inaccuracy of the geoid model.

The height difference between both points in the European Vertical Reference System realization EVRF2007 is 15.03 m. The accuracy is in cm, because the European Coordinate Reference Database gives the results in these values. Comparing height difference between both points in Baltic Normal Height System 1977 (database information) and the European Vertical Reference System realization EVRF2007, the results are almost the same. The reason for the same result is almost equal height difference between Baltic Normal Height System and the European Vertical Reference System realization EVRF2007 for the territory of Latvia and Lithuania in this area. It can be also seen in table 1 where the height difference of point 718 located in the same area in both systems in both countries is just 0.5 mm.

4. Conclusions and proposals

1. The height difference between height systems using the transformation formula is not a constant value but changes in SE-NW direction and depends on point location (coordinates).
2. This varying height difference does not have significant influence on the practical land surveying works;
3. The differences between Latvian - Estonian and Latvian - Lithuanian EVRS height systems are in the range of -33.9 to +16.9 mm.
4. United EVRS height system for the common Baltic space is advisable.
5. The change from Baltic Normal Height System 1977 to EVRS system in the Baltic countries will deny the possibility of direct comparison of the Earth's vertical movement dynamics and comparison of the prior levelling results.
6. After the implementation of the EVRS height system, it is desirable to give height values for each current geodetic sign in Baltic Normal Height System 1977.

5. References

[1] Celms A, Kronbergs M, Cintina V and Baumane V 2013 Precision of Latvia Leveling Network nodal point Height *Civil Engineering'13: Proc. of fourth Int. Scientific Conf. (Jelgava, LLU)* part I, vol 4, pp 310–317

[2] Celms A, Bimane I and Reke I 2014 European vertical reference system in Baltic countries *Baltic Surveying* **1** 49–55

[3] Directive 2007/2/EC of the European Parliament and of the Council of 14 March 2007 establishing an Infrastructure for Spatial Information in the European Community (INSPIRE) [Available]: http://eur-lex.europa.eu/LexUriServ/LexUriServ.do?uri=OJ:L:2007:108:0001:0014:en:PDF

[4] Description of national Coordinate Reference Systems of European Countries [Available]: http://www.crs-geo.eu/nn_159882/crseu/EN/CRS__Description/crs-national__node.html?__nnn=true

[5] National Coordinate Reference Systems: European Coordinate Reference Systems Database [Available]: http://www.crs-geo.eu/nn_124226/crseu/EN/CRS__Description/crs-national__node.html?__nnn=true

[6] Project: Amendment in regulation of Cabinet of Ministers No.879. [Available]: http://tap.mk.gov.lv/mk/tap/?pid=40335465

[7] Valsts augstumu izejas līmeņa noteikšana 2009 (Riga Technical University, Department of Geomatics) Scientific research project p 63

Earth's Surface Displacements from the GPS Time Series

D Haritonova[1,2,3], **J Balodis**[1], **I Janpaule**[1] **and K Morozova**[1]
[1] University of Latvia, Institute of Geodesy and Geoinformatics, 19 Raina Blvd., LV-1586, Riga, Latvia
[2] Riga Technical University, Faculty of Civil Engineering, Kalku Str. 1, Riga, LV-1658, Latvia

E-mail: diana.haritonova@rtu.lv

Abstract. The GPS observations of both Latvian permanent GNSS networks – EUPOS®-Riga and LatPos, have been collected for a period of 8 years – from 2007 to 2014. Local surface displacements have been derived from the obtained coordinate time series eliminating different impact sources. The Bernese software is used for data processing. The EUREF Permanent Network (EPN) stations in the surroundings of Latvia are selected as fiducial stations. The results have shown a positive tendency of vertical displacements in the western part of Latvia – station heights are increasing, and negative velocities are observed in the central and eastern parts. Station vertical velocities are ranging in diapason of 4 mm/year. In the case of horizontal displacements, site velocities are up to 1 mm/year and mostly oriented to the south. The comparison of the obtained results with data from the deformation model NKG_RF03vel has been made. Additionally, the purpose of this study is to analyse GPS time series obtained using two different data processing strategies: Precise Point Positioning (PPP) and estimation of station coordinates relatively to the positions of fiducial stations also known as Differential GNSS.

1. Introduction

The vertical and horizontal displacements of the Earth's surface can be measured to a high degree of precision using Global Navigation Satellite Systems (GNSS). In Latvia two permanent GNSS networks – LatPos and EUPOS®-Riga, are operating providing continuous observations. Time series of GNSS station positions from both networks are developed at the Institute of Geodesy and Geoinformatics of the University of Latvia (LU GGI) applying scientific post-processing software package.

Local surface displacements obtained from the Latvian GNSS station position time series are under discussion in this study. As Latvia is located in the region, which is under the effect of the ongoing

[3] To whom any correspondence should be addressed.

IOP Conf. Series: Materials Science and Engineering **96** (2015) 012039 doi:10.1088/1757-899X/96/1/012039

relaxation of the Earth in response to past ice mass loss, the comparison of the obtained results with the data from the deformation model NKG_RF03vel is presented.

Additionally, the purpose of this study is to analyse GPS time series obtained using two different data processing strategies: PPP and DGNSS, as well as to compare site velocities obtained with these approaches.

2. Participation in EUPOS® ECC

The EUPOS® initiative is an international expert group of public organisations coming from the field of geodesy, geodetic survey and cadastre. Partners from Central and Eastern Europe work on the provision of compatible spatial reference infrastructures by using the Global Navigation Satellite Systems GPS, GLONASS and as soon as it is available GALILEO by operating Differential GNSS EUPOS® reference station services [1].

In order to further strengthen the national EUPOS®/ETRS89 realizations and ensure the homogeneity of the national EUPOS® services, the FÖMI Satellite Geodetic Observatory (SGO) proposed the idea to establish the EUPOS® Combination Centre (ECC) at the EUPOS® Conference 2010. The main task of the ECC is the combination of the weekly national EUPOS® SINEX solutions into a single weekly EUPOS® solution. This activity is analogous to the EPN data analysis and combination approach, and the EUPOS® combined solution may be regarded as the densification of the EPN, namely, further densification of the actual ETRS89 realization. The EPN analysis standards should also be followed, therefore ECC can only handle SINEX solutions computed by a scientific post-processing software package. The detailed description of the analysis guidelines can be found in [2].

The Institute of Geodesy and Geoinformatics calculates and submits to ECC the Latvian SINEX solutions regularly. Currently ECC activities are very successful and the study area covers almost the entire Europe [3]. The result of ECC activities is depicted in figure 1.

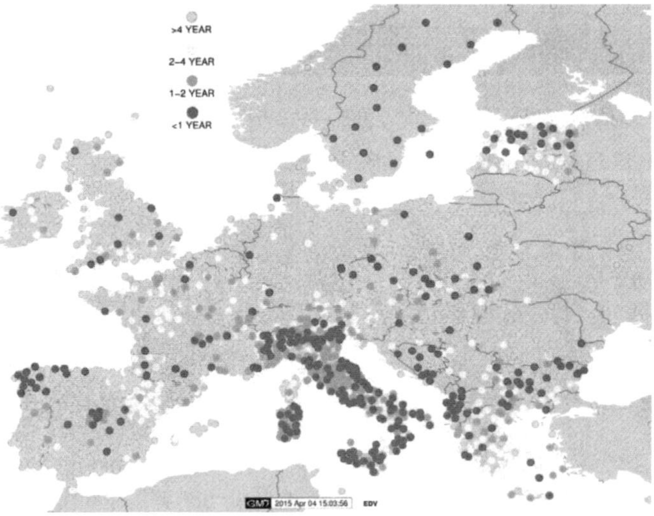

Figure 1. GNSS station observation availability for the EUPOS® combined solution shown with time intervals (status in April 2015).

3. Latvian GNSS stations

Within the framework of EUPOS® regional development project, two GNSS permanent station networks have been developed in Latvia – LatPos [4] and EUPOS®-Riga [5], which have been operating since 2006.

At present time LatPos network includes 24 permanent stations as shown in figure 2. Some of the initial stations have been renamed and removed within relatively small areas during whole period of operation, thus causing discontinuities in site position time series.

EUPOS®-Riga network consists of 5 stations, which are located in Riga and its surroundings. This network was modified once in 2012, when two stations were carried to Riga surroundings. EUPOS®-Riga stations are shown in figure 3. Location of station RIGA from IGS and EPN networks is shown as well.

Figure 2. Actual LatPos network with 24 continuously operating GPS/GNSS stations.

Figure 3. EUPOS®-Riga and IGS/EPN station RIGA.

EUPOS®-Riga and LatPos networks are primarily geodetic reference networks established for surveying and navigation purposes in the territory of Latvia. But according to the worldwide experience and trends in space geodesy it is commonly accepted to use GNSS stations for studies of geophysical processes. The vertical and horizontal displacements of the Earth's surface can be measured to a high degree of precision using GNSS.

3.1. Network solution and data processing

Time series of GNSS station positions of both EUPOS®-Riga and LatPos networks have been computed by Bernese GPS software version 5.0 developed at the University of Bern, Switzerland [6]. Station coordinates have been obtained for 8-year long observation period – from 2007 to 2014.

IGS precise orbits, Earth orientation and clock products from International GNSS Service (IGS) data base with CODE final ionosphere product are used. The data are corrected due to solid Earth tide and ocean tide loading effect by the online loading provider by H-G Scherneck. Using Differential GNSS processing strategy together with Precise Point Positioning approach applied for initial coordinate determination, site coordinates were computed in IGS05/08 frames, i.e. in IGS08 since GPS week 1632 – introduction of the new IGS08 frame [7]. Used reference stations are from EPN (some of them belong to IGS network) in the surroundings of Latvia.

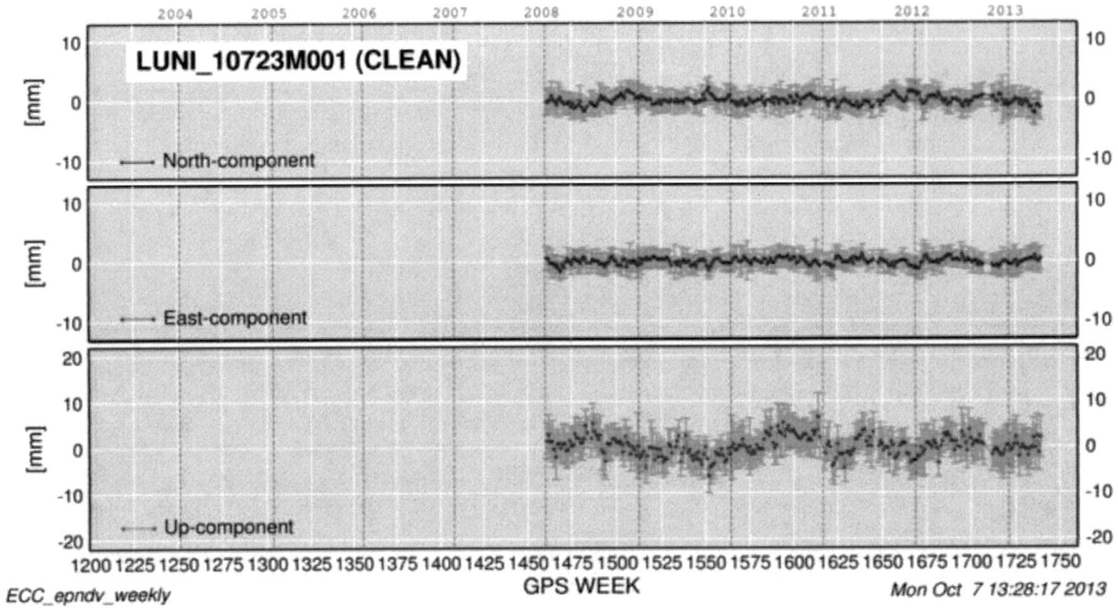

Figure 4. North, East and Up compo-nents of LUNI station position from the EUPOS® cumulative weekly solution.

The obtained daily solutions of Latvian GNSS station positions are composed to weekly SINEX solutions and are submitted to ECC for further combination into a single weekly EUPOS® solution. For example, figure 4 shows LUNI station position time series in North, East and Up components from the EUPOS® cumulative solution.

3.2. Results and discussion

Final velocity components derived from the minimum constrained cumulative solution till GPS week 1830 and expressed in the ETRF2000 frame are given in table 1. Bold font in column "Station name" of table 1 is used to mark operating GNSS stations, which are shown in figures 2 and 3.

Velocity fields are depicted in figures 5 and 6. Figure 5 shows station vertical movements, and figure 6 represents horizontal movements.

The obtained results have shown a positive tendency of vertical displacements in the western part of Latvia – station heights are increasing, and negative velocities are observed in the central and eastern parts. Station vertical velocities are ranging from -3.3 to +0.5 mm/year. This data range could be considerably reduced eliminating vertical velocities of four stations, which exceed -1.9 mm/year. As can be seen in figure 5, these velocities stand out against the background of velocity field. This could be caused by site-specific displacements. After outstanding data elimination, the resulting range of vertical velocities for the territory of Latvia is 2 mm/year; it is from -1.5 to +0.5 mm/year. In the case of horizontal movements, site velocities are ranging from 0.2 to 1.0 mm/year, excluding three stations, and oriented to the south.

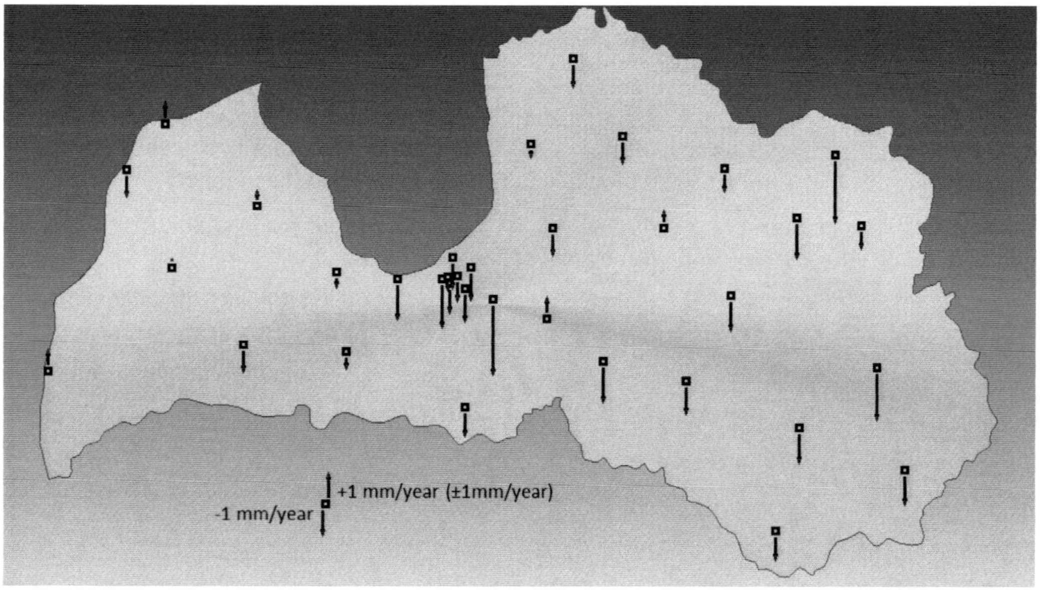

Figure 5. Cumulative vertical movements from EUPOS® observation set (status in April 2015).

Figure 6. Cumulative horizontal movements from EUPOS® observation set (status in April 2015).

Table 1. Latvian GNSS station velocities in North, East and Up components from the minimum constrained cumulative solution till GPS week 1830 expressed in the ETRF2000 frame and velocities from the deformation model NKG_RF03vel. Positive values of horizontal displacement are obtained using North and East component values. Additional marks: station name in bolt - operating GNSS station, values in bold – maximum and minimum values for Up component's velocities, underlined values – excluded values from final data ranges.

Network	GNSS station name	ECC cumulative solution, released on 4 April 2015 by the EPN RF coordinator, mm/year				Deformation model NKG_RF03vel, mm/year			
		North	East	Horizontal	Up	North	East	Horizontal	Up
LatPos	AIZK	-0.22	0.04	0.2	**-1.54**	-0.47	-0.01	0.5	0.48
	ALUK	-0.12	-0.17	0.2	-2.86	-0.57	0.07	0.6	0.73
	BALV	-0.81	0.29	0.9	-0.56	-0.52	0.03	0.5	0.54
	BAUS	-0.33	-0.09	0.3	-0.93	-0.46	-0.02	0.5	0.49
	DAGD	-0.45	-0.05	0.5	-1.18	-0.37	-0.10	0.4	**0.04**
	DAU1	-0.40	-0.16	0.4	-0.90	-0.34	-0.12	0.4	**0.04**
	DOB1	-0.30	-0.10	0.3	-0.33	-0.52	0.02	0.5	0.73
	GULB	-0.57	-0.03	0.6	-1.46	-0.54	0.04	0.5	0.62
	IRBE	-0.49	-0.28	0.6	0.50	-0.73	0.15	0.7	**1.72**
	JEK1	-0.65	-0.34	0.7	-1.07	-0.45	-0.03	0.5	0.37
	KUL1	-1.02	-0.24	1.0	0.00	-0.62	0.07	0.6	1.18
	LIMB	-0.52	0.38	0.6	-0.05	-0.64	0.12	0.7	1.14
	LIE1	-0.41	-0.16	0.4	0.48	-0.54	-0.01	0.5	0.99
	LODE	0.10	-0.98	1.0	0.30	-0.56	0.06	0.6	0.74
	LVRD	-0.60	-0.06	0.6	**0.52**	-0.51	0.01	0.5	0.62
	MADO	-0.66	-0.15	0.7	-1.19	-0.50	0.01	0.5	0.51
	MAZS	-2.38	0.11	2.4	-0.90	-0.69	0.17	0.7	1.37
	OJAR	-0.88	-1.14	1.4	-0.90	-0.56	0.05	0.6	0.81
	PLSM	-0.82	0.29	0.9	-0.59	-0.59	0.08	0.6	0.83
	PREI	-0.28	-0.55	0.6	-1.22	-0.40	-0.07	0.4	0.19
	REZ1	-0.51	-0.41	0.7	-2.07	-0.43	-0.05	0.4	0.24
	SIGU	-0.50	0.02	0.5	-0.79	-0.58	0.07	0.6	0.85
	SLD1	-0.43	-0.21	0.5	-0.77	-0.54	0.02	0.5	0.86
	TALS	-0.63	-0.10	0.6	0.18	-0.65	0.10	0.7	1.29
	TKMS	-0.20	-0.18	0.3	-0.18	-0.59	0.07	0.6	0.97
	VAL1	-0.74	-0.23	0.8	-0.82	-0.63	0.11	0.6	1.04
	VENT	-0.60	-0.17	0.6	-0.75	-0.70	0.12	0.7	1.62
EUPOS®-Riga	ANNI	-1.13	0.10	1.1	-1.90	-0.56	0.05	0.6	0.83
	KREI	-0.42	-0.09	0.4	-0.97	-0.58	0.06	0.6	0.88
	LUNI	-0.63	0.17	0.7	-0.68	-0.56	0.05	0.6	0.82
	MASK	-0.35	-0.46	0.6	-1.09	-0.55	0.04	0.6	0.78
	SALP	-0.65	-0.29	0.7	-3.29	-0.54	0.03	0.5	0.73
	VAIV	-0.38	0.31	0.5	-1.52	-0.57	0.06	0.6	0.89
	VANG	-0.24	0.13	0.3	-1.06	-0.56	0.06	0.6	0.83
C / F	**RIGA**	-0.43	-0.18	0.5	-0.11	-0.56	0.05	0.6	0.83

The present deformations due to the ongoing relaxation of the Earth in response to past ice mass loss – Glacial Isostatic Adjustment (GIA), have the rates ~10 mm/year in the vertical direction in the northern Scandinavia [8]. In the horizontal direction the displacements are smaller, about 0-2 mm/year [9]. In Latvia, the postglacial rebound is relatively small.

The velocity field from the GIA model (presented in [9]) is transformed to the GPS-derived velocity field [10], thus, the horizontal velocity field describes horizontal displacements relative to stable Eurasia as defined by the ITRS2000 and its rotation pole for Eurasia [11]. North and East components from this field are given in table 1. According to it, horizontal velocities of the Latvian GNSS stations range from 0.4 to 0.7 mm/year with the orientation to the south.

In the case of vertical displacements, according to model NKG2005LU(ABS) [12], maximum is about +1.7 mm/year, and minimum – close to 0 mm/year.

As obtained velocity fields (for horizontal and vertical displacements) derived from GPS observations and velocity fields from the Glacial Isostatic Adjustment (GIA) model are expressed in different reference frames, only the ratio of velocity ranges is given: for the horizontal displacement it is 2.7 (0.8/0.3), and for the vertical displacement – 1.2 (2.0/1.7).

4. Site velocities from PPP and DGNSS solution

The Bernese software estimates station coordinates relative to coordinates of a known station. This processing strategy is known as Differential GNSS, since it is the coordinate difference or the baseline between the two stations which is processed. The purpose of processing the baseline between two stations is that some of the errors can be reduced or eliminated.

Bernese software also contains processing strategy called Precise Point Positioning, which is able to give single station coordinates. PPP is very effective because coordinates for every station are processed individually. Hence, the coordinates do not depend on any reference station [13].

The Bernese software version 5.2 was used to process GPS data from the Latvian GNSS stations (table 2) applying two mentioned approaches. Station daily coordinates were computed in IGS08 (ITRF08) frame with respect to the EPN stations, as well as without referencing, and then have been transformed to ETRF2000 according to [14].

Residual position time series in Up component for the year 2014 are depicted in figures 7 and 8 representing three station data from EUPOS®-Riga, LatPos and IGS/EPN networks. Figure 7 shows time series of the Up component obtained from the PPP solution, and figure 8 – from the DGNSS solution.

In the case of DGNSS solution, Up components which are out of the diapason of ±15 mm were eliminated. Corresponding day coordinates were used for PPP solution; in this case coordinates which are in range of ±30 mm are used for linear trend determination (plot scale has not been changed for the better representation of data quality). Outstanding values could be observed from the PPP solution mainly during the winter time. Applying DGNSS approach, amplitude of station Up components essentially decreases, but also mentioned outliers could be seen. The winter time usually correlates with snow coverage of some GNSS antennae [15].

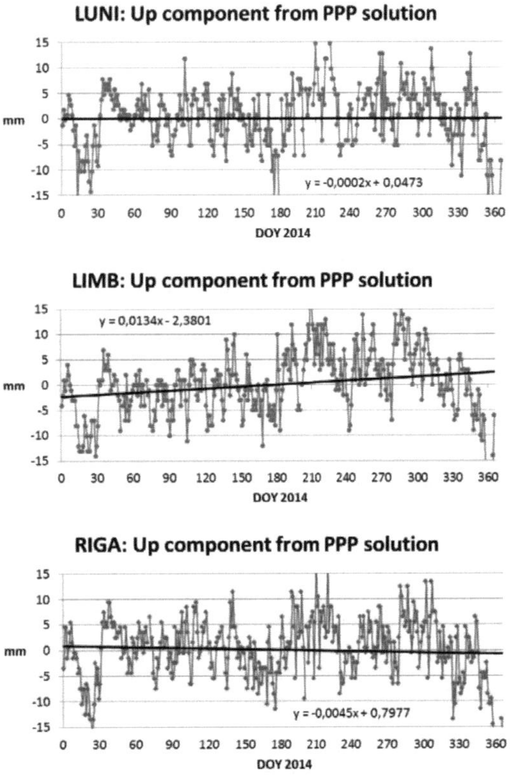

Figure 7. Up component and its trend of EUPOS®-Riga station (upper), LatPos station (middle) and IGS/EPN station RIGA (bottom) obtained from the PPP solution.

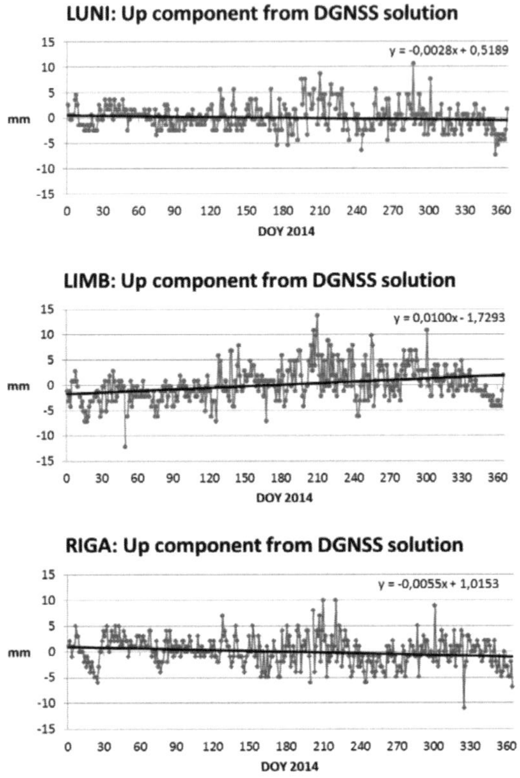

Figure 8. Up component and its trend of EUPOS®-Riga station (upper), LatPos station (middle) and IGS/EPN station RIGA (bottom) from the Differential GNSS solution.

Site velocities obtained from these time series and their differences are summarized for 24 Latvian GNSS stations in table 2. About one-third of all stations have the same velocities or velocities which more or less coincide between DGNSS and PPP solution. Half of all stations have velocity differences which are close to 1 mm/year. The maximum difference between site velocities obtained with two different processing approaches is 1.7 mm/year.

Table 2. Latvian GNSS station position velocities (mm/year) in the Up component obtained using two GPS data processing approaches, and velocity differences. Only observations of the year 2014 are used.

Station name	BAUS	BALV	DAGD	DAU1	DOB1	IRBE	JEK1	KUL1	LIMB	LIPJ	LODE	LUNI
DGNSS	-0.2	-1.2	+1.9	-4.2	-0.8	-1.6	-0.3	-1.8	+3.7	-0.7	-0.4	-1.0
PPP	+0.7	-2.1	+3.1	-3.0	-0.5	-0.6	+0.9	-1.9	+4.9	-0.4	-1.4	-0.1
Diff.	0.9	-0.9	1.2	1.2	0.3	1.0	1.2	-0.1	1.2	0.3	-1.0	0.9
Station name	LVRD	MADO	MAZS	OJAR	PLSM	PREI	SIGU	SLD1	TALS	TKMS	VAL1	RIGA
DGNSS	-1.0	-0.8	+0.5	-1.4	-0.8	+0.2	+0.4	-2.3	-0.1	-0.4	+1.4	-2.0
PPP	-2.3	-0.5	-0.2	-2.7	+0.9	0.0	+0.4	-2.3	-0.7	+0.5	+1.2	-1.6
Diff.	-1.3	0.3	-0.7	-1.3	1.7	-0.2	0.0	0.0	-0.6	0.9	-0.2	0.4

5. Conclusions

EUPOS®-Riga and LatPos station displacements obtained from continuous GPS observations have been summarized for the period of 8 years – from 2007 to 2014. The range of station velocities in vertical direction (after outlier elimination) for the territory of Latvia is 2 mm/year, it is from -1.5 to +0.5 mm/year. Horizontal velocity range is 0.8 mm/year – from 0.2 to 1.0 mm/year.

The comparison of the obtained results with data from the deformation model NKG_RF03vel shows the ratio of 1.2 between ranges of site vertical velocities. Horizontal velocities due to postglacial rebound are from 0.4 to 0.7 mm/year. Ratio of velocity ranges for the horizontal displacement is 2.7, but both fields have horizontal velocities up to 1 mm and the orientation to the south.

Latvian GNSS station positions have been obtained for one year observations applying PPP and DGNSS approaches. Outstanding Up components could be observed during winter time period in both solutions, but applying DGNSS processing, amplitude of station Up components is two times smaller. About one-third of all stations have the same velocities or velocities which more or less coincide between DGNSS and PPP solution. The maximum difference is 1.7 mm/year.

6. References

[1] Rosenthal G 2008 European position determination system status and activities *3rd Meeting of the Int. Committee on Global Navigation Satellite Systems (Pasadena, 8-12 December 2008)*

[2] EPN Coordination Group and the EPN Central Bureau 2013 Guidelines for the EPN Analysis Centres *EPN Guidelines*

[3] Kenyeres A 2014 EUPOS Combination centre now extended as EPN densification *EUPOS ISC Meeting (Latvia, Riga, 6-7 May 2014)*

[4] Abele M, Balodis J, Balodis K, Normanda M, Janpaule I, Mitrofanovs I, Rubans A, Silabriedis G and Zarins A 2008 Implementation of new positioning system in Riga *Int. Symp. EUREF (Belgium, Brussels, 18-20 June 2008)*

[5] Zvirgzds J 2007 Geodetic measurements using GPS base station system LatPos *Geomatics* **11** 81–89

[6] Beutler G et al 2007 *User Manual of the Bernese GPS Software Version 5.0* (Astronomical Institute, University of Bern)

[7] Rebischung P 2011 Upcoming switch to IGS08/igs08.atx IGSMAIL-6354 [Available]: http://igscb.jpl.nasa.gov/pipermail/igsmail/2011/006346.html
[8] Kierulf H P, Steffen H, Simpson M J R, Lidberg M, Wu P and Wang H 2014 A GPS velocity field for Fennoscandia and a consistent comparison to glacial isostatic adjustment models *J. Geophys. Res. Solid Earth* **119** 6613–29
[9] Milne G A, Davis J L, Mitrovica J X, Scherneck H-G, Johansson J M, Vermeer M and Koivula H 2001 Space-geodetic constraints on glacial isostatic adjustments in fennoscandia *Science* **291** 2381–85
[10] Lidberg M 2004 *Motions in the Geodetic Reference Frame - GPS Observations* Thesis for the degree of Licentiate of engineering Chalmers
[11] Altamimi Z, Sillard P and Boucher C 2003 The impact of a No-Net-Rotation condition on ITRF2000 *Geophysical Research Letters* **30** (2) 1064
[12] Ågren J and Svensson R 2006 Land uplift model and system definition used for the RH 2000 adjustment of the baltic levelling ring *NKG General Assembly* Copenhagen
[13] Khan S A 2005 *Surface Deformations Analyzed using GPS Time Series* Thesis for PhD degree Tscherning C C and Knudsen P
[14] Boucher C and Altamimi Z 2011 *Memo: Specifications for Reference Frame Fixing in the Analysis of a EUREF GPS Campaign* [Available]: http://etrs89.ensg.ign.fr/memo-V8.pdf
[15] Kenyeres A and Bruyninx C 2009 Noise and periodic terms in the EPN time series *Geodetic Reference Frames IAG Symposia* **134** 143–148

Acknowledgments
We would like to express our gratitude to Izolde Jumare for her support in data management and processing.

IOP Conf. Series: Materials Science and Engineering **96** (2015) 012040 doi:10.1088/1757-899X/96/1/012040

Analysis of Land Cover Change in a Coastal Area using Remotely Sensed Data

I Jaunzeme[1,2], M Kaļinka[1], M Reiniks[1] and J Kaminskis[2]
[1]Riga Technical University, Faculty of Civil Engineering, Department of Geomatics, Kalku Str. 1, Riga, LV-1658, Latvia
[2]ETH Zurich, Zurich, Switzerland

E-mail: inese.jaunzeme@rtu.lv

Abstract. Coastal area monitoring is a significant task in the national development and environmental protection. Study area of this work is the Baltic Sea Region, particularly focusing on the land cover changes in the coastal area from Cape Kolka to the Latvian-Lithuanian border. The aim of this research is to estimate and illustrate different examples of monitoring and mapping land cover changes in the coastal area using remotely sensed data – orthophoto, multispectral data and radar data. The results of the research include vector maps created from satellite images and comparison between different land cover value identification methods.

1. Introduction

Latvia's sea coast is approximately 500 km long, the coastal zone during the last decades has been subjected to notable changes [1, 2].

The aim of this research is to estimate and illustrate different examples of monitoring and mapping land cover changes in the coastal area using remotely sensed data – orthophoto, multispectral data and high resolution infrared images. Research was done using vegetation indices and urban territories. It is new for the territory of Latvia to use high-resolution infrared images. Infrared images come together with traditional orthophoto. The quality of the Landsat images depends on cloud density. In the territory of Latvia, the cloud cover is observed for approximately 80% days of the year.

Color infrared remote sensing and imagery are based on the fact that each type of land cover absorbs a particular portion of the electromagnetic spectrum and reflects the remaining portion, which can be recorded with a passive imaging system.

[2] To whom any correspondence should be addressed.

Determining how the object responds to the infrared light (absorbs, transmits, or reflects) one can explore such land cover conditions, which are undetectable on color imagery, as stressed vegetation, moist areas in the fields or plants.

Moderate resolution Landsat Thematic Mapper (TM) images have been used in various studies for land cover change in coastal area detection, such as automatic thresholding technique using vegetation indices [3, 4, 5]. Another method is satellite image classification to identify land cover classes like water and non-water, forest, man-made objects. Water identification is comparatively simple, because corresponding spectral class usually is well separated from other land cover classes.

Land cover classes, which we have detected, are bare soil, water, coastline, urban areas, cultivated land – agriculture, forests.

Landsat data are implemented in GIS and used for land use and land cover change detection in coastal areas, they can also be used in modeling of long term changes [8, 9, 10]. Nowadays Geographical Information Systems (GIS) can provide a suitable platform for data processing, analysis, update and retrieval.

2. Material and methods

2.1. Study area description

Research area of this work is the coastal area from Cape Kolka to the Latvian-Lithuanian border. The study area is situated next to the Baltic Sea and is subjected to the impact of west winds. The study particularly focused on the land cover change in the coastal area in 3 places (see figure 1):

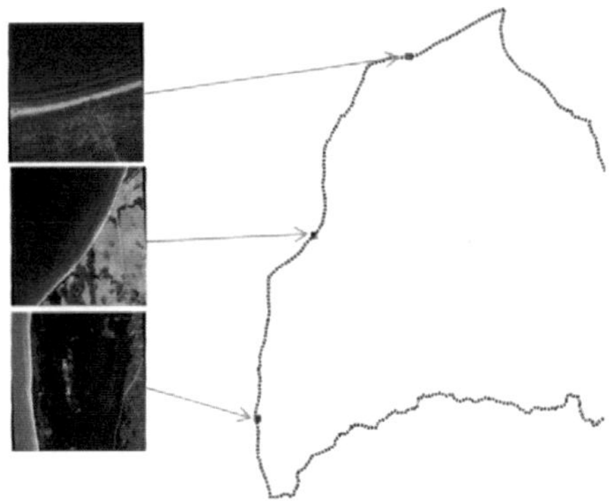

Figure 1. Study area and data.

The areas are selected on the basis of the type of shore.

2.2. Data

The research was carried out using 4 Landsat 5 Thematic Mapper (TM) and one Landsat 8 Operational Land Imager (OLI) images corresponding to the 189-020 scene, it covered the period from July 1986 to July 2014. The TM and OLI images possess 30-meter spatial resolution with multispectral coverage ranging from visible lights to the middle infrared lights of the electromagnetic spectrum. We did not use the TM and TIRS (Thermal Infrared Sensor) thermal band data in this study.

All images used for the evaluation of the methodology were downloaded from the United States Geological Survey (USGS) Global Visualization Viewer (Glovis) website as Level 1T data with precision and terrain correction, and selected comparatively in cloud-free conditions. Images were orthorectified using second order polynomial and nearest neighbor method.

Preparing the processing algorithms and processing the data, the deference in satellite sensors was taken into account.

Bandpass wavelengths for Landsat 8 OLI and TIRS sensor, compared to Landsat 5 TM sensor:

Table 1. Bandpass wavelengths Landsat 5 and Landsat 8.

Bands	Landsat 5 Wavelength (micrometers)	Landsat 8 Wavelength (micrometers)
Coastal aerosol	-	0.43 - 0.45
Blue	0.45-0.52	0.45 - 0.51
Green	0.52-0.60	0.53 - 0.59
Red	0.63-0.69	0.64 - 0.67
Near Infrared (NIR)	0.76-0.90	0.85 - 0.88
SWIR 1/MIR	1.55-1.75	1.57 - 1.65
SWIR 2	2.09-2.35	2.11 - 2.29
Panchromatic	0.52-0.90	0.50 - 0.68
Cirrus	-	1.36 - 1.38
Thermal Infrared 1	10.40-12.50	10.60 - 11.19
Thermal Infrared (TIRS) 2		11.50 - 12.51

Figure 3. Landsat 8 compared images.

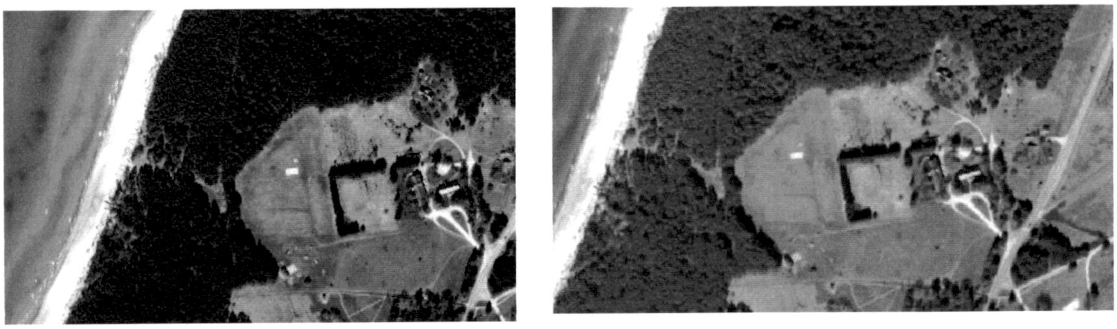

Figure 4. Ortophoto and infrared images.

Remote sensing image processing was performed using ERDAS Imagine, Integraph Geomedia and GRASS GIS.

The results compared with the reference data were obtained from the Latvian Geospatial Information Agency. As a reference, we used orthophotos and infrared images from research area.

2.3. Methodology

In order to determine land cover area changes in the coastal area, three remote sensing derived indices have been used: Normalized Difference Vegetation Index (NDVI), Modified Normalized Difference Water Index (MNDWI) and Normalized Difference Built-up Index (NDBI).

Using vegetation indices, we followed a four-step methodology to detect changes in the coastal area and create 6 land cover class vector maps. The first step was extracting the study area from a satellite image. Vegetation indices were calculated for the extracted area and then thresholded to binarize each image. A threshold binarization was used to separate the specific land cover classes. The threshold was determined on the basis of experiments for each vegetation index, and also on the knowledge of the common range of specific features. Finally, we classified individual changes.

The NDVI [6, 7] is the most commonly used spectral vegetation index, calculated by the formula:

$$NDVI = \frac{NIR - red}{NIR + red} \tag{1}$$

where NIR is a near infrared band, and red is a red band.

The value of this index ranges from -1 to 1. The common range for water is from 0.1 to -1. Soils generally exhibit a near-infrared spectral reflectance somewhat larger than the red, and thus tend to also generate rather small positive NDVI values (0.1 to 0.2). The common range for the forestry area is 0.6 – 0.8.

An urban and built-up area from data was detected using Normalized Difference Built-up Index (NDBI).

$$NDBI = \frac{MIR - NIR}{MIR + NIR} \tag{2}$$

IOP Conf. Series: Materials Science and Engineering **96** (2015) 012040 doi:10.1088/1757-899X/96/1/012040

where MIR is the middle infrared band.

Coastline and water objects (rivers, lakes etc.) are effectively mapped, detected through arithmetic manipulation of Modified Normalized Difference Water Index (MNDWI). The MNDWI is derived from the NDWI using the middle infrared instead of near-infrared [5, 11].

$$MNDWI = \frac{green - MIR}{green + MIR} \tag{3}$$

where green is the green band.

To correct the influence of soil brightness when vegetative cover is low Soil-Adjusted Vegetation Index (SAVI) was used:

$$SAVI = \frac{NIR - red}{(NIR + red + L)} * (1 + L) \tag{3}$$

where L is soil brightness correction factor band. The value of L varies by the amount or cover of green vegetation.

3. Results

The method to combine different data proved to be relevant for the territory of Latvia. The combined images from Landsat images and Ortophoto images including the infrared images were used. To analyze different periods, we obtained the results of changes of the vegetation near the coastline. Using NDVI and supervised classifications for infrared images, we facilitate classification.

The coastline changes should be analyzed in the long term. A long period of observations allows detecting visible changes and geometric changes.

To ensure higher precision of the images in the shorter time period, infrared images that give precise information of the changes for geometry and visible information should be used. It is possible to get visible information from high precision orthophotos and geometrical information – from the combination of the image bands. The change of the vegetation for 10 years period was 5 %. Coastline changes depended on water level - 2 % amplitude.

4. Conclusion

Clouds can significantly reduce the quality of the results, especially using the green band for the processing. So more advanced methods should be used for eliminating cloud shadows before creating water masks to avoid false water area detection.

Geometric corrections are an important step before processing Landsat satellite images to ensure valid results.

At the same time, the limitation in spatial, spectral, and radiometric resolution in the remotely sensed data is an important figure affecting the estimation performance.

The technique produces vector files of land cover classes which can be analyzed using GIS to estimate rates of change over relatively long time periods or can be used for modeling long term changes.

IOP Conf. Series: Materials Science and Engineering **96** (2015) 012040 doi:10.1088/1757-899X/96/1/012040

5. References

[1] Eberhards G, Grine I, Lapinskis J, Purgalis I, Saltupe B and Torklere A 2009 Changes in Latvia's seacoast (1935–2007) *Baltica* **22** (1) 11–22

[2] Ulsts V 1998 *Baltijas jūras Latvijas krasta zona* ed. Brangulis A J (Rīga: Valsts ģeoloģijas dienests) p 96

[3] Kuleli T, Guneroglu A, Karsli F and Dihkan M 2011 Automatic detection of shoreline change on coastal Ramsar wetlands of Turkey *Ocean Engineering* **38** (10) pp 1141–49

[4] McFeeters S K 1996 The use of the Normalized Difference Water Index (NDWI) in the delineation of open water features *Int. J. of Remote Sensing* **17** (7) 1425–32

[5] Xu H 2006 Modification of normalised difference water index (NDWI) to enhance open water features in remotely sensed imagery *Int. J. of Remote Sensing* **27** (14) 3025-33, 0143–61

[6] Rouse J W, Haas R H, Shell J A, Deering D W and Harlan J C 1974 *Monitoring the Vernal Advancement of Retrogradation of Natural Vegetation* Final Report, Type III, NASA/GSFC, Greenbelt, MD p 371

[7] Lillesand T M, Kiefer R W and Chipman J W 2004 *Remote Sensing and Image Interpretation* (United States of America: John Wiley & Sons) p 763

[8] Muthusamy S, Rosario Arunkumar X, Naveen Raj T, Lakshumanam C and Jayaprakash M L 2010 Use and Land Cover Changes Detection Using Multitemporal Satellite Data, Cuddalore Coastal Zone, Se-Coast of India *Int. J. of Geomatics and Geosciences* **1** (3) 610 – 619

[9] Wang Y 2005 Involving Geospatial Information in the Analysis of Land-Cover Change along the Tanzania Coast *Coastal Management* **33** 87–99

[10] Reis S 2008 Analyzing LandUse/Land Cover Changes Using Remote Sensing and GIS in Rize, North-East Turkey *Sensors* **8** 6188–6202

[11] Mwakapuja F, Liwa E and Kashaigili J 2013 Usage of Indices for Extraction of Built-up Areas and Vegetation Features from Landsat TM Image: A Case of Dar Es Salaam and Kisarawe Peri-Urban Areas, Tanzania *Int. J. of Agriculture and Forestry* **3** (7) 273–283

IOP Conf. Series: Materials Science and Engineering **96** (2015) 012041 doi:10.1088/1757-899X/96/1/012041

Accuracy of Orthomosaic Generated by Different Methods in Example of UAV Platform MUST Q

N Liba[1,2] **and J Berg-Jürgens**[1]
[1]Estonian University of Life Sciences, Institute of Forestry and Rural Engineering, Tartu, Estonia

E-mail: natalja.liba@emu.ee

Abstract. Development of photogrammetry has reached a new level due to the use of unmanned aerial vehicles (UAV). In Estonia, the main areas of use of UAVs are monitoring overhead power lines for energy companies and fields in agriculture, and estimating the use of stockpile in mining. The project was carried out by the order of the City of Tartu for future road construction. In this research, automation of UAV platform MUST Q aerial image processing and reduction of time spent on the use of ground control points (GCP) is studied. For that two projects were created with software Pix4D. First one was processed automatically without GCP. Second one did use GCP, but all the processing was done automatically.
As the result of the project, two orthomosaics with the pixel size of 5 cm were composed. Projects allowed ensuring accuracy limit of three times of the pixel size. The project that turned out to be the most accurate was the one using ground control points to do the levelling, which remained within the error limit allowed and the accuracy of the orthomosaic was 0.132 m. The project that didn't use ground control points had the accuracy of 1.417 m.

1. Introducion

Development of photogrammetry has reached a new level due to wide use of unmanned aerial vehicles (UAV). Wide use and availability of UAVs contribute to many fields, like environmental surveillance, agriculture [1], science [2], military and entertainment. On these territories, it is possible to act faster and reduce risks to human well-being.

Furthermore, software designed to process photographic materials made by UAVs has taken greats strides toward making it easier to use and to reduce the need for human involvement [3; 4]. At the same time, operational simplicity makes it harder to estimate the quality and precision of the measurements taken. Different fields of study need different levels of precision and on many instances less is sufficient. In this paper we compare the accuracy of orthophotos created automatically with the

---

[2] To whom any correspondence should be addressed.

ones created using manual assistance. Differences between the two separate projects will be determined by RMSE.

Former research with similar study has not given any acceptable results for automation orthomosaic generation [5; 6; 7]. But different technology packages provide different accuracy and UAV with different sensors piloting the aerial vehicle could lead to the higher accuracy [8; 9].

2. Materials and method

The object of the study that took place on 17 November, 2014, was the phase one aerial survey of Tartu using UAV flowing 1.8 kilometres over boroughs of Ihaste and Annelinn. The project was carried out in collaboration with the City of Tartu, Estonian Mapping Centre, ELI OÜ and Estonian University of Life Sciences. Its goal was to collect up-to-date data for the city planning department of Tartu. Territory that was measured is seen in figure 1.

Flight plan was plotted using Mission Planner (freeware) in which flying altitude, area and flight direction was inserted and the program loaded ground based control station. Flight took place 180 meters from the ground and total of 625 photos were collected (GSD 4.7 cm).

Figure 1. Schematic of partial areal-survey with ground control points of Tartu, concentrating on boroughs of Ihaste and Annelinn.

Measurements were taken using RTK method with receivers Trimble R6 and 5800 GNSS. Ground control points were located using minimal number of satellites and taking into consideration ground obstructions in the surrounding area. The base station of Estonian University of Life Sciences was used as the base station.

In the survey taken on 17 November, 30cm x 30cm plastic marking sheets were used. Survey on 20 November used easily recognisable ground points from nature and areal-photos.

For the areal-photography UAV type of plane named MUST Q was used (figure 2). MUST Q is an unmanned plane developed by ELU OÜ, which is able to take areal-photos and carry out laser scanning.

Figure 2. Unmanned plane MUST Q.

MUST Q is able to fly autonomously on autopilot due to the on-board control system. That system allows, when specific flight plans are necessary or obstacles occur, to repeat the pre-programmed flight plans. However, it is also possible to give commands to the plane manually and track the flight using live video feed.

MUST Q's empty weight is 2.5 kg, it has length of 65 cm and wingspan of 200 cm. The fuselage is made of expanded polypropylene aka EPP, because it is durable and inexpensive to replace when damaged. The airplane is driven by on-board brushless DC electric motor that is powered by a battery, which lasts about 90 minutes and has an operation range up to 10 km. Aircraft speed is from 10 to 25 m/s.

Figure 3. Ground control station used to guide and monitor MUST Q.

Ground Control Station consists of a control module based on Panasonic notebook and antenna, which receives and sends signals from and to the plane (figure 3).

On board an airplane there is a compact camera Panasonic DMC - GX1 (figure 4), used to send out a video feed. Plus there is a 3D Robotics GPS and IMU. All measurements of the sensors are synchronized with GPS that allows all the areal-data to be localized.

Figure 4. Compact camera Panasonic DMC - GX1.

Parameters of the camera Panasonic DMC-GX1 are given in table 1.

Table 1. Panasonic DMC-GX1 parameters.

Resolution	4612 x 3468
Sensor size	3.75 µm x 3.75µm
Focal lenght	19.159 mm
Weight	0.318 kg

The data was processed using photogrammetry-software Pix4D, which is capable to create a geo-referenced 2D orthomosaic and 3D point-cloud based on the taken photos. The software has three processing levels (figure 5):

- Initial processing, where it is possible to select an area of orthomosaic with cloud points and add a ground control points, thereby performing areal-triangulation.
- Point Cloud Densification – this will create a 3D point cloud.
- DSM and orthomosaic generation – creates the model of the ground and orthomosaic.

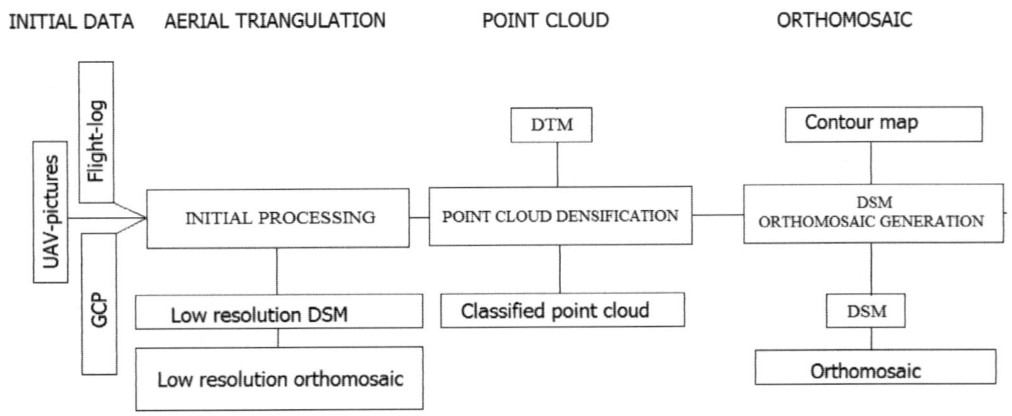

Figure 5. Example of the workflow of Pix4D of UAV.

Flight logs of aerial photos taken by the platform MUST Q are shown in figure 6. The logs reflect all data that are associated with the plane in the air – the condition of the battery, engine, different sensors and the readings, plus the times when photos are taken and GPS measurements.

```
GPS, 3, 121966200, 1819, 10, 1.30, 58.3462781, 26.7669005, 0.05, 38.01, 0.03, 311.60, -0.01, 513232
ATT, 513257, 1.21, 6.07, 316.61, 0.01, 0.00
IMU, 513257, 0.0006890818, -0.0003485335, -0.002389171, 1.073766, -0.1952583, -10.00044
ATT, 513355, 1.20, 6.06, 316.60, 0.01, 0.00
IMU, 513355, 0.00334963, -0.001412753, -0.001324952, 1.06419, -0.2197989, -9.993255
```

Figure 6. Extract from the flight logs of platform MUST Q.

The log files contain all the needed parameters measured by the sensors during the flight, which will allow automatically after the completion of aerial triangulation to set up locations and positions of the photos taken.

The accuracy control and the analysis of orthophotos were performed on the basis of two projects – automatic and semi-manual. Geometric accuracy assessment was measured by calculating the RMSE by comparing it with checkpoints measured from nature and on interactive orthomosaic. Table 2 shows the descriptions of the two projects.

Table 2. Creating orthophotomosaic in different projects (+ manual, - automatic).

Stage	Changeable stage	Project Automatic	 Semi-automatic
		Manual intervention	
Initial processing	Flight log	+	+
	GCP	-	+
	Tie points	-	-
Elevation model - DSM	Point cloud	-	-
Orthomosaic		-	-

The automatic method was conducted as automatically as possible. In case of Pix4D software and current project, images and the log file were downloaded and orthomosaic was created without interfering. To pass the areal triangulation, flight log was imported, which was then used to geo-reference areal images. This was used to automatically obtain tie points and level the block adjustment.

In this project, a point cloud was created by triangulating areal block (figure 7a), in which case it was determined in what parts of the areal-photo points were created. The given work considered a half of the image area. In addition, density of the points was set – one point for each of the four pixels. After that, minimum number of images where points can be seen was set – on four images.

Different filters were used to create elevation model DSM. Noise filter was used, which finds corrupted points and reduces their height to medium altitude of surrounding points. In addition, a smoothing filter was used, in this project it is a sharp filter, which tries to portray all objects as close to reality as possible. The pixel size of the orthomosaic (figure 7c) was determined – how precisely it is displayed and correlated.

In case of semi-manual method, leveling was done by some manual intervention. In this method, images and log files were loaded and then ground control points imported and measured. To pass the areal triangulation, nine GCP were used. Points were measured by L-EST 97 coordinate system. To create point-cloud, the parameters analogous to the automatic project were used. After generating the point-cloud (figure 7b) orthomosaic was created (figure 7d), the parameters of which were similar to the automatic project.

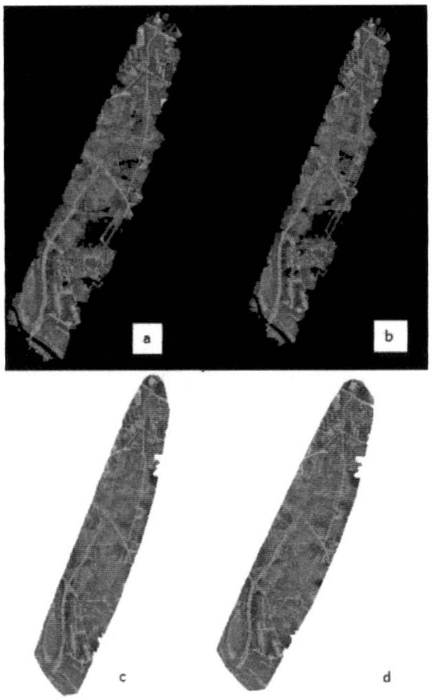

Figure 7. a) point cloud generated by the automatic project; b) point cloud generated by semi-manual project; c) orthomosaic created by the automated method; d) orthomosaic created by semi-manual method.

IOP Conf. Series: Materials Science and Engineering **96** (2015) 012041 doi:10.1088/1757-899X/96/1/012041

3. Results

Geometric accuracy assessment of automatic and manual intervention was done on the basis of 10 check points. RMSE were measured and compared to checkpoints measured in nature and interactively on orthophotos.

The results are given in figures 8 and 9.

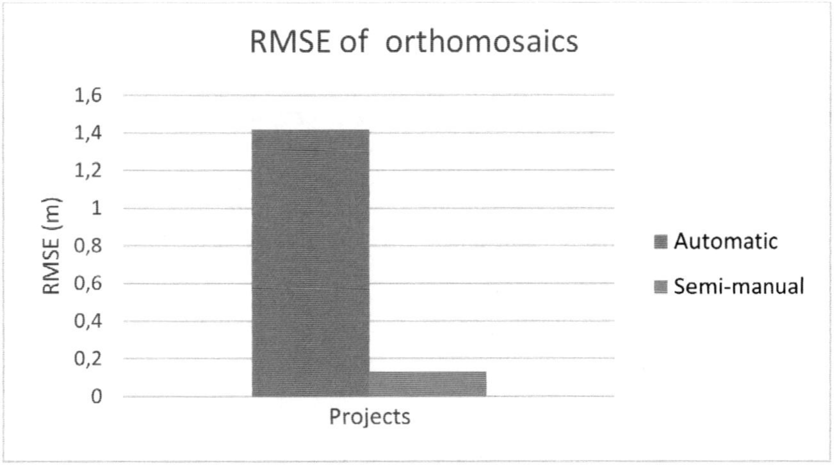

Figure 8. RMSE of orthomosaic.

The projects allowed accuracy limit that was three times of the size of the pixel, in this case it was 15 cm. RMSE of the semi-manual project was 0.132 m, which is within the limit. Accuracy of the automatic project was 1.417 m, which exceeds the allowable limit.

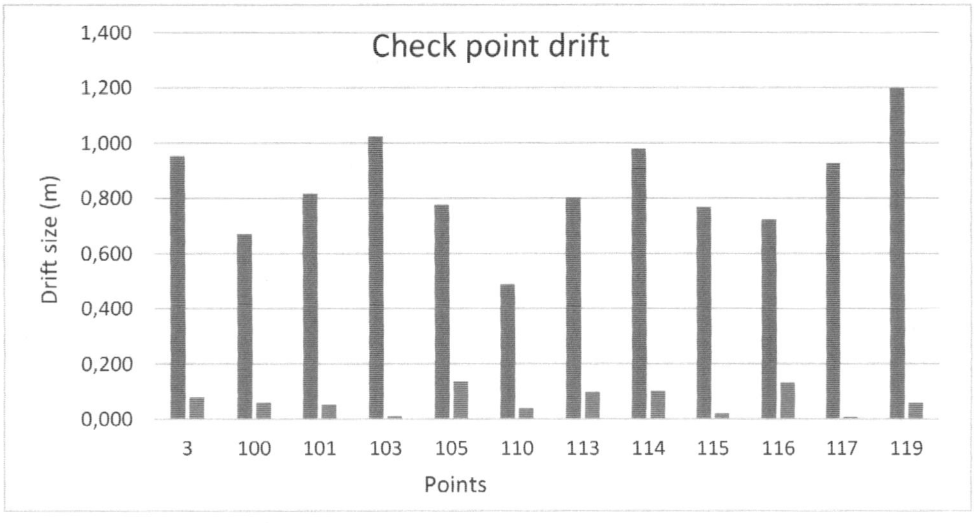

Figure 9. Drift values of the check points.

Figure 9 shows that the smallest differential between the X - axis direction of the automatic draft was 0.358 m, and 0.283 m in the direction of the Y-axis. Semi-manual project had the smallest differential X-axis of 0.000 m and 0.007 m in the direction of Y-axis.
Semi-manual project had the largest displacement of 0.138 m and the lowest of 0.013 m. Automatic project had the largest and smallest shifts of 1.200 and 0.488 m respectively. In the manual project all of the points of the project were within normal limits, but automatic project exceeded all of them.

4. Conclusions

As the result of the work, two orthomosaic by pixel size of 5 cm were obtained. As previously mentioned projects permissible accuracy limit was three times of the pixel size, or 0.15 m. When geometric accuracy of the orthomosaics was measured, we found that RMSE of the semi-manual project was 0.132 and that of automatic project 1.417.

The semi-manual project was within permitted limits and the automatic was not, which means that the use of ground control points is important and without doing it, it is not yet possible to create sufficiently accurate orthophotomosaics using UAVs.

Based on this paper it is recommended that further study of fully manual project is done, where tie points and elevation-model modification are used in addition to the ground control points.

5. References

[1] Gómez-Candón D, De Castro A I and Lopez-Granados F 2014 The accuracy of mosaics from unmanned aerial vehicle (UAV) imagery for precision agriculture purposes in wheat *Precision Agriculture* **15** (1) 44–56

[2] Chiabrando F, Nex F, Piatti D and Rinaudo F 2011 UAV and RPV systems for photogrammetric surveys in archaelogical areas: two tests in the Piedmont region. *J. of Archaeological Science* **38** 697–710

[3] Beyeler A, Floreano D, Fua P, Gervaix F, Kung O, Strecha C and Zufferey J-C 2011 The accuracy of automatic photogrammetric techniques on ultra-light UAV imagery *Int. Archives of the Photogrammetry, Remote Sensing and Spatial Information Sciences* ISPRS ICWG, Zürich pp 57–63

[4] Eisenbeiß H 2009 UAV Photogrammetry (dissertation) (Zürichi Tehnikainstituut: Zürich)

[5] Kerle N, Malet J-P, Niethammer U, Rothmund S and Stumpf A 2013 Image-based mapping of surface fissures for the investigation of landslide dynamics *Geomorphology* **186** 12–27

[6] Pagliari D, Pinto L and Sona G 2014 Experimental analysis of different software packages for orientation and digital surface modelling from UAV images *Earth Sci Inform.* **7** 97–107

[7] Panissod F, Strecha C, Tracol M and Vallet J 2011 Photogrammetric performance of an ultra light weight swinglet "UAV" *Int. Archives of the Photogrammetry, Remote Sensing and Spatial Information Sciences* ISPRS ICWG, Zürich, 36–42

[8] Nonami F, Kendoul F, Suzuki S, Wang W and Nakazawa D 2010 *Autonomous Flying Robots* (New York: Springer) p 348

[9] Raag R 2010 Telemeetriaüsteem taktikalisele mehitamata õhusõidukile (Master thesis, Estonian, Tallinna Tehnikaülikool)

IOP Conf. Series: Materials Science and Engineering **96** (2015) 012042 doi:10.1088/1757-899X/96/1/012042

The Near Real Time Ionospheric Model of Latvia

M Kaļinka[1,2,3], J Zvirgzds[1,2], D Dobelis[2], E Lazdāns[2] and M Reiniks[2]
[1]University of Latvia, Institute of Geodesy and Geoinformatics, Riga, Latvia
[2]Riga Technical University, Faculty of Civil Engineering, Department of Geomatics, Kalku Str. 1, Riga, LV-1658, Latvia

E-mail: maris.kalinka@rtu.lv

Abstract. A highly accurate ionosphere model is necessary to enable a fast and reliable coordinate determination with GNSS in real time. It is a partially ionized atmospheric region ranging up to 1,000 km height, affected by spatial variations, space weather, seasonal and solar cycle dependence. New approaches and algorithms of modelling techniques are sought to provide better solutions in the territory of Latvia. Ionospheric TEC value has large differences in Western Latvia and Eastern Latvia. Actual ionospheric map should be calculated and delivered to the surveyors near real time and published on the WEB. Delivering actual map to rover GNSS devices in a field will provide the surveyors with ionospheric conditions and allow choosing best time for surveying and making geodetic measurements with higher accuracy and reliability.

1. Introduction

The Global Navigation Satellite System (GNSS) has become an important tool for monitoring and predicting the variations in the earth's ionosphere based on the total electron content (TEC). There are international services providing global and regional ionospheric products derived from GNSS data processing. To achieve more detailed results for small areas, the activity of ionosphere can be processed and modelled using Continuously Operating Reference Networks (CORS) data. CORS network in Latvia (LatPos) has been fully operational since year 2005. The post-processing data for research of ionospheric activity during last solar cycle is fully available from up to 25 permanent GNSS base stations. The data in RINEX format from LatPos base stations are used to develop techniques and algorithms to improve local ionosphere modelling.

In this study, the local ionosphere modelling from LatPos data is presented. The achieved results are compared to the global ionospheric products derived from IGS.

[3] To whom any correspondence should be addressed.

To provide ionosphere activity related information to LatPos users, the data flow chart of calculations and ionosphere modelling in near real time are introduced.

2. Sets of the data

Continuously operating reference network system LatPos data have been used in calculations for the territory of Latvia. In the time span from 2005 to 2013, the positions of the antennas have been changed and new hardware installed. The network also has new stations installed. Base station "Ojars" was installed in 2001. The LatPos network schema is shown in figure 1.

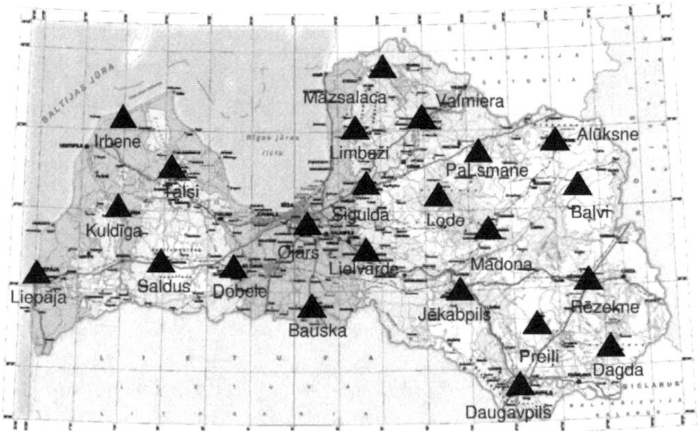

Figure 1. Continuously Operating Reference Network System LatPos – positions and names of base stations.

At the maximum of the solar activity, the LatPos network showed a lot of problems for precise measurements. The high ionosphere activity was reached from 10 o'clock GPS time. LatPos network has a set of data to analyze the ionosphere changes at the regional and local level. Table 1 shows the GNSS base stations used for modelling the ionosphere models.

To show the influence of solar activity on GNSS data processing, periods of minimum and maximum values of solar activity have been used. As solar activity cycle is 11 Years, 24 hour data sets were selected on 1 August 2006 and 1 August 2012. Base station coordinates are shown in table 2.

Table 1. LatPos station names and codes.

2006.08.01. GPS Day 213		2012.08.01. GPS Day 214	
Station Name	**Station code**	**Station Name**	**Station code**
VENTSPILS	VENT	IRBENE	IRBE
LIEPAJA	LIEP	LIEPAJA1	LIE1
DOBELE	DOBE	DOBELE1	DOB1
OJARS	OJAR	OJARS	OJAR
VALMIERA	VALM	MAZSALACA	MSLC
BALVI	BALV	BALVI	BALV
MADONA	MADO	MADONA	MADO
DAUGAVPILS	DAUG	DAUGAVPILS1	DAU1

Data model calculated using 8 base stations is shown in table 1. The network of the LatPos has been upgraded in the period from 2006-2012 and positions of the base stations have been changed. This does not affects data processing as the base stations provide information about the entire country and global ionosphere parameters of the ionosphere change in minimal amplitude. Base stations and calculated vectors are shown in figure 2.

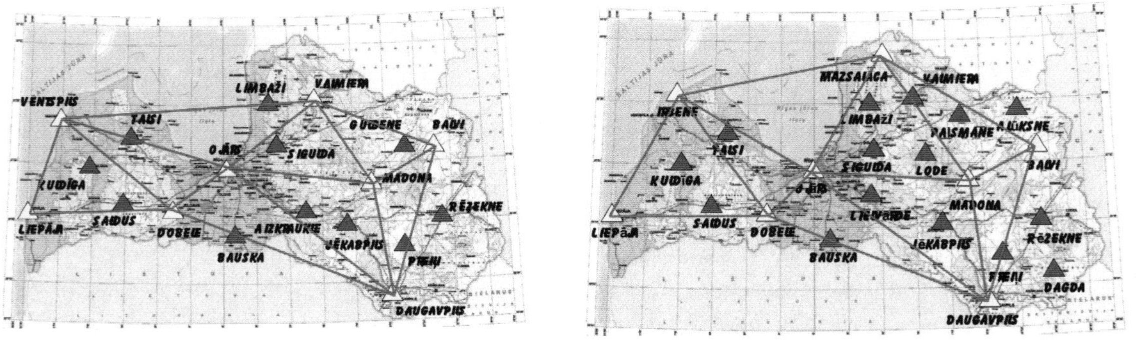

Figure 2. LatPos base stations in year 2006 (left) and 2012 (right) and the calculated vectors.

Table 2. LatPos base station coordinates.

NUM	STATION NAME	X (M)	Y (M)	Z (M)
1	IRBENE	3183614.9330	1276707.5630	5359315.0420
2	BALVI	3084535.0560	1589675.9260	5333791.7030
3	DAUGAVPILS1	3209600.8390	1601536.4170	5256389.4850
4	DOBELE1	3229567.8250	1389764.1370	5303835.6430
5	LIEPAJA1	3293067.5830	1265205.9660	5295955.7890
6	MADONA	3136049.8430	1544577.1080	5317122.7360
7	MAZSALACA	3081539.1520	1440527.8530	5377339.2580
8	OJARS	3185444.6290	1423322.9900	5321411.2010
9	DAUGAVPILS	3209659.7949	1601282.2846	5256433.8291
10	DOBELE	3230387.1544	1389950.5810	5303282.7935
11	LIEPAJA	3293151.6277	1264543.6873	5296068.1552
12	VALMIERA	3099939.8934	1472346.3504	5358234.0646
13	VENTSPILS	3204596.0472	1265426.8193	5349535.1459

3. Methods of calculation

Bernese GPS V5.0 software with two modeling principles – the local and regional model – was used for calculating the ionosphere models.

Local TEC model can be calculated if one or more double frequency stations have been used. The model can be described with the following algorithm:

$$E(\beta, s) = \sum_{n=0}^{n_{max}} \sum_{m=0}^{m_{max}} E_{nm}\,(\beta - \beta_0)^n (s - s_0)^m,$$

where

n_{max}, m_{max} are the maximum degrees of two-dimensional Taylor series expansion in latitude β and longitude s,

E_{nm} are the (unknown) TEC coefficients of the Taylor series, i.e., the local ionosphere model parameters to be estimated, and

β_0, s_0 are the coordinates of the origin of the development.

β is the geographic latitude of the intersection point of the line receiver-satelite with the ionospheric layer and s is the sun-fixed longitude of the ionospheric pierce point (or sub-ionospheric point). s is related to the local solar time (LT) according to

$$s = LT - \pi \approx UT + \lambda - \pi$$

UT is Universal Time and λ denotes the geographical longitude of the sub-ionospheric point. For satellites at elevation angles 15/20 with widely different azimuth, these sub-ionospheric points can be separated by up to 3000/2000 kilometers. Nevertheless, the representation of algorithm for local ionosphere models not well suited for regional or even global applications because of limitations in the (β, s)-space.

Regional ionosphere model was used as a global model, including the geometrical parameters of the region. The mathematical algorithm for calculating the global ionosphere model is:

$$E(\beta, s) = \sum_{n=0}^{n_{max}} \sum_{m=0}^{n} \widetilde{P_{nm}}\,(sin\beta)(a_{nm} cos\; ms + b_{nm} \sin ms),$$

where

n_{max} is the maximum degree of the spherical harmonic expansion,

$\widetilde{P_{nm}} = \Lambda(n, m) P_{nm}$ are the normalized associated Legendre functions of degree n and order m, based on normalization function $\Lambda(n, m)$ and Legendre functions P_{nm},and

a_{nm}, b_{nm} are the (unknown) TEC coefficients of the spherical harmonics, i.e., the global ionosphere model parameters to be estimated.

4. Results

The ionosphere models from stations mentioned in the previous sections are plotted in the following figures. For each base station, local ionosphere model was calculated each hour to get daily changes of VTEC. Changes of VTEC on 1 August 2006 and 1 August 2012 from several base stations can be seen in figure 3. The values of CODE processed global ionosphere model for Europe are also added to see the correlation between them.

IOP Conf. Series: Materials Science and Engineering **96** (2015) 012042 doi:10.1088/1757-899X/96/1/012042

In figure 4 and figure 5, the visualization of the regional ionosphere model over the territory of Latvia can be seen. The activity of ionosphere was visualized in two different epochs – at low activity (top of each figure) and high activity (bottom of each figure). The values in these figures are taken form IONEX ionosphere exchange format files calculated with Bernese GPS software v.5.0.

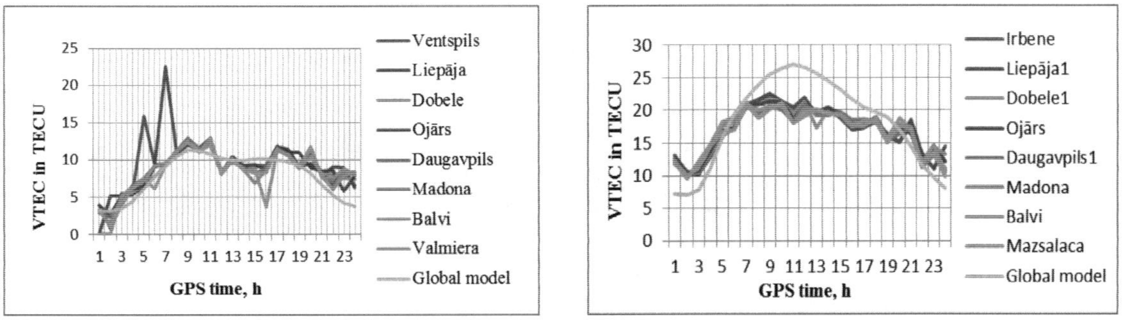

Figure 3. Changes of VTEC values at several LatPos base stations on 1 Aug 2006 (left) and on 1 Aug 2012 (right).

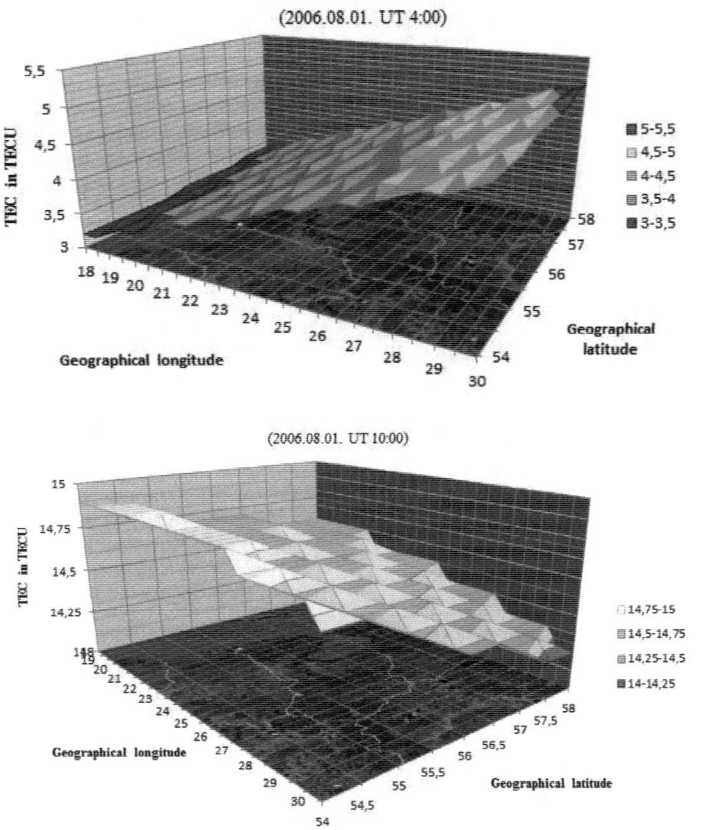

Figure 4. Regional ionosphere model visualization over the territory of Latvia on 1 Aug 2006.

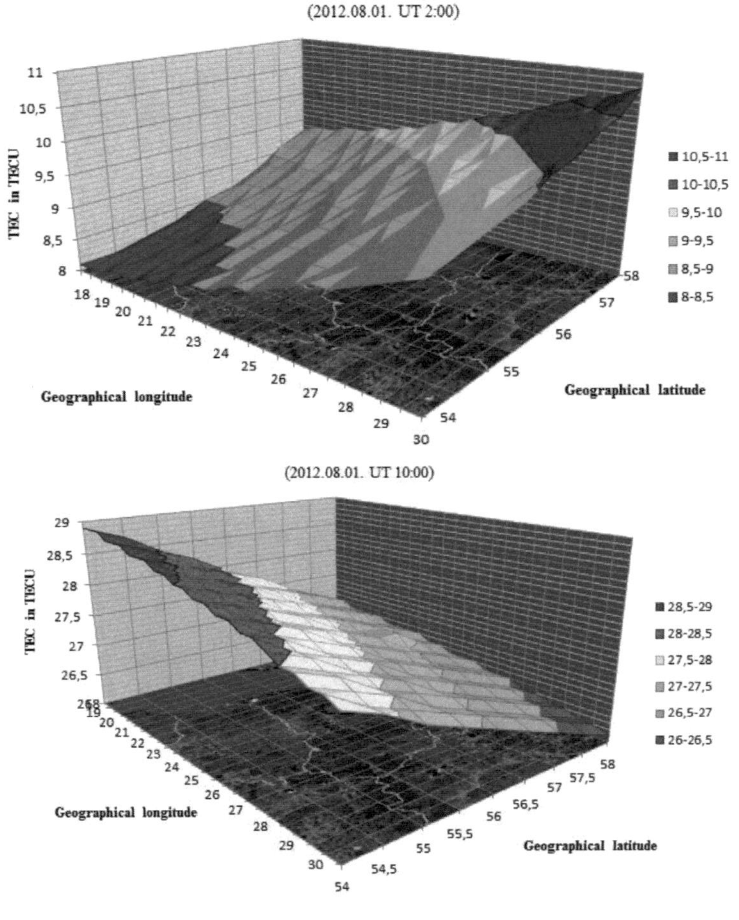

Figure 5. Regional ionosphere model visualization over the territory of Latvia on 1 Aug 2012.

5. Ionosphere modelling in near real time

Information about ionosphere activity today and in the next years will be very important for CORS users. If the CORS users can get information about daily ionosphere changes, they can better plan the schedule for RTK measurements. Furthermore, possibility to check the actual status of ionosphere in near real time is very helpful to detect problems and validate the measurements in problem cases.

Due to the facts mentioned above, the research on ionosphere modelling in near real time mode is still going on. The principal data flow and processing chart are introduced in figure 6. The introduced data flow chart includes many processing modules starting from GNSS data conversation to standard RINEX format, followed with pre-processing part of GNSS data (done with open source software BNC.EXE, teqc.exe), in the next step the ionosphere is calculated with Bernese GPS software, and later the data are stored in the database to accumulate ASCII data for ionosphere visualizations in WEB or mobile applications.

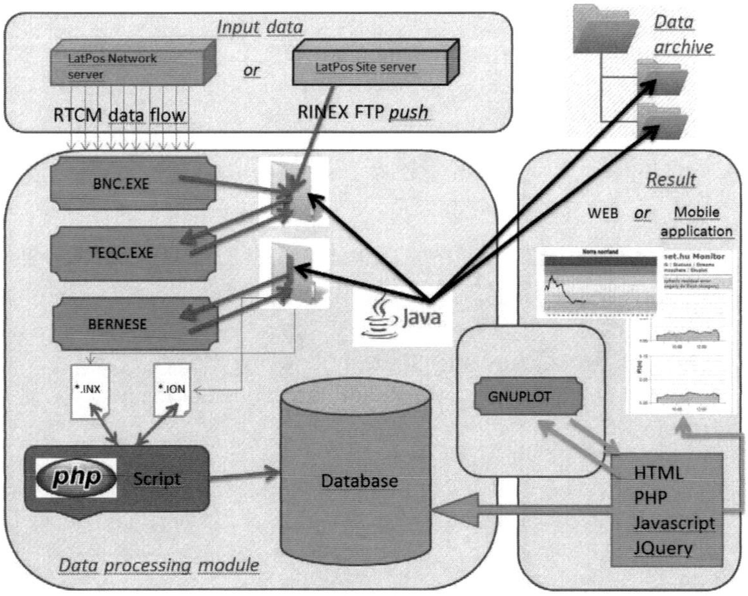

Figure 6. Data flow and processing chart of ionosphere modelling in near real time.

6. Conclusion

Differences in ionosphere activity on relatively small areas can be determined if ionosphere is modelled from GNSS data of the local area CORS networks. It can be noticed that TEC values change during the day period and depending on geographical location.

Ionosphere modelling in near real time and data delivery to the users for information purposes and/or calculations is important due to high solar activity.

7. References

[1] Hernández-Pajares M et al 2011 The Ionosphere: effects, GPS modelling and benefits for space geodetic techniques *J. of Geodesy* **85** (12) 887–907

[2] Schaer S, Beutler G and Rothacher M Mapping and predicting the ionosphere (Switzerland, University of Berne) CH3012 p 12

[3] Schaer S and Gurtner W 1998 IONEX: The IONosphere Map Exchange Format Version 1 ESA/ESOC (Darmstadt, Germany) p15

[4] Gao Y and Liu Z Z 2002 Precise Ionosphere Modeling Using Regional GPS Network Data *J. of Global Positioning Systems* **1** (1) 18–24

Effect of Silica Fume on two-stage Concrete Strength

H S Abdelgader[1,2] **and A S El-Baden**[1]
[1]Tripoli University, Faculty of Engineering, Civil Engineering Dept., Tripoli, Libya

E-mail: hakimsa@poczta.onet.pl

Abstract. Two-stage concrete (TSC) is an innovative concrete that does not require vibration for placing and compaction. TSC is a simple concept; it is made using the same basic constituents as traditional concrete: cement, coarse aggregate, sand and water as well as mineral and chemical admixtures. As its name suggests, it is produced through a two-stage process. Firstly washed coarse aggregate is placed into the formwork in-situ. Later a specifically designed self compacting grout is introduced into the form from the lowest point under gravity pressure to fill the voids, cementing the aggregate into a monolith. The hardened concrete is dense, homogeneous and has in general improved engineering properties and durability. This paper presents the results from a research work attempt to study the effect of silica fume (SF) and superplasticizers admixtures (SP) on compressive and tensile strength of TSC using various combinations of water to cement ratio (w/c) and cement to sand ratio (c/s). Thirty six concrete mixes with different grout constituents were tested. From each mix twenty four standard cylinder samples of size (150mmx300mm) of concrete containing crushed aggregate were produced. The tested samples were made from combinations of w/c equal to: 0.45, 0.55 and 0.85, and three c/s of values: 0.5, 1 and 1.5. Silica fume was added at a dosage of 6% of weight of cement, while superplasticizer was added at a dosage of 2% of cement weight. Results indicated that both tensile and compressive strength of TSC can be statistically derived as a function of w/c and c/s with good correlation coefficients. The basic principle of traditional concrete, which says that an increase in water/cement ratio will lead to a reduction in compressive strength, was shown to hold true for TSC specimens tested. Using a combination of both silica fume and superplasticisers caused a significant increase in strength relative to control mixes.

1. Introduction

Placement of concrete in the places that are difficult to reach, such as concrete repairing or underwater applications, is one of the challenging tasks for construction engineering, due to the risk of segregation and bleeding or washout of concrete constituents. Thus, the quality of concrete is highly influenced by its placement technique [1-3]. The goal when applying any of the placement techniques is to place

[2] To whom any correspondence should be addressed.

concrete into the formwork with minimum segregation, minimum honeycombing and maximum possible homogeneity of the concrete constituents, in order to achieve a concrete with high performance and durability properties. Concrete time dependent deformations and consequently cracking have been a growing concern. This has been the case in modern concrete work due to the use of higher content of mineral and chemical admixtures, low water-to-cement ratio and extended exposure to severe conditions during service; all of which may contribute to increasing volume instability in concrete. For example, minimizing of shrinkage can be ensured by the usage of adequate mix proportions, thorough curing and the use of shrinkage compensating cement [4, 5]. It was attested that using non-traditional concrete in engineering applications has been considered as an efficient solution to overcome challenges of limitations of the use of normal conventional concrete. Such new types of concrete, which have been developed and produced, are completely different from the conventional concrete in the methods of mixing, handling, pouring, consolidation, behaviors, cost, etc. Based on the technology of ready-mix self-compacting concrete (SCC), a new type of non-conventional concrete has been introduced and named as: two-stage concrete (TSC) or slurry infiltrated concrete (SICON) [3]. As the name would suggest, TSC is produced through a two-stage process. Firstly washed coarse aggregate is placed in the designated destination. Later a specifically designed grout is introduced into the form from the lowest point under gravity pressure to fill the voids, cementing the aggregate into a monolith [3], as presented in figure 1. As concrete technology develops, basic grouts for TSC can be modified to produce structural concrete more effectively. Such grouts may be modified chemically by using suitable combinations of mineral admixtures, such as silica fume (SF) and superplasticizers, or mechanically by using a specially designed high-speed mixer. Silica fume imparts very good improvement to rheological, mechanical and chemical properties. It improves the durability of the concrete by reinforcing the microstructure through filler effect and thus reduces segregation and bleeding [6, 7].

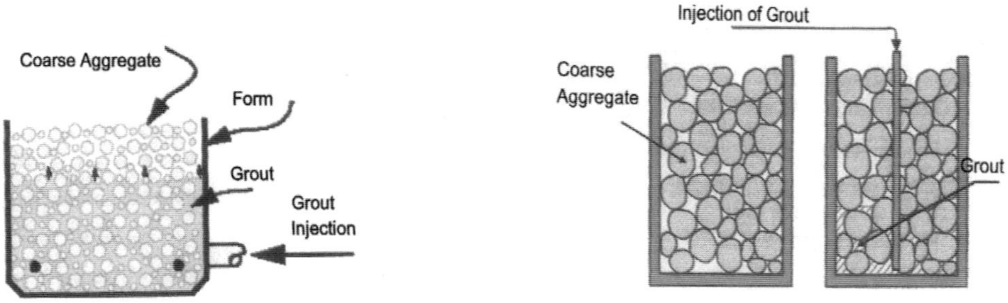

Figure 1. Two stage concrete basic technique.

This paper presents the results from a research work attempt to study the effect of silica fume (SF) and superplasticizers admixtures (SP) on compressive and tensile strength of TSC using various combinations of water to cement ratio (w/c) and sand to cement ratio (c/s). Thirty six concrete mixes with different grout constituents were tested. From each mix twenty four standard cylinder samples of size (150mmx300mm) of concrete containing crushed aggregate were produced. The tested samples were made from combinations of w/c equal to: 0.45, 0.55 and 0.85, and three c/s of values: 0.5, 1 and 1.5. Silica fume was added at a dosage of 6% of weight of cement, while superplasticizer was added at a dosage of 2% of cement weight.

2. Experimental program and discussion

2.1. Selection of materials

Materials, mix proportions and sample preparation procedures will be presented in this section as follows.

Cement. Throughout the research, ordinary Portland cement (Type I) conforming to ASTM-C150 [8] was used. Different quality control laboratory tests were conducted and the results are tabulated in table 1.

Water. Ordinary potable water available in the laboratory was used.

Fine Aggregate. The fine aggregate used in the manufacture of grout was natural beach sand with specific gravity 2.68 and maximum size 2 mm, conforming to ASTM-C33 [9]. The grading of the fine aggregate obtained from the sieve analysis is shown in figure 2.

Coarse Aggregate. The coarse aggregate used was crushed dolomite limestone of specific gravity 2.68, crushing value of 18.83%, abrasion value of 23.81%, and absorption value of 1.66%. Two aggregates of maximum size 19 mm and 10 mm were combined together to obtain the best combined aggregate grading as shown in figure 2.

Table1. Properties of cement.

Test Name	Result	Specifications Limitations
Standard Consistency	28%	32-27%
Initial setting Time	2:25 hr	≥ 45 Min.
Final Setting Time	3:40 hr	≤ 10 hrs
Soundness	1 mm	≤ 10 mm
Compressive strength (3 days)	22 MPa	≥ 21 MPa
Compressive strength (7 days)	30 MPa	≥ 24 MPa
Compressive strength (28 days)	40 MPa	≥ 39 MPa

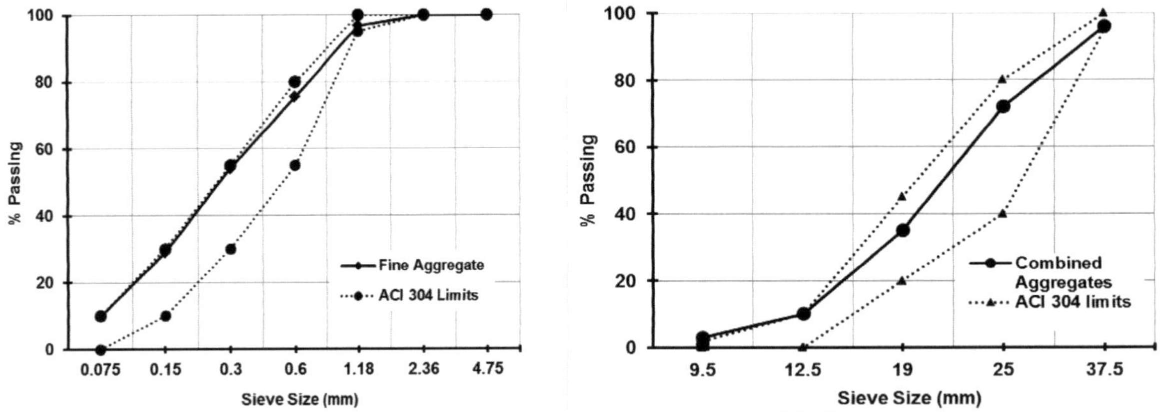

(a) Fine aggregate (b) Coarse aggregate

Figure 2. Fine aggregate and coarse aggregate sieve analysis grading.

Chemical Admixture. A viscocrete based superpalsticizer (SP) imported from SIKA-company is the only kind of admixture used in this research. It is a liquid material with a chemical composition consisting mainly of naphthalene-formaldehyde derivative, trade name 'SikaMent-163'. Superplasticizer was added at dosage rate of 2% by weight of cement. The superplasticizer dosage was mixed with the designated water and added after dry mixing of the mix constituents was completed.

Mineral Admixture. Silica fume (SF) powder containing 97% SiO_2 and having a specific surface area of about 20,000 m^2/kg was the only mineral admixture used in this research. The silica fume was used on dry basis as an additive at a rate of 6% by weight of cement.

2.2. Grout mix preparation

The selection of water-cement-sand ratios is more critical in TSC because the amounts of sand and water control the pumpability and propagation of grout, therefore, an essential requirement should be met in the production of TSC [5]. American Concrete Institute (ACI) method was used during this research to obtain the self-compacted grout mixtures proportions, followed by adjustments to the mix constituents to characterize the mixes for filling ability, passing ability and resistance to segregation [5]. The void content of the preplaced aggregate was 47 % and the bulk density – 1430 kg/m^3, based on a previous preliminary investigations conducted by the authors and others [5,7]. A total of thirty six mixes divided into four groups of concrete mixes were designed and implemented. Each group comprised a combination of nine mixes having the following constant definitions: 1 -The w/c ratio of: 0.45, 0.55 and 0.85. 2- c/s ratio of: 0.5, 1, 1.5. The first group was designed and implemented without using any chemical or mineral additives and was denoted in the text as (No Admixture). Superplasticizers were only used in the second group as a chemical additive at a dosage rate of 2% by weight of cement, denoted in the text as (SP). Silica fume was only used in the third group as a mineral additive at a dosage rate of 6% by weight of cement, denoted in this paper as (SF). A combination of both silica fume and superplasticizer was used in the fourth group with the same dosages as in the other groups and denoted in the text as (SP+SF). Grout preparation was accomplished by combining ingredients with an electric mixer for about two minutes in a dry basis then followed by other three minutes of wet mixing to achieve the desired grout uniformity and consistency after adding water and the other specified additives. Table 2 presents the contents of all the TSC mixes investigated.

Table 2. Grout mixture proportions.

Mix No.	w/c	c/s	Cement	Sand	Water
				(kg/m^3)	
1		1/0.5	524.093	262.046	235.841
2	0.45	1//1	438.367	438.367	197.841
3		1/1.5	376.740	565.115	169.533
4		1/0.5	474.370	237.185	260.903
5	0.55	1//1	403.032	403.032	221.667
6		1/1.5	350.345	525.518	192.689
7		1/0.5	369.268	184.634	313.877
8	0.85	1//1	324.550	324.550	275.867
9		1/1.5	289.492	434.238	246.068

IOP Conf. Series: Materials Science and Engineering **96** (2015) 012043 doi:10.1088/1757-899X/96/1/012043

2.3. Methods of placement and testing
The placement of TSC was experimentally modelled as shown in figure 1. One injecting polyvinyl chloride (PVC) pipe 700 mm high and 20 mm in diameter was vertically inserted till about 20 mm from the bottom of the mould (150mmx300mm). The aggregate was cast slowly and gradually, being distributed as uniformly as possible inside the mould. The aggregate completely filled the entire mould and then the prepared grout mix was pumped directly through the injecting tube by gravity. After one day of casting all specimens were removed from their moulds and stored in water curing tanks in the laboratory climate. The experimental program consisted of a series of unconfined compression tests and split tensile tests on the samples prepared from different grout mixtures after 7, 28, 90 and 120 days of water curing in the laboratory climate. Cylindrical specimens of size 150 x 300 mm were prepared in the same manner and tested. On the other hand, grout specimens were tested in compression using standard cubes of size 100x100x100 mm. Nine specimens from each mix were prepared and tested in unconfined compression and split tension at 7, 28, 90 and 120 days, at a rate of three samples for each kind of tests. Unconfined compression tests on TSC cylinders were tested in accordance with standards (ASTM-C873) [10] using a universal hydraulic testing machine with a capacity of 2200 kilo newton (KN) and a rate of loading of approximately 7 KN per minute. The samples were subjected to loading till failure. Splitting tensile tests were also conducted on three specimens of each concrete according to the procedures outlined in (ASTM-C496) [11].

3. Analysis and discussion

3.1. Compressive strength test results
Table 3 reports compressive strength results at different ages for the tested TSC mixtures using various ratios of w/c, c/s and different chemical and mineral admixtures. Results indicted generally that the TSC mixtures exhibited comparable compressive strength trend with respect to the type of admixtures. Generally, at the same age and w/c ratio, it was observed that increasing c/s would reduce slightly the strength of TSC, which is obviously related to the reduction in flowability and passability of the grout, which leads to production of TSC microstructure with honeycombed structure with partial binding of the aggregate skeleton. Usage of SP enhances the flowability and the passability of the grout and leads to improved strength results. Moreover, using SF alone or in combination with SP exhibited better compressive strength results compared to the mixes with no admixtures or with SP only, and this could be attributed to the following: a) Higher fluidity of grout using SP, which enables the grout to fill all the voids between aggregate particles; b) Pozzolanic activity of SF causes acceleration and increase in the hydration rate and products respectively, which in turn leads to enhancement of particle packing density of concrete microstructure. For example, at c/s ratio of 1:0.5 using a combination of SF and SF causes the 28 days strength to be increased by an average percentage of 40% for all w/c ratios considered. On the other hand, increasing the c/s for the same w/c causes an average percentage reduction in strength of 12% at 28 days, which is not a significant value.

Table 3. Compressive strength results of TSC.

w/c	Type of admixture	Curing Period (days) 7	28	120	Curing Period (days) 7	28	120	Curing Period (days) 7	28	120
		c/s=1: 0.5			c/s=1:1			c/s=1:1.5		
0.45	SF	16.79	18.49	26.59	16.60	17.35	25.65	13.58	16.22	25.09
	SP	12.07	18.67	20.18	13.00	17.35	19.05	15.40	15.47	17.17
	SF+SP	17.35	25.65	28.29	16.79	22.63	26.03	13.95	22.26	25.65
	No admixtures	11.13	15.28	20.37	11.13	14.15	18.11	9.24	13.00	16.03
0.55	SF	13.29	15.85	24.89	11.88	15.10	20.56	10.75	13.58	19.81
	SP	14.34	16.79	20.75	12.64	16.03	20.56	11.32	14.52	18.49
	SF+SP	13.20	21.32	26.88	12.83	19.24	21.50	11.70	15.00	21.87
	No admixtures	8.67	12.45	15.85	8.49	9.05	13.58	7.73	8.49	12.07
0.85	SF	7.36	10.56	13.58	5.28	9.43	12.07	4.90	9.43	11.69
	SP	6.98	13.20	16.22	8.11	11.51	15.47	8.86	10.94	14.52
	SF+SP	9.05	14.71	16.98	9.81	11.70	15.85	10.19	11.32	15.47
	No admixtures	6.60	8.29	11.69	4.53	7.36	10.19	3.96	5.66	9.81

According to the design algorithm presented in [4, 5], a relationship between compressive strength for (fc`) and w/c ratio has been assumed and presented in Equation1, as the following:

$$fc` = A + B*(w/c) + C*(w/c)^{D} + E*(c/s) \quad (1)$$

where fc` represents the estimated compressive strength of TSC, w/c is the water-to-cement ratio and c/s is the cement-to-sand ratio. Table 4 shows the values of the regression coefficients.

Table 4. Regression coefficients for equation 1.

Type of admixtures	Regression Constants A	B	C	D	E	Correlation Coefficients
SF	23.838	-19.641	7.022	-0.815	-2.715	0.975
SP	-110.44	-172.456	291.909	0.418	-3.823	0.947
SF+SP	8.73	7.265	2.399	-2.456	-2.395	0.977
No admixtures	7.34	5.278	1.427	-2.647	-3.357	0.979

3.2. Tensile strength test results

The splitting tensile strength is commonly used to evaluate the tensile strength of TSC. As shown in table 5, the TSC tensile strength followed a similar trend to that of the compressive strength. The splitting tensile strength at 7 days was low, especially for the mixtures without admixture and then its rate of development increased with time. For instance, the 7-days tensile strength for the mixture incorporating no admixtures was 20 % lower than that of the SF and SP combination mixture. Generally, the higher the TSC compressive strength, the greater the TSC tensile strength was, in agreement with the previous studies [5, 12, 13].

Table 5. Tensile strength results of TSC.

w/c	Type of admixture	Curing Period (days) 7	28	120	Curing Period (days) 7	28	120	Curing Period (days) 7	28	120
		c/s=1: 0.5			c/s=1:1			c/s=1:1.5		
0.45	SF	2.075	2.428	2.785	1.933	2.357	2.738	1.789	1.982	2.547
	SP	1.873	2.405	2.783	1.655	1.950	2.542	1.137	1.693	2.167
	SF+SP	2.452	2.972	3.772	1.982	2.765	2.207	1.796	2.363	3.111
	No admixtures	1.745	1.837	2.547	1.604	1.673	2.406	1.507	1.603	1.984
0.55	SF	1.620	1.626	2.359	1.677	1.653	2.078	1.430	1.510	1.698
	SP	1.666	2.213	2.266	1.636	2.071	2.165	1.415	1.688	2.075
	SF+SP	1.716	2.122	2.265	1.691	1.839	2.170	1.603	1.608	2.124
	No admixtures	1.424	1.522	1.768	1.383	1.518	1.628	1.180	1.190	1.579
0.85	SF	1.226	1.508	1.603	1.037	1.340	1.536	0.713	1.320	1.483
	SP	0.852	1.450	2.452	1.084	1.320	1.549	1.179	1.320	1.601
	SF+SP	0.943	1.674	2.304	1.226	1.406	1.699	0.943	1.264	1.625
	No admixtures	0.884	1.461	1.529	0.870	1.320	1.622	0.578	1.226	1.447

3.3. Compressive and tensile strength relationships

The results of this investigation show that there is a good correlation between the compressive strength and tensile strength of TSC. As the compressive strength increased, the tensile strength was also found to increase in the same manner. In the present work, however, the following formula presented in Equation 2 was developed by regression analysis to relate tensile strength (ft) to compressive strength (fc`). The graphical presentation of this relationship is shown in figure 3.

$$ft=A+B*(fc`)+C*(fc`)^{D} \tag{2}$$

where ft is tensile strength and fc` is compressive strength. Table 6 shows the values of the regression coefficients.

Table 6. Regression coefficients for equation 2.

Type of admixtures	Regression constants A	B	C	D	Correlation Coefficients
SF	-6.048	0.223	26.158	-0.674	0.943
SP	-78.020	-0.003	75.050	0.022	0.901
SF+SP	66.074	0.584	-52.448	0.122	0.907
No admixtures	-4.197	0.249	24.023	-0.871	0.936

IOP Conf. Series: Materials Science and Engineering **96** (2015) 012043 doi:10.1088/1757-899X/96/1/012043

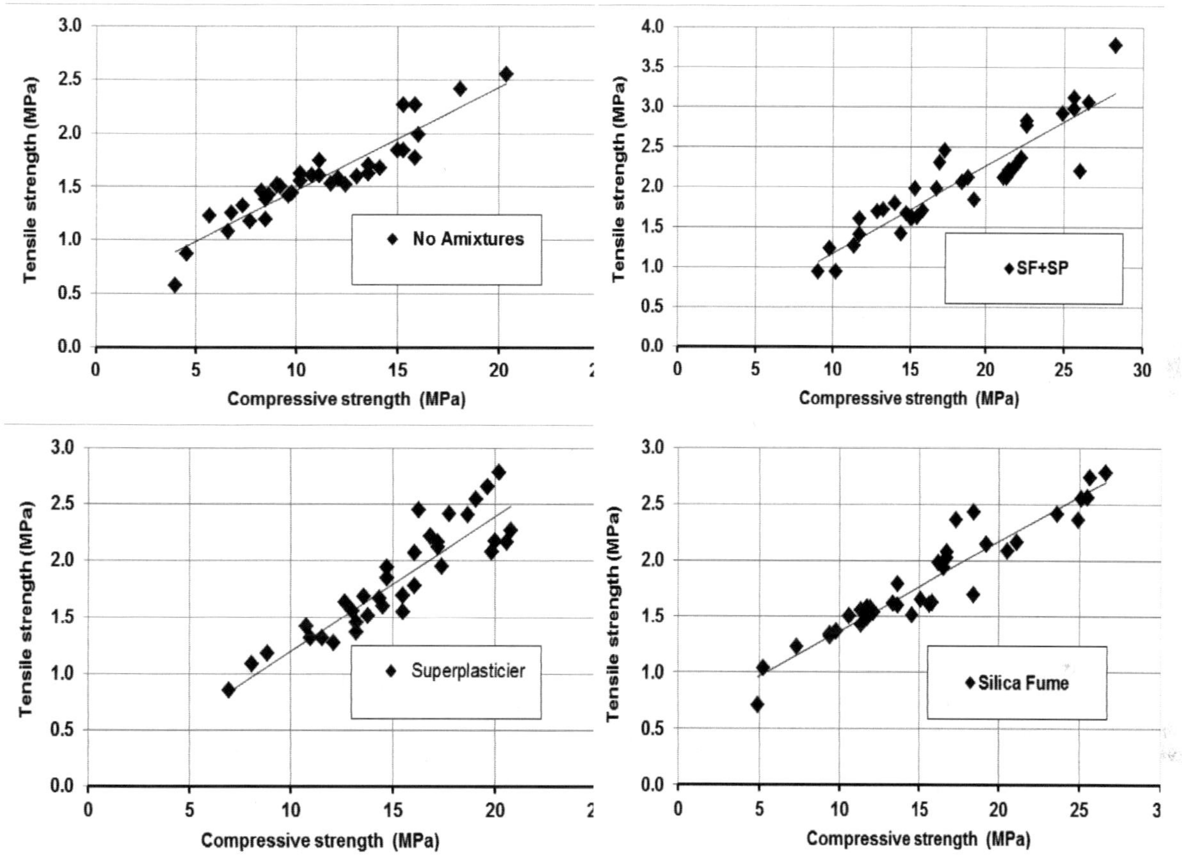

Figure 3. Relationship between compressive strength and tensile strength with and without admixtures.

4. Conclusions

1. As the method of placement in TSC is entirely different from that of NC, using chemical admixtures is necessary to satisfy the requirement for the pumping ability of grout. The superplasticizers were found to be the most suitable admixture in TSC as it enhances the flowability and the passability of the grout and leads to better strength results.
2. The compressive strength of TSC was tested with and without admixture for all grout proportions. On the basis of the results, a correlation between compressive strength and grout proportions was statistically derived.
3. The splitting tensile strength of TSC was investigated for all grout proportions. On the basis of these results, a relationship between tensile and compressive strength of TSC has been statistically derived.
4. The fractured specimens of TSC showed that a large proportion of failures occurred by cracking through the coarse aggregate particles.
5. SF causes acceleration and increase in the hydration rate and products respectively, which in turn leads to enhancing of particle packing density of concrete microstructure.

6. Using SF alone or in combination with SP exhibited better compressive strength results compared to other mixes and this could be attributed the pozzolanic activity of SF.
7. The authors believe that there are many aspects of the failure mechanism of TSC that require clarification through further theoretical and experimental studies. Among them the most notable feature is micro-cracking.

5. References

[1] Neville A 1981 *Properties of Concrete* Fourth edition (London: Pitman Books Limited)
[2] Neville A and Brooks J 1987 *Concrete Technology* First edition (London: Longman Scientific and Technical)
[3] ACI Committee 304 2005 *Guide for the Uuse of Preplaced Aggregate Concrete for Structural and Mass Concrete Applications* (USA, Michigan: American Concrete Institute, Farmington Hills) pp 21–24
[4] Abdelgader H 1999 How to design concrete produced by a two-stage concreting method *J. of Cem. and Conc. Research* **29** (3) 331–337
[5] Abdelgader H and Elgalhud A 2008 Effect of grout proportion on strength of two-stage concrete *Structural Conc. J.* **3** 163–170
[6] Kin J 1959 Concrete by intrusion grouting *Handbook of Heavy Construction* Second edition (New York: McGraw-Hill)
[7] Abdelgader H and Górski J 2003 Stress-strain relations and modulus of elasticity of two-stage concrete *ASCE J. of Mater. in Civil Engin.* **4** 251–255
[8] ASTM-C150 1994 Specification for portland cement, American society for testing and materials (Philadelphia, Pennsylvania)
[9] ASTM C33 1997 Specification for concrete aggregates (Philadelphia, Pennsylvania)
[10] ASTM C873 2010 Standard test method for compressive strength of concrete cylinders cast in place in cylinder molds, American society for testing and materials (Philadelphia, Pennsylvania)
[11] ASTM-C496/C496M 2004 Standard test method for splitting tensile strength of cylindrical concrete specimen, American society for testing and materials (Philadelphia, Pennsylvania)
[12] Najjar M, Soliman A and Nehdi M 2014 Two-stage concrete made with single, binary and ternary binders *Mater. and Struct. Conc. J.* **62** 47–58
[13] Najjar M, Soliman A and Nehdi M 2014 Critical overview of two-stage concrete: properties and application, and building Materials *Mater. and Struct. Conc. J.* **62** 47–58

Synthesis and Characterization of Reactive Powder Concrete for its Application on Thermal Insulation Panels

V Chozas[1], Í Larraza[1,4], J Vera-Agullo[1], N Williams-Portal[2], U Mueller[2], N Da Silva[2] and M Flansbjer[3]
[1]ACCIONA Infrastructure SA, Technological Centre, Valportillo II 8, Alcobendas, Spain
[2]CBI Cement and Concrete Research Institute, 501 15 Borås, Sweden
[3]SP Technical Research Institute of Sweden, 501 15 Borås, Sweden

E-mail: inigo.larraza.alvarez.ext@acciona.com

Abstract. This paper describes the synthesis and characterization of a set of textile reinforced reactive powder concrete (RPC) mixes that have been prepared in the framework of the SESBE project which aims to develop façade panels for the building envelope. In order to reduce the environmental impact, high concentration of type I and II mineral additions were added to the mixtures (up to 40% of cement replacement). The mechanical properties of the materials were analysed showing high values of compression strength thus indicating no disadvantages in the compression mechanical performance (~140 MPa) and modulus of elasticity. In order to enable the use of these materials in building applications, textile reinforcement was introduced by incorporating layers of carbon fibre grids into the RPC matrix. The flexural performance of these samples was analysed showing high strength values and suitability for their further utilization.

1. Introduction

There is a vastly growing demand for increased energy efficiency of the buildings we live and work in, which is one of the great challenges of the European construction sector, including materials and processes. The EU has set ambitious energy-saving goals: to achieve a reduction 20% of the total energy consumption, a reduction of 20% of Green House Gases (below 2005 level) and for 2050 to ensure that most buildings and districts will be energy-neutral and have zero CO_2 emissions (Energy Policy set by the European Council in March 2007).

This work has been developed within the SESBE project, which aims to develop façade elements (sandwich and half sandwich) with high insulation ability and at the same time a reduced thickness in

[4] To whom any correspondence should be addressed.

conjunction with superior mechanical and durability properties. Reactive powder concrete (RPC) has been chosen as the material for the outer layer of such façade elements. RPC is a cement-based material characterized by the property that its microstructure is optimized by a precise gradation of all particles in the mix to yield maximum compactness and a greatly reduced water/powder ratio, usually below 0.2. The theory of particle packing has been previously investigated in many studies [1,2] and it is based on using specific components which complement each other to form an evenly distributed particle size curve which reaches from particles ≤ 0.1 µm to ≥ 1 mm. The optimization of the amount of the single binder and aggregate components will reduce the w/b ratio, the amount of super plasticizer and will enhance the workability of the fresh RPC mix.

It uses extensively the pozzolanic properties of highly refined silica fume and other mineral additions. The superior properties are achieved by the optimization of the binder chemistry and by strongly reducing the binder porosity, in particular capillary porosity. In this context RPC should be more appropriately seen as a cold ceramic rather than a concrete [3]. Due to these properties, it is increasingly being utilized also for external façade cladding thus enabling a considerable reduction in the thickness of concrete elements. However, commercial RPC formulations on the market have drawbacks, the present work has been aimed to investigate and mitigate such barriers. In terms of sustainability due to their high cement content and heat curing which is often applied to increase final strength and material density. In this work, supplementary cementitious materials (SCMs) such as F class fly ash (FA) and ground granulated blast furnace slag (GGBS) have been used in order to partially replace the commercial cement. Moreover, in this work the RPC materials have been cured at room temperature.

Its toughness makes RPC a fairly brittle material, in order to improve this behaviour, fibre and textile reinforcement is typically incorporated in panels made of RPC. In this way, the reinforcement will prevent sudden collapse of the panel upon first cracking and would be activated under increased strain. In the present work, in order to enable a significant reduction of thickness and reduce the brittleness of these elements, panels of reactive powder concrete reinforced with carbon fibre textile grids have been developed. The composite behaviour of RPC reinforced by textiles has been investigated by uniaxial tensile and four-point bending tests.

The presented work has been carried out at two locations: ACCIONA in Spain and CBI in Sweden, in order to use local materials, being especially interesting in the case of SCMs which properties vary depending on the procedure. In total, three different mix formulations were designed and tested in terms of workability and mechanical properties. The matrices were reinforced with carbon fibre textile grids primarily to enhance the ductility and tensile strength of the material.

2. Materials and methods

This chapter describes the materials used for the synthesis of the RPC mixes, the mixing procedure and the analytic methods for their characterization.

2.1. Materials

Three RPC matrices were prepared with commercially available materials. The proportions of the components used in each of the cases have been collected in table 1. In total, three final mixes were designed, synthesized and characterized, two by CBI (C1 and C2) and one by ACCIONA (A1).

Table 1. Materials used for the RPC mix design. All the amounts are given in wt% except the water/cement ratio.

Component	C1	C2	A1
Cement[a]	31	27	22
Silica Fume	3.1	2.7	1.1
SCM[b]	6.7	9.2	15.7
Aggregates + quartz filler[c]	51.6	53.3	52.5
Water	6.3	6.3	6.6
Superplasticizer	1.3	1.3	1.9
w/c	0.23	0.27	0.30

[a]For C1 and C2, CEM II/A-V 52.5 cement was used. For A1 CEM I 52.5R was used.
[b]F class FA was used in both cases, for A1 GGBS was also used (GGBS/FA 2.34)
[c]C1 and C2 were prepared with quartz filler and aggregates (0/0.2, 0.1/0.3, and 0.2/1 mm), A1 was prepared with quartz powder (0/100 μm) and aggregates (0/600 μm and 0/2 mm)

The mix design was optimized by a particle packing model according to Andreasen [4] calculated with the software Emma by ELKEM [5]. The mix design was adjusted to locally available materials in each location. Mix A1 had ground granulated blast furnace slag as the main cement replacement.
As it has been previously mentioned, reinforced RPC was also explored in this work. Carbon fibre textile grids were incorporated in the RPC matrices and their flexural behaviour was investigated by means of four-point bending tests. Two alternatives consisting of one or two layers of a 2D carbon grid (figure 1) have been analysed. Additionally, a 3D carbon grid composed of two layers interconnected by a low-modulus polyester (PET) spacer with a thickness of 12 mm (figure 2) was used. Initial works pertaining to the 3D carbon reinforcement have been initiated and will be published in a future paper. The grids are made of carbon rovings consisting of fibre bundles of 24K (thousands of fibres), the surfaces of which were initially coated by approximately 15 wt-% styrene-butadiene resin (SBR). The mesh sizes are indicated in figures 1 and 2 for both types of grids. In this work, the reinforced RPC panels have been denoted as RC1, RC2 and RA1, whereby R stands for reinforced followed by the matrix type. The inclusion of an additional polymer coating to the carbon grid surface was also investigated and is denoted with "-C".

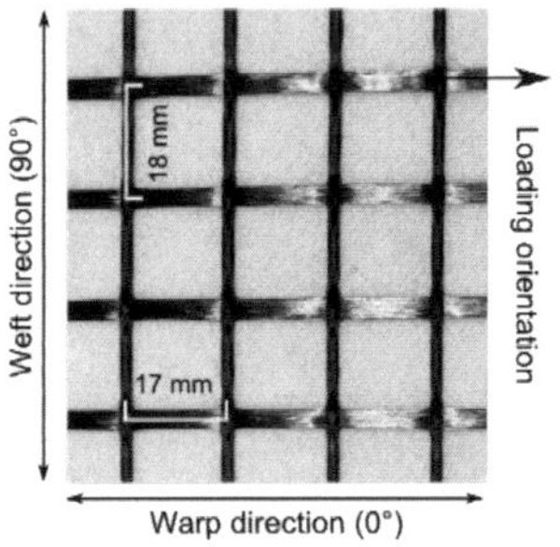

Figure 1. 2D carbon fibre textile grid.

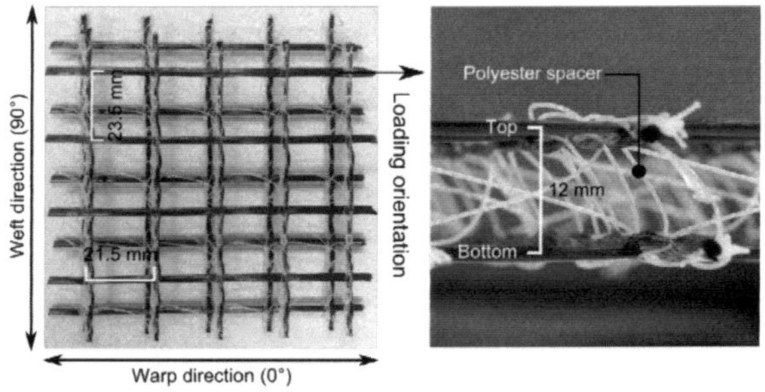

Figure 2. 3D carbon fibre textile grid.

2.2. Methods

2.2.1. Mix protocol.
In all cases, the RPC mixes were prepared according to the following procedure: firstly, the cement was homogeneously mixed with the supplementary cementitious materials (fly ash and blast furnace slag). Then, 75% of the previously mixed solution containing water and the superplasticizer additive was added and mixed for 4 minutes. After that, the aggregates, quartz powder and silica fume were incorporated into the mixture and all the components were mixed for 5 minutes. Finally, the remaining (25%) water/superplasticizer solution was added and mixed for 5 more minutes. All the mixtures were prepared in 50 L planetary concrete mixers and the mixing rate was set at 60 rpm. The rheology of the RPC fresh mortar matrixes was analyzed by a flow table test according to the European standard EN 1015-3. The result of the test consisted of the maximum diameter of the mortar mass measured after 60 s.

2.2.2. Preparation of the carbon grid reinforced RPC samples.
The samples were cast in wood molds. The form consisted of a panel shape in the size of 700 x 100 x 20 mm^3, see figure 3. In the case of the 2D carbon grid, 2 layers were included in each sample to attempt to match the geometry and placement of the 3D grid. In samples with the 3D grid, only one layer was used as the 3D grid already incorporates two carbon textile layers. In order to position the textile grids properly within the samples, they were cut a little larger and fixed by the formwork sidewalls. Thereafter, the RPC paste was poured into the molds penetrating through the mesh and allowed to harden at room temperature, as per figure 4. The specimens were demolded after 24 hours and cured in a wet room at 20 °C and 96% relative humidity.

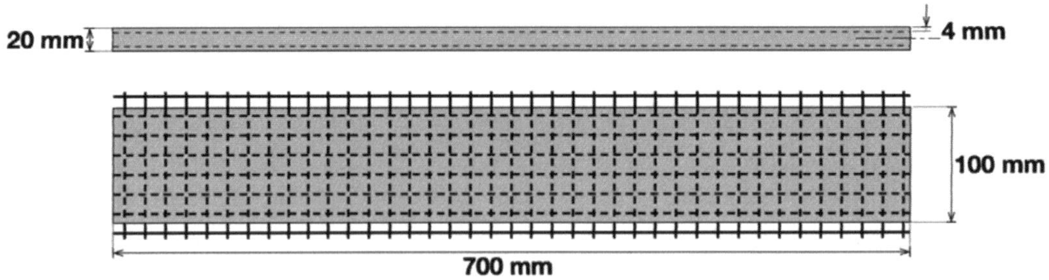

Figure 3. Schematic representation of the reinforced RPC panels.

Figure 4. Photography of the RPC being cast with two 2D carbon grids as reinforcement.

2.2.3. Measurements.
The mechanical properties of the RPC materials with and without reinforcement were investigated by analysing different parameters. Compressive strength was measured on prismatic and cubic unreinforced specimens for screening purpose (160 x 40 x 40 mm^3 and 100x 100 x 100 mm, respectively) and on cylinders (Ø 54 x 100 mm and Ø 150 x 300 mm). Tensile strength, modulus of elasticity as well as Poisson's ratio were also measured in the latter specimens. The compressive strength and the elastic properties were measured according to standards EN 12390-3 and EN 12390-13, respectively. The uniaxial tensile test was performed according to the RILEM TC 232-TDT draft recommendations for textile reinforced concrete. Deformations in the tensile test were measured by a Digital Image Correlation System (DIC), while deflections in the flexural tests were measured by standard displacement transducers at midspan. The flexural behaviour of the reinforced RPC panels was measured using four-point bending tests according to standard EN 12390-5.

3. Results and discussion

3.1. Characterization of the RPC matrices

3.1.1. Rheology.
After blending the mixes, the mixes behaved as self-compacting and bleeding was not observed. The slump test was carried out on fresh concrete according to EN 12350-2. The maximum diameter reached was 65 to 70 cm (C1 and C2) and 90 cm (A1), respectively. The slump flow values evidence the suitability of the synthesized RPC pastes for their potential use at larger scale.

3.1.2. Mechanical characterization of the unreinforced RPC matrices.
The mechanical properties of the RPC samples were evaluated by measuring the compressive and tensile strength, elastic modulus, ultimate strain and Poisson's ratio. The values of these parameters, measured at 28 days of age, are summarized in table 2. The results show that the differences between the mixes are not significant. However, mix C1 showed a slightly higher value of compressive strength, attributed to the higher proportion of cement in its composition. The elevated value of 50

GPa for the E-modulus and the post peak behaviour indicate the extreme brittleness of the material with a sudden and almost explosive failure.

Table 2. Results of the mechanical properties of unreinforced RPC matrices tested at 28 days of age.

	Compressive strength (MPa)	E-modulus (GPa)	Ultimate strain (‰)	Poisson ratio	Tensile strength (MPa)
C1	147.2 (2.3)	49.7 (1.7)	3.9 (0.2)	0.22 (0.02)	5.1 (0.5)
C2	135.7(4.3)	49.9 (2.4)	4.1 (0.3)	0.22 (0.01)	3.9 (0.7)
A1	135.0 (2.4)	48.3 (0.1)		0.21 (0.02)	

The development of the compressive strength of the three different final RPC mixes is shown in figure 5. All three strength curves showed a similar course. After 28 and 56 days of age, respectively, the differences in strength were only marginal for all three mixes. However, mix C2 showed a lower strength within the first 24 h of hydration but still well above 20 MPa. It is important to note that after only a short time period compressive strength was significant. Initial setting time was determined for C1 at 5.5 h and C2 at 7 h. The slower setting and lower initial strength of mix C2 was attributed to the higher content of fly ash in the mix.

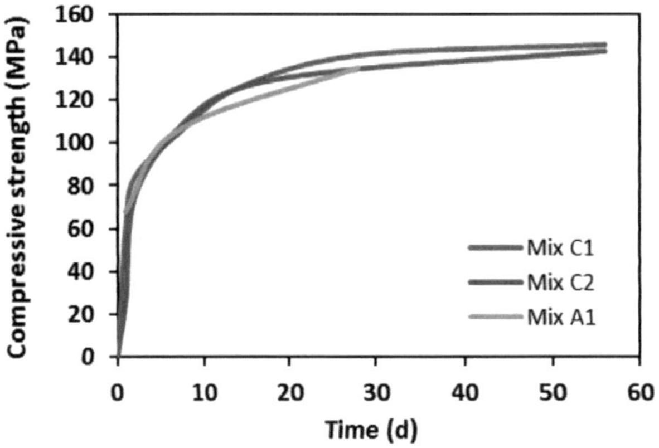

Figure 5. Development of compressive strength for three different mixes.

3.1.3. Characterization of the reinforced RPC materials.

The RPC mixes with one and two layers of textile reinforcement grid were tested. Figures 6 and 7 show the load versus deflection of the specimens composed of A1 and C1 matrices reinforced with one layer and two layers of carbon grid, respectively. In both cases, during the initial stage it was possible to appreciate the contribution of the RPC matrix until its brittle failure, reaching load values between 1000 and 1400 N. In the case of RA1 specimens, after the first load drop, it was possible to observe the contribution of the reinforcement, represented by an increment of the load/deflection curve, reaching a value close to that shown in the failure peak. After this stage, the specimens showed a gradual loss of their flexural capacity until a final failure of the reinforcement occurred after large deflections. In the case of RC1, the contribution of the reinforcement and the so-called increase in flexural capacity after first cracking of the matrix was not observed and the test was terminated due to the continued pull-out of the reinforcement. It is thought that the observed behaviour for RC1 could result from poor bond at the interface of the reinforcement grid and matrix, which should be further investigated.

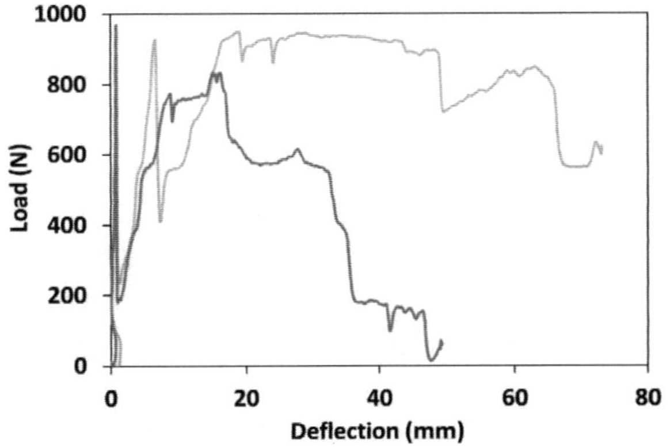

Figure 6. Flexural test results of RA1 with one layer of carbon grid reinforcement (lines correspond to independent tested specimens).

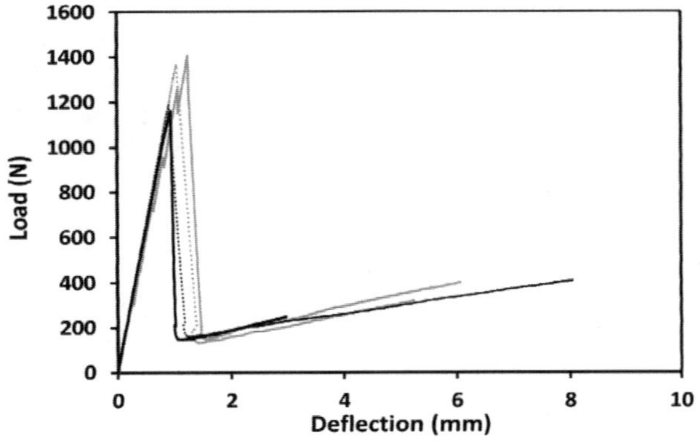

Figure 7. Flexural test results of RC1 with two layers of carbon grid reinforcement (lines correspond to independent tested specimens).

Figure 8 depicts the flexural characterization of the RC1-C specimens reinforced with two layers of carbon grid additionally treated by the aforementioned polymer coating. It can be seen that the three specimens showed a rather similar behaviour, however particularly the initial pre-cracking stiffness differed slightly due to material variability. The first peak corresponds to the contribution of the stiff RPC matrix followed by a series of multiple cracks and so-called crack stabilization, which is a favourable behaviour for this type of reinforced matrix. Compared to RC1 (figure 7), the first cracking load was slightly lower despite the fact that the same matrix is applied in RC1-C. The failure load was also lower than in the case of RC1, which could be explained by the fact that multiple cracking took place causing the matrix to subsequently lose stiffness. In this case, the RPC matrix did not undergo a brittle failure and it was observed that far from losing its flexural capacity, the additionally coated reinforcement led to a continuous increment of the flexural strength to around 80-100% with respect to the unreinforced matrix. It is worth highlighting that the scatter of the results is relatively low, compared to the typical scatter values characteristic of the discontinued fibre reinforced concrete materials.

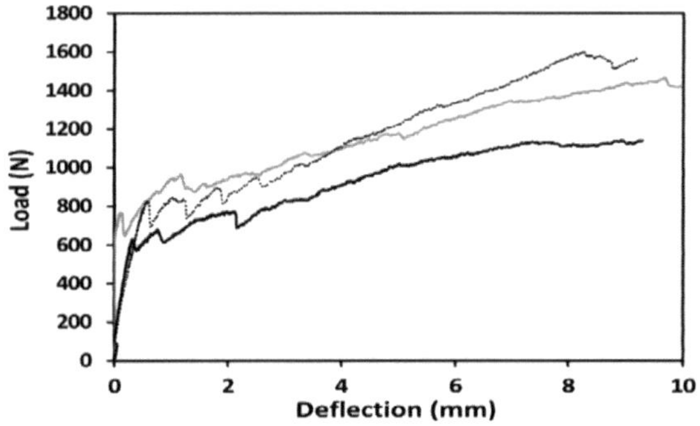

Figure 8. Flexural test of RC1-C with two layers of additionally coated carbon grid reinforcement (lines correspond to independent tested specimens).

4. Conclusions

The presented work has been developed in the framework of a project which has as an ultimate goal to produce panels for the building envelope and improve energy efficiency of the buildings in Europe. For this application, RPC has been chosen as the material to form the external layer of the panels and three sets of RPC mixes have been designed, prepared and characterized at two different locations (CBI and ACCIONA). In order to reduce the environmental impact of the materials, high amounts of mineral additions have been used to replace the cement.

The mechanical analyses evidenced the superior mechanical properties of the RPC matrices, characterized by high values of compressive strength (135-147 MPa) and modulus of elasticity (~50 GPa). In order to counteract the brittleness of the RPC and improve the flexural behaviour, textile reinforcement has been used in the form of one or two layers of carbon fibre textile grids. The flexural behaviour of the resulting panels was studied by four-point bending tests revealing the positive contribution of the reinforcement. This effect was especially evidenced when using two layers with additional coating, whereby the brittle first cracking was followed by multiple cracking and an increment of the flexural load resistance of about 80-100% with respect to the value of the unreinforced matrix.

5. References

[1] De Larrard F and Sedran T 1994 *Cement and Concrete Research* **24** 997
[2] Teichmann T and Schmid M 2004 *Int. Symposium on Ultra High Performance Concrete* (Kassel: Kassel University Press) p 313
[3] Richard P and Cheyrezy M 1995 *Cement and Concrete Research* **25** (7) 1501
[4] Andreasen A H M 1930 *Kolloid-Zeitschrift* **50** 217
[5] [Available]: https://www.elkem.com/silicon-materials/high-performance-concrete

Acknowledgments
This project has received funding from the European Community's Seventh Framework Programme NMP.2013-1 under Grant agreement no: 608950

IOP Conf. Series: Materials Science and Engineering **96** (2015) 012045 doi:10.1088/1757-899X/96/1/012045

Plastometry for the Self-Compacting Concrete Mixes

V Ā Lapsa[1] A Krasnikovs[1], V Lusis[1] and A Lukasenoks[2,3]
[1]Riga Technical University, Institute of Building Production, Faculty of Civil Engineering, RTU Concrete Mechanics Laboratory, Kalku Str. 1, Riga, LV-1658, Latvia
[2]Institute of Mechanics, RTU Concrete Mechanics Laboratory, Riga Technical University, Ezermalas Str. 6, LV-1006, Riga, Latvia

E-mail: Arturs.Lukasenoks@rtu.lv

Abstract. Operative determination of consistence of self-compacting concrete mixes at plant or in construction conditions is an important problem in building practice. The Abram's cone, the Vebe's device, the U-box siphon, L-box or funnel tests are used in solving this problem. However, these field methods are targeted at determination of some indirect parameters of such very complicated paste-like material like concrete mix. They are not physical characteristics suitable for the rheological calculations of the coherence between the stress and strains, flow characteristics and the reaction of the concrete mix in different technological processes. A conical plastometer having higher precision and less sensitive to the inaccuracy of the tests in construction condition has been elaborated at the Concrete Mechanics Laboratory of RTU. In addition, a new method was elaborated for the calculation of plasticity limit τ_0 taking into account the buoyancy force of the liquid or non-liquid concrete mix. In the present investigation rheological test of the concrete mix by use the plastometer and the method mentioned earlier was conducted for different self-compacting and not self-compacting concrete mixes.

1. Introduction

Abram's cone outflow diameter measurement [1], Vebe's device, U-box siphon [2], L-box [3] or funnel tests are usually used to operatively determine rheological properties of self-compacting concrete at site [3].

In Abram's cone method, metallic cone is filled with the concrete mix, then by lifting the cone its content is poured on metallic sheet and outflow diameter is measured (see figure 1 a and b). The time when cone outflow reaches diameter 500 mm is also sometimes measured.

[3] To whom any correspondence should be addressed.

The Vebe consistometer testing method is similar to simple Abram's cone test method for determination of the workability of concrete, but it has the advantage of a mechanized action. After lifting the slump cone, the concrete undergoes vibration to determine its slump (se figure 2).

U-box siphon test method apparatus consists of a vessel that is divided by a middle wall into two sections. An opening with a moving gate is placed between the two compartments. Reinforcement bars are installed at the gate with centre-to-centre spacing of 50 mm. Bars create spacing of 35 mm. The left hand section is filled with concrete, then the gate is lifted. Concrete flows upwards into the other section. The height of the concrete in both sections is measured (see figure 3).

(a)

(b)

Figure 1. a) Abram's cone; b) Abram's cone outflow view (for its diameter measurement).

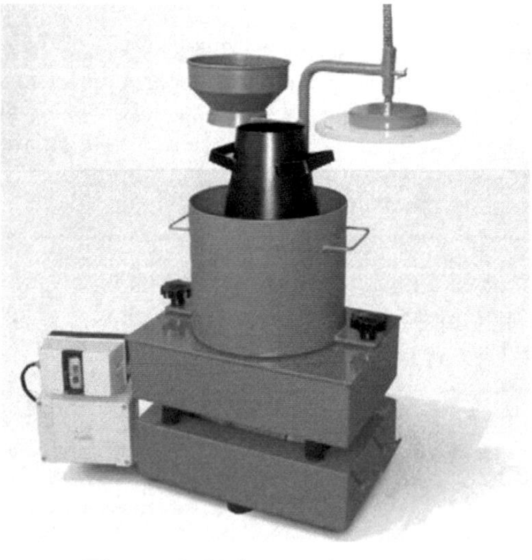

Figure 2. Vebe consistometer.

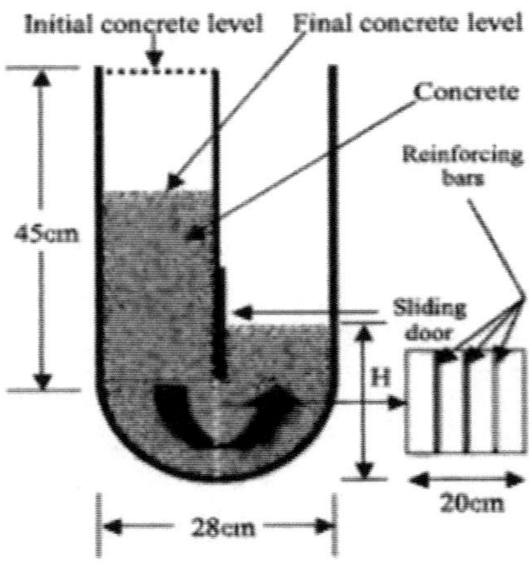

Figure 3. U -box siphon.

L – box test method apparatus consists of a rectangular L-shaped section box. The box consists of a vertical and horizontal section separated by a vertical gate. In front of the gate reinforcement bars are placed. The vertical section is filled with concrete, and then the gate is lifted. Concrete flows into the horizontal section. The height of the concrete at the end of the horizontal section is expressed as a proportion of that remaining in the vertical section (see figure 4).

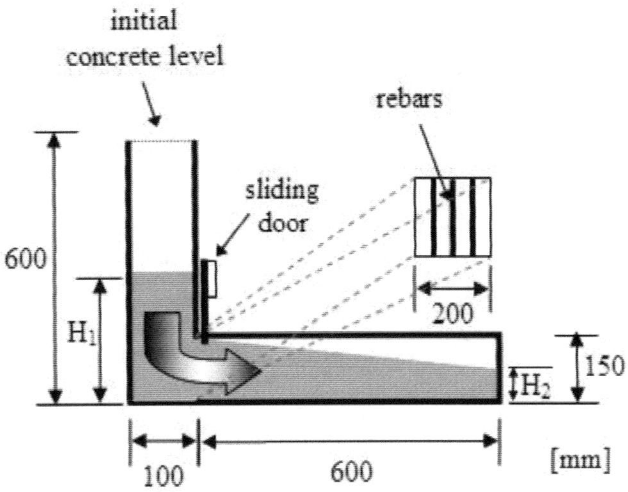

Figure 4. L-box.

V-funnel test method apparatus consists of V - shaped funnel and horizontal gate and the bottom of the funnel. The funnel is filled with concrete and horizotal gate is opened. The time that takes concrete to flow through the apparatus is measured (see figure 5). This method is extended sometimes by letting concrete rest in the funnel for 5 minutes and then test is performed. By this modification self-compacting concrete stability can be determined.

Figure 5. V-funnel test.

However, looking at such very complicated paste-like material like concrete mix, all these field methods can be used only for determination of some conditional characteristics. They are the slump of the conical sample or the outflow spot diameter in the Abram's cone method, degree of self-leveling in siphon, in the U-box method, it is time of flow-out in the funnel method or the Vebe time. Abram's cone method where spread diameter is measured can indicate yield stress level and measuring T500 (time when spread diameter reaches 500 mm) provides data on plastic viscosity. Similar to V- funnel and L-box apparatus, they indicate concrete's form filling ability and obstacle passing ability. Unfortunately, all these values are not physical characteristics suitable to determine rheological behavior of stress and strains of the liquid concrete mix, which is necessary for concrete flow characterization in different technological processes.

Rheological properties of concrete are described by Bingham visco-plastic model. Bingham flow is characterized by plastic limit or yield stress and viscosity coefficient η_{pl} as it is shown in equation 1:

$$\tau = \tau_0 + \eta_{pl}\dot{\gamma} \qquad (1)$$

Where: τ – shear stress;

τ_0 – plastic limit or yield stress;

η_{pl} –viscosity coefficient;

$\dot{\gamma}$ – shear rate.

2. Problem statement and the proposed solution

The main rheological parameter, which distinguishes conventional concrete mixes from self-compacting mixes, is the Bingham plasticity limit τ_0 . For the usual paste-like mixes, it is larger than zero, but for the liquid-like self-compacting concrete, it must be $\tau_0 = 0$ in all situations.

In our investigation, a conical plastometer was used for plasticity limit τ_0 measurement of a cement mortar. Traditional view is that this device and method are suitable for small volume paste-like liquid samples with high τ_0 [4] values and are not suitable for determining τ_0 in big volume structures using liquid mixes having $\tau_0 = 0$.

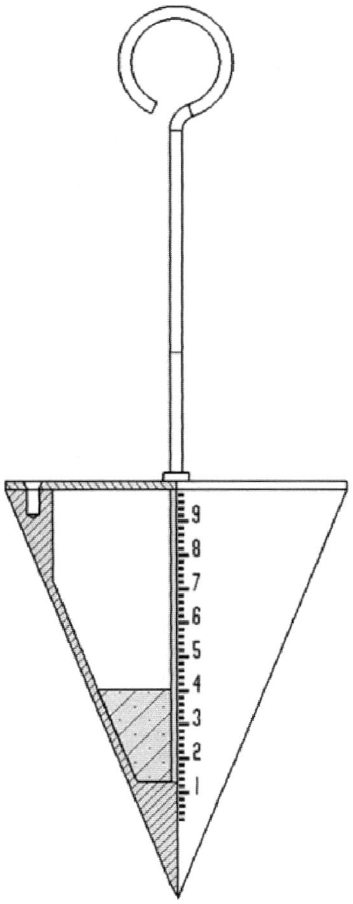

Figure 6. Conical plastometer.

IOP Conf. Series: Materials Science and Engineering **96** (2015) 012045 doi:10.1088/1757-899X/96/1/012045

For the field tests, where concrete volumes are big and the used self-compacting mixes have a liquid-like rheological properties, the known method of determining the properties is not precise enough, because the conical inductor is too small (15 mm) and the buoyancy force of the liquid traditionally is also not taken into consideration.

Based on above-mentioned facts, in the Concrete Mechanics Laboratory of RTU a conical plastometer was elaborated, which has higher precision and is less sensitive to inaccuracy of the test (what is possible in field conditions). A new method was also elaborated for the calculation of plastic limit τ_0, taking into account the buoyancy force of the liquid concrete:

$$\tau_0 = \frac{k}{h^2}(F_\Sigma - F_{Arh}) = \frac{k}{h^2}\left[F_\Sigma - \frac{\pi h^3}{3} \cdot \rho_c tan(\alpha)\right] \quad (2)$$

where F_Σ –the total axial force acting on the indenter;
F_{Arh} –the buoyancy force;
h –the immersion depth of the indentor into the concrete mix;
k –the coefficient dependent on the angle α of the pointed end of the indentor, if $2\alpha = 45^0, k = 0.416$;
ρ_c –the density of the concrete mix.

In figure 6 it is possible to see how the conical plastomer looks like, in figure 7 graphs calculated according to formula (2) for concrete mixes having different density are shown.

Conical plastometer was used for determining plasticity limit τ_0 of different self-compacting concrete mixes depending on time, counting it from mix preparation moment. All self-compacting concrete mixes were made with CEMEX CEM I 42,5 N cement. Measurement graphs are shown in figure 3. The actual duration of the test was 5 hours from the moment of adding water to the mix. In all cases, densities of concrete mixes were practically similar – for mix A it was 2.32 t/m^3, for mix B – 2.30 t/m^3 and for mix C it was 2.29 t/m^3. Simultaneously Abram's cone tests were done for all mixes. Some differences for the three mixes were found in the standard Abram's cone outflow diameters – for mix A it was Ø 620 mm, for mix B – Ø 670 mm, for mix C – Ø 720 mm.

Taking into account relatively small difference of the tested mix densities ρ, the starting depth of the conical plastometer was equal ~ 58 mm, and value $F_\Sigma = F_{Arh} = 48.2\ kN$ was obtained for the force. That means that all three tested mixes at the testing moment had plastic limit $\tau_0 = 0$ and they can be characterized as viscous liquids.

However, with time plastic limit values for all tested mixes changed from $\tau_0 = 0$ and increased at different rate depending on particular mix composition. Kinetics of these processes is shown in figure 8. The early coagulation structure, which distinguishes the paste as a medium from the liquid in mix A starts to form approximately after 55 minutes from the mixing moment end. This short time is very important. It means that for reaching the maximal hardening potential of these real mixes, the construction technologist must have a short time to drive and to pour and compact this mix into the formwork.

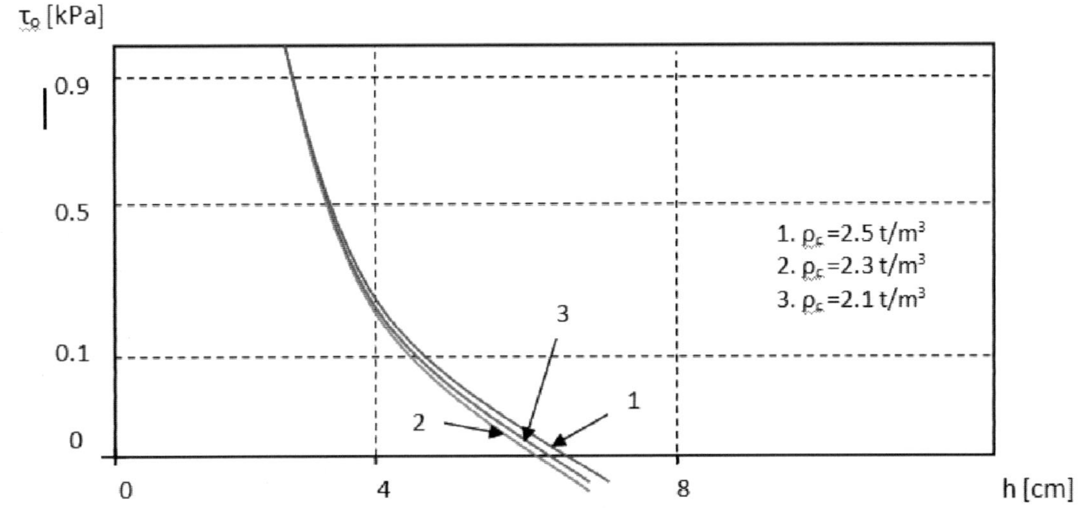

Figure 7. The immersion depth of the plastometer h depending of the concrete mix plasticity τ_o. ρ_c is the concrete mix density.

Every technological mechanical operation with the mix is done after the said term: transportation, pumping, pouring in the formwork and compacting of the mix will more or less destroy the early created and fresh coagulation structure of the cement mineral hydrates. In the later hardening time, the mentioned microdefects formed in the early phase will not overgrow and the hardened building structure will not obtain the potential strength.

As it is evident from figure 3, the critical term when $F_{\Sigma} = F_{Arh}$ for the mix B is 1.75 h, but for the mix C it is 2.5 h.

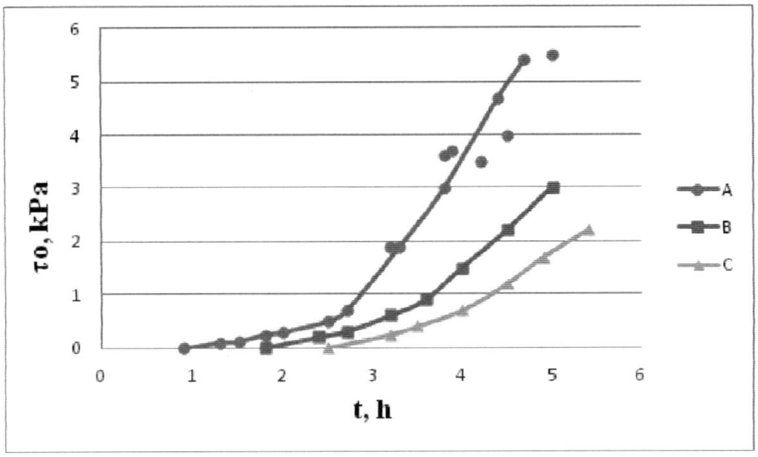

Figure 8. Plasticity limit τ_0 growth kinetics for three different self-compacting concrete mixes A, B and C.

3. Conclusion

The proposed rheological test in site conditions is simpler and easier than the U-box siphon, L-box or funnel tests [1, 2, 3]. It is also easily repeatable and easy to use directly in the mixer or in the formwork.

In such a way, the use of the conical plastometer and the mentioned method account for the buoyancy force $F_{Arh,}$ it is also possible to answer to the following questions:

- Is the tested mix self-compactible or not?
- If yes, $\tau_0 = 0$, but if not so ($\tau_0 > 0$) we can calculate the real value of the mix plasticity limit (or the yield stress) τ_0 using the conical plastometer.
- Technologists can follow these measurements for further determination of the time of appearance and increase of plasticity limit $\tau_0 > 0$ and for further kinetics investigation at early stage cement hydration product setting and formation forming new microstructure in the concrete.
- Taking into account term F_Σ , when $F_\Sigma = F_{Arh}$, while microstructure coagulation of the cement mineral hydrates has not yet started to occur, before starting further technological operations, we get a possibility to utilize all technological potential of the real mix for obtaining maximum strength characteristics as well as other qualities of the used self-compacting concrete.

4. References

[1] Zoran J Grdic, Despotovic I and Toplicic-Curcic G 2008 *Properties of Self-Compacting Concrete with Different Types of Additives* Facta Universitatis, Serie Architecture and Civil Engineering, vol.6 (2) 2008 pp 173–177

[2] Yang F 2004 *Self-Consolidating Concrete* CE 241: Concrete technology, Spring 2004, Report No. 1, March 9

[3] Okamura H and Ouchi M 2003 Self-compacting concrete *J. of Advanced Concrete Technology* **1** (1) 5–15

[4] Khaidapova D D and Pestonova E A 2007 Strength of interparticle bonds in soil pastes and aggregates *Eurasian Soil Science* **40** (11) 1187–92

[5] Lapsa V A, Krasnikovs A and Lusis V 2012 *Rheological Testing Method for Paste-like Materials* Latvian Patent No. 14530 April 2, 2012

IOP Conf. Series: Materials Science and Engineering **96** (2015) 012046 doi:10.1088/1757-899X/96/1/012046

Influence of Carbon Nanotubes on the Structure Formation of Cement Matrix

S Petrunin[1,3], **V Vaganov**[2], **V Reshetniak**[2] **and L Zakrevskaya**[2]
[1]Limited Liability Company "Macromer", Vladimir, Russia
[2]Vladimir State University, Vladimir, Russia

E-mail: ser-petru@yandex.ru

Abstract. The potential of application of CNTs as a reinforcing agent in cement composites is governed by their unique mechanical and electronic properties. The analysis of concrete strength changes under CNTs introduction shows non-uniformity and sometimes inconsistency of results. Due to the fact that CNTs influence the hydration kinetics, structure and phase composition of concrete, an idea concerning the importance of interaction between the surface of CNTs and hydrate ions formed by the dissolution of the clinker phases has been suggested. In this paper, the theoretical and experimental study of interaction between hydrate ions and CNTs surface is discussed. Reference nanotubes and nanotubes functionalized by carboxylic groups are used in this research. Phase composition was determined by X-Ray analysis according to the Rietveld method. It was found that the presence of oxygen-containing functional groups on CNTs surface leads to intensification of the hydration process and increase in concentration of C-S-H gel from 65.9% to 74.4%. Special attention is usually paid to interactions between Ca^{2+} ions and CNTs, because the hardening rate and structure of cement stone are determined by principle of Ca^{2+} localization in the solution. In this paper the possible binding mechanisms are discussed. Based on the experimental results, the hypothesis regarding the formation of cement composite structure for different CNTs surface functionalizations is considered. According to this hypothesis, the CNTs act as the centers of crystallization for hydration products contributing to the acceleration of hydration, increase of the concentration of C-S-H gel and strength improvement of CNTs based composites.

1. Introduction

Carbon nanotubes (CNTs) were considered as a promising filler for Portland cement composites by multiple authors [1-3]. Modification of concrete matrix by CNTs can increase the strength characteristics from 25% to 250%. The significant experimental scatter of mechanical properties may indicate high potential for technological optimization. Nevertheless, due to misunderstanding of the

[3] To whom any correspondence should be addressed.

strengthening mechanism, the problem is still unresolved. Properties of cement composites depend on surface chemistry of CNTs. Modification of cement system with functionalized CNTs contributes to a substantial increase in concrete strength compared with pure CNTs. Microstructural investigation demonstrated the existence of functionalized CNTs covered with hydration products, whereas hydration products on the surface of pure CNTs are absent. Functionalized CNTs can interact with hydration products of cement, leading to a decrease in porosity and, as a result, to an increase in composite strength. CNTs influence the hydration kinetics, structure and phase composition of concrete that is why an idea concerning the importance of interaction between the surface of CNTs and hydrate ions formed by the dissolution of the clinker phases has been suggested.

Special attention is usually paid to interactions between Ca^{2+} ions and CNTs, because the hardening rate and structure of cement stone are determined by principle of Ca^{2+} localization in the solution. As it is known [4], the interaction between Ca^{2+} ions and aromatic hydrocarbons is usually realized by polarization mechanism. Due to its high dielectric constant, water makes such interactions much weaker, thus the influence of interaction on the ion distribution should not be strong [4, 5]. On the other hand, as multi-walled CNTs chemically and by the electron structure are much closer to graphene than to aromatic hydrocarbon [6, 7], the differences in the interaction mechanisms (between Ca^{2+} ions and multi-walled CNTs and Ca^{2+} ions and aromatic hydrocarbons) can be expected. That is the reason why the interaction between Ca^{2+} ions and graphene surface is considered in the paper. The interaction was studied theoretically on the DFT GGA level using the plane wave method, which is realized in the Quantum Espresso program [8]. Calculations showed formation of the metallic type bonds between Ca^{2+} ions and graphene sheet, with the binding energy 0.75 eV.

2. Experimental program

2.1. Materials

Portland cement CEM I 42.5B supplied by JSC "Mordovcement," standard quartz sand with a fineness modulus of 1.5 and multi-walled carbon nanotubes produced by CVD technology at Vladimir State University (Russia) were used in this research. The CNTs' concentration in the composite was 0.05% of the cement mass. Polycarboxylate (PCE) superplasticizer P-11 supplied by JSC "Macromer" was used to reduce the sedimentation and facilitate the dispersion of CNTs in water.

2.2. Preparation of concrete samples

All cement composites had the component ratio of 0.3:1:1 for water, cement, and sand, respectively. The CNTs were mixed with water and the polycarboxylate superplasticizer and sonicated at 20 kHz using the ultrasound processor (IL100-6/1-1) at maximal power (700 W) for 60 minutes. Cement and sand were added into a planetary mixer. Then the required amount of water with superplasticizer and carbon nanotubes was added into the mixer. The entire mixture was stirred for 7 minutes. After this procedure, the composite mixture was placed into the metal molds with dimensions of 40×40×160 mm and compacted by vibration. The composite specimens were demolded after 24 hours of initial hardening and were stored in lime water until the testing age.

2.3. CNTs functionalization

CNTs functionalization by oxygen groups was carried out to improve the hydrophilic properties and the interaction between the surface of CNTs and the material matrix. The functionalization was carried out as follows: CNTs were placed in a flask and backed up with the mix of concentrated HNO_3 and

IOP Conf. Series: Materials Science and Engineering **96** (2015) 012046 doi:10.1088/1757-899X/96/1/012046

H_2SO_4 at a ratio of one-to-five. The mix was heated to 70℃ at continuous hashing for 3 hours. During this process, the nickel catalyst particles left from the synthesis were oxidized with gas liberation. The solution was then cooled, diluted with distilled water and filtered. The filtered CNTs were washed out and dried. The assessment of functionalized groups occurring on the CNTs surface was made with the help of infrared spectroscopy.

Coagulation resistance of CNT dispersions depends on the character of charge distribution on nanotubes surface formed by the absorption of ions from the solution and the dissociation of the CNTs surface. Inoculating of oxygen-containing groups promotes the redistribution of surface charges and improves the hydrophilic properties of CNTs.

2.4. Research methods

The phase composition was defined on the Brucker D8 Advance diffractometer using the Rietveld method.

The supercell of the graphene with 50 carbon atoms (5x5 unit cells) was considered. In the cell one Ca^{2+} ion was inserted (figure 1). To reduce the effect of the interaction between neighbour graphene sheets, the length of the orthogonal to the graphene sheet translation vector was applied equal to 4 nm. Calculations showed that the change of the cell energy due to increase of the vector length is not sufficient.

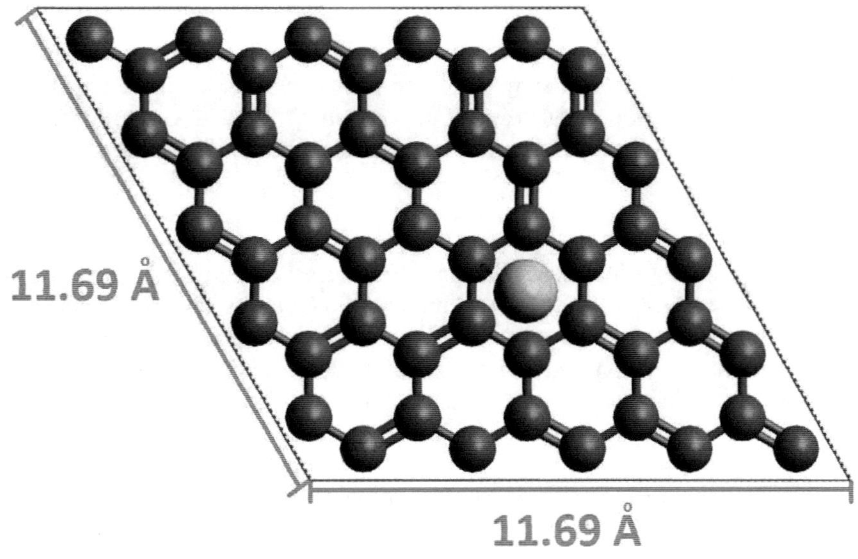

Figure 1. The structure of the supercell of the Ca^{2+} - graphene complex.

The calculation of the electron wave functions was performed using plane wave basis set and periodic boundary conditions on the cell boundaries. The Quantum Espresso program [8] and DFT method were used. The PBESol exchange-correlation functional [9], and relativistic Hartwigsen-Goedecker-Hutter pseudopotentials [10] were applied. The 9x9 Monkhorst-Pack k-point grid [11] was used.

3. Results and discussion

3.1. The X-ray analysis of hydrates

The changes of C-S-H due to the introduction of CNTs are identified. The resulting changes in phase composition of composites are demonstrated in table 1. The increase in the amorphous phase corresponds to the increase of cement hydration, promoting acceleration of hydration and an increase in the strength of a material.

Table 1. Phase Composition.

Minerals	Reference	CNTs	Functionalized CNTs
C_3S	8.9 %	8.1 %	6.5 %
C_2S	4.4 %	4.1 %	3.7 %
C_4AF	3.2 %	2.8 %	2.6 %
$Ca(OH)_2$	9.2 %	7.3 %	7.7 %
$CaCO_3$	6.7 %	5.6 %	4.9 %
Ettringite	0.3 %	0.3 %	0.2 %
C-S-H	65.9 %	71.9 %	74.4 %

The amorphous phase in the reference specimen is 65.9%. The increase of the amorphous phase to 74.4% was recorded for the composite with the functionalized CNTs. The character of the amorphous phase development corresponds to a tendency of compressive change at the age of 1 day. The composite with functionalized CNTs contains the greatest volume of C-S-H gel and possesses higher strength, whereas composites modified by pure CNTs have lower C-S-H gel concentration and weaker strength characteristics.

3.2. DFT calculation

The relaxation shows that there is a local energy minimum corresponding to the structure with Ca^{2+} ion position on the symmetry axis of the hexagon formed by the carbon atoms. The distances between Ca^{2+} ion and 6 closes neighbour C atoms are equal to 0.259 nm. The Löwdin population analysis shows charge transfer between Ca^{2+} ion and C atoms: the charge of the Ca ion due to the interaction became equal to +1.33e. Due to the interaction, the 3d orbital of the Ca ion becomes partially occupied and overlaps with the π-orbital of the graphene, which is formed by the 2p orbitals of the neighbour C atoms. The occupancy of the 3d orbital of the Ca ion is equal to 0.4. The hybridization of the electronic shells leads to the binding between Ca ion and graphene sheet by the exchange mechanism.

The map of the charge density distribution in the Ca^{2+} - graphene complex is shown in figure 2. The positive values of the charge density corresponds to the electron density and the negative values – to the hole density. In figure 2a the electron density map in the orthogonal to graphene sheet plane is shown (the light bar at the bottom of the figure corresponds to the carbon atoms of the graphene sheet, the bright spot in the centre – Ca ion). In figure 2b the charge density in the same plane is represented. The charge density was calculated by the subtraction of the electron density of the separate fragments (Ca^{2+} and graphene) from the electron density of the complex. In figure 2c the charge density map in the plane, parallel to the graphene sheet and equidistant to the both Ca ion and graphene plane is shown. It is seen that there is charge transfer between the Ca ion and C atoms, there is hybridisation of the electron shells, and that the formed bonds are nonlocal.

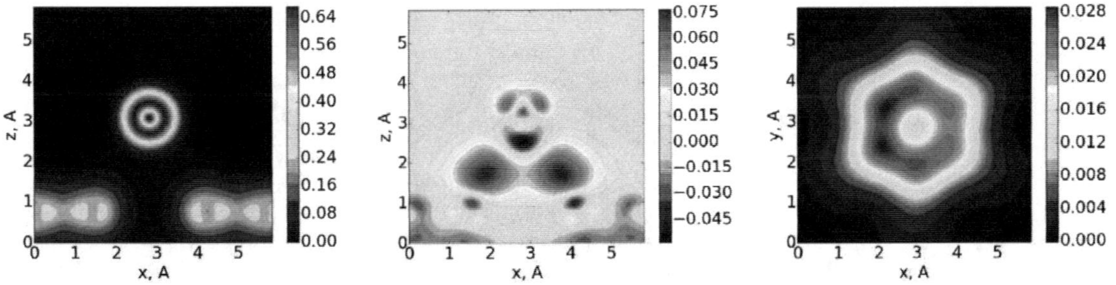

(a) (b) (c)

Figure 2. The charge density distribution in the orthogonal (a, b) and the parallel (c) planes.
a – electron density for all electrons, considered in the calculations;
b, c – charge density: the residual of the electron density of the complex and the separate fragments (Ca^{2+} and graphene).

Due to charge transfer, the C atoms became positively charged, and there is Coulomb repulsion between Ca ion and graphene sheet, which is competing with the metallic binding. Due to the transportability of the electrons in the graphene, the charge distribution in the sheet is uniform. It should be noted that the effect may lead to the dependence of the binding energy on the Ca ions concentration.

To calculate the binding energy between Ca ion and graphene, the potential energy scan was performed. The distance between the Ca and graphene plane was changed in the range from 0.19 nm to 0.51 nm. The calculations showed the nonmonotonic character of potential curve, which is shown in figure 3. There is a local maximum of the energy, which corresponds to the distance between Ca and graphene 0.35 nm. The energy in the point of the local maximum was set to be zero. The maximum appears due to the charge redistribution between fragments and the competing between Coulomb repulsion and metallic binding. Further increase of the distance decreases electron shells hybridisation and leads to the formation of the Ca^{+} ion and charged graphene sheet.

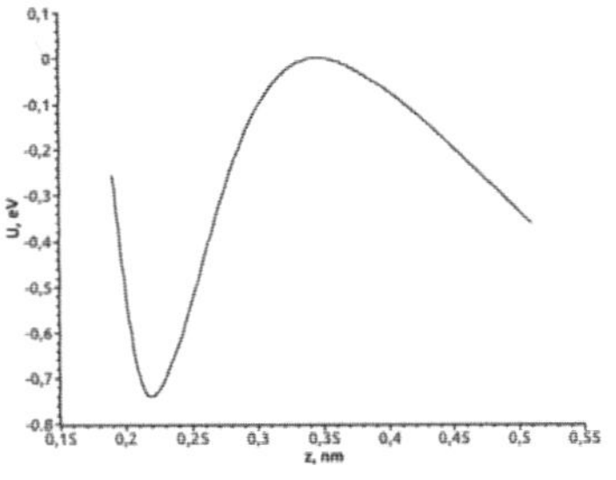

Figure 3. Potential energy dependence on the distance between Ca ion and graphene sheet.

As contrasted with the Coulomb interaction, which is vastly going down due to solvation with the solvent with high dielectric constant, the metallic bond is less susceptible to the solvation and may impact structure formation in the concrete significantly.

3.3. Hypothesis related to composite structure formation

In the process of cement hydration, the OH-, Ca^{2+}, CaOH- and $H_2SiO_4^{2-}$ ions are deposited on the surface of the cement grains and form a layer consisting of C-S-H gel with embedded $Ca(OH)_2$ crystals. With time, the thickness of this layer increases and its ionic permeability decreases. The formation of C-S-H gel and $Ca(OH)_2$ takes place when Ca^{2+} concentration exceeds the saturation limit by factor of 1.5–2.

The intensification of hydration in the system with CNTs is due to an additional sorption of Ca^{2+} ions on the surface of the CNTs. In this case, the hydration is accelerated, since the formation of C-S-H and $Ca(OH)_2$ crystals occurs in the zones supersaturated by Ca^{2+} ions. The fragmentary formation of hydration products on the CNTs surfaces contributes to the improvement of ion exchange between the solid and liquid phases and results in acceleration of new phase formation. The presence of oxygen-containing groups on the surfaces of the CNTs leads to a more active sorption of Ca^{2+} ions and higher rates of hardening when functionalized CNTs are used. According to this hypothesis, the CNTs act as the centers of crystallization for hydration products contributing to the acceleration of hydration, increase of the concentration of C-S-H gel and strength improvement of CNTs based composites.

4. Conclusions

Quantum chemical modeling showed formation of the metallic type bonds between Ca^{2+} ions and graphene sheet, with the binding energy 0.75 eV. The metallic bond is less susceptible to the solvation and may impact structure formation in the concrete significantly. The presence of oxygen-containing functional groups on CNTs surface leads to intensification of the hydration process and increase in concentration of C-S-H gel from 65.9% to 74.4%. Based on the experimental results, the hypothesis of formation of cement composite structure for CNTs is described. According to this hypothesis, the CNTs act as the centers of crystallization for hydration products contributing to the acceleration of hydration, increase of the concentration of C-S-H gel and strength improvement of CNTs based composites.

5. References

[1] Sanchez F and Sobolev K 2010 Nanotechnology in concrete – a review *Construction and Building Materials* **24** 2060–71
[2] Konsta-Gdoutos M S, Metaxa Z S and Shah S P 2010 Muli-scale mechanical and fracture characteristics and early-age strain capacity of high performance carbon nanotube/cement nanocomposites *Cement and Concrete Composite* **32** 110–115
[3] Li G Y, Wang P M and Zhao X 2005 Mechanical behavior and microstructure of cement composites incorporating surface-treated multi-walled carbon nanotubes *Carbon* **43** 1239–45
[4] Harder S 2010 *Chem. Rev.* – 2010 – 110, No 7 p 3852
[5] Ma J C and Dougherty D A 1997 *Chem. Rev.* – 1997 – 97, No 5 p 1303
[6] Bhupesh Ch 2009 *Synthesis and Electron Transport in Known Chirality Single Wall Carbon Nanotubes* Ph. D. Thesis, Columbia university
[7] Nizam R, Rizvi S M A and Azam A 2011 *Int. J. of Science and Technology* **4** 153
[8] Giannozzi P, Baroni S et al 2009 *J. Phys. Condens. Matter.* **39** 395502
[9] Perdew J P, Ruzsinszky A, Csonka G I et al 2008 *Phys. Rev. Lett.* **100** 136406-1:4
[10] Hartwigsen C, Goedecker S and Hutter J 1998 *Phys. Rev. B.* **58** 3641
[11] Monkhorst H J and Pack J D 1976 *Phys. Rev. B.* **13** 5188

IOP Conf. Series: Materials Science and Engineering **96** (2015) 012047 doi:10.1088/1757-899X/96/1/012047

Fibre-Reinforced Adhesive for Structure Anchoring

J Barnat[1, 2] and M Bajer[1]
[1]Brno University of Technology, Faculty of Civil Engineering, Veveri 331/95, 602 00 Brno, Czech Republic

E-mail: barnat.j@fce.vutbr.cz

Abstract. The topic of this paper is the glue-concrete interface of bonded anchors loaded by tension force. The paper is closely focused on bond strength experiments using high strength concrete up to class C50/60 or higher together with pure epoxy resin and fibre-reinforced resin. The goal of this research is to find the limits of the effective use of such glue types in high performance concrete, and also to verify the most commonly used design methods for bonded anchors. The presented research includes experimental analysis of the glue-concrete interface and the influence of its parameters on anchor behaviour. The presented analysis shows some problems of the 'separated failure modes' approach and also presents experimentally verified bond strength values obtained for the currently most widespread glue types. Results of fibre reinforced epoxy resin are also presented in this paper.

1. Introduction

The base line of this analysis consequents from one of the typical failure types of bonded anchor loaded by static tension force. According to widely spread ETAG documents [1] and also new EC standard [2], the failure type implied by adhesive compound is in general defined by bond strength value.The characteristic value of anchor resistance corresponding to this failure type is given as (1).

$$N_{Rk,o} = \pi \cdot d \cdot h_{ef} \cdot \tau_{Rk} \qquad (1)$$

In equation (1),h_{ef} is the effective anchorage length, τ_{Rk} is the characteristic bond strength, and d is the anchor bolt diameter.

The bond strength is defined as the peak value of tangential stress on the defined contact area in most cases. Contact area is given by effective anchoring length and diameter. In principle, the correct diameter should be used according to the failure position. However, it is very difficult to determine the correct failure mode for specific anchoring conditions. Using the anchor bolt diameter leads to complete neglecting of the thickness of the adhesive layer, figure 1.

[2] To whom any correspondence should be addressed.

This is an acceptable model of problem simplification because layer thickness for most bonded anchors systems is equal to several millimetres (according to anchor bolt diameter 1 or 2 mm). The bond strength as a material parameter defined in this way does not represent real material characteristic in many cases [3]. If the failure on the steel anchor bolt and adhesive compound occurs, the tangential stress on the contact area is not a bond stress (or bond strength in the peak value) but rather the shear strength of the adhesive itself. It is possible to assume that there is no adhesion failure on concrete-glue contact either.

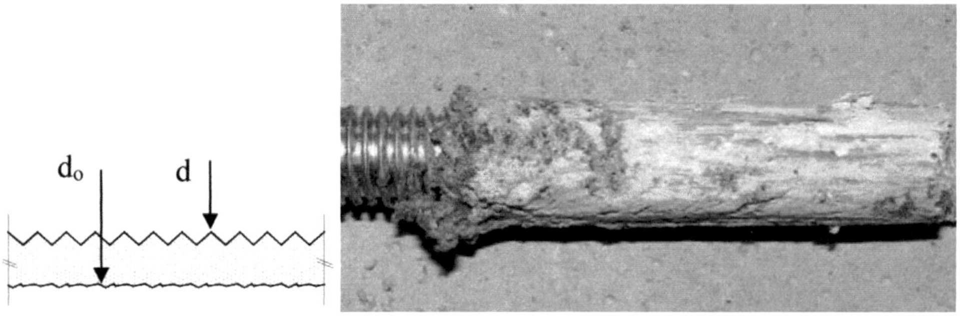

Figure 1. Bond failure close to glue-concrete contact defined by drill diameter d_o.

1.1. High-strength concrete effect

The surface of a drilled hole (at least in the usual case of hammer drilling procedure) is very rough/bumpy. Failure on this contact means the failure of concrete or glue, or both materials. The weaker from both materials affects the failure on themajor scale. This simple assumption means that anchor resistance cannot be increased by the use of high strength concrete without the use of high performance adhesive as well.

Problems connected with the combined concrete-bond failure mode are in detail described in [4]. The following relation (2) for the tensile resistance of an anchor experiencing the combined concrete-bond failure mode has been published in [4].

$$N_{Rk} = \pi \cdot \tau_{Rk} \cdot d_o \cdot h_{ef} \cdot 0.74\left(1 - e^{-1.5\frac{f_{cc,150,k}}{\tau_{Rk}}}\right) \tag{2}$$

In equation (2),h_{ef} is the effective anchorage length,$f_{cc,150,k}$is the characteristic compressive concrete strengthassessed for 150 mm cubes,τ_{Rk} is the characteristic bond strength, andd_0 is the diameter of the drilled hole.

Equation (2) was determined from experimental results (pull-out tests and bond quality tests) for the currently most widely used industrial glues based on vinyl-urethane resin, polymer cement mixtures and epoxy resin. All tests were conducted using common concrete ranging in class from C15/20 to C37/45 [5].

Figure 2 shows the results obtained by equation (2) for different bond strength values in relation to concrete strength. According to the previous experimental results for the bonds provided by currently used glues [4], [5], such bond strengths range from 15 to 20 MPa. This bond strength value, together with the standard anchorage length, which is about nine times the anchor diameter, ensures the effective utilization of materials with a concrete class of up to C40/50 and a steel bolt quality of 8.8.

For anchoring in high-performance concrete without using extended anchorage lengths, it is necessary to also use high strength steel such as 10.9 or 12.9 or different materials for the anchor bolts, for example, composite fibres. However, the bond strength provided by the glue remains the limiting parameter for this kind of anchor application.

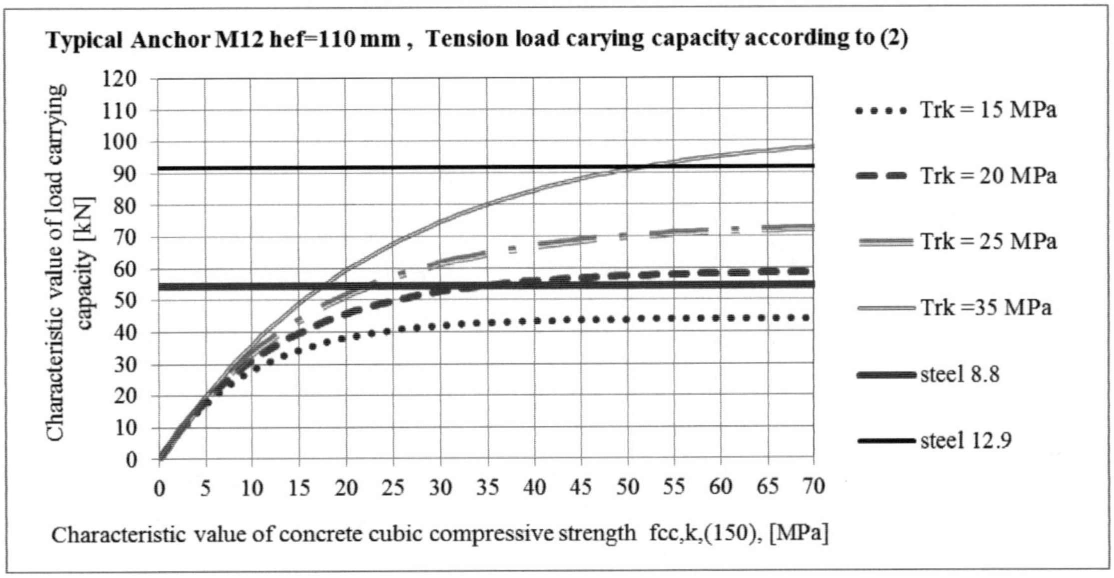

Figure 2. Tension load carrying capacity according to (2).

2. Fibre reinforced epoxy resin

One of the ideas is to reinforce the glue layer by carbon or glass fibres. The size of the contact area (and also the stress value) is influenced by the diameter. When the glue strength is lower in comparison to concrete, failure "a" will be located around the anchor bolt diameter d. By improvement of glue strength the failure moves towards the concrete contact around the drilled opening diameter d_o – failure "b". Both the area of contact and the bond strength increase, figure 3.

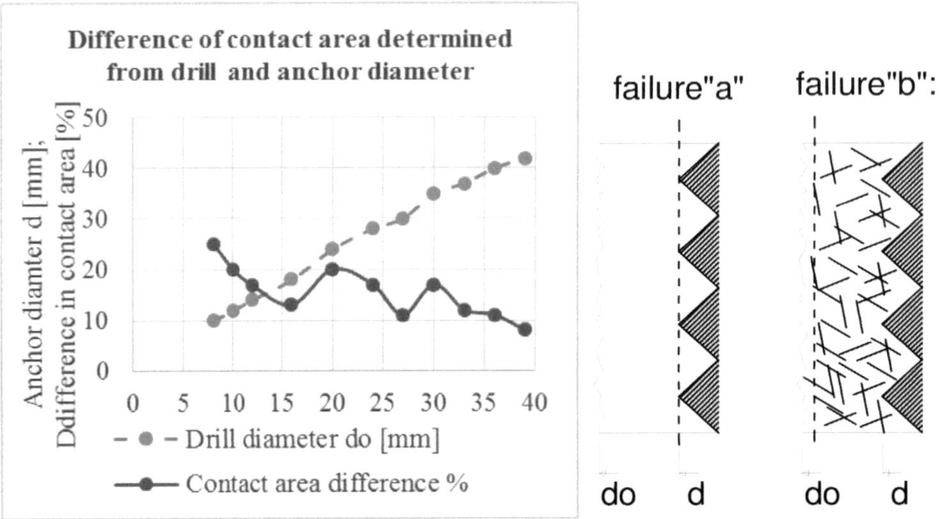

Figure 3. Differences of contact are determined from drill and anchor diameters.

This improvement is limited by the difference between diameters. The theoretical value of percentage improvement for common anchor diameters is summarised in table 1.

Table 1. Typical values of anchor and drill diameters for epoxy resin anchor types.

Anchor diameter d [mm]	**8**	**10**	**12**	**16**	**20**	**24**	**27**	**30**	**33**	**36**	**39**
Drill diameter do [mm]	10	12	14	18	24	28	30	35	37	40	42
Contact area difference %	25	20	17	13	20	17	11	17	12	11	8

2.1. Experiment

The main principle of the bond quality test used (as depicted in figure 4) [1] is to restrict bond failure to the anchor only. Load is applied to the anchor bolt by the loading mechanism, which is itself supported by the concrete in the immediate vicinity of the installed anchor. Experiments are comprehensively described in [4] and [5], where the results for several types of adhesives used in concrete C20/25 are also presented. The main result is the experimental study adverting to values of bond strength of the currently used glues. These are usually in the range between 15-25 MPa. This is also enough for concrete up to C50/60 class.

Glues based on epoxy resin usually give the best values of bond strength. The disadvantage of this type of glues is the long hardening time at normal temperatures (it takes several hours to reach significant values of strength and several days to reach full strength). The glue used in the following experiments was also based on epoxy resin but it was produced with an aim to reach maximum shear strength [5].

Two sets of test were performed. Anchors in the first set of experiments were installed in high strength concrete with mean value of compressive cubic strength 75.8 MPa, which corresponds to class C60/75. The concrete specimens were prepared in the shape of a cube with edge length 200 mm. The predetermined length of the contact was 40 mm. The steel rebar M12 12.9 was used as an anchor bolt. Concrete specimens were substituted by steel specimens in the second set. Another metric thread was performed in the specimen opening to face the anchor bolt surface. This configuration ensures the shear failure of glue itself. Therefore, the shear strength of adhesive can be determined. Specimens with plain epoxy adhesive are marked as N and specimens with fibre reinforced epoxy resin are marked as C.

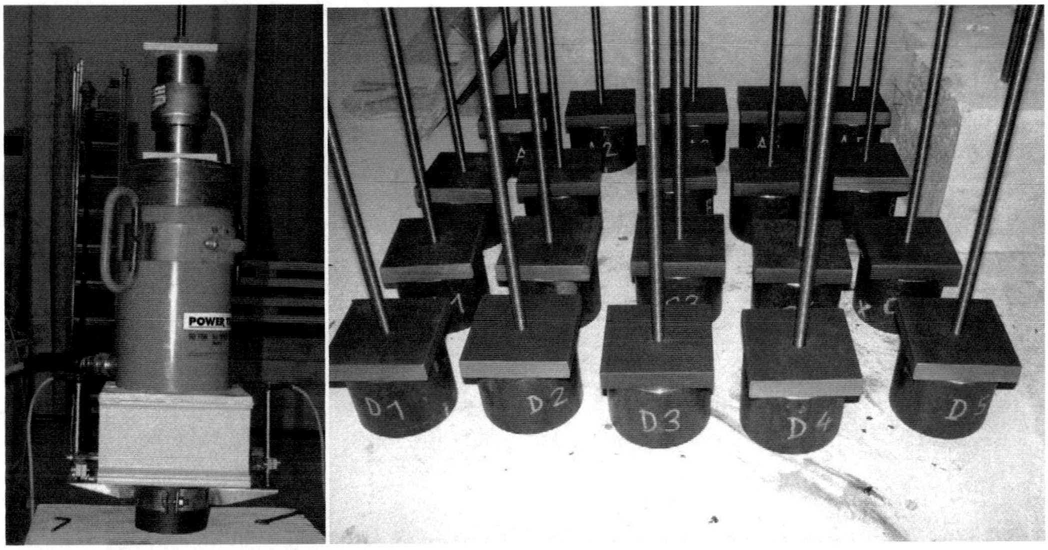

Figure 4. Experiment configuration.

In all tests the anchor was loaded by constantly rising tension force using hydraulic press with central opening, see figure. 4. The speed of loading was 2 kN per second. The applied force and the vertical deflection of the anchor bolt were also constantly measured. The temperature during the installing of the anchor, hardening time and testing was around 20°C. During specimen preparation, the standard cleaning procedure was used [6]. That means cleaning by air blowing (hand pump) and brushing with steel brush.

2.2. Results

Figure 5 shows both specimens after failure. The difference between failure positions is obvious. Bond failure for specimens with epoxy glue reinforced by carbon fibres occurred on the concrete glue contact or, more precisely, close to the area defined by drill diameter d_o.

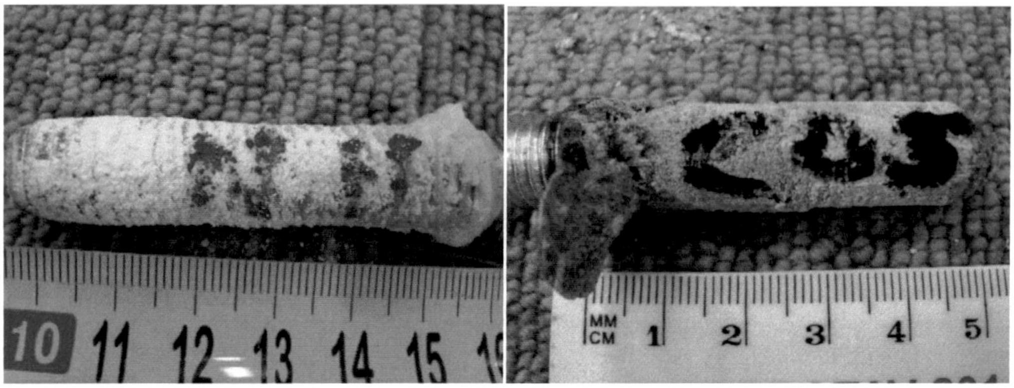

Figure 5. Experiment configuration – plain epoxy (left); reinforced epoxy (right).

Figure 6 shows a load deformation diagram from experiments in high strength concrete for three tests of each type of glue.

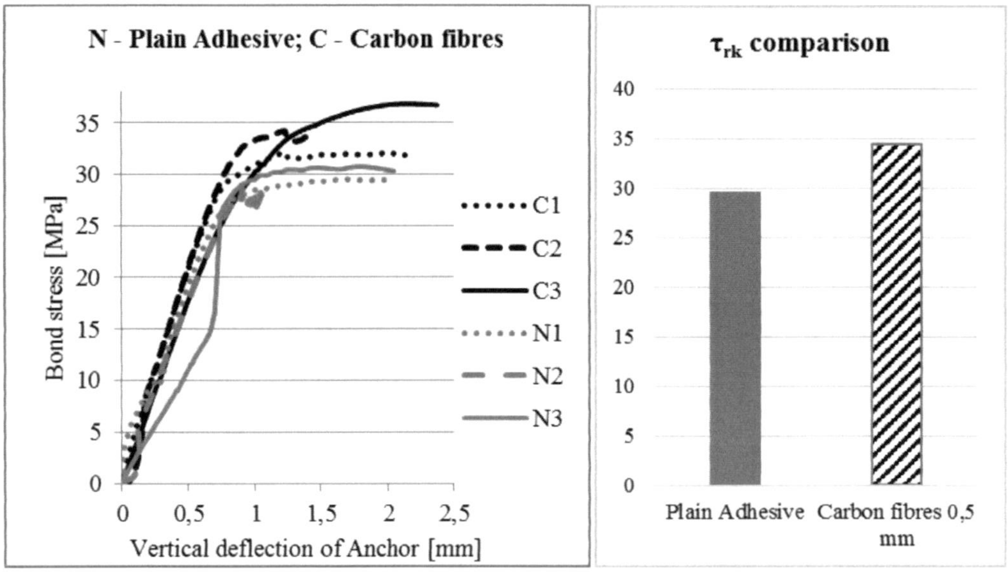

Figure 6. Experimental results; load deformation diagram.

Figure 6 also shows the comparison of mean values of bond strength determined from these experiments.

The difficult part here is to determine the amount of fibres which should be added to the compound. The reinforcement ratio has to be analysed in further research. There is a problem with the viscosity of the final adhesive compound.

Raising the amount of fibres leads to the loss (decrease) of viscosity. As the first approximation, the reinforcement ratio was set to 10% of the weight of the plain epoxy resin. Carbon fibres were melted to size lower than 0.5 mm. Material characteristics of the fibres were Young modulus 230 GPa and

tension strength 3600 MPa. The epoxy adhesive consists of two components which have to be mixed together in specific ratio. Fibres were first mixed with one component (not the activator) by accu-screwdriver by 600 rpm, for one minute. After this the fibres looked well-mixed in the compound.
The procedure of mixing used for the presented starting sets of specimens is not ideal, because it cannot prevent the formation of air spaces.In further experiments, the use of vacuum mixing is planned.

3. Summary and conclusion

Only a small number of tests were performed for such kind of fibre reinforced epoxy resin glue. In spite of that, there is a positive influence of carbon fibre reinforcement. This type of reinforcement enhanced the bond strength by 10% or more. The difference in the tension load carrying capacity for failure on the steel-glue contact and the failure on the glue-concrete contact cannot reach the theoretical value described in table 1 only by the useof fibre reinforcement. This is because fibre reinforcement is not active on the contact area defined by diameter d_0. However, it is obvious that bond connection of adhesive anchors can be improved in this way. Larger amount of small fibres has a greater effect on the adhesive compound shear strength. The correct amount of fibres, their size, and mixing procedure are the issues for further research activities.

This kind of adhesive anchor system (threaded rod + adhesive) can remain the simplest anchoring solution also for high performance concrete with compressive strength over 60 MPa. According to the presented results, reinforcing of currently used industrial adhesives is an effective solution.

4. References

[1] *ETAG 001 Anchoring Design Methods* 2011 EOTA-European Organisation for Technical Approvals, Avenue des Arts 40, Kunstlaan, 1040 Brusel
[2] ČSN P CEN/TS 1992-4-5 2009 *Design of Fastenings for Use in Concrete – Part 4 – 5: Post-Installed Fasteners – Chemical Systems*, CEN, 2 Avenue Marnix 17, B-1000 Brusel
[3] Hradil P and Kala J 2014 Analysis of the shear failure of a reinforced concrete wall *Applied Mech. and Mater.* **621** 124–129
[4] Bajer M and Barnat J 2012 The glue-concrete interface of bonded anchors *Constr. and Build. Mat.* **34** (9) 267–274
[5] Barnat J, Bajer M and Vyhnankova M 2012 Bond strength of chemical anchor in high-strength concrete *Proc. Engin.* **40** 38–43
[6] Cook R A, Bishop M C, Hagedoorn H S, Sikes D, Richardson D S, Adams T L and De Zee C T 1994 Adhesive bonded anchors: Bond properties and effects of in service and installation conditions. *Report No. 94-2A* (University of Florida, Department of Civil Engineering, College of Engineering, Gainsville)

Acknowledgements
This paper has been worked out under the project No. LO1408 "AdMaS UP - Advanced Materials, Structures and Technologies", supported by Ministry of Education, Youth and Sports under the "National Sustainability Programme I" and with the financial support of project TA03010680.

IOP Conf. Series: Materials Science and Engineering **96** (2015) 012048 doi:10.1088/1757-899X/96/1/012048

Glued Joint Behavior of Ribs for Wood-Based Composite Plates

G Frolovs[1,2], **K Rocens**[1] **and J Sliseris**[1]
[1]Riga Technical University, Faculty of Building and Civil Engineering, Kalku Str. 1, Riga, LV-1658, Latvia

E-mail girts.frolovs@rtu.lv

Abstract This article presents experimental investigations of composite sandwich plywood plates with cell type core and their connections between skin layers of birch plywood and a core of straight and curved plywood honeycomb-type ribs. This shape of core ribs provides several improvements for these plates in the manufacturing process as well as improves the mechanical properties of plywood plates. This specific form of ribs allows simplifying the manufacturing of these plates although it should be detailed and improved. The most typical cases (series of specimens) were compared to the results obtained from FEM (ANSYS) simulations. All thicknesses of elements are chosen according to plywood supplier assortment. Standard birch plywood (Riga Ply) plates were used - three layer plywood was chosen for skin elements (Surfaces) and three or five layer plywood was chosen for edge elements. Different bond pressures were taken to compare their influence on joint strength and stiffness.

1. Introduction

The consumption of composite materials has increased over the last few decades [1]. But on the other hand, it would be important to use wood based materials as much as possible and economically viable, as it would satisfy the increased demand for high performance materials.

Wood based panels as well as other wood based structural elements become more significant in construction. Plywood is one of the most common wood based secondary (transformed) materials and its consumption (and production) has been increasing over the past decades, especially in Asia [2], [3]. These plates could be used in civil engineering as covering plates, as separating elements or some furniture elements where they are subjected to bending.

To investigate the plate's stiffness normal stress σ occurs in the structure and is determinant for bending stiffness, but since the height of the plate increases and the material in the middle layer of the plate is reduced, shear stresses τ and the influence of shear deformation increases. The material in the

[2] To whom any correspondence should be addressed.

middle layer (plate's core) is reduced but it should take mainly the shear stresses becoming the weakest point since the height of the plate increases. The influence of shear strength increases becoming more important in determining strength and stiffness of the plate. As a result, the joint strength between the skins and the core becomes even more important. Therefore, the properties of glued joint (both strength and stiffness) should be investigated.

For economically grounded design it is important to choose the skin layer, core layer and bond material with defined properties to provide the most use of their properties to ensure efficient use. It is advisable to provide that the material breaks in the wood section not at the glued joint in a way that the performance of the skin material is at its maximum.

In literature, the recommended bond pressure for wood materials and sandwich constructions is 3.5 – 10.5 kg/cm^2 (50 – 150 psi) [4] to provide the optimal bond strength and stiffness.

2. Methodology and scope

Serviceability limit state for these composite sandwich plates with the cell type core is mostly determined in the same way as for most standard sandwich plates. Although when the height of the plate increases, the influence of a joint between the skins and the core increases and becomes determinant. In this paper, the joints between the surfaces of the skins and the edges of ribs were analyzed.

2.1. Bending tests

EN 789 [5] is most widely used standard for determination of properties of wood based panels. Standard sets out the guidelines for choosing the dimensions and measurement requirements for determination of properties.

For the case study, the plate with dimensions of 1.2 m (1.1 m span) in longitudinal direction and 0.3 m width was chosen. It has been proved that dimensions and other geometrical parameters have an influence on plate's stiffness and specific stiffness [6]. The plate was loaded with uniformly distributed load with intensity of 1 kN/m^2.

2.1.1. ANSYS Simulations.

The ANSYS finite element model was chosen for calculation of these plates. For numerical simulation of the bending tests, a FEM commercial code ANSYS v.15 (2014) was applied. A parametrical model of the panel was created with variable cross section parameters and bending loading set up options (according to Eurocode 5). The finite element mesh size was chosen 1/5 of the panel height. All plates (upper, lower and waved core plates) were defined by using ANSYS 4-node shell element SHELL 181.

In each step, there was a new numerical model made and calculated and the height of the plate in the first step span was changed; the same was done for different span length.

2.2. Shear tests

Shear properties should be determined for these plates as the decreased strength and stiffness of the core part. Stronger skin materials can resists larger loads than their design load, but the weakest point could be the bond between skins and the core materials. So the shear strength becomes determinant in ribbed sandwich plates, especially at the bond between the ribs and skin layer of the plywood.

2.2.1. Shear test standards.
Several shear standards exist to determine the shear properties of composite sandwiches and glued joints [7, 8, 9, 10, 11, and 12]. Single lap joint is mostly used for determination of glued joint shear properties, although some problems remain like eccentricities or complicated joints when designing a joint with cut-ins. In this case, the double lap joint was chosen ASTM D3528-96 [13] that provides two possibilities of double lap glued joint to be used on tension machines with grips for shear tests.

2.2.2. Specimen for shear test.
The provided methods of testing shear properties in standard tests mostly consider just glued overlap bond between two surfaces, but there is poor information on edge-to-surface joints.

For determination of shear strength, extra type of specimens with double overlap was chosen. The lap joint consists of surface elements 4x25x55 and edge elements 4(6,5)x25x75 with overlap 25mm on both sides and a gap 5 mm between edge elements (figure 1).

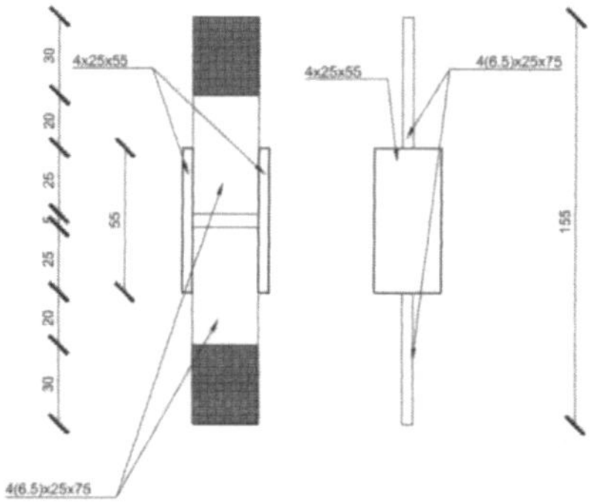

Figure 1. Shear specimen with dimensions.

All parts of specimen were formed with circular saws and taken direct without any additional processing and glued in conditions of relative moisture 6-8%.

Sample size could be calculated from the given formulas in ASTM E122-00 [14]

$$n = (3\,V_0/e)^2 \tag{1}$$

Number 3 is the recommended value for determination of some value, but it could be reduced when the precision requirements of the samples are reduced. As a result, value 3 is decreased to 1.96 and coefficient of variation is expected around 7% and the value e is assumed as 5%. So the sample size in this case is determined with the previously given formula (1) and the calculated value of 7.5 and assumed as 7 specimens in one sample.

A total of 63 glued plywood double lap shear glued connections were tested using shear standard tension test method. The specimens are labeled with PW-symbol-number. PW stands for Plywood Joint, the second symbol means the orientation of edge plywood elements and the number means the thickness of edge plywood elements. Orientation of outer layer fibers of skin parts are in the same direction as the applied tension load but for the edge elements labeled as (II) for outer layer fibers – in the same direction and (⊥) for transversal direction to the applied tension load.

In ANSYS Simulations for shear, the properties and element types were chosen the same as for bending simulations.

2.3. DIC method

Digital Image Correlation method was investigated for several specimens from each sample series. The method based on fixing pixel positions and analyzing their position in loading process strain can be determined by applying a correlation algorithm [15]. In this experiment, strain was measured in the middle of the specimen and close to the glued joint to determine the difference in particular points or eccentricity of the joint.

2.4. Material properties

In this study, the analyzed plywood was designed from standard birch values [16] assumed as one ply properties and applied to each layer of plywood composite element.

Polyvinyl acetate glue PVA D3 [17] was chosen for all glued joint parts. This type of glue provides water resistant joints according to EN 204 [18]. This type of glue is used for gluing of wooden joints in structural as well as carpenter works. It is suitable for hot press and high frequency drying applications in that way increasing the efficiency of the manufacturing process. In this experiment, only the edge part was covered with glue and then the surface part was placed on this glue covered edge and after that pressed with various pressures as defined in this research. The processing temperature was 20°C as recommended. As the hardening of the glue is dependent on the temperature, material, porosity and humidity, since the hot press was not used, the recommended hardening and pressure were bonded for at least 24 h.

3. Bending tests

Normal stresses in plate skins (figure 2) and shear stresses in the joint at the ribs (figure 3) and deformations [6] [19] were analyzed for plates with cell type hollow ribs. The results of serviceability limit state (SLS) show that for plates up to 1/25 from plates' height the plate's skin is the most loaded.

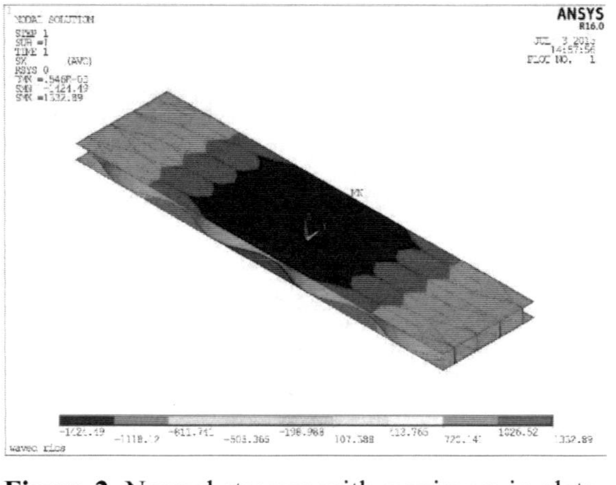

Figure 2. Normal stresses with maximum in plate skins for the plate with the cell type hollow core.

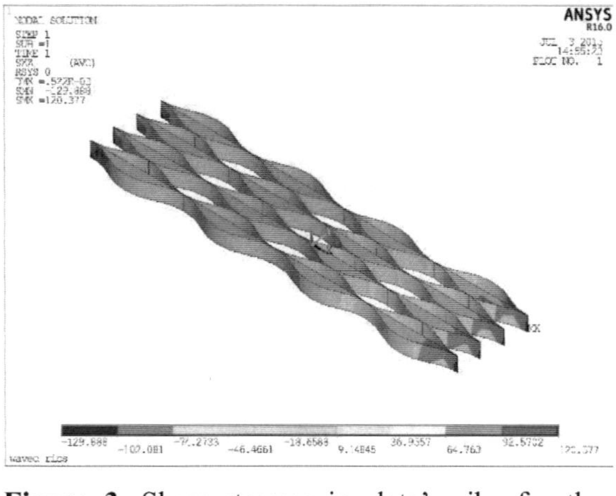

Figure 3. Shear stresses in plate's ribs for the plate with the cell type hollow core.

By increasing plate's height to span ratio up to 1/5, the ultimate limit state for plates skins are reached before the SLS (figure 4). The glue layer of plate skin-to-core still has a reserve of about 200%. But if thicker or other material plate skins were taken, this joint could become the weakest point.

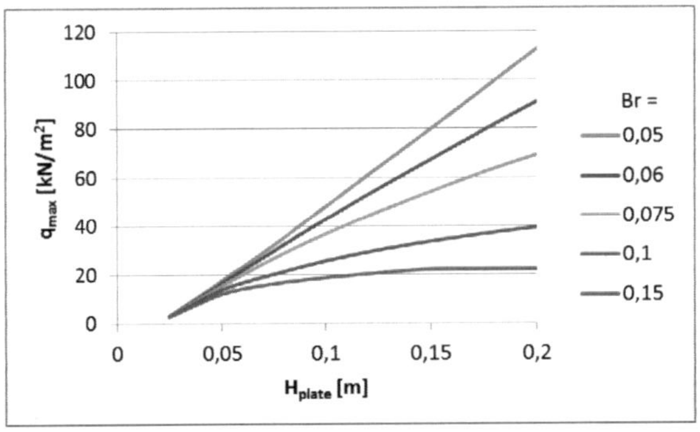

Figure 4. Maximal load for various heights of plate H_{plate} and widths of rib B_r.

4. Shear tests

4.1. Shear strength

The shear test specimen chosen in chapter 2.2 was simulated with ANSYS software code with the real dimensions of specimens for comparison of deformations and ultimate load.

4.1.1. FEM results.

As the linear properties of materials were assumed, ANSYS model was used for determination of shear stresses and it showed that shear stresses in the edge elements were distributed eccentrically (at the edge part that is closer to the center of the surface part they are the smallest (near to 0) but at the edge of the surface element they are the largest and for 1 kN tension load they are 4.0 N/mm^2 (figure 5), although for normal stresses, respectively, for 1 kN are 1.81 N/mm^2 and 18.81 N/mm^2 (figure 6) so in this case the difference is almost 10 fold.

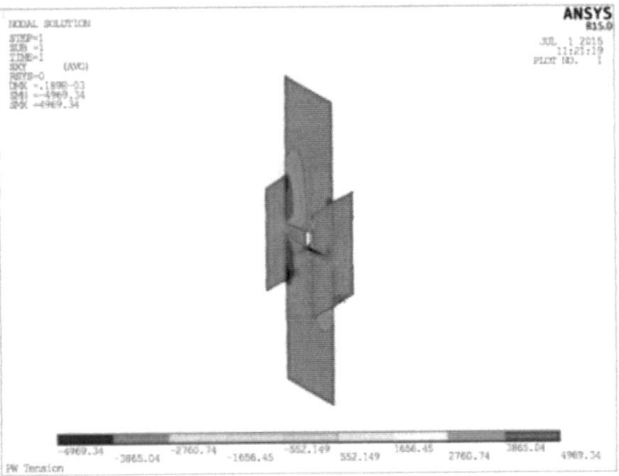

Figure 5. Non-uniform distribution of shear stresses along the joint.

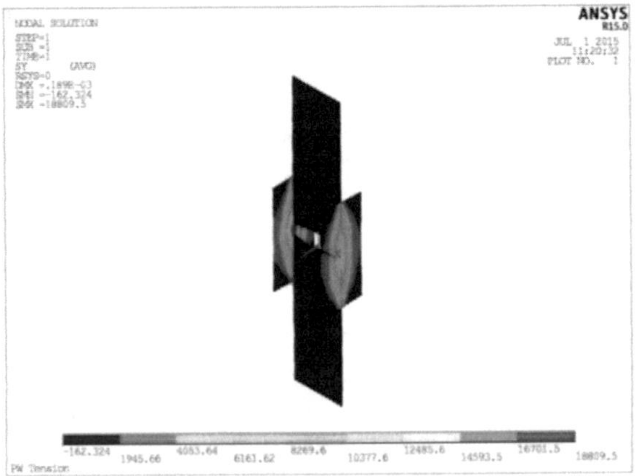

Figure 6. Non-uniform distribution of normal stresses along the joint.

4.1.2. Experimental investigations.

The samples previously calculated and analyzed with ANSYS were tested. As mentioned above, nine samples were made with different bond pressure, edge plywood orientation and thickness. The specimens and test equipment are shown in figure 7.

Figure 7. Specimen placed in the testing machine (Instron 3000).

In figure 8; figure 10; figure 12 the ultimate load for different level of bond pressure was achieved. The load-deformation plots for all specimens of the samples with most common bond pressure 0.5 N/mm^2 are shown in figure 9; figure 11; figure 13.

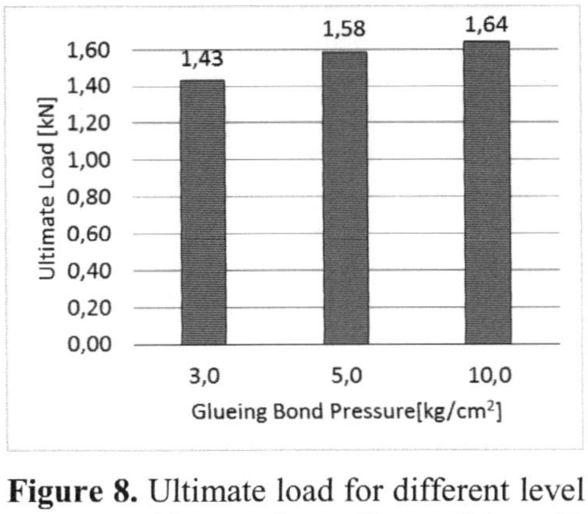

Figure 8. Ultimate load for different level of pressure with outer layer fibers of the middle layer (4.0 mm) in longitudinal direction.

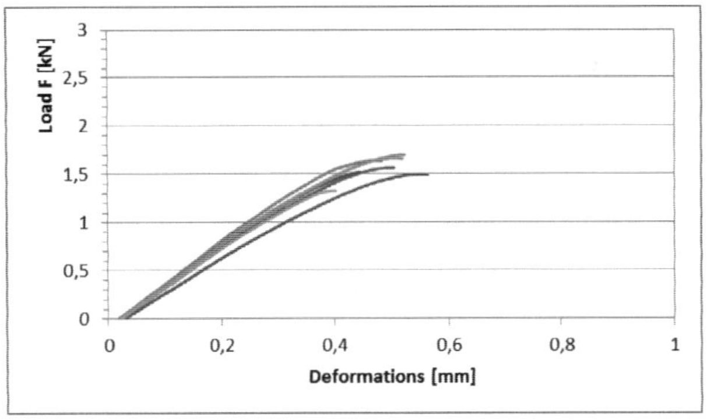

Figure 9. Load-deformation plot for specimens with bond pressure 5.0 kg/cm^2 and outer layer fibers of the middle layer (4.0 mm) in longitudinal direction.

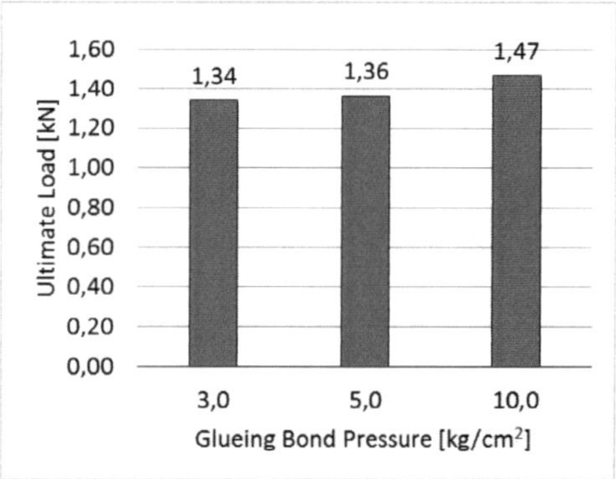

Figure 10. Ultimate load for different level of pressure with outer layer fibers of the middle layer (4.0 mm) in transversal direction.

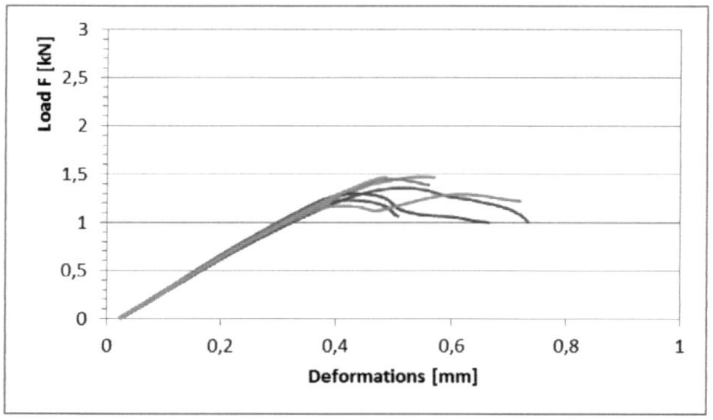

Figure 11. Load-deformation plot for specimens with bond pressure 5.0 kg/cm^2 and outer layer fibers of the middle layer (4.0 mm) in transversal direction.

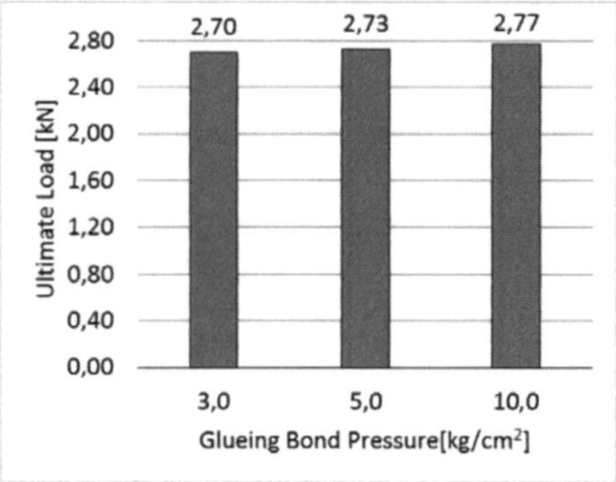

Figure 12. Ultimate load for different level of pressure with outer layer fibers of the middle layer (6.5 mm) in longitudinal direction.

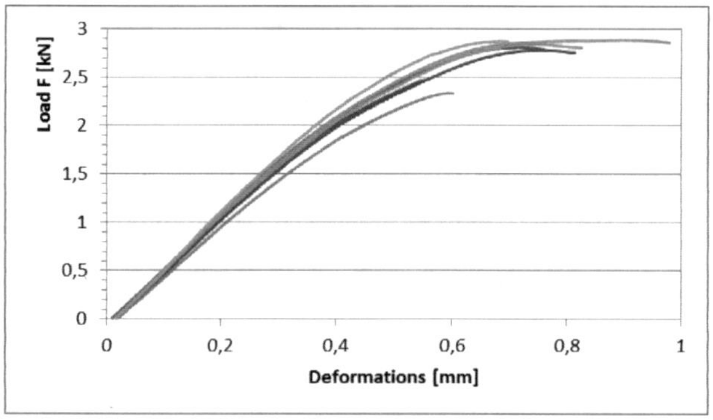

Figure 13. Load-deformation plot for specimens with bond pressure 5.0 kg/cm^2 and outer layer fibers of the middle layer (6.5 mm) in longitudinal direction.

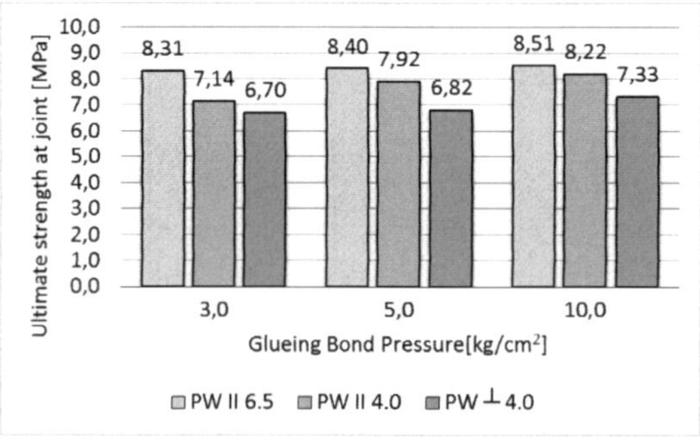

Figure 14. Ultimate strengths for different levels of pressure for all samples.

The main dimensions of specimens were shown in figure 1. Detailed specifications of the tested specimens are summarized in table 1. Results show that by increasing bond pressure from 0.3 N/mm^2 to 0.5 N/mm^2 the ultimate strength increases by 11% and 1% for five layer plywood with outer ply fibers oriented in longitudinal direction and by 2% for three layer plywood when outer ply fibers oriented in transversal direction to skin ply and added tension load direction. But when the bond pressure is increased to 1.0 N/mm^2 the ultimate strength increases respectively by 15%; 2% and 9%. To summarize these results, it could be concluded that the ultimate strength is dependent on the layer count in transversal direction of the bonded surface. It means when the more layers are oriented in transversal direction, the increase in bond pressure is required to achieve tighter connection. The bond pressure is dependent on the layers in transversal direction to surface. For three layer plywood connection and five layer plywood connection with outer ply fibers directed in longitudinal direction it is advisable to increase bond pressure from 0.3 to 0.5 N/mm^2, but further increase of bond pressure is not necessary. Although for plywood with outer ply fibers in transversal direction to skin ply it is recommended to increase bond pressure to 1.0 N/mm^2. Apparently, for 5 layer plywood this difference is not as large as it was for 3 layer plywood as the ratio between parallel and transversal layers smoothens.

Table 1. Comparison of pressure provided at the joint (comparison of bond pressure).

	Length of a glued joint mm	Thickness of a PW T mm	Pres-sure at joint kg/cm^2	Ultimate strength N/mm^2	F_{max} kN	F_{max}/ $F_{max,teor}$	F_{max}/ $F_{max3.0}$	Standard Deviation	Coeff. of var.	Range
PW II 4.0	25	4.0	3.0	7.14	1.43	-11%		0.17	12.12	0.54
PW II 4.0	25	4.0	5.0	7.92	1.58	-1%	11%	0.13	8.14	0.20
PW II 4.0	25	4.0	10.0	8.22	1.64	3%	15%	0.11	6.45	0.26
PW II 6.5	25	6.5	3.0	8.31	2.70	4%		0.10	3.67	0.24
PW II 6.5	25	6.5	5.0	8.40	2.73	5%	1%	0.16	5.59	0.42
PW II 6.5	25	6.5	10.0	8.51	2.77	6%	2%	0.14	5.19	0.20
PW ⊥ 4.0	25	4.0	3.0	6.70	1.34	-16%		0.21	15.26	0.49
PW ⊥ 4.0	25	4.0	5.0	6.82	1.36	-15%	2%	0.10	7.08	0.24
PW ⊥ 4.0	25	4.0	10.0	7.33	1.47	-8%	9%	0.17	11.03	0.44

The average strength was 7.76 N/mm^2 for PW II 4.0; 8.41 N/mm^2 for PW II 6.5 and 6.95 N/mm^2 for PW ⊥ 4.0. As it is seen, the average values are very similar to the values that were achieved with bond pressure 0.5 N/mm^2.

For most of the specimens the failure was in the plywood material. Delamination of the outer layer of skin plywood or of the longitudinal fibers of edge plywood was accompanied with glue abruption in transversal layer fibers was more common for lower pressure (0.3 N/mm^2) specimen (figure 15) it can be concluded that there is no need to change the glue for these applications because the strength of the glue is sufficient.

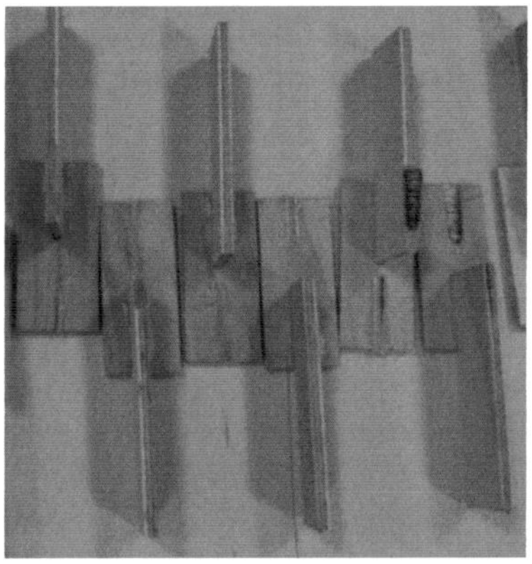

Figure 15. Different cases of failure mode.

4.2. Shear stiffness

Shear strength and stiffness at glued joints between core ribs forming plywood and skin forming plywood have been determined. As previously mentioned, this is the connection.

4.2.1. FEM results.

The plot of deformations in tension is shown in figure 16. The results show that deformations in the middle part are by 24% larger than they are in the parts close to the joint.

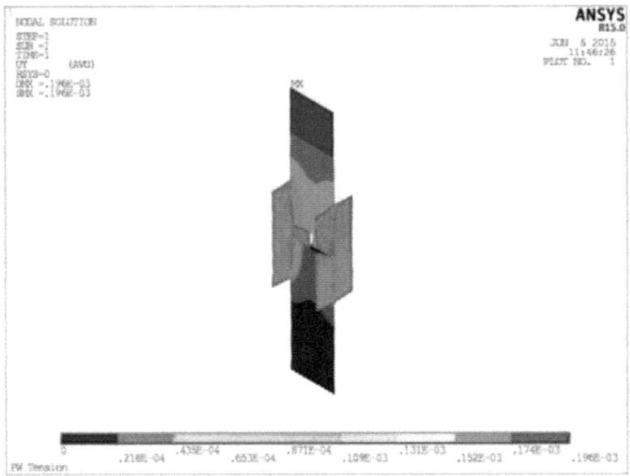

Figure 16. Shear specimen with dimensions.

4.2.2. Experimental investigations.

Shear stiffness was calculated from the measured load and deformations. For different bond pressure, the stiffness of the joint remains in 5 % level independent of bond pressure (table 2). The coefficient of variation is under 10% showing that these results are feasible to assume properties in modeling and determination of joints in design of composite ribbed cell type plates. The average stiffness is equal to 4.24 N/mm^2 for *PW II 4.0*; 3.25 N/mm^2 for *PW II 6.5* and 3.59 N/mm^2 for *PW* ⊥ *4.0*. In this case the stiffness properties are similar to strength properties, the average values are very similar to the values that were achieved with bond pressure 0.5 N/mm^2.

Table 2. Stiffness comparison of pressure provided at the joint.

	Pressure at joint kg/cm^2	Joint stiffness N/mm^2	Comparison to 0.3 N/mm^2	Standard Deviation	Coefficient of variation	Range	Average N/mm^2	Comparison to average
PW II 4.0	3,0	4.15		0.21	5.07	0.34		-2%
PW II 4.0	5,0	4.19	1%	0.25	5.91	0.76	4.24	-1%
PW II 4.0	10,0	4.38	6%	0.12	2.66	0.30		3%
PW II 6.5	3,0	3.15		0.09	2.62	0.26		-3%
PW II 6.5	5,0	3.28	4%	0.14	3.96	0.35	3.25	1%
PW II 6.5	10,0	3.32	5%	0.17	5.30	0.55		2%
PW ⊥ 4.0	3,0	3.53		0.13	3.72	0.36		-2%
PW ⊥ 4.0	5,0	3.60	2%	0.07	1.92	0.19	3.59	0%
PW ⊥ 4.0	10,0	3.65	3%	0.14	3.73	0.34		2%

4.2.3. DIC results.

Digital Image Correlation (DIC) tests were done for several experiment specimens to calibrate the load measurement and determine the deformation (and stress) distribution along the joint and its elements. It has been noticed that the strain in the middle part is by 24% on average (20-30%) larger than the strain in the parts close to joint. For one specimen of sample PWII4.0 0.5N/mm2 is shown in figure 17.

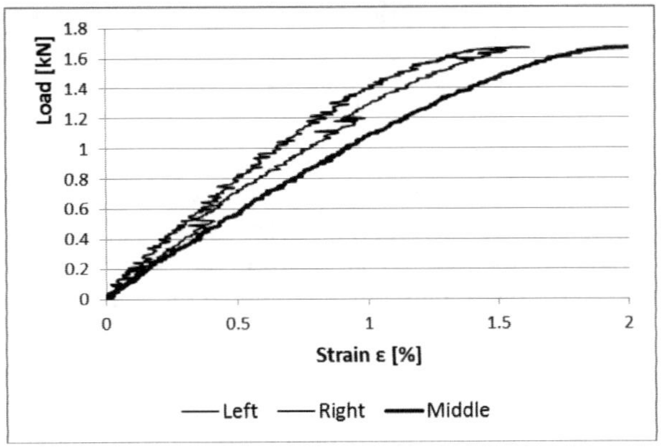

Figure 17. Deformations in the middle part of the joint and exactly at joint comparison.

5. Conclusions

In cases when more layers are oriented in transversal direction to longitudinal direction than other layers of the plywood surface that is glued, it is useful to increase the bond pressure comparing to traditionally used 5 kg/cm^2 (70 psi), although when more layers are oriented in longitudinal direction then it is practical enough to apply 5 kg/cm^2.

The obtained results from experimental investigations show that the achieved results for lower bond pressure less than (5 kg/cm^2) show approximately similar mean values of ultimate strength, although the dispersion of characteristic values increases.

6. References

[1] Pepke E 2010 *Global Wood Markets: Consumption, Production and Trade* [Available]: http://www.unece.org/fileadmin/DAM/timber/mis/presentations/PepkeGlobalWoodMkts050510.pdf
[2] *Vital Forest Graphics* 2009 [Available]: http://www.unep.org/vitalforest/Report/VFG_full_report.pdf
[3] *Production of Wood–based Panels* [Available]: http://eippcb.jrc.ec.europa.eu/reference/BREF/WBP_online.pdf
[4] Campbell F C 2010 *Structural Composite Materials* (USA: ASM International)
[5] EN789 Timber structures – test methods – determination of mechanical properties of wood-based panels
[6] Frolovs G, Rocens K and Sliseris J 2014 *Experimental Investigations of Composite Plywood Plates with Vertically Placed Waved Ribs* (Kaunas, Lithuania)
[7] *Adhesives Design Toolkit* [Available]: http://www.adhesivestoolkit.com/Docs/test/MECHANICAL%20TEST%20METHOD%201%20-%20Table%203.xtp
[8] Broughton W R and M R D 1997 *Review of Durability Test Methods and Standards for Assessing Long Term Performance of Adhesive Joints* (National Physical Laboratory, Teddington, Middlesex TW11 0LW)
[9] ASTM D3166-99 2012 Standard test method for fatigue properties of adhesives in shear by tension loading (Metal/Metal), ASTM International
[10] BS 5350-C5:2002 Methods of test for adhesives. Determination of bond strength in longitudinal shear for rigid adherents
[11] BS EN ISO 9664:1995 Adhesives. Test methods for fatigue properties of structural adhesives in tensile shear
[12] EN 1465:2009 Adhesives. Determination of tensile lap-shear strength of bonded assemblies
[13] ASTM D3528-96 2008 Standard test method for strength properties of double lap shear adhesive joints by tension loading, ASTM International
[14] ASTM E122-00, Standard practice for calculating sample size to estimate, with a specified tolerable error, the average for characteristic of a lot or process, ASTM International
[15] *Digital Image Correlation (DIC) Measurement Principle* [Available]: http://www.dantecdynamics.com/measurement-principles-of-dic
[16] Pereligin L and Ugolev B 1971 (L.M. U. B.N., Lesnaja promyshlennost') *Wood Science* p 288
[17] *PVA D3*, 22 June 2015 [Available]: http://www.polyline.lv/lv/docs/PDF/pva_d3_apraksts.pdf
[18] EN 204:2001 Classification of thermoplastic wood adhesives for non-structural applications
[19] Frolovs G, Rocēns K and Šliseris J 2015 Comparison of a Load Bearing Capacity for Composite Sandwich Plywood Plates *Proc. of the 10th Int. Scientific and Practical Conf. (Latvia, Rezekne, 18-20 June, 2015)* vol 1 (1) pp 39–45

Acknowledgments
The research leading to these results has received the funding from Latvia State Research Program under grant agreement "INNOVATIVE MATERIALS AND SMART TECHNOLOGIES FOR ENVIRONMENTAL SAFETY, IMATEH"

IOP Conf. Series: Materials Science and Engineering **96** (2015) 012049 doi:10.1088/1757-899X/96/1/012049

Validation of Finite Element Updated Models of the Developed Façade Scaffold Anchor

J Ilcik[1,4], V Arora[2] and J Dolejs[3]
[1]Faculty of Civil Engineering, CTU in Prague, Czech Republic
[2]Department of Technology an Innovation, SDU, DK
[3]Faculty of Civil Engineering, CTU in Prague, Czech Republic

E-mail: jiri.ilcik@fsv.cvut.cz

Abstract. The paper focuses on the updating of the finite element models of the newly developed facade scaffold anchor in the light of the experimental results. The experiments (force-displacement curves) have been carried out on the anchor. The façade scaffold anchor overcomes the problems arising in the process of joining them to the facades through the thermal insulation layers. Using the current methods, the wind load cannot be effectively transferred into the façade and scaffolding stability is decreased. Experimental results are presented and the finite element models of the anchor are developed using non-linear beam and solid elements. It has been observed that the predictions of finite element models that is force-displacement curve do not match with the experiment results. Subsequently, the finite element models of the developed anchor have been updated in the light of experimental results by using the parameter-based finite element model updating method. In case of an anchor, modelling of stiffness of the joints and values of the materials are expected to be dominant sources of inaccuracy in the FE model, assuming that the correct geometric parameters are known. After updating joint stiffness of the anchor joints and material properties, the finite element predictions match with experimental results. The outcomes show that there is a good correlation between the updated finite element models and the experimental data. The accuracy of the updated finite element models is demonstrated by overlaying force-displacement curves with the curve from the experiment, it can be concluded that the updated finite element models of the anchor accurately represent reality.

1. Introduction

To use a thermal insulation layer on the building façades is quite a common practice. But the current methods used for the fixing of scaffold anchors do not consider these thermal insulation layers. In the most common way, the distance provided by the insulation layer is crossed by the thick scaffold screw

[4] To whom any correspondence should be addressed.

as shown in figure 1. Under the influence of the wind load, the screw is being deformed [1]. Due to that fact, the long scaffold screw can hardly transmit the forces from the scaffold construction into the façade, which results in a deformation of the screw as well as the damage of the surrounding thermal layers; the stiffness of the whole scaffold system is considerably decreased. It should be also mentioned that for the safety reasons scaffolds systems are covered with nets or planks. However, the cladding increases horizontal force acting on the anchors under the impact of the wind load, which is crucial in most of accident cases [2, 3].

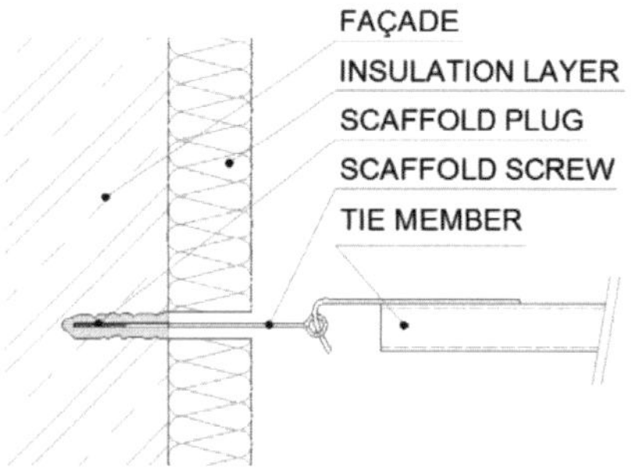

Figure 1. Vertical section of the fixed anchor – because of an insulation layer a long scaffold screw is used.

Figure 2. The fixing of an anchor to the insulated façade.

The scaffolds are generally weak systems, because of the loosened connections of components [4]. This results in instability of the anchor systems [5]. The stability of the scaffolding has been investigated in many papers [6, 7]. Scaffold constructions are generally divided into two wide groups based on their main bearing components [5]. First there are non-system tubes constructions, where the main components are mild steel tubes with diameter of 48.3 mm and 4 mm thickness. Tubes are connected together by couplers. In the European countries, these are still very frequently used because of their variability. Based on this principle, the bamboo scaffolds, which are very often used in the countries of East Asia, can be also added to this group. Second large group is represented by using prefabricated parts, mostly frames, in certain system configurations. This group can be also called proprietary scaffolds. In case of the non-system tubes constructions, the stability is provided only by the anchoring and bracing. For those reasons, the importance of the anchoring should not be underestimated.

It is necessary to develop a new scaffold system, which overcomes the problems connected with the existing scaffold anchors. These problems include damage of the façade thermal insulation layers surrounding the scaffold fixing and related decrease of scaffold stability.

2. The stability investigations and the basic principles of the new anchor

One of the most appropriate stability investigations has been provided by Dolejs in his thesis [1]. Dolejs analysed a typical non-system tubes scaffold construction with 12 storeys. The scaffolding has been modelled according to the netted cladding, see figure 3, and the values of loading and load combinations have been based on European standard EN12 811-1 [8]. The load class is set as 2, so the uniformly distributed load, which is part of the service loads, has the value of 1.50 kN/m^2. According to the standard, there are 4 loading combinations. First two represent the service condition while the second two represent the out of order condition. The 3rd combination is important, which represents the out of order condition with the maximum value of the wind load perpendicular to the façade. The anchors have been modelled as pinned supports according to the standard [8 and 9]. The scaffold tubes have been connected by semi-rigid joints respecting M – φ relation obtained by the previous experiment. Generally, the semi-rigid connections as well as initial imperfections affect the results [10]. Critical load coefficients for each combination have been computed. The values vary in the range from 1.68 to 2.74, while the value of the 3rd combination is 1.80.

2.1. The stability of the scaffolding using long screws in fixing

For the purpose of demonstrating the phenomenon of the long scaffold screw used in fixing (as show in figures 1 and 2), the previous Dolejs' model has been enhanced by adding these screws. The screw is added as a rod with 12 mm diameter and 300 mm in length. After that the stability analysis has been computed again. The critical load coefficient of the 3rd combination has the value of only 0.38, which is acceptable by considering the fact that most of the accidents happen during the strong wind conditions [2, 3]. The stability shape of this combination is shown in figure 4.

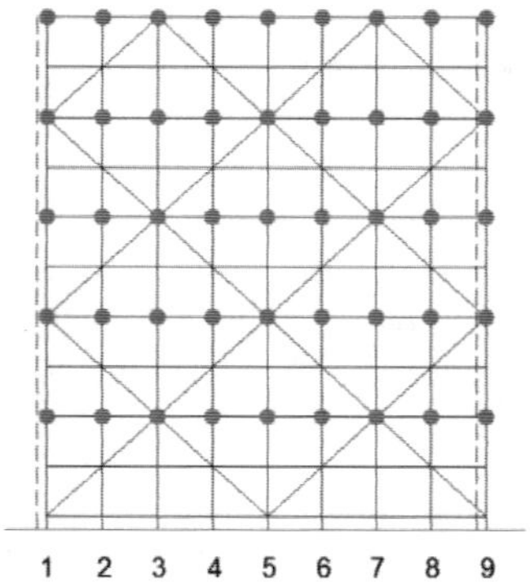

Figure 3. The front view on scaffolding – the anchor position pattern for the netted scaffolding. The dots represent anchors.

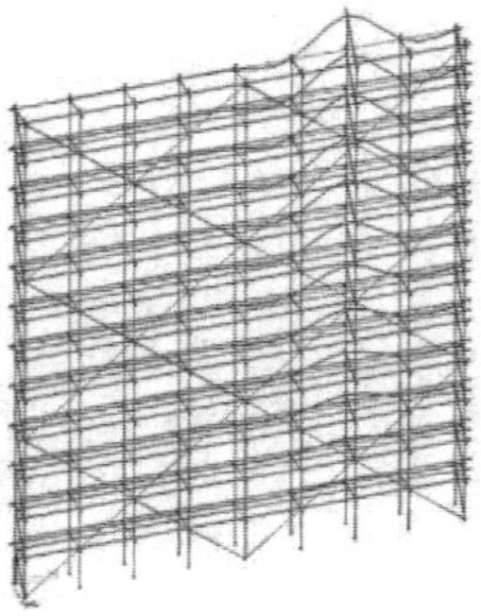

Figure 4. The first stability shape of the scaffolding fixed with long scaffold screws

2.2. The stability of the scaffolding using the proposed new anchor

For the purpose of increasing the scaffold stability, the parametrical study on the existing scaffold model has been carried out. The parameters include the position patterns of the anchors and the degree of freedom of supports.

It has been concluded that one of the most important parameters, which affect the final critical coefficient, is the support rigidity. Also it has been observed that even the semi-rigid support has a large influence on the critical constant value. Because of that it has been determined that for the next investigations, the new anchor will be considered as being rigid in the horizontal plane. For this type of support the optional position pattern has been developed as it is shown in figure 5.The final critical load coefficient has risen up to 3.24. The stability shape of this value is represented in figure 6. Also from this model the maximum values of the support forces have been computed by geometrical non-linear analysis with imperfection (GNIA) like in the case of the Dolejs' model [1].

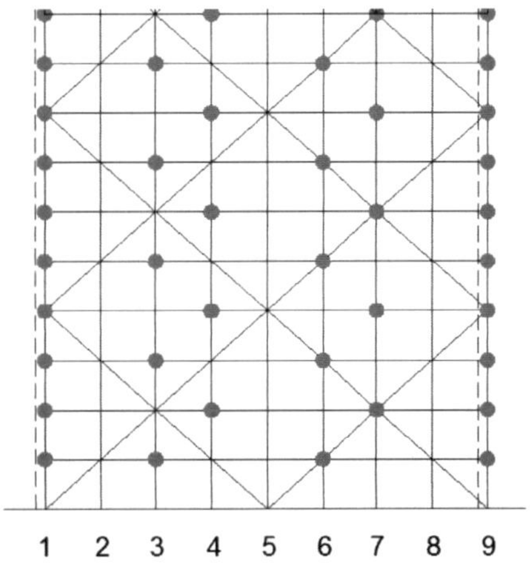

Figure 5. The developed position pattern for the new anchor.

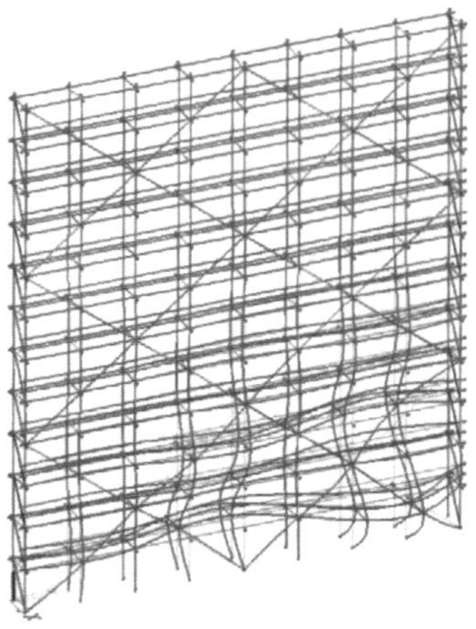

Figure 6. The first stability shape of the scaffolding using new anchors (supports are rigid in the horizontal plane).

2.3. The developed shape of the anchor

Considering the previous investigations, a new anchor shape has been developed and officially registered. The shape of the new anchor is termed Lever Anchor. The anchor consists of a rigid frame, which is connected to the facade in the horizontal way. The frame columns lead through the facade insulation layer. The frame is stiff enough to carry out the forces from the scaffold construction without deformations, so the insulation layer is secured. The anchor consists of several parts, because it has to provide variability during installation on the façade.

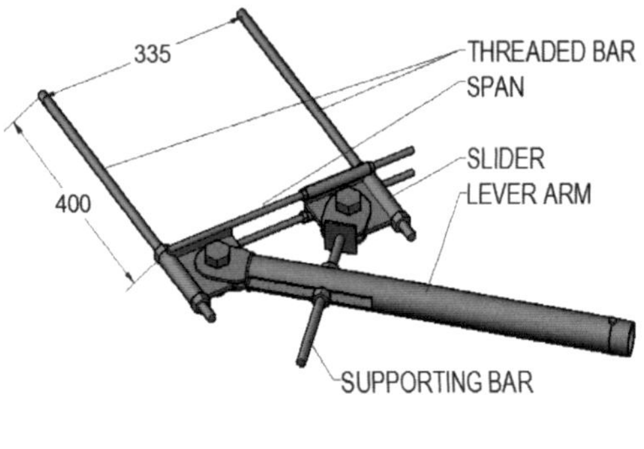

Figure 7. The developed new anchor shape – The Lever Anchor.

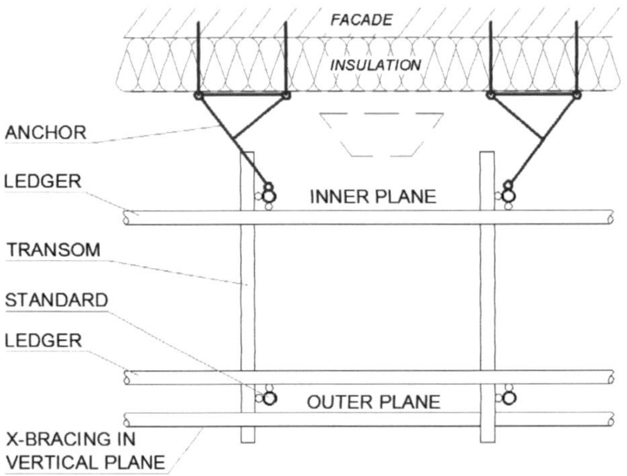

Figure 8. The plane view on the scaffolding with the anchoring by two Lever Anchors.

3. The experimental verification of the Lever Anchor

It is necessary to develop an accurate finite elements model, which will be used during the optimization. The accuracy of the finite element model is confirmed by the experimental verification. The prototype of the Lever Anchor has been created and an experiment has been carried out. The scheme diagram of the laboratory experiment is shown in figure 9.

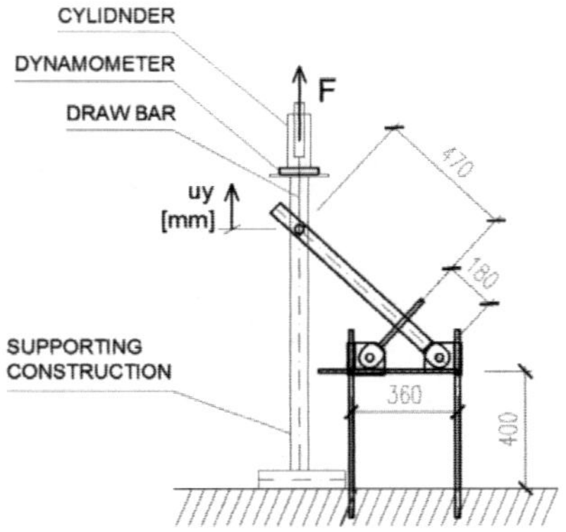

Figure 9. The scheme of the laboratory experiment.

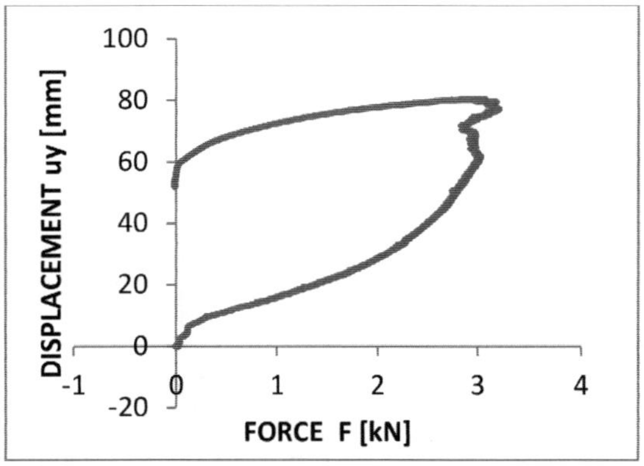

Figure 10. The force-displacement curve as a result of the laboratory experiment.

The anchor was loaded continually by tension force in vertical direction. The force magnitude rose from 0 to 3 kN. The displacement was measured at the end of the lever arm. The result was the force-displacement curve as shown in figure 10.

The force was provided by the single acting hollow plunger cylinder with the one-handed pump. The cylinder was situated on the top of the supporting steel construction made from two parallel columns with reversed T shape. The pressure was measured by tensometric dynamometer. The steel draw bar connected a hydraulic cylinder to the end of the lever arm. The values of vertical displacement were measured by the common absolute potentiometer sensor.

In the beginning, the member Span (the description of the anchor is shown on figure 7) reached the plastic behavior. After applying the force of approximately 3.0 kN, the first small cracks on the span's bars were detected, so the weakest part of the anchor was the Span member.

4. The finite element models of the tested anchor

Two accurate finite element models have been developed in the software Ansys. These models have difference in the used finite elements. The first is a 2D beam model and the second is a 3D solid model.

4.1. The initial 2D model

In the beam model all parts are connected with longitudinal and torsional stiffness. The stiffness values have been computed by using empiric equations. The real values of the joint stiffness are unknown and they are considered to be the source of error in the finite element model as well as the material properties. The defection profile is calculated using the applied force and subsequently plotted along with the experimental results. It can be observed from figure 11 that the finite element results do not match with the experimental results.

4.2. The initial 3D model

The solid final element model of the anchor has also been developed. It can be observed from figure 11 that the deflection curve does not match the experiment as well. Cross sections and material properties have been used the same as in the beam model. The only important difference is that in the solid model, the stiffness of the connections is provided by the rigidity of the solid bodies itself. Therefore, the material properties are the only source of errors.

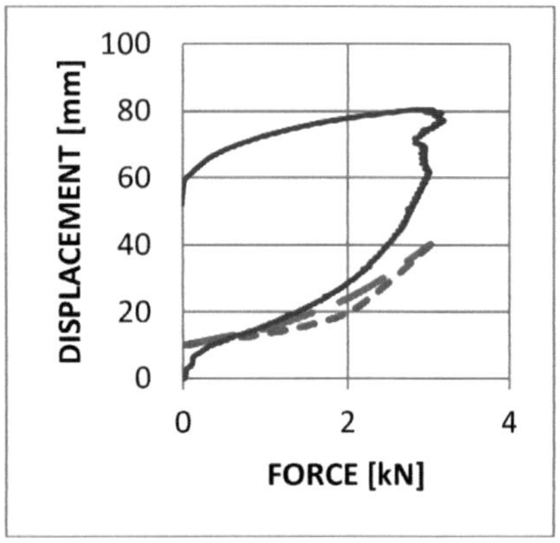

Figure 11. The force-displacement curves of the initial models. Solid line – experiment, Dot line – beam model, Dash line – solid model.

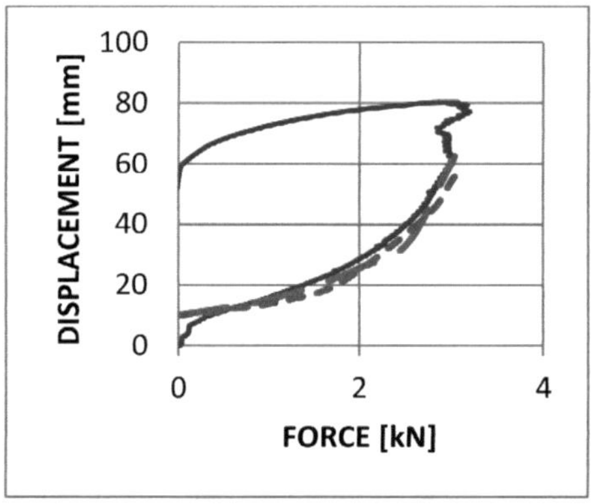

Figure 12. The force-displacement curves of the updated models. Solid line – experiment, Dot line – beam model, Dash line – solid model.

4.3. The accurate models

The choice of updating parameters on the basis of engineering judgment about the possible locations of modelling errors in a structure is one of the strategies to ensure that only meaningful corrections are made. In case of an anchor, modelling of stiffness of the joints and the values of the material properties are expected to be a dominant source of inaccuracy in the FE model, assuming that the correct geometric parameters are known. The values of stiffness of joints and values of the material parameters are updated in the light of experimental results. The same method has been used in many papers, see [11, 12]. In both models the parameter studies have been carried out, finally both force-displacement curves of the solid and beam model match the experimental curve well, as demonstrated in figure 12. For more detailed description, the updated parameters in both models follow.

In the updated 2D (beam) model, the important parameters include the torsional stiffness of the joints; see figure 13 and Table 1. The longitudinal stiffness in both axes (X, Y) has been left with no change to the initial model. Also the material yield strengths have been updated. The threated rod's Structural Steel 4.8 has been adjusted from 300 MPa to 275 MPa and in the case of the Structural Steel 8.8 the 580 MPa value has been decreased to 550 MPa. The plate's Structural Steel S235 has been left without change at the value 200 MPa.

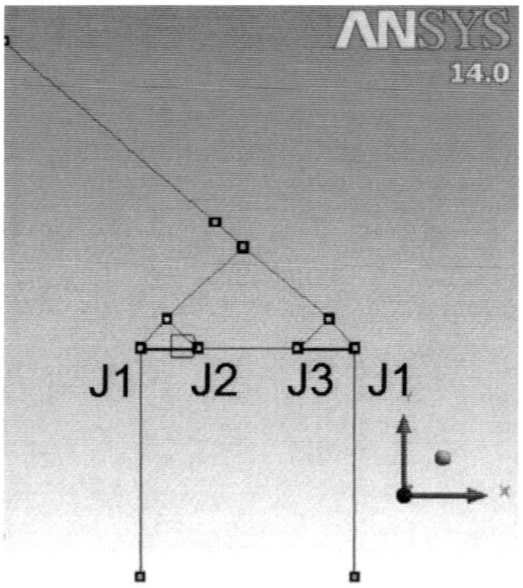

Figure 13. The description of the most important joints in 2D beam model according to the final force-displacement curve.

Table 1. The updated torsional parameters of the joints in [MNm/rad].

Name	**Initial**	**Updated**
J1	0.10	2.41
J2	0.01	0.34
J3	0.10	0.11

In the updated 3D (solid) model, only the material yield strengths have been updated. The threated rod's Structural Steel 4.8 has been lowered from 300 MPa to 260 MPa and Structural Steel 8.8 from 580 MPa value to 500 MPa value. The plate's Structural Steel S235 has also been left without change and stayed at the value 200 MPa.

5. The optimized anchor

Based on the accurate finite element models, the optimization process has been carried out. The dimensions have been adjusted for the safe transmitting of the support forces. These forces have been obtained from the scaffold construction, which has been previously shown in figures 5 and 6.

The final optimized anchor is shown in figures 14 and 15. The geometry of the anchor has been changed according to the standard [8] and some parts have been split up to ensure better manipulating and ease of the production process. The cross sections of all parts have been fortified. In these conditions, the Lever Anchor should be able to carry a pressure force up to 18 kN, which is directed to the end of the lever arm.

Subsequently, the anchor will be experimentally tested again. In this experiment the force will be placed in all 4 horizontal directions for the best description of the behaviour and for accurate specifying of the support semi-rigidity.

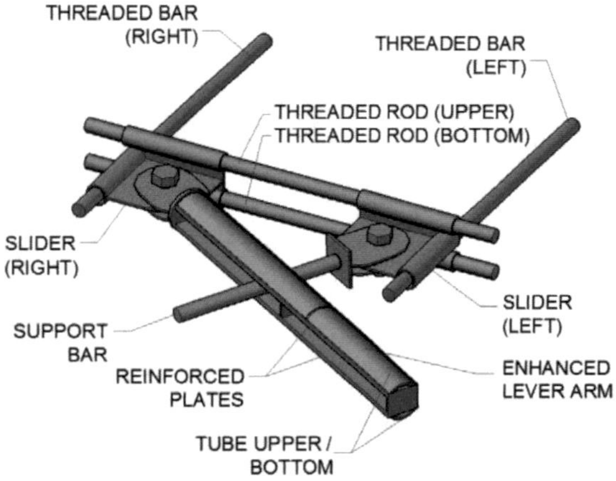

Figure 14. The description of the optimized Lever Anchor.

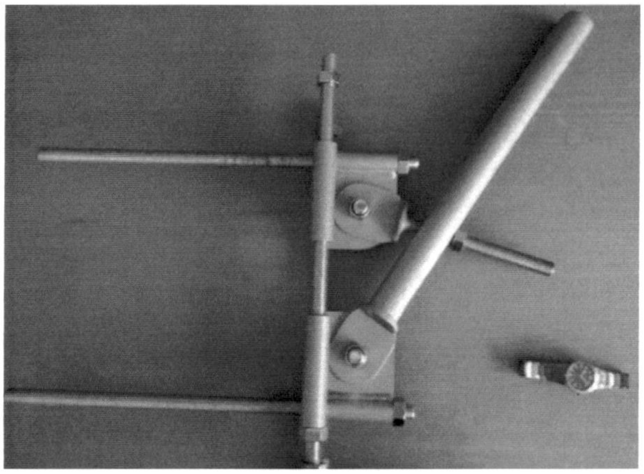

Figure 15. The created optimized Lever Anchor (with the wrist watch in the bottom right corner as a scale).

6. Conclusions

In this paper the new type of the scaffold anchor with the unique position pattern on the scaffolding as well as the finite elements models of that anchor have been developed. It has been observed that the finite element models do not match with the experimental results. After updating joint stiffness of the joints and the material properties of the anchor, the finite element prediction matches with experimental results. Subsequently, the result of the following optimization process is also presented. The paper is a part of the long-term project aimed at developing a new fixing system for the façade scaffolding.

7. References

[1] Dolejs J 2013 *Spatial Coactions of Facade Scaffolds Elements and Components* Habilitation thesis

[2] Wang F, Tamura Y and Yoshida A 2013 Wind loads on clad scaffolding with different geometries and building opening ratios *J. of Wind Engin. and Industrial Aerodynamics* **120** 37–50

[3] Wang F, Tamura Y and Yoshida A 2014 Interference effects of a neighboring building on wind loads on scaffolding *J. of Wind Engin. and Industrial Aerodynamics* **125** 1–12

[4] Chandrangsu T and Rasmussen K J R 2011 Investigation of geometric imperfections and joint stiffness of support scaffold systems *J. of Constructional Steel Research* **67** (4) 576–584

[5] Beale R G 2014 Scaffold research – A review *J. of Constructional Steel Research* **98** 188–200

[6] Liu H, Zhao Q, Wang X, Zhou T, Wang D, Liu J et al. 2010 Experimental and analytical studies on the stability of structural steel tube and coupler scaffolds without X-bracing *Engineering Structures* **32** (4) 1003–15

[7] Peng J-L, Wu C-W, Chan S-L, Huang C-H 2013 Experimental and numerical studies of practical system scaffolds *J. of Constructional Steel Research* **91** 64–75

[8] EN 12811-1 2003 *Temporary Works Equipment – Part 1: Scaffolds – Performance Requirements and General Design* (Brussels: European committed for standardization)

[9] Dolejs J, Picek Z, Skreta Z, Vlasak M and Vlasak S 2011 *Navrhovani konstrukci z leseni I.* (Prague: Czech Technical University, monograph)
[10] Prabhakaran U, Beale R G and Godley M H R 2011 Analysis of scaffolds with connections containing looseness *Computers & Structures* **89** (21-22) 1944–55
[11] Arora V, Singh S P and Kundra T K 2009 Damped FE model updating using complex updating parameters: Its use for dynamic design *J. of Sound and Vibration* 324 (1-2) 350–364
[12] Arora V, Singh SP and Kundra T K 2009 Finite element model updating with damping identification *J. of Sound and Vibration* **324** (3-5) 1111–23

Acknowledgements
The authors are grateful for the financial supports by the Czech Technical University (Project no. SGS13/168/OHK1/3T/11) and to the program ERASMUS+ (CZ PRAHA10). The technical support and supervision during experiments by Jonáš of HILTI is gratefully acknowledged as well as the support from the Experimental Centre of the Faculty of Civil Engineering, CTU in Prague, and Picek, PKL servis.

IOP Conf. Series: Materials Science and Engineering **96** (2015) 012050 doi:10.1088/1757-899X/96/1/012050

Slender Compressed Plate in Component Based Finite Element Model

M Kurejková[1,3], F Wald[1], J Kabeláč[2] and L Šabatka[2]
[1]Czech Technical University in Prague, Thákurova 7, 166 29 Czech Republic
[2]IDEA RS, U vodárny 2a, 616 00 Brno, Czech Republic

E-mail: marta.kurejkova@fsv.cvut.cz

Abstract. The paper presents an advance design model of a slender plate in the structural steel joint. Finite element methods and material models are described and design procedure for slender plates in numerical models of steel joints is proposed. The design procedure is demonstrated on examples. The results are verified with an analytical model according to European standards. A compressed beam with slender web and beam-to-column joint are studied by numerical analysis, buckling resistances are determined and results verified. The verification shows very good agreement.

1. Introduction

The analytical model represented by component method (CM), which is described in EN 1993-1-8, gives good prediction of behaviour for typical structural steel joints [1]. However, many components are limited by plate slenderness, because the design procedure is valid for the third class cross section and better, see [2] and [3]. Buckling of slender plates should be assessed in a separate step according to EN 1993-1-5 [4]. The code also contains recommendations for design by numerical modelling as an alternative to the analytical model. The use of numerical models is today widely spread in design practice and especially buckling analysis of plated structures gets special attention [5]. Finite element models (FEM) give very good prediction of global behaviour of joints, although the accuracy of the results is closely connected to the meshing, element types, interfaces and sub-modelling. When using FEM tools, special attention should be paid to the selection of software, modelling of material properties, the use of imperfections and modelling of boundary conditions and loads.

2. Finite element method in plate stability analysis

With respect to the ultimate limit state, the code provides five categories of numerical analysis with the following assumptions:

---

[3] To whom any correspondence should be addressed.

- Material and geometric linear;
- Material nonlinear and geometric linear -> plastic resistance;
- Material linear and geometric nonlinear -> buckling;
- Material linear and geometric nonlinear with imperfections;
- Material and geometric nonlinear with imperfections -> ultimate resistance.

Plastic resistance is determined by material nonlinear and geometric linear analysis. Material diagram is elastic with strain hardening and the structure is modelled with its theoretical geometry without imperfections. Ultimate limit state is reached by 5% plastic strain. The coefficient $\alpha_{ult,k}$ is obtained, where $\alpha_{ult,k}$ is the minimum load amplifier for the design loads to reach the characteristic value of the resistance of the most critical point.

Critical buckling modes are determined by material linear and geometric nonlinear analysis. The critical buckling factor α_{cr} is determined and stands for the load amplifier to reach the elastic critical load under complex stress field.

The load amplifiers are related with the non-dimensional plate slenderness, which is determined as follows:

$$\bar{\lambda} = \sqrt{\frac{\alpha_{ult}}{\alpha_{cr}}} \tag{1}$$

Reduction buckling factor ρ is calculated according to EN 1993-1-5 Annex B. Conservatively, the lowest value from longitudinal, transverse and shear stress is taken. Figure 1 shows the relation between plate slenderness and reduction buckling factor.

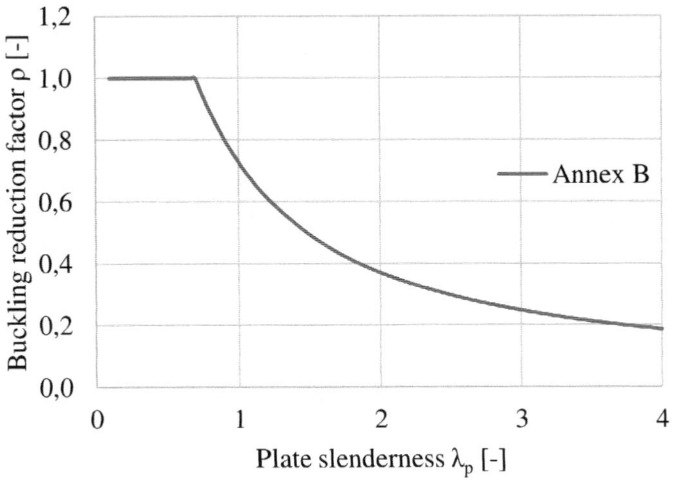

Figure 1. Buckling reduction factor ρ according to EN 1993-1-5 Annex B.

The verification is based on the von-Mises yield criterion and reduced stress method and sums up the load effects of normal and shear stresses. Buckling resistance is assessed as:

IOP Conf. Series: Materials Science and Engineering **96** (2015) 012050 doi:10.1088/1757-899X/96/1/012050

$$\frac{\alpha_{ult} \cdot \rho}{\gamma_{M1}} \geq 1 \tag{2}$$

where γ_{M1} is partial safety factor.

It should be noted that the first critical buckling mode in the joint may not be in terms of the previous condition crucial. More buckling modes need to be assessed in a complex joint, because they are related to different parts of the joint. The procedure is sufficiently general, robust and can be quite easily automated. Its advantages are in the advanced joint FEM analysis. In addition, it is part of the Eurocode standards. The design tool gives a quick overview of the joint behaviour, its critical components and allows fast stiffening to prevent instabilities.

The above described design procedure is used in component based finite element model (CBFEM), which combines the advantages of finite element models and the component method. The stress distribution in the joint is close to the real behaviour and the components bolt, weld and plate are assessed according to analytical models approved by experiments, see [6] and [7]. Compared to research finite element model (RFEM), which uses material and geometric nonlinear analysis with imperfections, CBFEM is time-saving in modelling, FEM analysis and evaluation of ultimate limit state. CBFEM is an innovative design method for steel joints with complex geometry and stress field, see [8]. Material model uses true stress-strain diagram, which is obtained by tensile tests, in RFEM and ideal plastic or elastic with strain hardening for design FEM, see figure 2.

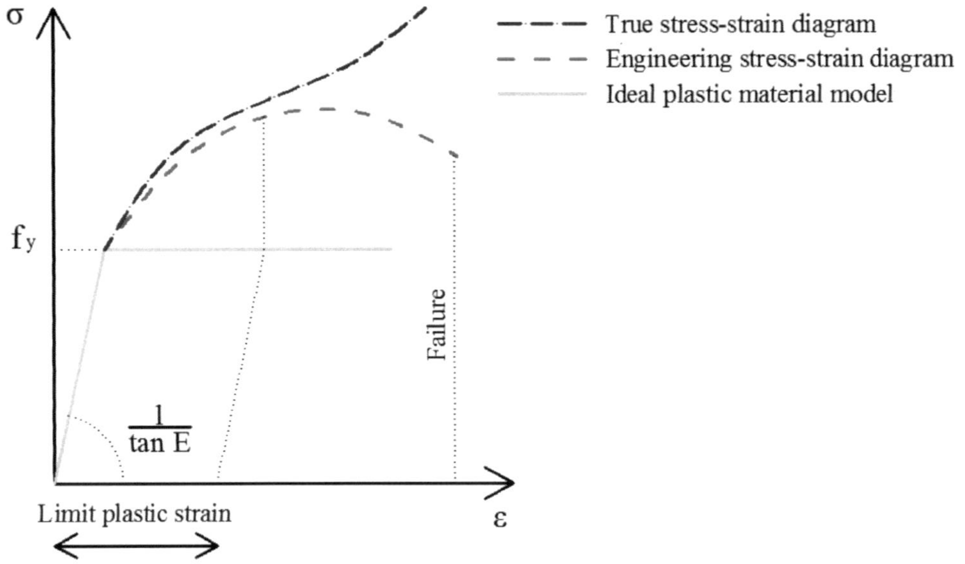

Figure 2. Material models of steel for research and design numerical models.

Validation and Verification (V&V) is a native process of computer based design, see [9], where validation means comparison of numerical models to experimental data and verification comparison of the numerical solution with the accurate analytical or numerical solution. Application of V&V to steel connections design is limited to a few published benchmark studies, see [10]. System response quality

(SRQ) contains a description of a selected joint, results of CM and CBFEM, comparison of resistances and Benchmark case to allow the user to check his results. In some cases the CBFEM method gives higher resistance. Advanced FEM model with shell elements validated on own experiments or experiments from literature is used in these cases to get proper results. CBFEM is approved by this procedure. The proposed procedure for slender plate in CBFEM model is based on Annex B in EN1993-1-5 and verified on two examples.

3. Verification of the compressed beam web

First verification example shows a welded beam loaded in compression. Four beams B2-B5 are studied and the ultimate resistances are calculated. The beam dimensions are fixed, only the web thickness t_w is changing between 2 and 5 mm. A plate 200x10 mm is used for flanges and a web height is set to 380 mm. The range of web's slenderness is set from 1 to 2.5. Yield strength of the beam is set to S235. Figure 3 shows the first buckling mode of the compressed beam.

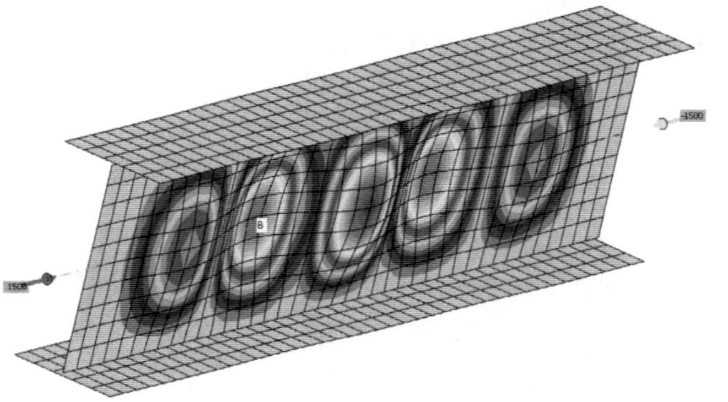

Figure 3. First buckling mode of the compressed beam.

A comparison of compressed beam's ultimate resistance for analytical calculation according to EN 1993-1-5 and CBFEM is shown in table 1. It is observed that the resistances are in good accordance with the CBFEM calculations.

Table 1. Ultimate resistance of compressed beam – Comparison.

	t_w	$\alpha_{cr,CBFEM}$	$\alpha_{ult,k,CBFEM}$	$\lambda_{p,CBFEM}$	F_{CBFEM} [kN]	F_{Rd} [kN]	$(F_{CBFEM} - F_{Rd})$ / F_{CBFEM} [%]
B2	2	0.12	0.77	2.53	338	330	2
B3	3	0.30	0.83	1.66	553	553	4
B4	4	0.56	0.89	1.26	777	758	2
B5	5	0.91	0.95	1.02	1016	1004	1

A comparison of ultimate resistances for the compressed beam is shown in figure 4. The vertical axis of the diagram shows the ultimate resistance by CBFEM model, while the horizontal axis shows the results of an analytical solution. An allowed deviation of 10% is marked by dotted lines. A very good agreement is observed in the diagrams.

IOP Conf. Series: Materials Science and Engineering **96** (2015) 012050 doi:10.1088/1757-899X/96/1/012050

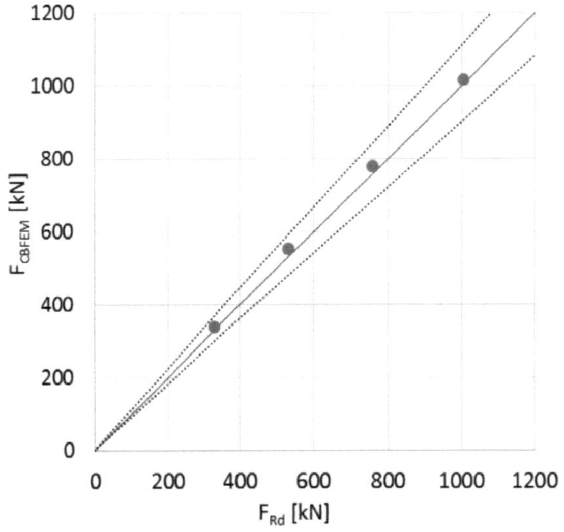

Figure 4. Comparison of compressed beam's ultimate resistances for CBFEM and CM.

4. Verification of the beam-to-column joint

The second verification study covers a beam-to-column joint. Four joints W2-W5 are studied and the ultimate resistances are calculated. Beams and columns are welded cross sections with flange 300x8 mm and web height 584 mm. Only the thickness of a column web panel t_w is changing from 2 to 5 mm in the verification study. The range of column web panel's slenderness is set from 0.8 to 2.0. The joint is loaded in bending and yield strength is set to S235. A plastification of the column web panel and the first buckling mode are shown in figure 5.

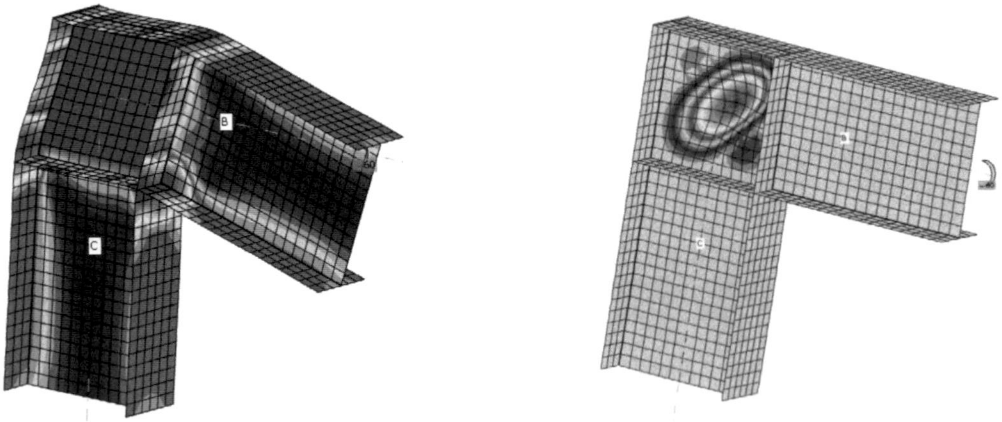

(a) (b)

Figure 5. Yielding in column web panel (a) and the first buckling mode.

A comparison of beam-to-column joint's ultimate resistance for analytical calculation according to EN 1993-1-5 and CBFEM is shown in table 2 and in figure 6. It can be found that the maximum deviation of ultimate resistances between CBFEM model and analytical solution is 7%, which means good agreement exists.

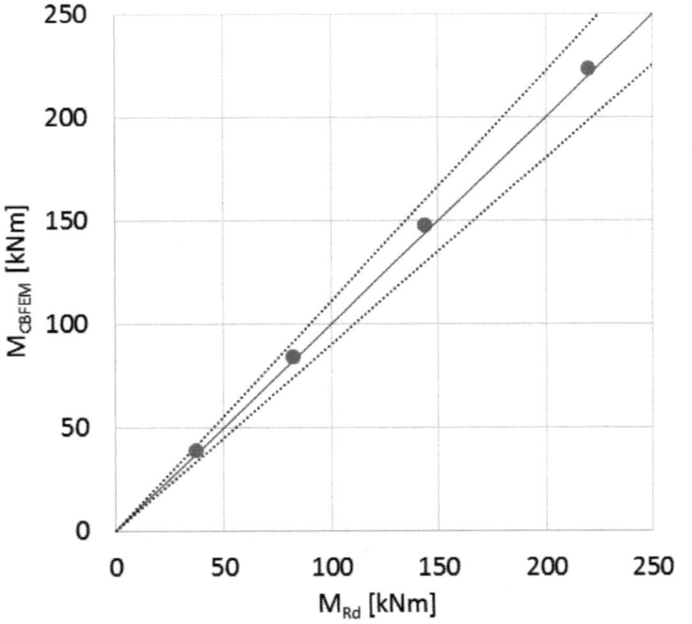

Figure 6. Comparison of beam-to-column joint's ultimate resistances for CBFEM and CM.

Table 2. Ultimate resistance of beam-to-column joints – Comparison.

	t_w	$\alpha_{cr,CBFEM}$	$\alpha_{ult,k,CBFEM}$	$\lambda_{p,CBFEM}$	M_{CBFEM} [kNm]	M_{Rd} [kNm]	$(M_{CBFEM} - M_{Rd})$ / M_{CBFEM} [%]
W2	2	0.10	0.41	2.02	39	36	7
W3	3	0.33	0.59	1.34	84	81	4
W4	4	0.77	0.79	1.01	148	141	4
W5	5	1.48	0.97	0.81	223	216	3

5. Conclusion

It is proposed to use the reduced stress method for the design finite element model (DFEM) or component based finite element model (CBFEM) of compressed plates in structural steel joints. The verification examples show that compressed plates could be designed in finite element models without applying imperfections. Essential part of the following research is the verification of buckling curve for non-regular plate shapes. Using the buckling curve instead of nonlinear analysis will reduce the calculation time. The design procedure covers local buckling of the compressed plates, shear buckling of a slender web panel and local buckling of a compressed plate between the bolts.

6. References

[1] Wald F, Kurejková M, Gödrich L, Martínek K, Šabatka L and Kabeláč J 2015 Simple and Advanced Models for Connection Design in Steel Structures *Int. Conf. on Advances in Civil and Environmental Engineering (Pulau Pinang, 28-30 July 2015)* pp 1 – 9

[2] Zoetemeijer P 1981 Report 6-81-15 Bolted connections with flush endplates and haunched beams (Delft: Delft University of Technology)

[3] Zoetemeijer P 1981 Report 6-81-23 Bolted beam to column knee connections with haunched beams (Delft: Delft University of Technology)

[4] EN 1993-1-5 2005 Eurocode 3: Design of steel structures – Part 1-5: Plated structural elements (Brusells: CEN)

[5] Beg D, Kuhlmann U, Davaine L and Braun B 2010 Design of Plated Structures (Berlin: ECCS)

[6] Kurejková M and Wald F 2014 Compressed Stiffeners in Structural Connections *Proc. Eurosteel 2014 (Napoli, 10-12 September 2014)* pp 317 – 318

[7] Gödrich L, Kurejková M and Wald F 2014 The Bolts and Compressed Plates Modelling *Proc. 12th Int. Conf. on Steel, Space and Composite Structures (Prague, 28-30 May 2014)* pp 215-224

[8] Wald F, Gödrich L, Šabatka L, Kabeláč J and Navrátil J 2014 Component Based Finite Element Model of Structural Connections *Proc. 12th Int. Conf. on Steel, Space and Composite Structures (Prague, 28-30 May 2014)* pp 337 – 344

[9] Wald F, Kwasniewski L, Gödrich L and Kurejková M 2014 Validation and verification procedures for connection design in steel structures *Proc. 12th Int. Conf. on Steel, Space and Composite Structures (Prague, 28-30 May 2014)* pp 111-120

[10] Bursi O S and Jaspart J P 1997 Benchmarks for finite element modelling of bolted steel connections *J. of Constr. Steel Res.* pp 17 – 42

Acknowledgments

This study was carried out with support from a grant by the Technology Agency of the Czech Republic within the MERLION project No. TA02010159.

IOP Conf. Series: Materials Science and Engineering **96** (2015) 012051 doi:10.1088/1757-899X/96/1/012051

Effect of Waste Plastic Shreds on Bond Resistance between Concrete and Steel Reinforcement

K B Osifala[1,4], M A Salau[2] and A A Adeniyi[3]
[1]Department of Civil Engineering, Yaba College of Technology, Yaba, Lagos, Nigeria
[2]University of Lagos, Department of Civil & Environmental Engineering, Yaba, Lagos, Nigeria
[3]Department of Civil Engineering, Yaba College of Technology, Yaba, Lagos, Nigeria

E-mail: kosifala@yahoo.com

Abstract: This paper investigates the effect of waste plastic shreds on steel-concrete bond. Forty RILEM test specimens with 16mm and 20mm diameter high-yield reinforcing bars were cast and tested. Fifteen specimens with 16mm and 20mm each were cast with the addition of waste plastic shreds at varying percentages of 1%, 1.5% and 2%; another ten RILEM specimens with 16mm and 20mm diameter bars at 0% of waste plastic shreds were cast as reference. Nine 150mm cubes, with three taken from each batch of various percentages of waste plastic shreds, were used to monitor the concrete strength. From the test results and analysis, the compressive strength of concrete was found to reduce with increased percentages of waste plastic shreds, while the waste plastic shreds material was found not to improve the bond resistance between concrete and steel. However, though lower than normal concrete, there was an increase in the bond resistance with increase in the percent of plastic shreds. The bond resistance of 16mm was also found to be higher than that of 20mm in all the specimens tested.

1. Introduction

The use of plastic has grown substantially all over the world in the recent years, and this has created huge quantities of plastic waste in the environment. Plastic waste cannot be dumped in landfills because of its bulk and slow degradation rate; incorporation of plastic waste into concrete if found suitable could be one of the solutions for disposing it.

The use of methylcellulose (0.4% to 0.8% by weight of cement) as an admixture in cement paste or concrete was reported in [1]. This was found to increase the shear bond strength with steel rebar, steel fibre or carbon fibre to the values attained by using latex (20% by weight of cement) as an admixture. The bond strength increased with increasing methylcellulose amount. The combined use of silica fume

[4] To whom any correspondence should be addressed.

IOP Conf. Series: Materials Science and Engineering **96** (2015) 012051 doi:10.1088/1757-899X/96/1/012051

(15% by weight of cement) and methylcellulose (0.4% by weight of cement) as admixtures was found to give concrete, apart from high bond strength to steel rebar, high tensile modulus, tensile ductility, flexural strength and flexural toughness.

Experimental investigations to study the suitability of the use of re-engineered plastics as fibres for road pavements were carried out in [2]. The properties studied include compressive strength, tensile strength, flexural strength under reversed cyclic loading, impact resistance, plastic shrinkage and abrasion resistance, etc. Efforts to compare it with steel fibres showed that the improvement of concrete properties at lower cost is obtained with re-engineered plastic shred reinforced concrete.

A laboratory experimental study carried out to utilize waste plastics (in the form of strips) obtained from milk pouches in the pavement construction was reported in [3]. Results of the study indicate that by adding plastic strips in the soil, shear strength, tensile strength and CBR values of the soil increased. In this study, plastic or polythene sheets having thickness of 0.5mm and which were made up of high density were used. Three types of plastic strips were used in the study to act as a reinforcing material. The first one was cut into 20mm x 40mm size, second one was 25mm x 50mm size and the third one was of 30mm x 60mm size. These plastic strips have innumerable advantageous properties like high tensile strength, low permeability etc. These plastic strips act as a good barrier to gases and liquids and are unaffected by cycles of wetting and drying. For all the strips used in the experimental work, an aspect ratio of 2 was maintained.

An investigation on the improvement of bond strength of epoxy coated bars using stirrups was reported in [4]. Three beam sizes (300mm x 230mm, 300mm x 200mm, and 300mm x 180mm) were used with 28mm, 20mm, 16mm diameter high yield bars, the stirrups used were 8mm and 10mm diameter high yield bars. It was found that stirrups improved the bond efficiency of epoxy coated bars, it was also reported that application of coating materials in sequence and as a mixture on coated bars improved bond strength. Epoxy was applied first and then tyrolin was later applied. The two materials were also mixed together and applied; the results gave higher bond efficiency. Simple lap spliced beams were used in a four point loading system. The two methods improved bond but application in sequence gave higher bond strength.

Polyethylene terephthalate and high density polypropylene (disposable glass) fibres were studied in [5]. The parameters investigated were workability, compressive strength and flexural peak load of concrete. The cubes of size 150 mm and slabs of size 740mm x740mm x 50mm were cast to determine the compressive strength and flexural peak load respectively. It was observed that slump and compaction factor was increased on addition of 1% of HDPP fibres and 2% of PET fibres. The compressive strength of cubes was increased by 4.2 % with 1% HDPP fibres and 5.63 % with 1% PET fibres. The flexural load before failure was also increased by using both the fibres. The bond strength was not studied.

Experimental work on compressive strength of cubes and cylinders, were used in [6] to study their behaviour under test. It was concluded that plastic strips have potential to act as a secondary reinforcement. From the above study, it was also concluded that the fibres made of recycled polyethylene terephthalate (PET) are appropriate for concrete reinforcement. The mixing ability of PET fibre is excellent and it is a promising material to reinforce the concrete. In view of this, attempt is being made to investigate the bond strength of such concrete.

The effect of increased cover on bond efficiency of coated reinforcing bars was studied in [7]. Increased concrete cover was found to increase bond efficiency of coated reinforcing bars, but the increase was not proportional to the additional cover thickness.

Pullout test was used in [8] to analyze the bond behaviour between different kinds of carbon and glass Fibre Reinforced Plastic (FRP) bars with different surface configurations and two concrete strengths. The result confirmed the trend of rebars with larger diameter developing lower bond strength, especially for high concrete strength series. An increase in bond strength and change in failure surface were observed when concrete compressive strength was increased.

The effects of fibre surface area on the bond properties of the fibre were studied in [9]. It was shown that significant increase in surface area can be achieved by changing the cross-sectional shape, which in turn leads to an increase in bond strength. Fibre crimping further increases the bond strength. More importantly, it was also found out that crimped macro-synthetic fibres with modified cross sections (cross, star, etc.) can attain a bond stress versus bond slip relationship similar to that of steel fibres, characterized by a steep and linear elastic frictional bond component, followed by a "bend-over point" and a subsequent parabolic increase to the maximum bond stress.

Bond behaviour between steel reinforcement and recycled concrete was studied in [10]. It was concluded that the bond strength decreases with the increase of the percentage of recycled coarse aggregate used. Transverse reinforcement was reported in [11] to improve the bond strength of coated bars and the ductility of the beams.

The effect of fibre content on the interfacial bond properties of ultra-high performance fibre reinforced concrete was investigated in [12]. Test results showed that 3% steel fibre by volume yielded the best performance in terms of compressive strength, elastic modulus, shrinkage behavior and interfacial bond strength. These parameters improved as the fibre content was increased up to 3 % volume. Flexural behaviours such as flexural strength, deflection, and crack mouth opening displacement at peak load had pseudo-linear relationships with the fibre content. Through inverse analysis, it was shown that fracture parameters including cohesive stress and fracture energy are significantly influenced by the fibre content: higher cohesive stress and fracture energy were achieved with higher fibre content.

This study is part of the continuing effort to investigate the bond properties of fibre-like materials with a view to generating data and improving knowledge on bond strength between concrete and steel, attempt to investigate the effect of waste plastic shreds on bond resistance between concrete and steel reinforcement, in order to see if it would improve bond. It is desirable and worthwhile.

2. Materials and methodology

2.1. Materials and samples preparation

The plastic waste was collected from a dump site, washed and shredded into fibre-like shape, figure 1. Ordinary Portland cement was used in this study. Fine aggregates used were river sand, while crushed granite of maximum nominal particle size of 12.5mm was used as coarse aggregate. The grading for the aggregate was done according to BS 1377 Parts 1 and 2.The uniformity coefficients (Cu) for the sand and granite were 5 and 5.4 respectively while their coefficients of curvature (Cc) were 1.36 and 1.0 respectively, figure 2. Potable water was used in producing the concrete mixture. The compressive mix ratio used was 1:2:4 (cement, sand and granite) with a water/cement ratio of 0.55. The waste plastic shredded material was added at varying percentages 0%, 1.0%, 1.5% and 2% by weight of concrete respectively.

Figure 1. Waste plastic shredded material.

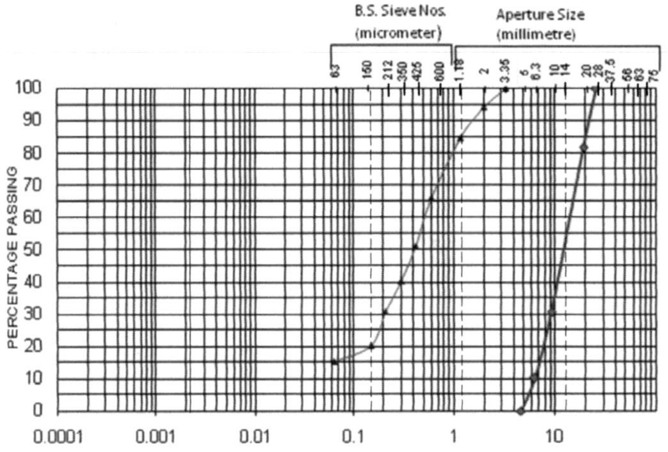

CLAY	*fine*	*medium*	*coarse*	*fine*	*medium*	*coarse*	*fine*	*medium*	*coarse*	COBBLES
	SILT			*SAND*			*GRAVEL*			

Figure 2. Grain size distribution for sharp sand and coarse aggregate.

2.2. Preparation of steel reinforcement

The reinforcements used were high yield steel rods (16mm and 20mm) which were cut into lengths of 1100mm each to allow for effective griping of the steel rod. The reinforcements were wire brushed and dusted. The cut length for 16mm and 20mm diameter bars was 1100mm (that is; 160mm within the cube, 50mm for the shorter end that resist pull and 890mm for the longer end where pull was applied). The encased region of the reinforcement as given by RILEM is 10D, where D is the diameter of bar. Half of this length (5D) was wrapped up with double layers of polythene to prevent bond between concrete and steel while the other half was left bare (unwrapped) in order to provide adhesion between the concrete and the reinforcement.

2.3. Testing of specimens

Before the commencement of the pull-out test, the RILEM specimens were kept under room temperature after the completion of the 28 days curing in water. Thereafter, each of the RILEM specimens and the strength cubes were weighed before they were finally taken to the machine for the pull-out and the compression testing.

In order to determine the bond resistance between concrete and steel reinforcement with waste plastic, the specimens were placed in the machine one after the other and it was carefully centralized between the upper and lower plates inside the frame, after which the tensile machine was loaded, and the reading at the time of pull out was recorded. The failure load was recorded for each of the specified variables, 0%, 1%, 1.5%, and 2% of each waste material (polypropylene plastic), including the reference. In the course of testing, at the point of failure, some of the RILEM specimens failed by splitting while some failed by slipping.

3. Results and discussion

3.1. Effect of waste plastic shredded material on compressive strength of concrete

The results of the compressive strength of 0% addition of waste plastic shreds material (polypropylene plastic) as compared to that of the 1%, 1.5% and 2% addition of waste plastic shreds is summarized in figure 3, table 1. For the normal concrete (0% plastic waste), the average cube strength is 27.48N/mm^2. For 1%, 1.5%, and 2% addition of plastic shreds, the observed average strengths are 23.85, 17.25 and 17.13N/mm^2 respectively. This reducing cube strength may be because the adhesion between polypropylene plastic and concrete depends largely upon the surface energy of the plastic. Metals in general have a high surface energy and are easier to bond with concrete, whereas plastics have a lower surface energy and are harder to bond with concrete. Hence, the decrease in compressive strength has been observed as waste shredded plastic content increases in the concrete mix.

IOP Conf. Series: Materials Science and Engineering **96** (2015) 012051 doi:10.1088/1757-899X/96/1/012051

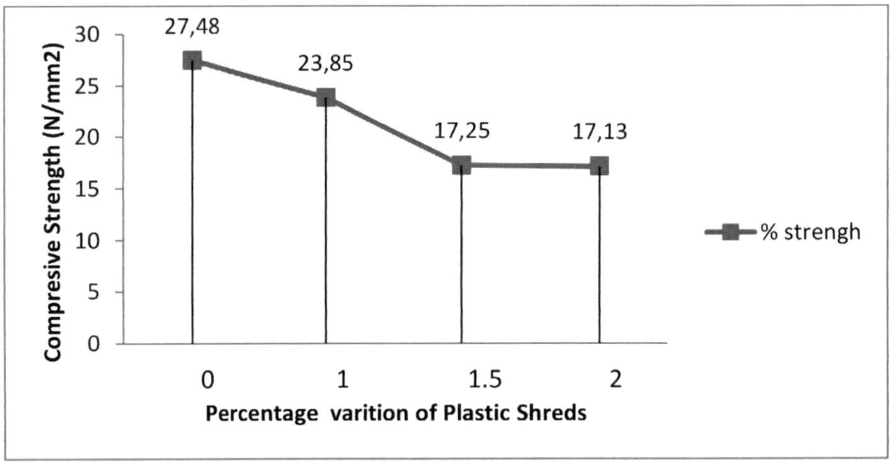

Figure 3. Compressive strength of concrete at 0%, 1%, 1.5% and 2% addition of waste plastic shreds (Polypropylene plastic).

Table 1. Compressive Strength (f_{cu}) of Concrete at 28 days with varying percentages of waste plastic shreds (150mmx150mmx150mm).

% of Plastic Shreds	Specimens ID	Maximum load (kN)	Cube Strength (N/mm^2)	Average Strength (N/mm^2)
	R01	680	30.00	
0	R02	520	23.11	27.48
	R03	660	29.33	
	P01	530	23.55	
1.0	P02	555	24.67	23.85
	P03	525	23.33	
	Q01	365	16.22	
1.5	Q02	400	17.77	17.25
	Q03	400	17.77	
	X01	295.5	13.13	
2.0	X02	455.5	20.24	17.13
	X03	406.0	18.02	

3.2. Crack pattern

Various crack patterns were recorded with different sizes of reinforcement (16mm and 20mm) at varying percentages of addition of waste plastic shreds material to the concrete mix. The RILEM concrete specimens with addition of plastic waste shredded material for 16mm and 20mm diameter steel reinforcing bars failed majorly by slipping, figure 4, while that for normal concrete (0% waste shredded inclusion) failed majorly by splitting, figure 5. Slip failure was observed, this may have been due to waste plastic shredded material finding its way across the path of the crack, and also the shredded plastics may have clustered around the steel reinforcement. The specimen tend to fail by slipping because the steel reinforcements slip off from the waste plastic shredded material due to the low energy surface of the plastic material and therefore the concrete did not bond to the steel reinforcement properly, hence slipping failure was observed during pull-out. The concrete with plastic shreds reached its maximum strength early due to the low friction between the plastic, concrete and the reinforcing bar.

Figure 4. RILEM specimen shreds with 1% waste shreds material failed by slipping.

Figure 5. RILEM specimen without plastic material failed by splitting.

3.3. Bond resistance of reinforced concrete with waste shredded plastic material

The results obtained from the tests gave the value of average bond stress of 1.61N/mm^2 for the specimens without addition of waste plastic shredded material (polypropylene plastic) with a few specimens failing by slipping, while the majority of the specimen failed by splitting. For the specimens with polypropylene plastic waste material the average bond stress of 0.85N/mm^2 was recorded at 1%, with all samples failing by slipping. The average bond stress of 1.1N/mm^2 at 1.5 % addition of waste plastic shredded material was recorded, with all the samples failing by slipping, the average bond stress of 1.43N/mm^2 at 2% addition of waste plastic shredded material was recorded, with all the samples also failing by slipping.

The results of samples cast with 20mm diameter bars gave the value of average bond stress of 0.79N/mm^2 at 1% addition of waste plastic shredded material, with all failing by slipping. The value of average bond stress of 1.09N/mm^2 at 1.5% addition of waste plastic shredded material (polypropylene plastic) was recorded with all the specimens failing by slipping, the value of average bond stress of 1.37N/mm^2 at 2% addition of waste plastic shredded was recorded with all the specimens failing by slipping, see figures 4 and 5, tables 2 and 3.

It was seen from the test results that the bigger the size of the specimens the lesser the bond strength. Samples with (200mm cube) 20mm diameter bars have bond resistance lesser than the samples (160mm cube) with 16mm diameter bars. This may be due to air pockets which settle below the steel bars during vibration which were with bigger diameter because of the bigger surface area. The graphs of bond stress for different percentages (0%, 1%, 1.5%, and 2%) are shown in figure 6.

Table 2. Bond Resistance and Crack Pattern of RILEM Specimens containing Plastic Shreds with16mm Diameter bar.

% Plastic Shreds	Specimen ID (160x160x160mm)	Bond Failure (kN)	Bond Resistance (N/mm^2)	Average Bond Resistance (N/mm^2)	Crack Pattern
0	R01	80.442	1.82	1.61	Split
	R02	86.328	1.95		Split
	R03	77.499	1.75		Split
	R04	51.012	1.15		Slip
	R05	60.822	1.38		Slip
1.0	S01	35.316	0.92	0.85	Slip
	S02	29.924	0.78		Slip
	S03	35.316	0.92		Slip
	S04	33.354	0.87		Slip
	S05	28.449	0.74		Slip
1.5	Y01	33.354	1.20	1.10	Slip
	Y02	31.392	1.13		Slip
	Y03	29.430	1.06		Slip
	Y04	26.487	0.96		Slip
	Y05	31.392	1.13		Slip
2.0	Z01	50.031	1.80	1.43	Slip
	Z02	40.221	1.44		Slip
	Z03	36.297	1.30		Slip
	Z04	39.240	1.41		Slip
	Z05	33.354	1.20		Slip

Table 3. Bond resistance and crack pattern of RILEM specimens. Containing plastic shreds with 20mm diameter bar (table 3. continues on next page).

% Plastic Shreds	Specimen ID (200x200x200mm)	Bond Failure (kN)	Bond Resistance (N/mm^2)	Average bond resistance (N/mm^2)	Crack pattern
0	A01	98.141	1.42	1.58	Split
	A02	108.701	1.57		Split
	A03	128.511	1.86		Split
	A04	110.500	1.60		Slip
	A05	98.716	1.43		Slip
1.0	B01	46.588	0.78	0.79	Slip
	B02	46.891	0.78		Slip
	B03	44.755	0.75		Slip
	B04	48.738	0.81		Slip
	B05	51.012	0.85		Slip

% Plastic Shreds	Specimen ID (200x200x200mm)	Bond Failure (kN)	Bond Resistance (N/mm^2)	Average bond resistance (N/mm^2)	Crack pattern
1.5	C01	51.012	1.18	1.09	Slip
	C02	45.773	1.06		Slip
	C03	42.183	0.97		Slip
	C04	46.20	1.06		Slip
	C05	50.869	1.17		Slip
2.0	D01	62.784	1.44	1.37	Slip
	D02	65.727	1.51		Slip
	D03	66.708	1.54		Slip
	D04	52.974	1.22		Slip
	D05	49.050	1.13		Slip

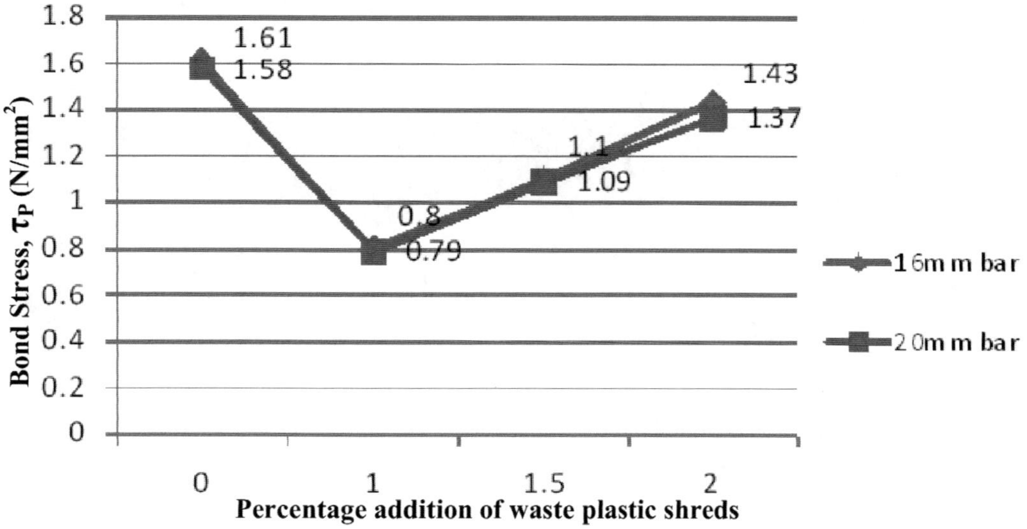

Figure 6. Bond resistance of RILEM specimens with waste plastic shreds.

With the failure load, the bond stress was determined using the following relationship in [13].

$$\tau_P = (0.00637F \times 25)\, 10^3 / Ø^2 \times f_{cu}, \tag{1}$$

Where:

τ_P = bond strength, (N/mm^2); F = Failure load, (kN);

$Ø^2$ = Bar diameter, (mm^2); f_{cu} = Compressive strength of concrete, (N/mm^2).

From the test results and analysis carried out, waste plastic shredded material incorporated in the concrete mix was found to immediately reduce the bond resistance between concrete and steel reinforcement when only 1% plastic shred was introduced. However, with increase in the percent of plastic shred, the bond resistance increased, when 1.5% and 2% of the plastic shred was added. This was due to the low force of adhesion between polypropylene plastic and concrete as adhesion depends largely upon the surface energy of the plastic, the surface energy of solid varies with its chemical make-up. Plastic has a lower surface energy and is harder to bond with concrete. Hence, low bond resistance was observed.

The bond resistance of waste plastic shred was also compared with that of other waste materials like steel, aluminium and wood [14], [15] and [16] to confirm the most effective material that can enhance the improvement of bond resistance, using 20mm steel reinforcing bars. The results are shown in figures 7.

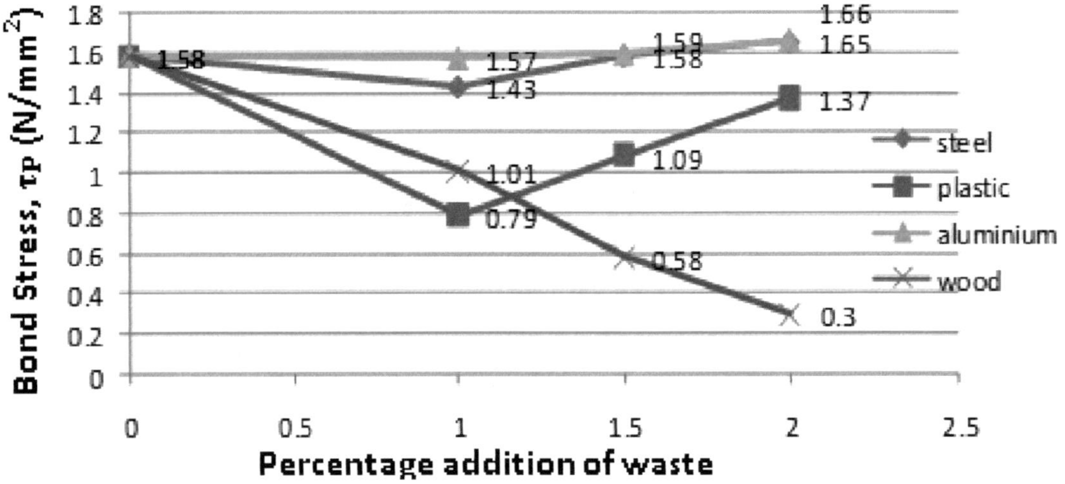

Figure 7. Comparison of bond resistance of various waste materials using 20mm bar.

4. Conclusions

Based on the results and the analyses carried out the following conclusions are drawn:

- It was observed that compressive strength of the concrete mix decreases as waste plastic shredded percentage increases.
- RILEM concrete specimens with addition of waste plastic shredded material for 16mm and 20mm diameter steel reinforcing bars failed majorly by slipping while that of 0% waste plastic shredded material failed majorly by splitting.
- Plastic shredded material does not at any percentage improve bond resistance between concrete and steel. However, bond resistance improved with increase in the percentage of plastic in the concrete from 1% addition, but was still lower at 2% when compared to concrete without any percentage of plastic shreds.

5. References

[1] Xuli F and Chung D D L 1998 Improving the bond strength of concrete to reinforcement by adding methylcellulose to concrete *ACI Materials J.* **95** (5) 601–608

[2] Laksmipathy M *et al* 2003 Use of reengineered plastic shreds as fibers in road pavements *Proc. of National Seminar on Furistics in Concrete and Construction Engineering* pp 3–5

[3] Chandrakaran S 2004 *Stabilization of Soils using Plastic Strips as Reinforcing Material* CE & CR pp 62–66

[4] Osifala K B 2005 *Improvement of Bond between Concrete and Coated Steel Reinforcement in Humid Environment* Ph.D. Thesis, Civil Engineering Department, University of Lagos p 235

[5] Venu M and Rao P N 2010 Effect of non-bio degradable waste in concrete slabs *Int. J. of Civil and Structural Engineering* **3** (1) 449–457

[6] Sarda V K, Dutta R K and Rajnish K C 2009 A study of compressive strength of fly ash mixed cement concrete reinforced with waste plastic strips *The Icfai University J. of Structural Engineering* **2** (1) 65–76

[7] Osifala K B and Akeju T A I 2010 Effect of increased cover on bond efficiency of coated reinforcing bars *Australian J. of Structural Engineering* **10** (2) 179–189

[8] Baena M 2010 *Study of Bond Behaviour between FRP Reinforcement and Concrete* Ph.D Thesis, University of Girona, Spain

[9] Choi O C, Hadje-Ghaffari H, Darwin D and McCabe S L 2012 Bond of Epoxy-Coated Reinforcement *Bar Parameters ACI Materials J.* **88** (2) 207–217

[10] Sindy S, Belen G, Javier E and Manuel H 2013 Bond behaviour between steel reinforcement and recycled concrete *J. of Materials and Structures* **47** 323–334

[11] Krishnamurthy and Gangolu 2013 Bond strength of epoxy coated bar splices confined with nominal lateral reinforcement *Central European J. of Engineering* **3** (1) 145–155

[12] Doo-Yeol Y, Hyun-Oh S, Jun-Mo Y and Young-Soo Y 2014 Material and bond properties of ultra-high performance fibre reinforced concrete with micro steel fibres *J. of Composite Engineering* **58** (3) 122–133

[13] RILEM 7-II-128 1994 RC6: *Bond Test of Reinforcing Steel 1* Pull-out Test. RILEM technical recommendations for the testing and use of construction materials, E & FN Spon, U.K. pp 102–105

[14] Osifala K B, Salau M A and Obiyomi T H 2013 *Improvement of Bond Resistance between Concrete and Steel Reinforcement using Steel Shavings* (Unpublished work) Civil Engineering Department, Yaba College of Technology, Lagos, Nigeria

[15] Osifala K B, Salau M A and Ikujebi A M 2013 *Improvement of Bond Resistance between Concrete and Steel Reinforcement Using Waste Aluminium Shavings* (Unpublished work), Civil Engineering Department, Yaba College of Technology, Lagos, Nigeria

[16] Osifala K B, Salau M A and Okonkwo C 2014 Improvement of Bond Resistance between Concrete and Steel Reinforcement Using Waste Wood Shavings *Proc. 9th Int. Conf. on Civil Engineering Design and Construction, Varna, Bulgaria*

IOP Conf. Series: Materials Science and Engineering **96** (2015) 012052 doi:10.1088/1757-899X/96/1/012052

The Behaviour of the Embedded Rail in Interaction with Bridges

P Ryjáček[1,3], Md M Howlader[1] and M Vokáč[2]
[1]Department of Steel and Timber Structures, Faculty of Civil Engineering, Thákurova 7, Prague 6, Czech Republic
[2] Klokner Institute, Czech Technical University, Šolínova 7, Prague 6, Czech Republic

E-mail: pavel.ryjacek@fsv.cvut.cz

Abstract. Since the introduction of the direct fastening system, Embedded Rail System (ERS) is probably one of the most spectacular and innovative developments in railway engineering. The development has become more rapid, especially in the last decade due to competitive advantages in comparison toother systems with respect to higher speed, cost effectiveness, environmental-material sustainability and others. But still there is a lack of specific interaction model, especially when it comes to bridges equipped with ERS system. Previous studies on ERS by Estzer Ludvigh[2] found out some important coefficients (vertical and longitudinal bedding coefficient) and compared the longitudinal resistance of ERS with flexible fastening systems. Other studies on ERS do not specifically deal with the evaluation of longitudinal resistance of such a system. With a view to establish the typical behaviour pattern of ERS, this paper is dedicated to find out the interaction of a specific Embedded Rail System with bridges. A small scale test was conducted on a sample of ERS in the laboratory under different combinations of vertical and longitudinal track loads and subsequently a Finite Element Model (FEM) was developed to simulate the test. The paper presents the FEA result validation with test results, the interaction pattern, longitudinal resistance of ERS track for both loaded and unloaded conditions, design load distribution for such systems and some specific influences of ERS on simply supported bridge systems.

1. Introduction

The primary component of an ERS in railway tracks is the embedding material used, which compensates for the use of traditional ballast along with the sleeper. Elastic embedment does the main work of providing the necessary longitudinal resistance against the wheel load along with better lateral stability, noise and vibration reduction, less influence of dynamic forces and a lot of other advantages [1, 2]. The properties of embedding material may change depending on manufacturer but the primary

[3] To whom any correspondence should be addressed.

objective fulfillment will be the same for all ERS. In this paper, the study is made on Edilon Sedra ERS (see figure 1), a widely known manufacturer of ERS. The numerical results found from the analysis are primarily Edilon Sedra embedding Corkelast VA-60 specific. Edilon Sedra Corkelast VA-60 is solvent and plasticizer free, two component, self-leveling casting elastomer system based on the state-of-the-art polyurethane resins with cork granulate and mineral fillers.

Other important features used in the system influencing track resistance are the use of hyperelastic resilient strip at the bottom interface of embedding, the steel angle and base plate bounding the embedding material and PVC pipes for space filling in the embedding.

An asymmetric steel I girder with a height of 300mm (see figure 1) was used with the ERS to represent a simply supported single span bridge system.

For connection, the top flange of the girder was directly bolted to the base plate and the steel angles of the ERS assembly. The bolts of 20mm diameter at 150mm spacing were used.

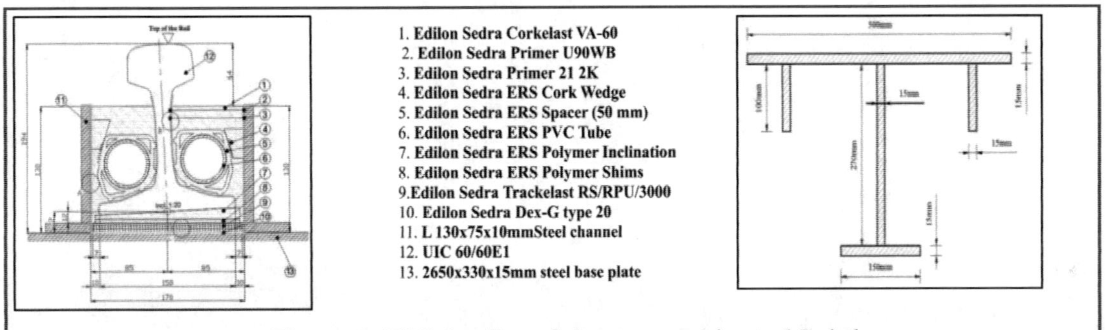

Figure 1. ERS details and the asymmetric steel I girder.

2. Laboratory Test for ERS Track

The static scheme of the whole assembly of ERS with Steel I girder is shown in figure 2. The idea was to impose longitudinal load on the rail only and to represent the ideal length of the rail i.e., the continuously welded rail (CWR) on the bridge deck. The test specimen was arranged for undergoing several stages of vertical and horizontal loading. The vertical load cell was positioned above the center of the girder to impose vertical load at the top of the rail and was supported by the steel beam and column. For the longitudinal load cell, the vertical support was provided by a steel cross beam at the rear end of the specimen. The horizontal support of the load cell was ensured by steel sections welded at the front end of the girder cross section. The final assembly of the test is shown in figure 3.

Four stages of vertical point load were considered up to maximum. And as the test was conducted on a single rail only, these were 0, 40, 80 and 125 kN respectively.

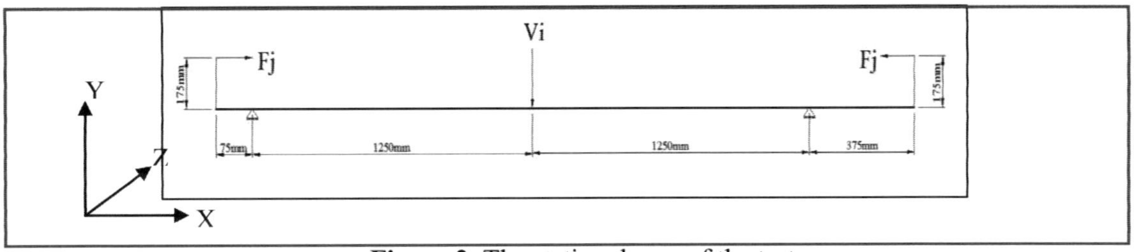

Figure 2. The static scheme of the test.

According to UIC-774R, the maximum longitudinal displacement for embedded rail is recommended as maximum 6mm for unloaded and 7mm for the loaded track [6]. Eurocode [7] has no recommendation on embedded rail. Hence, the maximum longitudinal load applied during the test was monitored according to the longitudinal displacement of the rail (not to exceed too much beyond 7mm) and it had never exceeded the maximum design longitudinal force value).

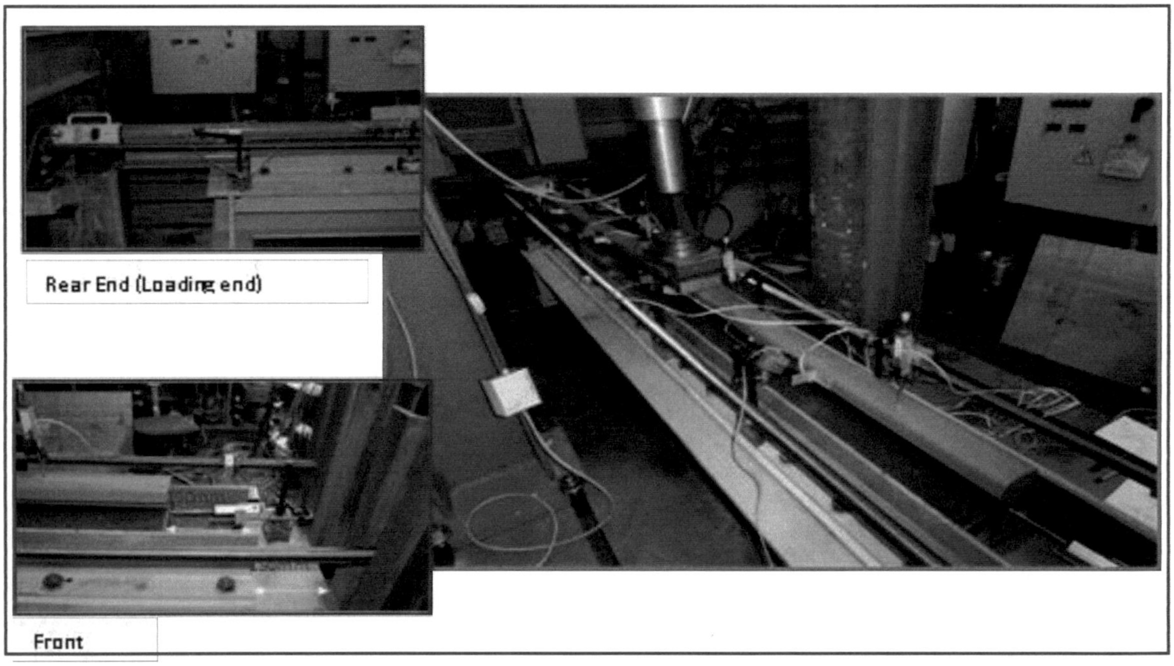

Figure 3. The final test assembly for the ERS on steel girder.

2.1. Test output

The relationship between the longitudinal resistance of the ERS in response to longitudinal displacement was found slightly curvilinear and can be approximated as a linear one, see figure 4.

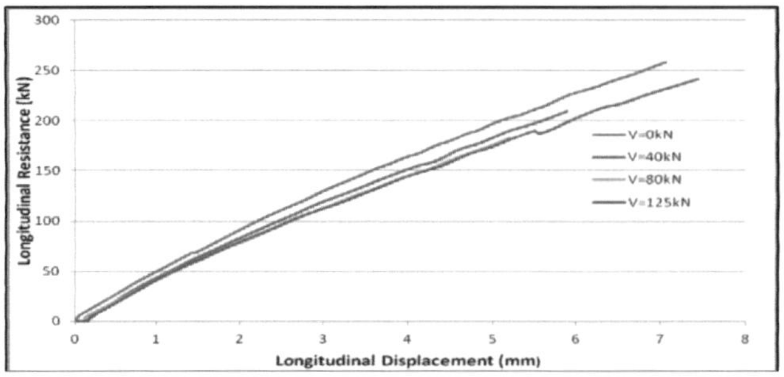

Figure 4. Longitudinal resistance of ERS against longitudinal displacement from the test data.

IOP Conf. Series: Materials Science and Engineering **96** (2015) 012052 doi:10.1088/1757-899X/96/1/012052

Although for the loaded track higher longitudinal resistance of track has been obtained as compared to the unloaded track in various studies of ballasted and non-ballasted track (UIC-774R & Euro Codes) [6,7], the test data obtained from the testing of ERS has shown slight decrease of longitudinal resistance with the higher vertical load combinations. Other than the longitudinal resistance and displacement data, the data were recorded for vertical displacements and horizontal stresses along the length of the rail. Stresses at the mid-span of the girder were also recorded.

3. Numerical FEM model Simulation and Validation

3.1. Numerical simulation

To accurately represent the real test specimen and to quantify the accuracy of the numerical solutions to experimental data, a 3-D finite element model was developed with the help of ANSYS APDL version 14.0. The model was composed of quadratic elements and thus hexahedral meshing was done. The lower order (SOLID185) for regular shapes and higher order (SOLID186) elements for irregular shapes elements were used.

One of the challenges was the choice of material for connecting different physical parts in the model. The adhesive materials used for contact between the embedding material and steel angle as well as the base plate have high adhesive strength (greater than 10 MPa and 35 MPa for steel angle and base plate respectively) that has been suggested to be enough to restrict any slippage between them by Edilon Sedra and the test observation attested that. However, the response from the laboratory test was recorded and the separation of physical parts of the arrangement was found only in two cases;

- Between the PVC pipe and the embedding material at the front and rear end.
- Between elastomer and the underlying steel girder at the rear end.

The two separations were modeled manually and the model was analyzed both for the glued and partially unglued regions, taking into consideration different length of separation. The main reason of such consideration was to identify possible influence of the separation on the results.

Both the embedding and the elastomer material show bimodular material properties[3]. Therefore, before picking the result from an analysis, a trial run was given to find out the tension and compression zone of these materials. Then the corresponding volumes were cut into parts to assign different material properties prior to the final run. Hyperelastic material model (Arruda-Boyce) was used for resilient strip in compression.

3.2. Validation

Mainly 3 parameters: longitudinal stress on rail, longitudinal displacement of rail and stresses on girder were used for the validation of the FEM. Longitudinal displacement values from FEA were found close to the experimental result for higher vertical load and showed lower displacement for the last load case (figure 5). The maximum deviation was 15.04% for unloaded condition and, most importantly, the pattern of displacement along the increment of vertical load was found similar.

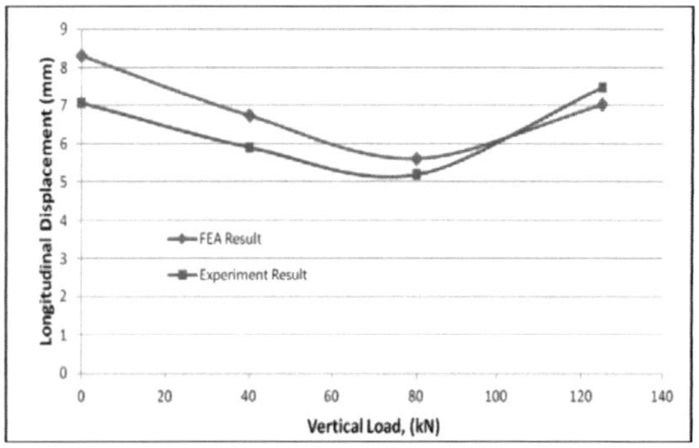

Figure 5. Longitudinal Displacement of Rail with the increment of vertical load.

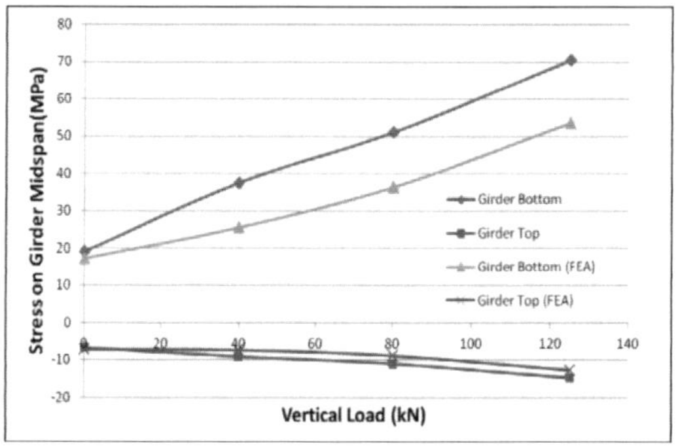

Figure 6. Comparison of stresses on girder.

Girder stresses in the top and bottom fibers in the mid span were found growing at similar pattern for both test and FEA. And for both tension and compression the results for FEA were found lower than test results (figure 6).The behavior of longitudinal stress in rail was found to be of similar pattern under the vertical load application zone (figure 7). This was considered more as a local behavior not impacting much the global behavior or quality of the whole experiment. It was found after a parametric study that higher Poisson's ratio of embedding material and shifting of the vertical load application point from the center can well describe the specific local variation of results.

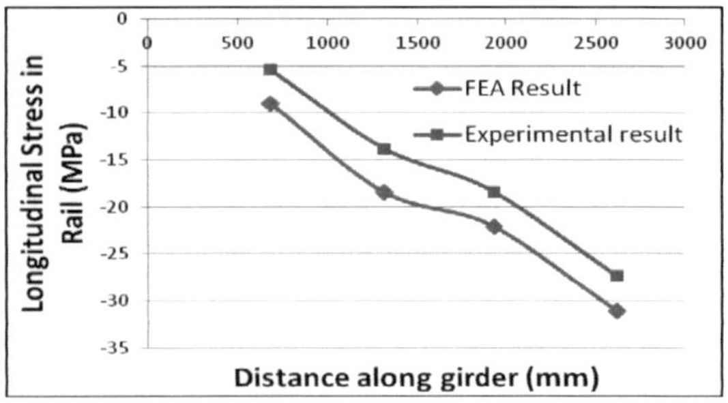

Combination 1: V=0 kN

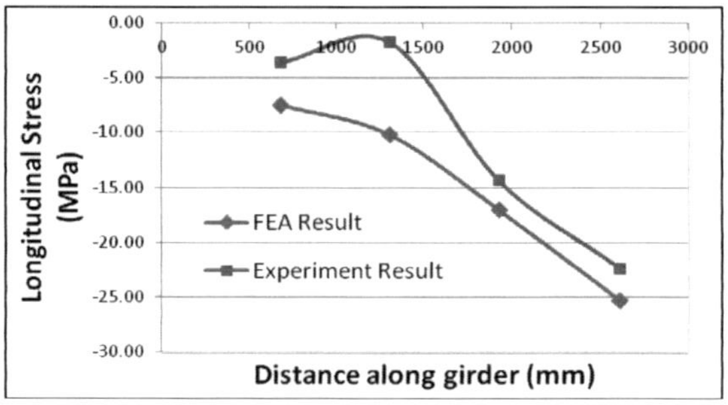

Combination 2: V=40 kN

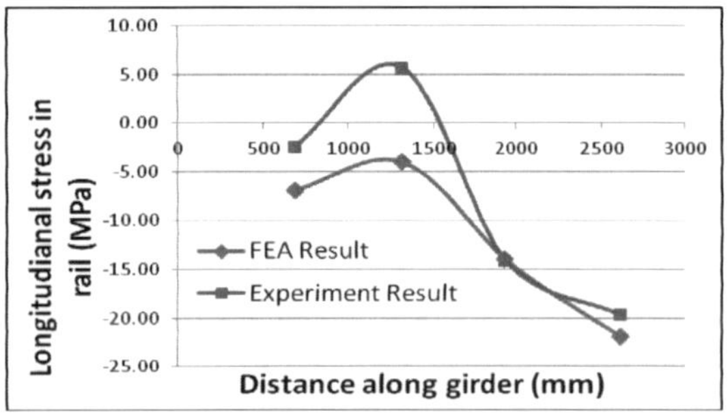

Combination 3: V=80 kN

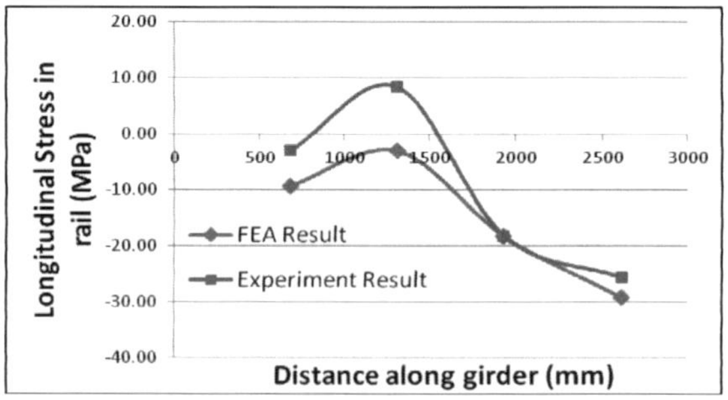

Combination 4: V=125 kN

Figure 7. Longitudinal Displacement of Rail with the increment of vertical load.

IOP Conf. Series: Materials Science and Engineering **96** (2015) 012052 doi:10.1088/1757-899X/96/1/012052

4. Evaluation of the longitudinal resistance of ERS track

4.1. Unloaded Track

The comparison of coupling relationship for 2190mm of embedding on a single rail was plotted first. Then converting the resistance for a 1 m track assembled with ERS system was suggested (figure 8). The result found in FEA is conservative in contrast to the experimental result for the unloaded track. Given the maximum allowable longitudinal displacement of 7mm in UIC-774-3R [7] for embedded rail, maximum longitudinal resistance have been found to be 200 kN for the unloaded track in FEA.

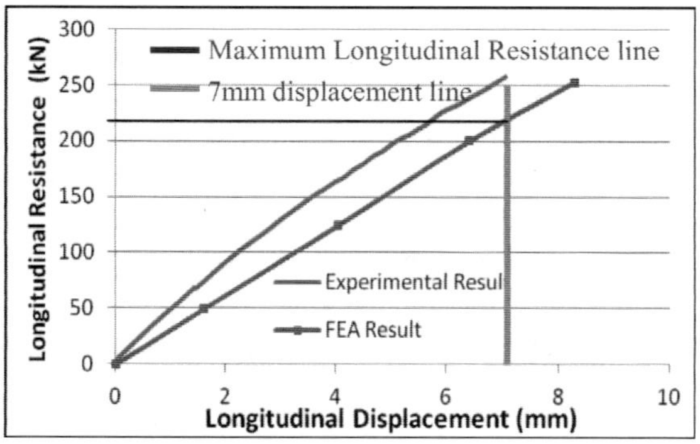

On single rail with 2190mm of ERS

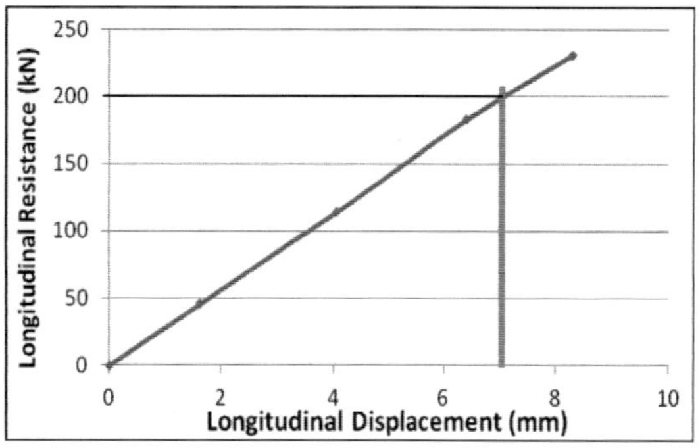

On 1m of ERS track

Figure 8. Longitudinal resistance of ERS track (unloaded condition).

4.2. Loaded Track

The result found in FEA is not conservative in contrast tothe experimental result for the loaded track (figure 9) and matches the recommendation of UIC-774-3R [6], see table1.

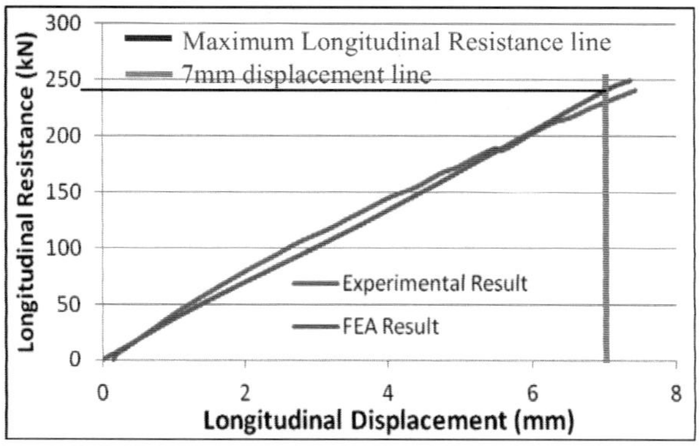

On single rail with 2190mm of ERS

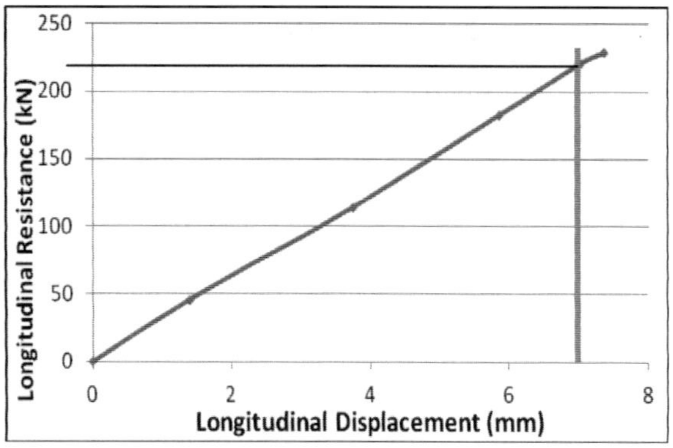

On 1m of ERS track

Figure 9. Longitudinal stress of ERS track (loaded condition).

Table 1. Comparison of resistance k value of track (UIC 774-3 vs FEA &experimental result).

	Unloaded track (kN/mm)	Loaded track(kN/mm)
UIC 774-3 Recommendation	13	19
FEA	28.57	31.42
Experiment Result	33.01	30.66

5. Evaluation of design stress on bridge deck

5.1. Distribution of vertical pressure along the length of ERS

The pressure under the sleepers in the ballasted track is widely known. However, the same information for ERS (essential for the analysis of the supporting structure)is not given anywhere. To find out the vertical pressure distribution beneath the ERS, FEA was done for vertical load 125 kN only and pressure distribution was found along the longitudinal central line of the girder. From the idea of effective width of stress distribution, an attempt was made to find out the general shape of this distribution. The peak pressure found was recorded as 1.38 MPa. The area under the pressure curve was calculated as 1792.44 N/mm.

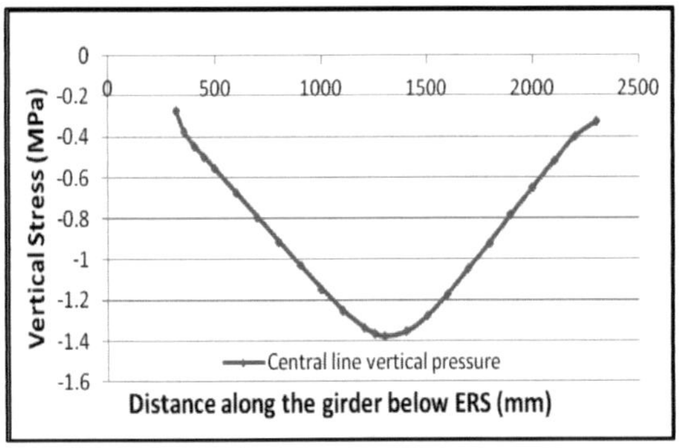

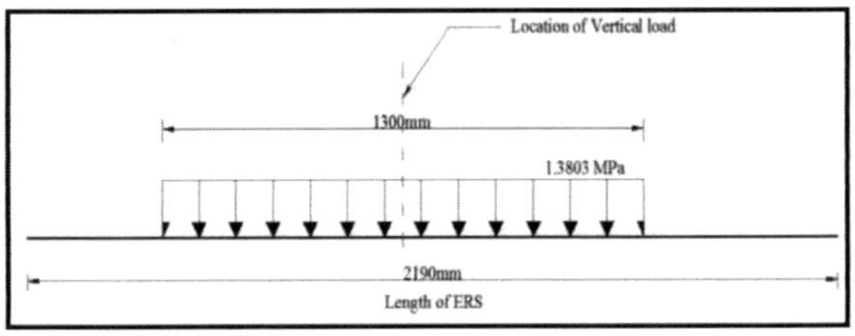

Figure 10. Vertical Stress distribution along the length of the girder (central line).

5.2. Transverse distribution of vertical pressure along length of ERS

The distribution of the vertical pressure in transverse direction was checked along the plane of peak stress as well as along two other intermediate planes. The vertical pressure in transverse direction can be also be idealized as distributed over an effective length. This information is important for the design of the bridge deck plate.

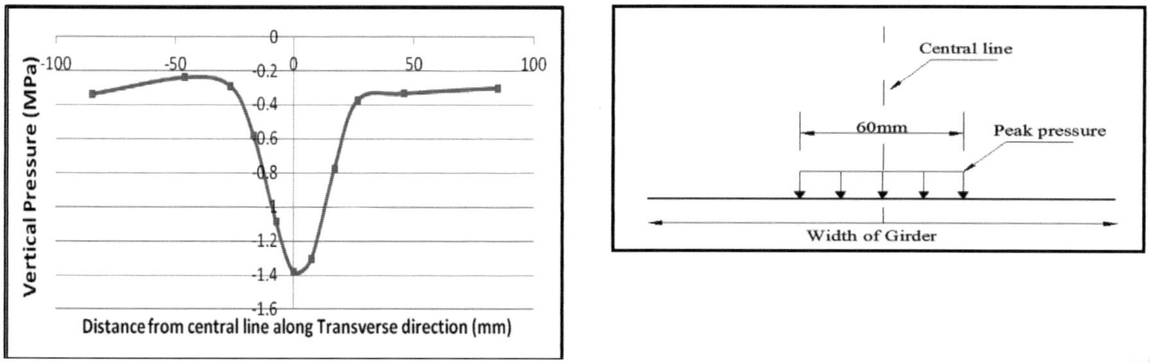

Figure 11. Vertical Stress distribution along transverse direction.

6. Effect of ERS on girder stress

The stiffness of the ERS system can have an impact on the stresses in the steel structure, due to its high stiffness. To analyze this effect, further analysis was performed with and without the ERS assembly for the loaded track condition. The reduction of girder stress was found for ERS application especially in the top fibers of the steel structure. On ERS track, the vertical point load is distributed over the length of the girder whereas for typical ballasted track the point load is applied on 3 adjacent points (to 3 adjacent sleepers) i.e., 50% load at the midpoint (directly under the wheel) and the rest equally divided to either side [7].

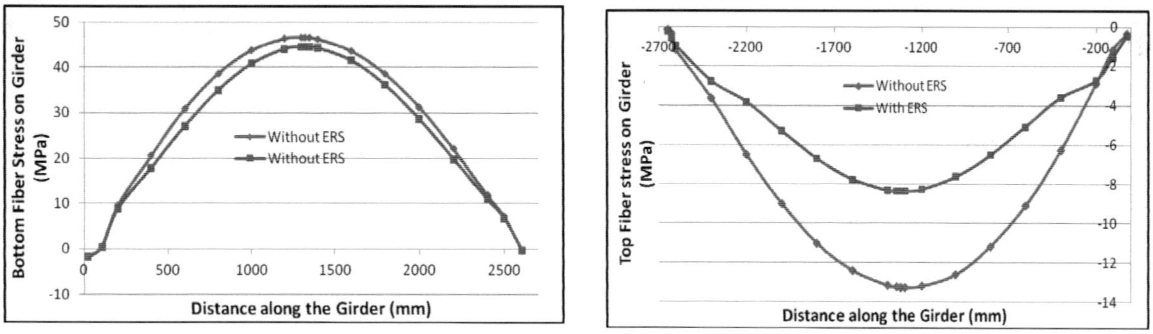

Figure 12. Reduction of girder stress due to ERS application.

7. Conclusion

The limitation to assigning bimodular material properties of embedding and elastomer material precisely (for highly irregular distribution of compression and tension elements) was verified in the analysis and little impact of the assumptions was identified.

It has been found that the resistance of the rail track is decreasing with the increment of vertical load while the same has been found increasing in FEA. This is a different and important conclusion, in contrast to the existing UIC recommendations. Also, the longitudinal resistance depends significantly on the speed of the loading. This is important for the proper analysis of the rail-bridge interaction actions like temperature and the brake/acceleration forces.

It is evident that gradual elastic softening of the material elasticity will lead to higher displacement under the same load and will influence the subsequent behavior of the other parts adjoin to that material. The last important conclusion is that the stiffness of the fastening is significantly higher than the value adopted in the UIC 774-3, which can lead to underestimation of the rail interaction stresses.

8. References

[1] Eszter L 2002 Elastic behavior of continuously embedded rail system *Periodica Politechnica Cer. Civ. Eng.* **46** 103-114

[2] Kormos Gy 2002 The Longitudinal Behavior of the Rail Embedded in Flexible Material *Periodica polytechnicaser. Civ. Eng.* **46** 115–124

[3] Jones R M 1977 Stress-strain relations for materials with different moduli in tension and compression *AIAA J.* **15** 16-23

[4] Yan B, Dai and Zhang H 2012 Beam-track interaction of high-speed railway bridge with ballast track *J. Cent. South Univ.* **19** 1447–53

[5] Lou P 2007 Finite element analysis for train–track–bridge interaction system *Arch. Appl. Mech.* **77** 707–728 (Berlin: Springer-Verlag Berlin Heidelberg)

[6] Union Internationale des Chemins de fer (UIC) 2001 *Track/bridge interaction, Recommendations for calculations* 774-3R

[7] European Committee for Standardization (CEN) 2003 *Actions on Structures; Traffic loads on bridges* Eurocode1, Part 2 (EN 1991-2)

Acknowledgement

Research reported in this paper was supported by Competence Centres Program of Technology Agency of the Czech Republic (TA CR), project Centre for Effective and Sustainable Transport Infrastructure (no. TE01020168). The authors greatly acknowledge the support and help from Edilon Sedra Inc.

Mechanical Properties of Heat Affected Zone of High Strength Steels

K Sefcikova[1,4], T Brtnik[1], J Dolejs[1,2], K Keltamaki[3] and R Topilla[3]
[1]Faculty of Civil Engineering, Department of Steel and Timber Structures, Czech Technical University in Prague, Czech Republic
[2] University Centre of Energy Efficient Buildings, Czech Technical University in Prague, Czech Republic
[3]Lapin AMK – Lapland University of Applied Science, Finland

E-mail: katerina.sefcikova@fsv.cvut.cz

Abstract. High Strength Steels became more popular as a construction material during last decade because of their increased availability and affordability. On the other hand, even though general use of Advanced High Strength Steels (AHSS) is expanding, the wide utilization is limited because of insufficient information about their behaviour in structures. The most widely used technique for joining steels is fusion welding. The welding process has an influence not only on the welded connection but on the area near this connection, the so-called heat affected zone, as well. For that reason it is very important to be able to determine the properties in the heat affected zone (HAZ). This area of investigation is being continuously developed in dependence on significant progress in material production, especially regarding new types of steels available. There are currently several types of AHSS on the world market. Two most widely used processes for AHSS production are Thermo-Mechanically Controlled Processing (TMCP) and Quenching in connection with Tempering. In the presented study, TMCP and QC steels grade S960 were investigated. The study is focused on the changes of strength, ductility, hardness and impact strength in heat affected zone based on the used amount of heat input.

1. Introduction

1.1. General

The research of welded connections of high strength steel started at Czech Technical University in Prague in cooperation with Kemi-Tornio University of Applied Sciences in 2012. The first research work was focused on using and evaluation of advanced techniques for the investigation of behaviour

[4] To whom any correspondence should be addressed.

of high strength steel [1]. These techniques including optical measurement in tensile tests, crack propagation study and an idea of using hydroforming machine as a bending press were presented.

Based on the conclusions of the above-mentioned research, the next part of the work is dealing with mechanical properties of heat affected zone of high strength steels. This paper focuses on strength of the heat affected zone and presents the first part of the conducted research dealing with mechanical properties of the heat affected zone.

Welded joints represent significant parts of some steel structures. Welded connections of common grade steels have been investigated and verified for many years; detailed rules for both welding technology process and structural engineering design therefore are clearly given in design standards [2] and other prescriptions. On the other hand, new or improved techniques of steel processing allow producing structural steel grades with tensile strength exceeding 1300 MPa. However, current European design codes for civil engineering permit usage of steel up to the grade S700 only and for some particular steel types only [3]. The main reason why new high strength steels cannot be used in the engineering praxis is the lack of knowledge about the behaviour of such steels in structures.

The main role of a structural weld is force transmission between two adjacent plates, which means that this connection must show clearly describable mechanical properties, ultimate load capacity and ductility especially. These characteristics are controlled not only by mechanical properties of base material but also considering a technology process and its procedures and quantities as, for instance, heat input, grade, diameter and type of welding consumable used, the welding method, number of passes (single or multilayer welds), preheating level, cooling rate, geometry of the weld and some other.

The welding process always causes changes in adjacent parts of base metal. This affected zone represents the weakest part of the connection in many cases. Therefore, it is necessary to determine the properties not only in the weld material (which in accordance with common rules is sufficiently resistant), but in the heat affected zone as well.

1.2. Influence of thermic conditions

It has been proven by many researchers [4] that resulting mechanical properties of HAZ are influenced especially by thermic conditions during the welding process. They are commonly described by heat input and preheating level, cooling rate and interpass temperature. According to the definition of heat input, a range for cooling rates ($t_{8/5,min}$; $t_{8/5,max}$) can be calculated. If the minimal preheating temperature and maximal interpass temperature are specified, so-called “working area” for the welding process can be plotted [5,6]. This area determines general requirements for the welding technology settings. The difference in the size of the working area for mild steel S355 and for high strength steel S960QL is displayed in figures 1 and 2. This basic comparison gives clear information about the main difference between welding process for mild and high strength steel: quite strict limitations for HSS make the welding much more difficult and do not allow an in-situ or manual workshop welding in many cases.

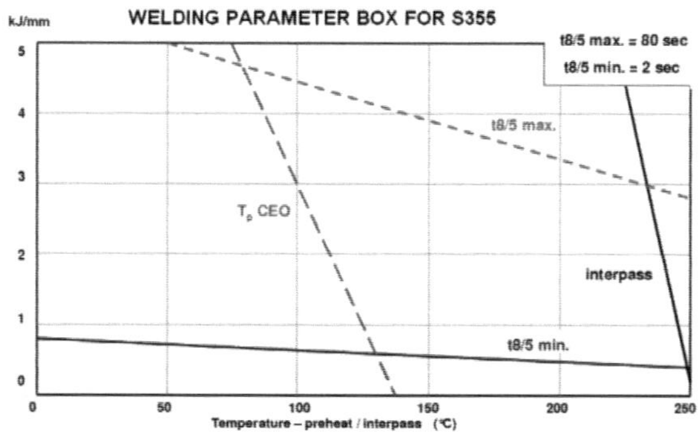

Figure1. Working area for S355.

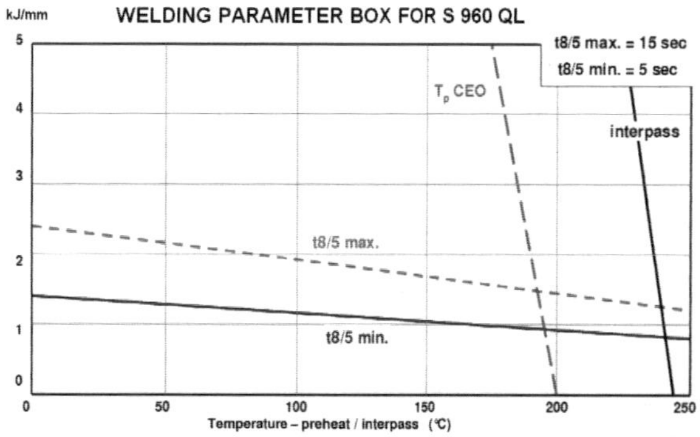

Figure 2. Working area for S960 QL.

2. Experimental program

2.1. Description

Thermomechanically rolled steel S960 MC and directly quenched steel S960 QC were used for determination of the impact of the thermic conditions on mechanical properties of the weld and heat affected zone. Matching and undermatching Stein Megafil flux-cored wires consumables and different heat inputs have been combined within the sample file.

2.2. Test configuration

22 sample plates were welded by combination of three different strength grades of welding consumables and three different heat inputs for the steel S960 MC and 13 sample plates for the steel S960 QC. The welding parameters are shown in table 2.

The sample plates made of the steel S960 MC were welded in the V-shape,the butt weld form and sample plates from the steel S960 QC – in the ½ V-shape.Each welded part of the plate (left and right) was provided by different rolling direction (parallel and perpendicular), see figure 3 and figure 4. Samples for tensile tests were tested without butt weld camber smoothing.Based on that it was possible to forecast that the failure of the tension test would appear in the heat affected zone,which is necessary to determine its mechanical properties.

From every welded sample plate 4-5 tensile samples, 9-10 V-Charpy impact strength and 2-3 samples for hardness measurement were cut by means of a high pressure water jet. Tensile samples and V-Charpy impact test samples from base material for longitudinal as well as for transversal rolling direction were taken as well. The quality of welding was observed using the tests of macrostructure. Due to these tests,it was possible to notice the areas of weld, HAZ and base metal, see figure 5.

Figure 3. Welded sample plate.

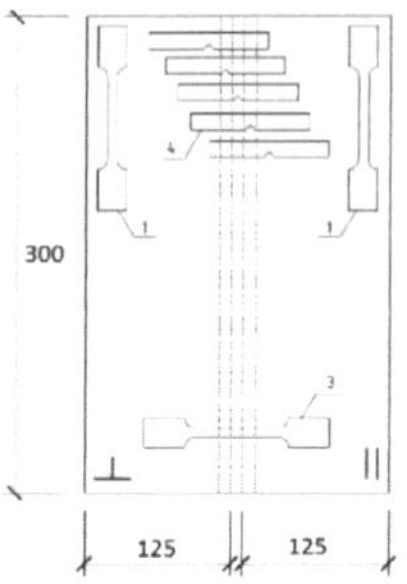

Figure 4. Scheme of cutting.

Table 1. Chemical composition of base metals.

Element	S960 MC	S960 QC
C	0.0870	0.0950
Si	0.2500	0.2000
Mn	1.8100	1.0600
P	0.0100	0.0090
S	0.0010	0.0031
N	0.0036	0.0058
Cr	0.0700	1.0900
Ni	0.2800	0.0590
Cu	0.0100	0.0180
Mo	0.4700	0.1200
Al	0.0400	0.0280
Nb	0.0030	0.0030
V	0.0200	0.0080
Ti	0.1900	0.0280
B	0.0011	0.0018

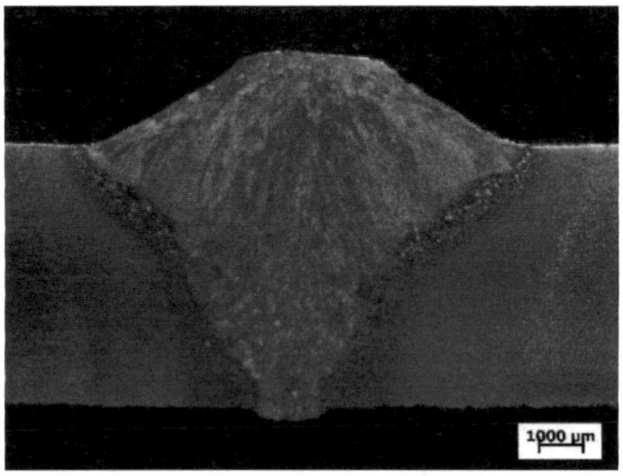

Figure 5. Macrostructure of butt weld.

Table 2. Welding parameters of samples.

S960 MC Sample	Electrode	Q [kJ/cm]	S960 QC Sample	Electrode	Q [kJ/cm]
1	MF 742	0.9	1	MF 742	1.1
2	MF 742	0.9	2	MF 742	1.1
3	MF 742	0.9	3	MF 742	1.1
4	MF 742	0.9	4	MF 742	1.1
5	MF 742	0.9	5	MF 807	1.1
6	MF 742	0.9	6	MF 807	1.1
7	MF 742	0.9	7	MF 807	1.1
8	MF 742	0.9	8	MF 807	1.1
9	MF 807	0.9	9	MF 1100	1.1
10	MF 807	0.9	10	MF 1100	1.1
11	MF 807	0.8	11	MF 1100	1.1
12	MF 807	0.8	12	MF 1100	1.1
13	MF 807	0.9	13	MF 1100	1.1
14	MF 807	0.9			
15	MF 807	0.8			
16	MF 807	0.8			
17	MF 1100	0.9			
18	MF 1100	1.0			
19	MF 1100	1.0			
20	MF 1100	1.0			
21	MF 1100	0.9			
22	MF 1100	0.9			

(green – parallel torolling direction; red – perpendicular to rolling direction).

2.3. *Test results*

The measured tension strengths of the heat affected zone material are shown in figures 6 - 14.

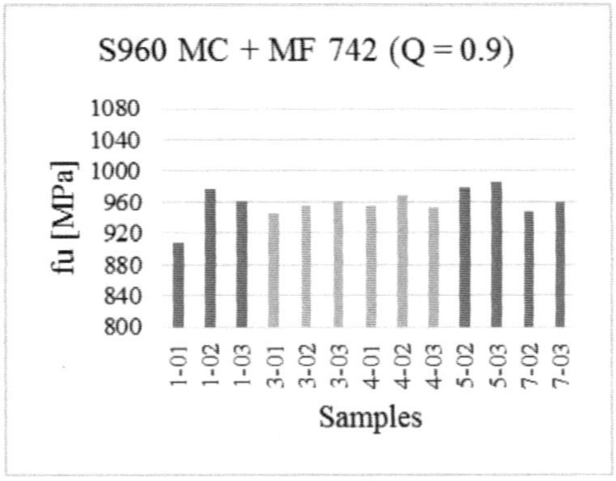

Figure 6. Strength of samples S960 MC.

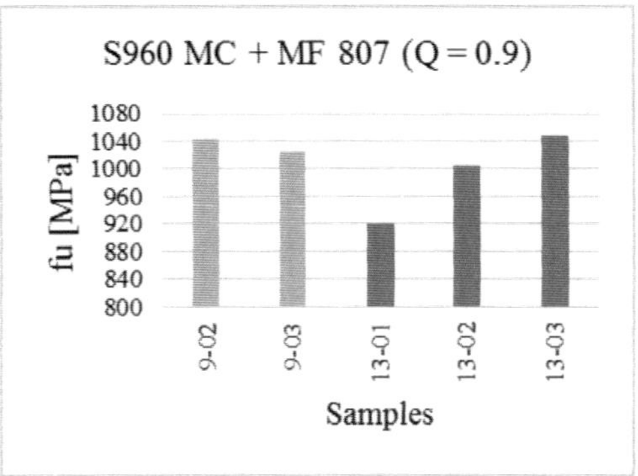

Figure 7. Strength of samples S960 MC.

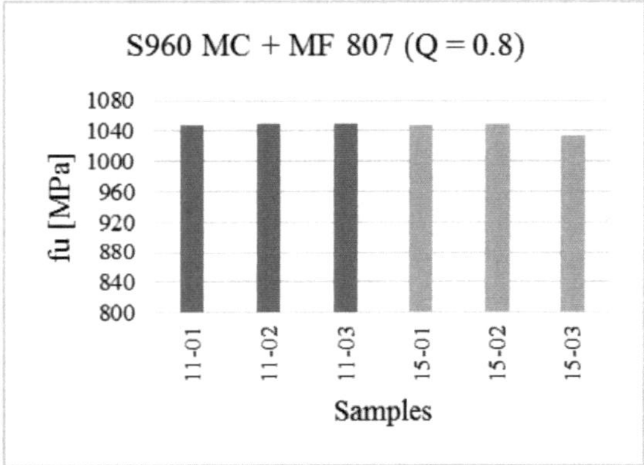

Figure 8. Strength of samples S960 MC.

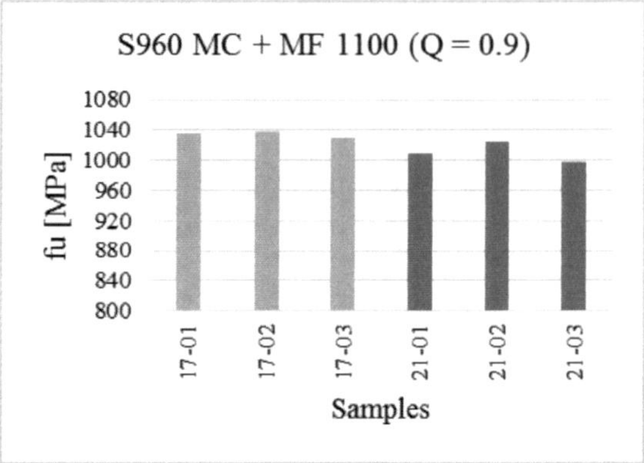

Figure 9. Strength of samples S960 MC.

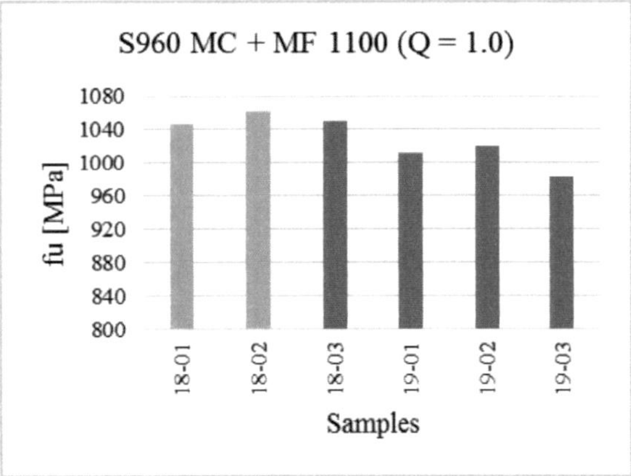

Figure 10. Strength of samples S960 QC.

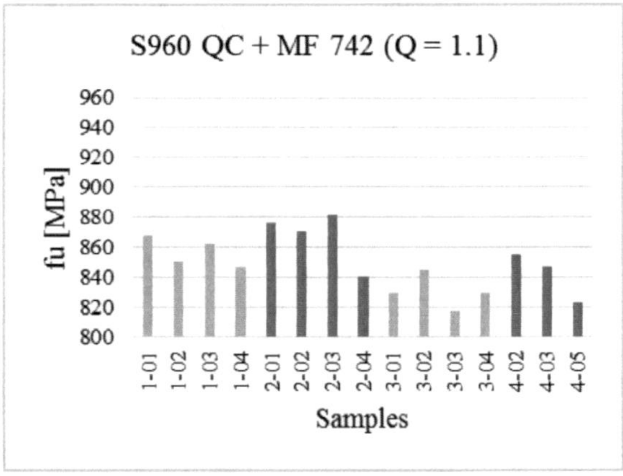

Figure 11. Strength of samples S960 QC.

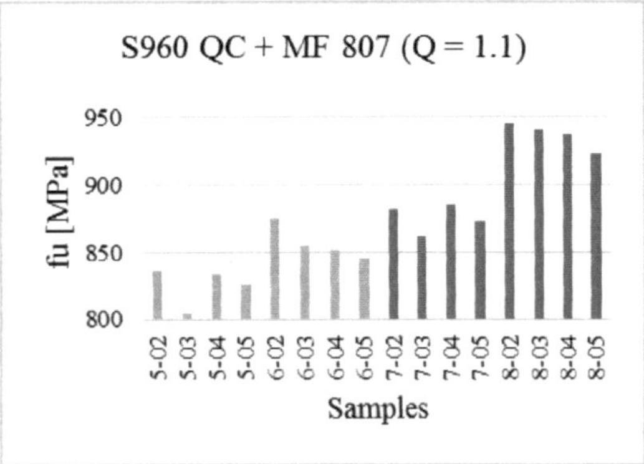

Figure 12. Strength of samples S960 QC.

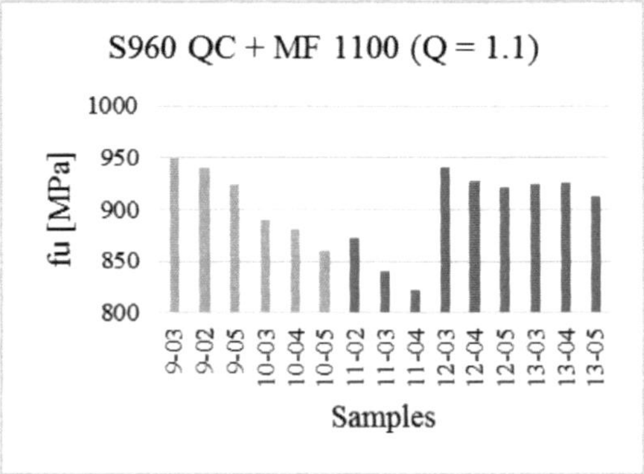

Figure 13. Strength of samples S960 QC.

A mean value of tension strength for all "MC" samples was 1016 MPa, for all "QC" samples – 876 MPa. As it was expected,theprincipal influence of filler metal grade on tension strength of HAZ was proven. The mean value of tension strength for all samples with MF742 filler metal grade was 804 MPa, MF 807 filler metal grade840 MPa and MF 1100 filler metal grade 957 MPa.

Different mechanical properties and inclination to material softening of heat affected zone is perceptible for direct quenched steel S960 QC. That phenomenon is caused by lower content of alloy components (as seenin table 1). Mean value of strength for MC with the heat input 0.8 was 1046 MPa, MC with the heat input 0.9 was 996 MPa, MC with the heat input 1.0 was 1029 MPa and QC with the heat input 1.1 was 875 MPa.

No clear correlation between tension strength and the rolling direction has been found. Figure 10 and figure 12 show that in some cases contrary results were obtained – QC steel with MF 807 shows significantly higher strength for "perpendicular" samples but the configuration MC + MF 1100 indicates an opposite trend.

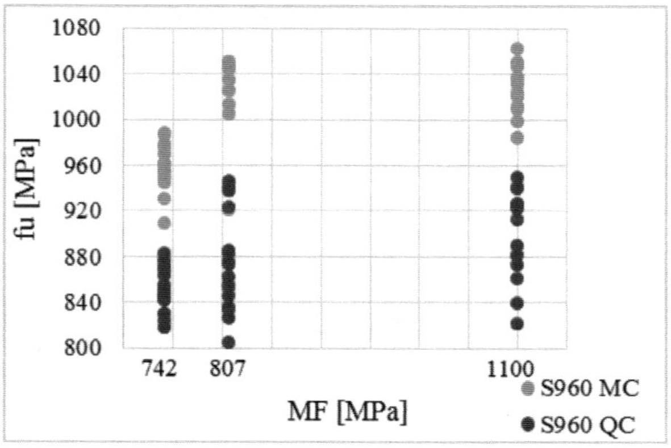

Figure 14. Strength of HAZ depending on MF.

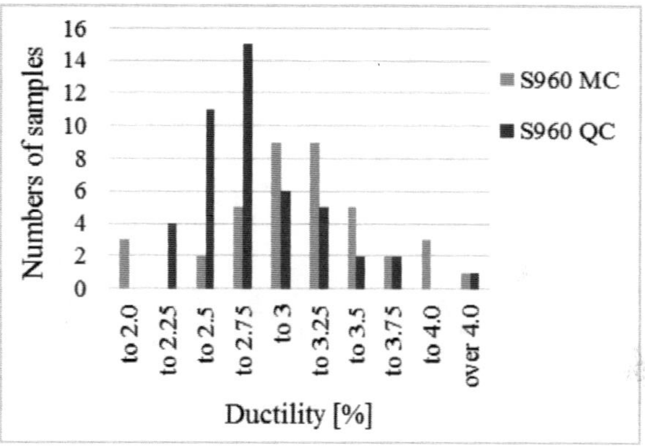

Figure 15. Ductility of HAZ.

Ductility of all specimen material was very low as it is demonstrated in figure 15. The mean values of ductility for steel S960 MC were 2.75 – 3.25 % and for steel S960 QC 2.5 – 2.75 %.

The V-Charpy tests results are not presented within this paper.

3. Conclusion and future work

All preliminary results, described within this paper, describe the mechanical properties in the HAZ. Both used materials (S960 MC and S960 QC) were welded using three different flux-cored wires and different heat inputs and the principal influence of filler metal grade on tension strength of HAZ was proven. Themean value of tension strength for all "MC" samples was 1016 MPa, for all "QC" samples – 876 MPa. Different mechanical properties and inclination to material softening of heat affected zone is perceptible for direct quenched steel S960 QC (this is caused by lower content of alloy components). However, direct correlation between tension strength and rolling direction was not validated. The follow-up research of our team is aimed at welded connection of I-shape hybrid girders.

4. References

[1] Virdi K and Tenhunen L 2011 *Proc. of the METNET Seminar 2011 in Aarhus* (HAMK University of Applies Sciences) pp 24–36

[2] EN 1993–1-8: Eurocode 3: *Design of Steel Structures – Part 1-8: Design of Joints*

[3] EN 1993–1-12: Eurocode 3: *Design of Steel Structures – Part 1-12: General – High Strength Steel*

[4] *IABSE: Use and Applications of High Performance Steels for Steel Structures* pp 99–110

[5] Gresnigt A M and Steenhuis C M 1997 *High Strength Steels – Construction Research Communications Limited* pp 31–41

[6] Davies A C *The Science and Practise of Welding, Welding Science and Technology* vol. 1

Acknowledgement

This work has been supported by the European Union, OP RDI project No. CZ.1.05/2.1.00/03.0091 – University Centre for Energy Efficient Buildings and by the SGS 15/042/OHK1/1T/11 project.

IOP Conf. Series: Materials Science and Engineering **96** (2015) 012054 doi:10.1088/1757-899X/96/1/012054

Design Methods for Load-bearing Elements from Cross-laminated Timber

A Vilguts[1] , D Serdjuks[1,3], and V Goremikins[2]
[1] Riga Technical University, Faculty of Civil Engineering, Institute of Structural Engineering and Reconstruction, Kalku Str.1, Riga, LV-1048, Latvia
[2] Czech Technical University in Prague, Department of Steel and Timber Structures, Thákurova 7, Prague CZ16629, Czech Republic

E-mail: dmitrijs.serdjuks@rtu.lv

Abstract. Cross-laminated timber is an environmentally friendly material, which possesses a decreased level of anisotropy in comparison with the solid and glued timber. Cross-laminated timber could be used for load-bearing walls and slabs of multi-storey timber buildings as well as decking structures of pedestrian and road bridges. Design methods of cross-laminated timber elements subjected to bending and compression with bending were considered. The presented methods were experimentally validated and verified by FEM. Two cross-laminated timber slabs were tested at the action of static load. Pine wood was chosen as a board's material. Freely supported beam with the span equal to 1.9 m, which was loaded by the uniformly distributed load, was a design scheme of the considered plates. The width of the plates was equal to 1 m. The considered cross-laminated timber plates were analysed by FEM method. The comparison of stresses acting in the edge fibres of the plate and the maximum vertical displacements shows that both considered methods can be used for engineering calculations. The difference between the results obtained experimentally and analytically is within the limits from 2 to 31%. The difference in results obtained by effective strength and stiffness and transformed sections methods was not significant.

1. Introduction

Cross-laminated timber (CLT) is a structural material, which at the present moment is widely used for several structural purposes. Cross-laminated timber possesses the mechanical properties, which enable to decrease structural cost and time of building erection in comparison with analogous structures made

[3] To whom any correspondence should be addressed.

of steel and reinforced concrete. Using of cross-laminated timber enables to obtain reliable load-bearing members and meet aesthetic and architectural requirements at the same time [1–3].

Multi - storey buildings of 4 to 10 storeys constructed entirely in timber, such as numerous structures recently constructed in Europe, are innovative and effective [4].

Buildings where timber is used as a structural material across the world are becoming more popular, because timber is one of the oldest and most traditional structural materials, which fulfills aesthetic requirements and is environmentally friendly. Timber is an efficient building material, not only regarding its mechanical properties, but also because it is a highly sustainable material considering all phases of the life cycle of timber structures: production, use and decommissioning [4].Cross-laminated timber is used for load-bearing walls and plates in one-storey and multi-storey timber buildings. The maximum number of storeys for existing buildings, which are made using of cross-laminated timber, is equal to nine. It is a residential building in London. The main vertical load-bearing elements of this building are external walls and the structural core, which are made of CLT. Horizontal load-bearing elements are CLT plates [2–7]. These load-bearing elements are subjected to compression with the bending and flexure.

(a) (b)

Figure 1. Structures with CLT load-bearing elements: a) model of residential building; b) pedestrian bridge with the CLT deck in Feldbach, Austria [4].

The structural solutions of multi-storey buildings with the load-bearing elements made of CLT can be divided in to three groups. These are the solution with the structural core and glulam columns at curtain walls, the solution with a structural core and internal walls with glulam columns at curtain wall and the solution with external walls and the structural core. The number of storeys can reach twelve for the first solution. But it rises up to twenty for the second and third solutions, correspondingly [8].

Cross-laminated timber plates are used for decking for pedestrian and road bridges [9–11]. The pedestrian bridge created in Feldbach, Austria, is an example of that kind of bridges (figure 1).

So, it can be concluded that CLT is widely used for the structural members subjected to flexure and compression with bending [9]. CLT plates are used for structures of floors, roofs and load-bearing walls. Thus, the aim of this paper is to consider and analyse design methods of CLT elements subjected to flexure and compression with bending. Effective strength and stiffness methods must be compared and verified by the laboratory experiments and FEM to obtain the identified aim.

2. Design methods for cross-laminated timber elements

2.1. Design methods for cross-laminated timber elements subjected to flexure

Two following methods are used for designing of CLT structural members subjected to flexure [1, 14–15]: effective strength and stiffness method and transformed section method. Both methods are based on the well-known approach to the design of structural members subjected to flexure, which is described in [12]. The ultimate limit state (ULS), which includes checks of bending stresses and the check of shear stresses, as well as serviceability limit state (SLS) must be checked for the CLT structural member subjected to flexure [12].

Bending stresses acting in the CLT plates are determined based on the recommendations of [14]. The stresses acting in the middle layers of CLT plates can be neglected in the case when the fibers of these layers are oriented perpendicular to the direction of the fibers of the outer layer.

Let us consider peculiarities of effective strength and stiffness method. In accordance with the effective strength and stiffness method, maximum value of bending stresses acting in the edge fibers of the outer layers of CLT panels must be determined by the equation:

$$\sigma_{edge,d} = \frac{M_{\max,d}}{K_{CLT}} \cdot \frac{a_{CLT}}{2} \cdot E_{i=5} \tag{1}$$

where $M_{max,d}$ – design value of maximum bending moment; a CLT – CLT plates height; K_{CLT} – effective stiffness of CLT plate; $E_{i=5}$ – modulus of elasticity of each layer in longitudinal direction.

Effective stiffness of CLT plate KCLT can be determined by equation (2):

$$K_{CLT} = \sum_{i-1}^{n}(J_i \cdot E_i) + \sum_{i=1}^{n}(A_i \cdot e_i^2 \cdot E_i) = (EI)_{ef} = E_0 \cdot \frac{h^3 \cdot a_{CLT}}{12} \cdot k_i \tag{2}$$

where, E_i, A_i – modulus of elasticity and area of cross-section of a separate layer; I_i – moment of inertia of the separate layer relative to its own main axis; E_0 – modulus of elasticity of timber in longitudinal direction; e_i –the distance from the middle plane of the element to the middle of the *i*-th layer; h – total thickness of the plate; k_i – composition factor which depends on certain loading conditions.

Composition factor is also used for determination of shear stresses and maximum vertical displacements taking into account bending and shear deformations. Equation (1) is written for the CLT plate, which consists of five layers. So, checks of ultimate limit state (ULS) and serviceability limit state (SLS) must be conducted by taking effective stiffness of CLT plate into account.

Let us consider the transformed cross-section method [13]. The transformed cross-section method is joined with the replacement of real cross-section of the element by the equivalent transformed cross-section. This method can be used in the case when fibers of each second layer of CLT are oriented perpendicular to the fibers direction of the first layer. Transformation of cross-section is based on the relation of modulus of elasticity of the layers in longitudinal direction:

$$n = \frac{E_{90}}{E_0} \tag{3}$$

where E_0 – modulus of elasticity of timber in longitudinal direction ; E_{90} – modulus of elasticity of timber in transversal direction.

The width of the layer whose fibers are oriented in transversal direction must be multiplied by the relation of moduli of elasticity.

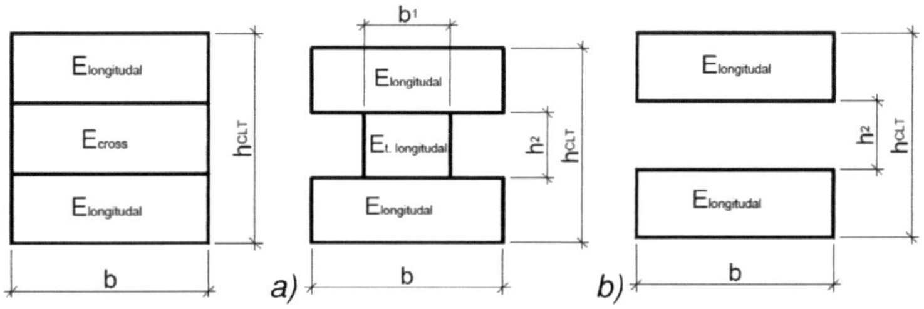

Figure 2. Transformation of cross-section: *a*) – middle layer is taken into account; *b*) – middle layer is not taken into account.

The obtained transformed double-tee cross-section then is considered as glued homogenous cross-section. Checks of ultimate limit state (ULS) and serviceability limit state (SLS) must be conducted based on the recommendations of [12].

The filling of the glued joints between the boards of the layer(s) with orientation of fibers in transversal direction is a very significant moment. If middle layer(s) can not be considered as a monolithic one, this layer can not be taken into account in the transformed section.

Verification of the transformed section and effective strength and stiffness methods analytically and by experiment is discussed below.

2.2. Design method for cross-laminated timber elements subjected to compression with bending

Transformed section method is considered for cross-laminated timber elements, subjected to compression with bending. The method is based on the well-known approach to the designing of structural members subjected to compression with bending, which is described in [12]. The method is divided into two sub-cases depending on the dominating internal force. The first sub-case takes place when compression internal force is dominating and condition (4) is satisfied.

$$\sigma_{c,0,d} \geq \sigma_{m,d} \tag{4}$$

where $\sigma_{c,0,d}$ and $\sigma_{m,d}$ are the normal stresses which were determined for the transformed cross-section of the cross-laminated timber element due to compression force and bending moment, correspondingly.

Stability of cross-laminated timber element can be checked by equations (6.23) and (6.24) [12] in the most common case, when relative slenderness of considered load-bearing elements exceeds the value 0.3. Otherwise strength of cross-laminated timber element must be checked by equations (6.19) and (6.20) [12]. The first sub-case can take place for load bearing walls where axial force is dominating.

The second sub-case takes place when bending moment is the dominating internal force and condition (4) is not satisfied. Stability of cross-laminated timber element can be checked by equation (6.35) in [12].

$$\left(\frac{\sigma_{m,d}}{k_{crit}\cdot f_{m,d}}\right)^2+\frac{\sigma_{c,0,d}}{k_c\cdot f_{c,0,d}}\leq 1 \qquad (5)$$

where $\sigma_{c,0,d}$ and $\sigma_{m,d}$ –design bending stress, which is determined for the transformed cross-section of cross-laminated timber element due to compression force and bending moment, $f_{m,d}$ and $f_{c,0,d}$ –design bending and compressive strengths parallel to grain; k_{crit} and k_c are the factors, which take into account the reduced bending and compression strengths.

All components of equation (5) must be determined for the transformed cross-section of the cross-laminated timber element. The second sub-case can occur in the horizontal load-bearing elements such as floor or roof plates, when non-sufficient axial forces took place. Checks of ultimate limit state (ULS) and serviceability limit state (SLS) must be conducted based on the recommendations of [12].

3. Verification of design methods of cross-laminated timber element subjected to flexure by experiment and FEM

3.1. General approach

The experiment has been carried out to verify the accuracy of the design method for current structural member, which is subjected to flexure. Two CLT plates with the length and width equal to 2 and 1 m, correspondingly, and thickness of 95 mm were considered. Both plates were formed by three layers of boards. Thicknesses of external and internal layers of boards are equal to 25 and 45 mm, correspondingly. Pine wood with strength class C18 [15–17] was chosen as a base material. Dimensions of the board's cross-sections for outer and middle layers were equal to 25x50 and 45x195 mm, correspondingly. All layers were joined together by the polyurethane glue under pressure in 400kg/m^2. Both plates were freely supported by the short sides and loaded by the uniformly distributed load. This static scheme is widely used for CLT plates in practice. A span of freely supported plates was equal to 1.9 m (figure 3).

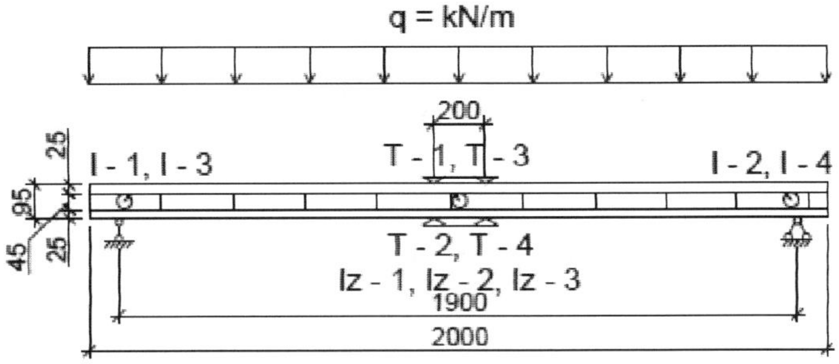

Figure 3. Design scheme and apparatus placement for CLT plates.

Both specimens were statically loaded by pieces of steel with approximate weight of 20 kg each, which were uniformly distributed by the plate surfaces. Intensities of uniformly distributed loads changed within the limits from 1 to 7.5 kN/m^2 with the step equal to 0.5 or 1.0 kN/m^2. The maximum intensity of the applied load is equal to the half from the design load-bearing capacity of the considered plates. The loading stage, when the plate was loaded by the uniformly distributed load with intensity of 7.5kN/m^2, is shown in figure 3. Maximum bending stresses acting in the edge fibers of the outer layers, maximum vertical displacements in the middle of the span and horizontal relative displacements of outer and middle layers of CLT plate were the main objectives of measurements. The local deformations in supports are not significant and were neglected.

Four strain gauges T-1, T-2, T-3, T-4, three deflectometers Iz – 1, Iz – 2, Iz – 3 and four indicators I – 1, I – 2, I – 3, I – 4 were used for this purpose (figure 3). Measurements by the apparatus were made in each stage of specimen loading. The parameters of the laboratory experiment and thicknesses of the slab layers were chosen in accordance with the literature recommendations [8, 15]. The second stage of the verification of the methodologies considered was joined with the calculation of maximum bending stresses, acting in the edge fibers of the outer layers, maximum vertical displacements in the middle of the span by the software ANSYSv14 and RFEM 5.0 [18–19]. CLT plate with dimensions 2x1 m and thickness of 95 mm was calculated using software ANSYSv14 and RFEM 5.0. Calculations of CLT plate using ANSYSv14 and REFM 5.0 are based on mechanics of laminated materials. The target of the calculation is verification of the results obtained by the reduced cross-section method and effective strength and stiffness method. The dependence between stress and strain for the considered CLT panel can be described by the generalized Hooke's law, which is written for the orthotropic model. The results obtained by the FEM software RFEM 5.0 and ANSYS v14 for the CLT plate with dimensions 2x1 m and thickness of 95 mm are given in figures 4 and 5.

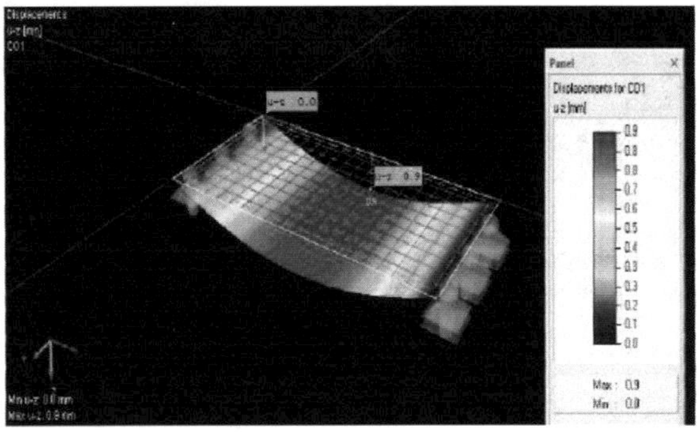

Figure 4. Maximum vertical displacements of CLT plate, which were determined by the software RFEM 5.0.

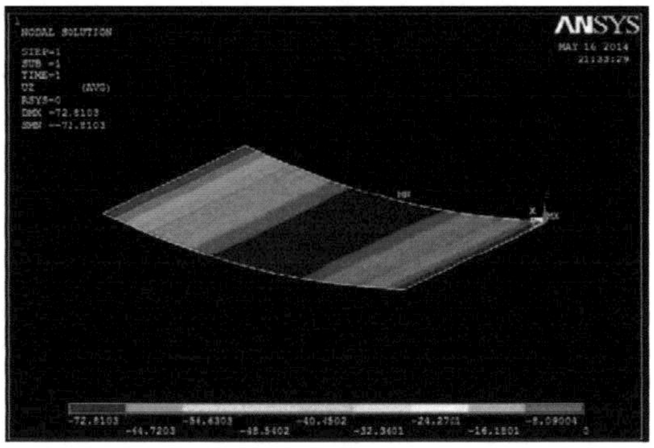

Figure 5. Maximum vertical displacements of CLT plate, which were determined by the software ANSYSv14.

The results, which were obtained for the considered CLT plate by the transformed section and effective strength and stiffness method, physical experiment and software RFEM 5.0 and ANSYS v14, were compared in the next chapter of this study.

3.2. Analysis of design methods of CLT elements subjected to flexure

The verification of the transformed section method and effective strength and stiffness method was carried out by the comparison of maximum bending stresses acting in the edge fibers of the outer layers, maximum vertical displacements in the middle of the span and horizontal relative displacements of the outer and middle layers of CLT plate, which were determined by these methods with the results of physical experiment and results obtained using RFEM 5.0 and ANSYS v14. The difference between the results obtained by the transformed section method and effective strength and stiffness method was insufficient. So, further these results will be mentioned as the results obtained by the design methods. The maximum differences between the results obtained by the design methods and physical experiment are the following:

- maximum bending stresses acting in the edge fibers – 22%;
- horizontal relative displacements of the outer and middle layers of CLT plate – 17%;
- maximum vertical displacements in the middle of the span – 31%.

The maximum differences between the results obtained by the design methods and software REFM and ANSYS v14 are the following:

- maximum bending stresses acting in the edge fibers – 10%;
- horizontal relative displacements of the outer and middle layers of CLT plate – 7%;
- maximum vertical displacements in the middle of the span – 3%.

The differences between the results obtained by the design methods and physical experiment can be explained by the deviation from the technological requirements during producing of both specimens. Thus, the necessary pressure during gluing of the CLT panels must be at least 600kN/m^2, but in reality it was 33% less and, probably, the necessary quality of glue joints was not provided.

The dependence of strains in the edge fibers of CLT plates as a function from the load intensity is shown in figure 6. The results obtained by the considered methods are designated as EC 5. The dependence of maximum vertical displacements in the middle of the span of CLT plates as a function of the load intensity is shown in figure 7.

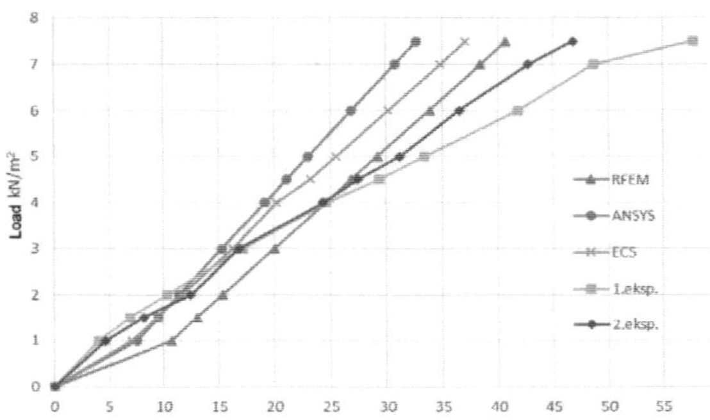

Figure 6. The dependence of strains in the edge fibers of CLT plates as a function of the load intensity.

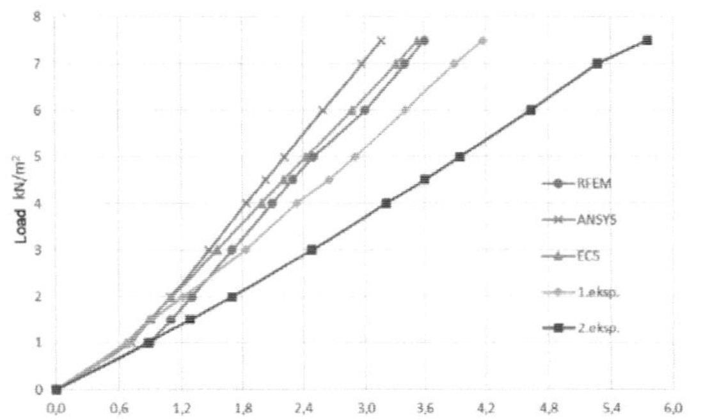

Figure 7. The dependence of maximum vertical displacements in the middle of the span of CLT plates as a function of the load intensity.

4. Verification of the transformed sections method for cross-laminated timber element subjected to compression with bending by FEM

The cross-laminated timber plate, which was described in details in chapter 3.1, was considered. A freely supported beam with the span equal to 1.9 m, which was loaded by the uniformly distributed load and axial force, was considered as a design scheme. The intensity of uniformly distributed load was equal to 7.5kN/m^2. The value of axial force was equal to 70 and 150kN. The values of axial force were taken to consider both probable cases for the cross-laminated timber elements subjected to compression with bending. When the value of axial forces is equal to 150kN, the compressive normal stresses are dominating. When the value of axial forces is equal to 70kN, the bending normal stresses are dominating.

The plate was modeled by the software ANSYS v15 using layered shell elements SHELL181 and orthotropic material properties. In case of dominating bending stresses, the maximum obtained stress by FEM is equal to 3.82 MPa (figure 8 (a)). The maximum stress, which was obtained by the transformed section method, is equal to 3.86 MPa. In case of dominating compressive stresses, the maximal obtained stress by FEM is equal to 5.32 MPa (figure 8 (b)); the stress obtained by the transformed section method is equal to 5.41 MPa.

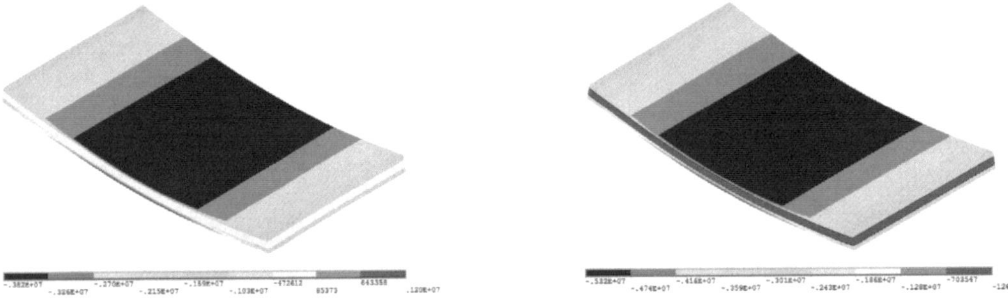

(a) (b)

Figure 8. Stress distribution in the plate: (a) in case of dominating bending stresses; (b) in case of dominating compression stresses.

Comparison of the results obtained by the transformed section method and FEM, which was performed by the software ANSYS v15, indicates that the difference between the obtained results does not exceed 1.7 % for the cases with dominating compressive and bending normal stresses. It allows us to make a conclusion that the transformed section methods can be used to predict behaviour of cross-laminated timber elements subjected to compression with bending.

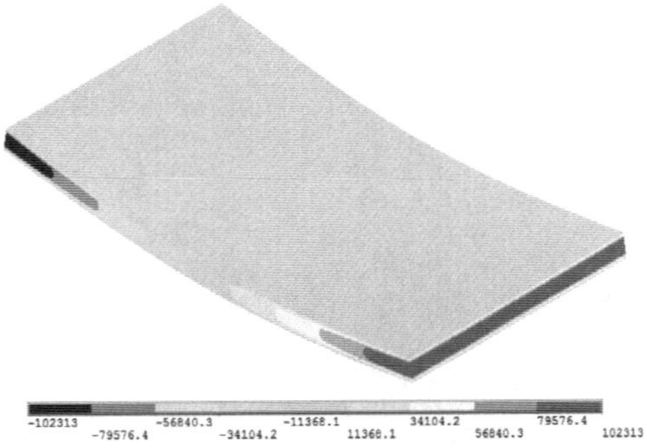

Figure 9. Shear stress distribution in the plate.

Distribution of shear stresses obtained by FEM is shown in figure 9. The level of shear stresses is much lower than shear strength. Shear stresses must be calculated using the transformed section method as for elements subjected to flexure.

5. Conclusions

Design methods of cross-laminated timber elements subjected to flexure and compression with bending were analyzed. The transformed sections and effective strength and stiffness methods were checked analytically and experimentally for cross-laminated timber panels. It was stated that the maximum differences between the results obtained by the transformed section method, physical experiment and software RFEM 5.0 and ANSYS v14 were equal to 31 and 10%, respectively, for the CLT panels subjected to flexure. The maximum differences between the results obtained by the transformed section method and software ANSYS v15 does not exceed 2% for the CLT panels subjected to compression with bending for the cases of dominating compressive and bending stresses. Therefore, the transformed sections and effective strength and stiffness methods enable to describe the behaviour of CLT elements subjected to flexure and compression with bending with the relative accuracy.

6. References

[1] Downing B and Spickler K 2013 Timber tower research project (Chicago: Softwood Lumber Board)

[2] Dujic B, Strus K, Zarnic R and Ceccotti A 2007 Prediction of dynamic response of a 7 – storey massive wooden building tested on a shaking table (Slovenia: World Connection on Timber Engineering)

[3] Ryan E S 2011 Interlocking cross – laminated timber: alternative use of waste wood in design and construction (Utah: University of Utah)

[4] Simon V E 2011 Medium rise timber buildings in the Netherlands. Master Thesis (Delft: Delft University of Technology)

[5] Ashtari S 2012 In – plane stiffness of cross – laminated timber floors. Master Thesis (Vancouver: The University of British Columbia)

[6] Sokolov V A 2014 Probabilistic Analysis of Intermediate Floor Steel and Wooden Structures in

the Old Urban Development Building *Applied Mech. and Mat.* **633** 1040 – 48
[7] Semenov K V and Kononova M J 2013 Konstrukciji iz dereva i plastmass (Sankt-Peterburg: Izdatelstvo Politehnicheskogo Universiteta)
[8] ETA-12/0281 2012 NORITEC X–LAM cross laminated timber (CLT) – solid wood slab element to be used as structural elements in buildings (Vienna: Noritec Holzindustrie GmbH)
[9] Harch B J L 2010 The investigation into the optimisation of cross laminated timber panels for use in the Australia building industry (Australia: Queensland University of Technology)
[10] ETA – 06/0138 2006 KLH solid wood slabs: Solid wood slab element to be used as structural elements in buildings (Österreichisches Institut für Bautechnik: KLH Massivholz GmbH)
[11] Miller J F 2009 Design and analysis of mechanically laminated timber beams using shear keys (Michigan: Michigan technological University)
[12] EN 1995–1–1 2004 Design of timber structures – General – Common rules and rules for buildings European Committee for Standardisation
[13] Porteous J and Ross P 2013 Designer's guide to Eurocode 5: Design of timber buildings EN 1995–1–1 (Bristol: ICE Publishing)
[14] Z–9.1–559 2012 CLT – Cross laminated timber (Graz: Graz University of Technology)
[15] EN 16351:2011–11 2012 Timber structures – cross laminated timber – Requirements (Vienna: Austrian Standards Institute)
[16] Reinhard B 2012 Production and Technology of Cross Laminated Timber (CLT): A state-of-the-art Report (Graz: Graz University of Technology)
[17] EN 391:2001–10 2012 Glued laminated timber – Delamination test of glue lines (Vienna: Austrian Standards Institute)
[18] Dlubal Software GmbH 2013 RF-LAMINATE Calculation and Design of Laminate Surfaces Program description (Tiefenbach, Germany: Dlubal Software GmbH)
[19] ANSYS Inc 2013 ANSYS Mechanical APDL Element Reference (USA: ANSYS Inc)

Acknowledgement
The research leading to these results has received the funding from Latvia state research programme under grant agreement "INNOVATIVE MATERIALS AND SMART TECHNOLOGIES FOR ENVIRONMENTAL SAFETY, IMATEH", project ID 1854, task 3.

Behaviour of Steel Arch Stabilized by a Textile Membrane

O Svoboda[1,2] **and J Machacek**[1]
[1]Czech Technical University in Prague, Faculty of Civil Engineering, Czech Republic

E-mail: ondrej.svoboda.4@fsv.cvut.cz

Abstract. Behaviour of the slender steel arch supporting textile membranes in a membrane structure with respect to in-plane and out-of plane stability is investigated in the paper. In the last decades the textile membranes have been widely used to cover both common and exclusive structures due to progress in new membrane materials with eminent properties. Nevertheless, complex analysis of such membranes in interaction with steel structure (carbon/stainless steel perimeter or supporting elements) is rather demanding, even with specialized software. Laboratory model of a large membrane structure simulating a shelter roof of a concert stage was tested and the resulting stress/deflection values are presented. The model of a reasonable size was provided with prestressed membrane of PVC coated polyester fabric Ferrari® Précontraint 702S and tested under various loadings. The supporting steel structure consisted of two steel arch tubes from S355 grade steel and perimeter prestressed cables. The stability behaviour of the inner tube was the primary interest of the investigation. The SOFiSTiK software was used to analyse the structural behaviour in 3D. Numerical non-linear analysis of deflections and internal forces of the structure under symmetrical and asymmetrical loadings covers various membrane prestressing and specific boundary conditions. The numerical results are validated using test results. Finally, the preliminary recommendations for appropriate numerical modelling and stability design of the supporting structure are presented.

1. Introduction

Tensioned Fabric Structures (TSF) are becoming more and more popular not only because of their visual attractiveness but also because of their ability to be a fully load-bearing component in a complex structure. Various membrane types are available for both single layer double curved shapes (hypars, cones, barrels) and pneumatic constructions (inflatable cushions, air supported pneumatic structures). The PVC coated polyester fabric (e.g. Précontraint FERRARI®) seems to be a rational choice for common single layer structures with lifetime up to 20 years, acceptable costs and with good joining possibilities (welding, sewing). When longer lifetime is needed, a glass fabric coated by PTFE (Teflon), silicon rubber or titanium dioxide could be used but at higher expenses. Rather expensive but

[2] To whom any correspondence should be addressed.

IOP Conf. Series: Materials Science and Engineering **96** (2015) 012055 doi:10.1088/1757-899X/96/1/012055

excellent material is expanded PTFE coated 2 sides by fluoropolymer film (TENARA®), joint by welding. The membranes may also be thermally insulated using Nanogel Aerogel™ and 2 sides coated with PTFE (in result translucent, with total thickness 9 mm, CABOT Cor.). Plastic foils of low thickness 50÷500 μm are used mainly as inflatable cushions, nowadays predominantly from ETFE (TEXLON®) or THV materials.

Whereas the geometrically and materially non-linear analysis with imperfections (GMNIA) is necessary for proper designing and material properties of the used membranes are essential inputs into the software, various models of material behavior have been established. The most sophisticated model with 15 input parameters obtained from laboratory tests was proposed by Kato *et al.* [1]. This model demonstrates excellent agreement with the test data but because of its complexity and time requirements is rather unsuitable for practical utilization. A non-linear material model with 5 parameters based on experimental results and depending on load ratios in warp and fill direction was proposed by Galliot and Luchsinger [2], [3]. Model parameters are: warp and fill Young's modules for load ratio 1:1, the change in warp and fill Young's modules and the Poisson's ration. Recently a simplified stress-strain model for coated plain-weave fabrics was developed by Pargana and Leitao [4]. The model consists of three nonlinear elements to model the yarns and an isotropic plate to model the coating.

Outstanding and unique structures using either common or sophisticated textile fabrics were designed in last few decades which indicate new ways and directions of novel structure design. Allianz Arena in Munich (design by Herzog & de Meuron, 2005), SkySong ASU in Scotsdale (FTL Design Engineering Studio, 2009) or Union Station in Denver (Skidmore, Owings & Merrill LLP (SOM), 2014) are just some examples of excellent projects realized in the last decade.

For these structures it is necessary to use supporting components that are as slender as possible to follow the concept of the delicate, light and attractive structure with the membrane surface and to avoid any visual intervention of supporting steelworks into the membrane area. While membrane surface is exclusively tensioned, supporting construction is most often exposed to compression or bending. This type of loading, in combination with slender elements, results in stability problems and design must be done with respect to these effects. Different situations occur, when membrane surface is joined with the supporting steel structure continually. In this case the membrane represents a spring support for the supporting structure, the critical length of individual elements is changing, and both parts of the entire system cannot be investigated separately but as one complex structure using proper software package, which allows integrated modeling and computing (e.g. EASY [5], FORTEN [6], SOFiSTiK [7], Rhino Membrane [8], NDN [9], etc.). Detailing of membrane structures and erection methods are well described by Seidel [10] and the basic design is discussed in the European Design Guide [11].

In this paper a significant stabilizing effect of membranes with respect to supporting steelwork is presented based on the data obtained from testing of a structure representing a reduced concert stage structure with two supporting arches. Influence of various membrane prestressing on the structure stabilization is analyzed and results are validated with tests.

2. Description of the model under experimental investigation

2.1. Geometry and material properties of the used components

Model of a reasonable size representing a roofing of a concert stage was designed, built and tested in the lab of the Faculty of Civil Engineering of CTU in Prague. The final shape of the tested configuration resulted from several analytical models examined using Formfinder software [12], which provides user friendly design interface and various possibility of shape designing. Floor projection of the model is *LxB* 4500x2250 [mm], with height $H = 1200$ [mm] (figure 1). From 8 configurations of loading (5 on arch alone, i.e. without membrane and 3 on the complex structure with membrane) two were used as representative. Both symmetrical and asymmetrical loading were applied on the stand-alone arch and on the arch connected with the membrane.

Inner arch, which was of primary interest within the investigation, was incurved from a hot formed CHS profile Ø 26.9x3.2 [mm] made of steel with yield strength (based on coupon tests) $f_y = 475$ MPa, ultimate strength $f_u = 595$ MPa and elongation 27.1%. Young's modulus of elasticity was considered in accord with Eurocode 3 as $E = 210$ GPa. Outer arch, incurved from hot formed CHS profile of Ø 88.9.9x3.2 [mm] cross section, proved to have yield strength $f_y = 460$ MPa and ultimate strength $f_u = 574$ MPa. Young's modulus was also considered as $E = 210$ GPa.

The membrane employed was Précontraint FERRARI® 702S with opaque surface (weight 830 g/m^2). Basically the material is polyester scrim coated on both sides with liquid PVC and PVDF topcoat, weldable for joining. The fabric, developed and patented by the French manufacturing group Serge Ferrari, is unique due to pre-stressing of the base fabric before and during the coating operations. The operation ensures in both warp and fill directions similar elongation behavior and minimum creep. The thick coating also supplies the membrane with longevity and extraordinary resistance against soiling. Concerning material characteristics, the biaxial test performed by Lab BLUM Stuttgart [13] is available. Both warp and fill braking strength was considered as $s_{ult} \approx 56$ kN/m, while working stress $s_{max} = s_{ult}/5 \approx 11.2$ kN/m to exclude tearing. The test prestressing was designed as approx. $s_p \approx 0.4$ kN/m. However, the material due to its structure is non-homogeneous, orthotropic (warp and fill directions) and non-linear. Based on the material model investigated by Galliot and Luchsinger [2] [3] dealing with five parameters, i.e. warp and fill Young's modules for 1:1 load ratio, the change in warp and fill Young's moduli and the Poisson's ratio, the mechanical properties for FERRARI® 702 used in the FE analysis of the test were established as 635.3 kN/m, 661.9 kN/m, 295.0 kN/m, 168.5 kN/m and 0.196, respectively. The formulas for orthotropic linear elastic behavior are given in [2] and appropriate shear modulus 14.3 MPa in [3].

The membrane was joined to the outer arch using riveted aluminum keder profile while to the inner tube via alternating pockets. Common 7x7 wire peripheral rope from CarlStahl Company with diameter of 6 mm in curved cuff fastened the membrane to the corner plates and anchors.

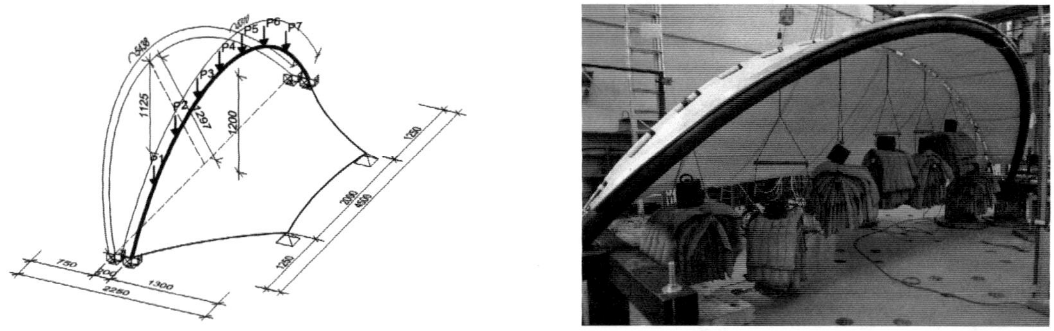

Figure 1. Geometry (left) and view of the loaded model at the laboratory (right).

2.2. Loading schemes and test results

The investigated arch was fitted with transducers (electrical potentiometers) in vertical (V), transverse (H) and longitudinal (L) directions to measure deflections. In supports and middle of the arch 12 strain gauges were placed always in four mutually perpendicular positions (figure 2).

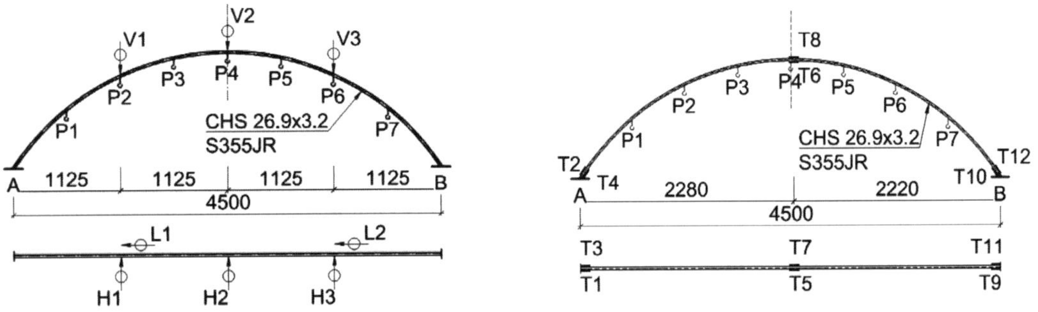

Figure 2. Position of transducers (left), strain gauges (right) and loading (P).

The loading schemes were performed with respect to the inner investigated arch in two phases. First, the arch standing alone was tested, loaded with calibrated pouches with steel pellets. Second, the membrane was assembled and prestressed by tightening of the perimeter ropes. The aim was to achieve a uniform prestress. Control measurement with strain gauges placed on the surface of the membrane was carried out to check a prestress uniformity. However, due to non-homogeneous structure of the membrane (perpendicular net of yarns and additional coating layer), the measurement was informative and real value of prestress could not be established. After membrane was assembled, the complete membrane structure was loaded in the same way as the arch standing alone. Sum of the total loading in individual loading schemes is shown in diagrams (figure 3) for both symmetrical and asymmetrical loading. Loadings were terminated when maximal permissible deflection of transducers or yield strength of steel was achieved.

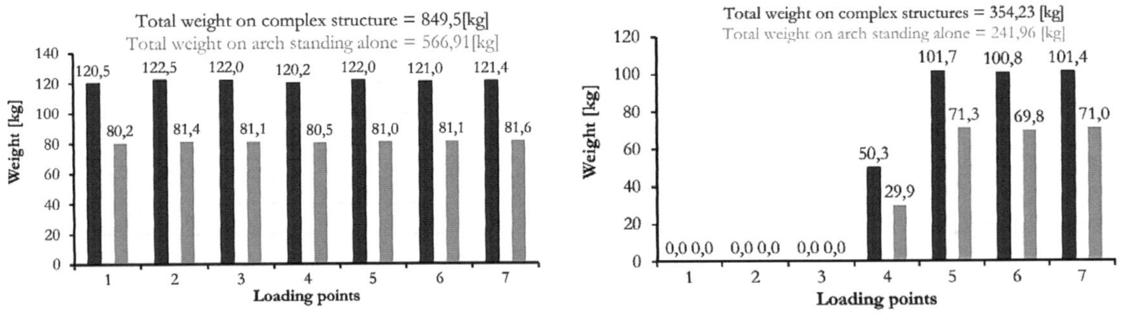

Figure 3. Maximal symmetrical (left) and asymmetrical (right) loading imposed on the complete membrane structure and the stand alone arch (without membrane).

The process of sagging of the inner arch under increasing loading in vertical direction is shown in figure 4. Unloading path is not included and in general it was fully elastic. Arch standing alone collapsed with out-of-plane buckling mode under total loading approaching $F_0 = 5.5$ kN (transverse deflection at the top of the arch was $\eta_0 = 79.7$ mm and vertical one $\delta_0 = 35.1$ mm) while the complex structure with the membrane resisted till the total load of $F_M = 8.3$ kN, with approaching the arch yield stresses. However, even at this level of loading the transverse deflection oscillated around the initial unloaded form.

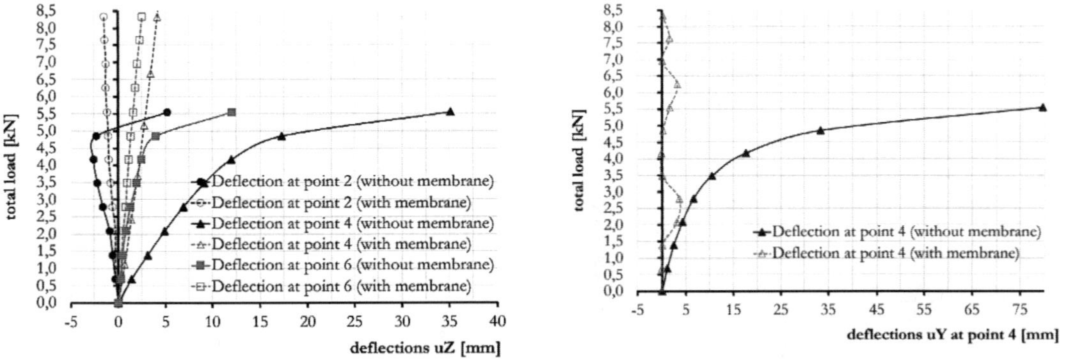

Figure 4. Comparisons of vertical (left) and horizontal (right) deflections under symmetrical loading of the inner arch standing alone and with the membrane.

The influence of the textile membrane on the slender steel tube behavior is even more significant under asymmetrical loading. Vertical deflection at load level $F_0 = 2.37$ kN of the arch alone at point P6 was $\delta_0 = 41.3$ mm, while the deflection at the same loading $F_M = 2.37$ kN of the arch with membrane $\delta_M = 18.5$ mm. At maximal loading of the structure with membrane $F_{M,max} = 3.47$ kN the vertical deflection was $\delta_{M,max} = 30.4$ mm and horizontal one $\eta_{M,max} = 1.7$ mm.

3. Numerical analysis

Software package SOFiSTiK was chosen for a detailed numerical analysis of both the arch standing alone and complete structure with the membrane. SOFiSTiK enables complete setting of all material parameters of the used membranes including their pretension in standard units [kN/m’]. For correct analysis, it is necessary to set proper material characteristics of all used materials, especially for orthotropic material of the used membrane Précontraint FERRARI® 702S.

SOFiSTiK is compatible with another software package AutoCAD, where the exact geometry with initial deformations was determined. Initial deformations were measured in the laboratory on the model just before first loading and were found to be around 5 mm in horizontal direction at the top of the arch and around +/-10 mm in vertical direction at the quarters of the arch. Initial shape of the membrane surface was generated automatically by the AutoCAD as a surface with minimal area. This approximation is sufficiently accurate for further computing in SOFiSTiK, to which the model from AutoCAD was exported. For each level of membrane prestress, exact value of cable pretension was calculated based on formula (1) [14].

$$S_S = R \cdot S_M \quad (1)$$

Where, S_S is a tension force in a rope,

R is a diameter of the rope,

S_M is prestress in membrane perpendicular to the rope [kN/m′].

Formfinding and all other calculations were done using geometrically non-linear analysis with Newton-Raphson method [15] and also using the theory of large deformation (in Germany called third order theory) which is necessary for the calculation of the membrane structure.

3.1. Comparison of numerical analysis and test results

Primary goal of the laboratory testing was to validate the numerical analysis using SOFiSTiK software. Subsequent work would head towards further modeling of different types of structures and a parametrical study. Comparison of the test results and the analysis of arch alone and the complex structure using geometrically nonlinear analysis with imperfections (GNIA) are shown in figure 5 and figure 6, respectively. The measured initial deflections were introduced in the analysis.

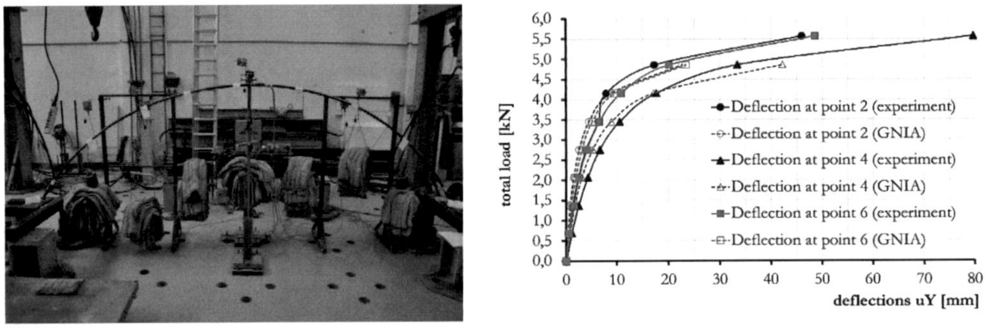

Figure 5. Symmetrical loading of the isolated arch, i.e. arch alone (left). Comparison of experimental horizontal deflections and numerical GNIA (right).

Excellent agreement between numerical analysis and test data is obtained for the arch standing alone. Agreement is presented by the diagram with symmetrical loading showing a transverse deflection uY out of the arch plane. Till sixth loading step, values from numerical analysis are lower than those from test. After sixth loading step, the material hardening of the real structure is noticeable and gives rather higher value. Under total load of 4.85 kN, the transverse test deflection of the middle of the arch was 33.4 mm and 42.2 mm at numerical analysis. Vertical deformation at the same point and at the same load level was 17.3 mm during experiment, respectively 13.1 mm in numerical analysis.

Table 1. Comparison of experimental vertical deflections and numerical ones with various pretension.

experiment		Pret. 0.2 kN/m		Pret. 0.4 kN/m		Pret. 0.6 kN/m		Pret. 0.8 kN/m		Pret. 1.0 kN/m	
Total load	Defl.	Defl.	Exp./ GNIA	Defl.	Exp./ GNIA	Defl.	Exp./ GNIA	Defl.	Exp./ GNIA	Defl.	Exp./ GNIA
[kN]	[mm]	[mm]	[%]	[mm]	[%]	[mm]	[%]	[mm]	[%]	[mm]	[%]
0.00	0.00	0.00	0%	0.00	0%	0.00	0%	0.00	0%	0.00	0%
2.08	3.69	3.43	-8%	3.22	-15%	3.10	-19%	3.02	-22%	2.97	-24%
4.17	8.33	8.14	-2%	7.60	-10%	7.27	-15%	7.05	-18%	6.91	-21%
6.25	13.86	13.50	-3%	12.50	-11%	11.90	-16%	11.50	-20%	11.30	-23%
8.33	**20.77**	20.10	-3%	**18.70**	**-11%**	17.60	-18%	17.00	-22%	16.50	-26%

Several levels of membrane pretension were investigated using software SOFiSTiK at this step as shown in the diagram (figure 6, right). It is obvious that level of pretension is a crucial parameter for the arch stability and resistance capacity. Higher pretension in membrane logically means higher stability and smaller deflection of the investigated arch. It should be noted that the real pretension during the testing was due to alternating "pockets" connection of the membrane and arch rather non-uniform and slacked. This is reflected in numerical analysis with pretension close to zero (see the value of 0.2 kN/m in table 1.) which is near but still lower than the experimental one. From measured values of stresses in the membrane during test a real value of membrane pretension was assessed roughly as 0.4 kN/m′. Nevertheless, taking all mentioned notes into account, the received results from the numerical analysis are sufficient for justification of the described SOFiSTiK model to be used in planned wider parametric studies and investigations.

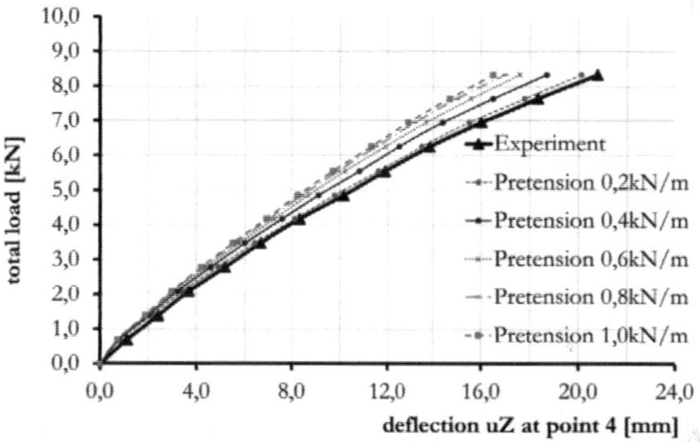

Figure 6. Symmetrical loading of complex structure, i.e. with membrane (left). Comparison of experimental vertical deflections and numerical ones with various pretension (right).

4. Conclusion

The paper deals with the detailed description of an experiment concerning an assembly of a textile membrane with supporting steelwork. Relevant values of deflection, stresses and pretension of the membrane and behavior of the supporting inner arch are presented and compared. The results confirmed the enormous influence of the textile membrane on stability of the slender arch which can be considered in practical design and should be involved in the complex analysis. Separate modeling of the membranes and supported steelwork is not recommended and in large structures leads to incorrect design and finally can lead to the collapse of the whole structure.

Based on the obtained data, software SOFiSTiK is a credible software package for further investigations and parametrical studies of crucial parameters of membrane structures.

5. References

[1] Kato S, Yoshino T and Minami H 1999 Formulation of constitutive equations for fabric membranes based on the concept of fabric lattice model *Engineering Structures* 691–708

[2] Galliot C and Luchsinger R H 2009 A simple model describing the non-linear biaxial tensile behaviour of PVC-coated polyester fabrics for use in finite element analysis *Composite Structures* 438–47

[3] Galliot C and Luchsinger R H 2009 Non-linear properties of PVC-coated fabrics used in

tensairity structures *ICCM17 (UK, Edinburgh, 27-31 July 2009)* p 10
[4] Pargana J B and Leitão V M A 2015 A simplified stress–strain model for coated plain-weave fabrics used in Tensioned Fabric Structures *Engineering* 439–50
[5] Technet gmbh Berlin-Stuttgart [Available]: http://www.technet-gmbh.com
[6] IxForten 4000 [Available]: http://www.ixforten.com/
[7] SOFiSTiK 2014 [Available]: http://www.sofistik.de/
[8] Rhino Membrane [Available]: http://www.membranes24.com/rhino-membrane.html
[9] Membrane NDN software [Available]: http://www.ndnsoftware.com/
[10] Seidel M 2009 *Tensile Surface Structures. A Practical Guide to Cable and Membrane Construction* Ernst & Sohn, Verlag für Architektur und technische Wissenschaften GmbH & Co. KG
[11] Foster B and Mollaert M 2004 European design guide for tensile surface structures *Tensinet* p 354
[12] Formfinder Software GmbH, Wien [Available]: http://www.formfinder.at/main/software/
[13] Laboratorium BLUM Stuttgart 2005 Report on biaxial tests Précontraint 702 [Available]: http://arc-ae.net/ARC1010-Pisco/
[14] Ziegler R and Wagner W 2001 Von der Formfindung zum Membrantragwerk *Bauen mit Textilien* **4** 7–17
[15] SOFiSTiK AG 2014 *Manual Book – Program ASE* Oberschleissheim, Germany

Acknowledgements
All laboratory works were performed at the Central Laboratory of the Faculty of Civil Engineering of CTU in Prague. Thanks are due to the laboratory technicians and technical managers.
The supply of fully formed steel tubes by company EXCON Cor. Prague including delivery to laboratory is highly appreciated.
Thanks are due to company Archtex Ltd. Prague for help with preparation and assembly of the textile membrane.
Financial support of the CTU in Prague grants SGS15/041/OHK1/1T/11 and grant of the Czech Grant Agency GACR No. No. 103/13/25781S is gratefully acknowledged.

IOP Conf. Series: Materials Science and Engineering **96** (2015) 012056 doi:10.1088/1757-899X/96/1/012056

Behaviour of Mechanically Laminated CLT Members

P Kuklík[1,2] **and L Velebil**[1,2]
[1] Czech Technical University in Prague, University Centre for Energy Efficient Buildings, Třinecká 1024, 273 43 Buštěhrad, Czech Republic

E-mail: kuklik@fsv.cvut.cz, lukas.velebil@fsv.cvut.cz

Abstract. Cross laminated timber (CLT) is one of the structural building systems based on the lamination of multiple layers, where each layer is oriented perpendicularly to each other. Recent requirements are placed to develop an alternative process based on the mechanical lamination of the layers, which is of particular interest to our research group at the University Centre for Energy Efficient Buildings. The goal is to develop and verify the behaviour of mechanically laminated CLT wall panels exposed to shear stresses in the plane. The shear resistance of mechanically jointed CLT is ensured by connecting the layers by screws. The paper deals with the experimental analysis focused on the determination of the torsional stiffness and the slip modulus of crossing areas for different numbers of orthogonally connected layers. The results of the experiments were compared with the current analytical model.

1. Introduction

Cross laminated timber (CLT) is one of the structural building systems based on the lamination of multiple layers, where each layer is oriented perpendicularly to each other. CLT is a plate element that consists of several board layers oriented in an angle of 90° to each other. The most mature method of connecting the layers is by sticking them with glue. Recent requirements are fully addressed to develop an alternative process based on the mechanical lamination of the layers omitting the glue. This technology could be cheaper and more accessible for producers. Currently, no detailed procedure for determining the buckling resistance and shear resistance of mechanically jointed CLT panels is described in the codes. Modifications of the procedure [1], [2] were used to simplify an analytical approach. Whereas buckling resistance of CLT wall panels loaded in the plane only depends on the effective bending stiffness EI_{eff} for the interaction of the layers oriented in the direction of loading, their shear resistance depends on the torsional stiffness of the connections between the boards. The connections of wooden structures are usually considered as articulated because fasteners in the joints

[2] To whom any correspondence should be addressed.

are distributed on a small area and the possibility of the torsional moment distribution to the fasteners is limited. Rigid joints are, therefore, used more as corner joints for three articulated frames or cantilevered timber columns. Mechanically jointed CLT shear walls use a large number of locally rigid connections. The orthogonal orientation of the layers and their connection with screws causes the behaviour different from the current design approach that determines the shear resistance of wood-based walls. The shear deformation of a CLT wall panel is caused by shear deformation of the boards as well as shear deformation of the crossing areas. Whereas shear deformation of the boards depends on the shear modulus of used wood, shear deformation of the crossing areas depends on the stiffness of the connection by mechanical fasteners that can be described by the slip modulus of fasteners K_{ser} and torsional stiffness of joints $K_{r,ser}$. The boards have significantly greater shear stiffness than their connection by fasteners. The rotation of the joints and shear deformation of the entire panel are, therefore, formed mainly by torsional displacement of fasteners, see figure 1.

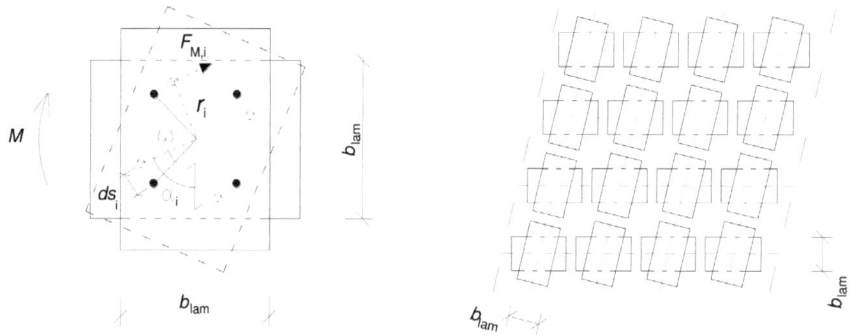

Figure 1. Torsion shear in the joint of boards (left) and in the entire panel (right).

The aim of this paper is to determine slip modulus K_{ser} and torsional stiffness $K_{r,ser}$ for different numbers of orthogonally connected layers and to compare experimental results with the current analytical model.

2. Previous research on mechanically jointed CLT structures

Wall panels with mechanically jointed layers were tested to determine the characteristic values of buckling resistance and shear wall resistance.

2.1. Evaluation of buckling resistance

The test set-up for the evaluation of buckling resistance is shown in figure 2. Two series of wall panels were tested: one series for a 3-layered panel (3x 27 mm) and one series for a 5-layered panel (5x27 mm). Each series consisted of three specimens with dimensions of 2200x1000 mm. The load-time diagram of a 5-layered panel is shown in figure 3.

Figure 2. Testing wall panel buckling resistance.

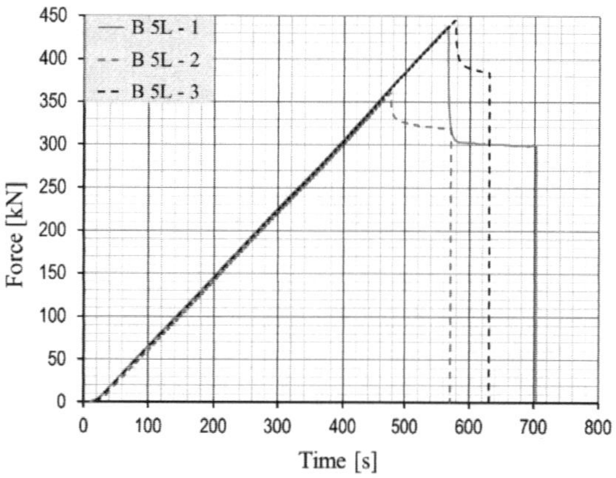

Figure 3. Testing procedure of a 5-layered wall panel.

2.2. Evaluation of shear wall resistance

The test set-up for the evaluation of shear wall resistance is shown in figure 4. Horizontal displacements were measured to evaluate the load-displacement diagrams. Both tested series for a 3-layered panel and for a 5-layered panel consisted of 5 identical specimens with dimensions of 3000x2500 mm. Horizontal deformation to the horizontal force diagram for a 5-layered panel is shown in figure 5.

Figure 4. Testing shear wall panel resistance.

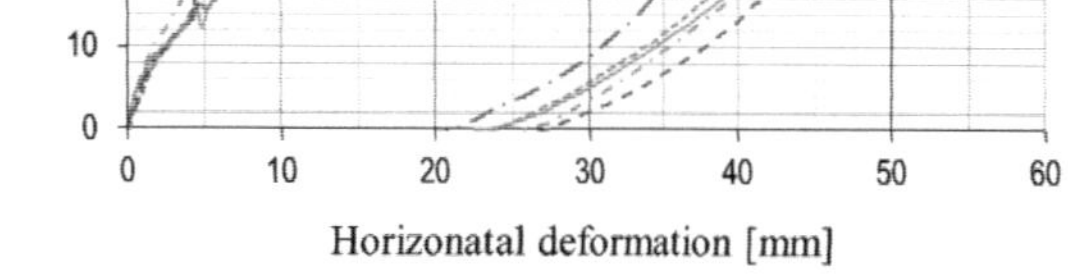

Figure 5. Horizontal deformation to horizontal force diagram for a 5-layered wall panel.

3. Tested specimens

Three series of specimens were tested; the series of 2, 3, and 5-layered specimens (series A, B, and C, respectively). Series A consists of 3 identical specimens, series B and C consist of 2 identical specimens. All layers of the specimens were made of spruce boards, strength class of C24 with dimensions of 170 x 27 x 500 mm, connected together with threaded galvanized screws 5/80 mm. Two

IOP Conf. Series: Materials Science and Engineering **96** (2015) 012056 doi:10.1088/1757-899X/96/1/012056

screws were used in each board's crossing for connection of the layer. 5-layered specimens in series C were created from 3-layered ones, where additional 2 layers were connected with the same screws, but in the diagonal position. This means that no screw goes through all layers in those specimens. A scheme of the samples used in the experiments is shown in figure 6.

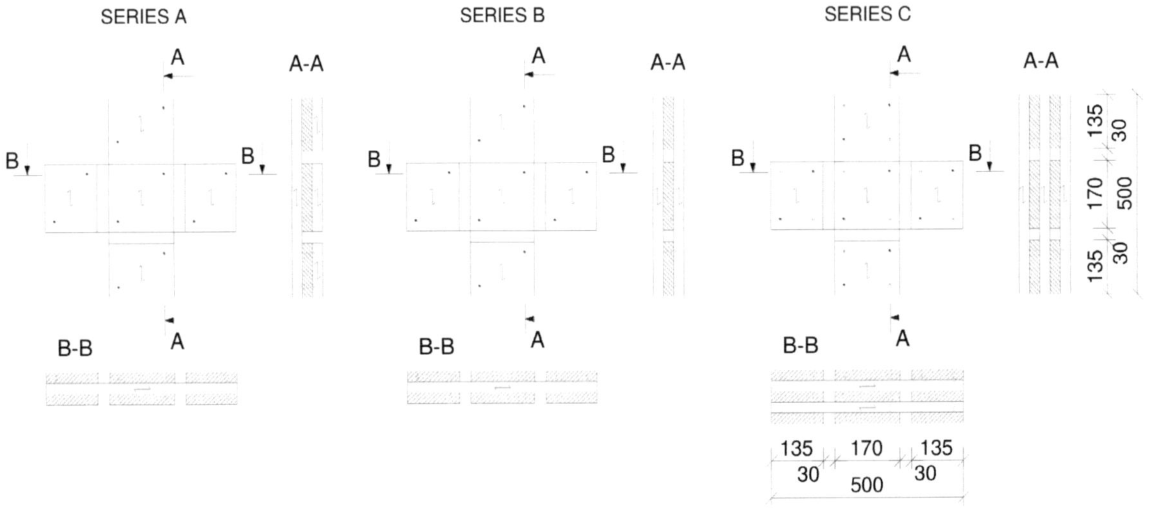

Figure 6. Scheme of the samples used.

4. Experimental setup

The specimens were inserted into a steel device during the test. This device consisted of two arms which were connected by a pin, and thus they could mutually rotate. Sufficient rigidity of the steel device causes uniform loading for fasteners in the timber joint. Therefore, there was only stress due to the torsional moment at the joint, not due to shear forces. Torsional moments which caused the rotation of joints were determined from loading forces at a distance of 250 mm from the centre of the joint. Vertical shifts of horizontal layers at both sides of the joint were measured by two inductive sensors mounted on the front surface of the specimens. Rotations of the joint were calculated from the average values of these shifts. The joints were loaded at a rate of 0.1 mm/s and the loading was continued until the joint failure. The slip modulus and torsional stiffness of the joints were determined by conducting a test setup according to figure 7 using the requirements and assumptions of [3]. The test device was based on a configuration that uses the Flaig and Meyer approach to determine the slip modulus of connections between crosswise bonded boards [4].

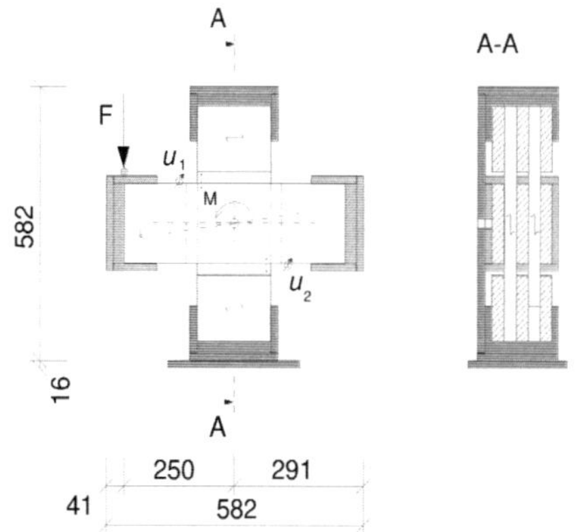

Figure 7. Test set-up to determine the slip modulus and torsional stiffness [4].

5. Results

Torsional shear tests of CLT joints have been performed to evaluate the rotation of joints in torsional moment diagrams. The evaluated characteristics with average values for each series of specimens are shown in figure 8. As one can see, the differences in the connection rigidity between the boards correspond to the differences in the number of shear planes in the joints. The increase in initial stiffness was the largest in series 5L and the smallest in series 2L, as expected. After this initial steep increase in stiffness, the curves begin to be flatter, due to the embedment of wood grain by the screw, see figure 9.

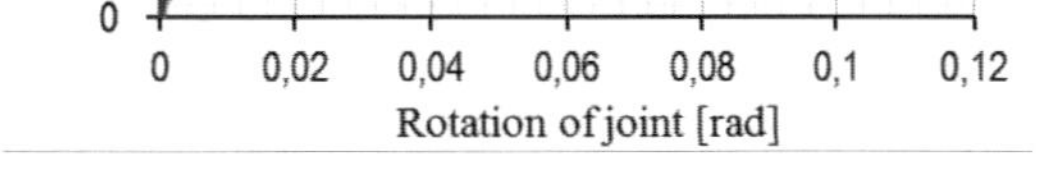

Figure 8. Rotation of joints to the torsional moment diagram.

Figure 9. Failure of the joint due to wood grain embedment.

The orthotropic properties of wood and the comparison of strength in compression perpendicular and parallel to grain indicate a significant influence of the angle between the direction of fasteners' loading and the direction of wood fibres on the resistance in the connection of two perpendicular layers. From this assumption, different values of the slip modulus for orthogonal boards may be obtained, as presented by Ohashi and Sakamoto [5]. The analytical model referred to in [1] contemplated the same value of the slip modulus for screws in both fibre directions, as one can see in equation 1. Slip modulus $K_{ser,i,cal}$ is dependent on the diameter of fasteners d and on the density of wood material ρ_m.

$$K_{ser,i,cal} = \rho_m{}^{1,5} \cdot \frac{d}{23} \tag{1}$$

Torsional stiffness $K_{r,ser,cal}$ can be determined from equation 2 using the slip modulus values.

$$K_{r,ser,cal} = K_{ser,i,cal} \cdot r_i{}^2 \cdot n \cdot m \tag{2}$$

where r_i is the distance of fasteners from the centre of rotation, n is the number of fasteners and m is the number of shear planes in joints. Torsional moments $M_{max,cal}$ were calculated from the characteristic load-carrying capacity for the used screws. Slip modulus $K_{ser,i,exp}$ was determined based on experiments for one shear plane of the fasteners according to equation 3, considering the condition of linear behaviour of the fasteners in the joint.

$$K_{ser,i,exp} = \frac{M_{04}}{\frac{4}{3} \cdot \varphi_{04-01} \cdot r_i{}^2 \cdot n \cdot m} \tag{3}$$

where M_{04} is the torsional moment at 40 % of the ultimate load, $\varphi_{04\text{-}01}$ is the mutual rotation between orthogonal layers that was determined from the values, related to the difference between 10 % and 40 % of the ultimate load. Torsional stiffness was determined according to equation 4 from the ratio of torsional moment and the appropriate rotation of the joint.

$$K_{r,ser,exp} = \frac{M_{04}}{\frac{4}{3} \cdot \varphi_{04-01}} \tag{4}$$

Experimental results and their comparison with the analytical model are given in table 1.

Table 1. Comparison between the analytical model and experiments.

	Calculation			Experiment		
Specimen	$M_{max,cal}$ [kNm]	$K_{ser,i,cal}$ [kN/m]	$K_{r,ser,cal}$ [kNm/rad]	$M_{max,exp}$ [kNm]	$K_{ser,i,exp}$ [kN/m]	$K_{r,ser,exp}$ [kNm/rad]
2L - I	0.52	1871.18	27.03	0.42	1431.81	20.69
2L - II				0.47	719.18	10.39
2L - III				0.47	591.97	8.55
3L - I	0.63		54.07	0.62	1297.33	37.49
3L - II				0.64	1444.48	41.75
5L - I	1.25		108.14	1.38	2152.84	124.43
5L - II				1.42	1652.00	95.49

IOP Conf. Series: Materials Science and Engineering **96** (2015) 012056 doi:10.1088/1757-899X/96/1/012056

6. Conclusion

Mechanically jointed CLT panels have a potential to be widely used for multi-storey buildings. Tests of the buckling resistance and shear wall resistance for wall structures were performed. Currently, no detailed analytical procedure for determining shear resistance is described in the codes; therefore, the mechanical behaviour of layers' connections in shear walls is investigated and described in this paper. Ongoing research is focused on an experimental program comprising different variations of mechanical layer connections.

The comparison of experimental results within each series of samples indicates a large dependence of the fasteners' slip modulus and torsional stiffness of joints on the material properties of wood.
The data obtained from the experiments show about 29.1 % less value of the fasteners' slip modulus $K_{ser,i}$ and about 51.1 % (2L series), respectively 26.7 % (3L series) smaller torsional stiffness of joints $K_{r,ser}$, than the values determined by calculation. The average measured values of torsional stiffness in the joints of 5L series are about 1.7 % greater, than the value determined by calculation.
By comparing the obtained values, it is obvious that the current analytical procedure may be used for the calculation of the bearing capacity of mechanically connected CLT elements, but it does not provide a sufficient level of accuracy for the calculation of stiffness in torsional displacement of boards.

Further sets of displacement tests are being prepared. On the basis of this experimental program, suitable possibilities will be considered for further research into shear wall structures.

7. References

[1] Comité Européen de Normalisation EN 1995-1-1 2006 Eurocode 5: Design of timber structures - Part 1-1: General - Common rules and rules for buildings (Bruxelles: Belgium)

[2] Racher P 1995 *Timber Engineering - STEP 1, C16 - Rigid joints in torsioned* ed H J Blass (The Netherlands, Centrum Hout) chapter C16 pp C16/1-C16/11

[3] Comité Européen de Normalisation - EN 26891 1991 Timber structures. Joints made with mechanical fasteners. General Principles for the determination of strength and deformation characteristics (Bruxelles: Belgium)

[4] Flaig M and Meyer N 2014 A new test configuration to determine the slip modulus of connections between crosswise bonded boards *Proc. COST FP1004 – Enhance Mechanical Properties of Timber, Engineered Wood Products and Tmber Structures (Prague, CR, 21-23 May 2014)* ed Schober K U (University of Bath) pp 77 – 84

[5] Ohashi Y and Sakamoto I 1989 Study on laminated timber moment resisting joint *Proc. of the 2nd Pacific Timber Engineering Conference (Aucktand, Australia)* vol. 2 pp 37 – 42

Acknowledgement

This work has been supported by the European Union, OP RDI project No. CZ.1.05/2.1.00/03.0091, University Centre for Energy Efficient Buildings and by the SGS13/169/OHK1/3T/11 project, Composite Structures at Normal Temperature and in Fire Situations.

IOP Conf. Series: Materials Science and Engineering **96** (2015) 012057 doi:10.1088/1757-899X/96/1/012057

Inspection of Construction Works According to Polish Construction Law

A Czemplik[1]
Wroclaw University of Technology,
Wybrzeże Wyspiańskiego 27, 50-370 Wroclaw, Poland

E-mail: andrzej.czemplik@pwr.edu.pl

Abstract. Construction regulations are still different in many European countries, even though the European Union directives have unified many acts for construction works and construction products in the member countries. The scheme of the construction process presented in the paper could be valid for most countries, despite of detailed regulations of legal systems. The number of construction regulations to be followed in order to get the Construction Permit in Poland is rather big, so the time between the start of the investment process and the day when the Construction Permit is issued could be several months. Only licensed professional engineers can play the role of site managers, site inspectors and designers, registered for the given construction project. Duties and responsibilities (civil liability) of these engineers are strictly defined by regulations. The obligatory inspection of construction works should be executed by the licensed site inspectors. Moreover, the works can be incidentally inspected by Authority, banks, insurance companies or designers. Foreign designers and foreign site engineers in order to be allowed by respective Authority to play official roles on Polish construction sites should present documents proving that they can do the same jobs in their countries as per regulations obligatory there.

1. Polish construction law

According to the European Attractiveness Survey 2014 prepared by Ernst&Young, Poland is the most attractive country for investment in Central and Eastern Europe. At the end of 2013, Foreign Direct Investments in Poland totaled at EUR 160.5 bn [12]. Great number of new construction projects have been performed in the recent years. So, international investors like to be familiar with all the fundamental roles of managing the construction projects in Poland. The objective of this paper is to show the basic roles of construction project inspection as per Polish regulations, presented from project management perspective. It is worth to notice that there is a big diversity with regard to

[1] To whom any correspondence should be addressed.

construction project supervision regulations in the EU countries. According to the World Bank Group ranking, in 2015 Poland has been classified 137 (out of 189 economies ranked) in the category "Dealing with Construction Permits" [4]. Fortunately, the number of construction regulations obligatory in Poland is decreasing year by year. On the other hand, organizational scheme of the construction project as normally performed in Poland is similar to the schemes in other countries. Its simplified proposal is presented in figure 1. As the effect of the big number of various obligatory documents and reports to be elaborated and approved by respective authorities in order to submit the application for the Building Permit, the whole project preparation phase usually takes several months.

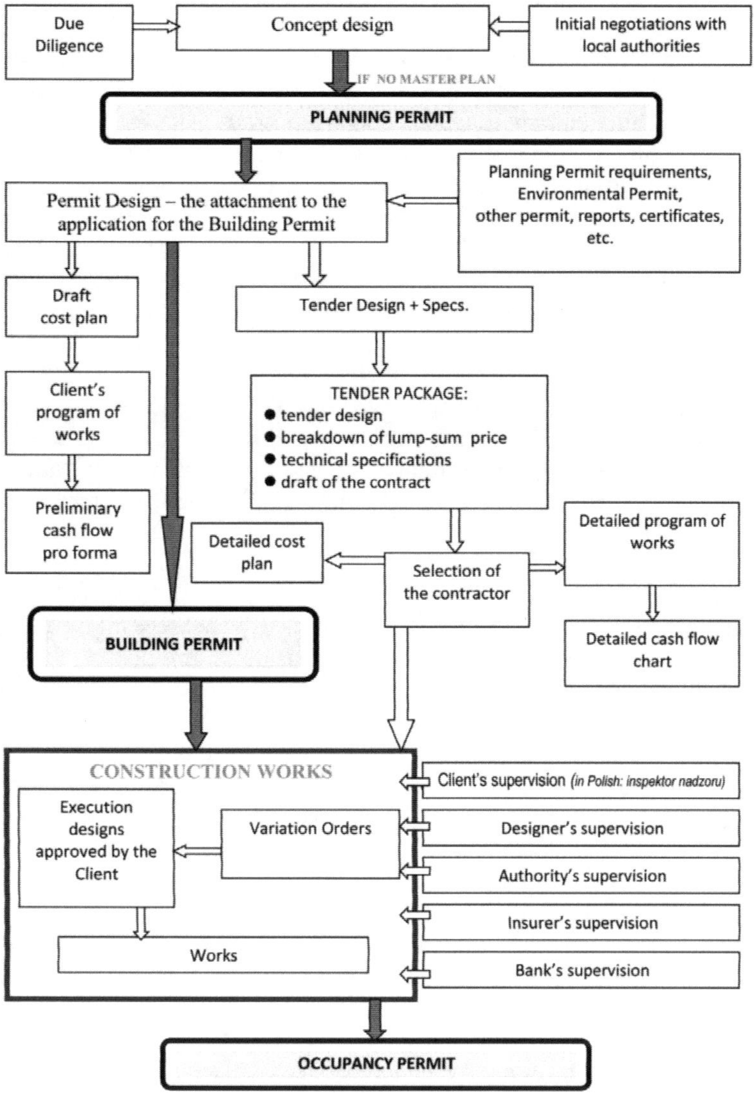

Figure 1. Simplified organizational scheme of the construction project.

In general, the Client prior to submitting the application for the Building Permit should collect a lot of approvals and other respective administration permits. During the whole preparation phase as well as during the construction phase, the Client should fulfil not only the Construction Law regulations, but also regulations of other laws regarding the environment, energy, wastes, public utilities, fire protection, etc. All in all, the number of formalities, approvals, and reports required makes the process really time consuming to a considerable extend. Clients used to order all the Building Permit formalities and the required documentation with an experienced architect or engineer who respectively authorized by the Client acted as the leading designer as well as the client representative. Despite time-consuming dealing with construction permits, the increase of new construction investments is considerable in Poland [14].

Very specific element of Polish Construction Law concept is civil liability of key construction process participants, such as the site managers, designers, and site inspectors, for their site job performance [1, 7]. The mentioned individuals must be respectively licensed as Professional Engineers by the Polish Chamber of Civil Engineers. So, in the light of Construction Law, any failures, accidents, use of wrong materials, design mistakes, etc. it is not corporate responsibility of the contracting company, but rather civil responsibility of the respective individual licensed Professional Engineers, officially registered as paying duties for a given project. The contracting companies face contractual responsibility correspondingly to their contracts for the construction works. The Building Permit design for medium and large size of construction projects must be elaborated by licensed designers (architect, structural, electrical, HVAC, etc.). Construction works of such projects should be managed by a licensed site manager and should be supervised by licensed site inspectors. Very common practice is that the site manager is employed by the Contractor while the site inspector – by the Client, however, other scenarios are not excluded as per the Construction Law. However, the key factor is that the mentioned licensed professionals do perform their jobs fully independently and they are responsible as individuals for their construction site decisions, even if they are employed by the contractor or the client.

2. Inspection of construction works

Construction inspections in public and private projects have been investigated by many authors in several directions, like planning and coordinating of inspection activities [6, 10], automation of the processes supporting inspection [2, 5] or developing models and systems of inspection process to avoid defective construction [3, 13]. Other research projects are those focused on detailed inspection actions required for a particular type of works, like concrete works, piling, etc. [8]. Polish construction regulations do not present any restrictions for using the newest research achievements in planning, automation or modelling of the inspection process. According to regulations, the inspection team should secure full agreement of everyday construction works with construction standards and design specifications.

In Poland, the main obligatory inspection of construction works during the whole project duration is performed by the licensed site inspectors who inspect civil, electrical, HVAC/mechanical and road works- depending on the type of their licenses. They are registered in the local Authority by the Client as site inspectors of the given construction project. Their main duty is to check systematically the works before these works become covered by the next phase works. So, for example, each reinforcement should be checked before the concrete-mix is placed into the formworks. Also, site inspectors should check if all the construction materials used by the Contractor comply with the project specifications and with the Polish Construction Materials regulations presenting conditions for use of products in construction (like certificates, declarations of performance, etc.). Finally, the inspectors check if all the works are being executed as per the Permit Design, quality standards and as per the obligatory regulations of the Construction Law. Schematic scope of main duties of site inspectors is presented in figure 2. The site inspectors represent the interests of the Client and also they represent public interests checking if works follow national construction and safety regulations.

For some specific or complex projects, the Authority can also demand the additional inspection to be performed by designers. In case of such a demand, the designer's inspection is obligatory. But, also

Figure 2. The scheme of main obligations of site inspectors.

in case it is not obligatory by law, the designer's inspection can be always demanded by the Client. The objective of designer's inspection is to check if Permit Design is fully followed during all site works. Also, the designer inspecting the site advises in change order acceptance and makes detailed design sketches or additional design drawings, if needed.

Several Authority departments (fire, sanitary, labor, safety, etc.) are entitled to inspect the construction site based on general law regulations. These are incidental inspection visits, usually rather detailed though, and they may result in fines addressed to site managers or site inspectors. The Authority departments do not have obligations to visit each site, so many construction sites are never inspected by the Authority department representatives.

Intensity of insurance inspection activities depends on the worth of the insurance policy. Some of insurance companies employ experienced engineers for systematic inspection visits to the site in order to get confirmation about the risk allocated to the works performed. It is normal that the insurance company is interested to reduce risk of any future claims which could be received from the Client / Contractor. So, the duty of insurance inspection is to warn about any case of risk which was not considered while the insurance policy CAR/EAR was signed. Thus, the inspectors are focused on looking for opportunities to reduce any construction risks on site.

Construction investments are usually financed, at least partially, by bank credits. Banks send their inspectors every two weeks to the site to check if their money is invested as per the credit agreement. Bank inspectors are also called bank monitors in project management terminology. They are experienced engineers who are able to identify true cost of works completed to date of visit, with very good tolerance.

The Polish Chamber of Civil Engineers is entitled to approve the foreign engineer, who has no Polish license for site inspection in Poland. In general, the considered engineer must prove that she/he is entitled to do the equivalent inspection job in the home country, according to local regulations.

The official register book called the Site-Log-Book is the main document issued by the Authority for each construction project to record all important phases of the project. The licensed site manager, registered at the given project, is formally responsible for the Site-Log-Book. All comments, approval or rejection notes should be written in the S-L-B by site inspectors just after each inspection visit.

According to the data of General Office of Building Control of Poland, 193,172 construction permits were issued during the year 2014 in Poland, so this is also the number of commenced construction projects. In the same year, about 115,000 engineers licensed to all the fields, who play a role in the construction process, have been the members of the Polish Chamber of Civil Engineers [11]. So, it is a good balance between market needs and the number of licensed inspectors available, considering that the site inspector's job is usually not a full-time assignment and very often is performed by freelance engineers. So, one engineer can try to perform the duties of an inspector for several sites simultaneously. The average cost of the overall construction project site inspection including all needed inspectors is negotiable and nowadays constitutes about 1.5 – 3.0 % of the cost of all site works, depending on project complexity, duration, etc.

3. Conclusions

Construction process administrative procedures are subject to permanent improving legislation modifications to meet the development of the construction industry in Poland. Recommendations on improvements usually concern administrative procedures and not the concept of the site inspection practice, which has been well established in the last years [9]. The recent discussions have been conducted in Polish construction environment and also at the Polish Parliament bodies about changes needed in construction regulations. The debate has not contested the concept of site inspection. So, the construction site works should be systematically inspected by licensed, independent professional engineers, and members of the Polish Chamber of Civil Engineers. Each inspector is responsible for respective type of works (civil, electrical, HVAC/mechanical, road). Construction regulations secure good and efficient site practice, which help to ensure high quality standard of construction works. The obligation of construction site inspection in Poland executed by independent licensed engineers with civil liability for job performance is not a common European practice.

4. References

[1] Bielecka D and Bielecki M 2014 *The Construction Law. Prawo Budowlane* BECK
[2] Boukamp F and Akinci B 2007 Automated processing of construction specifications to support inspection and quality control *Automation in Construction* **17** 90–106
[3] Chien-Ho K and Shun-Chi L 2014 Enhancing submittal review and construction inspection in public projects *Automation in Construction* **44** 33–46

[4] *Ease of Doing Business Ranking 2015* World Bank Group webside [Available]: www.doingbusiness.org
[5] Gordon C, Akinci B and Garrett Jr J H 2008 Automated planning support for on-site construction inspection *Automation in Construction* **17** 705–718
[6] Gordon C, Akinci B and Garrett Jr J H 2009 Characterization and search of construction inspection plan spaces developed using a component-based planning approach *J. of Computing in Civil Engineering* **23** (4) 211–220
[7] Kapliński O 2008 Development and usefulness of planning techniques and decision–making foundations on the example of construction enterprises in Poland *Technological and Economic Development of Economy* **14** (4) 492–502
[8] Karlowski A and Paslawski J 2008 Monitoring of construction processes in the variable environment *Technological and Economic Development of Economy* **14** (4) 503–517
[9] Kietliński W and Janowska J 2015 *Proces inwestycyjny w budownictwie (Investment Process in Construction)* Wyd P W, Warsaw 2015
[10] Ochoa S F, Bravo G, Pino J A and Rodríguez-Covili J 2011 Coordinating loosely-coupled work in construction inspection activities *Group Decision & Negotiation* **20** 39–56
[11] *Polish Chamber of Civil Engineers* [Available]: piib.org.pl
[12] *Polish Information and Foreign Investment Agency* [Available]: www.paiz.gov.pl
[13] Sawhney A, Walsh K D, Bashford H H and Palaniappan S 2009 Impact of inspected buffers on production parameters of construction processes *J. of Construction Engineering and Management* **135** (4) 319–329
[14] *The Investment Potential Catalogue of Polish Cities* PAIiIZ, Warsaw 2013

IOP Conf. Series: Materials Science and Engineering **96** (2015) 012058 doi:10.1088/1757-899X/96/1/012058

Damage Identification in Beam Structure using Spatial Continuous Wavelet Transform

R Janeliukstis[1,3], S Rucevskis[1], M Wesolowski[2], A Kovalovs[1] and A Chate[1]
[1]Riga Technical University, Institute of Materials and Structures, Kalku Str. 1, Riga, LV-1658, Latvia
[2]Department of Structural Mechanics, Faculty of Civil Engineering, Environment and Geodesy, Koszalin University of Technology, Koszalin, Poland

E mail: Rims.Janeliukstis_1@rtu.lv

Abstract. In this paper the applicability of spatial continuous wavelet transform (CWT) technique for damage identification in the beam structure is analyzed by application of different types of wavelet functions and scaling factors. The proposed method uses exclusively mode shape data from the damaged structure. To examine limitations of the method and to ascertain its sensitivity to noisy experimental data, several sets of simulated data are analyzed. Simulated test cases include numerical mode shapes corrupted by different levels of random noise as well as mode shapes with different number of measurement points used for wavelet transform. A broad comparison of ability of different wavelet functions to detect and locate damage in beam structure is given. Effectiveness and robustness of the proposed algorithms are demonstrated experimentally on two aluminum beams containing single mill-cut damage. The modal frequencies and the corresponding mode shapes are obtained via finite element models for numerical simulations and by using a scanning laser vibrometer with PZT actuator as vibration excitation source for the experimental study.

1. Introduction

Modern engineering structures, including but not limited to civil buildings, bridges, dams, automotive and aerospace facilities have to maintain their integrity and functionality under severe environmental conditions. Their failure often leads to tragic consequences and therefore structures have to undergo regular inspections. Rigorous inspection of structures on a regular basis is usually a time-consuming and costly procedure; therefore non-destructive structural health monitoring (SHM) methods have become an important research area in civil, mechanical and aerospace engineering communities.

In the recent years, various vibration-based damage detection methods have been proposed for SHM. Many of them use various transformations of measured dynamic response of a structure. Dynamic responses, which in many cases can be easily obtained, offer damage information such as the

[3] To whom any correspondence should be addressed.

location and severity. These methods are based on the fact that dynamic characteristics, i.e., natural frequencies, mode shapes and modal damping, are directly related to the stiffness of the structure. Therefore, a change in natural frequencies or a change in mode shapes will indicate a loss of stiffness. Extensive literature reviews of the state-of-the-art in the methods for detecting, localizing, and characterizing damage by examining changes in the dynamic response of a structure can be found in [1,2]. Many studies [3-9] have shown that mode shapes and corresponding mode shape transformations are highly sensitive to damage and can be used for detecting and quantifying damage. One of the most promising techniques of processing dynamic response for damage identification is a wavelet transform.

Wavelet theory has been developed since the beginning of the 20th century by Haar, but it did not result in a unified theory. A current wavelet analysis as we-know-it was introduced by Grossman and Morlet in 1984. Later, in 1989 Mallat founded wavelet theory for signal decomposition [10]. Wavelet transform technique originated in the 1990's and was mainly used for signal singularity detection [10], signal denoising, image compression [11]. Later in the 1990's wavelet transform technique was used on vibrational data for damage detection for the first time by Newland to study the effects of building vibrations, caused by underground trains and road traffic [10, 12, 13].

Many researchers have been studying damage detection in beams using spatial wavelet transform technique [12-18]. Studies in [12] showed that local perturbations in beam deflection profile are induced by crack and amplified through use of wavelet transform. Researchers in [18] suggested that wavelet transform is also capable of detecting two cracks in beams. Interesting studies in [13] proposed to classify wavelet-based methods for damage detection into 3 groups: 1) variation of wavelet coefficients [14, 15], 2) local perturbation of wavelet coefficients in space domain, 3) reflected wave caused by local damage. The first of these methods is used to find the existence and severity of damage and is based on the fact that variation of wavelet transform coefficients is caused by change of modal parameters of the structure. The second method is used to localize the damage in structures by detection of irregularity of wavelet transform coefficients near the damage site. The third method is used to measure the severity as well as location of the damage. It is based on the analysis of the wave reflected from local damage in the structure. Work similar to the third method was done in [19]. Authors suggested wavelet transform technique for signal processing to distinguish between the changes in wave propagation, caused by local discontinuities and cracks in beams and rods.

In this paper, the applicability of spatial continuous wavelet transform (CWT) technique for damage identification in the beam structure is analyzed by application of different types of wavelet functions and scaling factors. To examine limitations of the method and to ascertain the sensitivity of the method to noisy experimental data, several sets of numerically simulated data were tested. A broad comparison of ability of different wavelet functions to locate damage in beam structure is given. Effectiveness and robustness of the proposed algorithms are demonstrated experimentally on two aluminum beams containing single mill-cut damage.

2. Continuous wavelet transform

Wavelets are special functions, which are generally used to analyze different signals in time domain. Wavelet transform is simply a mathematical method to transform the original signal into a different domain where additional data analysis becomes possible, therefore damage-affected signal portion is revealed. Spatial analysis is obtained by simply replacing time with a spatial coordinate, say, x. This gives rise to signal $f(x)$ [12]. If $\psi(x)$ is a wavelet function, also called *mother wavelet*, which, in general, is complex-valued, then the set of wavelet family functions $\psi_{s,t}(x)$ is created by *shifting* (*translating*) (parameter t) and *scaling* (parameter s) the $\psi(x)$ [14, 20].

$$\psi_{s,t}(x) = |s|^{-1/2}\psi\left((x-t)s^{-1}\right) \tag{1}$$

Parameter s is essential in noise reduction [16], and it is a real number. If $0 < s < 1$, the function is expanded, if $s > 1$, it is compressed [21]. In time domain s represents the signal's frequency content and t – location of wavelet in time.

2.1. Benefits with respect to Fourier Transform

In contrast to Fourier transform (FT), wavelet function has time locality as well as frequency locality [20]. If N is the data to be transformed, the required number of operations for CWT with orthogonal wavelets mentioned in [20] is $O(N)$ instead of $O(N\cdot\ln(N))$ for fast FT, which leads to much faster data transformation.

2.2. Wavelet properties

Compact support – wavelet function's average value is 0 [14, 20] and it has a finite length [14], thus

$$\int_{-\infty}^{\infty}\psi_{s,t}(x)dx = 0 \tag{2}$$

therefore wavelet function has at least some oscillations [20]. Another wavelet property is continuity. Continuous wavelet transform (CWT) is given by [14, 20]

$$W_{s,t} = \int_{-\infty}^{\infty} f(x)|s|^{-1/2}\psi^{*}\left((x-t)s^{-1}\right)dx = \int_{-\infty}^{\infty} f(x)\psi_{s,t}{}^{*}(x)dx \tag{3}$$

where asterisk denotes complex conjugation. We will deal only with real-valued wavelets, which are used in case of low wavelength signals [21]. Using this formula, CWT coefficients are calculated, which are extremely sensitive to any discontinuities and singularities of signal $f(x)$, therefore damage due to a sudden loss of stiffness can be detected in mode shapes that have large amplitude wavelet coefficients. It is a damaged location. This perturbation of wavelet coefficients due to this damage is clearer in the finest scales of the CWT [10].

3. Damage detection algorithm

CWT for mode shapes is depicted as follows (recall equation (3))

$$DI_i^n = W_{i\ s,t}^n = \int_L w_i^n \times \psi_{s,t}(x)dx \tag{4}$$

where L is the length of the beam, w^n is transverse displacement of the structure, n is a mode number, i is number of grid point in x direction, respectively and $\psi_{s,t}(x)$ is a particular wavelet function, which was chosen to be Daubechies, Gaussian and Symlet of orders 2, 4 and 8 (Db2, Db4, Db8, Gs2, Gs4, Gs8), Coiflet of orders 2 and 4 (Cf2, Cf4), Morlet (Morl), Meyer, Haar, and Mexican hat (Mex). DI_i^n is a damage index for each mode.

In practice, experimentally measured mode shapes are inevitably corrupted by measurement noise causing local perturbations into the mode shape, which can lead to false peaks in damage index profiles. These peaks could be mistakenly interpreted as damage or they could mask the peaks induced by real damage and lead to false or missed detection of damage. To overcome this problem, it is proposed to summarize results for all modes. The summarized damage index then is defined as the average summation of damage indices for all modes N, normalized with respect to the largest value of each mode

$$DI_i = \frac{1}{N} \times \sum_{n=1}^{N} \frac{DI_i^n}{DI_{i,\max}^n} \tag{6}$$

4. Numerical simulation

To validate the effectiveness of the proposed algorithm, numerical modal analysis based on finite element (FE) method was conducted by using the commercial FE software ANSYS. An aluminum beam (Beam 1) with single mill-cut damage is examined. The geometrical configuration of the beam is shown in figure 5. Mill-cut damage with a depth of 2 mm and width of 50 mm is introduced at a distance of 750 mm from one edge of the beam. FE model of the beam consists of 2D beam elements. Each node has 3 degrees of freedom, namely translations along the X and Y axes and rotation along the Z axis. Beam 1 was constructed by means of 125 equal length elements (i = 126 nodes). The elastic material properties are taken as follows: Young's modulus E = 69 GPa, Poisson's ratio v = 0.31, and the mass density ρ = 2708 kg/m^3. For the healthy beam, a constant flexural stiffness EI is assumed for all elements, while the damaged beam is modelled by reducing the flexural stiffness of the selected elements. Reduction of flexural stiffness is achieved by decreasing the thickness of elements in the damaged region of the beam. The modal frequencies and corresponding mode shapes for the first 15 flexural modes were calculated.

In order to compare the sensitivity of the algorithm to noisy experimental data, an artificial white Gaussian noise was added to the numerically simulated mode shapes

$$w_n'^i = w_n^i \cdot (1 + \delta \times (2r - 1)) \tag{7}$$

where prime indicates noisy mode shapes, δ = 0 %, $1 \cdot 10^{-3}$ %, $1 \cdot 10^{-2}$ %, $1 \cdot 10^{-1}$ %, 1 %, 5 % is the level of random noise and r are uniformly distributed random values in the range (0, 1).

In order to study situations from real life when it is either not possible or not needed to equip the structure with many sensors, different numbers of data points were examined – simulated data with 149 entries was divided by p = 1, 2, 5 and 10. Based on these considerations, the following test cases were studied (table 1):

Table 1. Simulated cases for damage detection.

Noise level	reduction of used data			
(%)	$p = 1$	$p = 2$	$p = 5$	$p = 10$
$\delta = 0$	Case 1.1.	Case 1.2.	Case 1.3.	Case 1.4.
$\delta = 0.001$	Case 2.1.	Case 2.2.	Case 2.3.	Case 2.4.
$\delta = 0.01$	Case 3.1.	Case 3.2.	Case 3.3.	Case 3.4.
$\delta = 0.1$	Case 4.1.	Case 4.2.	Case 4.3.	Case 4.4.
$\delta = 1$	Case 5.1.	Case 5.2.	Case 5.3.	Case 5.4.
$\delta = 5$	Case 6.1.	Case 6.2.	Case 6.3.	Case 6.4.

4.1. Results and discussion

According to [22], the damage indices, determined for each element are then standardized and a concept of statistical hypothesis testing is applied to classify damaged and healthy elements and to localize damage depending on the pre-defined damage threshold value. In hypothesis testing, the null hypothesis and alternate hypothesis are defined as H_0 (element i of the structure is healthy); H_1 (element i of the structure is damaged), respectively. In order to test the hypothesis, the summarized damage index, given in equation (6), is standardized

$$Z_i = \frac{DI_i - \mu_{DI}}{\sigma_{DI}} \tag{8}$$

where μ_{DI} and σ_{DI} are mean value and standard deviation of damage indices in equation (6), respectively. The decision for the localization of damage is established based on the level of significance used in the hypothesis test, which can be determined from a pre-assigned classification criterion: choose H_0 if $Z_i < C_r$ or choose H_1 if $Z_i \geq C_r$, where C_r is a threshold value. Typical values of C_r, widely used in literature are 1.28, 2 and 3 for 90 %, 95 % and 99 % confidence levels for the presence of damage.

To quantify the reliability of wavelets to identify damage location, a new parameter, called „*damage estimate reliability*" (DER) is introduced and calculated as follows:
The whole interval along the axis of the beam (x axis) is split into 3 parts:
the first part is the one in which $x < 750$ mm (part a), the second part is 750 mm $< x <$ 800 mm (part b) and the third part is $x >$ 800 mm (part c).
In each of these parts SDI values from equation (8) of a respective wavelet are summed and divided by the number of data points in this particular interval, giving average amplitude of SDI ($\overline{Z_i}$) in a respective part. DER is equal to average SDI in middle part (the damaged area or part b in our case) divided by average SDI in all parts combined. It is expressed in percent as:

$$DER_i = \frac{\overline{Z_i}(b)}{\overline{Z_i}(a) + \overline{Z_i}(b) + \overline{Z_i}(c)} \times 100\% \tag{9}$$

By calculating average DER for each wavelet over all the cases, the best and the worst wavelets were determined (table 2). The best wavelets in each case are shown in boldface, whereas the worst wavelets are underlined.

Table 2. DER calculation results (in %) for Beam 1 with CBC on numerical data for the chosen wavelets at different noise levels and the number of data points used (bold – best, underlined – worst).

	Cf2	*Cf4*	*Db2*	*Db4*	*Db8*	*Gs2*	*Gs4*	*Gs8*	*Haar*	*Meyer*	*Mex*	*Morl*	*Sym4*	*Sym8*
1.1.	84.6	80.0	80.9	83.9	85.3	77.5	86.9	80.1	**87.7**	72.3	<u>72.2</u>	80.9	85.6	79.4
1.2.	85.3	83.8	80.1	79.5	68.8	70.2	82.5	84.0	86.5	79.3	<u>65.0</u>	83.5	**86.9**	84.1
1.3.	77.6	76.6	76.7	<u>30.2</u>	41.8	**83.3**	81.3	77.0	54.5	74.0	81.7	75.7	79.0	77.0
1.4.	69.4	67.8	59.9	27.3	46.7	**73.5**	69.3	65.9	<u>24.1</u>	67.8	72.4	66.2	65.9	69.8
<1.>	79.2	77.1	74.4	<u>55.2</u>	60.6	76.1	**80.0**	76.7	63.2	73.4	72.8	76.6	79.4	77.6
2.1.	84.6	79.9	80.9	83.9	85.3	77.5	86.9	80.0	**87.7**	72.3	<u>72.2</u>	80.8	85.6	79.3
2.2.	85.3	83.8	80.1	79.5	68.8	70.2	82.5	84.0	86.5	79.3	65.0	83.4	**86.9**	84.1
2.3.	77.6	76.6	76.7	30.2	41.8	83.3	81.3	77.0	54.5	74.0	81.7	75.7	79.0	77.0
2.4.	69.4	67.8	59.9	27.3	46.7	73.5	69.3	65.9	24.1	67.8	72.4	66.2	65.9	69.8
<2.>	79.2	77.0	74.4	<u>55.2</u>	60.6	76.1	**80.0**	76.7	63.2	73.4	72.8	76.5	79.4	77.5
3.1.	84.4	80.4	80.8	83.8	85.4	77.5	86.9	80.3	87.7	72.4	72.2	81.1	85.7	80.0
3.2.	85.3	83.7	80.1	79.4	68.8	70.2	82.5	83.9	86.5	79.3	65.0	83.1	86.9	84.0
3.3.	77.6	76.5	76.7	30.2	41.5	83.3	81.3	77.0	54.5	73.9	81.7	75.6	79.0	77.0
3.4.	69.4	67.8	59.9	27.3	46.8	73.5	69.3	65.9	24.1	67.9	72.4	66.3	65.9	69.8
<3.>	79.2	77.1	74.4	<u>55.2</u>	60.6	76.1	**80.0**	76.8	63.2	73.4	72.8	76.5	79.4	77.7
4.1.	84.6	76.5	79.8	83.5	84.2	77.6	87.6	77.3	87.8	71.8	72.2	76.5	85.3	76.1
4.2.	84.5	82.8	79.9	78.9	65.4	70.2	82.4	83.1	86.5	79.0	64.9	82.4	86.3	83.3
4.3.	77.5	76.4	76.5	30.2	41.2	83.3	81.3	76.9	54.3	73.6	81.7	75.0	78.8	76.9
4.4.	69.3	67.5	59.8	27.7	46.3	73.5	69.3	65.5	24.1	67.9	72.4	66.1	66.0	69.5
<4.>	79.0	75.8	74.0	<u>55.1</u>	59.3	76.2	**80.1**	75.7	63.2	73.1	72.8	75.0	79.1	76.4
5.1.	82.8	75.7	79.7	69.8	74.9	76.6	85.8	76.3	88.3	77.3	72.0	72.4	84.1	75.3
5.2.	84.8	83.8	77.0	79.5	67.5	68.6	81.0	83.5	86.4	79.2	64.4	84.3	86.4	84.2
5.3.	79.0	78.6	77.0	23.4	48.0	83.3	81.6	78.9	54.5	75.7	81.7	77.0	79.3	78.5
5.4.	68.1	64.6	60.0	31.8	40.3	73.5	68.6	61.7	23.1	66.5	72.7	64.9	66.2	66.7
<5.>	78.7	75.7	73.4	<u>51.1</u>	57.7	75.5	**79.3**	75.1	63.1	74.7	72.7	74.7	79.0	76.2
6.1.	50.1	34.2	65.9	40.8	33.3	74.7	72.6	33.9	87.8	34.7	71.7	35.0	55.6	34.5
6.2.	81.0	81.2	75.0	76.4	57.3	65.0	76.9	81.4	86.3	76.0	64.2	75.4	83.5	81.6
6.3.	79.0	80.7	74.4	23.6	47.8	83.4	80.9	80.7	53.9	77.0	81.6	79.4	78.7	80.8
6.4.	63.3	63.1	59.8	39.2	46.0	74.3	69.3	60.4	19.3	59.6	72.0	56.5	63.1	66.0
<6.>	68.4	64.8	68.8	<u>45.0</u>	46.1	74.3	**74.9**	64.1	61.8	61.8	72.4	61.6	70.2	65.7
Aver	77.3	74.6	73.2	<u>52.8</u>	57.5	75.7	**79.0**	74.2	62.9	71.6	72.7	73.5	77.7	75.2
Δ (±)	3.9	3.7	3.7	2.7	2.9	3.8	4.0	3.7	3.1	3.6	3.6	3.7	3.9	3.8

- The best wavelet over all of the analyzed cases turned out to be Gs4 with an average DER value of 79 %. It is important to note that peaks at random positions with relatively large amplitudes appeared in cases, corresponding to noise level of 5 %. Also different results were obtained, performing the analysis over several repetitions, which was due to the random nature of noise. Therefore a relative error of 5 % was attributed to average values of DER, depicting that these values are changing, conducting the experiment several times (table 2). As a result, the wavelets with best average DER that lies in the interval of 75 % - 83 % (79 % ± 4 %), apart from Gs4, are also all other wavelets, except for Db4, Db8 and Haar.

IOP Conf. Series: Materials Science and Engineering **96** (2015) 012058 doi:10.1088/1757-899X/96/1/012058

- Haar wavelet dominated the damage detection at larger data sets. It was tightly located in the zone of damage and even increasing noise levels did not affect it significantly. As the number of the examined data points was reduced, Haar wavelet shifted to the larger x values – just out of the damaged zone.
- Db2 and Sym2 are equal at capturing damaged location, therefore Sym2 was not included in table 2. Noticeable tendency for Db4 is to deteriorate the damage localization effect as data set shrinks, which also makes it the overall least valuable wavelet in damage detection.
- Only finer scaling parameters s are appropriate at damage localization ($s = 2$) – at larger s values wavelets are performing poorly ($s = 32$) (not shown in this study).

SDI, using the best and worst wavelets – Gs2 and Db4, respectively, was plotted for original data amount and noise levels 0.1 % and 5 % (figures 1 – 4). The location and size of the introduced damage is indicated by red dotted lines.

As one can see in these figures, data set with more points is more sensitive to noise – there are many peaks with random amplitude due to noise, scattered all over the x axis in figure 1 (right), but, as data amount shrinks, cases with 0.1 % and 5 % noise levels look practically the same (this is especially true for case 6.4. in figure 4 (right)). Also, as one might expect, the reduction of the examined data points deteriorates damage identification. This is true by looking at these figures – for original data the best and worst wavelets both manage to capture that location of damage and as the examined data are reduced, the worst wavelet shifts out of the damage zone, and also the largest peaks for both wavelets become less localized in the zone of interest.

The conclusion is that there is a trade-off between sensitivity to noise and damage detection accuracy. We suggest that the best case corresponds to data, reduced by a factor of 2 (figure 2). Damage localization is almost as good as for original data and corruption due to noise is far less. As data is reduced further, damage localization worsens.

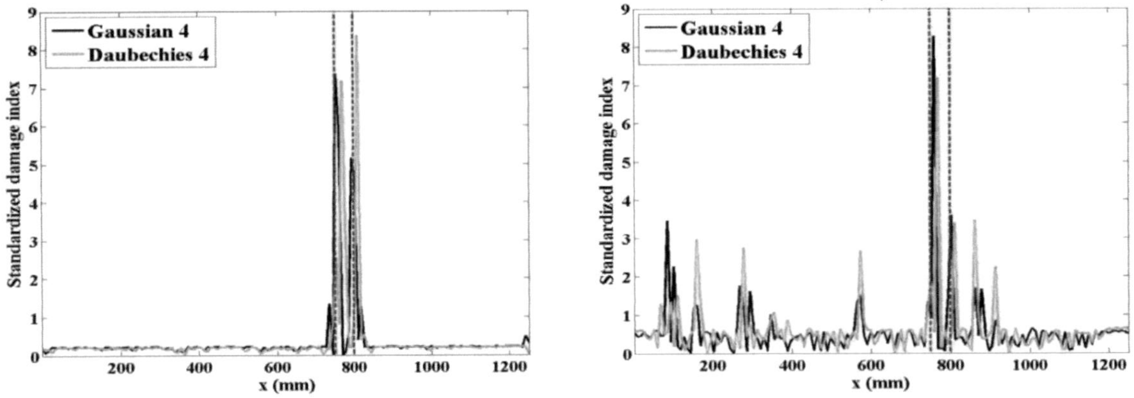

Figure 1. Beam 1, CBC, SDI. left: Case 4.1., right: Case 6.1. (numerical data).

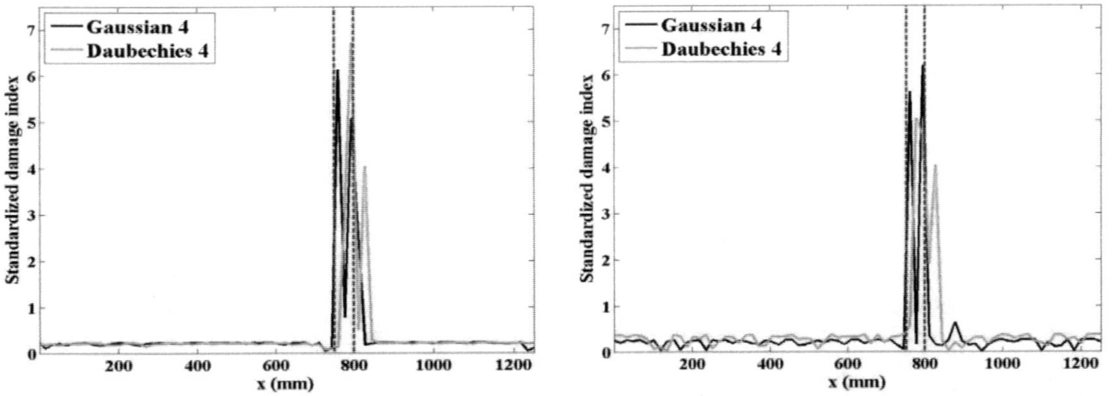

Figure 2. Beam 1, CBC, SDI. left: Case 4.2., right: Case 6.2. (numerical data).

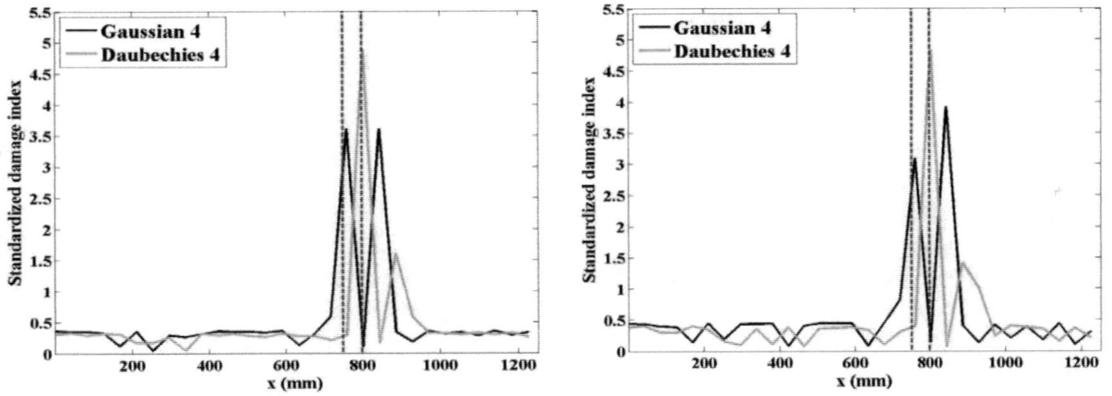

Figure 3. Beam 1, CBC, SDI. left: Case 4.3., right: Case 6.3. (numerical data).

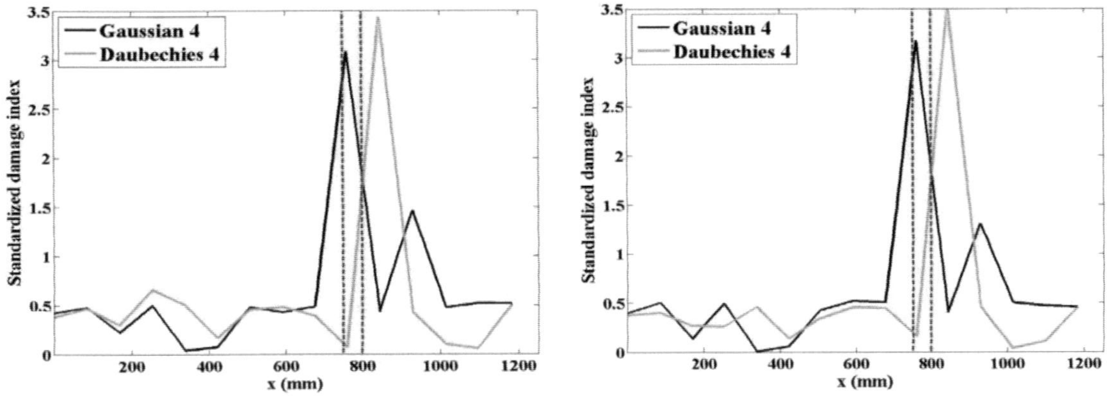

Figure 4. Beam 1, CBC, SDI. left: Case 4.4., right: Case 6.4. (numerical data).

5. Experiment

Specimens under test were 2 aluminum beams with the following dimensions – $length_1$ = 1250 mm, $length_2$ = 1500 mm, $width_{1,2}$ = 50 mm, $height_{1,2}$ = 5 mm. A single mill-cut damage ($width_1$ – 50 mm, $width_2$ = 100 mm, $depth_{1,2}$ – 2 mm) was introduced in a location at 750 mm from one edge of a *Beam 1* and at 950 mm from one edge of *Beam 2* (figure 5 (a)).

The modal frequencies and corresponding mode shapes of the beams are measured by a POLYTEC PSV-400-B scanning laser vibrometer. General experiment set-up consists of a PSV-I-400 LR optical scanning head equipped with a highly sensitive vibrometer sensor (OFV-505), OFV-5000 controller, PSV-E-400 junction box, a Bruel & Kjaer type 2732 amplifier, and a computer system with a data acquisition board and PSV software (figure 5 (b)).

PSV system requires defining the outer edges and geometry of the object and setting up a scanning grid, which consists of a number of points where vibrations are measured. To match the finite element model, 126 and 151 equally spaced scanning points are equally distributed to cover the area of *Beam 1* and *Beam 2*, respectively. Free-free (all edges free) boundary conditions (FBC) are simulated by suspending the beam with two thin threads under a frame. In order to simulate clamped-clamped (two ends fixed) boundary conditions (CBC) experimentally, two vices are used to fix the ends of the beam (10 mm from both sides) with the clamped torque of 20 Nm. The beams are excited by a periodic input chirp signal generated by the internal generator with a 1600 Hz bandwidth through a piezoelectric actuator discs. After the frequency response function and resonant frequencies have been obtained, the beam is excited by a periodic sine wave signal with a frequency corresponding to each resonant frequency in order to obtain the corresponding mode shape. When the measurement is performed in one point, the vibrometer automatically moves the laser beam to another point of the scan grid, measures the response using the Doppler principle and validates the measurement with the signal-to-noise ratio. The procedure is repeated until all scan points have been measured. The mode shapes are obtained by taking the Fast Fourier Transform of the response signal. The first 15 mode shapes were measured in total.

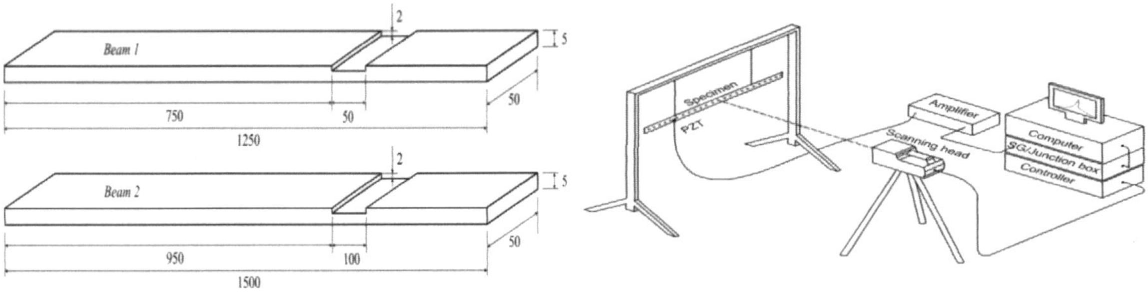

Figure 5. Left: geometry (in mm) of tested beams, right: experimental set-up.

5.1. Results and discussion

Experimental results for DER for both beams are shown in table 3. Differences between results for different boundary conditions were observed. While there was no correlation between best-worst wavelet sets of CBC case (Gs2-Morlet, respectively) and FBC case (Haar-Db2, respectively) for Beam 1, there was some correlation between different boundary conditions for Beam 2 in terms of best-worst wavelet, namely for both BC cases best wavelet was Mexican hat and worst was Cf2 (CBC) and Meyer (FBC).

Analysis for experimental data refers to case 1.1. for numerical simulations, therefore, for the purposes of objective comparison, only 1st row of table 2 should be compared to the experimental results. Note the fact that for both numerical simulation case 1.1 and the experimental results wavelet that performed the best in damage localization was Haar, only numerical simulation was done for Beam 1 with CBC, but the corresponding experimental result was obtained for FBC. As for Beam 2, Mexican hat performed the best, but this particular wavelet was the poorest in case 1.1. These opposite results were obtained for beams with only noticeable differences in their length, damage width and location.

Table 3. DER results (in %) on experimental data for both beams at different BC.

Wavelet	Beam 1		Beam 2	
	DER (CBC)	DER (FBC)	DER (CBC)	DER (FBC)
Coif2	41.72	37.57	22.36	32.21
Coif4	36.20	37.34	26.76	26.24
Db2	70.90	32.08	71.50	60.59
Db4	45.46	36.85	25.82	32.08
Db8	42.43	37.97	32.50	28.27
Gs2	**82.55**	58.95	80.21	69.26
Gs4	60.86	32.87	63.58	43.19
Gs8	38.02	37.91	28.70	25.67
Haar	69.51	**75.89**	66.09	55.65
Mexican hat	82.08	60.97	**80.58**	**69.87**
Meyer	38.92	36.86	39.68	24.83
Morlet	33.83	37.53	25.65	29.38
Sym4	43.85	37.63	32.04	29.35
Sym8	37.21	38.01	26.24	27.75

Based on table 3, results for the best (thick continuous black line) and the worst (thick continuous green line) wavelets were selected and plotted (figures 6-9). The best wavelets with the corresponding threshold are shown in the pictures to the right. Although the peak values occur at the pre-determined location, the large values also emerged at the boundaries of beams. The boundary distortion problem is caused by discontinuity of mode shapes at their ends due to boundary conditions. Commonly, two methods are used to reduce the boundary effect [5]. One method is to extend the mode shape data beyond the original boundary by cubic spline extrapolation based on points near the boundaries. The other method is simply to ignore those values near the boundaries by cutting them off or setting them to zeros. In this study the latter method is adapted – values of data points, corresponding to edges were set to zero for both beams. It should be noted that neither of both methods is capable of detecting damage close to boundaries, since both of them smooth out the information near the boundaries.

As one observes figures 6-9, it can be seen, that the algorithm works – CWT with the selected wavelet is an appropriate technique for damage detection – peaks with largest amplitudes are located between two red vertical lines (zone of damage). In figures with SDI threshold only, one peak remains in the zone of damage, indicating that statistical hypothesis approach (refer to equation (8)) is working.

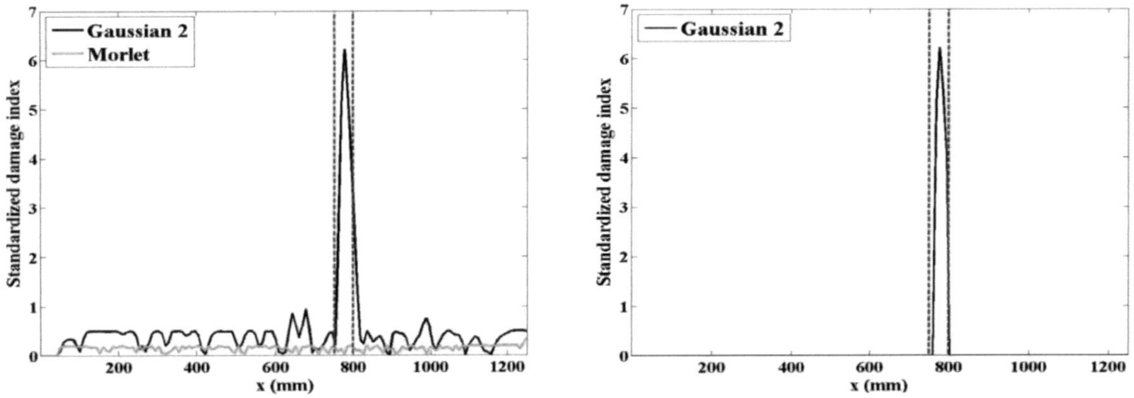

Figure 6. Left: SDI (Beam 1, CBC), right: SDI with threshold 3 for the best wavelet.

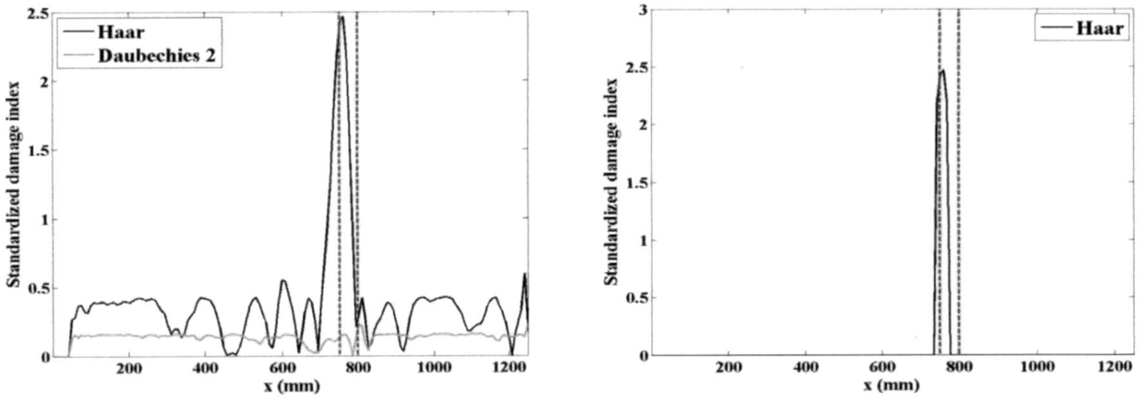

Figure 7. Left: SDI (Beam 1, FBC), right: SDI with threshold 2 for the best wavelet.

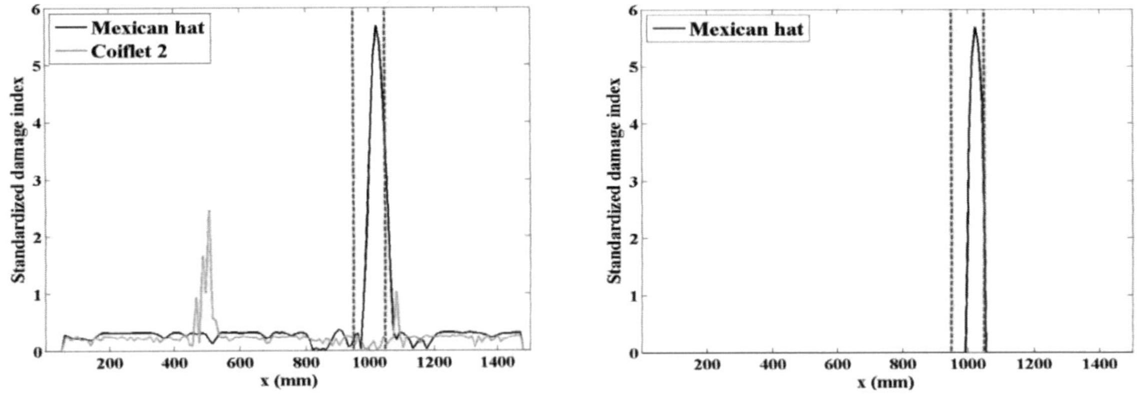

Figure 8. Left: SDI (Beam 2, CBC), right: SDI with threshold 3 for the best wavelet.

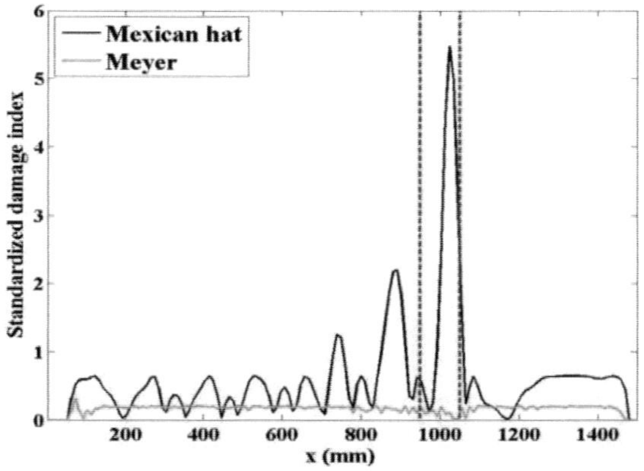

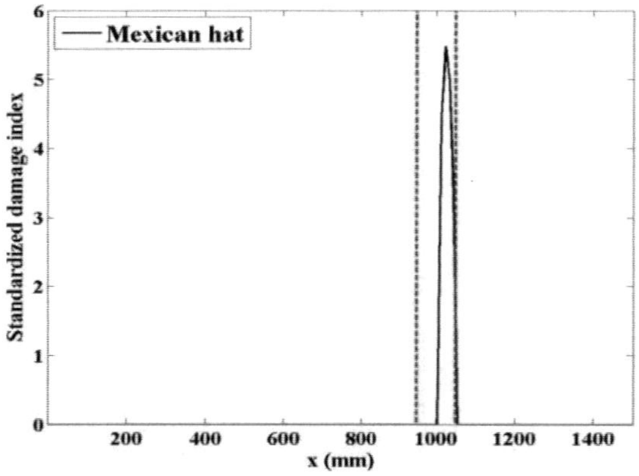

Figure 9. Left: SDI (Beam 2, FBC), right: SDI with threshold 3 for the best wavelet.

6. Conclusions

In this paper a numerical and experimental study on the applicability of mode shape based spatial continuous wavelet transform (CWT) technique for detection and localization of damage in beam structure is presented. The advantage of the proposed method is that it requires mode shape information only from the damaged state of the structure. Success of damage detection was demonstrated using statistical hypothesis approach, exploiting CWT. The obtained results show that the spatial CWT technique provides reliable information about the location of damage in two aluminum beams containing different length single mill-cut damage. To examine the limitations of the method, several sets of simulated data were analyzed. Overall, 15 different wavelets were tested at scale parameters 2, 8 and 32, of which scale 2 was suggested as the best. This analysis was performed at different noise levels, namely 0.001 %, 0.01 %, 0.1%, 1% and 5 % and different amount of data points, reduced by a factor of 2, 5 and 10.

Not all of the wavelets were equal at damage detection. Numerical simulation results were validated by experiment. Basing on the aforementioned results, the best and worst wavelets for both experimental and numerical cases are selected.

- At noise level 0 % and original data amount, Haar wavelet performed the best and it was only barely affected by increasing noise levels, on the other hand, data reduction affected its performance dramatically, so at the least amount of data Haar wavelet did not succeed in locating the damage. As for the experimental results, Haar wavelet, on the other hand, was the best at damage localization for Beam 1 with free – free boundary conditions (FBC).
- Gs4 was overall the best wavelet in terms of damage localization in numerical simulation, carried out for shorter beam – it possessed the best average DER, although several more wavelets came close. Its performance did deteriorate while working on fewer data points, but not significantly, and it performed satisfactory even at higher noise levels. As for experimental results, Mexican hat did the best job to locate the damage for longer beam for both clamped and free boundary conditions.
- Db4 was the worst wavelet for damage detection in terms of average DER for numerical simulation and also operated poorly on experimental data, however, the worst wavelets for experiment turned out to be Morlet (Beam 1, CBC), Db2 (Beam 1, FBC), Cf2 (Beam 2, CBC) and Meyer (Beam 2, FBC).

- As expected, higher noise levels corrupt data. However, as data amount is reduced, noise affects data to a lesser extent, meaning that "the more data points, the better" statement is not always true. On the other hand, data reduction deteriorates damage localization quality, therefore the optimal conditions (data amount-noise level) are suggested as investigated data decrease by a factor of 2, where due to fewer data points good damage localization still holds and is barely influenced by increasing noise levels.

Haar wavelet is suggested for damage detection if there are sufficient data points available even for noisy signals. If not Haar, then Mexican hat can also be used. On the other hand, Daubechies of order 4 is not recommended to use at all.

7. References

[1] Friswell M I 2007 Damage identification using inverse methods *Philos. Trans. R. Soc. A* **365** 393–310

[2] Fan W and Qiao P 2011 Vibration-based damage identification methods: a review and comparative study *Struct. Health Monitor.* **10** 83–111

[3] Wu D and Law S S 2004 Damage localization in plate structures from uniform load surface curvature *J. Sound Vib.* **276** 227–44

[4] Fan W and Qiao P 2009 A 2-D continuous wavelet transform of mode shape data for damage detection of plate structures *Int. J. Solids Struct.* **46** 4379–95

[5] Khiem N T and Tran H T 2014 A procedure for multiple crack identification in beam-like structures from natural vibration mode *J. Vibr. Control* **20** 1417–27

[6] Rucevskis S, Wesolowski M and Chate A 2009 Damage detection in laminated composite beam by using vibration data *J. Vibroeng.* **11** 363–73

[7] Rucevskis S, Wesolowski M and Chate A 2009 Vibration-based damage detection in a beam structure with multiple damage locations *Aviation* **13** 61–71

[8] Rucevskis S, Akishin P and Chate A 2015 Numerical and experimental study on application of mode shape curvature for damage detection in plate – like structures *Sol. St. Phen.* **220-221** 264–270

[9] Rucevskis S, Sumbatyan M A, Akishin P, Chate A 2015 Tikhonov's regularization approach in mode shape curvature analysis applied to damage detection *Mech. Res. Commun.* **65** 9–16

[10] Estrada R S 2008 Damage detection in bridges through vibration monitoring: evaluation and application, doctoral thesis, Universidade do Minho [Available]: http://hdl.handle.net/1822/9023

[11] Jerri A J 2011 *Introduction to Wavelets* First edition (Potsdam, New York: Sampling Publishing) pp 417–441

[12] Wu N and Wang Q 2011 Experimental studies on damage detection of beam structures with wavelet transform *Int. J. Eng. Sci.* **49** 253–261

[13] Kim H and Melhem H 2004 Damage detection of structures by wavelet analysis *Eng. Struct.* **26** 347–362

[14] Solis M, Algaba M and Galvin P 2013 Continuous wavelet analysis of mode shapes differences for damage detection *Mech. Syst. Signal Pr.* **40** 645 – 666

[15] Rucka M 2011 Damage detection in beams using wavelet transform on higher vibration modes *J. Theoret. Appl. Mech.* **49** (2) 399–317

[16] Gentile A and Messina A 2003 On the continuous wavelet transforms applied to discrete vibration data for detecting open cracks in damaged beams *Int. J Solids Struct.* **40** 295–215

[17] Douka E, Loutridis S and Trochidis A 2003 Crack identification in beams using wavelet analysis *Int. J Solids Struct.* **40** 3557–69
[18] Loutridis S, Douka E and Trochidis 2004 A Crack identification in double – cracked beams using wavelet analysis *J. Sound Vib.* **277** 1025–39
[19] Grabowska J, Palacz M and Krawczuk M 2008 Damage identification by wavelet analysis *Mech. Syst. Signal Pr.* **22** 1623–35
[20] Adeli H and Kim H 2009 *Wavelet-Based Vibration Control of Smart Buildings and Bridges* (Boca Raton, London, New York: CRC Press, Taylor and Francis Group)
[21] Weeks M 2011 *Digital Signal Processing Using Matlab and Wavelets* 2nd edition (Sudbury, Massachusetts: Jones and Bartlett Publishers)
[22] Bayissaa W L, Haritosa N and Thelandersson S 2008 Vibration-based structural damage identification using wavelet transform *Mech. Syst. Signal Pr.* **22** 1194–15

Acknowledgement

The research leading to these results has received the funding from Latvia State Research Programme under grant agreement "Innovative Materials and Smart Technologies for Environmental Safety, IMATEH".

Impact of External Heat Insulation on Drying Process of Autoclaved Aerated Concrete Masonry Constructions

S Rubene[1,2], **M Vilnitis**[1], **J Noviks**[1]
[1]Riga Technical University, Faculty of Civil Engineering, Department of Construction Technology, Kalku Str. 1, Riga, LV-1658, Latvia

E-mail: sanita.rubene@inbox.lv

Abstract. In the time when sustainable construction as well as cost saving on heating and cooling of buildings is one of the most important construction trends, it is important to acknowledge the possibilities of application of construction materials with high heat insulation parameters and the ways in which these parameters can be obtained. Autoclaved aerated concrete (AAC) is a load bearing construction material, which has high heat insulation parameters, although it has one significant disadvantage. If the AAC masonry construction has high moisture content, it loses its heat insulation properties. This is the reason why it is important to detect the humidity distribution throughout the cross section of the masonry elements in order to conduct the drying process of the AAC construction. Therefore, the question about non-destructive detection of humidity distribution throughout the cross section of the material arises. Humidity distribution throughout the cross section of AAC masonry constructions has a significant impact on its heat resistivity properties. Application of electrical impedance spectrometry (EIS) method for determination of humidity distribution throughout the cross section of AAC constructions has been a subject of research recently. The EIS method is an easily applicable non-destructive testing method for detection of the humidity distribution throughout the cross section of a construction. Research on the impact of the external heat insulation layer on the speed of humidity distribution changes is described in this paper.

1. Introduction

Autoclaved aerated concrete (AAC) is a load bearing construction material, which has high heat insulation parameters. However, as the AAC is a porous material, the moisture content of the material can reach high values and therefore the heat resistivity values of the material can decrease significantly. Such situations may lead to misinterpretation of the design data in the field of energy efficiency of the buildings.

[2] To whom any correspondence should be addressed.

Therefore, testing methods that allow credible determination of the moisture distribution throughout the cross section of the AAC masonry construction as well as moisture migration throughout the cross section of the construction have to be developed. The drying process of the masonry is a long process, therefore, it is preferable to use non-destructive testing methods because a number of measurement series have to be performed in order to provide credible data on the changes in moisture content throughout the cross section of the masonry construction.

In order to increase the heat resistivity properties of AAC masonry constructions, additional heat insulation layer from mineral wool or polystyrene is applied on the construction. Although the additional layer of insulation increases the heat resistive performance of the construction in design, it can result in quite opposite effect on construction site if it is applied on AAC construction with high moisture content. If AAC masonry blocks with high moisture content are covered with insulation layer, the moisture is locked in the masonry and the heat resistivity values drop significantly.

AAC masonry constructions are porous, composed of solid matrix and pores. In porous materials heat transfer is often coupled strongly with moisture transfer [1]. Accurate prediction of heat and moisture transfer in porous material is essential for optimization of building envelope with respect to energy consumption, hydrothermal performance, and indoor environment [2, 3]. It is important to detect not only the overall moisture content of the construction material but also to detect its distribution throughout the cross section of the respective element. Moisture distribution throughout the cross section of the element has a significant impact on its heat conductive properties.

2. Methods

Electric methods have become popular in characterizing different material properties. Electrical impedance spectrometry (EIS) is one of the most popular methods for detection of moisture content in construction materials. EIS can be applied through a number of different devices where each of them focuses on slightly different aspect of the material properties.

The possibility of application of this method on concrete based materials has been researched by McCarter and Garvin [4] and Rajabipour [5], Weiss et al. [6-8]. Barsoukov and McDonald [9] have introduced EIS as an optimal method for characterizing the electrical behavior of systems in which the overall system behavior is determined by a number of strongly coupled processes, each proceeding at a different rate. EIS can be applied for monitoring processes that take place in objects (e.g. changes caused by moisture in wet masonry sediments etc.), electrokinetic phenomena at boundaries or for describing basic ideas about the structure of an inter phase boundary (e.g. electrode/water or solid electrode to solid electrode boundary) [10-12].

The EIS is based on the periodic driving signal – the alternating signal. Its main measurement value is the impedance of AC circuit, which consists of electrodes (measurement device), and the solid electrolyte (in particular case AAC masonry construction).

For the particular research, a Z-meter III device (figure 1) with one pair of measurement probes (five channels in each probe) will be used for EIS measurements. For the measurements, the 0 channel measures the data closest to the external side of the wall construction.

This instrument has been verified in laboratory experiments and measurements on AAC constructions by the authors [13-18]. Correlations between the EIS measurement values on AAC masonry constructions and the moisture content of the respective AAC material samples have been established [18].

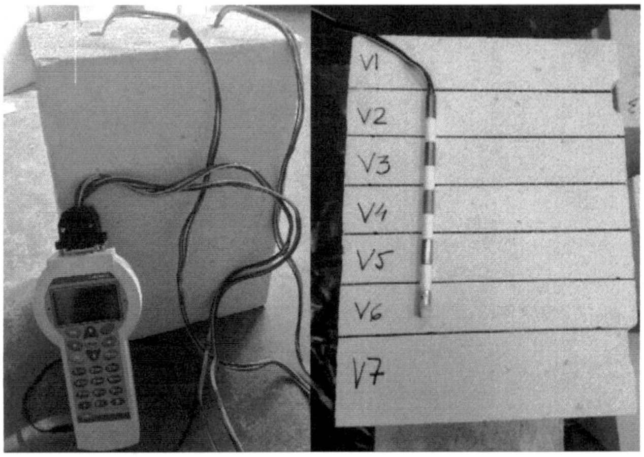

Figure 1. Z-meter III device with measurement probe.

Z-meter III device consists of an electronic and detachable measurement probes. The probes allow taking simultaneous EIS measurements in five different segments of the cross section in the sample. It allows determining the moisture content throughout the cross section of AAC masonry construction. This approach was used in the particular research to determine the differences in drying process of AAC masonry constructions with different heat insulation materials.

3. Description of the experiment

For the particular experiment three samples of AAC masonry wall construction were prepared with dimensions 1200x600x250mm (length x height x thickness). AAC with 375 kg/m^3 density was used [19] and an additional layer of heat insulation was attached to two samples. Most popular heat insulation materials which are usually used in Latvia were used – one of the samples was insulated with 100mm thick layer of mineral wool [20] (sample A) and the second sample was insulated with 100mm thick polystyrene insulation (sample B) [21]. The third sample was left without additional heat insulation layer in order to use it as a reference sample (sample C). Afterwards, both insulated samples and the third sample, which was left without any additional insulation, were covered with plaster from the external side of the insulation (e.g. the external side of the wall construction). Thus a model of construction phase when the cladding of the masonry construction had been finished and the external finishing of the wall had been finished was simulated (figure 2 (a) and (b)).

In each sample 17 bores were made in order to monitor the moisture migration in all directions of the sample by EIS. The monitoring of the drying process was performed for 12 weeks in laboratory conditions with air Rh in the range of 60-80% and the average temperature in the range from 18 to 25°C. Such conditions comply with the average weather conditions on construction site in Northern European summers when the building envelope is not fully closed and heating of the building has not started. The same drying conditions were maintained for all samples.

(a)

(b) (c)

Figure 2. (a) Sample constructions on stand; (b) Sample constructions on stand; (c) Measurement points on the sample.

4. Results

4.1. Initial moisture distribution throughout the cross section of the sample constructions

After the samples were constructed, the initial moisture distribution was determined by EIS. As the EIS measures the resistance values, the determination of the moisture content in the AAC material can be detected by application of correlation equations. Such correlation equation (1) between the electrical resistivity values of the respective AAC material and its moisture content has been developed by the authors [23] and tested in laboratory conditions.

$$y = -0{,}236\ln(x) + 3{,}1507 \quad (1)$$

The EIS measurement results provided information about the initial moisture distribution throughout the cross section of the sample constructions (figure 3 to figure 5)

The average moisture content of the AAC blocks at the beginning of the experiment was 25% of AAC dry mass. The moisture distribution throughout the cross section of all samples is even with the exception of channel 4, where the least moisture content is observed in all samples. This fact can be explained with the least impact of humid processes (e.g. finishing mortar on sample C or insulation installation mortar on samples A and B) on the internal side of the sample construction.

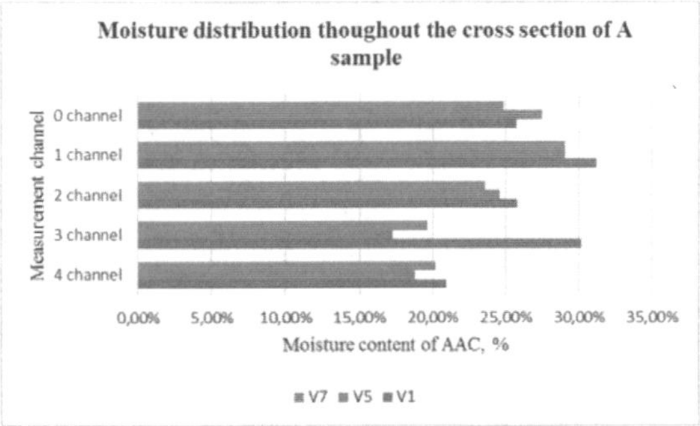

Figure 3. Moisture distribution throughout the cross section of sample A measurement points at the beginning of the experiment.

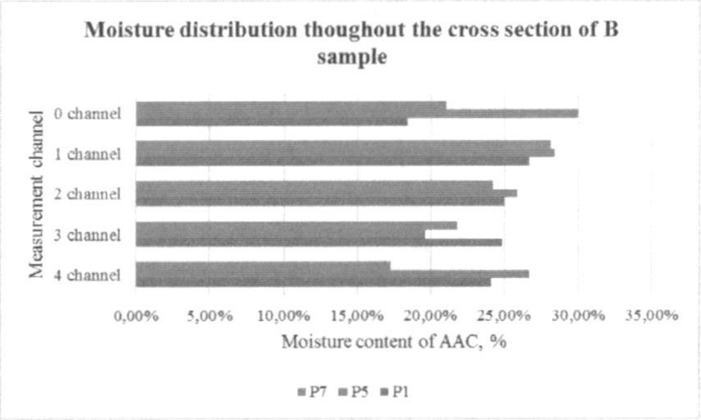

Figure 4. Moisture distribution throughout the cross section of sample B measurement points at the beginning of the experiment.

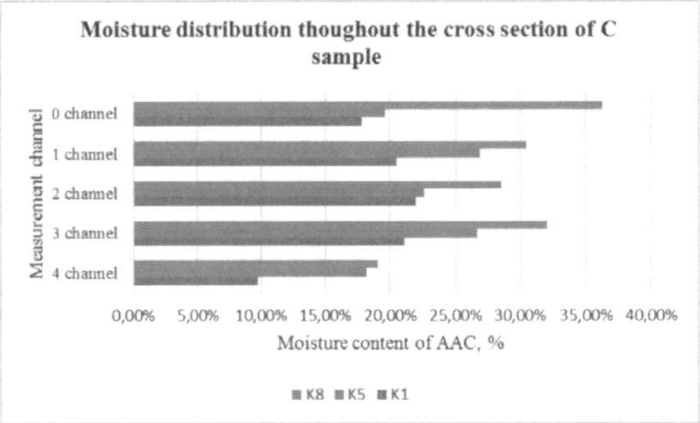

Figure 5. Moisture distribution throughout the cross section of sample C measurement points at the beginning of the experiment.

Further moisture distribution changes throughout the cross section of the samples were monitored by application of EIS measurements three times a week and certain dynamics of drying process were established for each sample construction. Figure 6 to figure 8 display the changes of moisture distribution throughout the cross section of the AAC samples with different external finishing. The obtained data were merged in one surface graph for each sample and the division of the data on x axis allows following the changes of moisture distribution throughout the cross section of the sample during the whole period of the experiment.

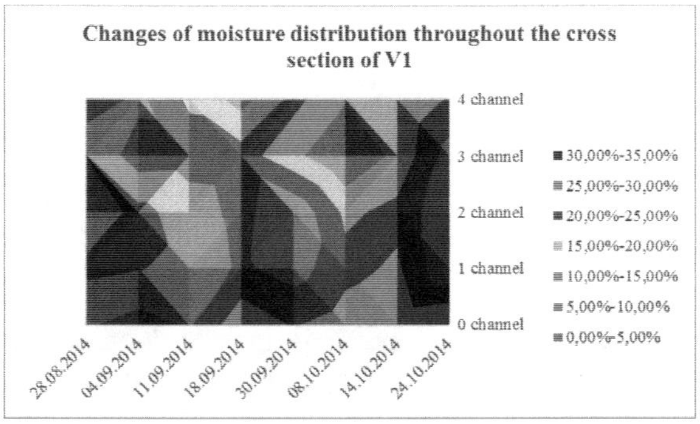

Figure 6. Changes of moisture distribution throughout the cross section V1 of sample A construction during the experiment.

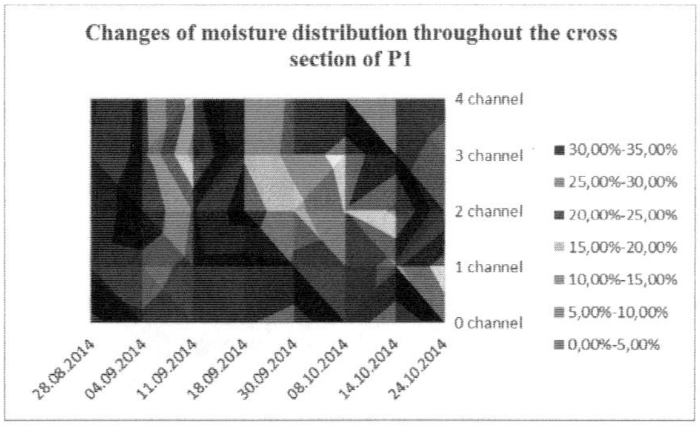

Figure 7. Changes of moisture distribution throughout the cross section P1 of sample B construction during the experiment.

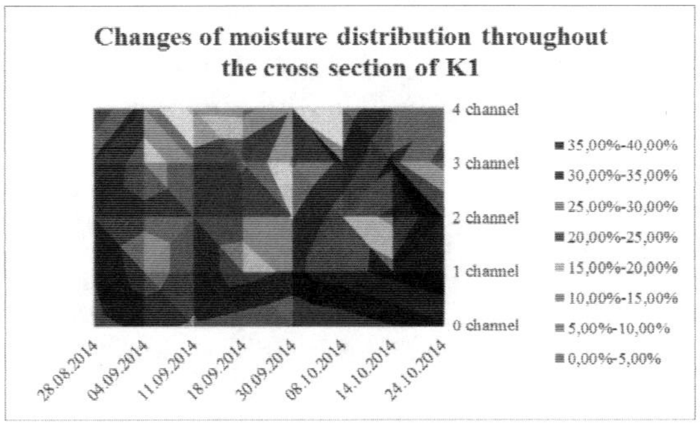

Figure 8. Changes of moisture distribution throughout the cross section K1 of sample C construction during the experiment.

The sample construction A displayed the fastest drying speed in comparison with sample constructions B and C. The slowest drying speed was observed for sample construction B. Therefore the assumption that the insulation material has significant influence on the drying speed of the wall construction has proved to be valid. Sample B attests that the vapor diffusion coefficient of the insulation materials has high impact on the drying speed of the AAC wall construction.

Table 1. Drying speed of the samples.

Sample	Average moisture content of the cross section at the beginning of the experiment, %	Average moisture content of the cross section at the end of the experiment, %	Drying speed (% of moisture lost/obtained in 12 weeks)
Sample A	26.82%	24.07%	-2.75%
Sample B	23.79%	24.04%	0.24%
Sample C	18.24%	29.24%	11.00%

The difference in drying speed of the respective samples varies by 12.75%. The results prove that the only type of insulation which allows decreasing AAC masonry construction's moisture rate is mineral wool as in sample A. Such performance depends on two factors. Firstly, the mineral wool insulation decreases the amount of moisture which is absorbed by the masonry construction and, secondly, allows the moisture to migrate through itself due to high value of the vapor diffusion coefficient. Sample B displayed a minimal increase of the average moisture content due to the fact that the low vapor diffusion coefficient of the polystyrene did not allow the moisture to migrate freely through the insulation layer and all moisture from the installation mortar was absorbed by the AAC masonry construction. The reference block (sample C) displayed the highest absorption of moisture which can be based on the amount of the moisture absorber from the finishing mortar and the fact that all surfaces of the reference construction were exposed to the external humidity of the premise while the other two samples had one covered surface. The fact that the changes of construction's moisture content are caused by the changes of the Rh rate in the premise has been proved by figures 7,8, which clearly indicate the increase of moisture rate in external layers of the construction starting from 30 September 2014, when the Rh rate of the premise increased as well.

5. Conclusions

The finishing and insulation material of the AAC masonry constructions have significant impact on the drying performance and therefore on the humidity distribution throughout the cross section of the wall construction. Furthermore, it influences the heat resistivity properties of the wall construction, therefore it should be considered during the design phase in order to avoid situations when buildings are not able to reach the designed insulation parameters.

The research proves that an insulation material with high vapor diffusion coefficient allows the construction to dry faster. Therefore, application of such materials is recommended for AAC masonry constructions in order to reach the designed insulation parameters as soon as possible after construction.

6. References

[1] Qin M H, Belarbi R, Aït-Mokhtar A et al 2009 Coupled heat and moisture transfer in multilayer building materials *Constr. Build. Mater.* **23** (2) 967–975

[2] Mendes N, Winkelmann F C, Lamberts R et al 2003 Moisture effects on conduction loads *Energy Build* **35** (7) 641–644

[3] Daniel A, Vries D E 1987 The theory of heat and moisture transfer in porous media revisited *Int. J. Heat Mass Transf.* **30** (7) 1343–50

[4] McCarter W J and Garvin S 1989 Dependence of electrical impedance of cement-based materials on their moisture condition *J. of Applied Physics Series D: Applied Physics* **22** (11) 1773–76

[5] Rajabipour F, Weiss J, Shane J, Mason T and Shah S 2005 A procedure to interpret electrical conductivity measurements in cover concrete during rewetting *J. of Materials in Civil Engineering* **17** (5) 586–594

[6] Weiss W J, Shane J D, Mieses A, Mason T O and Shah S 1999 Aspects of monitoring moisture changes using electrical impedance spectroscopy *Proc. of the second Symposium on Self-Desiccation and Its Importance in Concrete Technology (Lund, Sweden, 1999)*

[7] Schießl A, Weiss W J, Shane J D, Berke N S, Mason T O and Shah S P 2002 Assessing the moisture profile of drying concrete using impedance spectroscopy *Concrete Science and*

Engin. **2** 106–116
[8] Rajabipour F and Weiss J 2007 Electrical conductivity of drying cement paste *J. of Materials and Structures* **40** 1143–60
[9] Barsoukov E and McDonald R 2005 *Impedance Spectroscopy Theory, Experiment, and Applications* ed. John Wiley & Sons (USA: New York)
[10] Skramlik J and Novotny M 2008 One-dimensional moisture transport monitored by a non-destructive method *Int. J. of Ccomputers* **4** (2)
[11] Parilkova J et al 2011 Monitoring of changes in moisture content of the masonry due to microwave radiation using the EIS method *EUREKA 2011* Brno
[12] Parilkova J, Vesely J and Prochazka L E 2012 !4981 – Computerized measuring system for analysis of chosen characteristics and processes in porous environment by eis method, presentation of a project in 2012
[13] Rubene S and Vilnitis M 2013 Application of electrical impedance spectrometry for determination of moisture distribution in aerated concrete constructions *1st Conf. and Working Session EUREKA!7614 (Czech Republic, Brno, 2013)* pp 124–130
[14] Rubene S et al 2014 Determination of humidity level in aerated concrete constructions by non destructive testing methods *Proc. Innovative Materials, Structures and Technologies (Latvia, Riga, 2014)* pp 135–140
[15] Rubene S and Vilnitis M 2014 Monitoring of humidity distribution changes in aerated concrete masonry construction by EIS *Proc. second Conference and Working Session EUREKA!7614 (Czech Republic, Brno, 2014)*
[16] Rubene S, Vilnītis M and Noviks J 2014 Monitoring of the aerated concrete construction drying process by electrical impedance spectrometry *Proc. of fourth Int. Conf. Advanced Construction 2014 (Lithuania, Kaunas, 9-10 October, 2014)* (Kaunas: Kaunas University of Technology) pp 216–220
[17] Rubene S and Vilnītis M 2014 Correlation between EIS measurements and relative humidity distribution in aerated concrete masonry constructions *Recent Advances in Civil Engin. and Mechanics. Mathematics and Computers in Science and Engineering Series 35 (Italy, Florence, 22-24 November, 2014)* (Florence: WSEAS Press) pp 67–72
[18] Rubene S, Vilnītis M and Noviks J 2015 Impact of density and special features of manufacturing process on drying of autoclaved aerated concrete masonry blocks (Submitted for publication) *Proc. of tenth Int. Scientific Practical Conf. Environment. Technology. Resources (Latvia, Rezekne, 18 June, 2015)*
[19] [Available]: http://aeroc.lv/index.php?page=938&lang=lat&cnt=AEROC_Universal
[20] [Available]:http://www.rockwool.lv/izstr%C4%81d%C4%81jumi+un+risin%C4%81jumi/u/2011.product/2448/celtniecibas-izolacija/fasrock
[21] [Available]:http://tenapors.lv/lv/produkti/siltumizolacijas-materiali/sienas/

IOP Conf. Series: Materials Science and Engineering **96** (2015) 012060 doi:10.1088/1757-899X/96/1/012060

Wave Phase-Sensitive Transformation of 3d-Straining of Mechanical Fields

I N Smirnov[1,2,3] **and A A Speranskiy**[1]
[1]Advanced Vector Analytics Ltd, Riga, Latvia
[2]Lomonosov Moscow State University, Moscow, Russia

E-mail: Ilyas@ava-labs.com

Abstract. It is the area of research of oscillatory processes in elastic mechanical systems. Technical result of innovation is creation of spectral set of multidimensional images which reflect time-correlated three-dimensional vector parameters of metrological, and\or estimated, and\or design parameters of oscillations in mechanical systems. Reconstructed images of different dimensionality integrated in various combinations depending on their objective function can be used as homeostatic profile or cybernetic image of oscillatory processes in mechanical systems for an objective estimation of current operational conditions in real time. The innovation can be widely used to enhance the efficiency of monitoring and research of oscillation processes in mechanical systems (objects) in construction, mechanical engineering, acoustics, etc. Concept method of vector vibrometry based on application of vector 3D phase-sensitive vibro-transducers permits unique evaluation of real stressed-strained states of power aggregates and loaded constructions and opens fundamental innovation opportunities: conduct of continuous (on-line regime) reliable monitoring of turboagregates of electrical machines, compressor installations, bases, supports, pipe-lines and other objects subjected to damaging effect of vibrations; control of operational safety of technical systems at all the stages of life cycle including design, test production, tuning, testing, operational use, repairs and resource enlargement; creation of vibro-diagnostic systems of authentic non-destructive control of anisotropic characteristics of materials resistance of power aggregates and loaded constructions under outer effects and operational flaws. The described technology is revolutionary, universal and common for all branches of engineering industry and construction building objects.

1. Introduction

Vibration methods became of common usage in the sphere of test parameter measuring of mechanical systems. These methods are based on static models of experimental estimate of strength resource state according to two scalar jitter parameters such as amplitude and frequency. These elements are

[3] To whom any correspondence should be addressed.

IOP Conf. Series: Materials Science and Engineering **96** (2015) 012060 doi:10.1088/1757-899X/96/1/012060

emotional outer system indicators that reflect understanding of spatial deformation processes averaged and projected by scalar method. Monoscope monitoring provides only a capability to estimate systems and also it is entropy relative to insight. Because of it, the main reason of emergencies in critical infrastructure and of mass human presence is the lack of efficient equipment for monitoring of dynamic strength.

At the turn of XXI century in the context of state-of-the-practice of the idea to create vector phase-sensitive metrological means on the basis of computer technologies, there was a breakthrough opportunity of spatial updating of the outlined anisotropically hard dynamically-sized portraits. This opportunity allows estimating by direct method the deflected modes and vibration test parameters of the monitoring objects. The method, which is fundamental and flexible for all fields of machine building and construction industry, has been developed and implemented. This method represents innovative IT for environmental safety.

Deficiency of the known methods is that they do not allow reconstructing time-correlated multidimensional images of metrological and estimated spectral parameters of oscillatory processes in mechanical systems.

In order to effectively use the advantages of the substantial increase in informativity of the oscillatory measurements in mechanical systems attained through application of vibration accelerometers and 3D-receivers of mechanical oscillations, it is necessary to create multidimensional systematic interlinked spectral images of test parameters in oscillatory processes, which are adequate to physical and fundamental laws of researched and controlled natural phenomena and processes.

It would resolve the problem of enhancing the efficiency of research and evaluation of oscillatory processes in mechanical systems. Technical solution to the posed problem is the method of reconstruction of multiparameter images of oscillatory processes in mechanical systems. Creation of systematic interlinked multiparameter, multilevel, hierarchically organized cybernetic images of oscillatory processes (both superficial and volumetric) in mechanical systems is the main result of implementation of this innovation.

2. Description of the method

Solution to the posed problem is the method which implies creation of sets of three-dimensional elliptic images (hodographs) of linear vibration acceleration vectors and\or vibration speed vectors, and\or vibration displacement vectors in places where the sensors (3D-receivers) are installed on the monitored object based on synchronously measured spatial and dynamic (time-varying) components of test parameters' vectors, accumulation of spectral arrays consisting of the measured and estimated (including approximated) vector and scalar values in the specified range of mechanical oscillations frequencies, application of measured and estimated test parameters on 3D-model of monitored object, 3D-reconstruction of spectral sets of spatial elliptic hodographs of time-varying deformation vector parameters, defining the characteristic parameters of external forces and their relative positioning against measuring points through a correlation between values and direction of axes of deformation vector hodograph, as well as defining stress-strain states.

Metrological and estimated vector parameters are physically interlinked through tensor connection, these parameters are fundamental, integral and systemic and they form a stable aggregation of attributes which reflect dynamic homeostasis of monitored object adequately to the physical nature. An integrated reconstruction image of oscillatory processes in an object can be used in the course of further monitoring of the object.

IOP Conf. Series: Materials Science and Engineering **96** (2015) 012060 doi:10.1088/1757-899X/96/1/012060

Technical result is the creation of spectral set of multidimensional images which reflect time-correlated three-dimensional vector parameters of metrological, and\or estimated, and\or design parameters of oscillations in mechanical systems (Holospectrum). Reconstructed images of different dimensionality integrated in various combinations depending on their objective function can be used as a homeostatic profile or cybernetic image of oscillatory processes in mechanical systems for an objective estimation of the current operational conditions in real time.

The specified above technical result of the method of reconstruction of oscillatory process images is accomplished through the device which contains a space-time measurements module, module of synchronization of space-time measurement transformation, digital memory module, module of spectral processing of measured vector parameters, module of reconstruction of the measured parameters, module of standardized matrixes of tensor transformations, module of estimated test parameters of mechanical oscillations, module of reconstruction of object's 3D-model, module of reconstruction of multi-parameter images of oscillatory processes, module of visualization of the reconstructed parameters, registering module, and the module which organizes system interaction of all modules. Innovation can be widely used to enhance the efficiency of monitoring and research of oscillation processes in mechanical systems (objects) in construction, mechanical engineering, acoustics, etc.

The method is illustrated by figures 1 and 2 where the structure of measurement and buildup of test parameters of mechanical oscillation is schematically pictured. Reconstruction module of multi-parameter images of oscillatory processes in mechanical systems is displayed in figure 2. the method of reconstruction of multi-parameter images of oscillatory processes in mechanical systems is carried out as follows.

A spectrum of wave oscillations produced by external forces in the form of forces and moments' actions or by internal pathologies of structure's material, propagating on a contour surface and through the whole volume of the monitored object, influences sensing elements of the 3D-receivers of mechanical oscillations installed in the measuring points on a contour surface of an object. Metrological tool carries out the inverse tensor transformation of deformation spectrum acting upon it as well as carries out space-time measurement of superimposition of components of spectral set of metrological parameters in the Cartesian coordinate system of the receiver. Measured analog parameters are transformed synchronously into the digital form and are entered into a digital unit for further spectral processing (filtration) and vector-phase reconstruction of elliptic metrological parameters. In order to retain the linearity of interlinked metrological and estimated test parameters, it is necessary to standardize the measurements relative to maximum values defined by technical characteristics of metrological tools and estimated coefficients of transformation of physical quantities.

Multidimensional interlinked space-time set of the estimated and metrological parameters of oscillatory processes makes up an essence of object's cybernetic image and defines the degree of its adequacy to mechanical oscillations spectrum, produced in nature. Further, subject to the posed problem, the current values of test parameters are calculated based on spatial measurements, allowing evaluation of the dynamic condition (homeostasis) of an object. Inside the modules of visualization and registration of the reconstructed parameters both graphic vector images (including elliptic images) and scalar images are formed, allowing analyzing and effectively evaluating current homeostatic condition of an object.

Reliability of a cybernetic image of an object is supported by synchronous system interaction of all modules that generate, process, register, store and display sets of estimated metrological parameters of oscillatory processes.

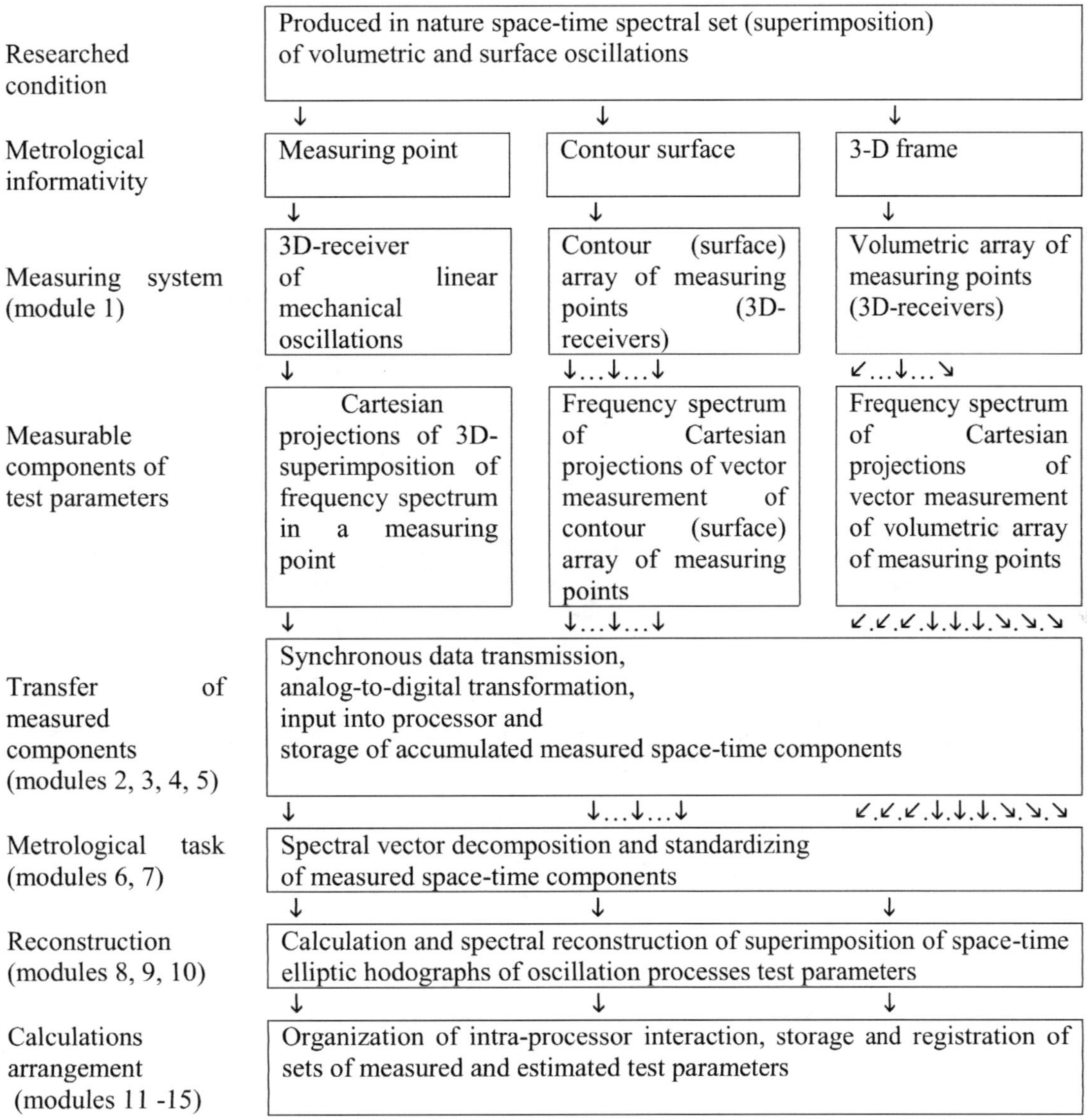

Figure 1. Structure of measuring and generating of test parameters in mechanical oscillations.

The project is delivered on the basis of such new IT as vector-phase method of multivariate spatial-temporal updating deformation fields via 3D-phasebalanced tensor – transformation of strength parameters of the monitoring objects. Software and hardware using vector vibration transducer are meant for estimate, supporting and safety systems in the building industry, in the sphere of fuel transfer, thermal and nuclear power industry, in air-space, heavy and machine industry, geological engineering, hydroacoustics and defectoscopy.

Comparative informativeness of scalar and vector-phase methods measurements of multivariate spatial-temporal updating deformation fields is presented in the picture (figure 3). Graphical symbols: red – block of spatial-temporal of instantaneous vectors of deformation fields, green – block of monoscope scalar projectures of instantaneous vectors.

Main objects of application of the method are dynamic aggregates and mechanisms, power static constructions, hydroacoustic systems, natural objects of anthropogenic type.

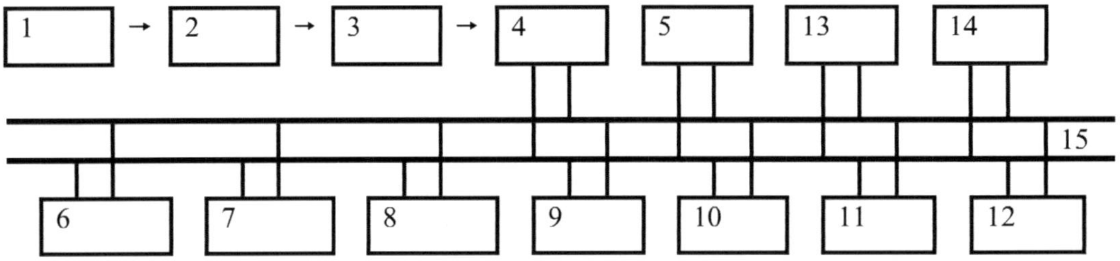

Figure 2. Reconstruction module of multi-parameter images of oscillatory processes in mechanical systems.

Designations in figure 2: 1 – 3D-receivers of mechanical oscillations, 2 – three-channel blocks of synchronous amplification and transformation of the charge, 3 – device/block of synchronous transfer of measured components of mechanical oscillations parameters, 4 – device/block of synchronous analog-digital conversion and input of components of mechanical oscillations parameters into processor, 5 – digital storage unit for measured components of mechanical oscillations parameters, 6 – module of spectral processing of the measured components of mechanical oscillations parameters, 7 – module of standardized matrixes of tensor transformations, 8 – module of spectral reconstruction of elliptic hodographs of test space-time parameters in measuring points, 9 – module of reconstruction of object's 3D-model, 10 – module of reconstruction of multi-parameter images of oscillatory processes, 11 – module of visualization of test parameters, 12 – registering module, 13 – module of synchronization of transformation of space-time measurements, 14 – module of organization of system interaction of all modules, 15 – information transfer and management bars.

3. Application examples

3.1. Comparison of informativity 1D and 3D methods

Vector vibroacoustic metrology method and laboratory production technology of vector vibration transducer is developed. Russian company «GAZprom» has developed, implemented and tested Vector Diagnostic Automated System (VDAS) of compressor station gas compressor units.

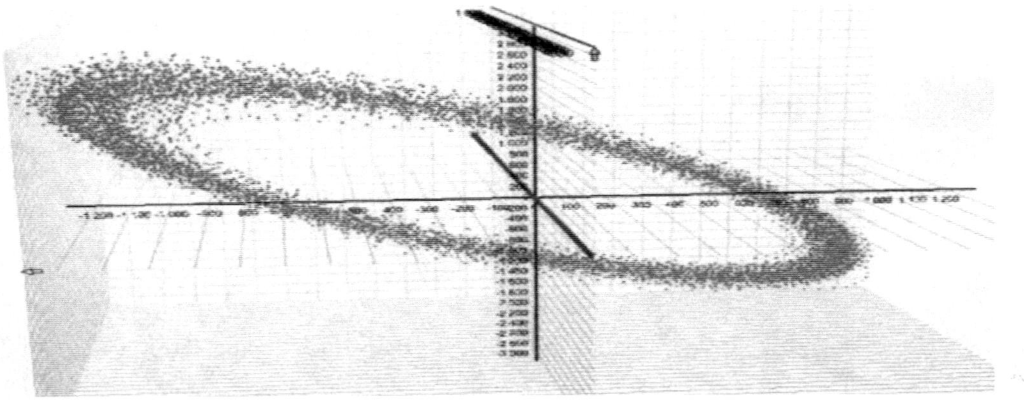

Figure 3. Comparative informativeness of scalar and vector-phase methods measurements (in m/s).

3.2. Static construction

Description of experiment: simple plywood construction 190x190x720 was placed on a vertical vibrating table. Frequency of the vibration was 50Hz. The platform was fixed on four dampers. Measurements were taken at different points of the structure (1-7 in figure 4).

There were two modes:

1- well fixed and dampened structure
2- one dampener removed forming an imbalance on one of the pillars

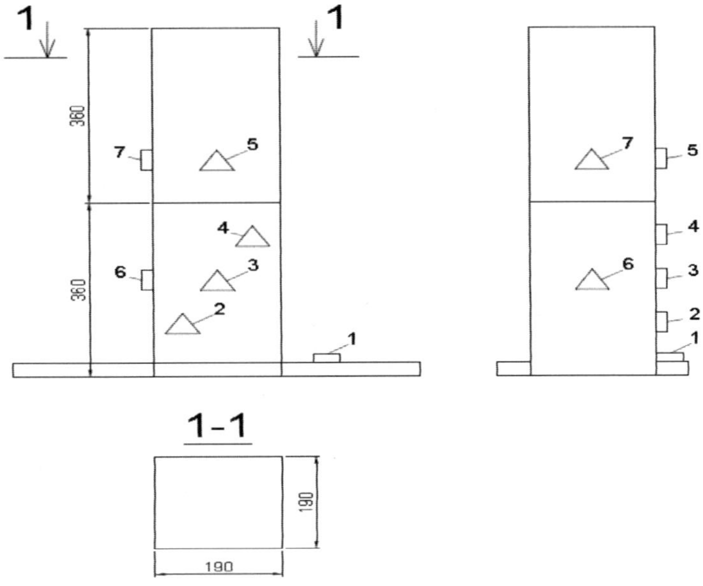

Figure 4. The structure used in the experiment; 1…7 – Measuring Point.

IOP Conf. Series: Materials Science and Engineering **96** (2015) 012060 doi:10.1088/1757-899X/96/1/012060

The presented data show that a perfectly balanced structure has its homeostatic holospectrum portrait in the band set elliptic trajectories having general directions of semi-axes, and a common plane. Adding an imbalance to the system at the same frequency band results in an unrelated set of elliptic trajectories, indicating more degrees of freedom, as well as an increase of the energy level of the structure. The energy level is characterized by the sum of areas of the resulting elliptic trajectories (Holospectrums). Depending on the task this situation can be analysed on deeper levels.

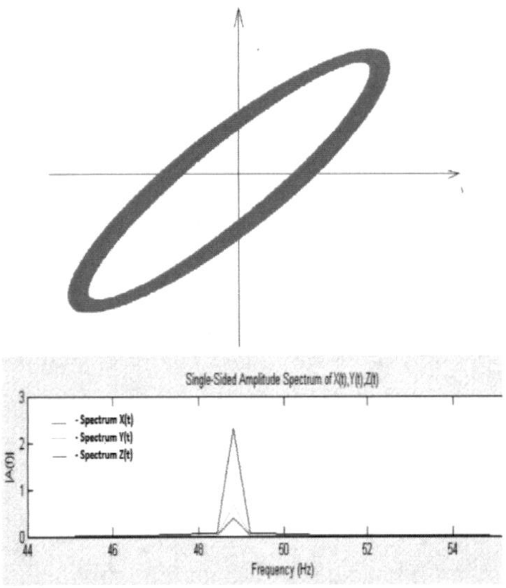

Figure 5. Box without mass.

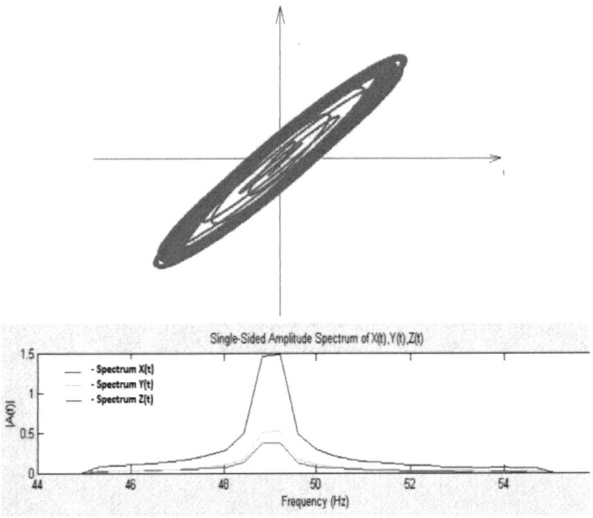

Figure 6. Box without mass, damper removed.

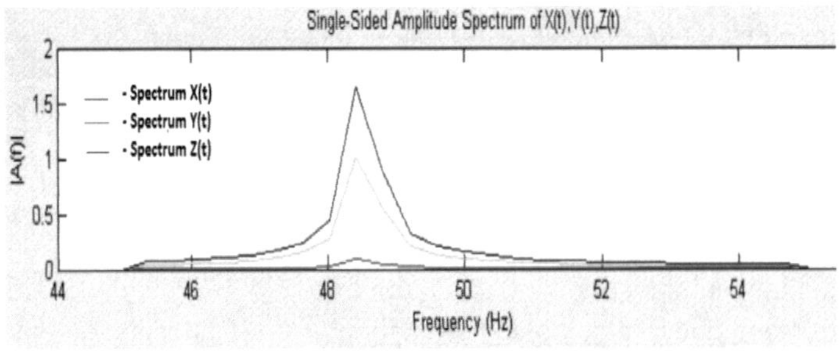

Figure 7. Box with mass, all dampers in stand.

4. Conclusion

In this article we have considered new approach to monitoring of buildings and constructions. A new approach to analysis of spectral data well-known as holospectrum approach was also considered. It demonstrates the simplicity of 3D data through elliptical vibrations portraits. Holospectrum analysis also allows us to estimate the energy fluctuations at the real physical measurement point. That in turn allows evaluating how the energy of external influences spread in the building and how much of it is stored in it. Thus, we can also evaluate the residual and accumulating stress and strain in buildings and metal structures.

5. References

[1] Zhilin P A A 1996 *New Approach to the Analysis of Free Rotations of Rigid Bodies* ZAMM Mech Z 76 p 4

[2] Qu L, Liu X, Peyronne G and Chen Y 1989 The holospectrum: A new method for rotor surveillance and diagnosis *Mechanical Systems and Signal Processing* **3** (3) 255–261

[3] Du* R and Yeung K 2004 Fuzzy transition probability: A new method for monitoring progressive faults, Part 1: the theory *Engineering Applications of Artificial* Intelligence **17** 457–467

[4] Braun S (eds) 2001 Encyclopedia of Vibration (USA: Academic Press) p 1685

[5] Strommen E N 2013 *Structural Dynamics* Springer p 521

[6] Levitskij D N, Maljutin D V and Speranskij A A 2006 *Novyj Tip Sredstv Izmerenija Vibrouskorenija dlja Dostovernogo Prostranstvennogo Monitoringa* (Izvestija VUZov: Neft' i gaz) No. 4 p 87–95

[7] Cernant A A, Speranskij A A, Kobjakov I B and Maljutin D V 2006 Vektornyj vibromonitoring – instrument obektnoj volnovoj tomografii v stroitel'stve, BST, No.12 p 52–64

Acknowledgments

This research was supported by grant RFBR project No. 14-01-31444 and Russian President Grant MK-7776.2015.1, NSH-4595.2014.1

IOP Conf. Series: Materials Science and Engineering **96** (2015) 012061 doi:10.1088/1757-899X/96/1/012061

Evaluation of Safety, Quality and Productivity in Construction

M A Usmen[1] **and M Vilnitis**[2,3]
[1]Wayne State University, Detroit, USA
[2]Riga Technical University, Kalku Str. 1, Riga, LV-1658, Latvia

E-mail: martins.vilnitis@rtu.lv

Abstract. This paper examines the success indicators of construction projects, safety, quality and productivity, in terms of their implications and impacts during and after construction. First safety is considered during construction with a focus on hazard identification and the prevention of occupational accidents and injuries on worksites. The legislation mandating safety programs, training and compliance with safety standards is presented and discussed. Consideration of safety at the design stage is emphasized. Building safety and the roles of building codes in prevention of structural failures are also covered in the paper together with factors affecting building failures and methods for their prevention. Quality is introduced in the paper from the perspective of modern total quality management. Concepts of quality management, quality control, quality assurance and Six Sigma and how they relate to building quality and structural integrity are discussed with examples. Finally, productivity concepts are presented with emphasis on effective project management to minimize loss of productivity, complimented by lean construction and lean Six Sigma principles. The paper concludes by synthesizing the relationships between safety, quality and productivity.

1. Introduction

Engineers, constructors and facility owners are concerned with the safety of buildings and constructed facilities because of the monetary losses and human tragedies resulting in personal injury, death and suffering. Safety issues may arise during construction due to failure to control the existing site hazards, including collapse of structures during the construction process. In addition, defects in constructed facilities usually lead to serviceability problems, while structural failures can cause catastrophic results. The effectiveness of quality management in all stages of a construction project has a significant impact on safety. It is the goal of the project managers to complete the job by assuring safety and quality, while closely controlling project schedule and cost. This means that

---

[3] To whom any correspondence should be addressed.

IOP Conf. Series: Materials Science and Engineering **96** (2015) 012061 doi:10.1088/1757-899X/96/1/012061

attention needs to be paid to productivity of the workforce. Unfortunately, knowledge on the relationship between safety, quality and productivity in construction is limited, and because of that there is wide spread assumption that time and expenditures associated with safety and quality in construction compromises productivity; i.e. they delay schedules and increase project costs [7].
This paper aims to provide background information on safety of construction and safety of structures (constructed facilities); describes the principles and tools of quality management and how they are applied to construction projects; covers the associated standards and codes; defines and discusses productivity concepts; and presents a synthesis on the relationship between safety, quality and productivity in construction projects.

2. Safety of construction and structures

Safety is defined as the state of being relatively free from harm, danger, injury or damage. Unsafe conditions in facilities under construction, or in constructed facilities, can be caused by the engineers' failure to develop and implement structurally and environmentally sound designs; performing construction in an unsafe manner on the jobsite; or allowing conditions during service to exceed the structural capacity of the facility. A safety hazard exists when such conditions prevail, and the result is an accident leading to jobsite injuries and fatalities during construction. Similarly, if the safety hazard exists in a structure in the form of an overload, it causes the facility in part or as a whole to become unstable a leading to collapse. Loss of lives and property damage are the common consequences. The role of the design and construction engineers is to anticipate and proactively prevent these happenings.

2.1 Occupational safety

Occupational safety deals with the identification and control of environmental and personal hazards in the work place. According to the US Bureau of Labor Statistics, the recordable injury rate was about 4.0 for every 100 full-time equivalent workers in the US construction industry in 2010, coupled by 774 fatalities in the same year. These numbers have improved significantly over the past few decades due to stricter enforcement of standards, industry training efforts, and advancement of technology. Certain problems have persisted, however; fatalities due to falls are still high (35 percent), followed by struck by and caught in between accidents (13 percent), and electrocution (9 percent) [6], [7].

Efforts to improve site safety are driven by humanitarian, ethical, legal and regulatory, and institutional and image concerns, but the biggest factor is the economic impact. Studies have shown that accidents cost significant sums of money because of increased workers compensation insurance premiums, which are based on the employers' safety records. So, improving the safety of contractors and subcontractors leads to measurable savings in project costs, which are passed on to the project owners (clients). The uninsured (hidden) costs of accidents that have to be borne by the contractors include clean-up, equipment repair, stand-by; work delays and rescheduling; post-accident extra safety supervision; regulatory an d civil fines; legal fees; administrative procedures, investigative reports; transportation to first aid/medical facility; wages paid to injured worker for time not worked; loss of crew efficiency; and training the new/substitute worker. The ratios of the uninsured costs to insured costs have been found to vary between 2 and 20 [7].

Safety hazards arise due to unsafe acts and unsafe conditions, usually in combination. Unsafe acts include unsafe use of tools and equipment; unauthorized operation of equipment; use of defective equipment; failure to use personal protective equipment; unsafe material handling; failure to follow safe procedures; poor housekeeping; and attitude problems (macho, horseplay, inattentive behavior, etc.). A partial list of unsafe conditions encompasses improper guarding (equipment, platform);

improper illumination, ventilation; hazardous substances (chemicals, explosives, etc.); poor site layout, housekeeping; improper dress; poor maintenance, lockout / tagout practices; and unsafe design and construction [6].

A systematic approach to minimizing and eliminating injuries and fatalities in the worksite is needed, so safety related losses can be controlled. The vehicle to accomplish this goal usually takes the form of a safety program or a safety plan, which can be applicable to project owners, designers (architect/engineer), general contractors and subcontractors. These are policies that can be implemented at the corporate level, or can be project-specific. U.S. federal legislation, through the Occupational safety and Health Administration (OSHA) provides guidelines for safety programs. The main elements emphasized are management commitment and employee involvement, worksite analysis, hazard prevention and control, and safety and health training.

Management commitment starts with the development and communication of a safety/health policy to all employees (signed by CEO) and pervades all operations through instilling accountability for safety, obeying safety rules, reviewing accident reports, conducting regular safety meetings, and involving employees, managers and supervisors in safety activities. Assigning responsibilities to coordinate safety activities, providing resources to accomplish the program goals, integrating safety into business practices (e.g., purchases, contracts, design and development), and recognizing employees for safe work practices are important elements.

Worksite analysis entails evaluating all workplace activities and processes for safety/health hazards, re-evaluating workplace activities when there are changes in processes, worksite conditions, materials and equipment; conducting on-site inspections to identify existing hazards and taking corrective actions; and establishing a hazard reporting system for employees to report unsafe conditions. This task includes investigating all accidents and near misses to determine root causes.

Hazard prevention and control is established by OSHA standards and implemented by training, and site inspections. Well known OSHA standards applicable to the construction industry (29 CFR Part 1926) include General Provisions (Subparts A through D); Personal Protective and Life Saving Equipment (E); Fire Protection and Prevention (F); Signs, Signals and Barricades (G); Materials Handling, Storage, Use and Disposal (H); Hand and Power Tools (I); Welding and Cutting (J); Electrical (K); Scaffolds (L); Fall Protection M); Cranes, Derricks, Hoists (N); Motor Vehicles, Mechanized Equipment (O); Excavations (P); Concrete and Masonry Construction (Q); Steel Erection (Subpart R); Underground Construction, Caissons, Cofferdams (S); Demolition (T); Blasting and the Use of Explosives (U); Power Transmission and Distribution (V); Rollover Protective Structures; Overhead Protection (W); Ladders (X); Commercial Diving Operations (Y); and Toxic and Hazardous Substances (Z). Also, 29 CFR PART 1910, which is applicable to General Industry covers topics applicable to construction, such as lock out/tag out, confined spaces, hazard communication, and so on.

A systematic way to prevent and control hazards is possible through the hierarchy of controls principle, which offers the following approach: 1. Eliminate the hazard (e.g., off site cutting of panel work); 2. Substitute the hazard (replace ladder with scissor lift); 3. Apply engineering controls (sensors and alarms, guardrails); 4. Administrative controls (job rotation, work instructions); and 5. Prescribe personal protective equipment (hard hats, goggles, respirators). Provisions of OSHA standards frequently utilize this principle.

Safety and health training is an integral part of a safety program. An effective training program includes needs assessment (planning); program development including training materials (printed, PowerPoint, video, interactive, demonstrations, etc.); program delivery (classroom lecture, field, online); and training effectiveness evaluation (surveys, pretest / posttest and observations) leading to program improvement. Evaluation can be performed at three levels: Level 1 – Trainee assessment of learning (feedback/perceptions); Level 2 – Measurement of learning (testing, observations); and Level 3 – Reflection in safety performance indicators (leading and lagging) [6].

Structural safety of a constructed facility relates to the understanding and elimination of the causal factors and prevention of failures (partial or total collapse). Some of the catastrophic failures in the US and elsewhere have been quite well documented [16] to [20]. For instance, in 1981, the walkway inside the Hyatt Regency Hotel in Kansas City collapsed during a social function resulting in 114 fatalities and 200 injuries. Inadequate connection detail between steel box beams and hanging rods caused the failure. When the original connection detail was altered by the steel fabricator, load was transferred from roof to the 4th floor creating an overload. The shop drawings had not been properly checked by the engineer of record allowing this to occur. The roof of the Algo Center Mall in Ontario, Canada experienced failure in 2012 due to an overload of a critical connection; this time corrosion due to poor waterproofing of the roof parking surface was the main cause. Two people died and 20 were injured. Forensic investigations revealed that there were construction defects (e.g. missing bolts) and no structural redundancy in design. The collapse occurred as a result of the failure of a simply supported beam. On 21 November, 2013 the roof of the Zolitude shopping centre in Riga, Latvia, collapsed resulting in the deaths of 54 people and injuries to another 41 people. Before the collapse, construction of the green roof of the shopping centre was in progress, as well as construction of underground car parking in the basement, intended for residents of the adjacent apartment building. There were plans for a layer of topsoil 20–30 centimetres thick on the roof; small recreational spots with benches, connected by cobblestone-paved paths, were planned for the resident of an apartment house that was part of the same complex [20]. Although the investigation is not yet complete, three possible causes are being set so far – 1) error in structural design, in particular bolt connection failure of two pieces roof trusses; 2) errors in building procedure and 3) construction of the green roof. On April, 2015 the State Police handed over the results (but not published it) of its massive criminal investigation to the prosecution, asking for charges to be brought against eight persons allegedly involved.

Eldukader and Ayyub [16] conducted a study on structural failures and found that 57% of structural failures resulted in collapse; 44% of failures occurred during construction; 51% of cases involved design errors; and 57% stemmed from construction errors. Structural designers were responsible for 48%; contractor's site staff were involved in 60%. A great majority (86%) of failures were reinforced concrete, while 34% were slabs and plates. Primary causes of failures were listed as inadequate construction methods; poor erection procedures; inadequate connection elements leading to unexpected load behavior; unclear contract information; and violation of contract. Secondary causes were highlighted as natural phenomena (disasters); environmental factors (climate); poor project controls; poor material and equipment usage; and poor workmanship. It was indicated that from a construction management perspective, inadequate planning and supervision, ineffective interpretation/evaluation of safety regulations, poorly defined work responsibilities, poor communications between project participants, insufficient work cooperation, and inadequate training of the workforce contributed to the failures. Engineering factors playing a role were poor site selection and site development, design deficiencies, construction issues, material deficiencies, and operational errors.

Under site selection and development challenges poor land-use planning, insufficient geotechnical surveys, and unnecessary exposure to natural hazards were mentioned. Errors in design concept, lack of structural redundancy, low safety factor, failure to consider a load, calculation errors, misuse of computer software, drafting errors, poor specifications development, and improper connection details were identified as main issues. Material defects covered were material inconsistencies, premature deterioration (including corrosion), poor choice of material, manufacturing or fabrication defects, improper substitutions, and mix-design problems. Finally, alterations to structure, change in use, inadequate maintenance, and negligent overloading were found to be among the critical operational matters contributing to structural failures.

General recommendations presented by Eldukader and Ayyub [16] for the prevention of construction and structural failures included education (including case studies), research, industry association events, failures database and attention to building codes.

2.1. Building codes

Building codes are developed by city, county and state building safety professionals with significant input from the design community and the construction industry. International Building Code (IBC) is currently used in most jurisdictions in the US, and it covers a wide range of topics such as building occupancy classifications, building heights and areas, interior finishes, foundation, wall, and roof construction, fire protection systems (sprinkler system requirements and design), materials used in construction, elevators and escalators, and means of egress. Besides the Building Code (IBC 2009), codes in common use are Mechanical Code (IMC 2009), Plumbing Code (IPC 2009), Fire Code (IBF 2009), Fuel Gas Code (IFGC 2009), Electric Code (NEC2002), Dwelling Code (IRC 2003), and Energy Code (IECC 2003). The primary aim is to establish provisions consistent with the scope of a building code that adequately protects public health, safety and welfare; and subject to the following provisions: (a) do not unnecessarily increase construction cost, (b) do not restrict the use of new materials, products or methods of construction, and (c) do not give preferential treatment to particular types or classes of materials, products or methods of construction.

Local or state government agencies administer and enforce the building code. They appoint building officials to render interpretations of the building code and to adopt policies and procedures in order to clarify the application of its provisions. The building officials perform the following duties: (a) They review applications and issue permits related to building construction work to implement the code; (b) They do all necessary building inspections; (c) They issue notices and orders to ensure compliance with the code; (d) They keep official records related to application, permit, inspections, and so on. Building officials will have the right to enter a building to enforce the code for ensuring building safety. In this process, they are relieved from personal liability for any damage to a person or property as a result of any act or omission in the discharge of official duties.

Construction documents submitted to building officials for obtaining permits include location and layout, site grading, location of utilities, plan and profile, standard construction notes, standard details, and specifications. A permit is not valid until the fees prescribed by law have been paid. A certificate of occupancy is issued after the building official inspects the building or structure and finds no violations of the provisions of the code or other laws that are enforced by the building department.

3. Construction quality management

3.1. Quality management concepts

Quality is associated with meeting project requirements. From a legal perspective, it relates to expertise, liability and standard of care; and from the aesthetics viewpoint, it means the fineness of visible features. When functionality is considered, the designer approaches it by ensuring workable plans and specifications; the owner focuses on project's timely and within budget completion, life cycle costs, and operations and maintenance; the constructors want plans and specifications adequate for estimating and bidding, as well as building the project with fair contract provisions, with timely response to change order requests and inspection reports. Public safety and health, and compliance with laws, regulations, codes and standards are the concern of the regulatory agencies [7].

Quality assurance (QA) and quality control (QC) are processes that are integrated and run together in quality management. QA encompasses all planned and systematic actions to provide adequate confidence that a structure, system or component will perform satisfactorily and conform to project requirements. QC, on the other hand, is a set of specific procedures to assess performance (against requirements) to ensure that the required standards of quality will be met. Generally QA is the responsibility of the owner or his/her designate, aiming to examine the QC methods in place to determine if the contractor is properly controlling work activities, making sure that the QC system and documentation are adequate, and deficiencies are corrected. QC, being a contractor responsibility, requires him/her to develop and maintain an effective QC system, perform all control activities and tests, and prepare acceptable documentation. A competent on-site representative with authority oversees the QC system, evaluating and controlling workmanship and methods to ensure quality is produced as defined in the contract.

Quality management includes processes and activities to establish quality policies, objectives, and responsibilities so that the project will satisfy the needs for which it was undertaken. Although QA/QC systems have existed for a long time, the modern thought of management of quality in a holistic way is relatively new (since 1980's). Total quality management (TQM) is a comprehensive and structured approach to organizational management that seeks to improve the quality of products and services through ongoing refinements in response to continuous feedback framework for implementing effective quality and productivity initiatives that can increase the profitability and competitiveness of organizations. The concept is also applicable to project management [1], [5].

There are many quantitative statistical methods that are employed in the implementation of TQM in quality measurement and improvement. The most commonly used ones are: histogram; Pareto analysis; scatter diagram; affinity diagram; control chart; flow chart; cause-and-effect (fishbone) diagram; and failure mode and effect analysis (FEMA) technique. An important tenet of TQM is doing things right the first time rather than relying on correcting errors. This is associated with the cost of quality, which is the sum of costs related to prevention of errors; appraisals costs for evaluating processes and outputs; failure costs tied with rework and error correction [16].

ISO 9000/9001 and Six Sigma are the two tools that systematically implement the TQM methodology. ISO is the family of standards related to quality management systems (QMS) and designed to help organizations ensure that they meet the needs of customers and other stakeholders. Six Sigma is a highly disciplined process that enables organizations deliver nearly perfect products and services through continued process improvements, where Sigma measures how far a given process deviates from perfection. Higher Sigma capability denotes better performance, the ultimate being 6 standard deviations from the mean which is being error free 99.9996 % of the time in a production situation. The approach to Six Sigma implementation is described by a series of steps involving Design, Measure, Analyze, Improve, and Control (DMIAC), each step involving the mentioned statistical tools [2], [3], [10].

IOP Conf. Series: Materials Science and Engineering **96** (2015) 012061 doi:10.1088/1757-899X/96/1/012061

3.2. Quality management challenge in construction

Construction is unique and different from manufacturing, in that structures have longer product life cycle; the owner usually influences production; project participants/stakeholders differ for each project, and each project is first-of-a-kind prototype (no mass production).This requires a unique quality management approach, incorporating relevant codes and permits, owner requirements, design provisions, testing and inspections, and project quality plans. Quality management in construction means making sure things are done according to plans, specifications and permit requirements. No comprehensive quality policy exists that is applicable to the entire industry or large segments thereof, with the exception of nuclear plants and interstate highways [7].

The owner, designer, constructors and regulatory bodies (building authority) must work together to inspect, document and correct deficiencies to manage quality. Contract documents (P&S) communicate design intent and provide technical information on materials, structural elements/assembly and quality requirements for construction and constructed facility construction (building, electrical, plumbing, fire, etc.) Codes present minimum requirements for quality, and construction permits are issued after the P&S review and inspections. Construction and building inspectors ensure that construction meets local and national building codes, zoning regulations, and contract specifications. There may be inspectors on a construction project working for owner, designer, regulatory bodies, or financial institutions.

The quality management plan for the project delineates the roles and responsibilities of project participants, required training and certification, QC/QA processes and systems; shop drawings and other submittals (e.g. samples, temporary structures, traffic plans), testing and inspections, record keeping, verification, correction , reporting and documentation protocols, and quality management audits [9].

4. Construction productivity and lean Six Sigma

Productivity is commonly expressed in terms of quantities produced (output) per employee hour of effort (input). The input can be man-hours, machine time, materials, investment, or capital, while output can be mass, area, volume or length. For construction, productivity can be measured by the number of hours required per item produced (output over input). Construction managers view productivity as the work accomplished per hour for a crew, or at the individual laborer level (e.g., cubic yard of concrete construction completed per crew hours or labour hours worked), which is essential for detailed estimates at the bidding phase. Productivity values can be expressed as an average for the whole project, so comparisons can be drawn within or between projects. A disadvantage is that lots of data may be needed for analysis [14].

Jobsite productivity is influenced by labor characteristics (skill, experience, training, leadership/motivation); project work conditions (job size and complexity, accessibility, equipment availability, breakdown, contractual agreements, culture and local climate); and non-productive activities (indirect labor for supporting construction activity, inclement weather, material shortage, absenteeism, strikes). Productivity is also influenced by design deficiencies, poor coordination, communication, supervision, and scheduling, and stakeholder involvement (e.g., scope changes). Application of lean management principles and tools can prevent productivity losses and improve efficiency. The focus in this approach is on eliminating non-value adding activities and components (waste). Types of wastes identified under the lean Six Sigma approach are: Defects, Overproduction, Waiting, Non-Utilized Talent, Transportation, Inventory, Motion, Extra-Processing (DOWNTIME). The lean tools are comprised of Sort, Straighten, Shine, Standardize, and Sustain (5S) [16].

IOP Conf. Series: Materials Science and Engineering **96** (2015) 012061 doi:10.1088/1757-899X/96/1/012061

5. Safety-quality-productivity synthesis

5.1. Lean construction safety

A recent industry survey indicates that there is synergy between lean production and safety management practices in construction [15]. Lean production is based on the Toyota Production System (TPS) to achieve best quality, lowest cost, shortest lead time, best safety and high morale. Lean production tools include Just-In-Time (delivering right items at right time and amount), autonomation ("jidoka" – never letting defective items go to next station; stopping workers and machines when defects are significant), production leveling (reducing variability and inconsistency), and continuous improvement (there is no best, there is always an opportunity to improve processes). These tools can be applied to improving safety management practices such as management commitment, staffing for safety, planning for safety, safety education/training, worker involvement, evaluation and recognition, subcontractor involvement, and accident investigation. Since the lean approach combines quality with productivity, and it can improve safety, one can conclude that safety, quality and productivity can be collectively improved [14], [15].

5.2. Construction safety vs. productivity

The question of how safety is related to productivity often comes up in decisions concerning how much attention and investment must be directed toward safety programs in the construction industry, especially in smaller and mid-sized firms. The following questions relevant to this topic can be considered by the managers.

1. Does safety have a negative or positive effect on construction productivity?
2. What is the significance of the relationship between safety and productivity?
3. Do the expenses associated with a safety program pay off in terms of cost savings for the project?
4. Do safety precautions (compliance) slow down construction work resulting in increased time and costs, and reduced productivity?
5. How can safety and productivity be simultaneously achieved?

Construction projects are expected to simultaneously achieve multiple goals concerning cost, schedule, quality, productivity and safety. These factors are related, while they may be in apparent conflict with each other. Because of pressures on cost, schedule, and productivity, workers may engage in unsafe behavior, believing that this enables them to be more productive. On the other hand, there is strong belief in the industry that no trade off exists between safety and productivity, and in fact they might be synergistic [8], [9], [17], [18], [20].

There is some evidence to support the notion of safety being negatively correlated with productivity. However, there is stronger support for the opposite; that is, safety and productivity are positively correlated, meaning there would be no loss of productivity due to taking safety measures or executing safety programs in a construction project. On the negative correlation side, there are reports in the safety literature that workers may ignore safety due to productivity concerns. In a recent study, safety audits conducted at 200 residential sites showed 59 % non-compliance, meaning widespread disregard for safety. It was found that this came about as a result of concern over taking time off work to learn new safety measures, fear of cost of compliance, and lack of knowledge (inadequate training). In another study, it was observed that managers believed there was not enough time to work safely and safety practices decreased productivity. Research on the impacts of schedule pressures on safety indeed showed increases in injury rate when such pressures are present. Another research study found that workers cut corners on safety to be more productive for fear of losing their jobs. In a study on roofers, it was observed that productivity dramatically decreased when fall arrest systems were used. Workers lost a lot of time adjusting their lanyards on steep roofs resulting in loss of productivity. In a more balanced study, it was concluded that there was a strong negative correlation between safety and productivity when new safety strategies were implemented; however, the correlation turned positive after some time elapsed in the project.

In support of positive correlation between safety and productivity, a U.K. study indicated that productivity losses were higher with safety violations than with preventive safety practices. Similar evidence in a different study suggested that safety management has a positive influence on productivity because of declining task performance by the entire crew in case of injuries. According to this study, damaged equipment and materials, time for additional record keeping, accident investigation and training also slow down production. According to Hinze's [17] distraction theory, workers are more productive when they perceive hazards to be minimal, than working with the perception of being in a high hazard environment. Productivity is compromised when distractions due to hazards turn the workers attention to mitigating safety risks. A recent study on manufacturing productivity vs. safety concluded that management should view safety and operating performance measures as being in concert with each other rather than competing entities. There is more scrap, rework and less employee involvement when safety performance is poor. Based on 18 case analyses, a study by Maudgalya et al. [18] revealed that several organizations realized a payback of their monetary investment in safety initiatives in less than a year. Following the implementation of workplace safety initiatives, organizations observed an average increase of 66% in productivity, 44% in quality, and 71 % in cost benefits. The study concluded that as a business case, safety can assist an organization in achieving the long-term benefit of operational sustainability. Finally, a study by McGraw [14] suggests that contractors are experiencing positive business outcomes from safety programs in terms of shortened schedules, budget/cost savings, increases in project return on investment (ROI), improved reputation, improved market share, and improved project quality.

6. Summary and conclusions

Given that some members and segments of the construction industry emphasize productivity over safety and quality, the main thrust of this study was to review how safety, quality and productivity affect each other, and whether they are in conflict with each other or they are synergistic. Safety was examined in the context of a construction project, as well as during service for constructed facilities. Factors affecting construction and building failures, preventive strategies, and the importance of quality management in securing safe performance for buildings, including the roles of building codes, inspectors and building officials were reviewed. Quality and productivity concepts were evaluated separately and in combination as related to the lean Sigma principles and tools, and this was extended into establishing the relationship between quality, productivity, and safety.

The following conclusions are drawn from this study:

1. Construction safety and productivity are positively correlated in many cases, meaning that investments made into the safety programs do not negatively impact productivity in construction, although some situations exist where safety precautions might slow down productivity.
2. Safety management and quality management are integrated, and they work together. Safe projects will support quality efforts.

3. Since the lean principles apply both to safety and quality in construction, and they play a significant role in increasing the productivity, there is synergy between these three facets of construction projects.

The information on the relationships between safety, quality and productivity in construction is quite dispersed and fragmented. Further research into safety, quality and productivity relationships is warranted, and is hereby recommended.

7. References

[1] Project Management Institute 2013 *A Guide to the Project Management Body of Knowledge* PMBOK Guide
[2] Hoyle D 2009 *ISO 9000 Quality Systems Handbook - updated for the ISO 9001:2008*
[3] US Army Corps of Engineers *Module 6: Quality Management for Construction Projects*
[4] Arditi D and Gunaydin H M 1996 Total quality management in the construction process *Int. J. of Project Management* **15** (4) 235–243
[5] ASCE 2012 *Quality in the Constructed Project: A Guide for Owners, Designers, and Constructors* ASCE Manual No. 73 third ed.
[6] Goetsch D L 2013 *Construction Safety & Health* 2nd ed. Pearson, ISBN:978-0-13-237469-30
[7] Usmen M 1994 *Construction Safety and Health for Civil Engineers* Instructional Module ASCE
[8] Dozzi S P and AbouRizk S M 1993 *Productivity in Construction* Institute for Research in Construction National Research Council, Canada
[9] Alarcon L 1997 *Lean Construction* Balkema Publishers
[10] Isa M and Usmen M 2014 Improving university facilities services using lean six sigma: a case study *Int. J. of Facilities Management*
[11] Hinze J 1996 The distraction theory of accident causation *Proc., Int. Conf. On Implementation of Safety and Health on Constr. Sites, CIB Working Commission W99: Safety and Health on Construction Sites* Alvez Diaz L M and Coble R J eds. Balkema, Rotterdam, The Netherlands, 357–384
[12] Maudgalya T, Genaidy A and Shell R 2008 Productivity-Quality-Costs-Safety: A sustained approach to competitive advantage–A systematic review of the National Safety Council's case studies in safety and productivity *Hum. Factors Ergon. Manuf. Serv. Ind.* **18** (2) 152–179
[13] Veltri A, Pagell M, Behm M and Das A 2007 A data-based evaluation of the relationship between occupational safety and operating performance *J. of Safety, Health and Env. Research* **4** (1)
[14] McGraw Hill Construction 2013 *Safety Management in the Construction Industry: Identifying Risks and Reducing Accidents to Improve Site Productivity and Project ROI* SmartMarket report
[15] Antillon E I, Alarcón L F, Hollowell M R and Molenaar K R 2012 A Research synthesis on the interface between lean construction and safety management *Proc. Construction Research Congress* ASCE, West Lafayette
[16] Eldukadir A and Ayyub B M 1991 Analysis of Recent U.S. Structural and Construction Failures *J. of Performance of Constructed Facilities, ASCE* **5** (1) 57–73
[17] Kaminetsky D 1991 *Design and Construction Failures: Lessons from Forensic Investigations* McGraw Hill
[18] Hyatt Regency Collapse [Available]: http://www.asce.org/PPLContent .aspx?id=12884909405

[19] Algo Centre Mall [Available]: https://failures.wikispaces.com/ Algo+Centre+Mall+Collapse
[20] Zolitūde shopping centre roof collapse [Available]: http://en.wikipedia.org/wiki/Zolit%C5%ABde_shopping_centre_roof_collapse

IOP Conf. Series: Materials Science and Engineering **96** (2015) 012062 doi:10.1088/1757-899X/96/1/012062

Strengthening under Load: Experimental and Numerical Research

M Vild[1,2] and M Bajer[1]
[1]Brno University of Technology, Faculty of Civil Engineering, Institute of Metal and Timber Structures, Veveří 331/95, 602 00, Brno, The Czech Republic

E-mail: vild.m@fce.vutbr.cz

Abstract. The paper presents experimental and numerical research of strengthening of columns under load using welded plates. Three sets of three columns each were tested. All columns were 3 m long. The load from loading cylinders was transmitted through knife-edge bearings, which ensured pinned boundary condition perpendicular to the weaker axis. Set (A) comprised columns with welded T shaped cross-section. Set (B) comprised columns with welded monosymmetric I shaped cross-section. Both sets (A) and (B) had been loaded monotonically until collapse occurred. Set (C) contained columns with T shaped cross-section with the same dimensions as the columns in set (A). The columns from set (C) were first loaded to 70 kN. The force was being held constant and the second flange was being welded to the web under load. After the welding process was finished and the specimen cooled, the column was loaded to failure. The average forces at collapse of column sets (A), (B) and (C) were 143kN, 308 kN and 323 kN, respectively. It was unexpected that the columns strengthened under load (C) had higher average resistance than the columns welded without preload (B). It could be caused by the residual stress and distortion caused by welding. The study includes the results of finite element models of the problem created in ANSYS software. The results from the experiments and numerical simulations were compared.

1. Introduction

Strengthening of steel members using welded plates is a common praxis because it is a fast and cheap method. Unloading of the structure may often be hardly feasible or economically inconvenient and then strengthening under load is carried out. Although there are several rules of thumb, current normative documents do not refer to this problem and no reliable analytical solution has been developed up to date. Most authors until 1990 [1, 2] allow only elastic distribution of stress in the strengthened member. Unterweger [3] used finite element study to determine ultimate plastic resistance of the member strengthened under load. Vild, Bajer [4] modified the method to be more

[2] To whom any correspondence should be addressed.

conservative and in accordance with current Eurocodes. Unfortunately, no method copes with residual stress and deflection caused by welding, which seems to have a significant impact on the buckling strength. Unterweger suggests reduction of ultimate load capacity by around 10 %. On the other hand, Rao and Tall [5] observed increase of ultimate load capacity.

The paper presents experimental and numerical research on strengthening of columns under load using welded plates. The experiments were conducted at the laboratory of the Institute of Metal and Timber Structures at Brno University of Technology in December 2014.

Figure 1. Specimen in the load set-up.

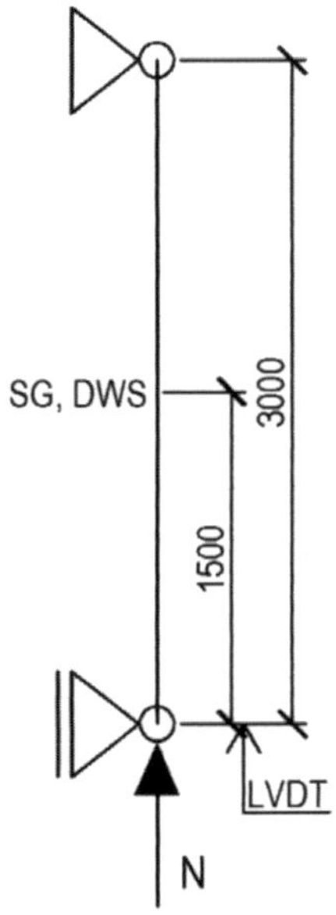

Figure 2. Loading scheme, LVDT, strain gauge (SG) and draw wire sensor (DWS) location.

2. Methods

Three sets of columns were selected for the research. The columns were subjected to compressive force (see figures 1 and 2). The cross-sections are presented in figure 3. All columns were welded with continuous weld with throat thickness 4 mm. Set (A) included columns with T shaped cross-section welded from the flange with dimensions 140 ×7.7 mm and the web with dimensions 200 × 6.3 mm. According to EN 1993-1-1 [6] classification of cross-sections, the columns of set (A) were class 4. Set (B) comprised the columns with monosymmetric I shaped cross-section. The flanges had dimensions 140 × 7.7 mm and 80 × 8.0 mm and the web 200 × 6.3 mm. This cross-section was class 2. Sets (A) and (B) were necessary for comparison of behaviour and resistance. Set (C) contained the columns with T shaped cross-section with the same dimensions as the columns in set (A). The columns from set (C) were first loaded to 70 kN and then strengthened under load with the second flange. The resulting cross-section was the same as of the columns from set (B).

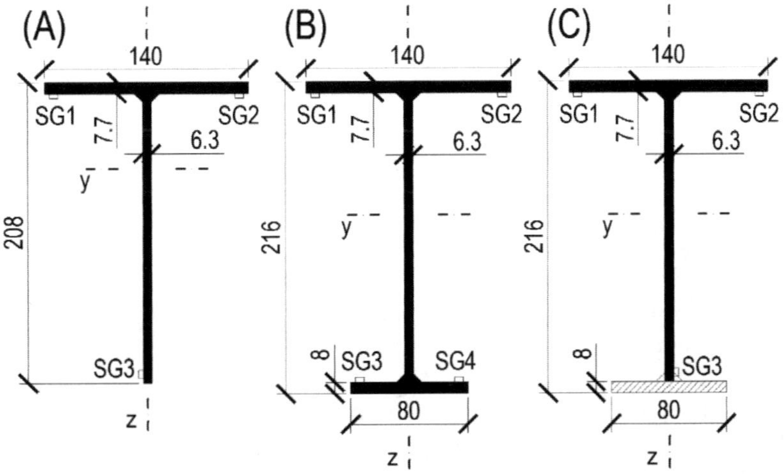

Figure 3. Column sets with dimensions and strain gauges (SG) position.

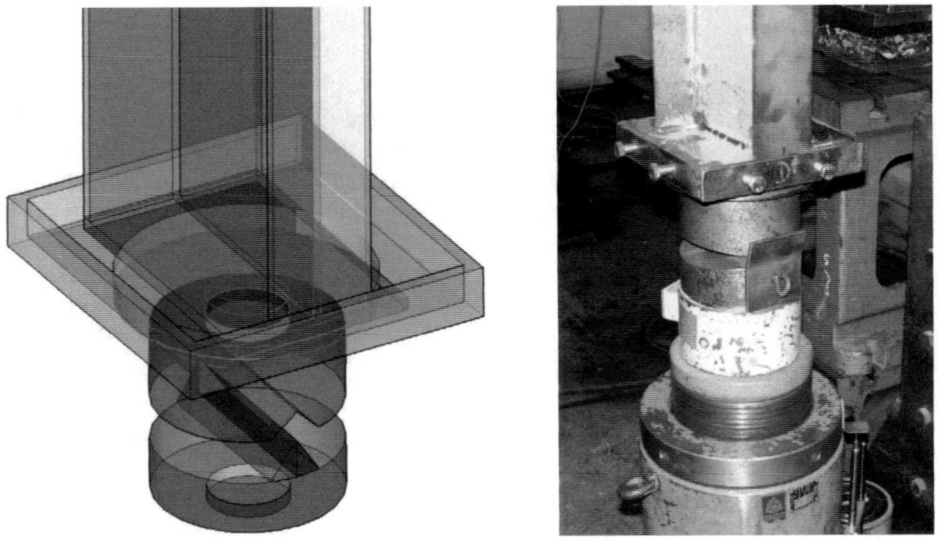

Figure 4. Scheme and photograph of the knife-edge boundary condition.

2.1. Analytical solution

The buckling resistance was determined by procedure from EN 1993 [6]. Length of the column plus knife-edge bearings thickness was used as buckling length in direction perpendicular to the weaker axis and half of this value in direction perpendicular to the stronger axis and in torsion. No additional bending moment was included. The buckling resistance of set (A) and set (B) is 159 kN and 269 kN, respectively.

Figure 5. Welding process and thermal image.

2.2. Experimental research

Each set comprised three columns for experimental research. All columns were 3 m long and loaded by loading cylinder. The load was transmitted through knife-edge bearings, which ensured pinned boundary condition perpendicular to the weaker axis and fixed perpendicular to the stronger axis (see figure 4). Both sets (A) and (B) had been loaded monotonically until collapse occurred. The columns from set (C) were first inserted into the loading set-up and loaded to 70 kN. The force was held roughly constant (± 5 kN) and the second flange with dimensions 80 × 8 mm was being welded to the web under the load. The shielded metal arc welding method was used. The force was rising by around 1 kN per 12 s while the steel was being heated by the welding process and dropped at the same rate while the process was paused and the column cooled. The temperature of the column was monitored with thermal imager (see figure 5). After the welding process was finished and the specimen cooled, the column was unloaded to 10 kN and then loaded to failure.

Strain gauges 1-LY11/6/350 were used (see figure 3). The vertical displacement and force at the bottom and the horizontal deflection and strains in the middle of the height were measured in case of all specimens. Tension coupon tests were conducted to determine steel mechanical properties. Four coupons were machined from webs and four from flanges. Precise thickness of plates was measured with callipers on several spots and the average value was used for analytical and numerical analysis. South well plot [7] was used to determine the initial imperfections.

2.3. Numerical modelling

Numerical simulations were performed to complement the values which have not been measured in the experiment. ANSYS software [8] was used for finite element analysis. SHELL 181 element type was used for all plates. The true stress strain diagrams obtained from the coupon tests were used for two material models – one for flanges and another one for webs. First, the ideal geometry was created and then it was updated with imperfections using Block Lanczos modal analysis. The first mode was selected with maximum deflection 5 mm, which is equal to 1/600 of the column length and roughly coincides with the results of Southwell plot. Pinned boundary conditions and node coupling in the longitudinal direction were used at the position of the knife-edge bearings. Then the elements were switched to SHELL 131 and two thermal analyses were conducted. The first was for the residual stress distribution of the base T shaped cross-section, the second was for deflections and residual stress caused by welding of the strengthening flange under load (see figure 6). Then the elements were changed again to structural and five consecutive static analyses were performed. First, the elements of the strengthening plate were deactivated and the residual stress for the base T shaped cross-section was implemented. Second, the column was loaded to 70 kN. Third, the elements of the strengthening plate were reactivated and residual stress and deflection caused by the welding process were implemented. Fourth, the column was unloaded to determine the change of initial imperfections of the column strengthened under load. Finally, the column was loaded with force until the solver stopped converging, meaning collapse was reached.

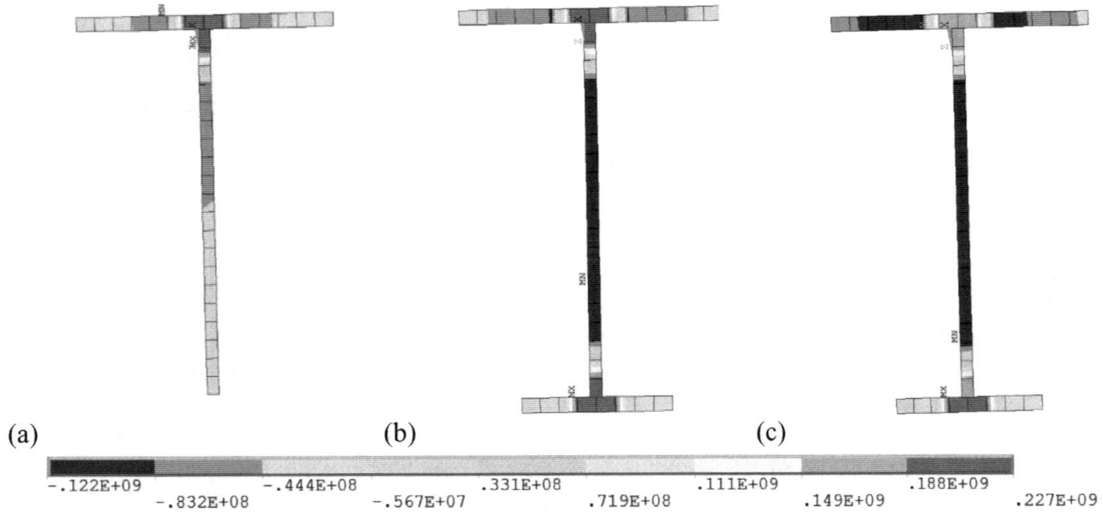

Figure 6. Residual stress from welding in longitudinal directionat the middle column height: set (a) – without load; set (b) – without load; set (c) – with preload 70 kN.

3. Results and discussion

3.1. Experimental research

The results of mechanical properties were averaged and summarized in table 1. The web steel seemed to be of slightly higher quality than the flange steel.

Table 1. Mechanical properties from coupon tests.

	E [GPa]	f_y [MPa]	f_u [MPa]
Web	211	334	480
Flange	203	310	457

Stress recalculated from strain gauges and deflectionsfrom draw wire sensors at the moment of collapse are plotted in figure 7, graphs of deflections dependent on axial force are shown in figure 9, note that deflections are in absolute values. All specimens failed by torsional-flexural buckling as expected (see figure 8) but the load, stress and deflection at collapse show great scatter.

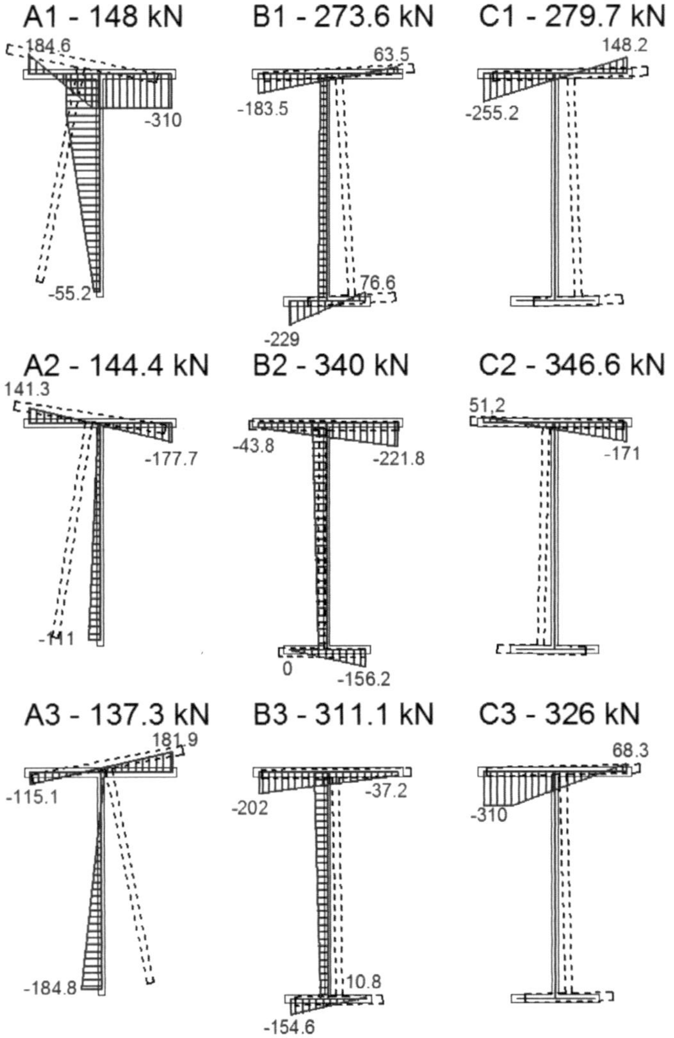

Figure 7. Stress recalculated from strain gauges and deflection at the moment of collapse.

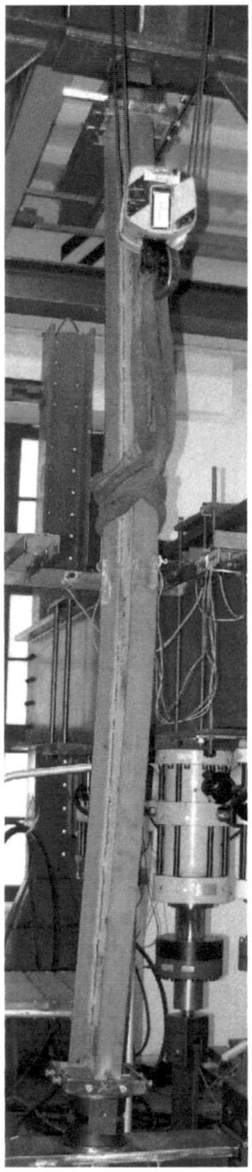

Figure 8. Collapsed column.

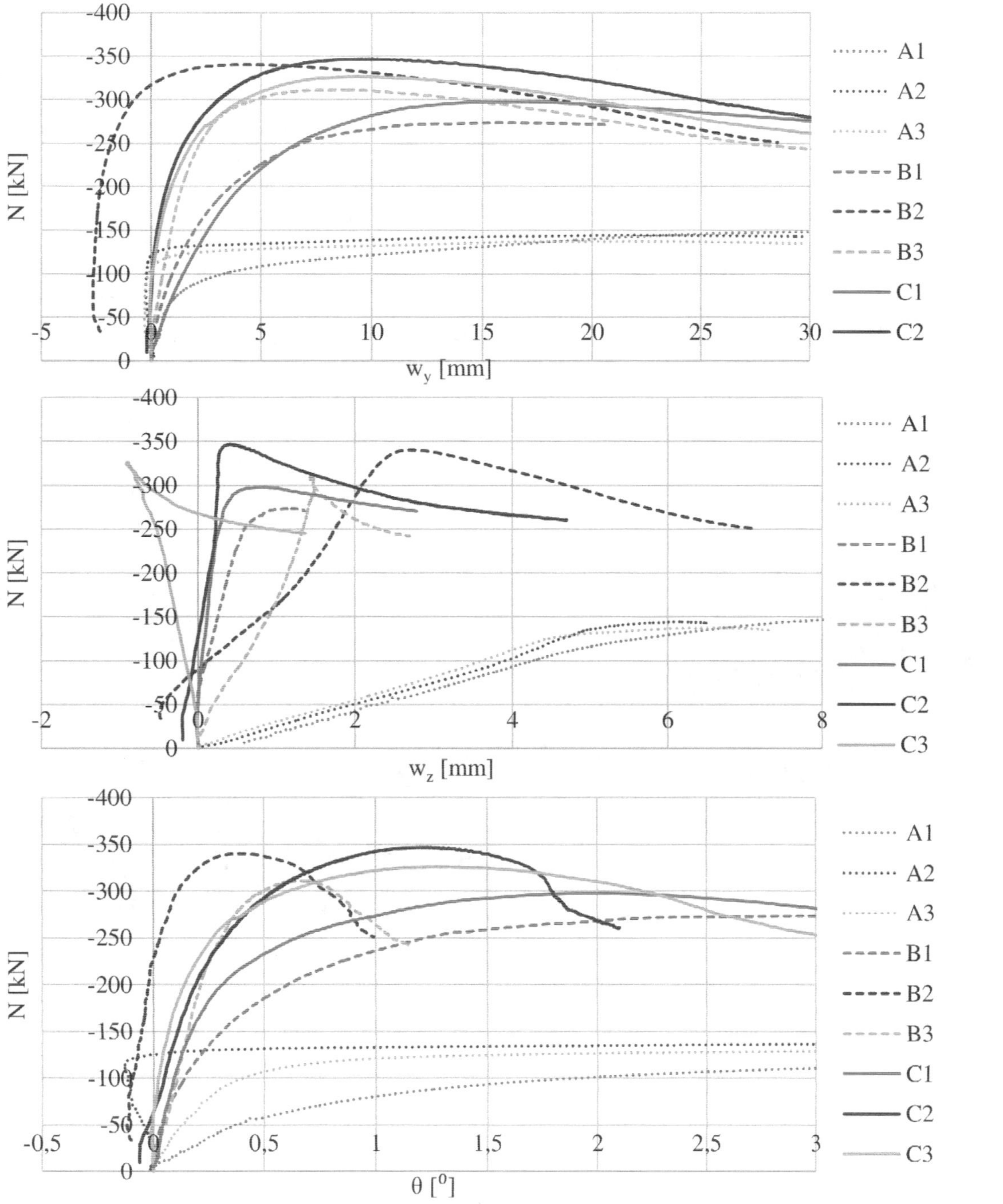

Figure 9. Graphs of deflections in the direction of stronger axis y and weaker axis z and rotation ϑ.

The average stress on the flange of the base T shaped cross-section (average of SG1 and SG2) at load 70 kNwas only -5 MPa (compression). The stress after the welding process even switched into tension to average value 18 MPa (tension). The stress on the web of the base T shaped cross-section (SG3) was -93 MPa, which is still a low value compared to the yield stress. The welding process near SG3 unfortunately destroyed the strain gauge but the weld clearly caused shrinkage. The specimens from set (C) deflected at the middle height by welding by average value -10.63 mm in *z* direction. The average ultimate load capacity of set (B) was 308 kN and of set (C) 323 kN. In the case of these 6 specimens, the resistance of the columnstrengthened under load (C) even increased compared to the column with I shaped cross-section welded without preload (B). This could be explained by the effect of residual stress and deflection caused by welding, which slightly returned the column to the original shape and caused the tension in the most compressed base section flange. This behaviour and increase of ultimate load capacity was also observed, for example, by Rao and Tall [5].

3.2. Numerical modelling

The numerical results coincide closely with the experiment. Von Mises stress at the moment of collapse is shown in figure 10. Deflections are plotted in figure 11; note that deflections are in absolute values and deflections caused by welding are set to zero in the case of experimental results, while in the numerical simulation all the processes – welding of T shaped cross-section, loading to 70 kN, welding of the strengthening flange, unloading and loading to failure – are plotted.

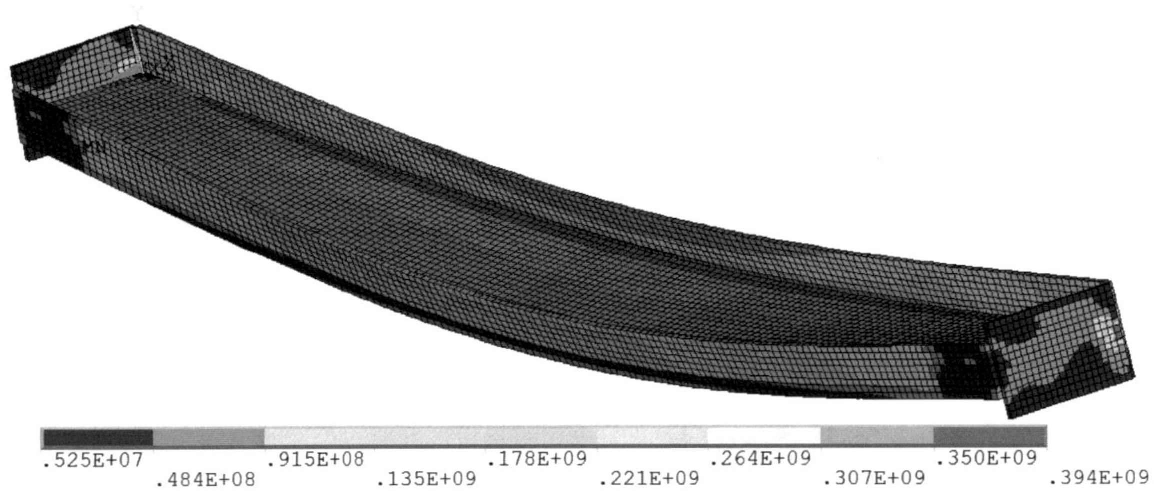

Figure 10. Von Mises stress distribution for the column from set (c) at failure (deflection scale 5).

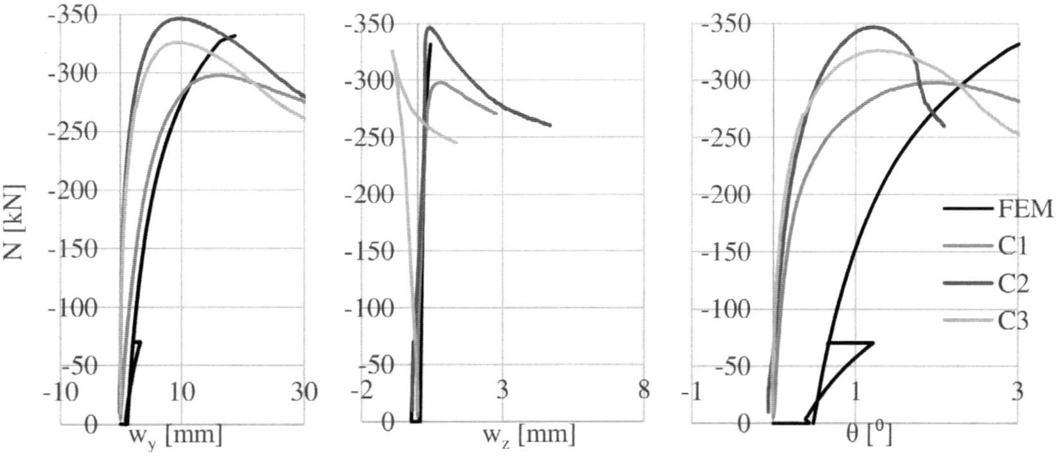

Figure 11. Comparison of experimental and numerical results of deflection.

Additional simulations were made to determine the effect of residual stress and preload on the ultimate load capacity of the column. The preload 70 kN decreased the ultimate load capacity by only 2.7 %, therefore, the increase in resistance observed from the experiment is not confirmed. The residual stress decreased the load capacity of the column with I shaped cross-section and the column withT shaped cross-section but the implementation of residual stress at the preload indeed showed increase of resistance by 6.3 %.

The sensitivity analysis was performed by Kala [9]. It showed that residual stress has the highest impact on columns with relative slenderness around 0.7. The relative slenderness of the specimens was higher (1.8 for T shaped and 1.5for I shaped cross-section). At this relative slenderness, axis curvature has the highest impact but residual stress is more determinant than, for example, material yield strength.

4. Conclusion

The experiments proved that strengthening of the column with T shaped cross-section under preload up to 50 % of the base section buckling resistance is safe and feasible. However, the force rises rapidly if the welding process is fast. Thus, temperature changes must be carefully monitored and stress changes considered especially if the column is less stiff than the surrounding structural components.

The numerical simulation serves to determine the values that had not been measured or are unclear in the experiment. The validated numerical model will serve for the behaviour analysis of members strengthened under load in future research in this area, which is in progress and other experiments have already been planned.

The results of the T shaped cross-section columns require further research. The buckling resistance with the consideration of additional moment caused by local buckling according to the Eurocode [6] (88.5 kN) is extremely conservative, but the resistance without additional moment (159 kN) is too high compared to the experiment.

5. References

[1] Spal L 1968 *Rekonstrukce Ocelových Konstrukcí* (Prague: Státní nakladatelství technické literatury) p 183
[2] Rircker T D 1988 Field welding to existing steel structures *Engin. J.* 16
[3] Unterweger H 1999 Ultimate load capacity of columns strengthened under load *Advances in Steel Struct.* 117–124
[4] Vild M and Bajer M 2015 Flexural buckling of columns strengthened under load *Advanced Mat. Research* 4
[5] Nagaraja Rao N R and Tall L 1963 Columns reinforced under load *Welding Research Council, Welding Research Supplement* 177–85
[6] ČSN EN 1993-1-1 Eurokód 3: Navrhování ocelových konstrukcí – Část 1-1: Obecná pravidla a pravidla pro pozemní stavby2011 (Prague: ČNI) p 94
[7] Southwell R V 1932 On the Analysis of Experimental Observations in Problems of Elastic Stability *Proc. Roy. Soc. London* Series A pp 601–616
[8] ANSYS® Academic Research, Release 15.0, Help System, ANSYS Parametric Design Language Guide, ANSYS, Inc.
[9] Kala Z and Kala J 2011 Sensitivity Analysis of Stability Problems of Steel Structures using Shell Finite Elements and Nonlinear Computation Methods ICNAAM *AIP Conf. Proc.* vol 1389 pp 1865–68

Acknowledgement

The financial support of projects FAST-J-15-2859, TA03010680 and No. LO1408 "AdMaS UP - Advanced Materials, Structures and Technologies", supported by Ministry of Education, Youth and Sports under the "National Sustainability Programme I" are gratefully acknowledged.

Road Safety Barriers, the Need and Influence on Road Traffic Accidents

Ž Butāns[1,3], **K A Gross**[1], **A Gridnevs**[2] **and E Karzubova**[2]
[1]Riga Technical University, Azenes Str.14-216, Riga, LV-1048, Latvia
[2]Ltd. Tilts, Spilves Str. 2a, Riga, Latvia LV-1055

E-mail: Zans.Butans@gmail.com

Abstract. Constantly increasing intensity of road traffic and the allowed speed limits seem to impose stronger requirements on road infrastructure and use of road safety systems. One of the ways to improve road safety is the use of road restraint systems. Road safety barriers allow not only reducing the number of road traffic accidents, but also lowering the severity of accidents. The paper provides information on the technical requirements of road safety barriers. Various types of road safety barriers and their selection criteria for different types of road sections are discussed.
The article views an example of a road traffic accident, which is also modelled by PC-Crash computer program. The given example reflects a road accident mechanism in case of a car-to-barrier collision, and provides information about the typical damage to the car and the barrier.
The paper describes an impact of the road safety barrier type and its presence on the road traffic accident mechanism. Implementation and maintenance costs of different barrier types are viewed. The article presents a discussion on the necessity to use road safety barriers, as well as their optimal choice.

1. Introduction

Along with the increasing road traffic intensity, the issue of road traffic safety is becoming more important requiring further attention from all the institutions. Road traffic accidents (RTA) are generally caused by vehicle diversion from the intended direction. If the driver loses attention, but the vehicle remains steerable, the presence of rumble strips along the centrelines and outer edges of undivided rural two-lane roads alerts the distracted, fatigued, or speeding motorist. This is an effective and relatively inexpensive installation system that can decrease the RTAs by approximately 14% [1]. However, if the vehicle has lost control, road safety barriers (hereinafter – barriers) play an important role in decreasing the RTA consequences provided that there are very few motorized two-wheel

[3] To whom any correspondence should be addressed.

vehicles using the roadway [2].

To decrease the number of RTAs and improve the overall traffic safety, a safety performance audit is required, which has proved to be an effective practice in enhancing road safety. Safety performance audit in China confirms that barrier systems, which are installed complying with the necessary requirements, improve traffic safety. It is also necessary to introduce more stringent requirements and improve standards [3].

The aim of this article is to show the necessity of barriers and their impact on the collision mechanism and collision consequences. A reference to all barrier types will lead to the optimal choice of barriers depending on their installation site.

2. Types of road safety barriers

Barriers are used to separate oncoming traffic, to isolate objects on the roadside (trees, buildings, water reservoirs and others) from possible interaction with the vehicle, and prevent animal access to the roadway. Based on their function, barriers are divided into two types – flexible (figure 1) and rigid systems [4] (figure 2). In some cases, designation of semi-rigid systems is also found. As a result of vehicle collision, flexible barriers undergo permanent deformation. Collision energy is absorbed by both the barrier (barrier deformation) and the vehicle (vehicle deformation). Flexible barriers are usually made of steel or other deformable material (for example, aluminium and wood). Rigid barriers undergo insignificant deformation as a result of vehicle collision. Collision energy is dissipated to vehicle deformation, movement of barrier elements, friction between the vehicle and the barrier, as well as lifting up the vehicle onto the barrier.

Figure 1. Flexible barrier.

Figure 2. Rigid barrier.

The choice of a barrier is primarily based on the technical requirements, but also on its appearance. In national parks or along scenic roads, it is desirable that barriers look as natural as possible and fit into the surroundings. Wood barriers blend into the overall landscape [5, 6], figure 3. Alternatively, composite material barriers may be chosen as a safe, energy absorbent, lighter-weight, easier to install, more durable and economically competitive alternative to existing steel guardrails [7].

Cable median barriers (see figure 4) are cost effective and flexible systems that cause less impact force than concrete barriers and are easier to maintain than the W-beam guardrails. However, there is a 3.6%chance of penetrating the cable median barriers [8], [9].

Figure 3. The barrier made using wooden elements.

Figure 4. Cable median barrier.

Figure 5. Portable water-filled barrier.

The most common barrier is the W-beam steel guardrail, with properties that can change depending on the height. Rigid systems – concrete barriers – are the second most commonly used type of barriers. These offer a temporary solution or are used where the road width and barrier deformation is limited. Portable water-filled road safety barriers as a semi-rigid barriermay also be used at roadwork sites (figure 5). The main advantage is the ease of transportation and installation [10].

2.1. Technical requirements of barriers

In case of a collision, the primary goal of a barrier is to restrain vehicles from an impending collision, and slowly dissipate the kinetic energy to the greatest possible barrier surface.This would prevent relatively sharp overloads (ride down acceleration) to the people in the vehicle. After collision with the barrier, the vehicle should glide along the barrier or move back to the driving lane until it stops. It is important to prevent the vehicle from hanging over the barrier.

On public roads in Europe, the barriers are installed in compliance with the European Standard EN 1317 [11]. In Latvia, barriers are installed in accordance with the Latvian Standard (LVS EN 1317). This is identical to the consolidated European Standard EN 1317 with three main criteria: a containment level, impact severity level and deformations, expressed by the working width W and the dynamic deflection D. The three criteria are based on test results, which confirm a barrier ability to absorb the vehicle impact load.

2.2. Barrier containment level

In case of a collision, the barrier should contain the vehicle. The barrier must not separate into pieces, be torn or break loose from the supports, and be thrown up on the roadway where it may collide with other vehicles or users of the roadway. Depending on the hazard level of the road, barriers will have different containment levels.

Table 1. Barrier containment level characteristics.

Containment level characteristics	Designation
Low angle containment	T1
	T2
	T3
Normal containment	N1
	N2
Higher containment	H1, L1*
	H2, L2*
	H3, L3*
Very high containment	H4a, L4a*
	H4b, L4b*

* Additional vehicle impact test at 110km/h

Barrier containment level may be anywhere from low to very high. At a low containment level, the barrier has to survive a collision from a 1,300kg automobile moving at 80km/h and a collision angle of 8°. At a very high containment level, the barrier has to endure a collision of up to 38,000kg truck at 65km/h and collision angle of 20°. The classification of barrier containment levels is provided in table 1.

The barrier containment level is determined from the permitted vehicle driving speed, traffic intensity, terrain and other criteria. For example, a normal containment level N2 is used on common road sections, but a high containment level H2 or very high H4a and H4b is used on bridges. A safety barrier will always satisfy the requirements of a lower level.

A low containment level is used where temporary safety barriers are required, such as sites for roadwork. If required, temporary barriers can also have higher containment levels.

2.3. Working width and dynamic deflection

When a vehicle collides with a barrier, part of the kinetic energy absorbed by the barrier goes to deforming and moving the barrier. Barrier deformation is determined from the dynamic deflection D and barrier working width W, figure 6. Working widths are classified in levels from W1 to W8, where W1 level allows a barrier movement up to 0.6m, whereas W8 level allows movement up to 3.5m.

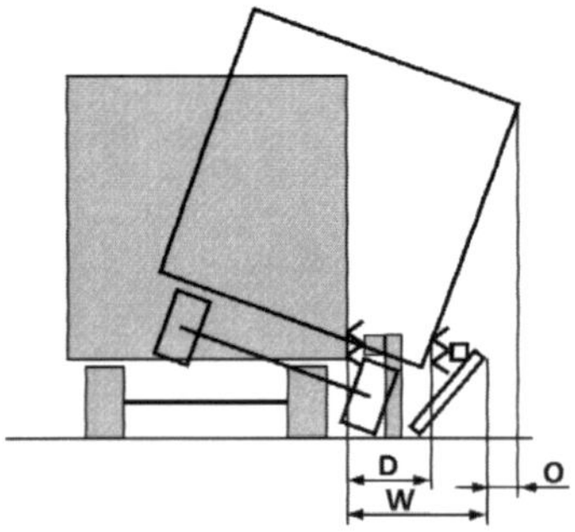

Figure 6. Schematic representation of barrier dynamic deflection – D, working width – W and vehicle overhang – O.

The installation of a barrier requires sufficient space on the side of the road, as defined by the deformation (W). Where barrier deformation is limited, the barrier must support the vehicle with an overhang (O), figure 6.

2.4. Impact severity level

The risk of injuryis mainly determined by the deceleration force of the driver and the passengers. Impact severity is assessed by ASI (acceleration severity index) and THIV (theoretical head impact velocity) indices. Impact severity is classified in three levels – A, B and C. Level A ensures a higher degree of safety than level B, which, in turn, is higher than level C. Level A and level B are unlikely to cause injury to the passengers. Level C impact severity will however cause significant injury and so barriers leading to undesirable outcomes are used as the last resort [4].

3. Barrier impact on road traffic accident mechanism

To visually show the effect of the barrier on the RTA mechanism, it is necessary to make reference to a collision; a BMW car and a SCANIA bus. In this case, the car collided into a barrier and had an initial impact with a bus. The maximum permitted driving speed was 70km/h. The car in the left lane (intended for higher speed traffic in Latvia)impacted a bus changing to the left lane. The right front corner of the car contacted the left rear corner of the bus. The car was deflected left until it contacted the barrier. After the collision with the barrier, the car continued movement forward turning until it stopped. Meanwhile, the bus continued traveling forward until it stopped. The sequence of event is shown in figures7-12.

The accident was modelled with a computer program PC-Crash (figure12). As measurements of friction coefficients and other parameters were not taken at the collision site, the friction coefficient between the driving lane and vehicle tyres was assumed to be 0.7, the friction coefficient of the asphalt pavement – 0.4, kinematic coefficient of restitution – 0.1. Energy spent on the car and bus deformation was determined with the comparison method using Dr. Melegh EES catalogue data [12]. The RTA mechanism was acquired from modelling, using the collision tracks, vehicle location, etc. It was determined that the initial car speed was ~145km/h, but the bus was moving at ~80km/h.

To demonstrate the impact of the barrier on the RTA mechanism by using the same data, modelling was performed, but this time without the barrier that separated the oncoming traffic.

From the second modelling case, it can be observed (figure13) that the car was hit left from its initial direction and continued to move left until it reached the opposite lane. After contact with the bus, the steering system of the car was damaged due to which it became uncontrollable. If there had not been the barrier installed in this road section, the car driver would have entered the opposite lane leading to more severe consequences.

Figures 7–11. RTA scene and vehicle photos.

It should be noted that for the given RTA mechanism, the barrier had to absorb a relatively small amount of energy from the car, which allowed the barrier system to easily meet the requirements – to divert the car back to the driving lane where it moved along the barrier until it stopped.

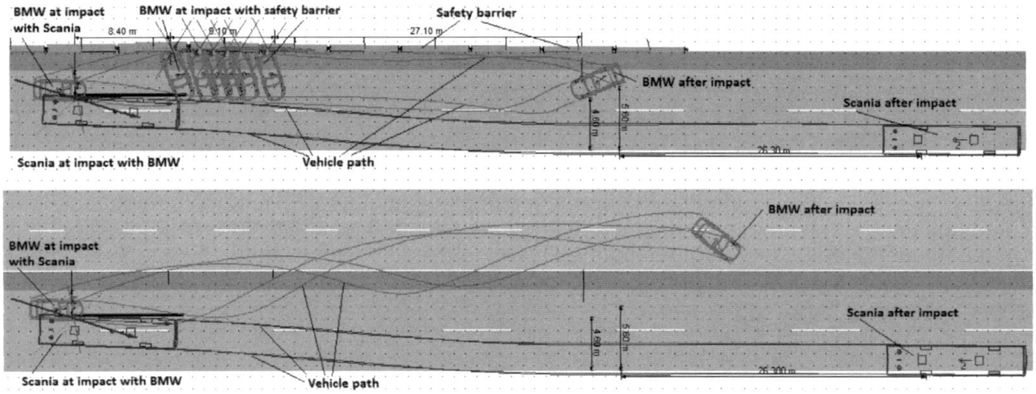

Figures 12–13. The RTA mechanism modelled by the PC-Crash program, with and without the barrier.

In many RTAs involving one vehicle, greater driving speeds are recorded. In such cases, barriers experience relatively large loads, and so the technical requirements of the barriers are a critical issue [3]. For W-beam guardrail systems, the soil is an important factor as barrier supports are fixed in it; the guardrail can fail to redirect the vehicle due to soft ground [13]. Lowering the height of the barrier by 40mm (for example, a renewed road surface with an additional asphalt layer) can still function to redirect pickup trucks and large SUVs, but a further lowering to 60mm can cause the vehicle to override the barrier [14].

4. Choice of barrier types, their economic feasibility

Although wooden barriers may be used in separate places in Latvia, the application of wooden elements should be assessed based on the lifetime of the structure and climate conditions. This type of structure may be chosen from the standpoint of the overall landscape, but cannot be justified as the most economically advantageous solution. Despite the fact that M.Borovinšek [5] indicated that in car to barrier Crash-test – "no load carrying parts of the barrier were separated from the barrier", in the photos it can be observed that relatively large pieces of wood separate from the barrier that can be dangerous to both people in the vehicle and other road users, especially, where there is a roadside footpath. It should be noted that standard EN 1317 binding for Latvia does not allow the barrier structure to break or separate at the moment of collision.

Despite wide use of composite materials in road construction (sign supports, lighting poles, highway delineators and small sign posts) [7], composites have not been used in barriers. Cable median barriers also cannot satisfactorily meet EN 1317 requirements; therefore, in the countries complying with this standard cable median barriers have almost no application.

In Latvia, flexibleW-beam guardrails (figure 1) and rigid concrete barriers (figure 2) are the most frequently used barriers. As both of these types of barriers could be the most used delimiting systems in the world, large amounts of resources are devoted to their studies and improvement of structure.

Effectiveness of W-beam guardrail depends on the location of a car contacting the barrier – whether the car collides with the barrier at a far distance between the posts, or in the vicinity of a post. With an aim to improve W-beam guardrail containment level and decrease barrier deformations, a barrier has been developed equipped with rubber dampers and lamellar shock absorbers [15]. It should be noted that this barrier system increases the costs as well as its total weight. To decrease the barrier costs and weight, Atahan [16] suggests simplifying W-beam guardrail system, using 3 instead of its 5 elements, gaining the system weight of 18.6 kg/m and ensuring containment level L1.

For a concrete barrier, the concrete composition is improved making it sufficiently resistant to impacts and at the same time crumbling at the collision moment, thus absorbing part of vehicle energy, as well as at the moment of barrier crumbling into fragments smaller than 4.75mm, its elements are not dangerous to other traffic participants [17]. Besides, recycled waste – tire rubber – is added to concrete, thus decreasing the barrier total weight and improving acceptable workability [18].

Safety criteria for barriers were developed primarily for cars and heavier vehicles rather than motorcycles, and both types of barriers – W-beam guardrails and concrete barriers – induce particularly severe motorcyclist injuries. In case of collisions with a barrier, a motorcyclist is 15 times more likely to be killed than people in a car. In Australia and New Zealand, the total length of W-beam barrier is 5,565km; whereas the total length of concrete barrier is 673km. Statistical data show that from 2001 to 2006, 73% of motorcyclist fatalities involved collisions with W-beams and 10% – collisions with concrete barriers. As the study does not include the data on collisions of motorcyclists with barriers where the motorcyclist survives, it is not possible to assess the impact of barrier structure on the collisions of motorcyclists with barriers [19], [20]. In reducing motorcyclist casualties, the structure of the barrier is more important than the energy absorption. It is important that a motorcyclist falling and gliding over the driving lane does not get caught by the barrier fixing posts. In this case, as the concrete barrier is a continuous system without gaps, it has an advantage in comparison with a W-beam barrier, where a motorcyclist gets caught on barrier fixing posts.

Gaps between barriers of different containment levels are not permitted. Figure 14 provides an example with two barrier structure types, where one barrier system is installed in a road section and the other system, with a higher containment level, is installed on a bridge. There is a gap between both barriers, which is not permissible from the standpoint of road traffic safety. After the reconstruction works, the barrier structure was made without the gaps combining the barriers of different containment levels, figure 15.

Installing a concrete barrier behind the W-beam guardrail, in such a case gaps between different barrier systems are also not permitted, and it is necessary to install a transition barrier system (figure 16). As a steel barrier system is usually replaced by a concrete barrier with a higher containment level delimiting a more dangerous road section, this transition system has to ensure optimal occupant impact ride-down acceleration, vehicle overhang over the barrier is also not permitted [21], [22]. In compliance with EN 1317, the height of both barriers is ~0.8m.

Figures14–15. Barrier fragment before and after road reconstruction works.

Figure 16. Transition barrier system between a W-beam and concrete barrier.

Transition to the barrier system without gaps allows significantly improving road traffic safety, in case of collision implementing a completely different collision mechanism, ensuring a higher safety level for people in the vehicle.

It should be noted that in case of collision both barrier systems (concrete barrier and steel barrier) have a relatively high difference in friction coefficients. Friction coefficient between the vehicle and concrete barriers varies in the range of 0.7-0.8, and the friction coefficient between the steel barrier and vehicle is 0.3 or less. This difference in friction coefficients can pose a substantial impact on the RTA mechanism. When a car collides with a steel barrier, collision of a more gliding nature can be observed, as a result of which the car is exposed to smaller rotation forces compared to collision with a concrete barrier.

Along with ensuring road traffic safety, costs of road construction and maintenance are considered. Barriers are designed in such a way that after a collision their damage would be minimal and the damaged barrier parts would be relatively quickly replaced investing minimum means. In accordance with the information provided by Ltd. Tilts, procurement and installation costs up to their containment level H1 are lower for steel barriers. However, taking into account the specified barrier lifetime (concrete – 35 years, steel barrier – 15 years), barrier system maintenance costs, as well as renewal of barriers after the potentially possible RTA, assuming that there is one collision in one year on average in a road section of 4km, concrete barrier system costs are lower for a period of 35 years than that of an analogue steel barrier.

5. Conclusions
Barriers for redirecting out-of-control vehicles on the road are undergoing several improvements. There is a need for both permanent and temporary road traffic barriers. The need for road resurfacing requires temporary barriers. They are made from natural materials where possible so that they blend into the surroundings more effectively. Regardless of the change in the barrier type and setting, further work needs to assess the impact of friction between various barriers and the car. This is believed to influence the movement of the car after collision.

6. References
[1] Persaud B N, Retting R A and Lyon C A 2004 *Accident Analysis and Prevention* **36** 1073–79
[2] Martin J L, Mintsa-Eya C and Goubel C 2013 *Accident Analysis and Prevention* **59** 443–451
[3] Wang Y G Chen K M Ci Y S Hu L W 2011 *Scientia Iranica A* **18** 1222–30
[4] Paeglitis A 2012 *Bridge Safety Barrier Guidelines* (Riga Technical University) pp 32
[5] Borovinšek M, Vesenjak M and Ren Z 2013 *Engin. Failure Analysis* **35** 625–635
[6] Davids W G, Botting J K and Peterson M 2006 *Constr. and Building Mater.* **20** 733–743
[7] Bank L C and Gentry T R 2001 *Composites Part:A* **32** 1329–38
[8] Fang H, Wang Q and Weggel D C 2015 *Advances in Engin. Software* **82** 1–13
[9] Wang Q, Fang H, Li N, Weggel D C and Wen G 2013 *Engin. Structures* **52** 240–256
[10] Thiyahuddin M I, Gu Y T, Thambiratnam D P and Thilakarathna H M 2014 *Int. J. of Impact Engin.* **72** 26–39
[11] EN 1317-1 to 1317-5 1998 *Road Restraint System* (European committee for standardization)
[12] Dr Melegh *Licensed Catalogue, Collection: CD-EES, 1998– 2005* (Hungary)
[13] Wu H and Thomson R 2007 *Int. J. of Impact Engin.* **34** 883–898
[14] NCAC 2007-T-003 Nov 2007
[15] Jiga G, Stamin S, Popovici D and Dinu G 2014 *Procedia Engineering* **69** 1191–00
[16] Atahan A O, Yucel A O and Erdem M M 2014 *Engineering Failure Analysis* **38** 25–37
[17] Shen W, Zhang C, Yang Z, Cao L, Yang Z and Tian B 2014 *Construction and Building Materials* **70** 394–398
[18] Elchalakani M 2015 *Structures* **1** 20–38
[19] Jama H H, Grebieta R H, Friswell R and Mclntosh A S 2011 *Accident Analysis and Prevention* **43** 652–660
[20] Gabauer D J Li X 2015 *Accident Analysis and Prevention* **77** 105–112
[21] Atahan A O and Cansiz O F 2005 *Finite Elements in Analysis and Design* **41** 371–396
[22] Ramli Sulong N H 2013 *Accident Analysis and Prevention* **59** 240–252

Acknowledgement
The research leading to these results has received the funding from Latvia State Research program under grant agreement "Innovative materials and smart technologies for environmental safety, IMATEH".

IOP Conf. Series: Materials Science and Engineering **96** (2015) 012064 doi:10.1088/1757-899X/96/1/012064

Analysis of Yearly Traffic Fluctuation on Latvian Highways

A Freimanis[1,2] **and A Paeglītis**[1]
[1]Riga Technical University, Faculty of Civil Engineering, Kalku Str. 1, Riga, LV-1658, Latvia

E-mail: andris.freimanis_1@edu.rtu.lv

Abstract. Average annual daily traffic and average annual truck traffic are two most used metrics for road management decisions. They are calculated from data gathered by continuous counting stations embedded in road pavement, manual counting sessions or mobile counting devices. Last two usually do not last longer than a couple of weeks so the information gathered is influenced by yearly traffic fluctuations.

Data containing a total of 8,186,871 vehicles or 1989 days from 4 WIM stations installed on highways in Latvia were used in this study. Each of the files was supposed to contain data from only 1 day and additional data were deleted. No other data cleaning steps were performed, which increased the number of vehicles as counting systems sometimes split vehicles into two. Weekly traffic and weekly truck traffic was normalized against respective average values. Each weekly value was then plotted against its number in a year for better visual perception. Weekly traffic amplitudes were used to assess differences between different locations and standard deviations for fluctuation comparison of truck and regular traffic at the same location.

Results show that truck traffic fluctuates more than regular traffic during a year, especially around holidays. Differences between counting locations were larger for regular traffic than truck traffic. These results show that average annual daily traffic could be influenced more if short term counting results are adjusted by factors derived from unsuitable continuous counting stations, but truck traffic is more influenced by the time of year in which counting is done.

1. Introduction

Traffic intensity is arguably the most widely used metric for road management decisions. Two different metrics are the most common – average annual daily traffic (AADT) and average annual truck traffic (AATT). AADT best describes traffic flow from traffic organization and planning viewpoint, AATT is best used to calculate necessary strength, modulus of elasticity and other road pavement parameters.

---

[2] To whom any correspondence should be addressed.

Such data are gathered at static vehicle counting stations embedded in the pavement or calculated from short term traffic counting sessions. Permanent counting stations are expensive to install and require constant maintenance during their lifetime, therefore, they are usually only installed on large highways. On smaller roads, counting sessions lasting from a couple of hours to few weeks long are used. Data from them are influenced by the time of a year/week/day the counting took place.

This influence can be corrected by calculating average daily traffic (ADT) and then using simple coefficients that increase or decrease it depending on the time of year. These coefficients have to be calculated from continuous counting station data that has been running for at least one whole year. Also traffic changes within a year can differ between different spots, therefore it is important to make sure that information from one place is suitable for drawing conclusions about another location.

This study uses data from 4 WIM stations in Latvia to assess the fluctuations in weekly traffic and weekly truck traffic depending on the time of year. Traffic recorded in all four places is then compared to see if data from one could be used to describe traffic in other places.

2. Related works

Few works about this topic can be found as this matter is usually presented in the so called "white papers". Latvian road design standards [1] presents a way to calculate ADT from both peak hour traffic and traffic counted from 3 PM to 7 PM. Both methods are used by road designers, however, neither gives AADT. And, as the following text will show, traffic varies greatly in a span of a single year.

Traffic patterns are studied in [2]. Authors show that difference between highest (August) and lowest (January) monthly traffic is about ten percent and attribute these fluctuations to seasonal migration. Similar fluctuations can be seen in the data used in this paper. However, they also note that traffic in metropolitan areas are studied. Results from less densely populated areas can be different [3]. States the need to correct the data from short-duration counting sessions and describes usage of seasonal adjustment factors. 12 factors are provided, one for each month. As following chapters will show, traffic fluctuates most around national holidays so one factor for each month might give misleading results if traffic is counted near these holidays, like in December between Christmas and New Year. It also notes that location of both counting places and differences between traffic volumes and roadway types must be taken into account.

Traffic density and its fluctuations are studied in [4]. While that is a different approach, daily traffic is easily calculable from density, therefore, similar approach is usable for assessment of traffic fluctuations. In the article coefficients are based on time series data and given for different times of day, days of week and months of year. The author also concludes that urban traffic is different from non-urban and coefficients derived from one should not be used to describe the other.

3. Available WIM data

Data from 4 WIM systems on highways A1, A3, A7 and P80 that use piezo-electric sensors were used. Stations were installed on two lane bi-directional highways so the data contained mixed traffic of both cars and trucks in the same lane, also vehicles going in the opposite of lane's direction were observed due to the same reason. Vehicles were automatically sorted into classes, based on their axle count, axle spacings and length. Records contained information about vehicle timestamps, length, wheelbase, gross vehicle weight (GVW), pavement temperature, axle loads and spacings. A total of 8,186,871 vehicles were recorded. Vehicle count, start and end date of recording period, and the number of days recorded on each of the highways are presented in table 1.

WIM systems stored each day's data in a single file. A simple check showed that 29 August 2014. records from A7 contained 5 days' worth of data. Additional vehicles were just duplicates from other days so the file was cleaned leaving only the vehicles from the correct date in it. This led to deletion of 57,151 vehicles. Files from 28 February2015 – 06 March2015 (A3), 30 March2015 (P80), 29 July 2013 (A1) were missing due to repairs.

Data cleaning usually uses a set of filters for various vehicle properties to select only those vehicles that authors are interested in, i.e. trucks, permit vehicles. etc. and delete the rest. In this case authors were interested in all of the vehicles as the variability in the amount of traffic is studied, so even if some records contained erroneous length or weight they were kept. We assumed that this would increase the amount of traffic slightly as split or duplicated vehicles would not be filtered, but the increase was considered insignificant.

Table 1. WIM data.

Highway	Number of vehicles	Start date	End date	Total days
A1	3,887,062	14.07.2012.	31.03.2015.	990
A3	1,351,800	05.09.2013.	31.03.2015.	566
A7	1,868,750	27.08.2014.	31.03.2015.	217
P80	1,079,259	27.08.2014.	31.03.2015.	216
$\sum$	**8,186,871**			**1989**

4. Methods

The authors studied yearly fluctuations of traffic. Two different cases were considered in this period – all vehicles (regular traffic) and only trucks (truck traffic), here defined as any vehicle with gross vehicle weight over 3,500 kg.

Fluctuation analysis was based on weekly periods for which average traffic and average truck traffic was calculated. In case some of the days were missing, the authors calculated values from the available data. Then traffic values were plotted against data from different highways. For better data comparison, the obtained values were normalized against average weekly traffic [5] and plotted with their respective week of the year.

Variability in truck and regular traffic was analyzed both between different sites and locally. Between different sites variability was defined as difference between maximum and minimum normalized values in the same week, thus at most 52 values were calculated per year. Calculated truck and regular traffic values were then compared. Locally standard deviations were calculated for data from each highway and each year. These were then compared to illustrate the differences between truck and regular traffic.

5. Results

Data splitting in weeks yielded 286 periods of data – 142 from A1, 82 from A3, 31 from both A7 and P80. Figure 1 shows normalized average traffic in these periods. A1 data has a huge jump each year from about middle of June to beginning of September, during this time several large festivals take place in a nearby town, which increases average traffic considerably. The authors found that 22 days with highest traffic (11,799 to 6,916 vehicles per day) were recorded within ± 2 days from these festivals. Also average daily traffic on A1 is the lowest amount among all four highways (3,938 vehicles per day), which increases the relative size of any deviation in the normalized data. Also highway A1 runs along Latvian coastline, so people going to the beach would also increase the traffic in summer.

Figure 2 shows that outside of few weeks in the middle of summer normalized traffic is quite similar from one site to another. Maximum and average differences between maximum and minimum weekly traffic values are shown in table 2. In addition, traffic outside of summer months follows the same pattern on all four highways: a gradual fall in traffic during autumn to winter (weeks 35 to 49), a rise in traffic around Christmas (weeks 49 to 51) and a sharp fall in the last week of a year followed by a strong rise till the middle of January (week 2) and a gradual rise till at least the end of March. It is hard to say anything about traffic patterns during the rest of the year as the data are available from only 2 sites.

Truck traffic was split into weekly periods and normalized average weekly truck traffic was calculated as described in the previous chapter. Figure 3 shows the plotted results. The same sudden drop as in regular traffic near New Year was observed. This drop, however, is more severe for truck traffic (truck traffic drops to around 40% of yearly average traffic, but regular traffic drops only to about 55% of yearly average). Plotted graphs show bottom in week 25 or 26 that is not present in regular traffic. Midsummer's Eve, which is a major holiday in Latvia, occurs during week 25 or 26. This would explain a drop in truck traffic as fewer businesses operate during this time, but regular traffic stays the same as people are still driving during holidays.

Figure 3 illustrates that truck traffic is less variable than regular traffic from site to site. Table 2 shows calculated average and maximum amplitudes between regular and truck traffic, they are twice as large for regular traffic if compared year by year.

Data shows that regular traffic is less fluctuating locally on all highways except A1. Table 3 displays standard deviation calculated for each year's data from each site. Comparison tells that in most cases truck traffic locally varies around 1.5 times more than regular traffic.

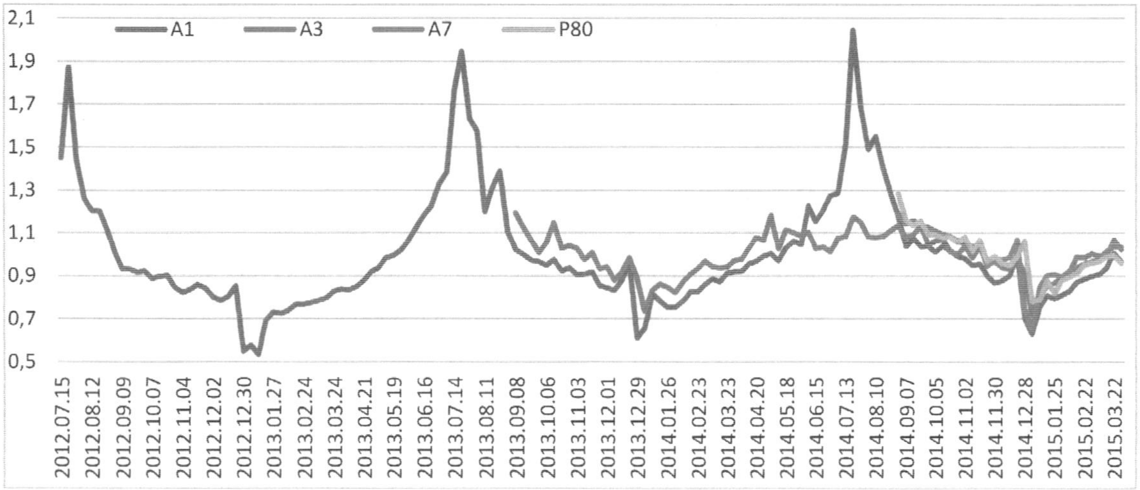

Figure 1. Weekly traffic variability on Latvian highways.

IOP Conf. Series: Materials Science and Engineering **96** (2015) 012064 doi:10.1088/1757-899X/96/1/012064

Table 2. Average and maximum amplitudes between normalized weekly traffic and normalized weekly truck traffic.

	All years		2013		2014		2015	
	All	Trucks	All	Trucks	All	Trucks	All	Trucks
Maximum	0.8699	0.4839	0.2796	0.2529	0.8699	0.3175	0.1536	0.1346
Average	0.2486	0.1659	0.1043	0.1091	0.1484	0.0795	0.0979	0.0383

Table 3. Standard deviations of normalized weekly traffic and truck traffic.

Year	Highway	Truck traffic	Regular traffic	Difference
2012	A1	0.1365	0.2779	-0.1414
2013	A1	0.1426	0.2881	-0.1455
	A3	0.1630	0.0888	0.0742
2014	A1	0.1382	0.2611	-0.1229
	A3	0.1361	0.0973	0.0388
	A7	0.1606	0.1024	0.0582
	P80	0.1120	0.0844	0.0276
2015	A1	0.1799	0.0984	0.0815
	A3	0.1986	0.0995	0.0991
	A7	0.1739	0.1012	0.0727
	P80	0.1680	0.0733	0.0947

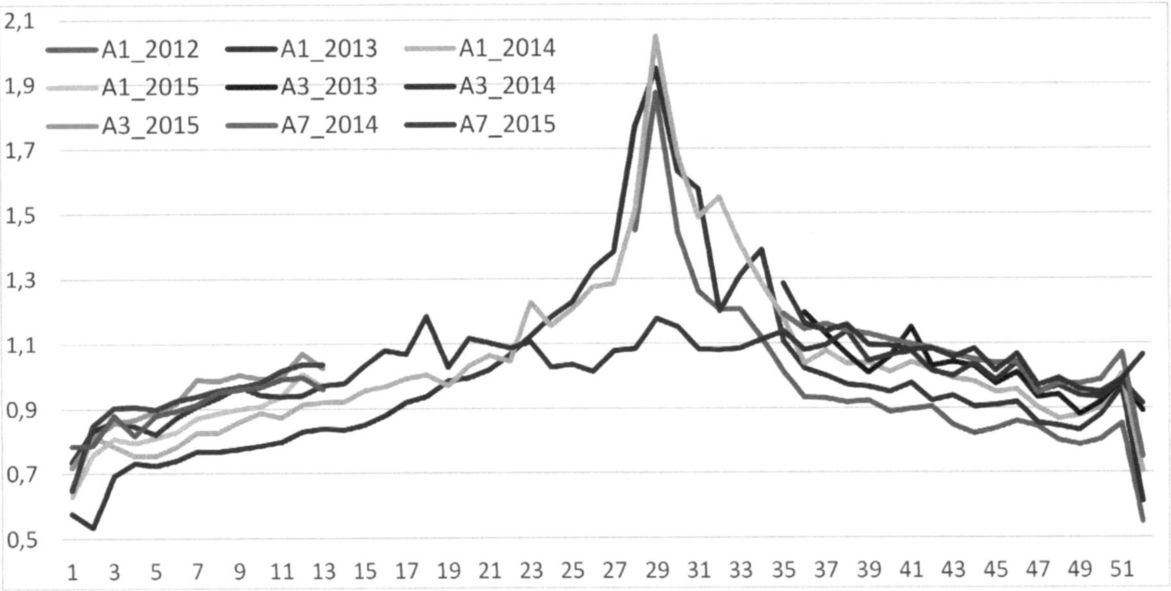

Figure 2. Normalized weekly traffic depending on the week of a year.

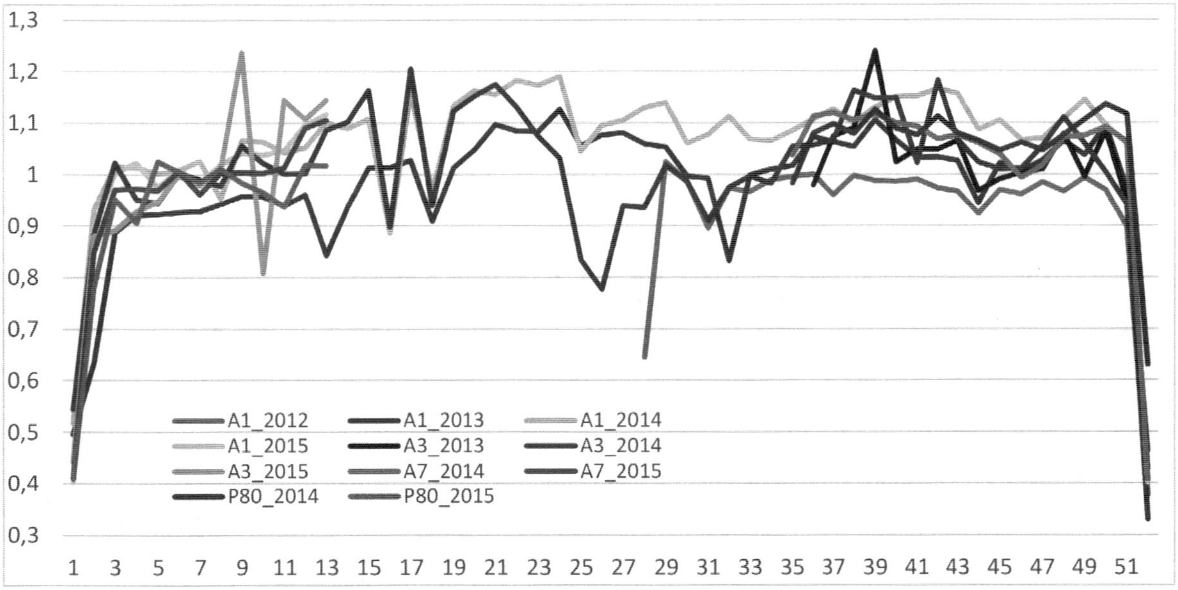

Figure 3. Normalized weekly truck traffic depending on the week of a year.

6. Conclusions and discussion

A study of weekly traffic and weekly truck traffic fluctuations using the data from 4 WIM stations (8,186,871 vehicles altogether) was done. Only two stations yielded the data for more than a year long period, which limited calculations and analysis itself.

Both truck and regular traffic show huge drops during the last and first two weeks of the year. The most likely explanation for this is that people use this time around Christmas and New Year for vacations and drive less. Also a drop in truck traffic is noticeable around Midsummer's Eve, which is a major holiday in Latvia, however, there is no drop in regular traffic. It could be caused by people having holidays and vacations during this time, thus driving less trucks, but the same or increased number (as there is no drop in traffic consisting of all vehicles) of cars. Traffic counting should not be done during these two lows as it might skew the data.

In highway's A1 case it is shown that local factors (festivals in nearby towns, close proximity to the beach) can severely increase traffic in summer. Local factors should definitely be considered if WIM data from such stations are used elsewhere.

Largest average amplitude of normalized traffic between two highways in one week was found to be around 25%, however, in this case the data for the entire year were used for calculations. If only single year's data is used then it drops to 15% and less. This goes to show that traffic from one site is not all that different from other sites. Thus these graphs and data from local vehicle counting could be used to predict the amount of traffic at a later date in some other location. If only truck traffic is considered then these average values are even lower (11% and less), so there is even less difference between sites if only truck traffic is considered.

Truck traffic is more variable locally than regular traffic, as shown in table 3. Less cyclicality is observable in truck traffic, however, each separate week differs more than separate weeks in regular traffic. If any predictions from the limited traffic counting data are made, engineers should adjust for potential low week. The exception for this rule is highway A1, this shows that local factors might have some influence.

The authors recommend that a larger study is carried out as the present analysis used the data from only 4 stations of which only 2 had more than one year of data available.

7. References

[1] LVS/STK/15 *Autoceļi. Ceļu projektēšanas noteikumi. Normālprofili* 2007
[2] Changjoo K, Yong-seuk P and Sunhee S 2008 Spatial and temporal analysis of urban traffic volume *ENSRI Int. User Conf. San Diego* (CA: ESRI)
[3] Washington State Depertment of Transportation 2010 *Guide for Interpreting Short Duration Traffic Count Reports* Washington State Depertment of Transportation
[4] Olma A 2009 *Coefficients of Traffic Density Fluctuations on the Urban Roads Network* Architecture Civil Engineering Environment pp 69–76
[5] Tuffery S 2011 *Data Mining and Statistics for Decision Making*

Acknowledgments

The authors would like to thank Latvian Road Administration for providing WIM data. The research leading to these results has received the funding from Latvia State Research Programme under grant agreement "Innovative materials and smart technologies for environmental safety, IMATEH".

Towards Practical Carbonation Prediction and Modelling for Service Life Design of Reinforced Concrete Structures

O S Ekolu[1]
[1]University of Johannesburg, Department of Civil Engineering Science,
Auckland Park 2006, Johannesburg, South Africa

E-mail: sekolu@uj.ac.za; sekolu@gmail.com

Abstract. Amongst the scientific community, the interest in durability of concrete structures has been high for quite a long time of over 40 years. Of the various causes of degradation of concrete structures, corrosion is the most widespread durability problem and carbonation is one of the two causes of steel reinforcement corrosion. While much scientific understanding has been gained from the numerous carbonation studies undertaken over the past years, it is still presently not possible to accurately predict carbonation and apply it in design of structures. This underscores the complex nature of the mechanisms as influenced by several interactive factors. Based on critical literature and some experience of the author, it is found that there still exist major challenges in establishing a mathematical constitutive relation for realistic carbonation prediction. While most current models employ permeability /diffusion as the main model property, analysis shows that the most practical material property would be compressive strength, which has a low coefficient of variation of 20% compared to 30 to 50% for permeability. This important characteristic of compressive strength, combined with its merit of simplicity and data availability at all stages of a structure's life, promote its potential use in modelling over permeability. By using compressive strength in carbonation prediction, the need for accelerated testing and permeability measurement can be avoided.
This paper attempts to examine the issues associated with carbonation prediction, which could underlie the current lack of a sound established prediction method. Suggestions are then made for possible employment of different or alternative approaches.

1. Introduction

Carbonation as a mechanism of deterioration has increasingly become an important worldwide problem in concrete structures. This is due to the fact that corrosion of steel reinforcement is the most widespread cause of deterioration in concrete structures. This raises the need for two sets of actions, consisting of: (a) durability specification, (b) service life design. Specification for durability requires understanding of the attack mechanisms and employment of measures that mitigate the attack process. Obviously, this requires knowledge of the attack process and the investigation of the material systems that are capable of mitigation the attack. This segment of understanding carbonation has been well-established by researchers over the past decades. For example, it is known that extenders generally tend to increase the progression of carbonation in concrete and that higher concrete strength reduces carbonation progression. With the present understanding of the carbonation process, durability specification for carbonation resistance is possible and is being practiced. But it is the development of a service life method of design for carbonation that has eluded researchers for quite a long time, despite the concerted efforts by the scientific community. A literature review by Parrot [1] and Sagues et al. [2] show that not only there are several factors that influence carbonation but synchronising the effects of all these factors into a prediction model comes with extremely intriguing tasks that continue to be investigated by the scientific community.

To date, a myriad of carbonation models have been proposed in the literature, which indicates the high level of interest and enthusiasm in carbonation studies. But it is often the case that prediction models developed on the basis of a particular data set tend to fail when applied to another independent data source. This paper attempts to identify the reasons for lack of practical carbonation prediction models and articulates on the important considerations in developing prediction models.

2. A future of carbonation

2.1. Historical perspective

Carbonation of concrete structures is a direct result of the increase in green house gas emissions in the Earth's atmosphere, which has emanated from the era of Industrial Revolution believed to have occurred from about 1750 to 1850. This event, which in terms of changes to human living, is often considered to be one of the most important eras in human history, was a period when production process shifted from the use of basic machines and hand tools to power-driven machinery which allows mass production. The emergence of technological advances over the hundred year period of Industrial Revolution did change lives dramatically as factories opened and workers congregated to locations of industrial production, which eventually turned into urban centres or cities. After 1860's a new drive was infused in what is often referred to as the Second Industrial Revolution, when scientists and engineers began to employ scientific approaches based on physics and chemistry in order to unlock industrial practices, which in turn led to further discoveries and has still continued to the present age.

To a great extent, the Industrial Revolution is hailed as the period of great transformation to improve living conditions, but along with its benefits it brought about huge problems which continue to this day. These include environmental hazards from toxic waste, atmospheric pollution from factory gas emissions, health problems for factory workers, diseases from waste disposal, poor sanitation, etc., which continue to be major modern challenges to the industry. The chain of success in industrialisation, however, is energy driven. Before the Industrial Revolution, human physical effort was the main energy source for machine tools and other forms of industrial work. This was replaced by steam engine power at the initial stages of the revolution, followed by the emergence of petroleum oil and electrical power in the 19th century and currently these remain the mainstream sources of industrial energy in the modern industry.

2.2. Future trends

Figure 1 shows that the growth in CO_2 emissions commenced from onset of the industrial revolution and intensified as the second wave of the revolution spread worldwide creating massive and more intense industrialisation as a means to economic sustainability of the growing populace. Another implication of CO_2 emission is climate change effect, believed to be the leading cause of global warming. In the reinforced concrete structures, CO_2 is responsible for steel corrosion which causes major repair and maintenance issues. As seen in figure 1, this trend of human co-existence with CO_2 emissions is set to continue for as long as industrialization exists in its current form of using fossil fuels. Even more crucial is the projected trend showing possible rapid or exponential increase in CO_2 emissions in the near future. While international efforts are presently being summoned to reduce CO_2 emission, it remains a major problem that the industry will contend with far into the future, provided fossil fuels remain the primary source of energy for industrial processes.

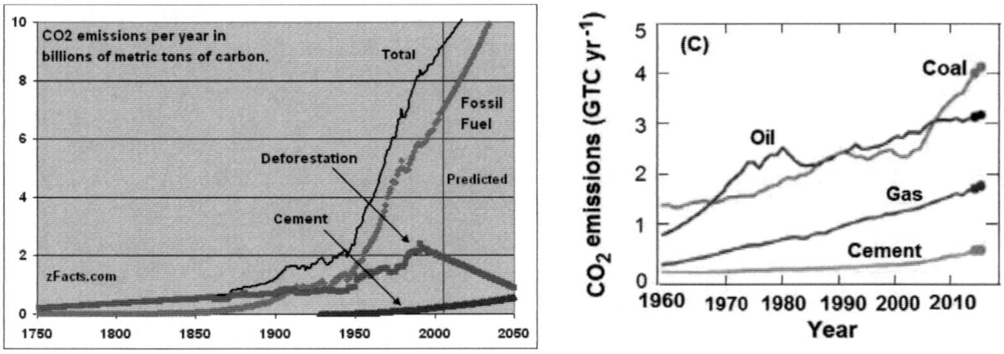

(a) (b)

Figure 1. Global trends in CO_2 emissions; (a) Traceability to Industrial Revolution; (b) Emission by fuel type [4-6].

3. Theoretical expression of carbonation

The fundamental law governing CO_2 diffusion into concrete is well-established to be the Fick's first law, written as

$$J = D.\frac{dC}{dx}$$

where J is the amount of ionic mass agent (CO_2 in this case) passing through a unit area (A) of a porous medium over a unit period of time (t), dC/dx is the concentration gradient between the surface and the interior, and constitutes the driving force of CO_2 ingress, D is the diffusion coefficient of the material i.e. concrete in this case, which is a property dependent on connectivity of the pore network in concrete. Application of this theoretical physical law to carbonation in concrete leads to the expression of CO_2 ingress into concrete as a square root function of time, given as [7,8]:

$x = K\sqrt{t}$, where K is the carbonation rate coefficient given as $K = \left(\frac{2D(C_1 - C_2)}{a}\right)^{1/2}$

D_c is CO_2 effective diffusion coefficient, $C_1 - C_2$ is concentration difference between CO_2 in air and carbonation front, a – alkaline content in the concrete. Models attempt to accurately predict the carbonation rate coefficient and to do so consistently for various types of concretes subject to different environmental exposure conditions. It is at this point where the complexity of developing or building prediction models unfolds.

IOP Conf. Series: Materials Science and Engineering **96** (2015) 012065 doi:10.1088/1757-899X/96/1/012065

4. Carbonation prediction and modelling approaches

Models attempt to predict carbonation in concrete by mathematically or empirically relating the progression of carbonation to a major property of the concrete material system. Theoretically, diffusion or permeability would naturally be the material property of interest in this regard. However, many models proposed on this basis have failed to provide consistent, sound and realistic carbonation prediction. Accordingly, researchers have come to realise that there are many other properties of concrete that relate strongly to carbonation. Through experimental studies, numerous forms of models have been proposed including empirical and mathematical models [1,2], kinetic or physicochemical models such as [10] that are based on chemical reaction determinants, statistical models such as [11], which typically employ multiple regression of various parameters to obtain predictive relationships, simulations and artificial neural networks such as [12], numerical models such as [13,14], which employ computational techniques that are heavily computer reliant. Of these model types, it is the empirical-mathematical models that often find preferred acceptance in design of engineering structures. Models of this kind have been developed in the past 50 years for creep and shrinkage phenomena and used in established international design codes such as ACI 209, BS 8110, RILEM B3, Wendner et al., 2013, CEB-FIP 1978, 1990; AS3600, SANS 10100 [15-22]. The empirical-mathematical models are admired for their universal veracity and realistic predictions with relatively less complexity as they tend to be of practical inclination. In carbonation studies, no such models have been fully developed and very few models, such as Duracrete [23], have been proposed as experimental models for practical carbonation prediction. In this paper, an attempt is made to highlight the potential sources of problems in the development of empirical-mathematical models for practical carbonation prediction.

In this work, selection of the empirical-mathematical models in the category of potential practical viability is made. However, in consideration of ease for practical application, it is essential to examine the parameters that are involved in the model function, i.e. whether they contain the merit of simplicity, whether the data for their use is easily available or can be easily acquired in a typical design project setting. These criteria lead to elimination of some models, such as those that require the use of mix composition, i.e. water-cement ratio (w/c) or cement content [1], since it is hardly possible to accurately determine these parameters in the hardened concrete. The CEB-FIP model [24], for example, is non-committal on recommending any specific performance property but suggests examples of using mix parameters (such as w/c, cement type etc) while recommending the use of experimental data and field testing to derive the carbonation rates. But the model also gives special consideration to the influence of exposure environment by incorporating the weather function. Following an extensive literature review, models that appeared to be oriented to these two criteria of empirical-mathematical functionality and potential practical applicability were selected as shown in table 1. These models currently represent arguably the foremost empirical-mathematical carbonation models proposed in the literature.

5. Discussion of problems in carbonation modelling

5.1. Selection of a suitable property of concrete for constitutive relations

It can be seen in table 1 that the two main properties of concrete that feature highly in mathematical functions to relate carbonation progression are permeability /diffusion or concrete strength. In fact, the majority of the models [8,10,23-26] use permeability /diffusion as their constitutive property. However, problems arise in considering its implementation, which renders its use highly questionable.

Table 1. Mathematical prediction models for carbonation, selected based on functional and material performance parameters.

Performance property	Prediction Model (d or A = carbonation depth)	Source
Permeability	$d = \frac{ak^{0.4}t_i^{\,n}}{c^{0.5}}$, k = permeability, c = CaO content, a = 64	Parrot [25]
CO_2 diffusion	$d = \frac{2C_o}{C_o + \frac{1-\epsilon}{\epsilon}Q}\sqrt{\frac{D_{eff}}{\pi}.t}$ C_o - Initial CO_2 concentration, D_{eff} – effective diffusivity coefficient, ϵ - porosity	Xu and Rodhe [26]
	$d = \sqrt{\frac{2D_c(C_1 - C_2)}{a}t}$ D_c – CO_2 effective diffusion coefficient, C_1– C_2 – concentration difference CO2 in air and carbonation front, a – alkaline content in the concrete	CEB [8]
	$d = \left(\frac{2[CO_2].D_{eCo2}]}{[CH]+3[CSH]}.t\right)^{1/2}$ Deff – effective diffusivity of CO_2, [CH], [CSH] are molar concentrations	Papadakis et al. [10]
	$A = \sqrt{2c_{s.co_2}.D_{ca}}$ $D_{ca} = k_{e,ca}.k_{c,ca}.D_{ca,o}.\left(\frac{t_o}{t+t_{ex}}\right)^{2n_{ca}}$ D_{ca} – carbonation rate from the laboratory test, $k_{e,ca}$ – environmental factor, $k_{c,ca}$ – curing factor, t_o – age of concrete during laboratory carbonation test, n_{ca} – age factor	DURACRETE /fib [23,24]
Concrete strength	$d = \frac{150ckd}{f_c}\sqrt{t}$ c – the parameter for CO_2 binding capacity, d – CO_2 surface concentration, f_c – compressive strength of concrete, k – the RH parameter	Bob [27,28]
	$d = C_{env}C_{air}a(f_{ck}+8)^b\sqrt{t}$ t – time of exposure (years), C_{env} – environmental coefficient, C_{air} – air content coefficient, f_{ck} – characteristic cube compressive strength, a,b – constants dependant on binder type	Hakkinen [29,30]
	$d = (A - B\sqrt{f_{28}}).\sqrt{t}$ A,B constants allows for different types of concrete materials and age effect on strength	Kokubu, and Nagatakis [31]
	$d = 680(f_{28}+25)^{-2.5} - 0.6$	de Fontenay [1,32]

IOP Conf. Series: Materials Science and Engineering **96** (2015) 012065 doi:10.1088/1757-899X/96/1/012065

Research has shown that permeability /diffusion property measurement is not sufficiently reproducible to be considered reliable. This also is the reason why there are no plausible standardized test methods for permeability that are universally recognised. The permeability property characteristically displays high variability, regardless of the kind of test method applied. This observation was highlighted by Hooton [33], who reported that the coefficient of variation (CV) for permeability varied from about 30% for oxygen permeability up to 50% for water permeability. The findings of Stanish et al. [34] reported CV values of 23.7% to 53.9% for oxygen permeability, which are in agreement with [33] as shown in table 2.

Table 2. Coefficient of variation for permeability or diffusion measurements.

Reference	Description	Coefficient of variation (%)
Hooton [33]	Water permeability; 36 tests x 3 replicates	51.9
Day et al.[33]	Oxygen permeability; 32 tests x 5 to 9 replicates	30.6
Hope and Malholtra [33]	Water permeability; 5 tests x 4 replicates	44.4
Stanish et al. [34]	Oxygen permeability	23.7 to 53.9
Imamoto et al. [35]	Air permeability; 16 tests	33.1 to 34.2

Table 3. Coefficient of variation for compressive strength measurements.

Reference	Description	Coefficient of variation (%)
Shimizu [36]	Built before 1961; 453 core tests, 52 buildings	26.8
	Built between 1961-65; 1769 core tests, 173 buildings	27.1
	Built between 1966-70; 2530 core tests, 260 buildings	19.8
	Built between 1971-75; 3398 core tests, 344 buildings	18.9
	Built between 1976-80; 2086 core tests, 244 buildings	15.6
	Built after 1980; 178 core tests, 21 buildings	15.6
Ekolu [37]	15 highway bridges and highways, 20 to 70 years of age	17.0
Other literature [29]	Recommended by various sources	20.0

IOP Conf. Series: Materials Science and Engineering **96** (2015) 012065 doi:10.1088/1757-899X/96/1/012065

Another concern of using permeability measurement property is the need for specialized test equipment. Permeability testing requires quite expensive equipment which includes the permeameter, conditioning chamber or oven, and pressure chamber. The expensive equipment also requires specialized training of technical staff. Altogether, these factors result in quite a major expense whether the testing is conducted in-house or through commercial laboratories. These issues can inhibit the potential use of permeability techniques in most developing countries, considering the lack of permeability equipment and skills for conducting the tests in these countries. In addition, the costs of installing such a facility may be unaffordable for typical or regular testing.

Conversely, compressive strength has a fairly low coefficient of variation, of about 20% as shown in table 3 showing reports from various literature sources.

Not only is compressive strength strongly related to carbonation as shown in figure 1, its data are also the most abundant set of test results typically acquired before and during construction. It is also simple to extract cores from existing structures for strength measurement. It can therefore be seen that the use of compressive strength in constitutional relations would provide significant advantages over the use of permeability. This is more important in the developing countries where the most basic tests that may be conducted in construction would be compressive strength tests.

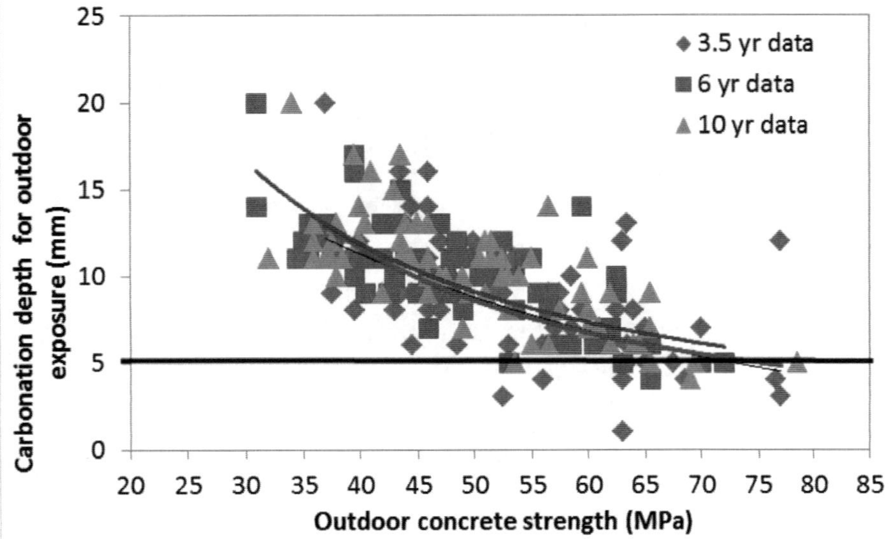

Figure 2. Relationship between compressive strength and carbonation (constructed from [38]).

5.2. Role of climatic regions and global geographical settings

It is often the case that the influence of climatic regions is not given much consideration in development of models. However, this factor plays a major role in considering practical model predictions. A crucial consideration is that most developing countries (Africa, Asia, South America) are located in the tropical regions, where carbonation tends pose a bigger problem than in the temperate climates (Europe, North America, Russia, etc.), where chloride attack is more dominant than carbonation. Accordingly, in the temperate climates, there may be sole carbonation ingress or combined carbonation/chloride attack in structures. Such models may need to give consideration to the synergy between the two mechanisms. In tropical countries, mainly sole carbonation progression

IOP Conf. Series: Materials Science and Engineering **96** (2015) 012065 doi:10.1088/1757-899X/96/1/012065

needs to be considered except along the coastal regions, where the risk of combined carbonation /chloride attack exists. The environmental factors are another key consideration. Whereas the relative humidity values in either region may vary from as low as 40%RH to as high as 80%RH, temperatures in the regions differ widely. While temperatures in the tropical regions are fairly steady at typically 18 to 30°C, those in the temperate regions may fall as low as -20°C in winter. Therefore, the models that may function properly in one climatic region may not directly apply to carbonation prediction in other climate types.

5.3. Accelerated tests versus natural exposure

Just about all diffusion-based models [8,10,23-26] depend on accelerated tests for determination of the diffusion coefficient to be applied in the model. However, there is always a problem that accelerated tests may not be able to accurately depict the variable field conditions and accordingly cannot accurately predict actual field behaviour of the structure under exposure to a natural phenomenon such as carbonation. Not only are the samples used in accelerated tests much smaller than the elements they represent in actual structures, they are also subjected to unnaturally severe test conditions such as high pressures during permeability or diffusion test, etc. At best, some kind of correlation may be established between accelerated tests and field results, but these correlations may change from one test set to another, making it difficult to obtain a universal relation. It is often the case that materials may fail accelerated tests but perform well under field conditions [39,40].

While there is a place for accelerated tests, specifically in studying behavioural characteristics from the theoretical perspective, it would be more appropriate to use carbonation data from natural carbonation exposure to assemble models for realistic carbonation prediction in structures. However, this option is immediately confronted with shortage or lack of data thereof, more so for long-term studies spanning several decades. This remains a challenge, although the approach of using the data from natural exposure is slowly gaining momentum [41].

6. Summary

The objective of this work was to identify some of the main problems affecting or hindering practical prediction of carbonation. While wide ranging approaches to carbonation prediction do exist, it seems that the development of a veracious constitutive relationship expressed mathematically remains a challenge. Many models allude to diffusion or permeability as the performance property to be used in a mathematical constitutive function, typically based on the theoretical Fick's law. Realistically, however, the high variability of permeability /diffusion test procedures makes the use of permeability data in any manner to be quite questionable. Besides this, accelerated tests are the only form of measurement that can be used to obtain permeability data, which typically do not truly represent realistic behaviour of the structure under natural exposure. In the developing countries, the use of permeability is far from practical, due to the high expense and skill required to conduct these tests.

It appears that the most realistic methods for practical prediction of carbonation for service life design should be empirical-mathematical models that can easily be incorporated into design codes. In employing these models, their applicability should be made realistic by adopting a material property which not only exhibits carbonation progression effectively but does so while maintaining practical viability of the model. Compressive strength was found to be potentially an effective material property, given its low variability of about 20%, availability and ease of data acquisition at any age of the structure.

Designers may need to distinguish the applicability of carbonation prediction models based on climatic regions, with tropical regions requiring mainly sole carbonation models while temperate climates may need to consider the synergy between carbonation and other deterioration processes, which might lead to more complex models. This can be done while allowing for different environmental factors in the specific regions.

7. References

[1] Parrot L J 1987 *A Review of Carbonation in Reinforced Concrete* (Cement and Concrete Association, BRE) p 42

[2] Sagues A A, Moreno E I, Morris W and Andrade C 1997 *Carbonation in Concrete and Effect on Steel Corrosion* Final Report, State Job No. 99700-3530-119, WPI 0510685, College of Engineering, University of South Floride, 4202 Fowler Avenue, tampa, Florida 33620-5350, p 299

[3] Hackett L 1992 *Industrial Revolution* [Available]: http://history-world.org/Industrial%20Intro.htm

[4] Zfacts CO_2: Global Carbon Dioxide Levels, Linked to Human Activity [Available]: http://zfacts.com/p/194.html

[5] Le Quere C et al 2014 Global carbon budget 2013, earth Syst. Sci. data 6 pp 235–263 [Available]: http://www.earth-syst-sci-data.net/6/235/2014/essd-6-235-2014.pdf

[6] In-Seok Y, Og˘uzhan C-opurog˘ lu, Ki-Bong P 2007 Effect of global climatic change on carbonation *Atmospheric Environment* **41** 7274–85

[7] Kropp J 1995 *Relations Between Transport Characteristics and Durability* Kropp J and Hilsdorf H K (eds), Performance cirteria for conrete durability (London E & FN: Spon)

[8] CEB 1997 New approach to durability design–an example for carbonation induced corrosion Schiessl P (eds), Bulletin vol. 238 Comite ′ Euro-International du Be ′ton, Lausanne

[9] Saettaa A V, Renato V, Vitaliani Renato V 2004 Experimental investigation and numerical modeling of carbonation process in reinforced concrete structures Part I: Theoretical formulation *Cement and Concrete Research* **34** 571 – 579

[10] Papadakis F and Vayenas 1992 Effect of composition, environmental factors and cement-lime mortar on concrete carbonation *Materials and Structures J.* **25** 293–304

[11] Silva A, Neves R and de Brito J 2014 Statistical modelling of carbonation in reinforced concrete *Cement & Concrete Composites* **50** 73–81

[12] Seung-Jun and Song Ha-Won 2010 Analysis of carbonation behaviour in concrete using neural network algorithm and carbonation modelling *Cement and Concrete Research* **40** 119–127

[13] Saetta A V and Vitaliani R V 2004 Experimental investigation and numerical modeling of carbonation process in reinforced concrete structures Part I: Theoretical formulation *Cement and Concrete Research* **34** 571–579

[14] Peter M A, Adrian M, Meier S A and Bohm M 2005 Modelling and simulation of concrete carbonation: competition of several carbonation reactions (Zentrum fur¨ Technomathematik, Universitet Bremen) p 39

[15] American Concrete Institute (ACI) ACI Committee 209, Subcommittee II. prediction of creep, shrinkage and temperature effects in concrete structures Report ACI 209 R92, (re-approved 1997)

[16] BS 8110 Structural use of Concrete 1997 Part 2, Code of Practice for Design and Construction, London, British Structural Institution 389 Chiswick High Road, London, W4 4AL, United Kingdom
[17] RILEM Model B3 1995 Creep and Shrinkage Model for Analysis and Design of Concrete structures: Model B3, RILEM Recommendations *Materials and Structures* **28** 357–365, 415– 430, 488–495 **29** 126
[18] Wendner R, Hubler M and Bažant Z 2013 The B4 Model for Multi-decade Creep and Shrinkage Prediction. Mechanics and Physics of Creep, Shrinkage, and Durability of Concrete pp 429–436
[19] CEB-FIP 1978 Comite' Europeen du Be'ton-Federation Internationale De la Precontrainte, International System of Unified Standard Code of Practice for Structures, vol 2. CEB-FIP Model Code for Concrete Structures (third ed.) p 56, 331–344
[20] CEB-FIP 1990, CEB, Bulletin d'Information n° 213/214–CEB-FIP Model code 1990, Telford T, London, UK, 1993
[21] AS 3600 Concrete structures, Standard Association of Australia, North Sydney 2009
[22] SANS 10100 Code of Practice for the Structural use of Concrete, Part 1, & 2, Part 1: Design, Part 2: Materials. Pretoria, South African Bureau of Standards 2000 limited
[23] Duracrete 2000, *General Guidelines for Durability Design and Redesign*, Report R15, in 'EU Brite-EuRam III project DuraCrete (BE95-1347): Probabilistic performance based durability design of concrete structures
[24] CEB-FIP 2010 Model Code 2010 First complete draft – vol 2, fib bulletin 56, Int. Federation for Structural Concrete (fib), Case Postale 88, CH-1015 Lausanne, Switzerland
[25] Parrott P J 1994 Design for avoiding damage due to carbonation-induced corrosion *Durability of Concrete 3th Int. Conf. (France Nice, 1994)* pp 283–298
[26] Xu and Rodhe Influence of alkali on carbonation of concrete *Durability of Building Materials and Components 7* vol 1 C. Sjostrom (Ed.) Spon
[27] Bob C 1999 Durability of concrete structures and specification *Proc. Int. Conf. Creating with Concrete Dundee* (ed. Dhir R K and McCar-thy M J) p 311–318
[28] Quillin K 2001 Modelling degradation processes affecting concrete, BRE Centre of concrete construction, CRC Ltd 151 Rosebery Avenue, London, ECIR 4GB
[29] Sarja and Vesiliari (eds) 1996 *Durability Design of Concrete Structures* Rilem Report 14 (Sarja A & Vesikari E eds) (UK: E & FN Spon) p 165
[30] Hakkinen T 1993 The influence of slag content on the microstructure, permeability and mechanical properties of concrete, Part 2, Technical and theoret-ical examinations *Cement and Concrete Research* **23** 518–530
[31] Kokubu M and Nagatakis S 1989 Carbonation of concrete with fly ash and corrosion of reinforcement in 20-years tests ACI. SP-114-14 pp 315–329
[32] LeSage de Fontenay C 1985 Effect of concrete admixture, composition and exposure on carbonation in Bahrain *Deterioration and Repairs. Baharain, Proc. VI* pp 467–483
[33] Hooton R D 1988 What is needed in a permeability test for evaluation of concrete quality *MRS Proc. Materials Research Society* vol 137 1989
[34] Stanish K, Alexander M G and Ballim Y 2004 *Durability index interlaboratory test results – statistical analysis of variance* Research Report, Departments of Civil Engineering, University of Cape Town /University of the Witwatersrand p 87

[35] Imamoto K, Shimozawa K, Nagayama M, Yamasaki J and Nimura S 2006 Evaluation of air-permeability of cover concrete by single chamber method *31st Conf. on Our World in Concrete and Structures (Singapore, 16 – 17 August 2006)* p 8

[36] Shimizu Y, Hirosawa M and Zhou J 2000 Statistical analysis of concrete strength in existing reinforced concrete buildings in Japan *12th World Conf. on Earthquake Engin. (WCEEE), (New Zeland, Auckland, 30 January - 4 February 2000)* p 8

[37] Ekolu S O 2010 Model validation and characteristics of the service life of Johannesburg concrete structures *Proc. of the National Symp. on Concrete for a Sustainable Environment, Concrete Society of Southern (Africa, Kempton Park, Johannesburg, Gauteng, 3–4 August 2010)* p 30–39

[38] Barker W R 1999 Final report on an investigation into the influence of fly ash on the durability of concrete, Division of Building and Construction Technology, CSIR, PO Box 395, 0001 Pretoria, South Africa, p 18 Also, Ash Resources Also, Ash Resources (pty) Ltd, PO Box 3017, Randburg 2125

[39] Crewdson M Outdoor weathering must verify accelerated testing, Q-Lab Weathering Research Service, 1005 SW 18th Avenue, Homestead, FL 33034-1725 U.S.A

[40] Tittelboom van K, De Nele B and Hooton R D 2013 Test Methods for Resistance of Concrete to Sulfate Attack – A Critical Review, Chapter 10 *Performance of Cement-Based Materials in Aggressive Aqueous Environments*, RILEM State-of-the-Art Reports vol 10 pp 251–288

[41] Tang L, Utgenannt P, Lindvall A and Boubitsas D 2010 Validation of models and test methods for assessment of durability of concrete structures in the road environment, 17-12-2010, CBI Betonginstitutet AB, 100 44 Stockholm, Besök Drottn Kristinas väg 26, 114 28 Stockholm

Acknowledgments

This work is based on the research supported in part by the National Research Foundation (NRF) of South Africa, Grant No. 96800. The authors wish to thank the NRF for financially supporting this research.

Load Bearing Innovative Construction from Glass

R Kalamar[1] **and M Eliášová**[1,2]
[1]Department of Steel and Timber Structures, Faculty of Civil Engineering
CTU in Prague, Czech Republic

E-mail: eliasova@fsv.cvut.cz

Abstract. Glass plays an exceptional role in the modern architecture due to the optical properties and transparency. Structural elements from glass like beams, facades and roofs are relatively frequent in common practice [1]. Although glass has significantly higher compressive strength in comparison with tensile strength, load bearing glass elements are relatively rare. This opens up new opportunities for application of glass in such structures as transparent columns loaded by the axial force. This paper summarizes the experimental results of the tests on glass columns loaded by centric pressure, which were performed in the laboratories of the CTU in Prague, Faculty of Civil Engineering. The first set of experiments was composed of three specimens in a reduced scale 1:2 to verify real behaviour of the specimens with enclosed hollow cross-section. The main goal of the experiment was to determine force at the first breakage and consequently the maximal force at the collapse of this element.

1. Introduction

Glass columns subjected to compression must have sufficient strength, stiffness and residual load bearing capacity to be able to transfer loads (e.g. dead load, snow load, wind load) and to be safe even after first cracks have appeared. Design methods for other materials cannot be taken into account and used for glass elements without any modifications because it is necessary to consider the effect of manufacturing tolerances (glass thickness), initial deformation, PVB layers for laminated glass, glass elastic behaviour without hardening effect, the effect of load duration, damage of the glass surface and the impact strength of glass, which is not necessary for other materials (e.g. steel).

Due to fragility – the main property of glass, it is necessary to keep in mind proper design and assembly of details. Any local irregularities can cause stress concentrations and thus the failure of glass. In case of glass columns, the most important detail is the place where the force is transmitted from the slab to the glass part of the column. Project design should also consider load transfer from

[2] To whom any correspondence should be addressed.

IOP Conf. Series: Materials Science and Engineering **96** (2015) 012066 doi:10.1088/1757-899X/96/1/012066

horizontal structures to the other parts of structures and prevent progressive collapse of the building or its part due to the collapse of a single glass column.

Recently, a set of experimental studies at research centres in Europe has been focused on the behaviour of glass columns with different cross section. Experimentally verified glass structures were used in the practice, such as a glass pavilion in Rheinbach, where all vertical and horizontal forces were transmitted by glass columns [2]. Laminated glass columns of simple rectangular cross-section section were used. Another example is the glass foyer at the headquarters of an industrial company Nordborg in Denmark, where the architects used cruciform shape for the glass columns which support the glass roof beams [3]. Experimental verification of glass columns was also performed at EPFL in Lausanne, where the research was focused on the loss of stability of glass elements loaded by compression and on compiling of basic buckling curves for different strength of glass [4].

2. Glass columns

2.1. Design of glass columns

Within the research of glass structures, which takes place at the CTU in Prague, the first set of experiments contains three test specimens. These specimens were prepared in a reduced scale 1:2 from float single layered glass which had been selected with regard to simpler behaviour under loading without effect of polymer interlayers. In the real case laminated tempered glass will be used for such kinds of structural load bearing elements.

The test specimens consisted of four float glass panes of thickness 6 mm with the length 1750 mm and the width of 150 mm, which were connected to each other at the corners into a square hollow cross-section by acrylate adhesive Sikafast 5211 - NT, see figure 1, figure 2.

Special attention was paid to the detail of the end parts of glass columns. At this point, it was necessary to provide a uniform transfer of normal force into the glass part of the column by the plastic pad made from polyamide (PA). This plastic pad was designed as a two-stage plate to prevent potential torsion deflection of the cross-section at the edges. Plastic pad was placed at the bottom and the top of the column between the glass part and special steel shoes.

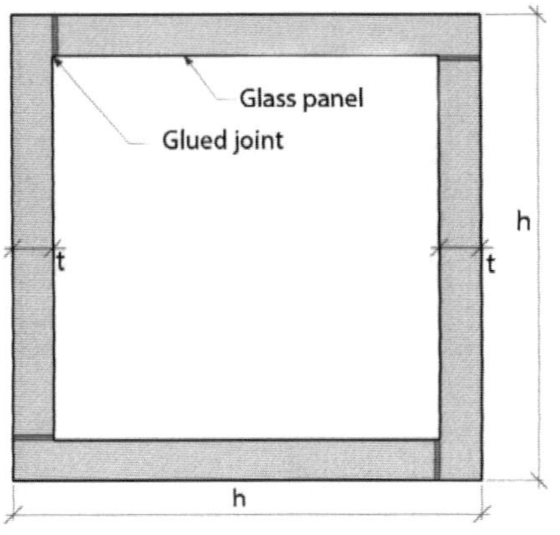

Figure 1. Square hollow cross-section.

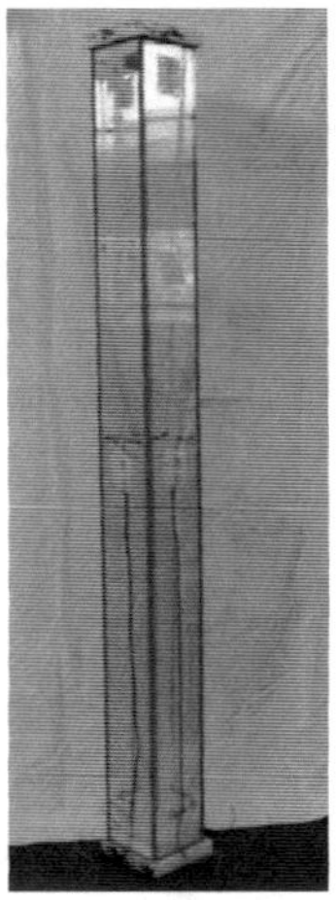

Figure 2. Test specimen.

2.2. Test set-up

All test specimens were continuously loaded by centric force at the speed of 50 N/s until the first crack appeared. After the first breakage of glass pane, the columns were loaded until they collapsed. Total of 16 strain gauges LY11-10/120 were used for indirect measurement of stress, they were installed at the three levels of the column, see figure 3. Horizontal mid-span deflections were also recorded during the test by four potentiometers, which were fitted on each panel.

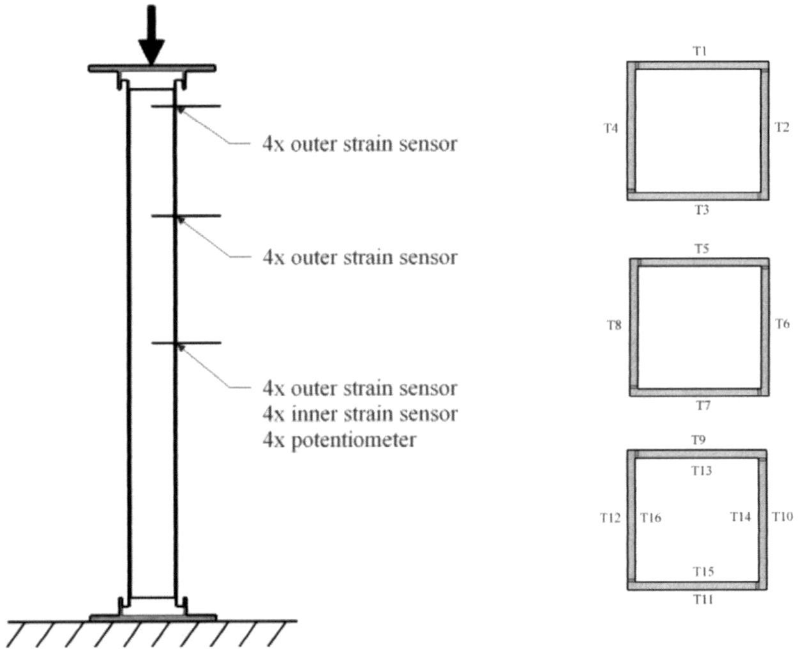

Figure 3. Position of the measuring sensors.

The load bearing capacity of the glass column could be defined as force F_{1st} in which the first crack has appeared but there is no total collapse of the column. The first cracks had been seen in all samples at the value of force $F_{1st} = 75$ kN. The first cracks appeared every time at the bottom of the column above the steel shoe and always had vertical direction. These cracks occurred due to the lateral tension. Thus, the column was still capable to transmit the normal force until it collapsed. We are talking about residual load bearing capacity. The residual load bearing capacity means the ability of partially broken column to transmit additional load without collapse. The residual load bearing capacity can be calculated as $F_{residual} = F_{collapse} - F_{1st}$, where $F_{collapse}$ is the ultimate loading force, see figure 4.

Columns were able to transfer additional load until they achieved forces in the range from $F_{collapse} = 85$ kN to $F_{collapse} = 168$ kN, i.e. columns had residual load bearing capacity in the range $F_{residual} = 13 - 93$ kN, see table 1.

Table 1. Experimental results.

Specimen	$N_{1st.}$ [kN]	$N_{collaps}$ [kN]	$F_{residual} = F_{collapse} - F_{1st.}$ [kN]
Specimen 1.1	72	117.5	45.5
Specimen 1.2	72	85	13
Specimen 1.3	75	168	93

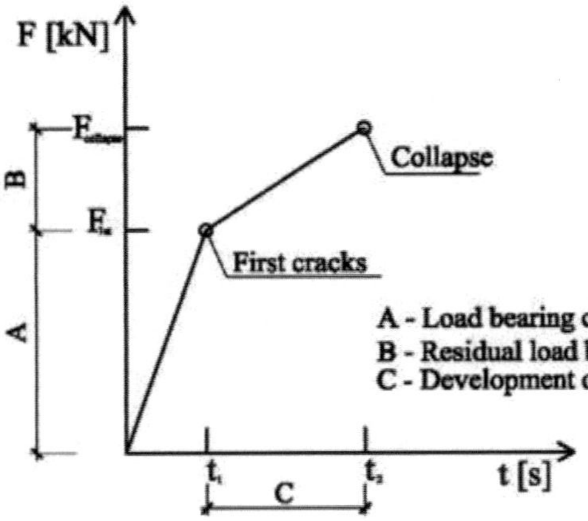

Figure 4. The basic behaviour of the glass column.

3. Analytical model

The real rod has lost its stability before the force reached the critical force N_{cr}. The behaviour of ideal compressed column with the length L can be described by a known differential equation (1). The column is simply supported and loaded by axial force N with initial imperfections w in the shape of a sinusoidal wave, figure 5.

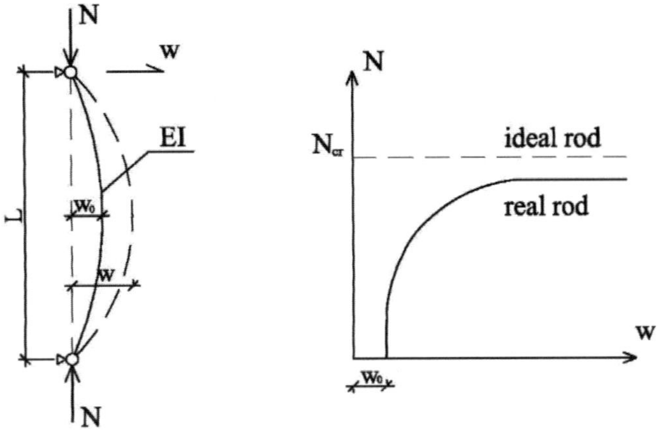

Figure 5. Eccentrically loaded compression rod with initial deformation [6].

$$EIw'' + Nw = 0 \tag{1}$$

For the real rod with initial geometric deflection w_0, the maximum bow imperfection in the middle of the length can be determined from the equation

$$w = \frac{w_0}{1 - \frac{N}{N_{cr}}} \tag{2}$$

The geometric deflection w_0 can be determined as the sum of material and structural imperfections according to the equation

$$w_0 = w_{0,1} + w_{0,2} \tag{3}$$

Material imperfection $w_{0,1}$ for float glass can be considered according to [5] as the value L/2500 and structural imperfection $w_{0,2}$ can be considered as L/400.

The critical Euler force for the ideal rod is equal

$$N_{cr} = \frac{\pi \cdot EI}{L^2} \tag{4}$$

where I is the moment of inertia.

Normal stresses in the middle of length and in the extreme fibers can be calculated by equation

$$\sigma = \frac{N}{A} \pm \frac{M}{W} = \frac{N}{A} \pm \frac{N \cdot w}{W} \tag{5}$$

where W is the section modulus.

The normal stress from the analytical model was determined according to the above mentioned equations and compared with the experimental data in the mid-span cross-section, see figure 6. Curves T_i - T_j represents the principal stress measured by stress strain gauges. Location and position of the stress strain gauges is shown in figure 3. The lines $AM_{,pressure}$, $AM_{,tensile}$ are average values calculated by equation (5).

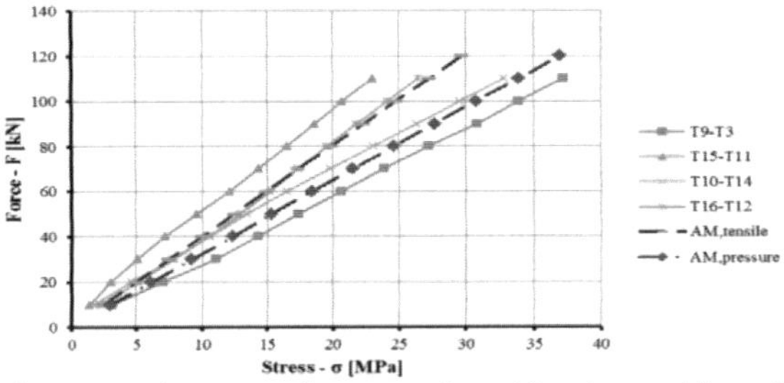

I Model, tensile – The value of compressive normal stress reduced by the positive effect of the bending moment.
I Model, pressure – The value of compressive normal stress increased by the effect of the bending moment.

Figure 6. Comparison of the analytical model with the experiments.

4. Numerical model

Program ANSYS ver. 14.5 was used for the numerical model. Each part of the column was created from volume elements SOLID. Glass was modeled as a linear isotropic material with Young's modulus E_{glass} = 70 GPa and Poisson coefficient 0.23. Plastic pad was defined as a linear isotropic material with Young's modulus E_{pad} = 3.5 GPa and Poisson coefficient of 0.39.

The adhesive was modeled by using multi-linear stress strain relation with Young's modulus E_{adh} = 0.26 GPa and Poisson coefficient 0.40. Stress strain graph was based on tensile tests of adhesive joints performed at CTU, [6]. Initial geometric imperfection for the numerical model was set as w_0 = L/400.

The load bearing capacity was chosen for comparison of the test results obtained from numerical model and real measured data. Deformed shape and compressive stress σ_z in the glass is demonstrated in figure 7, which is scaled for illustration by 1000:1. The maximum of the measured horizontal deflection in the middle of the column was $w_{y,real}$ = 0.22 mm. Deformation determined by the numerical model was by 25 % lower, $w_{num,y}$ = 0.175 mm. Comparison of compressive stress obtained from experimental results and numerical and analytical model is shown in table 2.

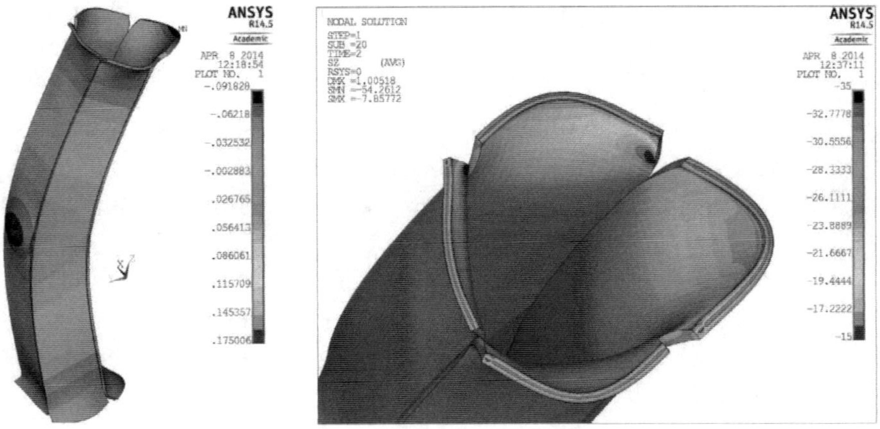

Figure 7. Deformed shape in the direction Uy (left); tension σz (right).

Table 2. Comparison of principal stress obtained from the experiments, numerical and analytical model.

Principal Stress	Analyt. model	Numer. model ANSYS 14.5	Specimen 1.1	Specimen 1.2	Specimen 1.3
	σ_{AN} [MPa]	σ_{NUM} [MPa]	σ_{exp} [MPa]	σ_{exp} [MPa]	σ_{exp} [MPa]
s_{max}	24.6	25.4	27.2	32.7	25.9
s_{min}	19.9	21.2	16.5	14.2	16.8

From table 2 it can be seen that test specimen 1.2 has different values of the principal stress than the others. It is due to higher assembling imperfection and the fact that during the test a part of the glass corner felt out. Thus, a local bending moment increased.

5. Conclusion

In the first phase of experimental research of the behaviour of a glass column with hollow square cross-section was verified. Columns were assembled from four float glass panels with thickness 6 mm. They were bonded together along the length by acrylate adhesive. Results showed sufficient load bearing capacity of the glass columns. The first breakage occurred approximately at a force $F_{1st} = 75$ kN when the first crack occurred above the bottom basement of the test specimen. The column was still able to carry the increasing load thought cracks started to develop. The collapse was observed at average force $F_{collapse} = 123.5$ kN. The detail of transferring axial forces from steel shoe through the plastic pad into the glass column has had a significant influence on the load bearing capacity. In all cases, the first cracks showed above the steel shoe due to the lateral tensile but the column was still be able to carry the increased load. The average residual load bearing capacity was $F = 51$ kN.

The next phase of the research will be experimental verification of real scale specimens made from laminated glass.

6. References

[1] Netušil M and Eliášová M 2015 Design and evaluation of bonded composite glass beams *Structures and Buildings Themed Issue on Structural Glass* **168** 490–499

[2] Wellershoff F and Sedlacek G 2003 Glass pavilion Rheinbach – Stability of glass columns *Glass Proc. Days* (Tampere, Finland) pp 316–318

[3] Nijsse R 2003 *Glass in Structures – Elements Concepts Design* (Basel, Berlin, Boston) Birkhäuser – Publishers for Architecture (Basel, Switzerland) pp 60–81

[4] Luible A and Crisinel M 2004 Buckling strength of glass elements in compression *Structural Engineering Int.* **14** 120–125

[5] Haldimann M, Luible A and Overend M 2008 *Structural Use of Glass* (Berne, Basel, Nottingham) IABSE-AIPC-IVBH (ETH Zürich, Switzerland) pp 107–129

[6] Netušil M and Eliášová M 2010 *Challenging Glass 2 – Conference on Architectural and Structural Applications of Glass* ed. Bos F et al (Delft university of Technology) Wöhrmann Print Service pp 269–276

Acknowledgement
This research was supported by the grant GAČR 14-17950S.

IOP Conf. Series: Materials Science and Engineering **96** (2015) 012067 doi:10.1088/1757-899X/96/1/012067

Dimensional Interval and Critical Sizes of Non-Legand Nano- and Microparticles

A N Kharkhardin[1], **V V Strokova**[1], **N I Kozhukhova**[1,2] **and V V Nelubova**[1]
[1]Department of Material Science and Material Technology, Belgorod State Technological University named after V.G. Shoukhov, 308012, Belgorod, Russia

E-mail: kozhuhovanata@yandex.ru

Abstract. The derivation of the equation for the calculation of the critical size of clusters and nanoparticles in the range of appearing of their unusual properties in comparison with the massive body as well as the equation for the calculation of the growth of spherical crystallites and aggregations is presented in the article.

1. Introduction

The main characteristic of a substance with granular structure is discreteness of corpuscular formations and the substance structure at different scale levels. Quantitative accumulation of the sources of the individual properties of the substance leads to a qualitative change in the system of its microparticles with the appearance of unusual properties and to formation of aggregations. Change in the material dispersity when transfering to macro-level in the process of nuclear clusters assembly of nano- and microparticles, or when shifting to micro-level during milling process with using 'sol-gel' technology brings about changing in the topological state of the disperse layer and appearance of unusual properties of nano- or microparticles. *Discreteness level (degree of fragmentation) of the substance with the appearance of the critical state of its disperse layer or with demonstration of unusual properties changing in its microparticles which are different of those of massive body can be considered a critical dispersity.* Metal nano- and microparticles raise a profound scientific interest taking into account their unusual physical and other properties that are not found in massive bodies.

When the number of atoms in a microparticle is equal or more than 13, packing density can be achieved if each atom is a surfacial one. When increasing the size of the microparticles, two types of atom packing take place: the inner one that is close to the crystal packing and external one that is a surfacial, loose in a near-wall layer. Surface atoms have lower coordination numbers than inner ones. Their electron shells are strongly polarized, their bonds are directed only inside of the particle. For such particles, more stable configurations occur with five bonds between atoms (five-order symmetry

[2] To whom any correspondence should be addressed.

axis), which is not valid for macrocrystals. There is a certain critical size of microparticles above which the pentagonal (icosahedral) structures become less stable than crystalline atom packing.

It is believed that the transformation from icosahedral to a crystalline structure is kinetically difficult to achieve due to the need to reorganize the structure and increase the number of atoms around. That explains saving of pentagonal shapes for relatively large (10–30 nm) metallic microparticles. However, a theoretically limited size of microparticles having a structure with stable pentagonal symmetry has not been found so far, because for this to be done, it is necessary to calculate free energy of huge number of isomers. For these particles, clear faceting characterizing crystals are still unstable and there is a tendency to form microparticles with spherical shape, where the corners and tips get round, and the edges are protruded. According to the data of Y. Petrov [2], in the range of 151–21100 the anomalies of physical properties are observed for a number of atoms in microparticles that results in transition region, where the center of microparticles gains more and more inner layers, where the package corresponds to the structure of the given compact metal. Metal ligand-free microparticles (cluster formations) have the following characteristics: surface and internal energy of a single cluster are comparable with number of atoms in a bulk. Structure of the cluster differs from the metal sample, where the packing density of crystal is absent. Interatomic distance is increased; melting temperature as well as Debye temperature of cluster formations is lower than that of massive metal, etc. Shape and structure of microparticles are in non-equilibrium state and correspond to the state, which is different from the minimal energy state. In this state, microparticles of metals, alloys and other simple mineral substances can exist.

It is known that as the consequence of spraying the metal vapor on a cold substrate (in cold helium) stable clusters with a magic number of atoms are formed. Ligand-free nanoparticles (cluster formation) of metals have been comprehensively studied. In such a way, for plumbum the most stable clusters consisting of 7, 10, 13, 17, and 19 atoms were obtained [3]. With 13 atoms in the clusters, the single-layer hexagonal and cubic close packing (η_1=0.7405) and icosahedral packing are formed (η_1=0.68834). In the first two packings the permanent groups of 3–5 atoms in the cluster are formed on the surface of the central atom. They contacts with each other and with the central atom, whereas in the icosahedral packing all 12 surface atoms with diameter *d* are separated from each other by an equal distance:

$$l = [(1/0.905)^{1/3} - 1]\, d \tag{1}$$

They only contact with the central atom forming 6 axes of 5th order. It is intermediate structure between crystalline inorganic solids and developing organic bodies that are typical of the most viruses. Mechanisms of interaction of isolated 'cold' atoms and formation of such structures are not established yet. One of the interesting results in this field is establishment of an important role of states of electron of the surfacial atoms in metal clusters.

These states are characterized by the lack of electron density in comparison with bulk states. In this regard, increasing of electron work function of the cluster in comparison with a large metal crystal is observed. In such a way, the difference for 13 atom cubooctahedral clusters in transition metal is 2 eV. In addition, for clusters the properties of bond unsaturability and orientation between the atoms are presented weakly or are completely absent. Up-to-date studies of small particles using computational and empiric methods (neutron scattering and slow electrons diffraction) found the existence of specific vibrations localized near the surface within a few atomic layers. The vibration spectrum is shifted to lower frequencies compared to the infinite lattice of microcrystals. The amplitude increases for surfacial atoms in particular [3]. In our opinion, the length of *p* and *d* electron orbitals in 'cold' atoms is reduced. Cluster assembly of 'cold' atoms and homogeneous crystalline (icosahedral) structures in inert medium and medium without charge carriers are based on a weakened quantum mechanical attraction and enhanced lateral Coulomb repulsion of the atoms according to two principles.

The first principle is the principle of close packing, it can be interpreted as a decrease of internal energy of cluster atoms or molecule collection in close packing as a result of attractive forces. Single 'hot' atoms or atoms in dense areas of fluctuations that come into contact with each other due to forces of attraction and in accordance with the quantum and mechanical compatibility of overlapping atom orbitals and permitted symmetry herewith in their translational stacking, tend to occupy the minimal volume and the densest packing fills in free space .

The second principle is the principle of Coulomb uniformity of distribution of the 'cold' atoms in nano-aggregations, it can be considered a result of Coulomb repulsion forces which are a homogenous distribution of atoms in a free assembly and free space around the central atom. Coulomb uniform distribution of atoms in the cluster formations leads to decrease in the density of their packing, the coordination number and the bond strength between them. Growth of metal nanoparticles according to the first principle of free holes filling in by single atoms formed by the surfacial atoms, as well as by increasing of density of their packing and coordination number, lateral or radial overlap of external p or d electron orbitals in accordance with the quantum and mechanical compatibility, formation of cuboctahedral or icosahedral structures in the presence of the central atom into the cluster or hollow clusters with formation of stray vacancies in there. The low packing density of surfacial atoms reduces bond strength between them as a result of Coulomb repulsion force and uniformity of their distribution in the surface layers. The growth of spherical crystals occurs when the packing density of atoms in the surface layers is equal to $0.4098^{2/3} \leq \eta \leq 0.64976^{2/3}$, where $\eta \leq 0.4098$ is the value indicating the impossibility of topological transition to a dense condensed state of substance [4]. Upon reaching the bulk packing density $\eta \geq 0.64976$, the quantum and mechanical interaction of atoms in compact packaging with high strength connection between them starts working. Quantum and mechanical compatibility of orbitals determines the type of symmetry and atomic structure with a minimum internal energy of the crystals. When putting the second layer of 6 atoms over, there is a possibility to reorganize the first layer with icosahedral structure of a cluster into a closest cubic packing density with formation of a stable octahedral shape. There are 19 atoms with a maximal coordinate number with up to 7 surfacial atoms in it.

$$\eta = V_0 / V = \left(19\pi d^3 / 6\right) / 0{,}4714 \cdot (3d)^3 = 0{,}7816$$

where V_0, V – an occupied volume of atoms and the volume of octahedra with the size of an edge of $3d$, excluding the volume of cone in the tips of the octahedron.

When reorganizing the first layer of balls (atoms) with icosahedral packing in a hexagonal packing for the formation of its first faceted shape of a cluster, obtained on the basis of packing, it is necessary to have 17 balls (atoms) of the size. A trigonal bipyramid herein formed consists of 30 balls, where the maximal coordination number is 9:6 balls. 6 balls are placed in the octahedral holes between 4 balls, 6 balls are in the wells between 3 balls; 5 balls are on the tips. Whereas for the formation of polyhedon with cuboctahedral and icosahedral shapes the number of balls of the second layer according to (5) with n=2 has to be equal to N=42. The total number of balls of the three-layer polyhedron is 147 and is calculated according to S.P. Gubin equation [5]:

$$N = 1 + \sum_{1}^{n} (10n^2 + 2) \tag{2}$$

IOP Conf. Series: Materials Science and Engineering **96** (2015) 012067 doi:10.1088/1757-899X/96/1/012067

The packing density of balls in a multilayer small polyhedron is unknown. It is assumed that it is η_1=0.7405. Then the diameter of the circumscribed sphere of cuboctahedron (η_1=0.7405) consisting of 147 balls, will be equal to $D \geq (147/\eta_1)^{1/3} d \geq 5.8d$ and (η_1=0.68834) is $D \geq (147/\eta_1)^{1/3} d \geq 6d$ for icosahedron. The packing density of balls in the circumscribed sphere with a single cuboctahedron and icosahedron $\eta_1=13d^3/(3d)^3$= 0.4815 and a hollow icosahedron when filling in pores in it, a ball's appropriate size would be 12.90 / 24.51 = 0.5264. In the absence of the central globe it is, respectively, the following:

$$\eta_1 = 12d^3/(3d)^3 = 12/27 = 0{,}4444$$

$$\eta_1 = 12d^3/(2{,}9d)^3 = 12/24{,}515 = 0{,}4895$$

Consequently, the low packing density of atoms in the clusters with icosahedral or cuboctahedral shape without the central atom is a condition for further growth, first of all, in the form of icosahedral, that is, spherical crystals with formation of vacancies in their cell or restructuring of a faceted crystal shape.

The study about the substance in the nano-scale range (1–100 nm) of its cluster formations, nano- and microparticles is related to the development of mathematical apparatus to describe them. The given size range of nano-objects is summarized according to empiric data, however, there is no theoretical justification depending on the size of an atom of simple substance and size of a unit of the crystal lattice of complex substances. The size range of micro- and nanoparticles when their unusual properties begin to appear is important when production and milling of minerals and inorganic raw materials as well as synthesis of geopolymer binders with non-hydration type of hardening and neocomposites occur [6, 7].

Regardless of the preparation method and structure of nanoparticles, the effects of their use in chemical and technological processes are promising to improve the quality of raw materials and construction composites during structure formation process. Major and fundamental aspect in the developing science about nanodispersed objects is the matter of dimension, structure and topological state of bulk and surface layers of nanoparticles, when their unusual properties appear, which differ from those of the massive body. It is believed that the unusual properties are associated with the relatively large proportion of surfacial atoms [8, 9]. It is known, when the formation of more loose atom packing in the surface layers than in the core of their clusters or crystal packing, the micro- and nanoparticles, generally rounded with atomicrough surface, are formed when cooling process. Under the conditions of supercooling of the molted mass, the crystalline shape becomes profitable. Packing density of atoms in atomic rough surface layers of micro- and nanoparticles can reach the critical value for the given substance.

Since the difference between the phases in critical point disappears, their specific surface free energy and the difference in thermodynamic (chemical) potentials go to zero. In this case, increasing in the size of nanoparticles is stopped for a while. As a result, the principle of close packing is observed when the interaction of atoms in nanoparticles occurs, starting from the deep of the layers. Direction of its action depends on the preparation conditions (formation) of the nanoparticles as well as on the size transition from Coulomb interaction to quantum-mechanical interaction. If it outgrows a structural-topological sphericity barrier, the growth of micro-crystals with faceted shape takes place.
Differential characteristics of microparticles are explained by different distribution of electron density in the surfacial layer of atoms compared to the bulk atoms, which manifested in size effects. The size effect is associated with near-wall factor of discrete systems, which is determined by different packing

density and number of atoms in the surfacial layer and in the bulk layer of nano- and microparticles. Parietal factor takes place at the macro level: the granular layer and in aggregations of microparticles in composite materials with granular filler and other materials with a granular structure. A critical size of micro- and nanoparticles, obtained by ultrasonication, dry or wet mechanical dispergation of granular materials and rocks can be determined by a certain size, when unusual changing of one or a complex of characteristics of the particles occurs in comparison with the massive solid body. It is well known that the decrease in particle size during milling process of rocks and materials leads to increasing in their specific surface. With the increase of surface curvature of such particles motivated by decrease of their size, a relative quantity of surfacial atoms of milling products has a tendency to growth and their packing density reduces because of reducing of coordination number and enhancing of Coulomb repulsion due to bigger bond distance between the atoms in the range of quantum-mechanical interaction. As surface elements of discreteness such as atoms and particles can be considered those atoms and particles, which located into two or more loose surfacial layers of clusters, microparticles aggregations or in near-wall layers of in a volume of dispersed or granular materials in a container. According to research data [10], the number of loose surfacial layers increases to 3–4 layers, when the container dimensions with the granular material in it decrease.

2. Experimental part and results

In order to solve one of the main issues of micro- and nanodispersed particles, it may be helpful to apply a simple research apparatus of applied science, that is, discrete topology [11] of random condensed systems, since a common factor for them is discreteness of the investigated subjects. Discrete topology of random condensed systems, having discreteness at a different dimensional scale, (i.e. macro- and microscale) uses mostly two basic characteristics – random packing density of elements of discreteness and their coordinate number that determines a topological condition of a system. Mathematical apparatus for the investigations is based on corpuscular-topological properties of discrete systems and basically consists of some fundamental principles [11]. The main principle is a relationship between discrete distributions of elements of discreteness (atoms, molecules, balls, nano-, micro- or macroparticles, etc.) in dimensional order with random close density in a system [11]:

$$\frac{d_n}{d_1} = \left(\frac{1}{10\eta_1(\sqrt{3}-1)^p} \right)^{\frac{mn}{n}} \tag{3}$$

or, at p = 3 and $m = 3$ - for random packing eq.3 takes the following form:

$$\frac{d_n}{d_1} = \left(\frac{2{,}549}{10\eta_1} \right)^n = (0{,}2549/\eta_1)^n = (\eta_{c1}/\eta_1)^n \tag{4}$$

where, d_1, d_2 – diameter of the biggest particles (balls), forming a packing and diameter of the particles (balls) with consistently smaller size, filling in the existing voids in the packing.

η_1 – random packing density of monodispersed layer of particles (balls, atoms – for elementary substances), atoms in a crystal cell ($\eta_1 \le 1$) – for complex substances;

m – class in the system of particle size distribution, determining discreteness in their size, $m = 0 \ldots 15 \ldots (n+1)$;

p – factor, determining short or long distance of elements of discreteness (granules, micro- or nanoparticles, atoms) in a system:

$p = 3$ – for random packing of mixed granules and particles; $p \leq 6$ – for contacted and free stayed solid particles (balls); $p > 6$ – for covering and penetrated to each other spherical particles.
The calculations made with formula 4 for critical radius of microparticles depending on packing density of metallic atoms in them are listed in table 1.

From eq. 3 at $m = 3$ and $d_n / d_1 = 1$, the following formula for random packing density of identical spherical particles can be obtained:

$$\eta_1 = \frac{1}{10(\sqrt{3}-1)^p} = \frac{0{,}1}{(\sqrt{3}-1)^p}$$

From the equation mentioned above, at $p = 3$, $p = 0$ and $p = 6$ we have the first $\eta_{c1} = 0.2549$ and the second $\eta_{c2} = 0.1$ critical packing density of elements of discreteness of a substance and the maximal theoretical random packing density $\eta_{c1} = 0.64976$.

From eq. 3 follows the equation for dimensional near-wall factor of discrete systems from the ratio of maximal voids (pores) size at $m = 3$, $n = 2$ in random packing ($p = 3$) to particles (balls) size ($m = 12$), which fill in all the voids or free volume in the system more compactly:

$$D \leq \left(\frac{10\eta_1}{2{,}549}\right)^3 \leq \left(\frac{\eta_1}{\eta_{c1}}\right)^3 \leq \left(\frac{\eta_1}{0{,}2549}\right)^3 \leq (3{,}923\eta_1)^3 \leq 60{,}377\eta_1^3 d \tag{5}$$

where D, d – critical size of atom aggregations (clusters, nano- and microparticles), particle aggregations of dispersed materials and the diameter of atoms in clusters, nano- or microparticles respectively for elementary substances and the most linear size of elementary crystal cell for complex substances. It is also a critical particle size for dispersed materials.

The results of calculation of critical radius for microparticles, depending on packing density of metal atoms in them using formula 4 are presented in table 1.

Table 1. Size of particles, obtained from liquid and crystal phases of metals.

Packing density of atoms in microparticles, η	Radius of microparticles, A (0,1 nm)					
	Zn	Pb	Sn	Au	Hg	Fe
	Radius of atom, Å (0,1 nm)					
	1,39	1,75	1,58	1,44	1,57	1,26
Crystal phase						
η = 0,7405	34,1	42,9	38,7	35,3	38,5	30,9
0,70323	29,2	36,7	33,2	30,2	33,0	26,5
0,68527	27,0	34,0	30,7	28,0	30,5	24,5
Liquid phase						
0,6370	21,7	27,3	24,7	22,5	24,5	19,7
0,63347	21,3	26,8	24,3	22,1	24,1	19,3
0,62634	20,6	26,0	23,4	21,4	33,3	18,7
0,60830	18,9	23,8	21,5	19,5	21,3	17,1
0,54100	13,3	16,7	15,1	13,8	15,0	12,0
0,49277	10,0	12,6	11,4	10,4	11,3	9,10
0,43406	6,96	8,76	7,91	7,21	7,86	6,31
0,4098	5,77	7,27	6,56	5,98	6,52	5,24
0,4040	5,53	6,97	6,29	5,73	6,25	5,01

A good correlation between the computed values of critical sizes for microparticles and empirical data can be taken as confirmation of formula 5. According to the empirical data given by A.S. Chizhik et al.[12], the radius of critical particles for the first four elements obtained from metallic molted mass is in the range of 20…30 Å. The data from the range has a good correlation with the calculated results using formula 5 for solid crystal and liquid phases of metals.

In compliance with eq. 5, the critical size of nanoparticles and number of atoms in forming aggregations from liquid phase depends on packing density of atoms in them. The critical size of dispersed particles from solid and liquid phase is not only a regular geometric parameter. According to the data given by S.L. Chizhik et al. [12], with decreasing of the size of metal microparticles, produced from molten mass, volume effect of super cooling disappears and latent heat gets equal to zero. At the critical diameter of microparticles, the difference between liquid and crystal condition comes out. The first class of phase transformation becomes continuous. The value of particle critical radius corresponds to the size in the field of short-range order in the liquid phase. Thus, the frozen density fluctuations in liquid at fast cooling are, to some extent, the nucleation centers of microparticles and crystal microphase. If the size of a microparticle dispersed from metal molted mass is extremely small, its volume consists of the compacted fluctuation zone. Depending on the size of microparticles cooled in the liquid metal phase, there is a probability of formation of different structures like loose amorphous and crystal structures. Critical size of nanoparticles with crystal structure produced from metal molted mass for the element presented in table 1 is in the range of 4…8.6 nm.

If the first critical random packing density of particles of dispersed materials is equal to $\eta_{c1} = \eta_1^{(9\ldots10{,}5)/3}$ where $= \eta_1 \leq 0.6403\ldots0.6498$ and a regular packing density of elements of discreteness of a substance is of $\eta_{c1} = \eta_1^{(11\ldots15)/3}$ [11], where $\eta_1 = 0.6802\ldots0.68834\ldots0.6981\ldots0.7405$, then the formula for dimensional range and for average critical size of clusters (n= 9…15) and spherical nanoparticles is the following:

$$D = 60{,}38\eta_1^{n/3}d\,, \qquad n = (0\text{-}3\text{-}6)\text{-}\mathbf{9\ldots12\ldots15} \tag{6}$$

where, n = 9–10.5 at $0.60377 \leq \eta_1 \leq 0.64976$ – for critical particle size of dispersed materials.
For the critical size of nano- and microparticles of crystal substances with different crystal cell:
n = 10.5 – 11.5 at $0.64976 \leq \eta_1 \leq 0.6802$ – volume centered;
n = 11.5 – 12 at $0.6802 \leq \eta_1 \leq 0.68834$ – icosahedral;
n = 12 – 12.4 at $0.68834 \leq \eta_1 \leq 0.6981$ – trigonal;
n = 12.4 – 15 at $0.6981 \leq \eta_1 \leq 0.7405$ – compacted hexagonal and face centered cubic cells.

The critical sizes of growing spherical new formations ($n \leq 0$), including faceted crystals in the interval of sphericity and potential reorganization of structure to their different unique form ($0 \leq n \leq 15$) can be taken from Eq. 4 as:

$$D = 60{,}38^{n/3}\eta_1^n d/\eta_s = \left(60{,}38^{1/3}\eta_1\right)^n d/\eta_s = \left(3{,}923\eta_1\right)^n d/\eta_s \tag{7}$$

where n = (0,1,2, **3)-6-9-12-15**….

$\eta_s = \eta_1^{1/9}$ – packing density of atoms in surface profile of interatomic layers. General equations for determination of the size of growing nano- and microparticles can be obtained from eq. 3 at $n = 2$, $p = 3$:

$$\frac{D}{d} = d_1 / \left(\frac{2{,}549}{10\eta_1}\right)^{m/3} d_1 = \left(\frac{10\eta_1}{2{,}549}\right)^{m/3} = (3{,}923\eta_1)^n, \quad n = m/3 \tag{8}$$

Table 2 presents the calculation results obtained using eq. 6, 7 for average and minimal critical size of spherical nanoparticles in the interval of potential reorganization of structure and appearance of their unusual properties in comparison with a massive solid body. Therein presented the dimensions of growing crystal new formations (6), which depend on packing density of elements of discreteness in them (metal atoms and other elementary substances), dimensional interval and an average critical size of spherical nanoparticles with icosahedral atom packing (η_1 = 0,68834).

Table 2. Dimensions of a growing crystal and the critical size of micro- and nanoparticles in the range when their unusual properties appear.

Index of power, *n-n*	Micro- and nanoparticle size $D \times d$ with packing density of structural elements in them, η_1				
	0,7405 (12,7)	at d = 0,3 nm	0,7405 (12,8)	0,6883 (12,7)	0,6403 (12,7)
1	2	3	4	5	6
0	**60 – 62**	**18 –19 nm**	**1,0 – 1,1**	**60 – 63**	**60,4 – 63,0**
3….1	45 – 46	13 – 14 nm	2,9 – 3,0	42 – 43	38,6 – 40
6 ….2	33 – 34	9,9 – 10 nm	8,4 – 8,7	28,6 – 29,8	24,8 – 26
9….3	**24,5 – 25**	**7,4 – 7,5 nm**	**24,5– 25,0**	**19,7 – 20,5**	**13,7 – 14,4****
12….4	18 – 19	5,4 –5,7 nm	71 – 73	**13,6– 14,2**	10,0 – 10,5
15….5	**13 – 14(12,6)***	**4,0 – 4,2 nm**	206 – 213	9,3– 9,7	6,5 – 6,8
18….6	9,9 – 10	**3,0 – 3,1 nm**	601 – 621	6,4 – 6,7	4,0 – 4,2
21….7	7,4 – 7,6	**2,2 – 2,2 nm**	1745 – 1804	4,4 – 4,6	2,7 – 2,8
24….8	5,5 – 5,7	**1,6 – 1,7 nm**	5071 – 5243	3,0 – 3,1	1,7 – 1,8
27….9	4,0 – 4,2	**1,2 – 1,3 nm**	growing	2,1 – 2,2	1,1 – 1,15
30….10	3,0 – 3,1	**0,9 – 0,93 nm**	growing	1,4 – 1,5	–

*– at n = 16/3, **– at n = 10/3

From table 2 and eq. 6, 7 it can be assumed that the maximal size of nanoparticles, when their unusual properties appear, corresponds to the value $n \geq 0$, at $n = 0$ will be equal to $D = 60 - 63d$. For example, at $d \approx$ 0.3 nm (for metals) we have D = 18 – 19 nm, and at η_1 = 0.7405 and n = 9;15 the value of D = 7.5; 4 nm. For dispersed materials at d = 0.1 μm and n = 0, the value of D = 6.0 – 6.3 μm and at η_1 = 0.64976– 0.6403 and n = (10.5 – 10)/3 respectively, the critical size of microparticles is equal to D = (14.0 – 14.4) d = 1.34 – 1.37 μm. From table 2 it follows that the average critical size of crystal nanoparticles is in the range of 7.5 – 19 nm, at the same time, the minimal critical size, when unusual changing of physical properties can appear is 4.2 nm. For microparticles of dispersed materials at η_1= 0.6403, the critical size is equal to 13.7 d and their maximal average dimensions are (60 – 63) d. The data of the critical size for nanoparticles of elementary substances (metals) are marked in table 2 with bold type.

For reorganization of the shape of growing spherical crystals into faceted crystals at appropriate thermodynamic conditions, the minimal critical size of elementary substances should be equal to (13 – 14)d, with the average critical size of 25 d and the maximal critical size of (60 – 62) d.

Since the size of monomer of a ion of silica (tetrahedron) d = 0.324 nm [13], the size of nanoparticles of freshly prepared sol at $\eta_c \leq 0.7405^{3...5}$, will be equal to 8.1 – 4.5 nm. Experience has shown that nanoparticles of silica and others rocks with minimal critical size about 8.1 – 4.5 nm form gel easily from liquid alkaline medium. The size of its stable aggregations at d = 8.1 nm equals to:

$$D_1 \leq 60{,}38\eta_1^{3...10/3} / \eta_1^{1/9} \cdot 8{,}1nm = 205...186nm$$

Critical size of quartz nanoparticles, having a size of crystal cell with given parameters a = 0.49 nm, b= 0.49 nm и c= 0.54 nm and $\eta_1 \leq 0.76$ equals:

$$D = 60{,}38 \cdot 0{,}76^{5...16/3} \cdot 0{,}54nm / 0{,}76^{1/9} = 8{,}5...7{,}8nm$$

Incorporating in eq. 7 the known values of regular packing densities of atoms for elementary substances and random packing of spherical particles, taking into account their packing densities in the surface of cross-section area of crystal formations with filled in atom voids upper and lower layers in the form of truncated ball segments ($\eta_s \approx \eta_1^{1/9}$) for different packing, we obtain:

for closest packing – $\eta_1 \leq 0.74048$, $D \leq 3^n d$

for icosahedral packing – $\eta_1 \leq 0.74048$, $D \leq 2{,}905^n d / \eta_s^{1/3}$

$\eta_1 \leq 0.68834$, $D \leq 2{,}70^n d / \eta_s^{3/4} \leq 2{,}70^n d / 0{,}9069^{3/4} \leq 2{,}90^n d$

for random packing – $\eta_1 \leq 0.64976$, $D \leq 2{,}549^n d / \eta_s^{1/3}$

(general packing) – $\eta_1 \leq 0.64029$, $D \leq 2{,}512^n d / \eta_s^{1/3}$

$\eta_1 \leq 0.63716$, $D \leq 2{,}50^n d / \eta_s^{1/3}$

$\eta_1 \leq 0.63405$, $D \leq 2{,}49^n d / \eta_s^{1/3}$

where $\eta_s^{1/3} = \left(\pi / 2\sqrt{3}\right)^{1/3} = 0{,}9069^{1/3} \approx 0{,}968, \eta_s^{1/3} \approx \eta_1^{1/9} = 0{,}74048^{1/9} \approx 0{,}9672$.

From the equation for icosahedral packing n = 1, η_s = 1 and $\eta_s^{3/4}$ = 0.9293 respectively, it follows that icosahedral atom aggregation (cluster) consists of one surface atom layer. Without central atom, the size of a central pore equals 0.9d that corresponds to the size of the central pore in an icosahedral packing of atoms, contacting each other. This way, vacancy defects form that gives an opportunity for growing of spherical aggregations. With the presence of the central atom in one-layered icosaherdral cluster, all 12 surface atoms contacting with it have an equal distance between each other.

Equations for dimension interval and for the average critical size (m = 0…1…2…3) of growing (n= ∞…0) clusters and spherical nanoparticles can be obtained from eq. 6, 7 in the following form:

$$\frac{(3{,}92\eta_1)^n d}{\eta^{1/9}} \leq D \leq \frac{(60{,}38\eta_{c1})^m d}{\eta^{1/9}} \tag{9}$$

$$\frac{1000\eta_1^n d}{\eta^{1/9}} \geq D \geq \frac{(60{,}38\eta_{c2})^m d}{\eta^{1/9}} \tag{10}$$

where $\eta_{c1} = \eta_1^{(11,5;\ 12;\ 12,4;\ 15)/3}$, $\eta_{c2} = \eta_1^{(18;\ 18,4;\ 19;\ 23)/3}$ at η_1 = 0.6802; 0.6883; 0.9681; 0.7405 respectively for systematic atoms packing (balls) in crystal structures, $\eta_{c1} = \eta_1^{(9;\ 10;\ 10,5)/3}$ and $\eta_{c2} = \eta_1^{(13,5;\ 15;\ 16)/3}$ – for random particles packing of dispersed materials at η_1 = 0.6038; 0.64029; 0.64976 respectively.

Both parts of equation 4 or 6, 7, including η_s take the cube and times $\pi/6$. After conversion we have a critical number of elements of substance discreteness for the corresponding packing:

$$N \le \frac{(3{,}923\eta_1)^n}{\eta^{1/3}},$$

$$N = \frac{60{,}38\eta_1^n}{\eta_1^{1/3}}$$

3. Conclusion

The obtained formulas 9, 10 give a number of elements of discreteness in spherical aggregations and do not determine their number in coordination polyhedrons such as cubooctahedrons and icosahedrons.

Equations 9, 10 describe ($m \ge 0$) critical sizes of elements of substance discreteness for the main level of discreteness, even at critical condition of a substance.

4. Reference

[1] Petrov Y I 1982 *Physics of Small Particles* Moscow p 359
[2] Petrov Y I 1986 *Clusters and Small Particles* Moscow p 367
[3] Kipnis A Y 1981 *Clusters in Chemistry. Znanie* Moscow No.7 p 64
[4] Kharkhardin A N and Kashibadze J 2009 Izvestia vuzov *Construction* **5** 109–113
[5] Gubin S P 1987 *Chemistry of Clusters. Basis of Classification and Structure* (Moscow: Nauka Press) p 263
[6] Strokova V V, Zhernovsky I V and Cherevatova A V 2011 *Nanosystems in Construction Material Science* Belgorod p 205
[7] Strokova V V, Zhernovsky I V and Fomenko Y V 2007 *J. Bulletin of BSTU* **4** 48–49
[8] Suzdalev I P 2006 *Nanotechnology: Physics and Chemistry of Nanoclusters, Nanostructures and Nanomaterials* (Moscow: ComKniga) p 592
[9] Wells A 1987 *Structural Inorganic Chemistry* First edition, Moscow p 178
[10] Aerov M E and Todes O M 1968 *Hydraulic and Thermal Basis of Apparatus Process with Stationary and Boiling Granular Layer* (S.-Peterburg) pp 32–36
[11] Kharkhardin A N 2011 *Structural Topology of Dispersed Materials* Belgorod p 288
[12] Chizhik S L et al 1985 *Experimental and Theoretical Physics J.* **88** (5) 1706–17
[13] Kukolev G V 1966 *Chemistry of Silica and Physical Chemistry of Silicates* Moscow p 59–60

Acknowledgement

The research work was accomplished under the financial support from the Ministry of Education and Science of the Russian Federation in the framework of the State Assignment using the equipment based on the High-Technology Center, BSTU named after V.G. Shoukhov.

Structural Engineering Managers – Innovation Challenges for their Skills

D Linkeschová[1,2] **and A Tichá**[1]
[1]BUT, Faculty of Civil Engineering, Institute of Structural Economics and Management, Veveří 331/95, Brno, 602 00 Czech Republic

E-mail: linkeschova.d@fce.vutbr.cz

Abstract. The profession of a structural engineer is highly responsible, because the consequences of a structural engineer's errors result not only in economic damage to the property and often irreversible damage to the environment, they can also lead to direct loss of lives. In the current turbulent, dynamically developing society the managerial methods of structural engineers should not stagnate at the level of the last century applications.
This paper deals with the challenges which the ongoing century poses to structural engineers and managers. It compares the results of research regarding the current state of managerial skills of structural engineers in Czech building companies to the defined skills of the 21st century's managers according to the global research programme ITL Research and according to the Vision for the Future of Structural Engineering, drawn up by Structural Engineering Institute – SEI ASCE.

1. Introduction

High responsibility borne by structural engineers results mainly from the far reaching and often even irreversible consequences of their errors, which may result in the loss to the environment, lives and health of people and their property. Professional building-technical skills are the natural requirement for managers in building companies and at construction sites. At present, they also include relevant language skills and good ability to manage information technologies. This knowledge, which managers acquire by education and practice, is examined by the Czech Chamber of Certified Engineers and Technicians Active in Construction at authorization testing and by the International Project Management Association (IPMA) in the case of project managers.

However, structural engineering managers must also have specific managerial qualities and skills, which enable them to cooperate in sequential teams, often at the international level. According to the

[2] To whom any correspondence should be addressed.

surveys carried out, these so-called soft skills are underestimated in the Czech Republic. Only building practice proves the ability or inability of a particular person to cope with these qualities and skills.

The survey shows that training in managerial skills in the companies are either totally missing or non-systematic, which is particularly characteristic of small and medium-sized companies. Issues required by legal regulations (occupational health and safety protection) and technical standards (ČSN – Czech National Standards) remain the focus of education and training of structural engineers and other professionals in the industry.

Similar problems are probably faced by structural engineers in other countries and also by engineers and managers in other professions (see new programmes [1] and professional materials [2]).

2. Methodology

The **aim of the current research** is to determine the current state of management skills in construction companies in the Czech Republic. In this article, surveys drawn up since 2010 to date have been used to find out the current state of their managerial skills. The target groups of these surveys include managers, site engineers and foremen on the one side and building workers in the position of subordinate personnel on the other side. The surveys examined in particular the situation in the building companies participating in smaller building contracts in which, in most cases, the leading and managing of the working crews fall under the competence of site foremen or site managers. The surveys are also based on the statistic data of persons working in civil engineering, which are published by the Czech Statistical Office at its website every year. [3]

The surveys are based on the questionnaire surveys and talks conducted with persons in civil engineering. The following qualities and skills were examined in particular: the skills of planning, organization, checking, decision-making, motivation, cooperation, communication, running a risk, quick response to situations, as well as the ability to correctly allocate financial resources.

The skills to be possessed by structural engineers in the future (until 2033) were taken from the available materials. "Engineers must become more visionary. They require more business skills and must be better leaders. Not only must they be better leaders, they must be perceived as leaders. Additional skill sets (oral and written communication, business strategies, and leadership techniques) can be incorporated with coursework already present in the university curriculum." [2]

3. Results

Today's building companies need productive management more than ever in the past. Reaching of the set goals, their timeliness and performance quality also depend on the managerial skills and approach of the superiors to their co-workers. Only the surveys that best characterize the requirements for the qualities and skills of structural engineering managers have been chosen from the surveys carried out to date.

3.1. Qualities and skills of structural engineering managers

The surveys carried out among employees and managers of building companies in 2012 were evaluated using the scoring method. The respondents assigned values of 1 to 7 points, where value 1 was quite insignificant and value 7 was ascribed to the most important quality or skill.

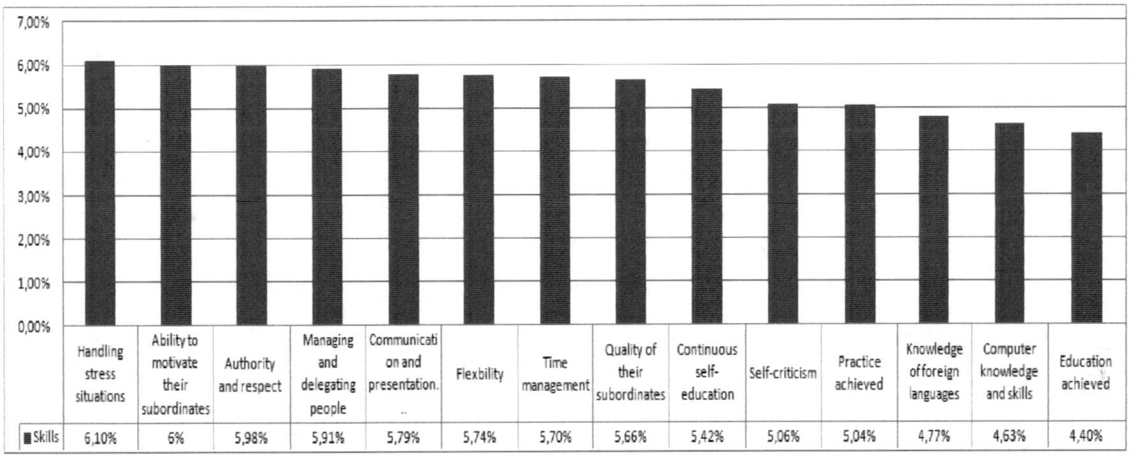

Figure 1. Required skills of structural engineering managers from the point of view of subordinate employees. [4, p. 44].

The graphs represent the average value of all received answers and individual elements in the graphs are arranged in the descending order by the average value. Employees have the following requirements for the qualities and skills of their managers (see figure 1).

From the point of view of managers in civil engineering, the situation looks as follows:

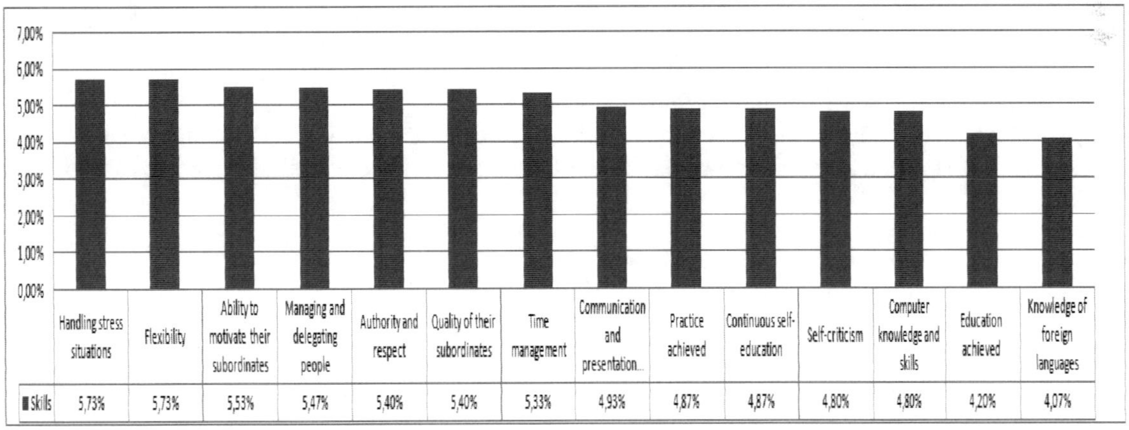

Figure 2. The required skills of structural engineering managers from the view of managerial staff. [4, p. 46].

The above-mentioned graph shows that the employees of building companies consider handling stress situations to be the most important factor for their managers. Value 7 was ascribed by 138 respondents, which is 53% of the answers, while 28% evaluated it by value 6, which is another 28%. The second is the ability to motivate subordinates (value 7 in 48% and 6 in 27% cases). The factor of authority and respect in subordinates ended up in the third place.

From the point of view of managing staff, resistance against stress along with flexibility are again in the forefront (value 6 in 47% and value 7 in 33% cases). Motivation, which the employees placed in the second place, appeared in the third place among the managing staff. Education along with language and computer skills are possibly considered a standard, but according to this survey they are assigned the least significant position. Apparently, the respondents expressed the opinion that there is no substantial difference between bachelor's and master's degree and thus additionally acquired knowledge is not so significant.

The results of the survey conducted in 2014 are shown in figure 3.

Figure 3. The required skills of structural engineering managers. [5, p. 52].

Communication skills (35%) are required most significantly. Professional knowledge (29%) is followed by the requirement for the ability to handle problematic situations (18%) and organizational skills (14%). It is interesting that knowledge of the language and computer skills again have negligible values. [5]

3.2. Roles and goals of structural engineering managers

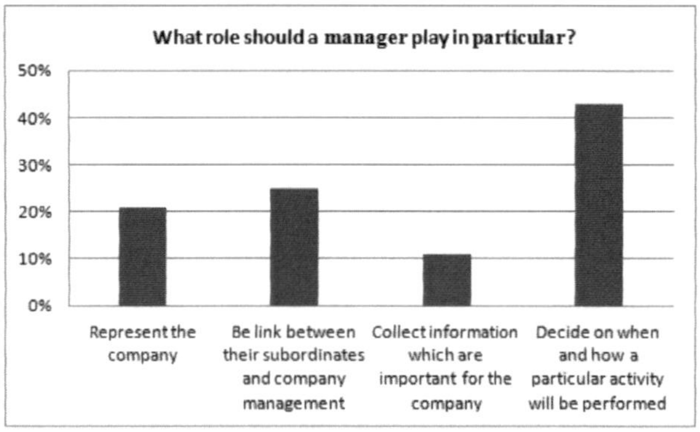

Figure 4. The required roles of structural engineering managers. [5 p. 53].

According to the respondents from the building companies, the most important role of managers is to decide on where and how a particular activity will be performed. This option was selected by 43% of the respondents. 25% pointed out that managers mainly have to be a connecting link between subordinates and company management. 21% of the respondents think that a manger primarily represents the company. According to 11% of the received answers, managers should mainly deal with obtaining information important for the company.

The declared goals of managers are shown in the figure below.

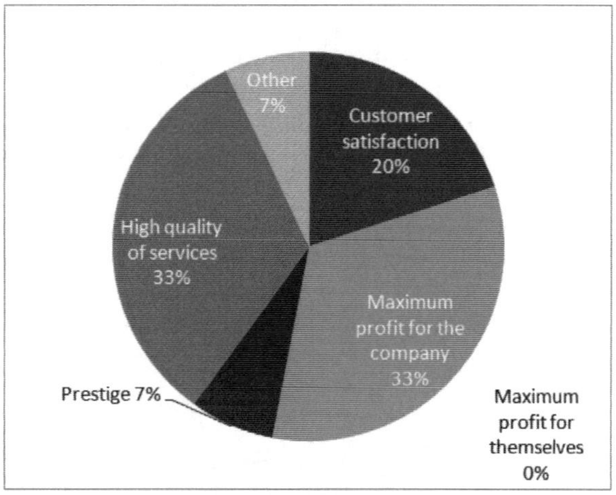

Figure 5. The required goals of structural engineering managers. [5, p. 55].

The main working goal of managers, in their opinion, is to achieve the maximum profit and, at the same time, to provide high quality of services. Such goals as creating an effective work team, satisfaction of employees, building the company goodwill, environmental impacts, etc. did not appear in this survey at all.

4. Discussion

The above-mentioned surveys conducted in the Czech Republic show that structural engineering managers often think that they meet the current demands for their profession. However, the amount of information and the complexity of solutions for upcoming problems increase and such situations occur more and more often and one cannot prepare for them in advance.

From the point of view of leading and managing personnel, "there is no significant difference between Human Resources Management and Personnel Management"[7].

More and more frequently not only companies, but also individuals need to immediately react to the changing conditions, they need to improvise. However, improvisation and the necessary flexibility are in no case substitutes for preparations and planning, they rather supplement them naturally.

The required skills and abilities of future managers of the 21st century are dealt with by the global research programme ITL Research. The necessary competences which will be added to the current demands include: collaboration, knowledge construction, self-regulation, real-world problem-solving and innovation, the use of ICT, skilled communication [1].

According to our opinion, also the ability to use both critical logical thinking and intuition, awareness of the consequences and sense, humour and control are necessary as the sources for creativeness and innovation. One must be aware that "leadership is the ability to lead the company beyond the current boundary of its economic horizons, give the company and especially the employees a vision for the future and react to the changes in global market demands or even create them. This way of leadership is often linked to magic terms and personal attributes, such as: "leader", "charisma", "mission" and "radiate" [6]. Obviously prerequisite are also ethical criteria (e.g. [8], [9]).

The Vision for the Future is drawn up according to the study of the Structural Engineering Institute. The expectations and Role of the Future Structural Engineer presents an opportunity, if not a mandate, to transform structural engineering education, training, and practice in ways that will foster an enduring and creative profession. Desired Outcome is to explore and define what the structural engineering profession is. The next generation of structural engineers need to have education, practical knowledge, and technology applications that should be developed throughout a career [2 p. 4]. Abilities and skills that must be developed are specified in the study. They include redefined technical skills, soft skills, advanced professional ability, creativity and curiosity. The study clearly shows that it is also necessary to set new mechanisms to enable the desired changes.

In the CEI survey, the respondents when asked about the next 10-30 years repeatedly identified technology, delivery methods, education, and globalization as the drivers of change; and current practitioners see these drivers as threats rather than opportunities [2. p. 40]. This also corresponds to the previously obtained conclusions of the surveys carried out in the Czech Republic. Final figures and statistical comparison of the state of managerial skills in our country and the world will be performed after the completion of the ongoing research in the country.

5. Conclusion

The above-mentioned surveys and studies show that the need of innovations in the work of structural engineers and managers is perceived as a necessity by personnel in civil engineering also in other countries. It is apparent that the requirements for structural engineering managers will significantly change within several years. The emphasis will be placed on the above-mentioned qualities and skills which are usually summarized under a uniform term "soft skills". "It means skills including communication and presentation skills - along with a full complement of the modern soft skills related to a high emotional intelligence quotient - such as an engaging personality, flexibility, adaptability, self-direction and self-control, and perseverance." [2. p. 18] According to that research, they are mainly communication and organizational skills. It is urgently needed to develop these skills at schools, as it is shown in figure 3, along with a strong ability to choose and unite subordinates and superiors into one efficient and cooperative team. Similar results have been obtained by the team under the leadership of the head of an international research team, Professor Ansgar Zerfass from the Leipzig University, who have conducted the European Chief Communication Officers Survey (ECCOS) [10]. 579 leaders of communication and public relations departments took part in the research in the European companies. As stated above, the profession of a structural engineer is a very responsible profession, because the consequences of structural engineers' errors are far-reaching. In the current society, it is no longer possible that the managerial methods of structural engineers stagnate. This is also confirmed by the previous results of the ongoing research in the Czech Republic that are listed in this article.

The profession of an engineer will still be an important profession. However, it will be necessary to be more aware of possible risks, economic and sustainable utilization of valuable resources, and the use of innovative technologies. It is also important to create inspiring structures and mechanisms which will allow and support the oncoming changes.

IOP Conf. Series: Materials Science and Engineering **96** (2015) 012068 doi:10.1088/1757-899X/96/1/012068

6. References

[1] ITL Research [Available]: http://www.itlresearch.com/images/stories/reports/21cld%20learning%20activity%20rubrics %202012.pdf

[2] SEI ASCE *Vision for the Future of Structural Engineering Zpracované Structural Engineering Institute* [Available]: http://www.asce.org/uploadedFiles/visionforthefuture.pdf

[3] Český statistický úřad [Available]: www.czso.cz

[4] Vostrčil M 2012 *Požadavky na Manažera Stavebního Podniku v Období Znalostní Ekonomiky* Brno No.68 p 17 příl. Bakalářská práce. Vysoké učení technické v Brně, Fakulta stavební, Ústav stavební ekonomiky a řízení.

[5] Málková D 2014 *Požadavky na Manažera Stavebního Podniku* Brno, příl. Bakalářská práce. Vysoké učení technické v Brně, Fakulta stavební, Ústav stavební ekonomiky a řízení p 66

[6] Linkeschová D 2014 The construction industry managers and the need for changes in their communication *The Conf. Proc. of People, Buildings and Environment 2014 (Kroměříž, Czech Republic)* p 266 [Available]: http://www.fce.vutbr.cz/ekr/PBE/Proceedings/2014/025_14162.pdf

[7] Savaneviciene A and Stankeviciute Z 2013Relabelling or new Approach: theoretical insights regarding Personnel Management and Human Resource Management *Engineering Economics* **24** (3) 236 [Available]: http://www.inzeko.ktu.lt/index.php/EE/article/view/1618

[8] Synthes. Global code czech. Září 2006 [Available]: http://us.synthes.com/CmsSynthes/compliance/pdf/?Global%20Code%20Czech.pdf.

[9] ABB. Kodex jednání ABB. Říjen 2010 [Available]: http://www.abb.cz/cawp/czabb016/acb8144b21436b0ac12576030037ae7d.aspx

[10] Zerfass A, Verčič D, Moreno Á, Verhoeven P, Tench R and Klewes J 2013 European Chief Communication Officers Survey 2013: Managing CEO positioning and international communication: Insights from interviews with corporate communication leaders (London and Brussels: Ketchum / EUPRERA) [Available]: http://www.ketchum.com/sites/default/files/insights/eccos_2013_report_-_ketchum_euprera.pdf

Shear Adhesive Connections for Glass Structures

K Machalická[1,3], I Horčičková[2] and M Eliášová[2]
[1]Czech Technical University in Prague, University Centre for Energy Efficient Buildings , Třinecká 1024, 273 43 Buštěhrad, Czech Republic
[2]Czech Technical University in Prague, Faculty of Civil Engineering, Thákurova 7, 166 29 Prague, Czech Republic

E-mail: klara.machalicka@fsv.cvut.cz

Abstract. Unique aesthetical properties of glass – not only transparency but also smooth, glossy and primarily reflective surface – give this material special importance in the contemporary architecture. In every structural application of glass it is necessary to solve the problem associated with connections between glass pane and other part from a different material or between two glass elements. Moreover, there are many types of hybrid structures that combine glass and different materials to achieve safe failure behaviour and high degree of transparency at the same time. Connection of brittle glass and reinforcing material is an essential part of these structures, where composite action between two parts is beneficially ensured by a glued joint. The current paper deals with the experimental analysis focused on the determination of mechanical characteristics of adhesives applied in planar connections under shear loading.

1. Introduction

Polymer materials of adhesives are no longer limited to the sealing function, they can also fulfill real load-bearing function [1], [2] [3]. Contrary to the traditional bolted connection, an adhesive joint provides many important advantages for brittle glass, such as uniform stress distribution along the larger connection area, the absence of drilling holes in the glass or possibility to join materials with different mechanical and thermal properties. The reliability of the bond is influenced by many factors, starting with selection of the appropriate adhesive for a particular joint. The shear strength of the connection may be also influenced by adhesion of glue to the substrate, cohesion of glue, joint dimensions and geometry, environmental factors (relative humidity, UV radiation, and temperature), thermal expansion coefficients of joining elements, duration of load, rate of load application, surface preparation and surface roughness, workmanship and curing [4].

[3] To whom any correspondence should be addressed.

IOP Conf. Series: Materials Science and Engineering **96** (2015) 012069 doi:10.1088/1757-899X/96/1/012069

The presented article deals with the experimental analysis focused on the determination of the mechanical characteristics of adhesives applied in planar connections under the shear loading. With regard to the fact that many types of adhesives are produced, six glues with different properties (polyurethanes, acrylates and UV-curing methacrylates) were selected for the research. The research also compares two similar adhesives (the older technology and the new one) produced by the same manufacturer in case of glass-to-steel connection. Specimens of all adhesives were prepared for various substrates and their surface treatment (smooth glass, roughened glass, steel, stainless steel and aluminum). As joint dimensions and thickness play a crucial role, adhesives were tested in several thicknesses according to the type of adhesive. Firstly, small-scale test specimens were carried out and evaluated. In the following stage, the most appropriate transparent UV-curing adhesive was selected for full-scale specimens of glass beam to assess the influence of size effect. Due to these several goals, the research provides comprehensive knowledge about the behavior of the chosen adhesives in shear connection under short-term loading.

2. Experimental program

Experimental analysis of the glued connections was performed in relation to hybrid steel-glass beams research [5], therefore adhesives were chosen in accordance with the application in glass-to-metal or glass-to-glass large planar connection loaded by shear.

Table 1. Shear tests setup.

Type of specimen	description	diagram
1st arrangement: glass-to-steel	<u>2 x float glass</u> 50 x 110 x 19 mm <u>2 x steel S235</u> 50 x 75 x 25 mm <u>Adhesives:</u> 1C-PU: SikaFlex-265+Booster Acrylate: SikaFast 5211 Acrylate: SikaFast 5211 N	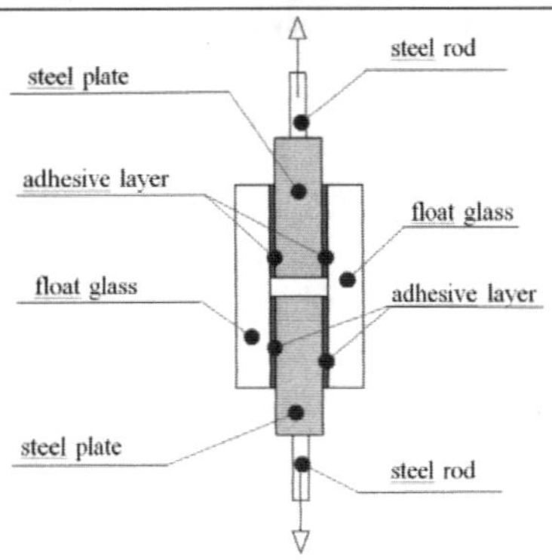
2nd arrangement: glass-to-steel	<u>2 x float glass</u>: 50 x 50 x 19 m <u>steel S235, stainless steel 1.4301, aluminium Al 6082</u> 290 x 100 x20 mm <u>Adhesives:</u> 1C-PU: SikaFlex-265+Booster 2C-PU: SikaForce 7550 Acrylate: SikaFast 5211	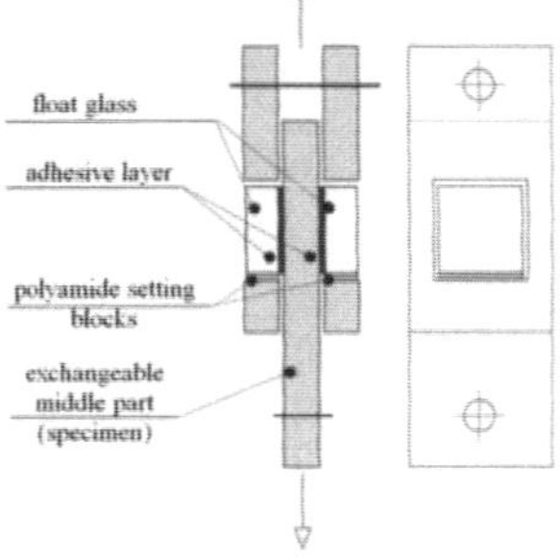

IOP Conf. Series: Materials Science and Engineering **96** (2015) 012069 doi:10.1088/1757-899X/96/1/012069

glass-to-glass	3 x float glass: 50 x 50 x 19 mm Adhesives: UV-curing acrylate: Ritelok UV50 UV-curing acrylate: Conloc 685	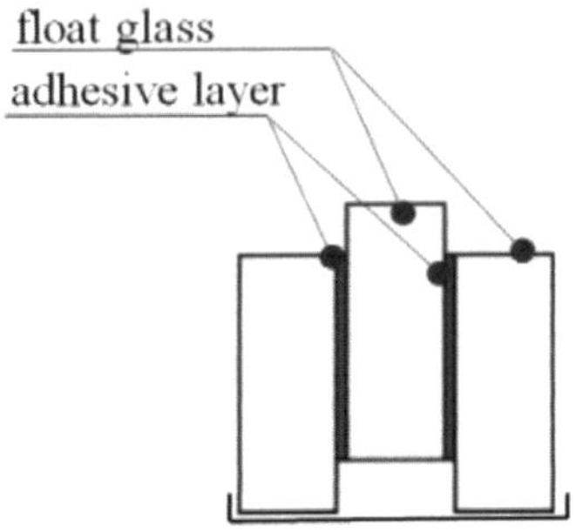

2.1. *Small scale shear test arrangement*

Since the important information about the glue is not only its ultimate strength but also its behaviour in the joint under loading, the research covers six types of adhesives with different mechanical properties – elastic adhesives (one-component and two-component polyurethanes), semi-rigid adhesives (two-component acrylic adhesives) and transparent rigid adhesives (UV-curing systems), which can be suitable in case of glass-to-glass bonding. Adhesives were applied in the test specimens; see Table 1, assembled from different materials to describe the influence on adhesion. The research covers glass-to-glass and glass-to-metal (steel, stainless steel and aluminum) joints with smooth or coarsened glass surfaces considering the adhesion improvement possibility.

The joint ability of flexible deformation is one of the most important conditions for the choice of adhesive for hybrid structures due to different thermal expansion coefficients of the connected materials. For the reason that flexible deformation corresponds to the adhesive layer thickness, glass-to-metal specimens were prepared in 1, 2, 3 and 4 mm thickness of the layer for polyurethanes and two-component acrylate glue.

Small-scale shear tests were performed according to three arrangements, see test setups in Table 1. The first setup was carried out according to the scheme in the first row in table 1. Specimens were prepared for three adhesives. Shear stress in the adhesive layer was reached by steel plates in tension. Tensile load was introduced to the setup by long steel rods. This arrangement is suitable for quick obtaining of glued connection stress-strain diagrams and was used only for glass-to-steel connections.

The second setup arrangement was used for detailed glass-to-metal joint analysis. Glues were applied in glass-to-steel, stainless steel and aluminum specimens, which create middle part of the setup. Glass for specimens was used both with smooth and roughened surface. The second arrangement allows introducing the tensile load by the pins positioned at both ends of the test device. Shear stress in adhesive layer was reached by tensile load in test device and middle sheet of metal (specimen). The main benefit of the rearranged test setup and usage of testing device with end-pins was the elimination of additional loading.

The third arrangement of small-scale shear tests was used on glass-to-glass connections with transparent UV-curing adhesives. Shear load in adhesive layer was introduced by compressive force applied to the middle glass element. At the bottom edge, lateral restraint against outspreading was applied.

3. Small scale shear tests results

Summarized stress-strain relationships were prepared for comparisons of joint types with various adhesives and glue thickness. The summarized diagrams show representative curves, which were chosen according to their approaching to the average values of results.

3.1. Polyurethane adhesives

Both tested PU-glues demonstrated similar behavior for all tested substrates, see stress-strain diagrams in figure 1. All specimens were broken by full cohesive failure mode. Both adhesives results proved lower strength values by 10% and two-component PU adhesive had also lower strain at break by 20 % in comparison with product data sheets.

3.2. Acrylate adhesives

Specimens glued by two-component acrylate adhesive (SikaFast 5211) with smooth glass surface reached shear strength values significantly between 5.5 and 7 MPa with shear strain at break 150-200%. Failure of these specimens started by adhesive mode at glue-glass interface but final failure was observed as cohesive at the places, where glue had sufficient adhesion to the glass, see figure 4. Contrary to this, specimens with roughened glass surface demonstrated satisfactory adhesion to the glass at whole joint area and they reached approx. 8 MPa shear strength, which is in conformity with the product data sheet. Summary of representative stress-strain relationships, see figure 2, provides comparison of various substrates in joints and different adhesive thicknesses as well. These results were obtained by usage of the second arrangement setup.

Stress-strain diagram illustrating the results for the first arrangement setup, see figure 3, shows comparison of adhesive SikaFast 5211 and SikaFast 5211 NT (new technology). It is noticeable that the new type of the adhesive (specimen 2) shows different behaviour than its predecessor (specimen 1). SikaFast 5211 NT reached significantly higher shear strength (7.5 MPa for 3 mm thick joint and the first type of arrangement, while originally the glue reached 4 MPa for the same joint and arrangement) and larger shear strain at break. The large initial stiffness in the joints of new adhesive did not change up to 3.5 MPa, although the older type of adhesive had the initial stiffness limited around 2 MPa. Cohesive failure with adhesion faults at glass-glue interface (similarly to its predecessor), see figure 5, or glass damage was typical for SikaFast 5211 NT specimens.

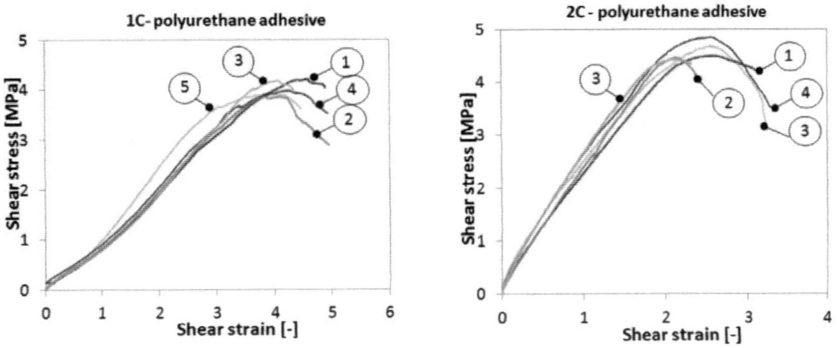

Figure 1. Shear stress-strain relationship – PU adhesives. 1 = steel + glass (3 mm), 2 = steel + glass (4 mm), 3 = steel + roughened glass, (3 mm), 4 = stainless steel + glass (3 mm), 5 = aluminium + glass (3 mm).

IOP Conf. Series: Materials Science and Engineering **96** (2015) 012069 doi:10.1088/1757-899X/96/1/012069

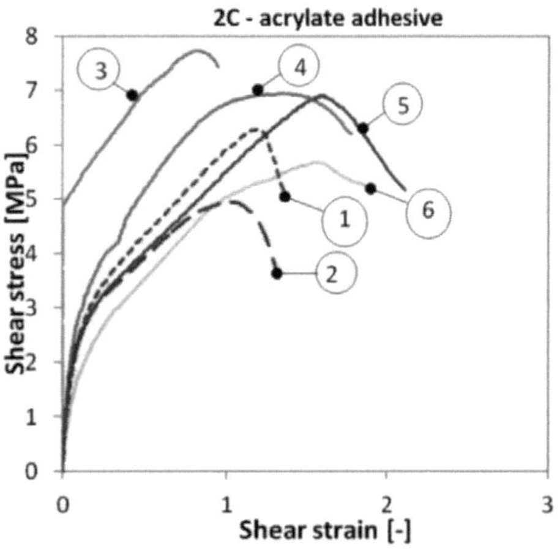

Figure 2. Shear stress-strain relationship – acrylate adhesive; 1= steel + glass (3 mm), 2 = steel + glass (4 mm), 3 = steel + roughened glass, (3 mm), 4 = stainless steel + glass (3 mm), 5 = stainless steel + glass (4 mm), 6 = aluminium + glass (3 mm).

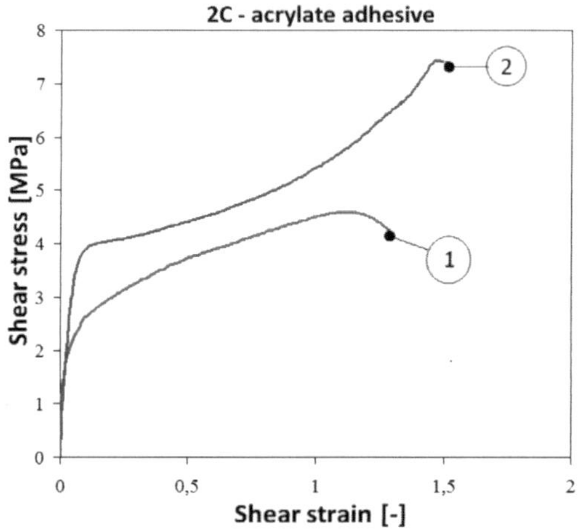

Figure 3. Comparison of acrylate SikaFast 5211 a SikaFast 5211 NT in glass-to-steel connection, first arrangement, 1 = SikaFast 5211 (3 mm), 2 = SikaFast 5211 NT (3 mm).

Figure 4. Typical adhesive-cohesive mode of failure for the specimens glued by SikaFast 5211.

Figure 5. Cohesive mode of failure for the specimen glued by SikaFast 5211 NT.

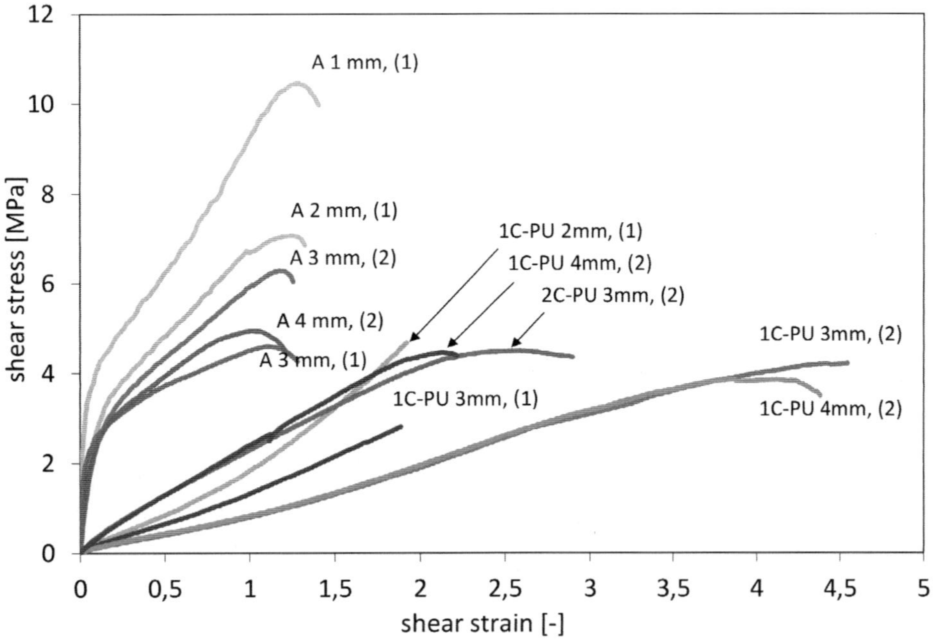

Figure 6. Influence of adhesive layer thickness on stress-strain relationship in glass-to-steel joint.

3.3. Adhesive layer thickness comparison

As the thickness of the adhesive layer is a very important factor affecting the mechanical properties of a joint, glass-to-steel specimens were produced in a 1, 2, 3 and 4 mm thickness of glue. The comparison of the results is shown in figure 6, where 1C-PU is one-component polyurethane with the Booster system (Sikaflex-265 Booster 20W), 2C-PU is two-component polyurethane (SikaForce-7550 L15) and A is acrylate adhesive (SikaFast 5211). It is noticeable that all glues show a different reaction to the change in thickness. Experiments proved the phenomenon that the strength of joint decreases with the increasing thickness of the adhesive layer simultaneously with the increase in elastic deformation and it was observed in semi-rigid acrylate adhesive more significantly than in flexible polyurethanes. The graph includes the results from the first (1) and the second (2) type of arrangement and it also provides the comparison of the results for the same connection tested in both arrangements (e.g. acrylate adhesive in 3 mm connections). It is evident that different results were mainly caused by embedded hinges in the second arrangement and by minimizing of additional loading.

3.4. UV-curing adhesives

Specimens bonded by UV-curing adhesives (RiteLok UV50, Conloc 685) showed similar behavior for both smooth and roughened glass surface. Shear strength reached was approx. 15 MPa with shear strain at break 100%; see figure 7, which is in accordance with the product data sheets. Failure was observed as a combination of cohesive break of glue together with rupture of glass.

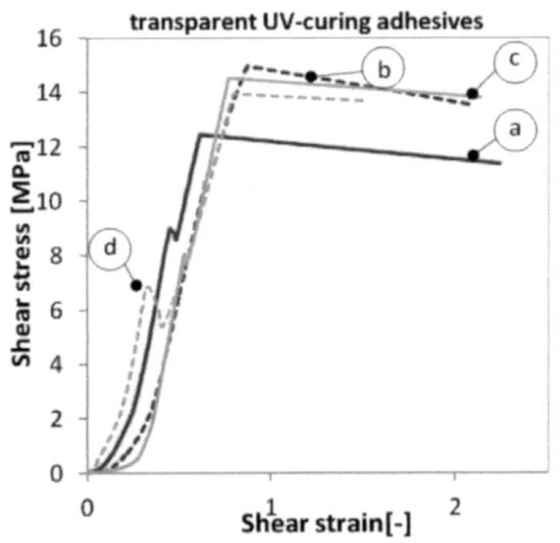

Figure 7. Shear stress-strain relationship – UV-curing adhesives; a = RiteLok UV50 (1 mm) smooth glass, b = RiteLok UV50 (1 mm) sand-blasted glass, c = Conloc 685 (1 mm) smooth glass, d = Conloc 685 (1 mm) sand-blasted glass.

4. Glued glass beam with I cross-section

The knowledge of the glued glass joint behaviour was verified in a real structural element – a fully transparent glass beam with I cross-section. The glued glass beam consists of a glass pane (web) that is double sided strengthened by rectangular glass profiles at its upper and lower edges. The connection between the rectangular glass strengthening profile and the web is adhesive bonded by Conloc 685. The span of the beam was 4.0 m. The beam was produced from heat-strengthened glass because of the high shear strength of the adhesive. Rectangular sheets of heat-strengthened glass may be manufactured in a width-to-length ratio of up to 1:10. For this reason, glass profiles at the edges had to be divided into segments. The whole beam geometry was developed so that no seam between two segments was in the same place of the flange from both sides.

For full-scale tests of a glued connection, two beams were prepared. The specimens were subjected to a 4-point bending test. In the test setup, each point load was at the distance 937 mm from supports to avoid seams in the flanges. Lateral restrictions were provided at supports and the mid-span. The beams were loaded at a rate of 50 N/s and loading was continued until the total collapse.

Full-scale tests proved the expected behaviour of the adhesive connection according to the previously performed small-scale tests. The measurements of strain gauges showed that the adhesive connection transferred stresses between cross-sectional parts very well, see table 2 representing stress transferring below the load introduction point at the bottom edge of the beam.

Figure 7. Glued glass beam.

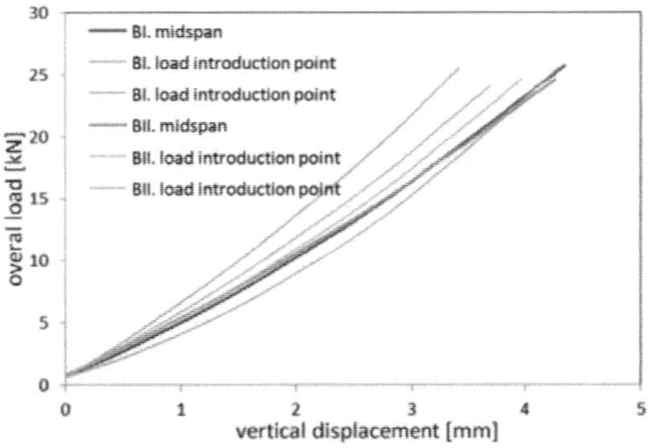

Figure 8. Load-displacement relationship for both beam specimens.

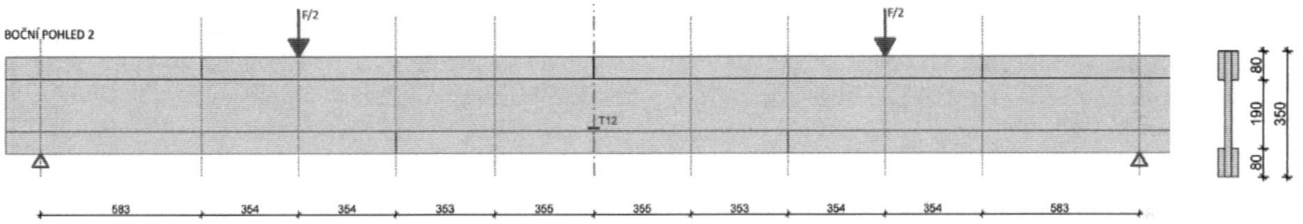

Figure 9. Glued glass beam experimental setup.

Table 2. Strain gauges measurements below the load introduction point

no. strain		description	bending stress [MPa]
T02		bottom flange segment	
	Beam 1		**11.2**
	Beam 2		**10.7**
T03		bottom edge of web	
	Beam 1		**11.1**
	Beam 2		**10.4**
T04		bottom flange segment	
	Beam 1		**10.6**
	Beam 2		**10.1**

Contrary to expectations, the collapse of both specimens occurred at relatively low loading and the failure started near the support by a shear crack, not in the middle of the span. The early failure was caused by several factors. Firstly, considering hand-made manufacture of the beams, there is a risk of stress concentrations near the bottom flange edge due to geometrical imperfections. Secondly, the asymmetrical seam arrangement of rectangular sections creating flanges caused additional transverse loading of the beam. As a result, at the seam locations, the upper compressed part of the cross-section with a seam was additionally bent to the web with the continuous rectangular section and the bottom tensile part of the cross-section was inflected to the interrupted rectangular flange section, thus both flanges were additionally bent in the same direction and the glued connection was loaded by peeling-off.

5. Conclusion

Sand-blasted glass surface can improve the shear strength, if the pure cohesive failure mode is not reached in the joint with a smooth glass surface and if the glue has a low viscosity. No decreasing strength values caused by sandblasting of the glass surface were observed because the cohesive strength of the adhesive was crucial in the glued joints. The influence of different metal materials in glued connections was not decisive.

The effect of thickness was noticeable for the semi-rigid 2-component acrylate adhesive (SikaFast 5211) – the strength of a glued connection was decreased while increasing the thickness of the adhesive layer, especially in the glass-steel type of connection. In flexible PU-adhesives, the influence of thickness was not noticeable.

Important difference was proved for behaviour of 2-component acrylate adhesives. The new type of acrylate glue (SikaFast 5211 NT) showed completely different behaviour under shear loading with significant, almost linear part of stress-strain diagram with high stiffness up to stress level of 3.5 MPa. For higher levels of loading, behaviour was changed and noticeable increase of strain was observed.

Behaviour of adhesive connections in glued glass beams as full-scale specimens corresponds to the knowledge of adhesive joints in small-scale test specimens. The joint between the glass web and the strengthening profiles (flanges) transferred stress very well.

Except for the above mentioned, it is important to say that long-term loading, cycling loading and ageing can significantly change the mechanical properties of glued joints and further research is necessary.

6. References

[1] Belis J, Van Hulle A, Callewaert D and Dispersyn J 2012 *Proc. of Challenging Glass 3 – Conf. on Architectural and Structural Applications of Glass (Delft University in Technology)* ed F Bos et al (Amsterodam: IOS Press) pp 177–186

[2] Weller B, Härth K, Tasche S and Unnewehr S 2009 *Detail Practise – Glass in Building, Principles, Applications, Examples* (Basel: Birkhauser) pp 68–71

[3] Abeln B, Richter C and Feldmann M 2014 *Proc. of Challenging Glass 4 – COST Action TU 0905 Final Conf.* ed Ch Louter et al. (Luiden: CRC Press/Balkema) pp 321–329

[4] Haldimann M, Luible A and Overend M 2008 *Structural Engineering Documents 10 - Structural Use of* (Zürich: IABSE-AIPC-IVBH) pp 152–163

[5] Netusil M, Eliasova M 2012 *Proc. of Challenging Glass 3 – Conf. on Architectural and Structural Applications of Glass (Delft University in Technology)* ed Bos F et al (Amsterodam: IOS Press) pp 715–724

Acknowledgement

This paper has been written with support by the European Union, OP RDI project No. CZ.1.05/2.1.00/03.0091 - University Centre for Energy Efficient Buildings and by the GAČR 14-17950S project.

Determination of Load Bearing Capacity for Spatial Joint with Steel Angle Brackets

P Sejkot[1,2,4], S Ormarsson[3], J Vessby[3] and P Kuklík[1,2]
[1]CTU in Prague, Department of Steel and Timer Structures, Thákurova 7, 166 29 Praha 6, CZ
[2]UCEEB of CTU in Prague, Třinecká 1024, 273 43 Buštěhrad, Czech Republic (CZ)
[3]Linnaeus University, Växjö, Sweden

E-mail: petr.sejkot@uceeb.cz

Abstract. The design of spatial connections in load bearing timber structures with steel angle brackets has insufficient support in the existing design standards. Therefore, research has been necessary to improve this state of the art. In the current paper an experimental study on two designs of angle brackets is presented and the results from full-scale experiments are compared to numerical and analytical computational models.

1. Introduction

Timber connections using thin-walled metal elements gradually supplant traditional carpentry joints. Their main advantage is that they do not significantly weaken the connected timber elements. Other advantages include the possibility of in-situ implementation or the possibility of direct connection of timber elements to steel and concrete structures. Their ductile behavior when subjected to load is another important advantage [1].

A steel angle bracket that is normally fastened to timber using annular ring nails (see figure 1) is the most common thin-walled metal connector.

Their main disadvantage is a very complex behavior when subjected to load. Connections using angle brackets are often loaded from various directions. The load distribution to the nails is not uniform and is not easily predictable because of high ductility of the metal plates. Moreover, deformed metal plates also cause prying of nails, i.e. loading them by an additional bending moment. So far any standardized calculation procedure to determine their load bearing capacities is missing and connections using metal works are mostly designed according to the producers' experimentally based catalogues.

[4] To whom any correspondence should be addressed.

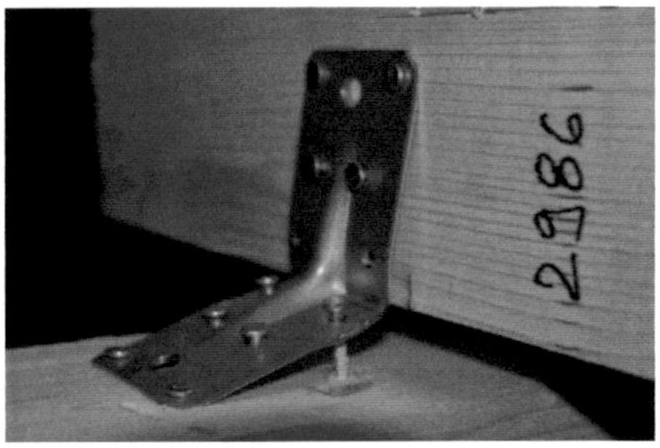

Figure 1. Spatial joint made by steel angle bracket after load-bearing test.

This situation is not very convenient for a designer to address, because they have only very limited possibilities of checking these values by using simplified analytical calculation models.

Developing new and reliable calculation models of connections using metal works is desired both by producers and designers. Producers will profit due to faster and cheaper development of their products. On the other hand, designers will be able to check their designs in a more precise way.

One of the most convenient ways of analyzing connections using metal works is to use the finite element method in 3D space.

2. Numerical simulations

The test set up used to determine the load bearing capacities of beam to beam connections made by nailed angle brackets with a rib has been simulated with a 3D-model in Abaqus CAE [2] (see figure 2). It aims to compare the numerical results with the experimental behavior presented in this paper.

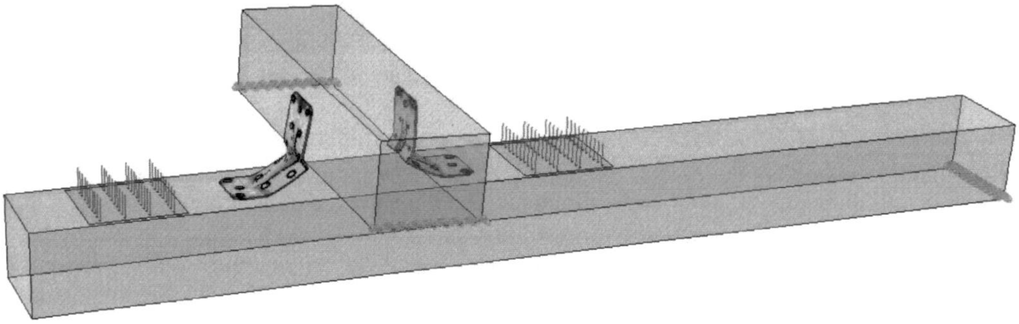

Figure 2. Geometry and boundary conditions of the experimental set-up studied.

Boundary conditions in the simulation were set according to the real test set-up, that is, the pinned end of the longer shoulder of the lower beam and both ends of the upper beam (see figure 2).

In the experiment, the connection was loaded by using load distribution device symmetrically transferring the external loading force to two areas on the lower beam of the connection (marked in figure 2).

Angle brackets are modelled as shell elements with attached thickness. The steel material used is considered as an elastic-plastic material with a yield stress σ_y= 280 MPa.

Timber beams are modelled as solid members. The wood material is assumed to be an orthotropic material with the elastic parameters E_l=9700 MPa, E_r=400 MPa, E_t=220 MPa, ν_{lr}=0.35, ν_{lt}=0.60, ν_{rt}=0.55, G_{lr}=400 MPa, G_{lt}=250 MPa, G_{rt}=25 MPa used in [3].

Nails were modelled as connector elements with properties based on the experimental data of annular ring nails, which were connecting the edges of the holes of angle brackets to the rigid bodies in timber beams. Axial behavior of the connectors was based on the data provided by a Czech nail producer Hašpl a.s. Lateral behavior was based on the data provided by Linnaeus University in Växjo, Sweden.

Although the provided data were not from experiments performed on the nails with exactly the same length and diameter as nails used in the connection, only the following observed essential behavior characteristics were adopted in the connection simulation: For the axial direction, the measured characteristic load bearing capacity of axially loaded nails was 75% of their mean load-bearing capacity, while the motion was only 67% of the motion in failure. For the lateral direction, the characteristic load bearing capacity of laterally loaded nails was 83% of their mean load-bearing capacity, while the motion was only 50% of the motion in failure (see figure 3).

According to the observed characteristics, simplified force-displacement diagrams of nail behaviour were created to be used in numerical simulations for both axial and lateral behaviour of nails. Characteristic load-bearing capacities of the nails used were calculated according to EC5 [4] and DIN 1052:2004 [5]. Mean values were created with respect to the ratios of mean to characteristic capacities of the tested nails. Displacements were adopted from the force-displacement curves of the tested nails (see figure 4).

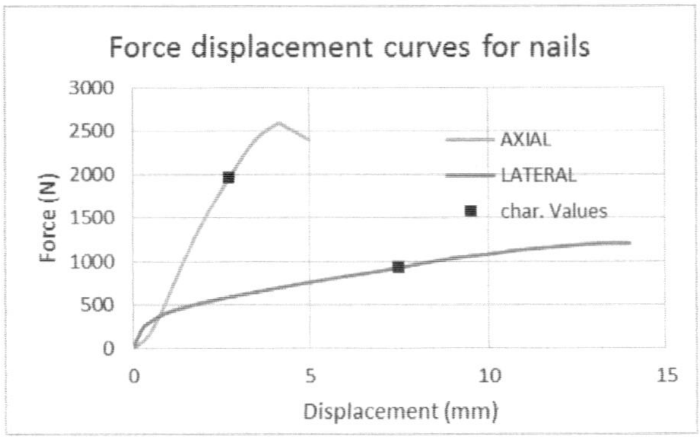

Figure 3. Force-displacement curves for nails (axially loaded: d=4.0 mm, l = 90 mm, laterally loaded: d=2.5 mm, l=40 mm).

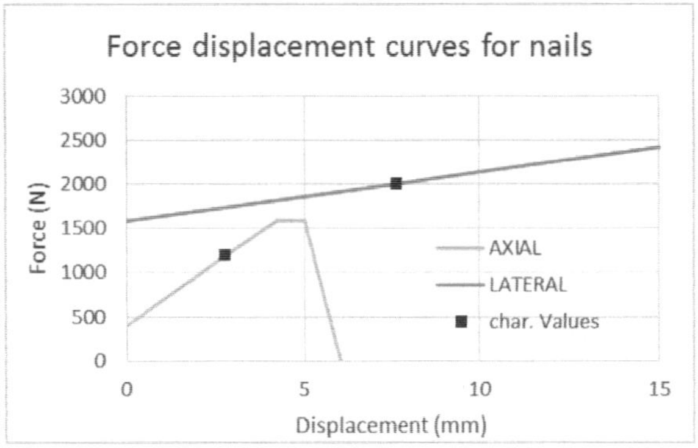

Figure 4. Idealized force displacement curves for the same nails as were used in full-scale experiments (d=4.0 mm, l=60 mm).

The nail behavior from figure 4 was implemented to numerical model and simulated force-displacement behavior was compared to the behavior of the tested connections.

The way used, when every nail is represented by a rigid body in the timber beam connected to the appropriate hole edge in the angle bracket by connector with predefined behavior, can simulate the following behaviors in the connection: Groups of rigid bodies simulate the behavior of the timber

blocks affected by groups of nails connecting the planes of angle brackets to beams. Each linear connector simulates both lateral and axial behavior of every single nail in the connection. And by using damage condition, lateral load reduces the maximum axial load which can be applied to the loaded nail.
Despite these simplifications, the behavior of the whole simulated connection with the applied external load quite well conforms to the behavior of the connections during the load test (see figure 5).

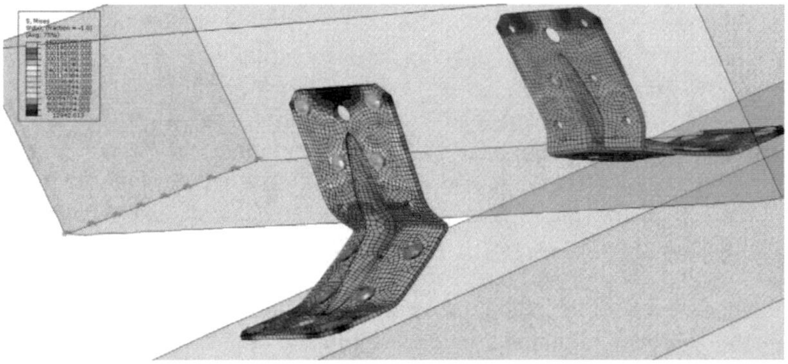

Figure 5. Contour plot of Von Mises stresses shown on the deformed geometry of the angle brackets in the connection.

Moreover, the benefit of simplification of these simulation models is that it is easy to create them and they can be adopted not only by researches but also by practicing engineers. Since these models have been experimentally verified, they can be used to calculate load carrying capacity for other load configurations.

3. Experiments

Full scale experiments on beam to beam connection (upper beam 100x100x400 mm, lower beam 100x100x1110 mm, see figure 2) made by two different versions of an angle bracket with a rib were performed. Ten specimens of the first angle bracket version were tested at the beginning of the research. Then the results were analyzed and geometry improvements were suggested. Finally, five specimens of the improved version of angle brackets were produced (by BOVA Březnice s.r.o.), then they were experimentally tested and their behavior was analyzed (see figure 6).

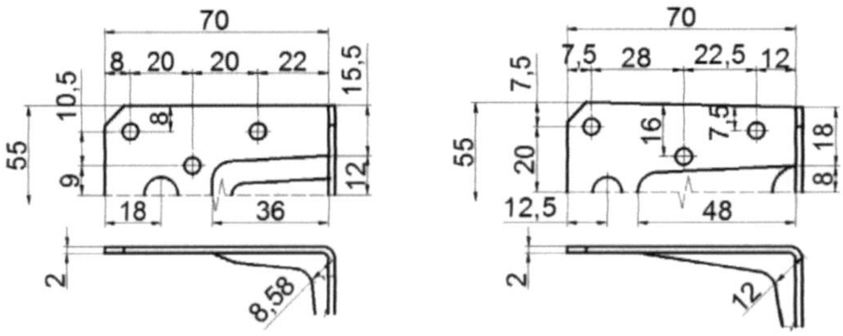

Figure 6. Angle bracket dimensions in millimetres (left – the original version, right – the improved version).

Both types of angle brackets were made of hot-dip galvanized steel sheet, 2 millimeters thick, classified as S280GD+Z275.

Both sets were loaded according to the analogical load diagrams based on the expected load-bearing capacities of tested angle brackets (i.e. 8.50 kN for the original one and 10.0 kN for the improved one). In the first 120 seconds of the test, the connection was loaded to the 40% of its expected load bearing capacity and left in that state for 30 seconds. Then, in 90 seconds the connection was uniformly unloaded to 10% of the expected load-bearing capacity and left in that state for 30 seconds. And finally, it was loaded till collapse by 3.33% of the expected load-bearing capacity per second.

Partial load bearing capacities were determined by two limits – collapse and maximal displacement of 15 mm. The force was applied by a hydraulic cylinder with a preset loading program in time. Displacement between the connected beams was sensed by a group of sensors. All input data were synchronized in the measuring center.

3.1. Evaluation procedure
The evaluation procedure combined four methods of assessment of the tested specimens.

Density was evaluated according to Eurocode 0 [6] with added reductions respecting timber joint specifics described in EOTA TR 16 [7] and ISO 8970 [8].

The final load-bearing capacity was evaluated according to EN 14358 [9] with the added reduction from EOTA TR 16 [7] taking into account the difference between the measured and standardized densities.

4. Analytical models of the connection
Two analytical models were compared with the experimental results. The first one was an elastic model inspired by research in Trento [10]. The second one was a plastic model, which considered the approach from EOTA TR 17 [11].

Both methods took into consideration the research conclusions from the research conducted by the university in Karlsruhe that annular ring nails of 4 mm in diameter behave in 2 mm thick steel plate in the same way as being in a thick plate [12] and that their axial resistance has squared relation to density [13].

4.1. Elastic model
The elastic model describes the behavior and force distribution before significant deformations occurred (see figure 7).

Figure 7. Deformed angle bracket during the elastic phase.

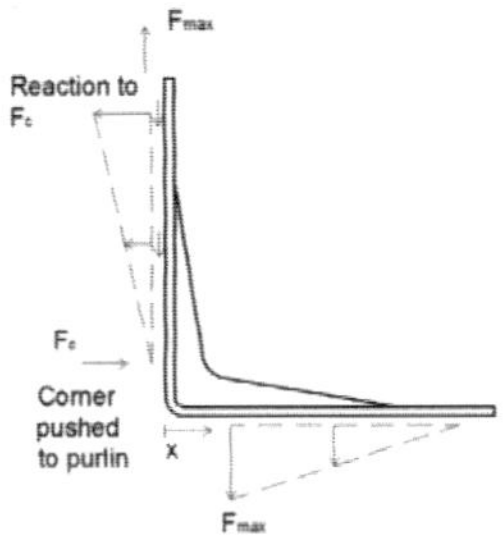

Figure 8. Schema of the elastic model.

The load-bearing capacity is determined according to the elastic model, where the resisting features are the corner of the angle bracket pushed to purlin and nails loaded both axially and laterally. In the vertical branch of the angle bracket, lateral load is uniformly distributed to all nails and withdrawal forces are linearly increasing from the corner to the edge of the angle bracket. The forces in the nails are balanced by the pushed corner of the angle bracket to the timber. The bending moment is transferred by a rib to the horizontal branch of the angle bracket. In the horizontal branch, the bending moment is balanced by linearly distributed withdrawal forces acting on the nails there (see figure 8).

4.2. Plastic model

The plastic model describes the behavior and force distribution just before the collapse (see figure 9). The static schema of this plastic analytical model is presented in figure 10.

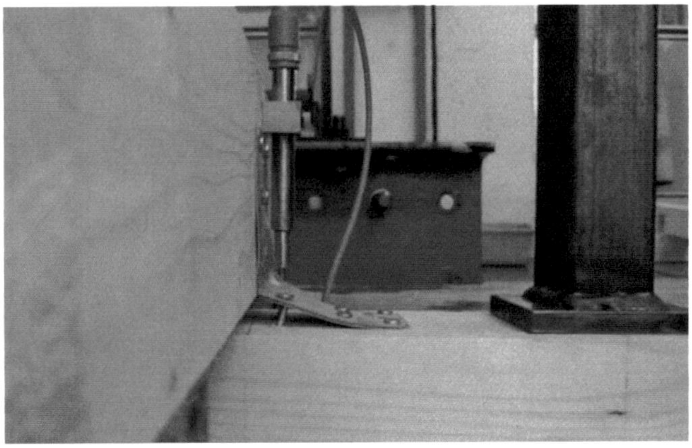

Figure 9. Deformed angle bracket during the plastic phase.

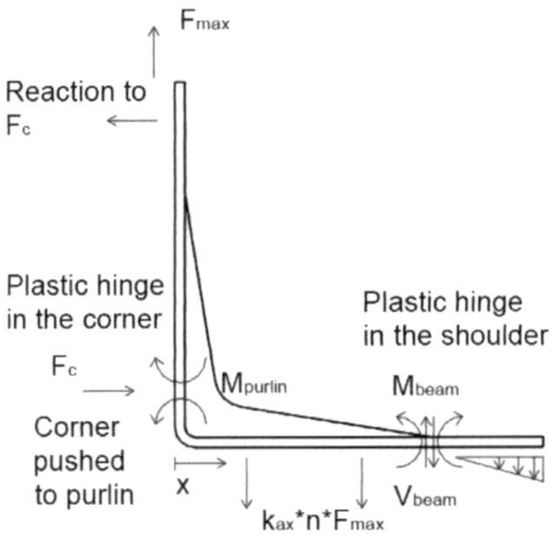

Figure 10. Schema of the plastic model.

The calculation is based on *"The example of a spatial joint of timber elements by steel angle bracket loaded in a direction that is opening to the angle"*. This example was published in the Technical Report 17 (TR 17) [11], which was released by the European Organisation for Technical Approvals (EOTA). The load bearing capacity is determined according to the model, where the resisting features are two plastic hinges and a group of nails in tension. The influence of pushing edge of the angle bracket is neglected. Moment capacities and places of plastic hinges are determined experimentally. Nail characteristics do not follow the Eurocode 5 [4] but rather the Danish timber codes.

5. Results and discussion

The experimentally determined load-bearing capacities of the connections were compared to the results from numerical simulations and analytical solutions.

Measured densities of tested angle brackets are presented in table 1.

Table 1. Evaluation of measured densities.

	Specimens	Mean density *(kg/m³)*	Standard deviation *(kg/m³)*	Characteristic density *(kg/m³)*
05-21 - original	10	447.9	32.1	386.3
05-21 - improved	5	424.2	23.6	369.1

Compared to table densities in EN 338 [14] of timber strength class C24 (ρ_{mean} = 420 kg/m^3, ρ_k = 350 kg/m^3), it is obvious that the used timber has a significantly better quality than is required for the appropriate strength class.

Table 2. Consideration of the natural variation of wood density with respect to measured load-carrying capacities.

	Mean capacity[a] *(kN)*	COV_ρ [b] *(%)*	COV_δ [c] *(%)*	EN ISO 8970 *(-)*	COV_R [d] *(%)*	k_{cov} [e] *(-)*
05-21 – original	12.324	7.16	12.41	Not satisfied	15.99	1.2881
05-21 – improved	17.492	5.57	13.01	Not satisfied	16.46	1.2652

[a] With transformed values to timber strength class C24
[b] COV_ρ is coefficient of variation of the measured wood density
[c] COV_δ is coefficient of variation of the measured or the modified load-carrying capacity
[d] COV_R is coefficient of variation of the load-carrying capacity considering full variation of the wood density
[e] k_{cov} is COV_R to COV_δ ratio used in Standard deviation calculation according to EOTA TR 16 [7]

In both sets, there were samples which did not meet density conditions from EN ISO 8970 [8]. This fact resulted in a high value of factor k_{cov} (see Table 2).

Load bearing capacities from the evaluated experiments were compared to the values calculated by analytical methods and by numerical modelling (see Table 3).

Table 3. Comparison of characteristic load-bearing capacities.

	Experiments *(kN)*	Analytically - elastic *(kN)*	Analytically - plastic *(kN)*	Numerically *(kN)*
05-21 – original	8.50	6.56	4.04	7.15
05-21 – improved	11.67	7.56	8.07	8.63

Load bearing capacity of each connection was experimentally determined either by collapse of the connection or by 15 mm displacement. It is noticeable that the improvement of the shape and nail positions of the angle bracket brought an over 30% improvement in the load bearing capacity.

Analytical elastic approach expected over 15% improvement of load bearing capacity of the connection and underestimated it. On the other hand, the plastic approach expected 50% improvement and overestimated it. From these results it is obvious that both methods are quite inaccurate.

The numerical approach gives more accurate results than the analytical approach. Characteristic values were determined by adding conditions to force displacement diagrams of nails (see figure 4) preventing nails from carrying higher force than their characteristic capacities.

Numerical simulations of the beam to beam connection also brought complete load-displacement curves for the studied connections using angle brackets. These results were compared to the experimentally determined force-displacement curves. The force represents the external load applied on the connection and displacement is measured considering opening of the gap between the beams in the connection.

In figure 11, mean behavior of the experimentally studied connection is compared to the simulated behavior of the connection using nail load-bearing capacities calculated considering the measured densities of timber used in each connection set.

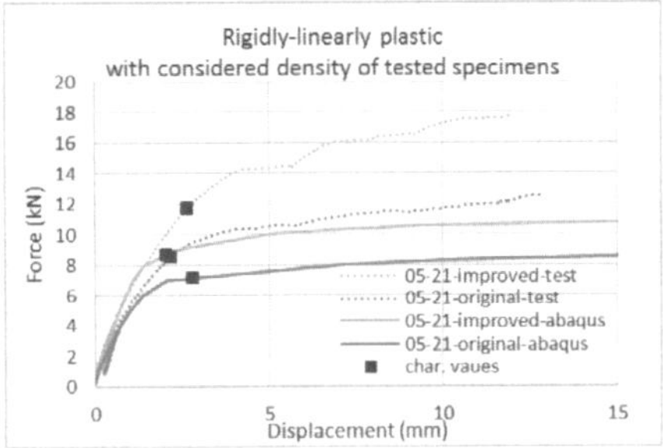

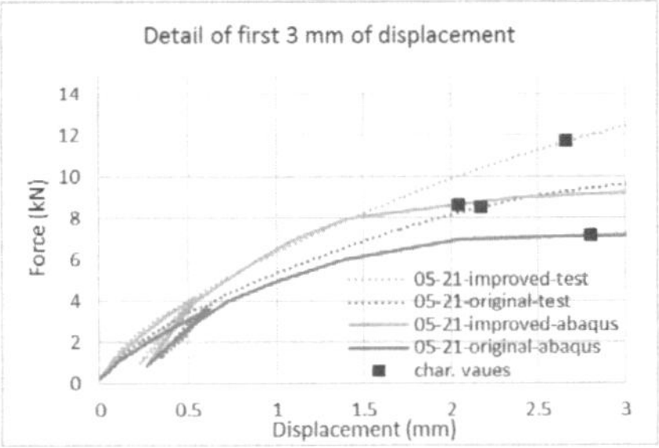

Figure 11. Mean force-displacement curves of opening of the gap between the connected beams under the applied external load. The model data used for nails are based on mean density of timber measured in tests.

As it is evident from the plots, both simulations match quite well with the experimentally studied behaviour until they started to behave nonlinearly. This quite high difference between mean load bearing capacities of the tested connections compared to the results from numerical simulations can be explained by neglecting of the prying of nails. It happens because angle bracket deformation induces a bending moment in the nails fixed in its holes. This bending moment increases the pressure of the threaded part of the nail on the timber and increases its withdrawal resistance.

This behaviour is different to the lateral load, which mainly increases stress at the not threaded part of the nail. This does not increase its withdrawal resistance, conversely, it reduces it due to the rope effect and the plastic motion.

6. Conclusions

The results presented show how to determine load bearing capacities of spatial connections using angle brackets in three different ways: by full-scale experiments according to the European Technical Reports, by an analytical method and by numerical modelling using simplified material and nail models. All determined load-bearing capacities were listed and compared.

Although the experiments are not dramatically expensive or time consuming, it is useful to have a cheaper, faster and reasonably reliable calculation procedure. The main benefit of the analytical solutions is that they are cheap. However, they are very conservative and their reliability is not well proved.

Numerical simulations provide more applicable results than the analytical solutions. Although numerical modelling of timber is still very challenging and numerical models were quite simplified, simulation proved to be a very useful tool for predicting the behavior of the whole connection and for optimizing the shapes of steel elements in the processed connections.

7. References

[1] Šťastný R 2005 *Tenkostěnné Ocelové Tvarované Elementy Používané v Dřevařském Inženýrství* doctoral thesis CTU in Prague Faculty of Civil Engineering

[2] Abaqus 6.14-1 documentation 2014 *Abaqus Analysis User's Manual* Simulia

[3] Ormarsson S, Dahlblom O and Petersson H 1999 A numerical study of the shape stability of sawn timber subjected to moisture variation Part 2: Simulation of drying board *Wood Science and Technology* **33** (5) 407–23

[4] CEN 2006 *EN 1995-1-1 Eurocode 5: Design of Timber structures – Part 1-1: General – Common Rules and Rules for Buildings*

[5] DIN 1052:2004: *Design of Timber Structures – General Rules and Rules for Buildings* 2004-088

[6] EN 1990 Eurocode 0: *Basis of Structural Design* Second edition

[7] European Organisation for Technical Approvals 2012 *EOTA TR16: Method of Testing Three-Dimensional Nailing Plates with Examples* Technical report 16

[8] CEN 2010 *EN ISO 8970: Timber Structures – Testing of Joints Made with Mechanical Fasteners – Requirmentsfor Wood Density*

[9] CEN 2007 *EN 14358: Timber Structures – Calculation of Characteristic 5-percentile Values and Acceptance Criteria for a Sample*

[10] Tomasi R and Sartori T 2013 Mechanical behaviour of connections between wood framed shear walls and foundations under monotonic and cyclic load *Construction and Building Materials* **44** 682–690

[11] European Organisation for Technical Approvals 2012 *EOTA TR 17: Worked Example Calculation of Characteristic Load-carrying Capacities of 90° Angle Bracket with a Rib*

[12] Ehlbeck J and Görlacher R 1982 *Tragfähigkeit von Balkenschuhen unter Zweiachsiger Beanspruchung* Forschungsbericht, Vesuchsanstalt für Stahl, Holz und Steine, Universität Karlsruhe

[13] Werner H and Siebert W 1991 Neue Untersuchungen mit Nägeln für den Holzbau *Holz als Roh- und Werkstoff* **49** (5) 191–198

[14] CEN 2003 *EN 338: Structural Timber – Strength Clases*

Acknowledgments

This work has been supported by the European Union, OP RDI project No. CZ.1.04/2.1.00/03.0091-University Centre for Energy Efficient Buildings.

Experimental research was also supported by the grant of CTU in Prague SGS/171/OHK1/3T/11. The authors would like to acknowledge the contribution of the staff of the Department of Building Technology at Linnaeus University, Växjö, and the Faculty of Civil Engineering at Czech Technical University in Prague and the financial support of Cost Action FP1004 through Short-term Scientific Mission grant.

Natural Oscillations of Welded Steel Beams in the Span Structures of Conveyor Galleries

V Volkova[1,2] and I Smolii[1]

[1]SHEI "National Mining University", Departament of Construction, Geotechnics and Geomechanics, 19 Karla Marksa Ave., Dnipropetrovsk 49000, Ukraine

E-mail: sebrle@bk.ru

Abstract. This article presents the results of numerical simulation of dynamic behaviour of welded metal beams in conveyor galleries. Investigations were carried out for the beams having symmetrical cross-sections and evenly-spaced transversal ribs with various spacing between the ribs in different beams. The regularities of changes in the vibration mode shapes depending on the beam geometric characteristics have been investigated.

1. Introduction

Belt conveyors have gained widespread acceptance for charging materials along the belt-conveyor trestles into the blast furnaces in the metallurgical coal-preparation and ore-dressing plants, for transporting extracted materials onto the surface in the open pit and underground mining, in the building and construction sector as well as in other industries. Belt conveyors are installed in adequate buildings and premises or in purpose-built galleries between these structures [1].

The experience gained in the 70-year period of industrial operation of the galleries in Ukraine proves that these structures belong to the class of facilities of the first criticality level. High damage rates in the galleries stem from the concurrent effects of a number of adverse factors, including dynamic loads [2].

Articulated and thin-walled beams, parallel-chord trusses, rectangular and circular cylindrical shells as well as other types of structural elements can be used as span bearing structures. From the point of view of cost saving and operating economy, the application of beam bearing structures is rational for the spans up to 30 m in length, while for the spans ranging from 24 m to 42 m it is recommended to use lattice girders as the load-bearing structures, particularly trusses. Shells may also be used as the load-bearing structures in the conveyor galleries with the spans exceeding 30 m in length [2]. They have higher reliability and require less operating expenditures as compared to trusses. Depending on

[2] To whom any correspondence should be addressed.

technological requirements, the span structures of conveyor galleries can be installed straight or inclined (10-15°) [6].

2. The object of research

The objects of this research are conveyor gallery span structures. These structures are the facilities of the first criticality level. With an aim to ensure the best reliability, strength and stability of span structures at all stages of their manufacturing, transportation and erection, the up-to-date Regulations and Standards of construction practices provide guidance to the selection of geometrical parameters and stiffness characteristics for the span structures.

In the real-world beam structural systems, a number of initial imperfections of the elements can be found, in particular, bent portions, curvature bending under transverse loads, and other defects. For this reason, the limit ratio of b_f/t_f as recommended in the Standards is slightly less than the value obtained on the basis of an idealized design scheme [3].

According to the State Standards and Norms for construction in Ukraine (DBN V.2.6-163: 2010), the conventional boundary slenderness parameters should be set as $\overline{\lambda_{uf}} = 0{,}36 + 0{,}1\bar{\lambda}$, for the beam chords and $\overline{\lambda_{uw}} = 1{,}30 + 0{,}15\bar{\lambda}^2$ for the web. In this case, the parameter $\bar{\lambda}$ is conventional slenderness of the structural framing, which is determined taking into account the overall stability of the system under the action of the central compression force [4].

Technological loads acting on the structural elements of conveyor galleries are dynamic in nature, due to induced vibrations of the conveyor belt transporting bulk material at various stages of operation. Due to this fact, investigations into dynamic characteristics of the span structures of conveyor galleries are of great significance.

Formerly, when calculating conveyor gallery structures, the most commonly used method was based on the flat cross-section hypothesis [10]. However, for dynamic behaviour of beams in the span structures, such approach is rather approximate. The models suggested by Timoshenko and Vlasov have also found a wide application [7].

3. The method of investigations

Simulation was performed for simply supported pin-ended beams. The results were obtained with the use of the SCAD software package and implementation of Lanczos method for dynamic analysis [8]. Rectangular finite element mesh was used in research. The finite elements represented the plates of 100×100 mm in size and 20 – 40 mm thick [12]. According to the standards, it is recommended to analyze from 3 to 5 principal modes of vibration. In order to avoid the building-up of errors in integration, simulation was carried out for 10 modes of vibration. In further investigations the first 5 modes were used. The results obtained were compared to available approximate solutions [9].

The computation was performed using the finite element method with the SCAD PC Programming Software. The numerical analysis was performed in the computation of the beams with the specified geometrical characteristics. The reduced values of slenderness of the beam webs varied within the range of $\overline{\lambda_w}$= 2.5 ÷ 13 and the slenderness ratios of the beam chords varied within the particular range of b_f/t_f = 10 ÷ 40.

The boundary conditions corresponded to a hinged structure. Dimensions of the beam cross sections were chosen in accordance with the recommendations in "The Handbook on Conveyor Gallery Designs", the Supplement to SNiP (Local Building Requirements) 2.09.03-85". The data were obtained for the I-section welded beam of the span length of 12 m. The web of beam was connected with the flanges via filled welds. Disposition, effective throat thickness and effective length of welds were chosen in accordance with the recommendations of [13]. The cross-sectional view of the beam is shown in figure 1.

The finite elements used in the simulations represented "four-node rectangular plates" (FE-analysis 13), i.e. the plate members subjected to bending and lying in the plane XOY. The finite elements of this kind are generally useful in case of uniform thickness materials. In each node of these finite elements, three degrees of freedom were defined, namely: w - vertical displacement (sagging) and also UX and UY – the angles of rotation about the axes X and Y, respectively [11]. The control points were chosen at equal intervals along the top and bottom flanges of the beam, as well as on the web. Displacements were controlled in the nodes of the finite elements. The numerical designation of the nodes for this computation scheme is shown in figure 2.

In the theory of the finite element method, a lot of attention is paid to the problem of convergence, i.e. the asymptotic behaviour of the accuracy estimates of the obtained approximate solution with the

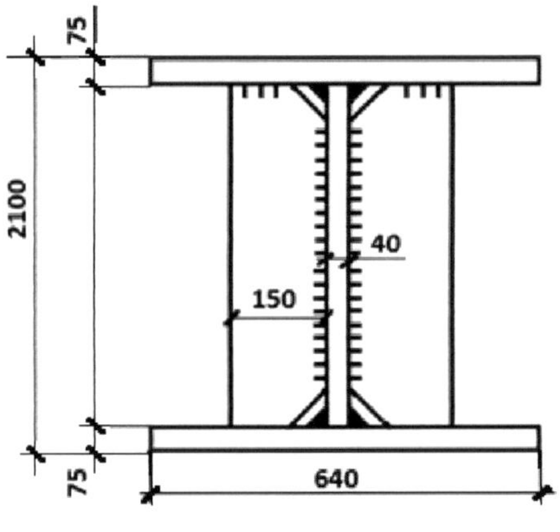

Figure 1. Cross-sectional view of the beam.

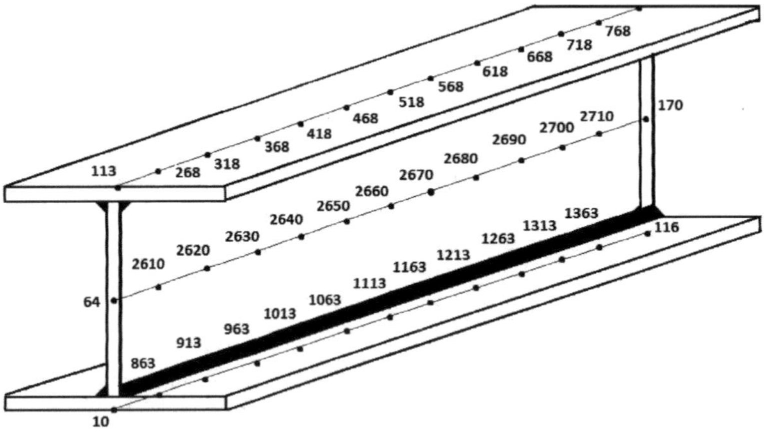

Figure 2. The numerical designation of the nodes for the computation scheme.

unlimited thickening of the finite element mesh. Proceeding from the results of the assessment of convergence of both the compatible finite elements and the incompatible ones, convergence of the given finite elements can be estimated. The spacings and density of the finite element mesh were chosen in accordance with the required accuracy of the approximate solution and also taking into account an uncertain error in the determination of stresses and deformations. According the estimates given in the paper [5], the rates of convergence of the computing procedures for FE No. 13 came to 2 parameters for displacements, 2 parameters for moments and 1 parameter for transverse forces.

4. Analysis of the results obtained

Based on the results of the calculation, the following regularities of changes in the vibration mode shapes were determined for various reduced flexural capacities of the beam web and beam chord. In the tested slenderness range of the web $\overline{\lambda_w} = 2.5 \div 5.5$ and that of the beam flanges $b_f/t_f = 10 \div 40$, the beams with the slenderness values of $\overline{\lambda_w} = 2.5$ and $\overline{\lambda_w} = 3.2$ belong to the beams with the rigid web and those with the slenderness value of $\overline{\lambda_w} = 5.5$ are the beams with the flexible web. The mode shapes for flanges are given in figure 3 ($\overline{\lambda_w} = 2.5$), figure 4 ($\overline{\lambda_w} = 3.2$), figure 5 ($\overline{\lambda_w} = 5.5$).

The first type of the vibration mode shapes in the beam flanges corresponded exclusively to node free mode shapes. The vibration phase was constant. A single shift of the vibration phase was revealed in the transition of the ratio b_f/t_f from value 10 to value 20. The effects of the slenderness

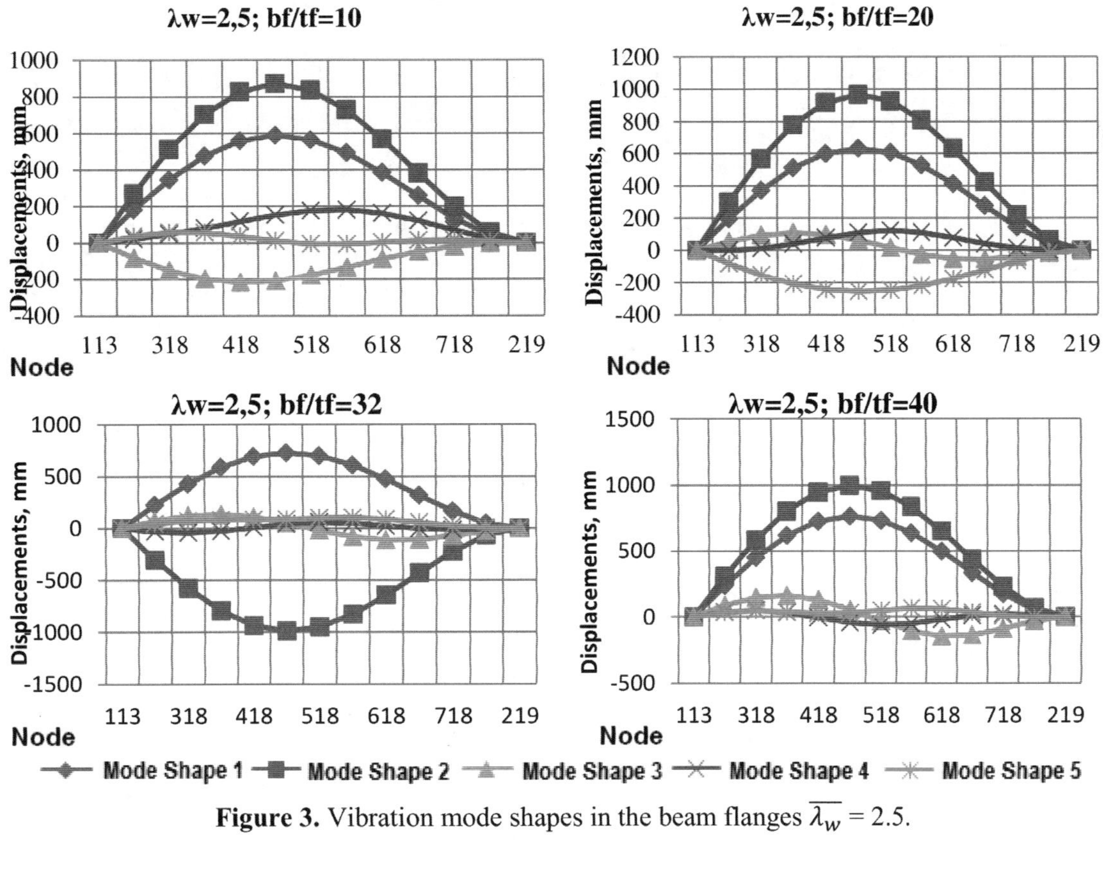

Figure 3. Vibration mode shapes in the beam flanges $\overline{\lambda_w} = 2.5$.

IOP Conf. Series: Materials Science and Engineering **96** (2015) 012071 doi:10.1088/1757-899X/96/1/012071

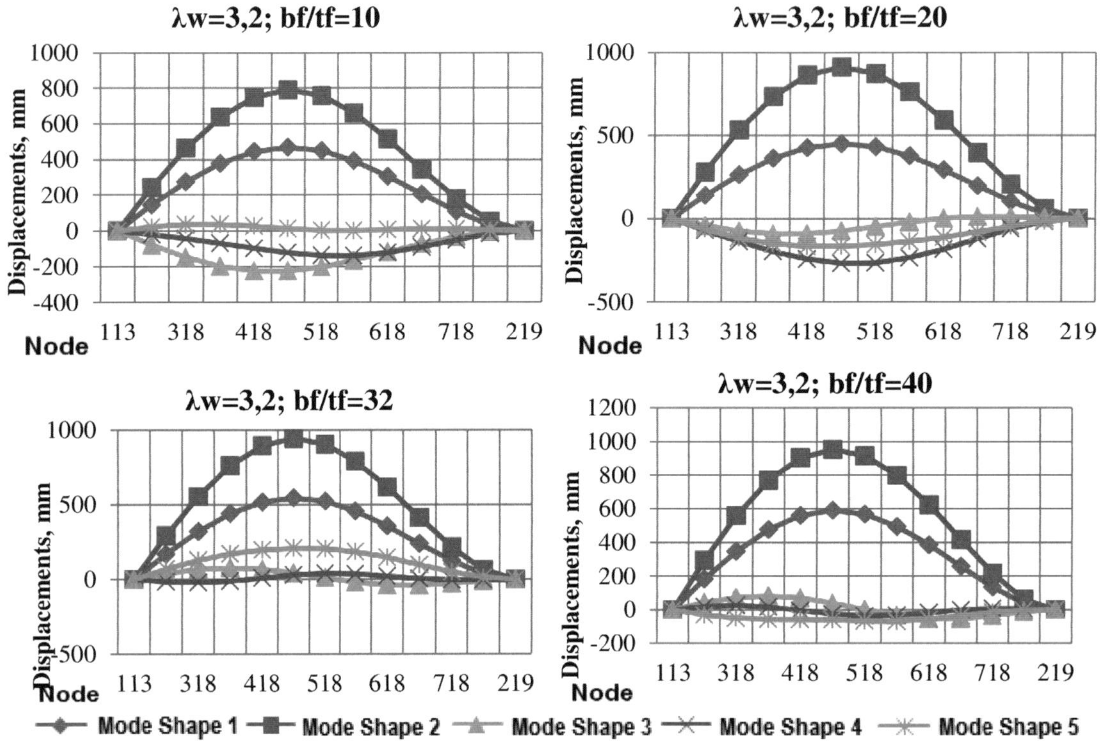

Figure 4. Vibration mode shapes in the beam flanges $\overline{\lambda_w}$ =3.2.

characteristics of the beam flanges and webs upon the vibration mode shapes of the second type are of the following nature. The beams demonstrated a node free mode shapes of vibrations. The shifts of the vibration phases were observed with the value of $\overline{\lambda_w}$ = 2.5 in the transition of the ratio b_f/t_f from value 20 to value 32 and from value 32 to value 40. The shifts of the vibration phase were detected with the value $\overline{\lambda_w}$ = 5.5 in the transition of the ratio b_f/t_f from value 10 to value 20. When the ratio b_f/t_f changed from value 20 to value 32, the transition from the node free vibration mode shapes to the single-node modes occurred.

The vibration mode shapes of the third type changed from the node free vibration mode shapes to the single-node ones in the transition of the ratio b_f/t_f from value 10 to value 20. In the transition of the ratio values of b_f/t_f from 20 to 32 with the parameter $\overline{\lambda_w}$ = 5.5, the node free vibration mode shapes transformed into the two-node modes.

In the fourth type of vibration mode shapes, with the web slenderness reduced value $\overline{\lambda_w}$ = 2.5, the transition of the vibrations from the node free modes into the single-node modes was observed when the ratio values of b_f/t_f changed from 10 to 20, but when the ratio values of b_f/t_f changed from 20 to 32, the single-node modes transformed into the two-node mode shapes. At the same time, the transition from the node free mode shapes to the two-node modes took place when the ratio values b_f/t_f changed from 20 to 32 and the value $\overline{\lambda_w}$ = 3.2. The node free mode shapes are characteristic only for the value $\overline{\lambda_w}$ = 5.5. Phase shifts were observed in both cases of transition of the ratio b_f/t_f from value 10 to value 20 and from value 20 to value 32.

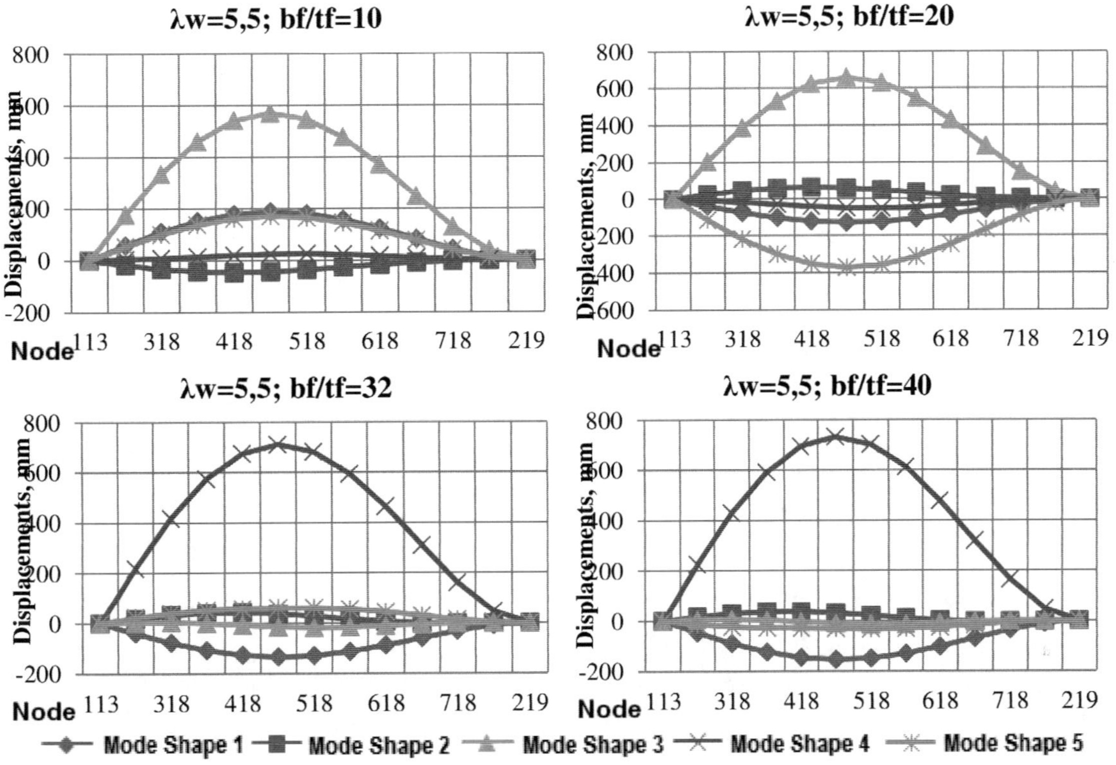

Figure 5. Vibration mode shapes in the beam flanges $\overline{\lambda_w}$ =5.5.

The following changes are peculiar to the fifth type of the vibration mode shapes. With $\overline{\lambda_w}$ = 2.5, the two-node mode shapes are typical for the value b_f/t_f =10, while the node free modes are characteristic of the ratios b_f/t_f = 20 and b_f/t_f = 32. Only the node free vibration modes were revealed with $\overline{\lambda_w}$ = 3.2 and $\overline{\lambda_w}$ = 5.5, at that, with the value $\overline{\lambda_w}$= 5.5, phase shifts took place in the whole range of changes of the ratio values b_f/t_f= 10 ÷ 40.

The following regularities were revealed in the vibrations of the beam web. The mode shapes for the web are given in figure 6 ($\overline{\lambda_w}$ = 2.5), figure 7 ($\overline{\lambda_w}$ = 3.2), figure 8 ($\overline{\lambda_w}$ = 5.5). The node free mode shapes corresponded to the first type of vibrations. With the value $\overline{\lambda_w}$ = 5.5, phase shifts in vibrations occurred in the transition of the ratio b_f/t_f from value 10 to value 20.

At the same time, the node free mode shapes represented the second type of vibration modes. The single-node mode shapes are characteristic of the value $\overline{\lambda_w}$ = 5.5, and phase shifts were detected when the ratios of b_f/t_f changed from value 10 to value 20.

The single-node mode shapes were characteristic of the third type of vibrations. Phase shifts were revealed with the value $\overline{\lambda_w}$=2.5 in the transition of the ratio b_f/t_f from value 10 to value 20, and with the values of $\overline{\lambda_w}$= 3.2 and $\overline{\lambda_w}$= 6, phase shifts took place in the transition of the ratio b_f/t_f from value 20 to value 32. With the value $\overline{\lambda_w}$ = 5.5, the transition from the node free mode shapes to the two-node vibration modes occurred when the ratio b_f/t_f changed from value 20 to value 32.

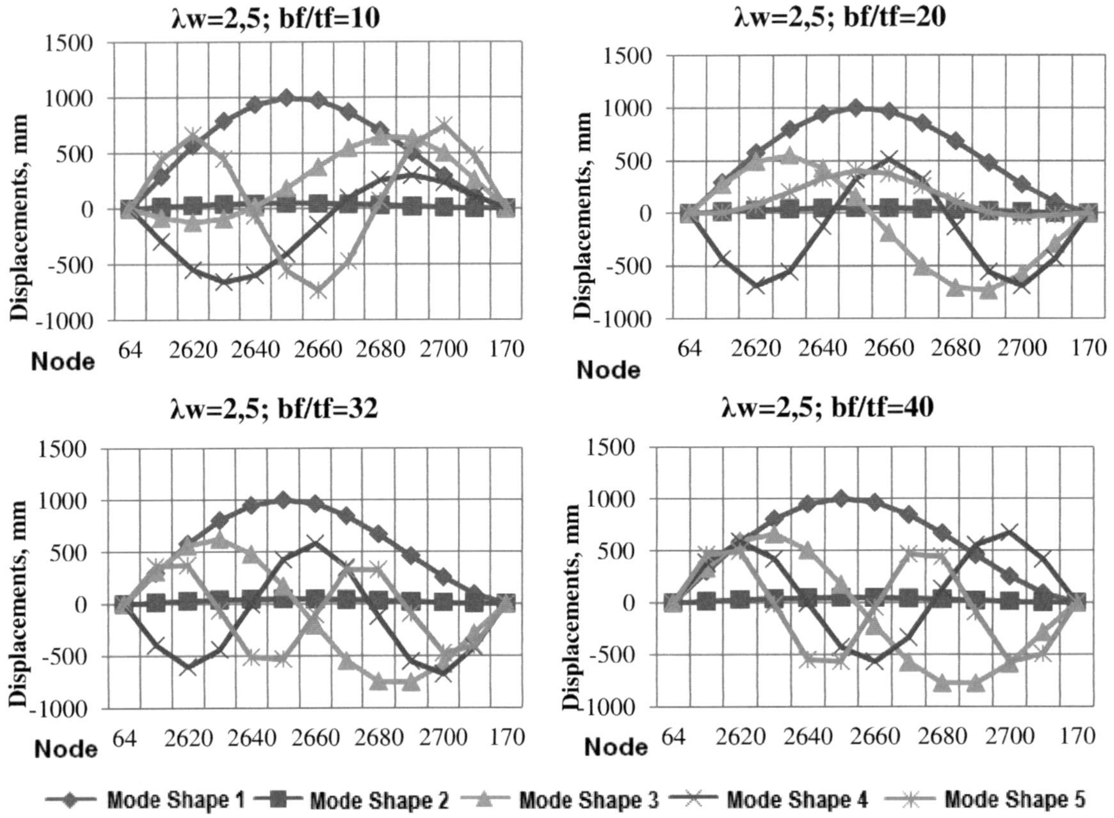

Figure 6. Vibration mode shapes in the beam web $\overline{\lambda_w}$ =2.5.

The transition of the vibrations from the single-node modes into the two-node modes was observed when the ratio b_f/t_f changed from value 10 to value 20. With the same values of $\overline{\lambda_w}$, phase shifts were revealed when the ratio b_f/t_f changed from value 32 to value 40. With the value $\overline{\lambda_w}$=5.5, the two-node vibration modes transformed into the node free modes in the transition of the ratio b_f/t_f from value 20 to value 32, and phase shifts were detected when the ratio b_f/t_f changed from value 10 to value 20.

The changes of the $\overline{\lambda_w}$ and b_f/t_f parameters have more complex effects on the fifth type of the beam web mode shapes. With the value of $\overline{\lambda_w}$= 2.5, during the transition of the ratio b_f/t_f from value 10 to value 20 and when the ratio b_f/t_f changes from value 20 to value 32, the single-node mode shapes passed into the two-node vibration modes; and when the values of the ratio b_f/t_f ran from 32 to 40, phase shifts were detected. At the same time, with the value of $\overline{\lambda_w}$= 3.2, the two-node vibration mode shapes evolve into the three-node modes when the ratio b_f/t_f was passing from value 20 to value 32, while the vibration phase shifts were observed in both cases – when the ratio values of b_f/t_f were going from 10 to 20 and from 32 to 40. With the value of $\overline{\lambda_w}$= 5.5, the three-node modes with the phase shifts were revealed in the whole range of the varying parameter b_f/t_f=10÷40.

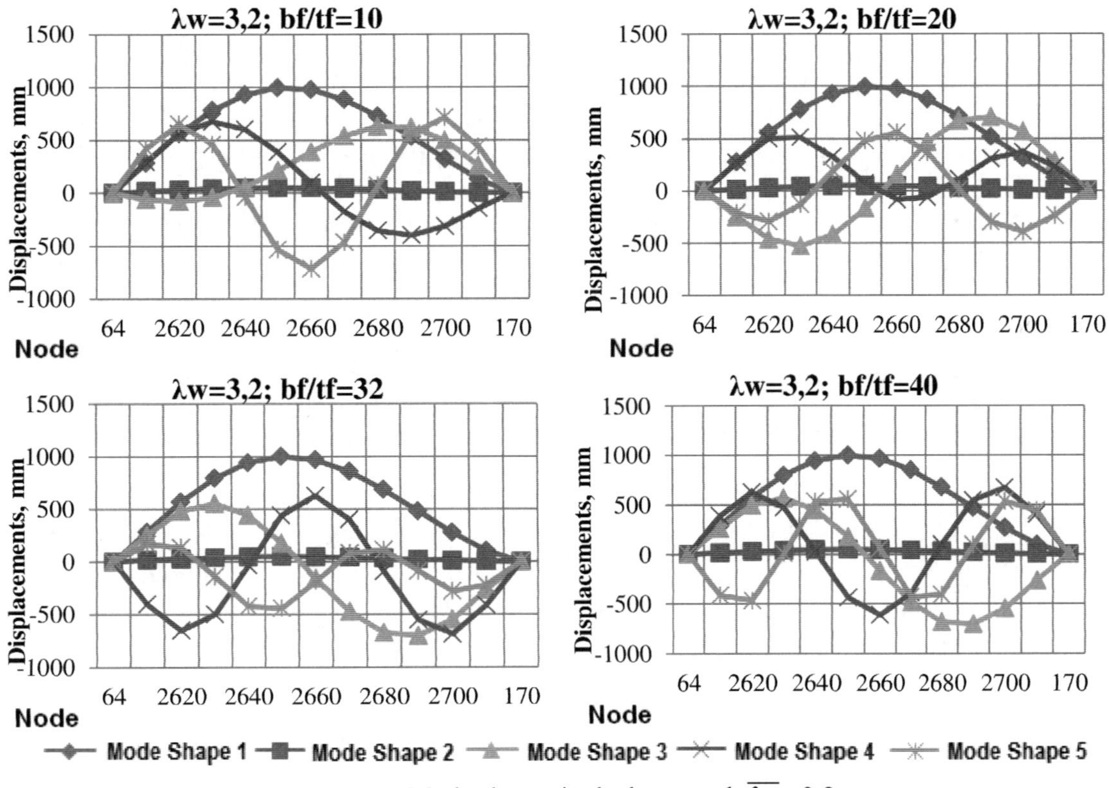

Figure 7. Mode shapes in the beam web $\overline{\lambda_w}$ =3.2.

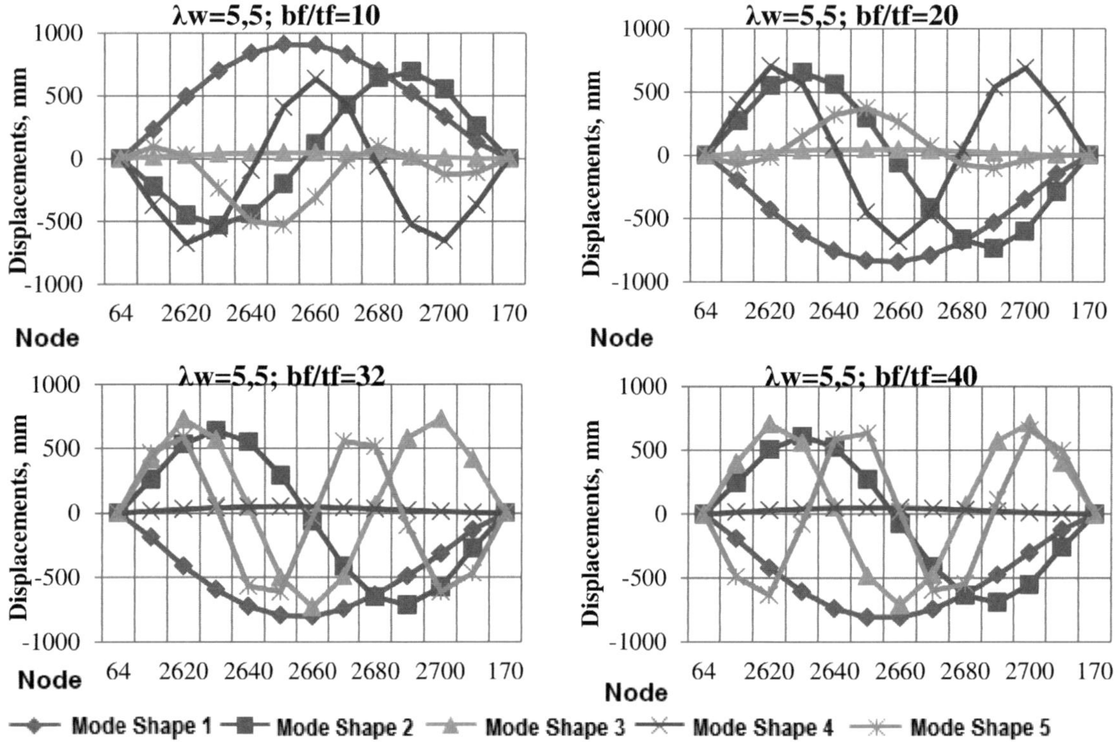

Figure 8. Mode shapes in the beam web $\overline{\lambda_w}$ =5.5.

5. Conclusions

Free oscillations of the system are a type of motion, which is realized in the system in the absence of any external influence or power impact from the outside. The results of the analysis of various modes of forced vibration mode shapes enable to reveal the internal dynamic characteristics. Based on these characteristics, it is feasible to predict the response of the structure to the influence of various external factors.

With the higher-level slenderness of the beam chords, the vibration amplitudes of the two lower-order mode shapes increase. The natural frequencies in the beams with the flexible webs decrease when the slenderness of the beam chord increases. The emergence of the two- and four-node vibration mode shapes was observed when the slenderness of the beam flange increased. With the higher-level slenderness of the beam webs and chords, the appearance of additional nodes in the mode shapes was revealed.

6. References

[1] Gordeyev V N, Lantukh-Lyashenko A I, Pashinsky V A, Perel'muter A V, Pichugin S F 2007 *Loads and Efects on Buildings and Structures* (Moscow: Publishing House of Association of Building Universities)

[2] Gorev V V 2002 *Metal Structures. Volume 3, Special Designs and Constructions* (Moscow: High School Publishing House)

[3] Gorev V V 2002 *Metal Structures. Volume 1, Structural Elements* (Moscow: High School Publishing House)

[4] *The State Standards and Norms for Construction in Ukraine (DBN V.2.6-163: 2010), Steel Structures (*Kiev: Minregionbud Publishing House of the Ministry of Regional Development, Construction and Housing and Communal Services of Ukraine)

[5] Karpilovsky V S, Kriksunov E Z 2003 *CAD System for Users (*Kyiv)

[6] *SNiP 2.09.03-85, Construction Rules and Regulations for Industrial Enterprises* (The USSR State Committee for Construction)

[7] Soltani M, Asgarian B, Mohri F 2014 *Finite Element Method for Stability and Free Vibration Analyses of Non-prismatic Thin-walled Beams* (Thin Walled Structures vol. 82) pp 245–261

[8] Fialko S 2007 The Lanczos method of displacements as applied for the seismic analysis of structures *CADmaster Magazine* **5** 102–105

[9] Perelmuter A 2014 *Discourse on Construction Mechanics Scientific Edition* (Moskow: SCAD Soft).

[10] Perelmuter A, Slivker V 2002 *Computational Models of Structures and Feasibility of their Analysis Scientific Edition* (Kiev: Stal Publishing House)

[11] Bazhenov V A, Perel'muter A V, Shishov OV 2009 *Structural Mechanics. Computer Technologies* (Kiev: Karavela Publishing House)

[12] Zinkevich O K 1975 *The Finite Element Method in Engineering* (Moscow: Mir Publishing House)

[13] *EN 1993-1-8 2005: Eurocode 3: Design of Steel Structures - Part 1-8: Design of Joints* (Authority: The European Union Per Regulation 305/2011, Directive 98/34/EC, Directive 2004/18/EC) pp 38–50

IOP Conf. Series: Materials Science and Engineering **96** (2015) 012072 doi:10.1088/1757-899X/96/1/012072

Monitoring Technical Conditions of Engineering Structures Using the Non-Linear Approach

V E Volkova[1]
SHEI National Mining University, Department of Construction, Televizionnaya Str.12 ap.48, Dnipropetrovsk, UA 49005, Ukraine

E-mail: drvev@mail.ru

Abstract. Conventional methods of monitoring technical condition are based on detection of damage in the structures of buildings or facilities during the entire period of their operation. In spite of considerable interest displayed to this issue and a significant number of publications, there is no unity of opinions. These methods differ from each other in the sets of values fixed for investigations, the techniques of their recording, transfer and further processing. Today's rules and regulations for structural designs expand the scope of application of the structures operating in the elastic-plastic stage. These damage-free structures originally display the non-linear properties and can be adequately described only by the non-linear models. This paper presents a method for determining the type and level of non-linearity from the structural oscillations data for monitoring the change in the health of structures. It is shown that a plot of acceleration against the magnitude of the displacement represents the restoring force of a structure. If the structure is damaged during a new striking motion, the phase trajectories in plane "acceleration-displacement" will deviate from its healthy signature.

1. Introduction

Expanding capacities of technological equipment, high speed rates and cargo-carrying capacities of the transport vehicles on the background of the increasing seismic activity have raised the significance of dynamic design calculations, which must ensure the maximal reliability of the systems. Either the theoretical and computational approach or the theoretical and experimental approach is generally accepted as a method of constructing dynamic models. The first approach is implemented at the stage of the structure design and consists in constructing theoretical mathematical models based on the application of the fundamental laws of the discrete and continual system mechanics which are expressed by analytical ratios. The second approach is based on the application of the methods of parametric and/or non-parametric identification. The models developed according to the results of the

[1] To whom any correspondence should be addressed.

IOP Conf. Series: Materials Science and Engineering **96** (2015) 012072 doi:10.1088/1757-899X/96/1/012072

experimental data have a number of advantages due to the fact that they more adequately depict the characteristics of the dynamic behaviour of the actual structure. Even though a significant error may be present in the models based on the experimental data, the results obtained with these models are generally of satisfactory approximation.

One of the ways of increasing the reliability of construction elements and buildings is monitoring of their technical condition. The main purpose of monitoring is to predict the possibility of damage, state of emergency or to establish emergency operation of the object of research. For this purpose, in the beginning, the reaction is fixed on a permanent or temporary, and maintenance, or test actions. In the next step, application of regression techniques by numerical simulation is performed to evaluate the performance of the object.

2. Methods of structural health monitoring

Conventional methods of monitoring technical condition are based on detection of damage in the structures of buildings or facilities during the entire period of their operation. In spite of considerable interest displayed to this issue and a significant number of publications, there is no unity of opinions [5 - 9, 11]. These methods differ from each other in the sets of values fixed for investigations, the techniques of their recording, transfer and further processing. However, with all their diversity, the structural health monitoring methods can be classified as based either on the frequency analysis or the time-dependent parameters [5-9, 11].

The damage identification methods, which are based on the frequency parameters and attribute the defects occurring in the structure to the deterioration of its structural stiffness, have found the wide-spread application. These methods use the finite element models for the estimation of the values of linear modal parameters, namely, the natural frequencies and the mode shapes, for the identification and, in some cases, even for location of the damage. It is assumed that the changes in the modal parameters of the structure during its life period are directly related to the occurrence of damage. These methods have been effective for the identification of substantial damages in the structures, but they are inapt in detecting the moment of the damage origination. It is worth noting that the changes in the frequency parameters and the mode shapes in the buildings and premises may be caused by the altered conditions of operation, and that restricts the application of the frequency-based method for the identification of significant damage.

The main shortcoming of all conventional methods [5-9, 11] consists in the fact that the same linear dynamic models are used in the analysis of the structure in its initial damage-free state as well as in the damaged condition, and that is contradictory with the basic principles of the physical process. The nonlinear effects in the above mentioned methods are averaged and divided evenly between the parameters of the individual modes.

The time-depended algorithms are based on the calculation of certain parameters over a predetermined period of time. These parameters include deformations (stress), buckling in the structures, building tilts, vibration amplitudes, etc., which are evaluated and compared to the reference values prescribed in the relevant construction Standards and Norms.

3. Damages and defects as the sources of nonlinear dynamic models in the building structures

Non-linearity of dynamic models in structures stems from geometrical non-linearity, the non-linear functioning of construction materials, non-linearity of dissipative forces. Today's rules and regulations for structural designs expand [12, 13] the scope of application of the structures operating in the elastic-

plastic stage. These damage-free structures originally display the non-linear properties and can be adequately described only by the non-linear models.
Cracks are one of the most common types of such damage; they can open and then close under operational loads. This type of damage includes: fatigue cracks, which occur near the bolt holes in truss joints; cracks caused by destruction of fragile structural materials; cracks resulting from significant deformation of the structure.

The response of the structures damaged with cracks to the repeated application of the operational loads is non-linear if during the recording of experimental data the cracks alternately open and close. In the contrary case, the response of the system is linear and the presence of cracks will only change the geometrical characteristics of the system. In a number of instances, it is difficult to simulate the actual dynamic behaviour as such, because any real structure has a well-defined specific pattern of crack locations.

Another common type of damage that leads to the non-linear response of the structures is the destruction of connections and fixtures. This category of defects includes not only the delaminating of the glued joints in the wooden structures but also in adhesion between concrete and reinforcement in the reinforced concrete structures or the destruction of bolt connections of the metal members. The emergence of structure-borne noise and thumps when uploading the structure with the operating loads is usually associated with a loss of bonds. The loss of bonds is one of the causes of crack development; moreover, cracking leads to the loss of bonds.

4. Selection of recorded values and experimental data processing

Traditionally, monitoring technical condition of structural elements in buildings and facilities is based on the application of measuring instruments for the survey of deformations and displacement of building structures and ground. A set of recording equipment, the points of location of individual sensors and the methods of data transmission are assigned individually. According to experimental data, the basic natural frequency ranges and Eigen vibration modes are determined. The next step is to calibrate the initial finite-element model. This approach has gained widespread acceptance. There are, however, a number of significant limitations to its successful application.

In particular, synchronization of the concurrent operation of sensors as well as discretization of the signals continuously coming from the sensors has significant impacts on the accuracy of the results obtained. As it is shown in research [4], the most efficient procedure consists in transferring the continuous signal $y_i(t)$ detected by the i- th sensor into its discrete form $y_i(i\Delta t)$at the discrete sampling rate of $f = (10...20) f_{\max}$. The maximum frequency value present in the signal $f_{\max}$ can be determined either by filtering high frequencies in the circuit connecting the sensor to an analogue-digital converter or according to the frequency parameters of the sensor. In each of these cases the following condition should be met:

$$A(f_{\max}) < 0.05 \max(A(f)), f > 0 \qquad (1)$$

where $A(f)$ is the module of the discrete Fourier transform for the transformation process $y_i(i\Delta t)$. When processing the results of the tests, which were conducted for the purposes of identification, it appeared that the value 200 Hz was sufficient frequency [4]. This limitation is in full compliance with the international standards and the Directive of the Cabinet of Ministers of Ukraine [14].

IOP Conf. Series: Materials Science and Engineering **96** (2015) 012072 doi:10.1088/1757-899X/96/1/012072

Another limitation consists in the simplicity of registration of the points of acceleration in the structure; nevertheless, in certain circumstances a number of individual points are inaccessible for registration of displacements.

The basic sources of noise are both a process of treatment of time processes and process of their receipt. Work of sensors depends on their quality and from different external factors taking place in the experiment. The range of noises, occurring at treatment of time processes, is quite wide. For suppression of noises there is a range of methods of rough-down of time processes. However, in every special case, a level is estimated by a specialist in the area of artificial perception in the manned regime. Obviously, that identification of noise requires the qualification of an expert from a researcher. So, inexact estimation of parameters of noise can result in the off-grade rough-down of time processes. In most cases, there is no *a priori* information about the character of noises, therefore the parameters of the method of rough-down of time processes sneak up by tuning.

It influences further process of treatment of time processes. As it is shown in [8], the underestimation of the level of multiplicative noise causes large breaks of the contours of the objects at the solution of the task of contour segmentation. It, in same queue, influences the error of recognition of these objects. Overextension of the level of additive Gausse noise conduces to breaking out of the contours of the objects, and that also affects the error of recognition. In filtration of time processes linear, non-linear and mixed methods are selected.

The use of the special receptions is required in non-linear methods, as the application of ordinary spectral or other linear filters can result in distortion of nonlinear structure of time row. Nonlinear filtration in contrast to linear does not lean against information in the frequency area with the purpose of distinction or clearing of useful constituent of time row from noise. Instead, nonlinear filters utilize the reconstructed phase space, in which the determined small-size geometrical inclusions are searched for, subsequently clearing them from noise components. The volume of information, which can be extracted from the finite set of points of a time row, depends on the properties of phase trajectories and on the properties of the dynamic model, generating these trajectories.

However, in the tasks of reconstruction of the time row, the properties of phase trajectories are *a priori* unknown; therefore doing some preliminary estimation of their parameters is impossible. The analysis shows that even at noised time process the numerical estimation of parameters of the model becomes rather inconvenient. The given fact is a signal to remove some percent of noise from the time process. It will allow investigating the dynamics of a mechanical system, instead of the dynamics of noise and, finally, receiving authentic numerical results.

5. The phase trajectories in the identification of dynamic models of building structures

Dynamic behaviour of mechanical systems is usually presented as oscillating processes in various graphic forms such as time processes, the Lissajous patterns and hodograph. Such patterns of presentation enable to determine the type of a process and to perform numerical estimations of its characteristics, but do not disclose any properties of the governing system. In contrast, classic phase trajectories have a range of advantages.

The study of the qualitative behaviour of a dynamic system implies the investigations into the behaviour of the trajectories in the phase space. The basic concepts of the theory of the qualitative study of dynamic systems were laid down by H. Poincaré. Many original fundamental contributions to the development of the qualitative research methods of dynamic systems were made by A.A. Andronov [1], E.A. Leontovich [1, 2], I.I. Gordon, A.M. Lyapunov, George Birkhoff [3]. The main objective of the classical theory of the qualitative research is to determine the dynamic properties of the systems without getting a closed form analytical solution. To meet these requirements, the phase trajectories on the plane $(y, \ddot{y})$ are commonly used.

The image on phase plane $(y, \dot{y})$ is a more vivid presentation because it depicts inharmonious oscillations particularly well. Each phase trajectory represents only one definite clearly defined motion. A disadvantage of phase trajectories consists in the fact that they do not provide for the immediate presentation of oscillating process in time. However, this drawback is compensated by a significant advantage. The geometric presentation of a single phase trajectory or a set of trajectories allows coming to important conclusions about the oscillation characteristics. It is, foremost, true with the oscillations, which are described with nonlinear differential equations. As is has been shown by the previous investigations, the expansion of a phase space by taking into account phase planes $(y, \ddot{y})$ and $(\dot{y}, \ddot{y})$ substantially promotes the efficiency in analyzing dynamic system behaviour. Hereby, we pass on to a three-dimensional phase space confined with three co-ordinate axes, i.e. displacement, velocity and acceleration. An interest taken in accelerations in dynamic systems is conditioned by the fact that these accelerations are more sensitive to high-frequency components in oscillating processes.

Of the main interest is the phase trajectory on the plane $(y, \ddot{y})$. This is motivated by the fact that the energy criteria on this trajectory are interpreted most accurately. In addition, the dependence $\ddot{y}(y)$ is antisymmetrical to the curve of elastic properties in relation to axis $\ddot{y}$. Particularly, phase trajectories $\ddot{y}(y)$ allow for identification of the type and degree of non-linearity of the system. It is known that accelerations of points are more sensitive towards deviations from harmonic vibrations [10].

Let us perform the analysis of free damped vibrations in the reinforced concrete beam, which are presented in figure 1. The time recordings $y(t)$ take the form of a monoharmonic process, which may represent the recordings of the displacements from a linear system as well as a non-linear system. The Poincare phase trajectories $(y, \dot{y})$ obtained in our experiments point to the existence of one stable equilibrium state in the structure. At the same time, the phase trajectories on the plane enable to reveal the nonlinearity of the elastic characteristic and also to determine its type, e.g., a 'soft characteristic' (see figure 2).

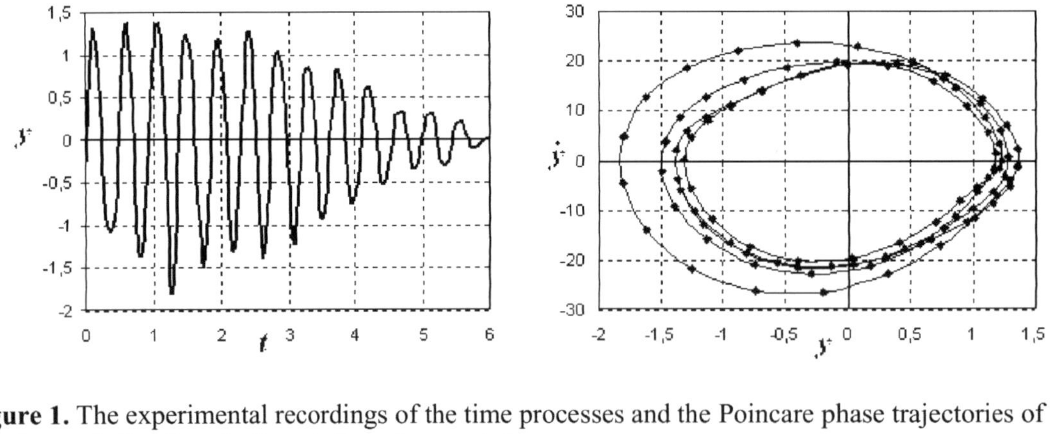

Figure 1. The experimental recordings of the time processes and the Poincare phase trajectories of free damped vibrations in the reinforced concrete beam.

A peculiar feature of this system is the frequency-versus-amplitude relationship of free vibrations. Thus, in the dynamic systems with the 'soft' characteristic of elastic force, the frequency of free vibrations decreases when the amplitude increases. The identification of the given nonlinear system with the linear ones is incorrect. With the simplified linear model neither the boundaries of the zones of fundamental-frequency resonant vibrations in the structure can be defined, nor the boundaries of the zones of subharmonic and ultraharmonic vibrations, nor the boundaries of the zones of amplitude multiplicity.

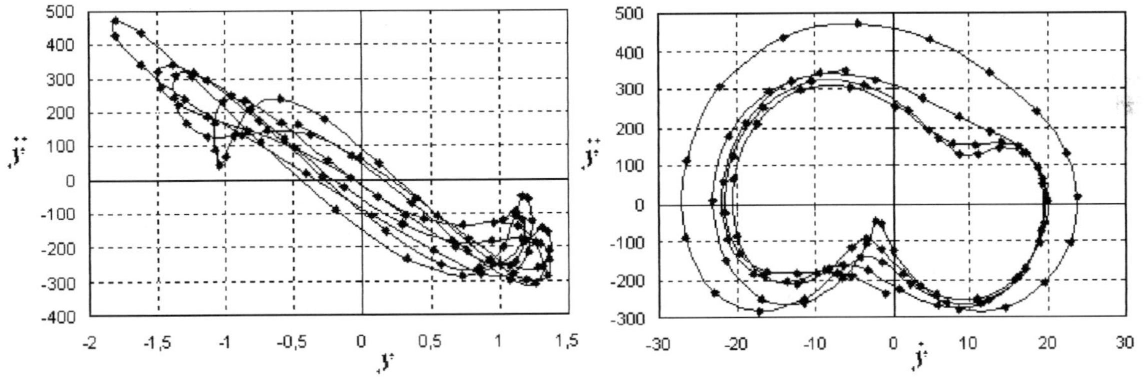

Figure 2. The phase trajectories of vibrations in the reinforced concrete beam.

The assessment of the properties of an actual structure is realizable only with a certain degree of approximation. In this regard, one of the most important tasks in the computer-based simulation of a load-bearing structure is to construct such mathematical model that enables the introduction of those parameters and property characteristics of the structure into the model, which will substantiate the results of the field experiments.

6. Conclusions

Most mechanical systems show non-linear features at certain parameters of outer excitement. Various sources of non-linear behaviour that may exist in engineering structures, for example, in the structures with bolted and pin joints, reinforced structures, damaged structural elements. Non-linearity is important in many structural dynamic applications that are of interest to engineers.

The identification of a non-linear model of mechanical systems experimental data represents a distinct challenge in view of the absence of superposition principle in non-linear dynamics. The problem discussed in the paper referred to the characterization of non-linearity. The method of nonparametric identification is suggested in the paper. The method is based on the usage of the information about accelerations, displacement, and also outer excitation. It uses the methods of the qualitative theory, and also regression by methods and approximating expressions of the elastic characteristic as functions from generalized co-ordinates. The capabilities of the suggested method are limited only to noise level, measuring error and length of process. The application of the suggested method is of interest and promising for engineering application.

7. References

[1] Andronov A A 1966 *The Qualitative Theory of Dynamic Systems of the Second Order* (Moscow: Nauka Publishing House)

[2] Bautin N N 1976 *Methods and Techniques of the Qualitative Study of Dynamical Systems on the Plane* (Moscow: Nauka Publishing House)

[3] George D and Birkhoff D 2002 *Dynamical Systems* (Moscow-Izhevsk: Regular and chaotic dynamics)

[4] Volkova V E 2011 The reconstruction of the phase trajectories of noisy dynamical systems *Theoretical Foundations of Civ. Engin.* **14** 323 – 328

[5] Kononenko V O and Plakhtienko N P 1976 *Methods of Identification of Nonlinear Mechanical Vibration Systems* (Kiev: Naukova Dumka Publishing House)

[6] Menshikov Yu L 2009 *The Identification of External Influences* (Dnepropetrovsk: Science and Education)

[7] Nikulchev E V 2007 *A Geometric Approach to the Non-linear Systems Modelling Based on Experimental Data* (Moscow: the Moscow State University of Printing Arts)

[8] Sushchev S P 2011 Monitoring technical condition of load-bearing structures in buildings and premises *Science and Safety* **1** 24 – 32

[9] Kerschen G 2006 Past, present and future of nonlinear system identification in structural dynamics *Mech. Syst. Signal Proc.* **20** (3) 505 – 592

[10] Volkova V E 2005 Qualitative theory and identification of dynamic system with one degree of *Int. J. Applied Mech.* **41** (6) 134 – 139

[11] Worden K 2001 *Nonlinearity in Structural Dynamics: Detection, Identification and Modeling* (New-York: Institute of Physics Publications)

[12] The State Standards and Norms for construction in Ukraine *(DBN B.2.6-98:2009), Concrete and Reinforced Concrete Structures. General provisions* 2011 (Kiew: Minregionbud Publishing House)

[13] The State Standards and Norms for construction in Ukraine *(DBN B.2.6-163:2010) The Designs of Buildings and Premises. Steel Structures* 2011 (Kiew: Minregionbud Publishing House)

[14] The Resolution of the Cabinet of Ministers of Ukraine *Measures on the Gradual Introduction in Ukraine of the European Union Directives on Sanitary, Ecological, Veterinary and Phytosanitary Regulations, and the International Standards* No.244 1997 (Kiew: Minregionbud Publishing House)

IOP Conf. Series: Materials Science and Engineering **96** (2015) 012073 doi:10.1088/1757-899X/96/1/012073

Cement Based Batteries and their Potential for Use in Low Power Operations

A Byrne[1,3], N Holmes[1] and B Norton[2]
[1]School of Civil and Structural Engineering, Dublin Institute of Technology, Bolton Street, Dublin 1, Ireland.
[2]Presidents Office, Dublin Institute of Technology, Grangegorman, Dublin 1, Ireland

E-mail: aimee.byrne@dit.ie

Abstract. This paper presents the development of an innovative cement-electrolyte battery for low power operations such as cathodic protection of reinforced concrete. A battery design was refined by altering different constituents and examining the open circuit voltage, resistor loaded current and lifespan. The final design consisted of a copper plate cathode, aluminium plate anode, and a cement electrolyte which included additives of carbon black, plasticiser, Alum salt and Epsom salt. A relationship between age, temperature and hydration of the cell and the current it produced was determined. It was found that sealing the battery using varnish increased the moisture retention and current output. Current was also found to increase with internal temperature of the electrolyte and connecting two cells in parallel further doubled or even tripled the current. Parallel-connected cells could sustain an average current of 0.35mA through a 10Ω resistor over two weeks of recording. The preliminary findings demonstrate that cement-based batteries can produce sufficient sustainable electrical outputs with the correct materials and arrangement of components. Work is ongoing to determine how these batteries can be recharged using photovoltaics which will further enhance their sustainability properties.

1. Introduction

Novel battery design is an area that can help ease humanity's dependence on fossil fuels. Research and development focuses on creating higher power storage, greater recharge capacity and extending the life of traditional batteries by adapting their components and materials. The electrolyte of a battery is an ionic conductor but an electronic insulator (resisting the movement of electrons) [1]. Liquid electrolytes are generally favoured due to the high mobility of ions and continuity of interface between electrode and electrolyte. The main issue with liquid electrolyte batteries is the use of toxic materials and their tendency to leak during use or after disposal. Solid electrolytes are not prone to leakage but their ionic conductivity tends to be less than their liquid counterparts and they are more costly. Some examples of solid electrolytes are polymers doped with ions [2-4] or ceramics with ions arranged to

[3] To whom any correspondence should be addressed.

allow movement [5-7]. Cement is an ionic conductor due to the solution that can be stored in, and travel through, its pores and micro-cracks. This facilitates its potential as a good electrolyte for novel cement battery designs. Cement-based batteries have been sparsely researched, therefore, there have not been many advances in making these batteries more efficient, powerful, long lasting or rechargeable. The work presented here follows on from previous research conducted by the authors in which the cement electrolyte design was refined for highest current (I) and longevity [8]. This paper presents a closer examination of the relationship between temperature and humidity inside the cell and the current output of the refined battery as well as methods of further increasing output by sealing the cells and connecting them in parallel.

Meng *et al.* [1] provided the initial proof of the concept that cement based batteries could be designed to provide a voltage and current output. The design consisted of electrode cement layers with active additives separated by a basic cement electrolyte as shown in figure 1. Examples of successful cement battery development tend to follow the same layered design [9, 10]. However, Burstein *et al.* [11] developed a battery with a steel cathode and an aluminium anode set into a concrete electrolyte which could provide a small current density. Similarly, Ouellette *et al.* [12] used probe type electrodes inside a cement-based electrolyte enclosed in a saltwater bath. Both of these batteries are closer to the form chosen for the research presented here and shown in figure 2, with cement only in the electrolyte.

The intended use of the cement batteries presented in this paper is for Impressed Current Cathodic Protection (ICCP) of steel reinforcement in concrete structures. ICCP is a method of protecting reinforcing steel in concrete from corrosion by connecting it to an inert, less noble, metal than steel and passes a low level of current through it using an external power source [13]. The recommended design current density is 20mA per m^2 of bar surface area [14] although many studies indicate that lower values are adequate [15-19]. Cathodic prevention, which is the provision of protective current before any corrosion has taken place, requires a lower current density of 2-5mA/m^2 [20]. Therefore, the testing regime incorporated a resistor load and focused on enhancing current output.

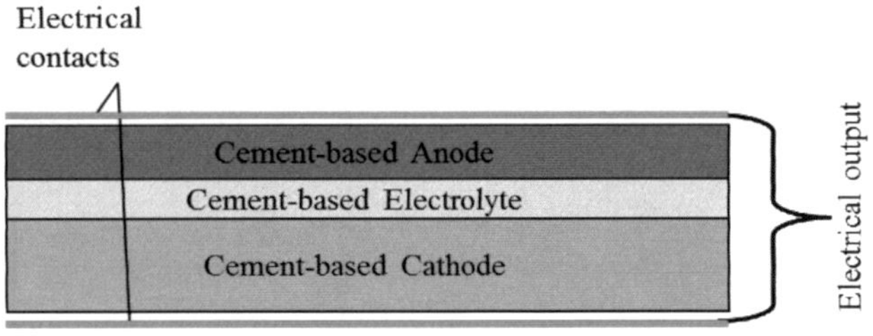

Figure 1. Layered style cement battery.

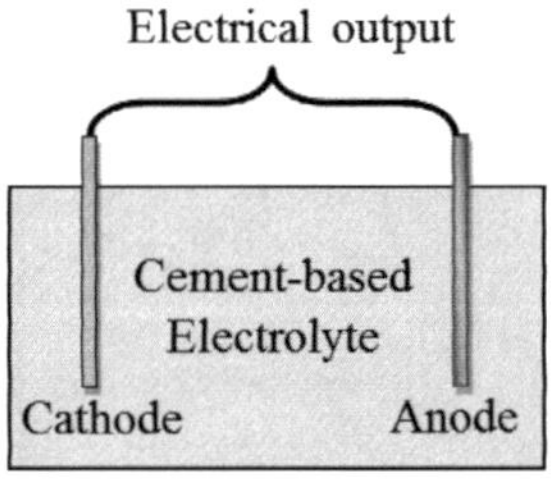

Figure 2. Probe style cement battery.

2. Battery design

2.1. Mix design

A standard form of battery was chosen (figure 2) and used to compare different electrolyte designs. The battery consisted of cement and water paste to form the electrolyte, a copper plate cathode and an aluminium plate anode. The effect of the water/cement ratio, additives and electrode spacing on current, voltage and lifespan were examined. Table 1 shows that optimal output could be achieved by designing high w/c ratios, the addition of carbon black, and adding salt for high current and lifespan. Further details of the design mixes, testing regime and outcomes are described by Holmes *et al.* [8].

Table 1. Summary of the impact of electrolyte constituents in refining the battery design.

Addition	Current (under 10Ω load)	Voltage (open circuit)	Lifespan
Increased w/c ratio	↑	=	=
Sand	=	=	=
Clay aggregate	=	=	=
Carbon black & plasticiser	↑	↑	↑
Increased electrode material	↑	=	=
Salt solution	↑	=	↑
Salt crystals	↑	=	↑
Waterglass	=	=	=
Closer electrodes	=	=	=

A final electrolyte design shown in table 2 produced an average continuous current output of 0.02mA for more than a month through a 10Ω resistor load and formed the battery design used for further development through sealing and connection in parallel as presented in this paper.

Table 2. Electrolyte constituent descriptions and proportions by weight.

Constituent	Proportion by weight (g per kg of total mix)	Description
Water	290.58	Mains supply tap water
Cement	594.94	CEM 1 (BS EN 197-1, 2000 [21])
AlKO8S2.12H2O (Alum salt)	47.28	>99% purity
MgSO4.7H2O (Epsom salt)	47.28	>99% purity
Carbon Black	9.85	Average size 30nm
Plasticiser	9.85	Sika VistoCrete 30HE

2.2. Battery cell preparation

The dry electrolyte materials shown in table 2 were weighed and passed through a 200μm sieve to remove any non-conforming lumps before being mixed with water and plasticiser and placed into 70 x 70 x 40mm plastic moulds to create the electrolyte. The electrode plates (60 x 30 x 0.5mm) were sanded and washed in a borax solution to remove any impurities or oxide layers and inserted into the wet electrolyte block protruding 5mm from the surface to facilitate connection to the resistor circuit. The batteries were then placed on a vibration table for 30 seconds to remove any remaining air. Two large blocks measuring 140 x 180 x 95mm were cast alongside using the same constituents to facilitate moisture and temperature testing over time.

In total, eight batteries and two large blocks were made as shown in figure 3. A k-type thermocouple was cast into the middle of each of the large blocks and two of the battery cells and allowed to cure under a polythene sheet for 24 hours. Four of the batteries and one of the blocks were then sealed using an acrylic based varnish suitable for cement surfaces, which was water and vapour proof. The remaining four batteries and single large block were unsealed.

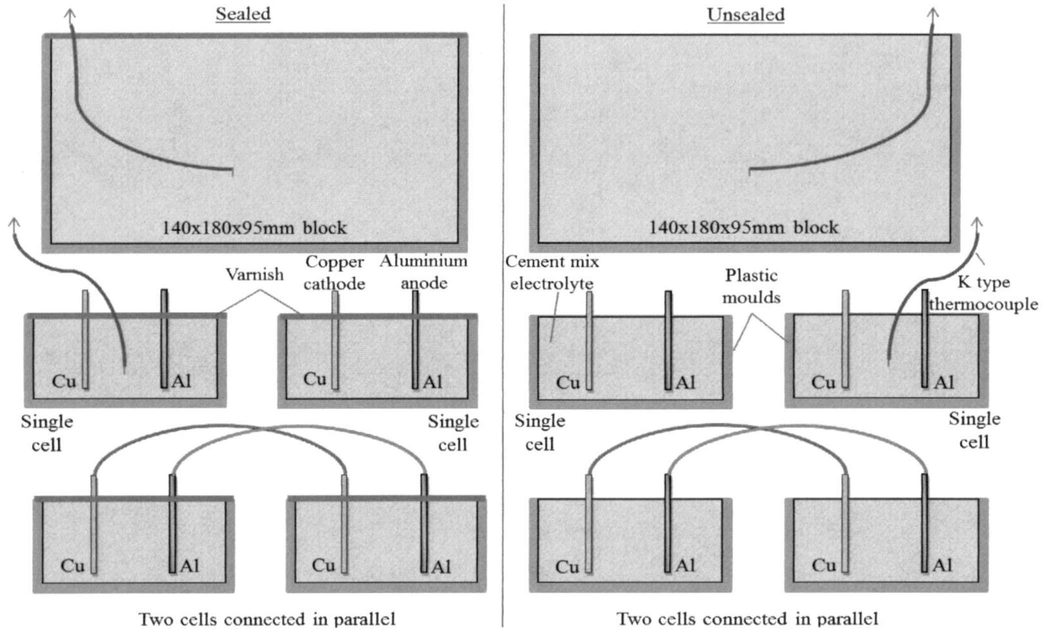

Figure 3. Schematic of sealed and unsealed batteries and blocks.

3. Recording

3.1. Resistor-loaded current

A 10Ω resistor was connected between the anode and cathode of each battery to act as a resistor load as per figure 4. Two unsealed and two sealed batteries were connected as such. The remaining four cells were joined in parallel as shown in figure 3 and connected to the resister as shown in figure 5.

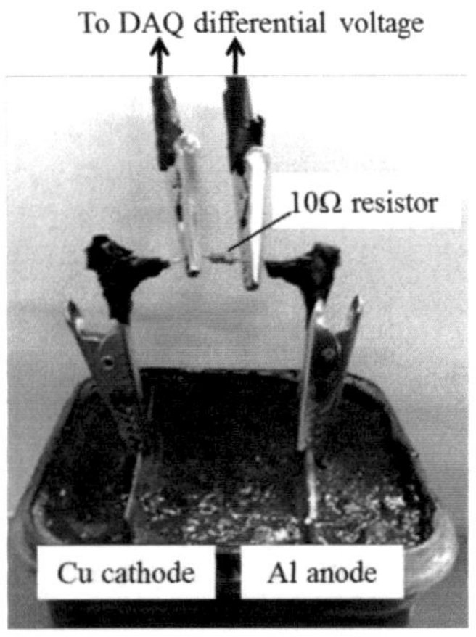

Figure 4. Differential voltage across a resistor for a single cell.

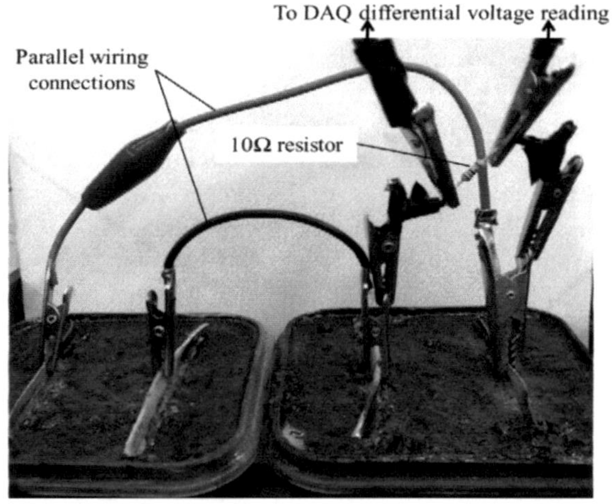

Figure 5. Differential voltage reading across a resistor for parallel cells.

A National Instruments differential data acquisition (DAQ) unit NI 9205 was used to record voltage across the resistors as shown in figures 4-6. Readings were taken every 10 seconds for the first two hours and every 10 minutes after that using a designed LabVIEW program. Current through the resistor was then obtained by dividing the voltage across it by the resistor value (10Ω). Temperature readings from the four thermocouples were recorded similarly using a NI 9211 DAQ unit. The final layout is displayed in figure 6.

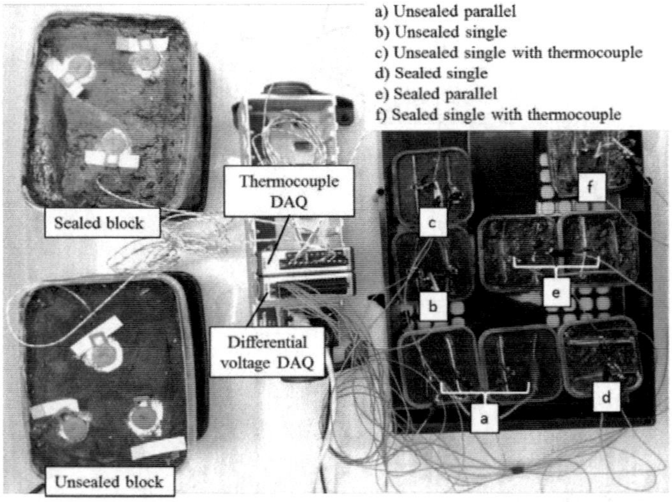

Figure 6. Setup showing DAQ for voltage recording and DAQ for thermocouple recording.

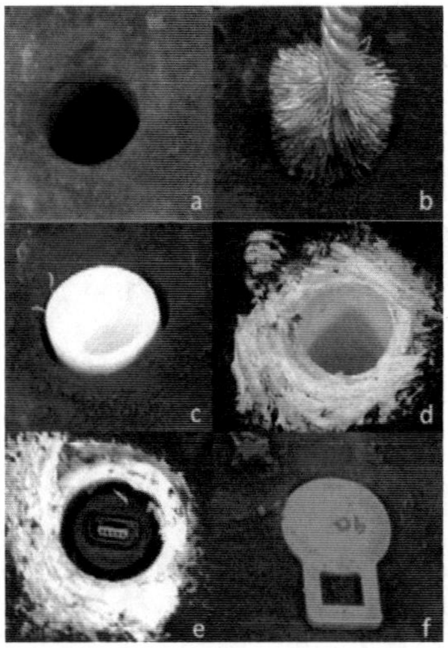

Figure 7. Steps in inserting relative humidity probes in concrete.

3.2. Internal relative humidity
Relative humidity probes were inserted into the two large blocks at depths of 40mm, 50mm and 65mm using Tramex CMEXPERT II Hygro I probes. As shown in figure 7, the holes were initially drilled into the hardened block 24 hours after casting and cleaned out. The plastic tubing provided was cut to length, inserted into the holes and sealed at the edge if required. Finally, the probe head was inserted and covered with the cap. The moisture meter could then be attached to the probe head at intervals to take recordings taken four times a day.

4. Results and discussion

4.1. Relationship with moisture
The impact of moisture retention on the discharge current of the cells was examined in two ways. Firstly, the difference in current between the sealed and unsealed cells indicates the impact of the increased moisture retention in the cells due to sealing. Secondly, the relative humidity probes that were used on the representative blocks provided an indication of the changing internal moisture condition in the sealed and unsealed cases over time.

IOP Conf. Series: Materials Science and Engineering **96** (2015) 012073 doi:10.1088/1757-899X/96/1/012073

All cells were made using the same constituents and proportions. As shown in figure 8 the sealed cells provided a resistor-loaded current that was on average 1.5 times greater than the unsealed batteries (0.066 mA unsealed and 0.101 mA sealed).

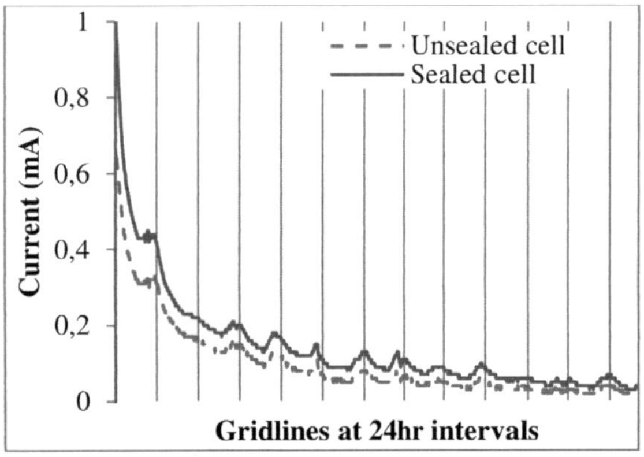

Figure 8. Discharge current through a 10Ω resistor for a sealed and unsealed battery.

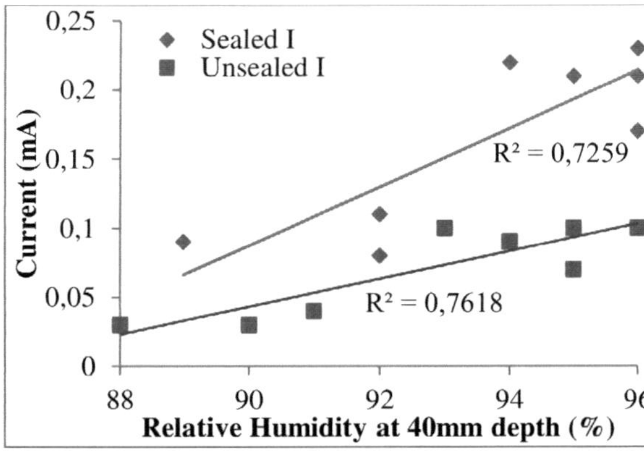

Figure 9. Relative humidity inside the blocks related to current discharged from the batteries.

Although moisture content of the cells could not be directly measured using the probe method due to size requirements, the blocks did provide an indication of the condition inside the cells. Figure 9 shows the difference between the moisture retention of the unsealed and sealed blocks. The sealed block retained on average 1% higher relative humidity than the unsealed case. The relationship between humidity and the current output from the associated cells indicates that there is a definite but not precise relationship between the two as shown by the R^2 values of around 0.7. For every 1% increase in relative humidity current was found to increase by 0.01mA for unsealed cells and 0.02mA for sealed cells. However, it is likely that during the first two weeks after casting only the area close to the surface of the large blocks will show marked differences between the sealed and unsealed cases whereas the minimum depth for probe testing was 40mm.

Over the lifespan of the battery the cells dry out as reflected by the relative humidity of the blocks. Also during this time reactants that cause the chemical reactions which create the current will be depleted and reaction products will build up on the electrodes. It is difficult to attribute the drop in current over time exactly. Further to the recorded change in relative humidity of the blocks over time, the sealed cells all showed higher current outputs than the air-dried cells consistently during the measurement period. Again this is due to both the enhanced moisture retention and the retention of reaction gases.

4.2. Relationship with temperature

The curing process naturally creates an increase in temperature as reactions take place. However, even after the peak of current during initial discharge (figure 8), small cyclical temperature variations with the environment showed a correlation with the current output. Internal temperature of the cells fluctuated cyclically with daily environmental conditions as shown in figure 10. It was observed that for every 1°C increase in temperature, the current output increased by 0.015 mA for unsealed cells and 0.028 mA for sealed cells, both with a high level of correlation with a 0.99 R^2 value. In batteries, higher temperatures are known to correlate with greater outputs as it is favourable for chemical reactions and improves electron or ion mobility reducing the cell's internal impedance [22].

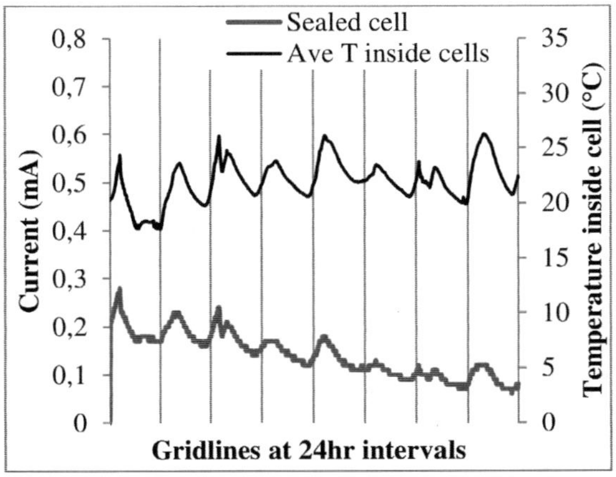

Figure 10. The relationship between temperature fluctuations within the cement electrolyte and resistor-loaded discharge current from the cells.

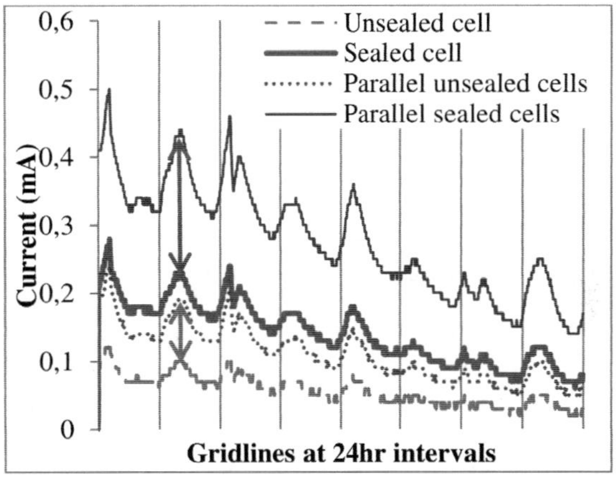

Figure 11. The resistor-loaded discharge current from single cells and two parallel-connected cells for the sealed and unsealed cases.

4.3. Connecting cells in parallel

As shown in figure 11, connecting two cells in parallel increased current output by 250% for unsealed cells to 0.16mA and 360% for sealed cells to 0.35mA. Parallel configurations are used to double the capacity while maintaining the same voltage. Capacity is the product of current multiplied by the number of hours it flows. In this case it can, therefore, be assumed that the additional current output is in sacrifice of some of the additional lifespan the cells could have otherwise expected had the current output only doubled.

5. Conclusions

This paper followed on from previous research into refining the design of a cement based battery [8] by advancing the understanding of the relationship between power output and conditions inside the cell. Impressed current cathodic protection limits the corrosion of a metal surface but requires an external direct current (DC) source. For reinforcement in steel the recommended design current density is $20mA/m^2$ [14]. The cement batteries presented in this paper were designed for the intention of use in cathodic protection systems and the discharge current through a 10Ω resistor was measured. Unsealed, air-dried cement batteries were able to achieve an average of 0.066mA current over the two week recording period. Sealing the batteries with weatherproof varnish increased this value to 0.101mA. Connecting sealed cells in parallel to increase capacity further increased the current output by 360%. Although the current value (0.35mA) is considerably lower than the required 20mA for a m^2 of reinforcement surface area, it is much closer to the $2\text{-}5mA/m^2$ required for cathodic prevention. Additionally, the resistivity of mild steel ($15x10^{-8}\Omega m$) reinforcement is considerably lower than the resistor used for these experiments (10Ω), therefore, future tests should use a resistor which is more representative of mild steel.

6. References

[1] Meng Q and Chung D D L 2010 Battery in the form of a cement-matrix composite *Cement and Concrete Composites* **32** 829–839

[2] Ashrafi R, Sahu D K, Kesharwani P, Ganjir M and Agrawal R C 2014 Ag+-ion conducting Nano-Composite Polymer Electrolytes (NCPEs): Synthesis, characterization and all-solid-battery studies *J. of Non-Crystalline Solids* **391** 91–95

[3] Sun B, Mindemark J, Edström K and Brandell D 2014 Polycarbonate-based solid polymer electrolytes for Li-ion batteries *Solid State Ionics* **262** 738–742

[4] Taib N U and Idris N H 2014 Plastic crystal–solid biopolymer electrolytes for rechargeable lithium batteries *J. of Membrane Science* **468** 149–154

[5] Shin B R, Nam Y J, Oh D Y, Kim D H, Kim J W and Jung Y S 2014 Comparative Study of TiS2/Li-In All-Solid-State Lithium Batteries Using Glass-Ceramic Li3PS4 and Li10GeP2S12 Solid Electrolytes *Electrochimica Acta* **146** 395–402

[6] Tatsumisago M, Takano R, Tadanaga K and Hayashi A 2014 Preparation of Li3BO3–Li2SO4 glass–ceramic electrolytes for all-oxide lithium batteries *J. of Power Sources* **270** 603–607

[7] Tatsumisago M and Hayashi A 2012 Superionic glasses and glass–ceramics in the Li2S–P2S5 system for all-solid-state lithium secondary batteries *Solid State Ionics* **225** 342–345

[8] Holmes N, Byrne A and Norton B Accepted for publication – 2015 First steps in developing cement-based batteries to power cathodic protection of embedded steel in concrete *Sustainable Design and Research (SDAR)*

[9] Rampradheep G S, Sivaraja M and Nivedha K 2012 Electricity generation from cement matrix incorporated with self-curing agent *Conf. Advances in Engineering, Science and Management (ICAESM)* pp 377–382

[10] Qiao G, Sun G, Li H and Ou J 2014 Heterogeneous tiny energy: An appealing opportunity to power wireless sensor motes in a corrosive environment *Applied Energy* **131** 87–96

[11] Burstein G T and Speckert E I P 2008 Developing a battery using set concrete as electrolyte *ECS Transactions* 42 ed pp 13–20

[12] Ouellette S A and Todd M D 2014 Cement Seawater Battery Energy Harvester for Marine Infrastructure Monitoring *Sensors J. IEEE* **14** 865–872

[13] Polder R B 1998 Cathodic protection of reinforced concrete structures in the Netherlands - Experience and developments: Cathodic protection of concrete - 10 years experience *Heron* **43** 3–14

[14] Polder R, Kranje A, Leggedoor J, Sajna A, Schuten G and Stipanovic I 2009 Guideline for smart cathodic protection of steel in concrete *Assessment and Rehabilitation of Central European Highway Structures*

[15] Glass G K, Hassanein A M and Buenfeld N R 2001 Cathodic protection afforded by an intermittent current applied to reinforced concrete *Corrosion Science* **43** 1111–31

[16] Glass G K and Chadwick J R 1994 An investigation into the mechanisms of protection afforded by a cathodic current and the implications for advances in the field of cathodic protection *Corrosion Science* **36** 2193–09

[17] Glass G K and Buenfeld N R 1995 On the current density required to protect steel in atmospherically exposed concrete structures *Corrosion Science* **37** 1643–46

[18] Koleva D A, Hu J, Fraaij A L A, Stroeven P, Boshkov N and van Breugel K 2006 Cathodic protection revisited: Impact on structural morphology sheds new light on its efficiency *Cement and Concrete Composites* **28** 696–706

[19] McArthur H, D'Arcy S and Barker J 1993 Cathodic protection by impressed DC currents for construction, maintenance and refurbishment in reinforced concrete *Construction and Building Materials* **7** 85–93
[20] Chess P M and Broomfield J P 2003 *Cathodic Protection of Steel in Concrete* Taylor & Francis
[21] British Standards Institution 2000 *BS EN 197-1 Cement: Composition, Specifications and Conformity Criteria for Common Cements London*
[22] Solanki C S 2013 *Solar photovoltaic technology and systems, a manual for Technicians, Trainers and Engineers* (Delhi: PHI Learning Private Limited)

Achnowledgements
This research was funded by Science Foundation Ireland's Technology Innovation Development Award (SFI TIDA).

IOP Conf. Series: Materials Science and Engineering **96** (2015) 012074 doi:10.1088/1757-899X/96/1/012074

Chemical Characterization of Lime-Based Binders in Historic Buildings of Latvia

I Kirilovica[1,2], L Gulbe[1], I Vitina[1] and S Igaune-Blumberga[1]
[1] Institute of Silicate Materials of Riga Technical University, 3/7 Paula Valdena Str., Riga, LV-1048, Latvia

E-mail: inta.kirilovica@rtu.lv

Abstract. The aim of this research is to investigate the chemical composition of stone materials of several local historic buildings with a purpose of elaboration of restoration strategy, including the choice of restoration materials. Most of the examined mortars are lime-based hydraulic mortars, characteristic of the architecture of $19^{th}/20^{th}$ century. Pure aerial lime binders show reduced compatibility with historic materials, that is why lime binders with pozzolan additive (cement) are an appropriate choice for restoration. In order to examine the changes of hydraulicity (i.e. the property of binders to harden when exposed to water) of perspective restoration binders, a series of blended lime–cement mixtures were synthesized with growing content of cement (up to 10 % by weight). A significant relationship between cement content and hydraulic properties has been shown.

1. Introduction

Principal factor of the sustainability of historic monuments is usage of compatible materials. In the conservation practice, both lime binders and cement binders often display incompatibility with historic mortars [1], [2], [3]. Cement binders have high content of soluble salts, low vapor permeability, high modulus of elasticity, while lime binders show low mechanical strength, slow hardening and low water resistance. To avoid these disadvantages, a wide range of materials and their combinations are recommended – hydraulic lime, Roman cement, lime with pozzolan additive etc. [4], [5], [6], [7], [8], [9]. It has been observed that substitution of a small percentage of lime by cement (below 10% in mass) does not modify the microstructure of the mortar and the mixture can be formulated in order to meet the requirements for conservation mortars [3]. Overall, the choice of material for each object must be based on the knowledge about the original material.

Scientific investigation of several mortars from local historic buildings (Vingrotaju Str. 1 (Riga), Daugavpils Fortress, the banks of the River Aleksupite (Kuldiga) etc. (total of 13 samples)) was

[2] To whom any correspondence should be addressed.

carried out at the laboratory of the Institute of Silicate Materials of Riga Technical University in order to characterize authentic materials and elaborate conservation plans. The results of analysis of the chemical composition of historic mortars allow indicating the original components of mortar, providing necessary information for the choice of the most compatible restoration materials.

Binders used for restoration are recommended to contain hydraulic components (associated with soluble Al_2O_3 and SiO_2). Matching composition and compatible properties to historic binders can be achieved by using lime-cement mixtures. A range of aerial lime – white cement binders were synthesized with growing cement content (up to 10 %); their composition was then determined.

2. Materials and methods

The historic mortar samples examined in this research are characterized in table 1.

Table 1. Location of samples of historical buildings.

Sample No.	Location	Sample characterization
1	Bank of the River Aleksupite in Kuldiga	Mortar with crushed brick particles
2	The right bank of the River Aleksupite in the niche near Pasta Street in Kuldiga	Mechanically strong mortar with crushed brick particles
3	Vault under the bridge of Skolas Street in Kuldiga	Mechanically strong mortar with crushed brick particles
4	Daugavpils Fortress, 1st coast lunette	Corroded mortar with high content of sulphates and chlorides
5	Daugavpils Fortress, 1st coast lunette	White disintegrated plaster with high content of sulphates and chlorides
6	Daugavpils Fortress, 1st coast lunette	Mechanically strong cement plaster on the surface of white lime plaster of decors
7	Vingrotaju Street 1 in Riga	White mechanically strong plaster
8	Vingrotaju Street 1 in Riga	White mechanically strong mortar for cladding bricks
9	Vingrotaju Street 1 in Riga	Disintegrated grey mortar joints corroded by sulphates
10	Belfry of Aleksander Nevsky Church on Brivibas Street 56 in Riga	White, old corroded mortar under cement layer
11	Professional school on Saules Street 15 in Ventspils	Greyish corroded plaster
12	Miesnieku Street 13 in Riga	Mechanically strong two layered part of a decor
13	Puzes Church near Ventspils	White calcitic lime plaster

The chemical composition of historic mortars as well as synthesized lime-cement binders was characterized through classical (wet) chemical analysis (according to EN 196-2 [10]). The mineralogical phases of mortar samples were determined by means of differential thermal and theromgravimetric (DTA/TG) analysis (SETARAM SETSYS Evolution – 1750) and X-Ray diffraction (XRD) analysis (Rigaku Ultima + with CuKα radiation at scanning interval 0-60° (2θ) and speed 2°/min). Classical chemical analysis comprised determination of loss of ignition, CaO, MgO, SiO_2, Al_2O_3, Fe_2O_3, Na_2O, K_2O and part insoluble in HCl. From the data of chemical analysis, the type of binder, its original components, modulus of hydraulicity and aggregate/binder ratio were determined. Size distribution of aggregate was analyzed after dissolving a sample in 10 % HCl, elution, drying and sieving [11].

For the synthesis of lime-cement binders with growing cement content (0, 2, 4, 8 and 10 %), mixtures of slaked lime ($Ca(OH)_2$) with sand (aggregate size 0-4 mm) and white Portland cement (CEM I 52,5 R) were used.

3. Results and discussion

By means of chemical analysis it was determined that most of the analyzed binders are based on dolomitic or calcitic lime with different levels of hydraulicity. The results of chemical and granulometric analysis are shown in table 2, table 3 and table 4.

The amount of hydraulic components is of great importance, providing mechanical strength and water resistance, and it is strongly connected with the grade of deterioration: historic binders with higher content of soluble SiO_2 and Al_2O_3 were found to be in a better condition and are mechanically stronger. It can be featured by using modulus of hydraulicity (m) (Equation 1) [12].

$$m = \frac{\%(CaO+MgO)}{\%(SiO_2+Al_2O_3+Fe_2O_3)} \quad (1)$$

The modulus of hydraulicity expresses the ratio between basic and acid oxides in mortar. If m>9, it is a pure lime binder, but if m<1.7 – the binder can be considered to be Roman cement. The modulus of hydraulicity of hydraulic lime is 1.7-9 [12].

All samples with m > 6 are observed as corroded and most of them also suffering from migrating salts and moisture.

Hydraulic nature of the examined mortars dated to 19th/early 20th century (or earlier) can be attributed to raw materials with high content of clay used in the binder production. The dehydroxylated clays acted as a pozzolan which imparts early strength to the mortar. In mortars with crushed brick additive, compounds of hydraulic type occur at the brick matrix interface [13].

Pure lime mortars without hydraulic additives are weaker attached to the original surface. As an example of unconformity between old and new (without hydraulic components) building materials, rendering of Puze Church (sample N° 13) can be mentioned (figure 1). The usage of the aerial lime binder with no hydraulic additives (m = 43.0) for the restoration of the façade has led to fast and intensive decay and pealing of exterior surfaces in the exposure to outdoor conditions.

Figure 1. Church of Puze, Latvia (sample N° 13).

Table 2. Results of chemical analysis of historical mortars, weight %.

Sample / Component	1	2	3	4	5	6	7	8	9	10	11	12	13	Accuracy, [± absol. %]
Loss of ignition at 1000°C	16.03	25.04	21.78	27.29	27.04	14.21	13.20	11.44	30.60	8.62	10.10	14.13	44.05	0.3
Sand and gravel	61.84	49.56	47.92	41.84	52.61	62.73	68.81	72.12	35.44	80.46	72.87	45.23	11.45	0.5
CaO	15.21	13.94	21.32	16.04	12.09	16.56	9.59	10.01	19.37	7.03	8.05	24.92	42.20	0.5
MgO	2.78	7.35	2.41	9.60	7.90	0.72	3.76	2.86	9.43	1.96	3.38	3.36	1.71	0.5
Soluble SiO_2	0.96	1.70	2.28	1.36	0.31	1.17	2.80	2.26	1.61	0.68	4.20	6.04	0.37	0.3
Al_2O_3	1.71	1.52	2.72	1.20	0.89	2.13	1.05	0.97	1.08	0.45	0.76	4.09	0.44	0.3
Fe_2O_3	0.96	0.50	1.08	0.65	0.33	0.53	0.34	0.28	0.56	0.19	0.38	1.07	0.21	0.1
Total	99.36	99.50	99.45	97.92	98.17	98.05	99.55	99.71	97.77	99.37	99.74	98.84	100.43	-
CaO/MgO	5.47	1.90	8.80	1.67	1.53	23.0	2.55	3.50	2.05	3.5	2.38	7.4	24.68	-
m (modulus of hydraulicity)	4.6	5.7	4.0	8.0	13.1	4.5	3.2	3.7	8.9	6.8	2.1	2.5	43.0	-
Aggregate/binder ratio	1:2	1:1.5	1:1.5	1:1.4	1:2.3	1:2.7	1:3.8	1:4.4	1:1	1:7	1:4	1:0.8	1:0.3	-

Table 3. Calculated parameters of the binder in historical mortars from chemical analysis, weight %.

Sample Component	1	2	3	4	5	6	7	8	9	10	11	12	13	Accuracy, [± absol. %]
CaO	68.8	54.9	70.4	51.9	51.8	71.8	53.4	60.9	57.0	64.4	47.1	61.4	90.3	0.5
MgO	12.6	28.4	7.9	31.0	33.8	3.1	20.9	17.4	27.8	17.9	19.8	8.3	3.8	0.5
Soluble SiO_2	4.5	6.7	7.5	4.4	1.3	6.1	15.6	13.7	4.7	6.2	24.5	14.9	0.8	0.5
Al_2O_3	7.7	6.0	8.9	3.9	3.8	9.2	5.8	5.9	3.2	4.1	4.4	10.1	1.0	0.5
Fe_2O_3	3.7	2.0	3.6	2.1	1.4	3.3	1.8	1.7	1.6	1.7	2.2	2.6	0.4	0.1
Hydraulic compounds [a]	15.9	14.7	20.0	16.4	6.5	18.6	23.2	21.3	9.5	12.1	31.1	27.6	2.2	-
Type of binder [b]	SH; CL	SH; DL	SH; CL	WH; DL	WH; DL	SH; C/L	SH; DL	SH; DL	WH; DL	WH; DL	SH; DL	SH; C	NH; CL	

[a] Hydraulic compounds = (SiO_2+ Al_2O_3+ Fe_2O_3).
[b] SH = strongly hydraulic; WH = weakly hydraulic; NH = non-hydraulic; CL = calcitic lime; DL = dolomitic lime; C/L = cement with lime addition; C = cement.

Table 4. Recalculated aggregate size distribution, weight %.

Particle size d, mm	1	2	3	4	5	6	7	8	9	10	11	12	13	Accuracy, [± absol. %]
> 0.7	26.1	38.8	28.2	44.0	32.8	9.5	30.2	Not investigated	31.2	11.2	Not investigated	14.8	Not investigated	0.5
0.5-0.7	44.6	31.2	32.7	28.8	31.1	48.0	36.2		48.9	60.1		45.1		0.5
0.2-0.5	16.8	15.7	15.8	10.8	15.5	24.5	17.5		11.9	23.4		21.3		0.5
0.16-0.2	3.1	5.3	6.2	5.3	6.9	5.3	8.5		3.8	2.4		6.7		0.5
<0.16	9.4	9.0	17.1	11.1	13.7	12.7	7.6		4.2	2.9		12.1		0.5
Total	100.0	100.0	100.0	100.0	100.0	100.0	100.0		100.0	100.0		100.0		0.5

In the selected historic mortar samples, XRD showed carbonates as the main crystalline phases formed by the carbonation of lime. In dolomitic mortars two forms of magnesium hydrogen carbonates were also detected – hydromagnesite ($4MgCO_3{\cdot}Mg(OH)_2{\cdot}4H_2O$) and nesquehonite ($MgCO_3{\cdot}3H_2O$) (figure 2). Calcium silicate hydrates (CSH) and calcium aluminate hydrates (CAH), which are among the main hydrated phases formed at room temperature after pozzolanic reaction, were not observable by means of XRD, which could be due to their decomposition caused by aging of the mortar as well as low crystallinity [14].

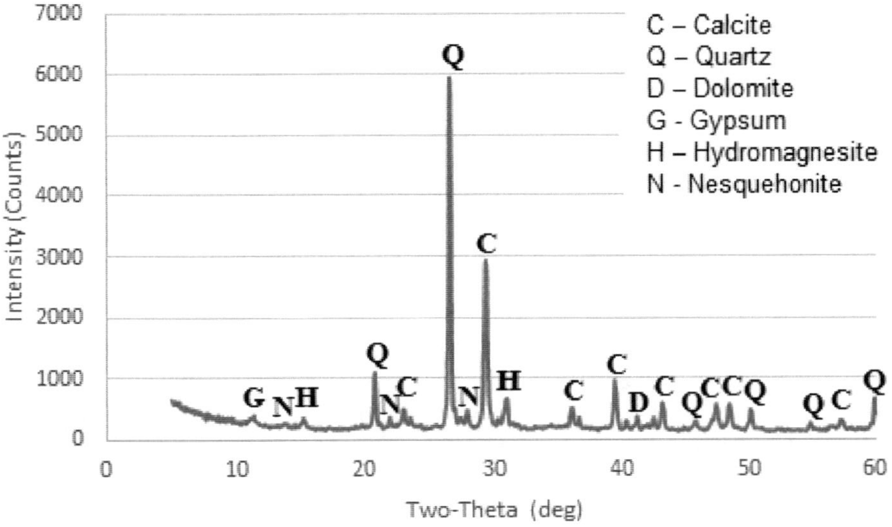

Figure 2. Example of XRD pattern of historic dolomitic mortar (sample N° 4).

Thermal analysis (DTA/TG) showed common basic thermo effects in the selected samples attributed to quartz inversion (573°C) and decomposition of carbonates (in figure 3 - calcite) (650-900°C). Dolomitic lime-based mortars show endothermic reaction at 350-450°C temperature attributed to dissociation of $Mg(OH)_2$ (figure 3). At the temperature range of 480-620°C the dissociation of uncarbonated portlandite ($Ca(OH)_2$) takes place [15]. Broad peaks indicate weight loss of the absorbed water in the temperature range from 60 to 180 °C and of the water bound to hydraulic components in the temperature range from 200 to 600 °C [16].

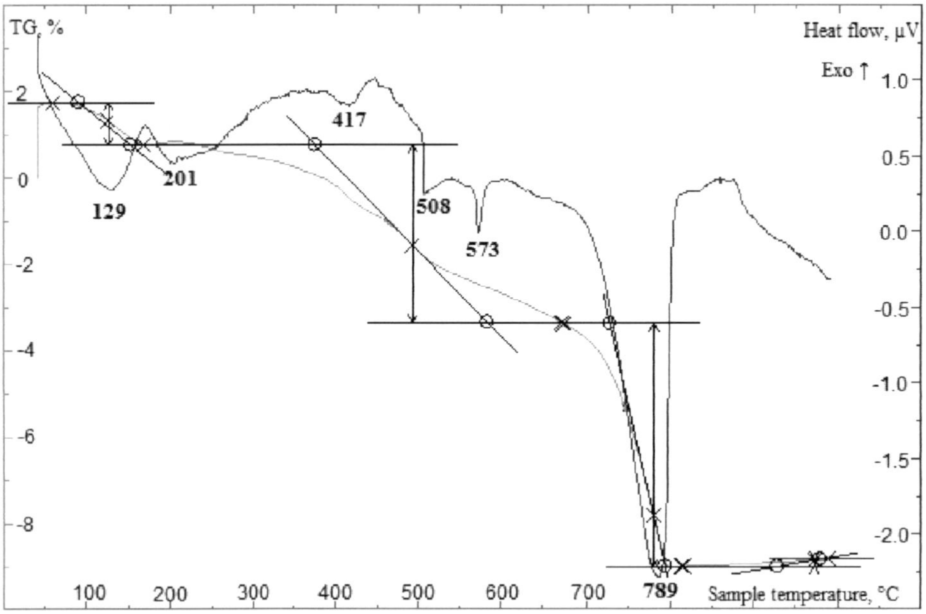

Figure 3. Example of DTA/TG curve of historic dolomitic mortar (sample N° 7).

The investigation of the change of the modulus of hydraulicity (m) of cement-lime mixtures approved that hydraulicity is contributed to the amount of added cement, allowing choosing the best matching composition for each object. In table 5, it can be seen that the modulus of hydraulicity of the lime binder without cement additive is very high (m=9.9) meaning low hydraulicity, but if cement content reaches 10 %, modulus decreases to 2.0. Beginning with 4 % cement additive, binder can be considered as strongly hydraulic. XRD data approve this coherence (figure 4) – the XRD patterns of the synthesized lime-cement mixtures after 2 months hardening show presence of cement mineral (calcium silicate – larnite ($2CaO{\cdot}SiO_2$)), if the content of cement in the binder is 4 % or higher. These minerals (calcium silicates and calcium aluminates) are responsible for hydraulic activity, forming hydrates while hardening. Hydrates of cement minerals – calcium silicate hydrates (CSH) and calcium aluminate hydartes (CAH) – are mostly amorphous [14] and for the determination of the presence of these phases such methods as DTA and SEM-EDS can be used.

The amount of soluble Al_2O_3 and SiO_2 (indicative of hydraulic components) is increasing with the content of cement in the binder (table 6).

Table 5. The results of chemical analysis of aerial lime-white cement mixtures, weight %.

Content of cement in mixture Component	0	2	4	8	10	Accuracy, [± absol. %]
Loss of ignition at 400°C	0.93	1.00	1.30	1.09	1.80	0.20
Loss of ignition at 1000°C	13.10	13.07	12.83	12.63	11.53	0.20
Insoluble residue in HCl (sand and gravel)	62.42	61.40	57.29	57.76	56.61	0.30
CaO	20.56	21.45	21.12	20.14	20.03	0.50
MgO	0.11	0.00	0.00	0.00	0.00	0.30
Soluble SiO_2	1.58	2.07	4.96	4.11	4.28	0.50
Soluble Al_2O_3	0.11	0.30	1.15	4.48	5.27	0.30
Fe_2O_3	0.39	0.41	0.46	0.47	0.53	0.10
Total	99.20	99.70	99.11	99.78	100.05	-
Binder to filler ratio	1:3.3	1:3.1	1:2.5	1:2.5	1:2.3	-
m (modulus of hydraulicity)	9.9	7.7	3.2	2.2	2.0	-
Type of binder	Weakly hydraulic binder	Weakly hydraulic binder	Strongly hydraulic binder	Strongly hydraulic binder	Strongly hydraulic binder	-

Table 6. Calculated composition of the binder according to chemical analysis, weight %.

Content of cement in mixture Component	0	2	4	8	10	Accuracy, [± absol. %]
CaO	87.5	86.8	78.1	69.7	66.5	0.8
MgO	0.5	0.0	0.0	0.0	0.0	0.0
Soluble SiO_2	6.7	8.4	15.3	15.1	16.2	0.8
Soluble Al_2O_3	0.5	1.2	4.0	14.2	16.5	0.8
Fe_2O_3	1.6	1.6	1.6	1.5	1.6	0.5

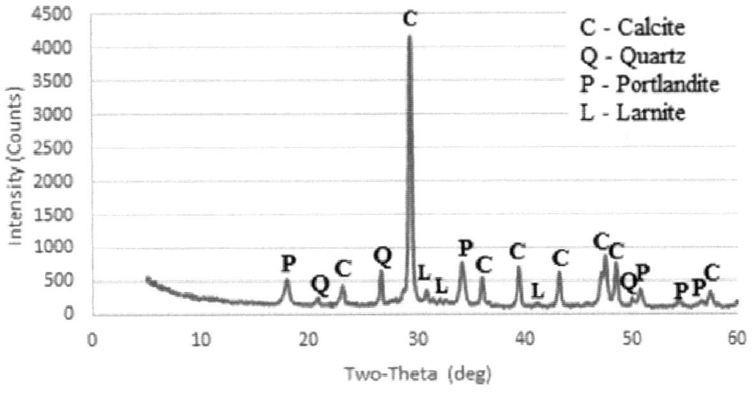

Figure 4. XRD pattern of aerial lime-white cement mixture with 4 % cement content.

Thermal analysis allows detecting hydraulic components (hydrated cement minerals), which are often hard to detect by XRD because of their poorly crystalline structure. Two endothermic peaks in the region 100-125 °C and 150–310 °C illustrate the loss of absorbed water and endothermic dehydration of pozzolanic hydrated products, calcium silicate hydrates and calcium aluminate hydrates, respectively (figure 5) [14]. As the cement-lime ratio increases in the mixtures, the mass loss attributed to total bound water of the pozzolanic reaction compounds also increases. Endothermic peak at 480°C can be attributed to dissociation of portlandite ($Ca(OH)_2$) and 650-900°C – to calcite ($CaCO_3$) [15]. Two endothermic minimums of calcite dissociation peak could occur due to 2 types of calcite origin in material: primary calcite – formed by the calcination of lime, and secondary – formed after the hydrolysis process of cement mineral alite.

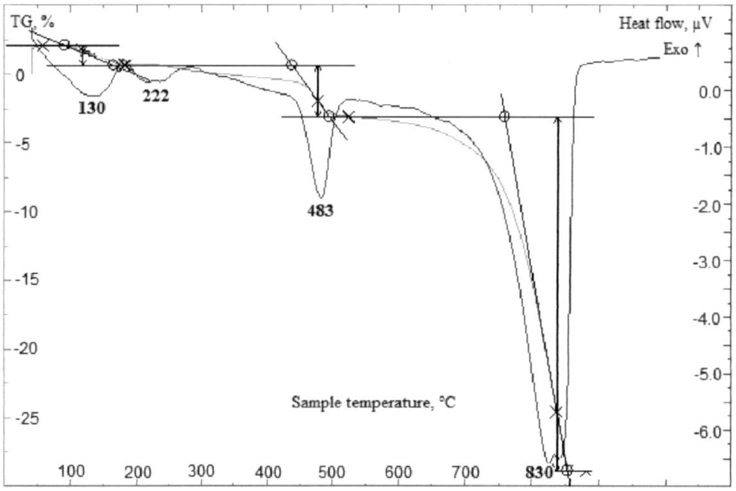

Figure 5. DTA curve of the mixture with 10 % of cement (after 2 months of hardening).

4. Conclusions

The choice of materials used for restoration purposes is based on the results of research on the historic materials. The basic aspects for the choice of the binder, filler and additives (if necessary) are the data obtained during visual observation, chemical and granulometric analysis, XRD and DTA/TG analysis.

Hydraulicity plays an important role in the durability of mortars. Most of the examined samples from historic buildings in Latvia are based on dolomitic or calcitic lime with hydraulic properties, achieved by pozzolan additives or natural admixture of clay.

Lime binders with Portland cement additive can be used in restoration as an appropriate material. The cement-lime mortars described in this study presented the increase of hydraulicity, which is needed in the restoration of historic masonries.

This study helps finding adequate proportions of pozzolan additive (white cement) in the restoration binders, in order to achieve desirable hydraulic nature of the binder, at the same time avoiding undesired changes of the hardened properties (for example, excessive improvement of the mechanical strength), which might cause incompatibility with the original materials.

5. References

[1] Silva B A, Ferreira Pinto A P and Gomes A 2014 *Constr. Build. Mater.* **72** 208–218

[2] Van Hees R P J, Binda L, Papayianni I and Toumbakari E 2005 Damage analysis as a step towards compatible repair mortars *RILEM TC COM 167: Characterisation of Old Mortars with Respect to their Repair* ed Groot C, Ashall G, Hughes J (Paris: RILEM Publications S.A.R.L.) pp 107–152

[3] Pacheco-Torgal F, Faria J and Jalali S 2012 *Constr. Build. Mater.* **30** 488–494

[4] Starinieri V, Hughes D C and Wilk D 2013 *Constr. Build. Mater.* **44** 192–199

[5] Maravelaki-Kalaitzaki P, Bakolas A, Karatasios I, Kilikoglou V 2005 *Cem. Concr. Res.* **35** 1577–86

[6] Kirilovica I, Krage L, Vitina I and Igaune-Blumberga S 2014 *Key Eng. Mater.* **604** 102–105

[7] Andrejkovičová S, Velosa A L, Ferraz E and Rocha F 2014 *Constr. Build. Mater.* **65** 132–139

[8] Matias G, Faria P and Torres I 2014 *Constr. Build. Mater.* **73** 125–136

[9] Lanas J, Pérez Bernal J L, Bello B A, Alvarez Galindo J I 2004 *Cem. Concr. Res.* **34** 2191–01

[10] LVS EN 196-2:2013 *Method of Testing Cement - Part 2: Chemical Analysis of Cement*

[11] Teutonico J M 1988 *A Laboratory Manual for Architectural Conservators* (Rome: ICCROM)

[12] Avots M, Eiduks J, et al 1968 *General Silicate Technology* (Riga: Zvaigzne)

[13] Moropoulou A, Bakolas A and Bisbikou K 1995 *Thermochim. Acta* **269-270** 779–795

[14] Arizzi A and Cultrone G 2012 *Constr. Build. Mater.* **31** 135–143

[15] Földvári M 2011 *Handbook of Thermogravimetric System of Mminerals and Its Use in Geological Practice* (Budapest: Geological Institute of Hungary)

[16] Ingo G M, Fragala I, Bultrini G, de Caro T, Riccucci C and Chiozzini G 2004 *Thermochim. Acta* **418** 53–60

IOP Conf. Series: Materials Science and Engineering **96** (2015) 012075 doi:10.1088/1757-899X/96/1/012075

Impact of Wetting/Oven-Drying Cycles on the Mechanical and Physical Properties of Birch Plywood

M Sooru[1], **K Kasepuu**[1], **R Kask**[1] **and H Lille**[1,2]
[1]Estonian University of Life Sciences, Institute of Forestry and Rural Engineering, Fr. R. Kreutzwaldi 5, 51014 Tartu, Estonia

E-mail: harri.lille@emu.ee

Abstract. The objective of this study was to explore some physical and mechanical properties and the dimensional stability of birch (Betula sp.) nine-ply veneers glued with phenol-formaldehyde (PF) after 10 cycles of soaking/oven-drying. The properties to be determined were bending strength (BS), modulus of elasticity in bending (MOE), Janka hardness (JH) and thickness swelling (TS), which were tested according to the European Standards (EN). An analytical equation was used for approximation of the change in the physical and mechanical properties of the samples depending on the number of cycles. It was shown that the values of the studied properties were affected most by the first soaking and drying cycles after which BS and MOE decreased continuously while the values of JH and TS stabilized. After 10 cycles the final values of BS, MOE, JH and TS accounted for 75-81%, 95%, 82% and 98.5% of the initial values, respectively.

1. Introduction

Plywood is an original engineered wood product made by bonding several softwood or hardwood plies. Plywood is a multi-used structural material in the manufacture of conventional construction (engineered flooring) and industrial products (even high-risk gasholder). The important factors that affect the physical and mechanical properties of plywood are tree species, density of wood, number of plies, type of adhesive, thickness of veneers and technological parameters of manufacturing [1,2,3,4].

Various test methods were applied in accordance with the European Standards (EN); bending strength (BS) and modulus of elasticity in bending (MOE) were determined according to the face grain orientation of the first plies; Janka hardness (JH) and thickness swelling (TS) were determined as well. Changes in BS, MOE, JH, TS at different soaking/oven-drying cycles were investigated in a Master's thesis of Sooru, Kasepuu [5, 6].

[2] To whom any correspondence should be addressed.

The basic method consists in soaking specimens in water at room temperature for 24 h, drying 48 h in a ventilated drying box at 65±1 °C for a certain number of cycles and testing (immediately after pre-treatment with cold water and after drying) using the computer-controlled mechanically actuated universal test machine INSTRON 3369. After these procedures the output of data and the analysis of the obtained results take place. The sensitivity of the measured data is studied and the expanded uncertainties of the computed mean values are presented.

An analytical expression was used to approximate the experimental data for the investigated BS, MOE and JH depending on the number of soaking/oven-drying cycles.

2. Experimental procedure and methods

Three nine-ply plywood panels, birch (*Betula sp*) veneers, of 3000×1500×12 mm^3, made by UPM-Kymmene Otepää, Estonia, were used. Engineered hardwood plywood panels are commonly glued with phenol - formaldehyde resin (PF) as the wood adhesive.

Test samples for MOE and MOR were cut from each of the panels in the parallel (major axis) and perpendicular (minor axis) directions to the face grain (parallel and perpendicular samples) [7]. Experiments were conducted in 11 series (minimum number of samples in a series being twelve). The samples were tested using the computer-controlled mechanically actuated universal testing machine Instron 3369. Deflection for calculating the modulus of elasticity was measured by an optical gauge (Advanced Video Extensometer 2663-821).

MOE and BS were found by three-point bending on the basis of dried samples and on the basis of samples pre-treated with cold water in accordance with the EN (figure 1) [8].

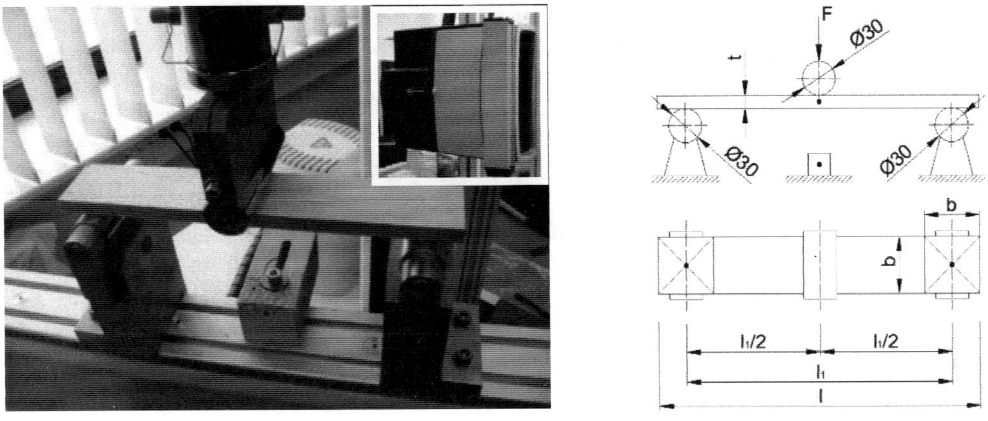

(a) (b)

Figure 1. (a) A photo of the three-point bending test and of the Advanced Video Extensometer 2663-821 in the upper right-hand corner; (b) schematic view of the three-point bending test; location of the testing area for determining static hardness and thickness swelling.

The BS was calculated by the following formula (see EN [8])

$$f_m = \frac{3F_{max} l_1}{2bt^2}, \tag{1}$$

where F_{max} is the maximum load, N; l_1 is the length between the supports, (240 mm); b is the width of the sample, (50±1 mm); t is the thickness of the sample, (12 mm).

MOE was determined by the following formula (see EN [8])

$$E_m = \frac{l_1^3 (F_2 - F_1)}{4bt^3 (a_2 - a_1)}, \tag{2}$$

where F_1 and F_2 are 10% and 40% of the maximum bending force, respectively; a_1, a_2 are deflections according to the loads F_1, F_2, respectively (see figure 2b).

Moisture content in the test samples after a soaking time of 24 h determined according to EN [9] was about 40%. The samples were dried (48h) to a moisture content of about 8% in a ventilated drying box at 65±1 °C. The JH was determined in the middle of the end area 50×50 mm^2 of the samples before the bending test in accordance with ISO Standard [10] (see figure 1b). All investigated properties were determined at both moisture content values.

Thickness swelling of the samples was determined according to EN [11] before the bending test at the cross section for which hardness was determined (see figure 1b).

Calculation of the uncertainty of the measurements was done in accordance with EN [7].

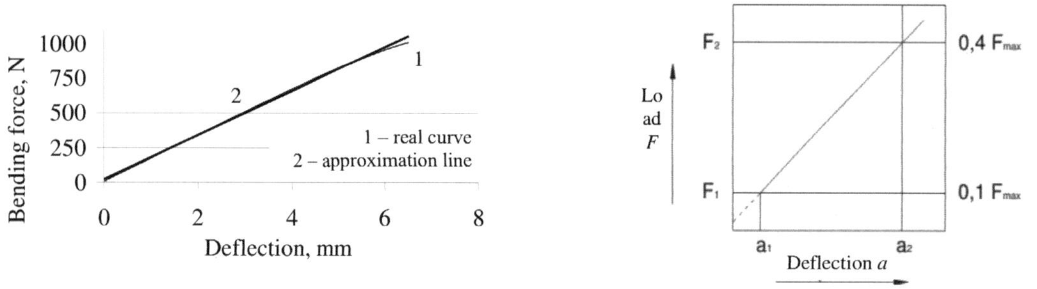

(a) (b)

Figure 2. (a) Dependence of deflection on bending force; (b) load-deflection curve within the range of elastic deformation [8].

The following linear-fractional function was used to approximate the obtained experimental data for the studied properties depending on the number of the soaking/oven-drying cycles [12]

$$Y(x) = \frac{d(Y_i - Y_f)}{cx + d} + Y_f, \tag{3}$$

where Y_i, Y_f are the calculated initial ($x = 0$) and final values of the studied properties, x is the number of cycles, and c and d are constants.

The formula allowed predicting to some degree the mechanical and physical properties of the specimens when their values after applying a small number of soaking/oven-drying cycles were known.

The initial and final values of the properties and constants should be determined so that the measured experimental data are approximated in the best way by minimizing the square of error (least squares regression). This problem was solved by using the program Mathcad 15.0 with the regression function *genfit(vx,vy,vg,F)*.

3. Results and discussion

The obtained values with the expanded uncertainty of BS, MOE, JH and TS are presented in figures 3a, 4a, 5a and 6a, respectively. All experimental data were approximated by formula 3. In figures 3b, 4b, 5b the mean values for one cycle are presented. The gradient of the studied properties was larger after the first soaking/oven-drying cycle. After the first soaking cycle, BS at moisture content 40% accounted for 50% and 63% of the values determined for the dry samples (8%) for the major axis and for the minor axis, respectively. The corresponding values for MOE were 57% and 61%, respectively. BS and MOE of the test samples decreased significantly after first soaking/oven-drying cycle, which is affected by the drying temperature and exposure time [13,14]. Also the first grain flaws appear in the surface of plywood. Thereafter BS and MOE decreased continuously with the increasing number of cycles, while JH and TS stabilized. After ten soaking/oven-drying cycles, the initial values of BS and MOE decreased significantly while the decrease was greater for MOE than for BS. The values of BS and MOE for the samples of major axis were clearly higher (ranking from 1.35 to 1.70) than those for the samples of minor axis. Evidently, this result is caused by the fact that the load direction of the load apparatus was perpendicular to the face grain direction of the samples of major axis, while it was parallel to the face grain direction of the samples of minor axis. In the face main grain flaws occurred after 2-3 soaking/oven-drying cycles. The flexural properties (MOE) of the samples of major axis were better than those of the samples of minor axis.

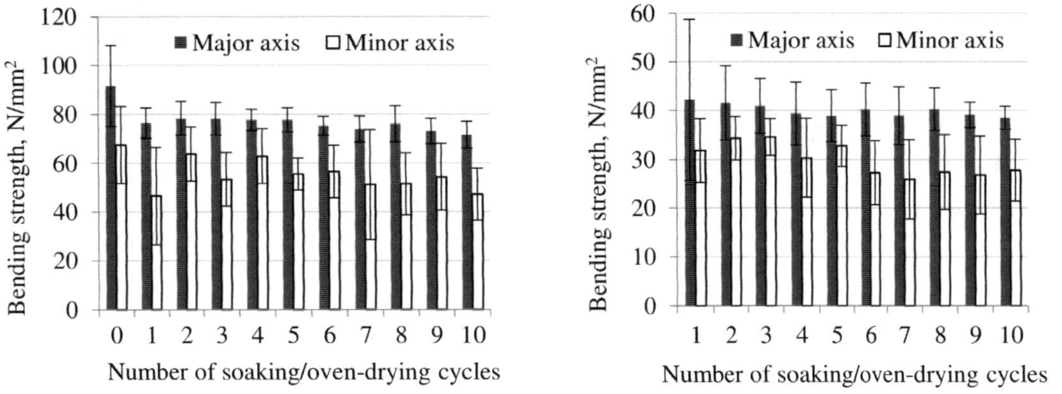

Figure 3. Dependence of BS on the number of soaking/oven-drying cycles: bar charts with expanded uncertainty.

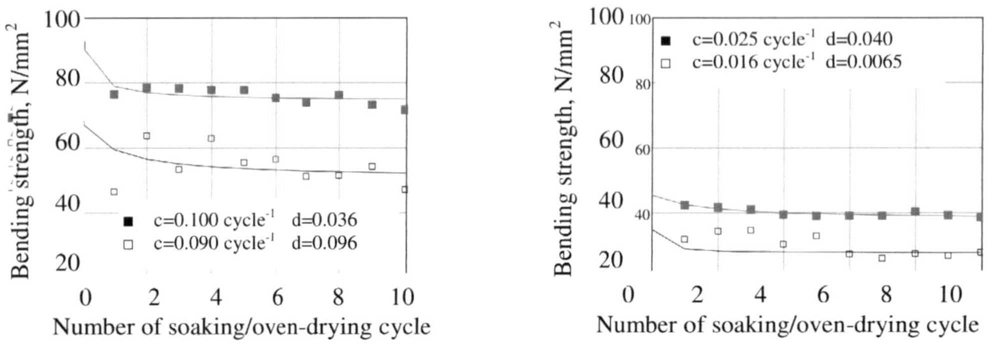

Figure 4. Dependence of BS on the number of soaking/oven-drying cycles: mean values of the experimental data, the curves of approximation, calculated initial and final values of BS at moisture content in test samples 8% - 40% were: $f_{m,i}$ = 91.6 N/mm^2 - $f_{m,i}$ = 45.4 N/mm^2 and $f_{m,f}$ = 74.3 N/mm^2 - $f_{m,f}$ = 37.8 N/mm^2; $f_{m,i}$ = 67.8 N/mm^2 - $f_{m,i}$ = 42.8 N/mm^2 and $f_{m,f}$ = 50.6 N/mm^2 - $f_{m,f}$ =19.6 N/mm^2, for the major axis and for the minor axis, respectively.

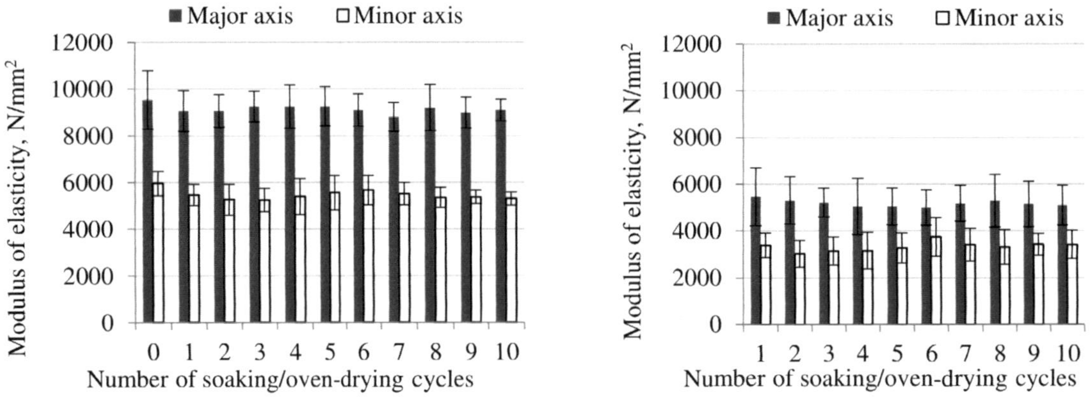

Figure 5. Dependence of MOE on the number of soaking/oven-drying cycles: a) bar charts with expanded uncertainty.

IOP Conf. Series: Materials Science and Engineering **96** (2015) 012075 doi:10.1088/1757-899X/96/1/012075

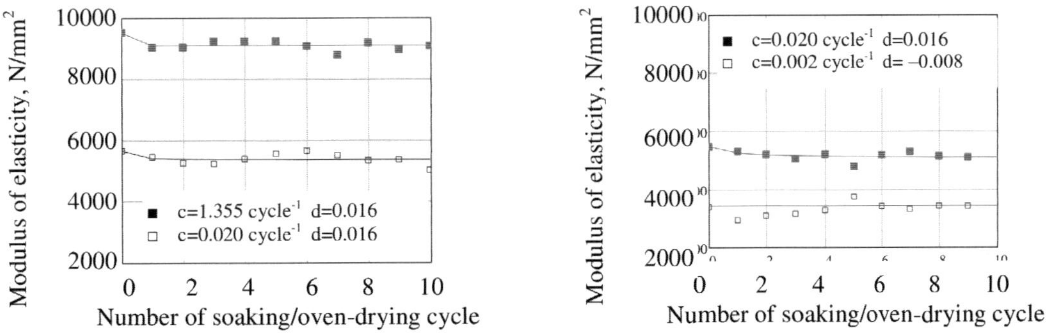

Figure 6. Dependence of MOE on the number of soaking/oven-drying cycles: a) bar chart with expanded uncertainty; b) mean values of the experimental data; the curves of approximation, calculated initial and final values of MOE at moisture content in test samples 8% - 40% were: $E_{m,i}$=9539 N/mm^2 - $E_{m,f}$ = 9105 N/mm^2 and $E_{m,i}$ = 5662 N/mm^2 - $E_{m,f}$ = 5369 N/mm^2; $E_{m,i}$= 5479 N/mm^2 - $E_{m,f}$ = 5071 N/mm^2 and $E_{m,i}$ = 3475 N/mm^2 - $E_{m,f}$ = 3370 N/mm^2, for the major axis and for the minor axis, respectively.

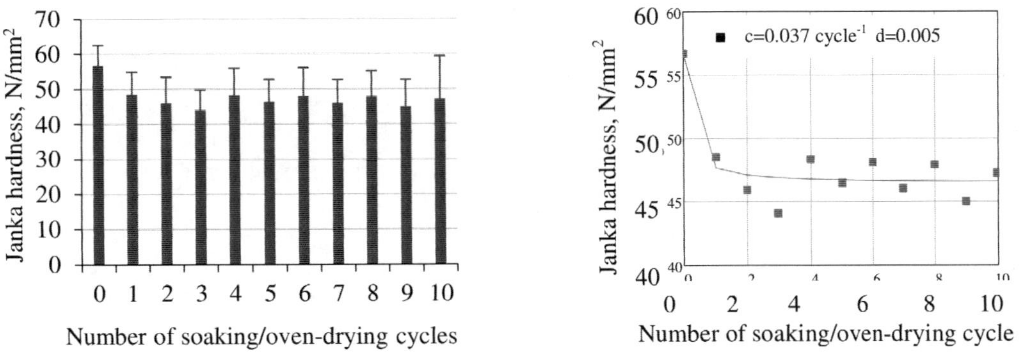

Figure 7. Dependence of JH on the number of soaking/oven-drying cycles: a) bar chart with expanded uncertainty; b) mean values of the experimental data; the curves of approximation, calculated initial and final values of JH at moisture content in samples were $H_{8c,i}$ = 56.7 N/mm^2 and $H_{8c,f}$ = 46.4 N/mm^2; c = 0.037; d = 0.005.

The experimental data of all properties fluctuated to a great extent and we can see that the proposed formula (3) approximated the experimental data satisfactorily.

The TS of the board did not change significantly after two soaking/oven-drying cycles, i.e. 8%. This has been observed in the case of wood. Swelling in wood takes place below the saturation point of fibre (about 30%) at which the total amount of water is present within cell wall [15] and, regardless of the number of soaking/oven-drying cycles, the dimensions do not change. Decrease of the dimensions of the test samples may be caused by the weight loss already after first soaking/oven-drying cycle [13]. This kind of a phenomenon was observed to a certain extent in our experiments in the case of BS, MOE, JH and TS.

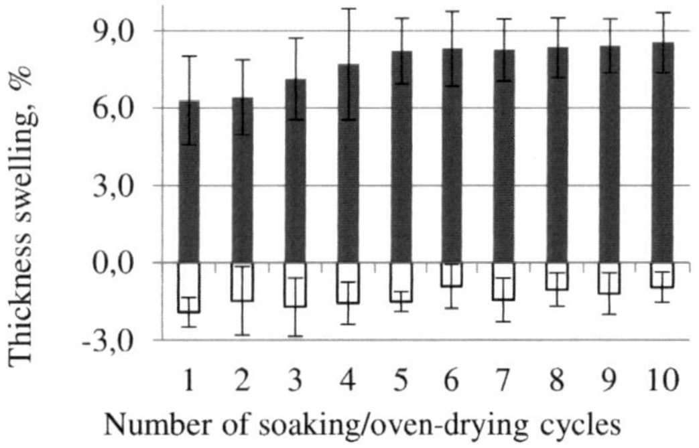

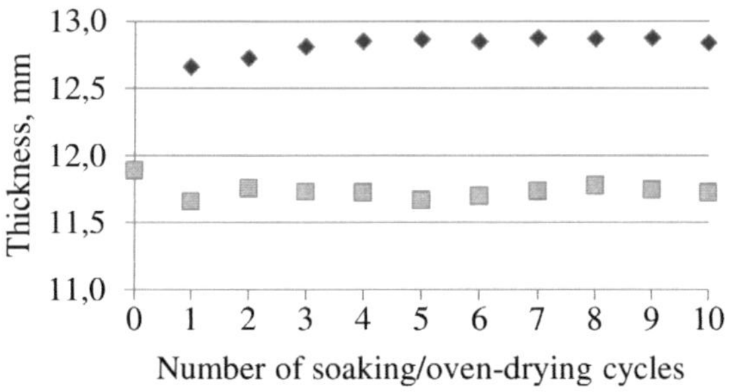

Figure 8. Dependence of TS on the number of soaking/oven-drying cycles: a) bar chart (light column – moisture content 8%, dark column – moisture content 40%), b) mean values of thickness.

4. Conclusions

The final mean values determined at a moisture content of 8% after ten soaking/oven-drying cycles were 95% for MOE (for both axes); 81% for BS for the major axis and 75% for the minor axis; 82% for JH and 98.5% for TS of the respective initial values.

After the first soaking (moisture content 40%), the values of BS accounted for 50% for the samples of the major axis and 63% for the samples of the minor axis; for MOE, 57% for the samples of the major axis and 61% for the samples of the minor axis and for TS, 106% of the values obtained for the dry samples (moisture content 8%).

The proposed analytical function approximated satisfactorily the experimental data of BS, MOE and JH, depending on the number of soaking/oven-drying cycles.

The presented analysis is restricted to the data obtained from the above-described experiments.

5. References

[1] Federation F F 2002 *Handbook of Finnish Plywood* (Kirjapaino Markprint Oy, Lahti, Finnland)

[2] Cosereanu C, Lica D, Curtu I, Lunguleasa A, Cismaru I, Brenci L and Fotin A 2010 *Proc. 7th Int. DAAAM Baltic Conf. Industrial Engineering* ed R Kyttner (Estonia /Tallinn: Tallinn University of Technology Press) p 417

[3] Lipinskis I and Spulle U 2011 *Drewno. Pr. Nauk. Donies. Komunik.* **185** 109

[4] Dieste A, Krause A, Bollmus S and Militz H 2008 *Holz Roh Werkst.* **66** 281

[5] Kasepuu K 2014 Effect of playwood soaking and drying cycles on mechanical and physical properties, Master Thesis, Estonian University of Life Sciences, Estonia/Tartu

[6] Sooru M 2015 Variation in mechanical and physical properties of birch plywood after soaking/oven-drying cycles, Master Thesis, Estonian University of Life Sciences, Estonia/Tartu

[7] EVS-EN 326-1:2002 *Wood-based panels. Sampling, cutting and inspection Part 1: Sampling and cutting of test pieces and expression of results* (Tallinn: Eesti Standardikeskus) p 14

[8] European Standard EN 310 1993b *Wood-based panels - Determination of modulus of elasticity in bending and of bending strength* (Brussels: European Committee for Standardization) p 11

[9] European Standard EN 322:1993 *Wood-based panels – Determination of moisture content* (Brussels: Central Secretariat) p 4

[10] ISO 3350:1975 *Wood-Determination of static hardness* (Switzerland: International Organization for Standardization) p 2
[11] EVS-EN 317:2000 *Particleboards and fibreboards - Determination of swelling in thickness after immersion in water* (Tallinn, Standardikeskus) p 12
[12] Lille H, Kõo J, Ryabchikov A, Reitsnik R, Veinthal R, Mikli V and Sergejev F 2013 *Key Engin. Mat. Engin. Mat. and Tribology* **527** 125
[13] Sinha A, Nairn A J and Gupta R 2010 *J. Wood Sci. Technol.* doi: 10.1007/s00226-010-0329-3
[14] Ayrilmis N, Buyuksari U and As N 2010 *J. Cold Regions Sci. Technol.* **63** 40
[15] Hiziroglu S Dimensional Changes in Wood. Oklahoma Cooperative Extension Service. [Available]: http://pods.dasnr.okstate.edu/docushare/dsweb/Get/Rendition-3624

Thermal Transmittance and the Embodied Energy of Timber Frame Lightweight Walls Insulated with Straw and Reed

M Miljan[1,2] **and J Miljan**[1]
[1]Estonian University of Life Sciences, Institute of Forestry and Rural Engineering, Department of Rural Building, Tartu, Kreutzwaldi 5, Estonia

E-mail: matis.miljan@emu.ee

Abstract. Sustainable energy use has become topical in the whole world. Energy gives us comfort we are used to. EU and national regulations determine energy efficiency of the buildings. This is one side of the problem – energy efficiency of houses during exploitation. But the other side is primary energy content of used materials and more rational use of resources during the whole life cycle of a building. The latter value constitutes about 8 – 20% from the whole energy content. Calculati[3]ons of energy efficiency of materials lead us to energy efficiency of insulation materials and to comparison of natural and industrial materials taking into account their thermal conductivity as well as their primary energy content.
Case study of the test house (built in 2012) insulated with straw bales gave the result that thermal transmittance of investigated straw bale walls was according to the minimum energy efficiency requirements set in Estonia U = 0.12 – 0.22 W/m^{2}K (for walls).

1. Introduction

World's demand for energy is increasing and greenhouse gas (GHG) emissions are rising. A big part of these processes is caused by building sector. In Europe 40% of the produced energy is consumed by building industry and 36% of GHG is produced by buildings [1]. In the USA homeowners spend an estimated 29% of their total energy-related expenditures on space heating alone, whereas commercial buildings spend more than 15% percent annually [2]. Berge had investigated how the embodied energy input of the building materials, usually about 85-95%, is distributed: a) direct energy consumption in extraction of raw materials and the production, b) secondary energy consumption invested in machinery, heating and lighting of the factory and the maintenance of the working environment, c) energy in transport of the necessary raw and basic materials depending on distance,

[2] To whom any correspondence should be addressed.

type of used fuel and method of transport [6]. Renewed data for calculating embodied energy of building materials is available in the database of the University of Bath [3, 5].

Using these data the calculations were done to find out the embodied energy of 1 m^2 of the five mostly used lightweight wall structures in Estonia for small residential houses. The U value of these walls was calculated as equal.

- Rendered lightweight block wall with 150 mm EPS insulation – 248 kWh/m^2;
- Timber cladding on lightweight block wall with 100 mm with mineral wool insulation – 223 kWh/m^2;
- Rendered 375 mm porous concrete block wall – 186 kWh/m^2;
- Hollow concrete block wall with stone cladding and with 100 mm mineral wool insulation – 173 kWh/m^2;
- Timber frame wall with timber cladding and with 100 mm + 50 mm mineral wool insulation – 139 kWh/m^2.

All wall structures in this case were done using industrial building materials [7]. The most energy effective was timber frame wall with timber cladding and mineral wool insulation energy content of which was 139 kWh per wall square meter.

The next step was to find out if any of the natural insulation materials is worth of use. Four different wall structure fragments were built into the window openings of the laboratory in the autumn of 2009 and measurements of temperature, RH and heat flow were done during winter. Results show that thermal transmittance of the wall fragments insulated with horizontally placed compressed reed, rendered with 50 mm of clay plaster inside and outside was U = 0.145 W/m^2K. Wall fragment built of straw bales and rendered with 50 mm of clay plaster from both sides has U = 0.182 W/m^2K and wall fragment built of reed bales and rendered with 50 mm of clay plaster from both sides has U = 0.148 W/m^2K [6, 8]. As reed turned out to be such a good insulation material, a small test house was built in 2010 to measure reed insulation properties more precisely and at the same time to determine the time and material consumption to calculate energy efficiency of the building. Walls of the test house had different structures: reed was placed loose and compressed in one wall horizontally and in one wall vertically. The third wall was built of handmade reed panels and the fourth wall was done of reed bales. Walls were 450 mm thick and rendered with 50 mm clay plaster from both sides. The measured thermal transmittance of walls was U = 0.207 – 0.383. The most heatproof wall was done of horizontally placed compressed reed and the worst was vertically compressed reed wall. [9] Later the calculations were done to determine the embodied energy of 1 m^2 of timber frame wall, insulated with reed (700 mm) and rendered with clay plaster, with U = 0.1 W/m^2K.

As the straw is the waste of agricultural activity, the next idea was to build a small house with straw bale walls to reduce the embodied energy content of insulation material. Calculated U = 0.092 W/m^2K and the respective thickness of the wall was 900 mm. The aim of this work was to determine the real thermal transmittance of the wall structure and to find the embodied energy content as well.

2. Methods

The office insulated with straw bales was built specially to carry out investigations.

2.1. Structure of the house with straw bale walls

The bearing structure of the house was done from straw bales, which were anchored together with metal bars pressed from upside down. Used bars were with 25 mm diameter, 2 m long and cc 1200 mm. The dimensions of wheat straw bales were: h = 1200 mm, w = 900 mm, l = 2200 mm, weight of a ball was 500 kg (figures 1, 2, 3). Straw bales were insulated from foundation using EPS plates (25 mm). On the top of the walls OSB plates were placed and on this layer timber beams (100 x 150 mm) were fixed for building the roof.

Figure 1. View of the straw bale office.

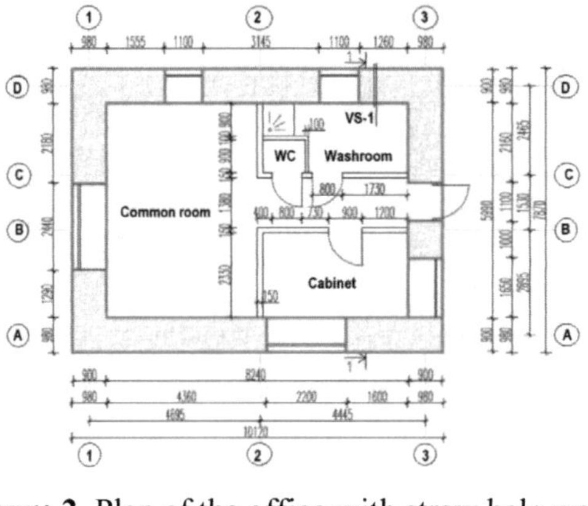

Figure 2. Plan of the office with straw bale walls.

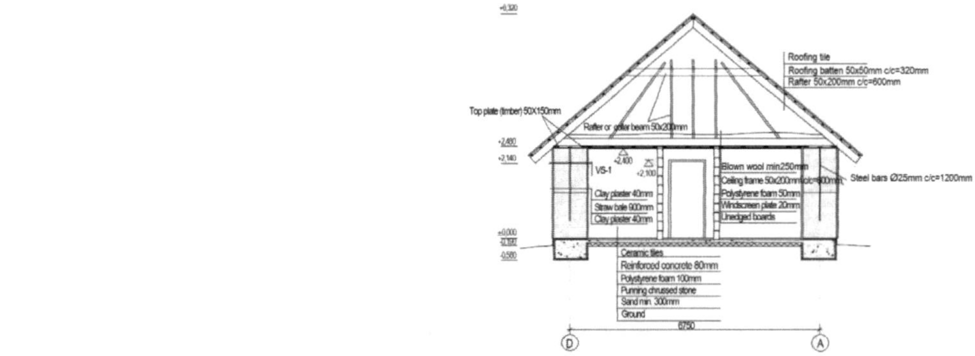

Figure 3. Cross-section of the office with straw bale walls.

To calculate the embodied energy content of one m^2 of the straw bale wall the amount of every used material was taken into account. To find the energy content of clay plaster the case study was done in the factory of Saviukumaja [4]. The following values were calculated separately: 1) mining of clay and sand, 2) transportation of sand and clay to the factory, 3) energy (electricity, fuels) consumption during preparation work: crushing, sieving, mixing renders, 4) energy consumption for sieving out non usable sand (stones 23%) (figure 5). Amount of the human work wasn't taken into account. Embodied energy of other used materials (including straw bales) was taken from the database of the University of Bath [3].

2.2. Measurements and placement of measuring devices
Measurements of thermal transmittance were done in non-steady environment beginning from February 2012 until February 2013. Data were recorded with 30 minutes intervals in Almemo and HOBO data recorders. Temperature and relative humidity were measured with several pairs of thermocouples, placed as shown in figure 4. Heat flow through wall was measured with heat flux plate.

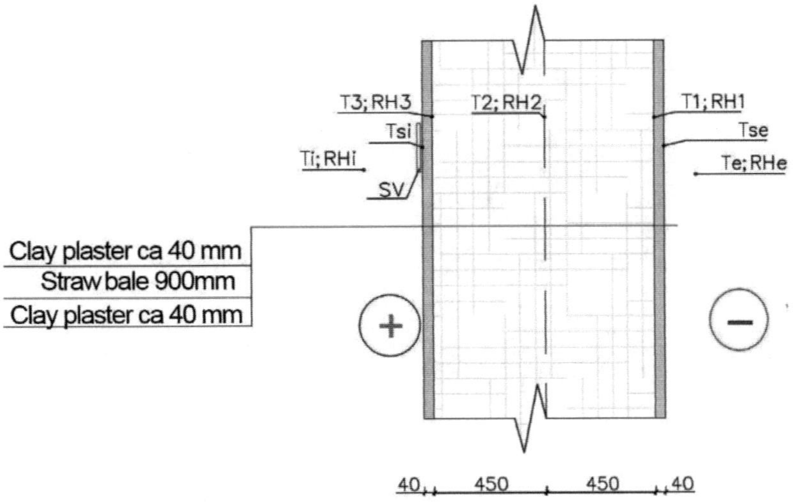

Figure 4. Placement of measuring devices (MD) on the cross-section of the wall.

2.3. Measured data and used equations
Next data were measured to achieve the set up aims.
Measured data:

T_e - outside temperature °C, MD: Hobo U12-006 with TMC6-HD
T_i - inside temperature °C, MD: Hobo U12-011
T_{si} - temperature on the wall outside °C, MD: FQA 019 C, datarec. Almemo 2890-9
T_1, T_2, T_3 - temperatures in different layers of the wall °C, MD: FHA 646 R
T_{se} - temperature on the wall inside °C, MD: Hobo U12-006 with TMC1-HD
RH_i - relative humidity in the room %, MD: Hobo U12-011
RH_e - relative humidity outside %, MD: Hobo U12-011

IOP Conf. Series: Materials Science and Engineering **96** (2015) 012076 doi:10.1088/1757-899X/96/1/012076

RH_1, RH_2, RH_3 - relative humidity in different layers of the wall %, MD: FHA 646 R
Heat flow (SV) - W/m^{2}K, MD: FQA 019 C, datarec. Almemo 2890-9

The minimum temperature during this year was -29 °C and maximum 34 °C. The mean room temperature was 21 °C and RH = 40.4%. The mean outside temperature was -5 °C and RH = 90.3%.

2.3. Results

During the year of investigation, the measured mean thermal transmittance of the straw bale wall was U = 0.125 W/m^{2}K which is in accordance with the requirements for the building walls stated in Estonia (0.12–0.22 W/(m^2·K) (for walls) [11]. The measured thermal transmittance differed from the calculated value which was U = 0.092 W/m^{2}K. One of the reasons might be the clay plaster, because the soft clay penetrates into the straw and therefore decreases the effective thermal insulation layer and the second reason is that the density of straw bales differs quite a lot. Calculations of the embodied energy of the straw bale wall are presented in table 1.

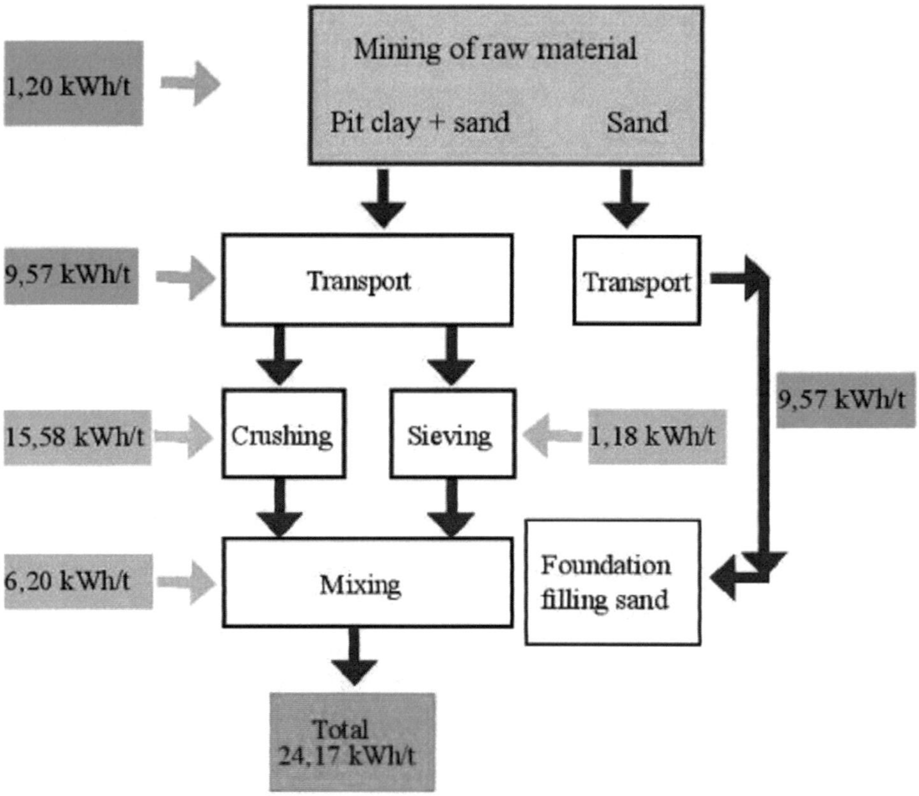

Figure 5. Embodied energy content of clay plaster (case study in the factory of Saviukumaja).

Table 1. Embodied energy content of the straw bale wall (case study at Tammistu).

Material	Energy content		Amount	Total
	MJ/kg	kWh/kg	kg	kWh
EPS	88.60[3]	24.61	16	394
Straw bales	0.24[3]	0.066	12600	840
Clay plaster (1:3)		0.09[2]	8000	193
Sand sieved off		0.04[2]	1380	16
Steel bars	8.80[3]	2.44	124	303
Particle board	9.50[3]	2.64	233	615
Timber	7.80[3]	2.17	535	1161
			Total	3522

From table 1 we can see that timber and particle boards are the main contributors to the total energy consumption. Embodied energy consumption of the straw bale wall was 55 kWh/m^2. [10]
For comparison the calculation of timber frame wall was done. The results are presented in table 2.

Table 2. Embodied energy content of the timber frame outer wall.

Material	Energy content MJ/kg	Amount kg	Total MJ	Total kWh
Timber cladding	7.80	905	7059	1976
Mineral wool 150 mm	16.6	270	4482	1255
Mineral wool 100+100	16.6	360	5976	1673
Wind proof plate	7.40	300	2220	621
Steam proof film	83.10	20	1662	465
Gypsum plate	6.75	1000	6750	1890
Frame timber	7.80	1461	11396	3190
		Total	39545	11070

Embodied energy consumption of the timber frame wall was 173 kWh/m^2. [10] Thermal transmittance of both walls was the same.

3. Conclusions

The results of presented research show that for evaluating the energy efficiency of a building it is not enough to assess only thermal transmittance of enclosures. One should take into account the embodied energy of the used materials. Calculations and measurements show that using natural insulation materials instead of industrial ones we may save about three times, as the embodied energy content of the compared walls was 55 kWh/m^2 against 173 kWh/m^2. These results are in the same order of magnitude as obtained in one of our previous investigations [5] where the embodied energy of reed insulated timber frame wall was 37 kWh/m^2 and that of timber frame wall insulated with mineral wool was 178 kWh/m^2. Minke [16] had also calculated embodied energy of different walls and the results are similar to ours. The straw bale wall with U = 0.11 W/m^{2}K has energy content 54 kWh/m^2 and embodied energy in timber framed wall insulated with rockwool was 175 kWh/m^2. Assuming that, we suggest that the energy consumption of used materials should also be taken into attention. On the basis of calculations done by measurements, there was no threat for condensate arising inside the wall [10].

4. References

[1] International Energy Agency 2013 FAQs: Renewable energy assessed in Jan 2015, Key World Energy Statistics accesssed in Jan 2015 [Available]: http://www.iea.org/aboutus/faqs/renweableenergy/
[2] U.S. Department of Energy 2011 *Buildings Energy Data Book* Expenditure data are from 2010
[3] Hammond G and Jones C 2011 ICE V2,0 [Available]: www.bath.ac.uk/mech-eng/sert/embodied
[4] Shvoerov A 2013 The impacts of using industrial and natural construction materials to embodied energy of building *Master Thesis of Rural Building of EULS* p 88
[5] Hammond G P and Jones C I 2012 Embodied energy and carbon in construction materials *Proc. of the Institution of Civil Engineers Energy* pp 87–98
[6] Berge B 2009 *The Ecology of Building Materials* second ed. Architectural Press p 427
[7] Miljan M and Miljan J 2013 Embodied energy in reed insulated structures *Compendium Reed and the Possibilities to use it EULS and TTU* pp 72–78
[8] Miljan M J and Miljan J 2013 Thermal transmittance of enclosures of the test house *Compendium Guidebook of Reed Business EULS* pp 38–40
[9] Miljan J and Kukka E 2009 Eestis enamkasutatud välispiirete primaarenergia sisaldus *Ehitaja J* **9** pp 72–75
[10] Miljan M J, Miljan M and Miljan J 2013 Thermal Conductivity of Walls Insulated with Natural Materials *Proc. of LUA Fac of Rural Engineering Civil Engineering'13 Part 1 vol.* 4 p 175
[11] Miljan M, Miljan M J, Miljan J, Akermann K and Karja K 2014 Thermal transmittance of reed-insulated walls in a purpose-built test house *J. in Internet Mires and Peat*
[12] Saar L 2013 A case study of hygdro-thermal properties of enclosures and indoor climate of the house built of straw bales *Master Thesis of Rural Building of EULS* p 82
[13] Minimum Requirements for Energy Performance Chapter 2 § 12 Requirements for the building envelope, State Gazette of Estonian Government 09.01.2015
[14] U.S. Department of Energy 2011 *Buildings Energy Data Book* Expenditure data are from 2010
[15] Minke G and Krick B 2009 *Handbuch Strohballenbau* p 139

Evaluation of Behaviours of Laminated Glass

L Sable[1,2] G Japins[1] and K Kalnins[1]
[1]Riga Technical University, Institute of Materials and Structures, Kalku Str. 1, Riga, LV-1658, Latvia

E-mail: lienesable@gmail.com

Abstract. Visual appearance of building facades and other load bearing structures, which now are part of modern architecture, is the reason why it is important to investigate in more detail the reliability of laminated glass for civil structures. Laminated glass in particular has become one of the trendy materials, for example Apple© stores have both load carrying capacity and transparent appearance. Glass has high mechanical strength and relatively medium density, however, the risk of sudden brittle failure like concrete or other ceramics determine relatively high conservatism in design practice of glass structures. This should be changed as consumer requirements evolve calling for a safe and reliable design methodology and corresponding building standards. A design methodology for glass and glass laminates should be urgently developed and included as a chapter in Eurocode. This paper presents initial experimental investigation of behaviour of simple glass sheets and laminated glass samples in 4-point bending test. The aim of the current research is to investigate laminated glass characteristic values and to verify the obtained experimental results with finite element method for glass and *EVA* material in line with future European Structural Design of Glass Components code.

1. Introduction

Transparency of glass and apparent lightness of design of glass structures are the reasons why in the 21st century glass is one of the most widely used building materials. However, brittle behaviour without plastic deformation is the negative characteristic that largely restricts application of glass as a material to be used in the load bearing structures.

Initially, glass as a material is obtained in the liquid form then cooled to the rigid state without crystallizing. Raw materials for glass production are 75% silicia (SIO_2), sodium oxide Na_2O from soda ash, lime CaO, and several minor additives. In the production process, all ingredients have been heated at the temperature higher than 1200^0C [1]. When the molten mass cools, glass gradually passes from being a liquid to a solid without – as it is normally the case with molten products – forming a

[2] To whom any correspondence should be addressed.

IOP Conf. Series: Materials Science and Engineering **96** (2015) 012077 doi:10.1088/1757-899X/96/1/012077

regular symmetrical or periodic crystal lattice. Glass is isotropic, i.e. its properties do not depend on direction or orientation [2].

The modulus of elasticity of glass is typically assumed to be 70 GPa, only about a third of that of steel but five times greater than hardwood [3]. Under loading glass shows linear stiffness behaviour where relative strain to the stress is fully linear up to brittle fracture with relatively small scatter between different glass treatment technologies. On the contrary, strength characteristics may depend on glass treatment type. For example, as a reference bending strength of annealed glass is $f_{ck,float} = 45\ N/mm^2$, heat strengthened glass – $f_{ck,heat\ strengthened\ glass} = 70\ N/mm^2$ and tempered glass – $f_{ck,thermally\ toughened\ glass} = 120\ N/mm^2$ [4/5].

Laminated glass is a glass unit that consists of at least two panes and one intermediate layer, whereby the panels are boned to the intermediate layer in the manufacturing process [6]. The materials used for the interlayers are *polyvinylbutyral* (PVB), *cast –in-place resin* (CIP), *ethylene vinyl acetate* (EVA) [3]. Full-surface bonding provides many opportunities to modify the mechanical and optical properties of glazing through the selection of the component layers, their sequence and thicknesses. Laminated safety glass remains as one piece when shattered and has an increased residual load-bearing capacity.

2. Materials and methods

2.1. Finite element method (FEM) – numerical solution

To determine the exact behaviour of glass at various loading stages and to verify the obtained test results, commercial finite element method engineering simulation software *ANSYS 11.00* was used. In the current research two different models were examined – single glass sheet and laminated glass model, which consisted of three sheets of glass and ethylene vinyl acetate (*EVA*) film between.

For the annealed single glass sheet model in *ANSYS* code the 4-node SHELL 181 element has been used. It is a four-noded element with six degrees of freedom at each node: translations in the *x*, *y*, and *z* directions, and rotations about the *x*, *y*, and *z* axes [6]. Mechanical parameters, which have been defined for the calculation, were density $\rho = 2500 kg/m^3$, Young Modulus $E = 70\ GPa$ and Poisson's ratio $\nu = 0.23$ [7].

For the laminated glass model in ANSYS code alternative element has been selected to incorporate fully nonlinear behaviour of polymer film, thus nonlinear layered structural 9-node *SHELL91* element has been selected. The mechanical properties of the tested glass specimens have been maintained, however, for (EVA) film stress and strain values obtained from test results and shown in figure 2 were used.

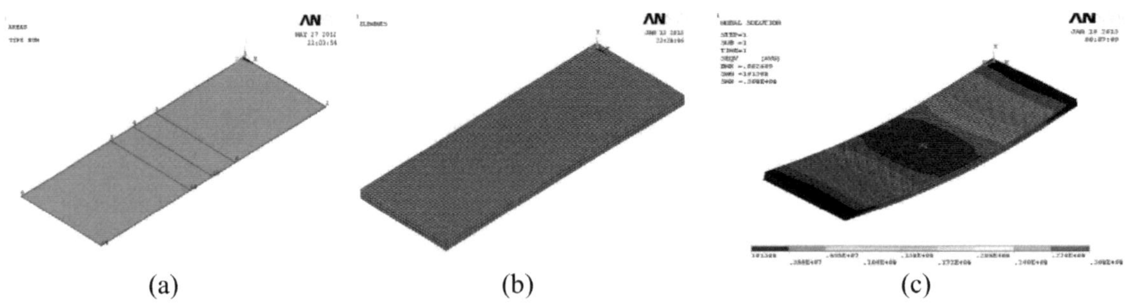

(a) (b) (c)

Figure 1. (a) Model geometry; (b) FE meshed structure; (c) deformed state.

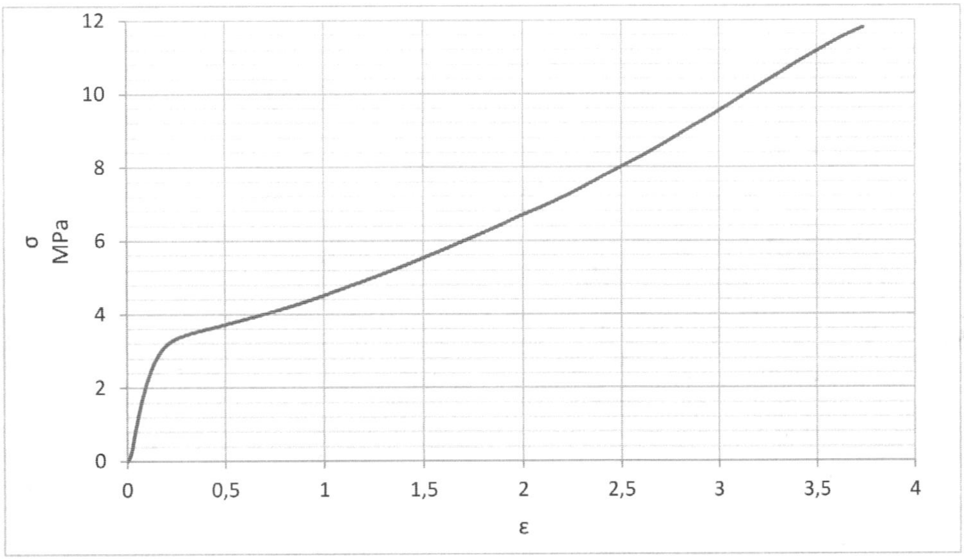

Figure 2. Stress strain curve of EVA polymer.

2.2. Analytical stress analysis

Bending stress can also be calculated with conventional analytical formulas, which formerly established values by testing. Single glass sheets stress calculation was made according to LVS EN 1288-3:2001 [8].

$$\sigma_{bB} = k\left[F_{\max}\frac{3(L_s - L_b)}{2bh^2} + \sigma_{bG}\right] \tag{1}$$

where b – width of the specimen; h – thickness of the specimen; L_s– distance between the center lines of the supporting rollers; L_b – distance between the centre lines of the bending rollers; y – central deflection of the specimen; k=k_e, – dimensionless factor as function of y/h (to determine the stress at the middle of the span k=1); σ_{bB} – bending strength; σ_{bG} – bending stress imposed by the self-weight of the specimen [8]. Bending stress imposed by the self-weight of the specimen is [8]:

$$\sigma_{bG} = \frac{3\rho g L_s^2}{4h}. \tag{2}$$

3. Materials and methods

3.1. Glass 4-point bending tests

Glass samples have been tested in 4-point bending according to LVS EN 1288-3:2001 [8] at Riga Technical University, Institute of Materials and Structures (*IMS*). For testing the *INSTRON 8802* testing machine (figure 3) was exploited. Universal test machine allows conducting tests in tension, compression, bending and fatigue, both statically and dynamically load up to the 250 kN capacity. Glass panel samples of size L_p= *1100 mm±5mm* long and b_p=*360mm±5mm* wide were cut from a single glass plate with a thickness of 8, 10, 12 mm (h_p). These were industrially cut on cutting machines and finished by grinding and polishing as required by industry standard for the best quality glass product. The spam length between supports has been assumed to be constant of 1000 mm and distance between the loading points set 300 mm, as shown in figure 3.

Specimens were divides in two groups for tests: simple annealed glass sheets with polishing edges ((glass thickness) mm_AP_(number)) and annealed laminated glass with two and three annealed glass sheets ((glass thickness)/EVA/(glass thickness)_number).

(a) (b)

Figure 3. (a) Single glass sheet sample test (b) laminated glass sample in bending test set up (INSTRON 8802).

4. Results and discussion

During the tests it was observed that the annealed glass comparing with the tempered glass has slightly lower bending strength, however, that does not mean that it cannot be exploited as a load bearing structural material. If annealed glass is part of a laminated glass structure then material properties are sufficient for load bearing strength requirements and acceptable by deflection limitations.

Annealed glass samples have been tested with polished edges, which basically eliminates any possibility of micro cracks. Bending strength of this type of glass samples resulted in load-bearing capacity increment of 1.5 times higher than the same sample without polished edges. As shown in figure 4, the sample collapsed directly beneath loading space and shards are with sharp edges.

(a)

(b)

Figure 4. Samples after testing: (a) sample in the testing machine, (b) cracks in glass.

Comparing test and numerical results with ANSYS software (table 1), one should note that specimens failed in deflection state in between 1/50 to 1/70 relative to the span length. On the other hand, the ultimate stress state observed during the tests ranged from 40 to 60 MPa. The test results correlate with linear analysis of the numerical procedure.

Table 1. Bending tests results.

Number of test	Failure load F_{max} [N]	Deflection u [mm]	Young's Modulus E [GPa]	Bending strength σ_{bB} [MPa]	Bending strength (ANSYS) σ_A [MPa]
8mm_AP_1	1188	20.3	74.5	56.43	55.9
8mm_AP_2	1137	19.1	74.5	54.11	53.3
10mm_AP_1	1 664	13.4	73.8	57.3	57.8
10mm_AP_2	1 497	13.3	73.8	51.7	52.1
10mm_AP_3	1 709	15.0	73.8	58.8	59.4
10mm_AP_4	2 031	17.9	73.8	66.5	70.5
10mm_AP_5	1 242	11.0	73.8	43.2	43.1
10mm_AP_6	1 609	14.1	73.8	55.5	55.9
12mm_AP_1	2935	14.1	72.6	60.9	61.4
12mm_AP_2	2904	14.0	72.6	60.3	60.7

The main aim of the laminated glass is to hold glass shards after initial brittle failure of outer plies. Permanent connection of two or more single pane glasses with sticky, elastic, highly tear-resistant foils (EVA) makes a multi-functional element from the glass, which can handle high static forces and constructive tasks in addition to its given transparency. The safety effect of the laminated safety glass is based on the extremely high tensile strength of the interlayer and its excellent adhesion to the adjacent glass surface [10]. In terms of mechanical stress such as shock, impact or influence from other forces breaking the glass, though, the fragments adhere to the interlayer so that the laminated safety glass will usually retain its stability under load [10].

The failure of a laminated glass sheet can be subdivided in five phases:

1. Elastic behavior of the glass plies/sheets.
2. The breakage of the first glass ply, however other plies are still intact; the interlayer is not damaged.
3. The breakage of the second glass ply; the interlayer stops behaving elastically.
4. The interlayer behaves plastically; the splinters are kept together by the interlayer.
5. The interlayer fails by reaching its failure strength or by cutting from the splinters [11].

Laminated glass test consisted of 4 series of samples (table 2). For finite element model, combination of two elements – shell and solid – was used. The same model was used by Sun et al. [12] for modelling the failure behaviour of windscreens. Solid element represents EVA interlayer, which use hyper elastic law.

Table 2. Laminated glass bending tests results.

Number of test	Thickness h_l [mm]	Failure load F_{max} [N]	Deflection u_l [mm]	Deflection (ANSYS) u_A [mm]
4/EVA/4- 1	8.42	1151	18.3	19.78
4/EVA/4- 2	8.39	1245	19.4	21.33
4/EVA/4- 3	8.26	1244	19.1	21.31
4/EVA/4- 4	8.31	843	13.1	14.71
4/EVA/4- 5	8.27	1446	22.7	24.64
4/EVA/4- 6	8.32	1206	19.1	20.69
4/EVA/4- 7	8.31	868	13.5	15.11
4/EVA/4- 8	8.41	979	15.3	16.95
5/EVA/5 -1	10.39	1706	14.2	14.95
5/EVA/5 -2	10.35	2263	18.6	19.65
5/EVA/5 -3	10.41	1726	14.4	15.11
5/EVA/5 -4	10.37	1380	11.4	12.19
5/EVA/5 -5	10.42	1914	15.8	16.70
5/EVA/5 -6	10.36	1735	14.4	15.19
5/EVA/5 -7	10.31	1373	11.3	12.13
5/EVA/5 -8	10.35	1361	11.1	12.02
6/EVA/6 -1	12.38	2375	11.7	11.99
6/EVA/6 -2	12.36	3259	16.2	16.32
6/EVA/6 -3	12.36	2775	13.3	13.95
6/EVA/6 -4	12.45	2034	9.9	10.32
6/EVA/6 -5	12.38	1401	6.8	7.22
6/EVA/6 -6	12.55	1622	8.0	8.30
6/EVA/6 -7	12.53	2438	11.9	12.31
4/EVA/4/EVA/4 -1	12.71	2076	10.4	9.98
4/EVA/4/EVA/4 -2	12.81	2426	12.1	11.61
4/EVA/4/EVA/4 -3	12.76	2730	13.7	13.02
4/EVA/4/EVA/4 -4	12.76	2200	11.2	10.55
4/EVA/4/EVA/4 -5	12.76	2473	12.6	11.83
4/EVA/4/EVA/4 -6	12.82	2706	13.7	12.91
4/EVA/4/EVA/4 -7	12.74	3235	16.6	15.37

The observed physical behaviour confirmed that glass laminate in 4-point bending test is perfectly linear (figure 5, 6). Figures 5 and 6 show that all laminate sample deformation levels up to 1/65 of the span numerically around 15 mm, however, it shoud be noted that ultimate load was varying. Two ply 4 mm glass laminate breaks at 1.1 kN where 5 mm at 1.9 kN and 6 mm at 2.2 kN hight load (figure 5). Three layer laminate has higher result and breaks at 2.5 kN (figure 6).

After testing and visual observation of the specimens, one can observed that for two layer laminate glass first breaks in the tension zone. However, for three ply laminate both two bottom layers break instantly and that means that for load bearing structures just two glass layers are designed.

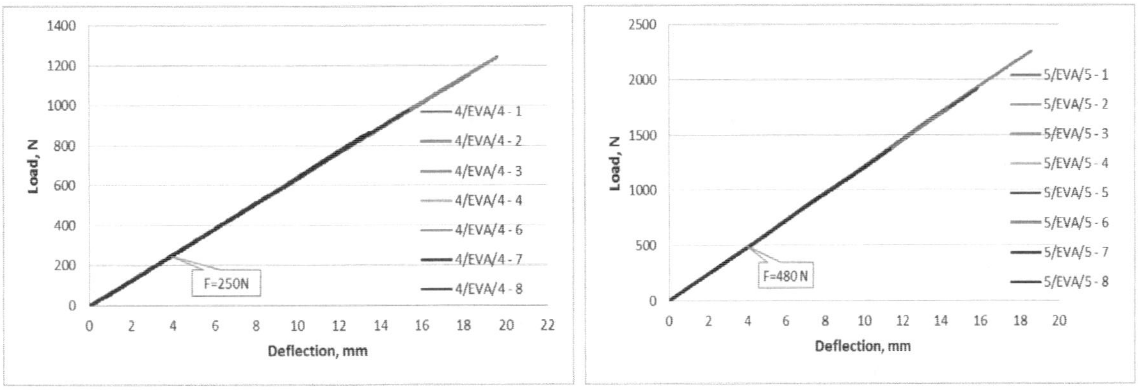

(a) (b)

Figure 5. Experimental data for two layer samples (a) *4/EVA/4* laminate, (b) *5/EVA/5* laminate.

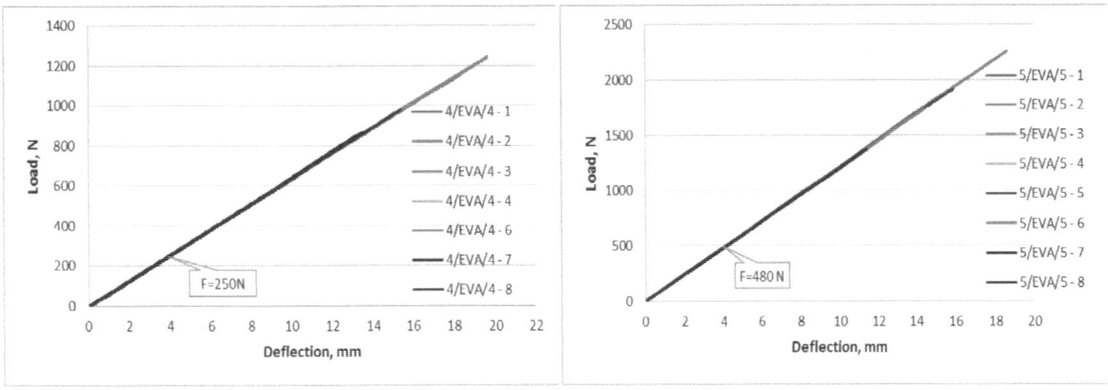

(a) (b)

Figure 6. Experimental data for two and three layer samples (a) *6/EVA/6* laminate, (b) *4/EVA/4/EVA/4* laminate.

5. Conclusions

In the present article it has been confirmed that the laminated glass has several advantages compared with the conventional glass sheets, for example, higher durability in bending and the interlayer insure that shards are kept together after the laminate collapse. A verification study confirmed that material properties implemented for nonlinear simulation of polymer films show good agreement with larger scale physical tests. One should note that different shell element types in commercial code ANSYS have been utilised for conventional glass sheet and laminate simulation. Test results confirmed that there is at least four times the safety margin if specimens are designed for bending according to deflection limit of 1/250 compared to the span length. More detailed evaluation is foreseen to estimate the reasons for relatively high strength value scatter.

6. References

[1] Button D et al 1993 *Glass in Buildings: A Guide to Modern Architectural Glass Performance* England p 372
[2] DIN EN 572 2004 *Glass in Building* (Berlin: Basiserzeugnisse aus Kalk-Natronglas)
[3] Wurm J 2007 *Glass Structures: Design and Construction of Self-Supporting Skins* Springer p 255
[4] EN 12150: *Glass in Building - Thermally Toughened Soda Lime Silicate Safety Glass*
[5] EN 1863: *Glass in Building - Heat Strengthened Soda Lime Silicate Glass Using Shell Elements in ANSYS Mechanical* ANSYS Technical Support [online][accessed on 10.05.2015.]
[6] Feldmann M, Kasper R et al 2014 *Guidance for European Structural Design of Glass Components* (Luxembourg: Publications Office of the European Union) p196
[7] LVS EN 1288-3:2001 *Glass in Building-Determination of the Bending Strength of Glass-Part3: Test with Specimen Supported at two Points (four point bending)* Brussels 2001
[8] Sable L and Kalnins K 2013 *Evaluation of Glass in Design of Load Bearing Structures* Riga Technical University
[9] *Glass Time* 2012 Technical manual GUARDIAN Europe S.a r.l., Dudelange, Luxembourg, September
[10] Larcher M, Solomos G, Casadei F and Gebbeken N 2011 Experimental and numerical investigations of laminated glass subjected to blast loading *Int. J. of Impact Engineering* **39** pp 42–50
[11] Sun D, Andrieux F, Ockewitz A, Klamser H and Hogenmüller J 2005 Modelling of the failure behaviour of windscreens and component tests *5th European LSDYNA Users Conf., 25e26 May, 2005*

IOP Conf. Series: Materials Science and Engineering **96** (2015) 012078 doi:10.1088/1757-899X/96/1/012078

Effects of Elevated Temperature on Concrete with Recycled Coarse Aggregates

M A Salau[1,2], O J Oseafiana[1] and T O Oyegoke[1]
[1]University of Lagos, Department of Civil and Environmental Engineering, Lagos State, Nigeria

E-mail: ajibsalau@yahoo.co.uk

Abstract. This paper discusses the effects of heating temperatures of 200°C, 400°C and 600°C each for 2 hours at a heating rate of 2.5°C/min on concrete with the content of Natural Coarse Aggregates (NCA) partially replaced with Recycled Coarse Aggregates (RCA), obtained from demolished building in the ratio of 0%, 15% and 30%.There was an initial drop in strength from 100°C to 200°C which is suspected to be due to the relatively weak interfacial bond between the RCA and the hardened paste within the concrete matrix;a gradual increase in strength continued from 200°C to 450°C and steady drop occurred again as it approached 600°C.With replacement proportion of 0%, 15% and 30% of NCA and exposure to peak temperature of 600°C, a relative concrete strength of 23.6MPa, 25.3MPa and 22.2MPa respectively can be achieved for 28 days curing age. Furthermore, RAC with 15% NCA replacement when exposed to optimum temperature of 450°C yielded high compressive strength comparable to that of control specimen (normal concrete). In addition, for all concrete samples only slight surface hairline cracks were noticed as the temperature approached 400°C. Thus, the RAC demonstrated behavior just like normal concrete and may be considered fit for structural use.

1. Introduction

Heavy Construction Industries such as manufacturing of foundries, furnaces, gas turbines are at the forefront of development in developing countries, like Nigeria, as they produce resources that are necessary for driving the economy. At the moment, only natural aggregates are used in construction for the needs of these industries, in spite of the fact that using recycled concrete aggregates reduces the need for virgin aggregates. Use of recycled aggregates reduces the environmental impact of the aggregate extraction process as well as enhances the greenhouse effect by absorption of large amount of carbon dioxide from the surrounding environment. The primary challenge with using recycled aggregates such as partial or total substitutes for coarse aggregates in concrete, as observed by Fady [1] is that their usability is a function of many uncontrollable factors such as the amount of impurities

[2] To whom any correspondence should be addressed.

e.g. glass, wood, brick etc., water content of the parent concrete and the quantity of natural aggregates to be replaced in the concrete matrix.

Katz [2] concluded that the concrete with recycled coarse aggregate in concrete are said to exhibit similar behaviour to that of Natural Aggregate Concrete (NAC) when exposed to temperatures, ranging between 25°C and 800°C if the replacement level is limited to between 0% and 40%, and done with hydrated old concrete. Salau et al [3] used Cassava Peel Ash (CPA) as partial replacement for cement in concrete which is exposed to relatively high temperature. It was shown that with 15% CPA replacement for Cement, the concrete can withstand up to 200°C without any negative effect on compressive strength. Salau and Oseafiana [4] focused on the thermal technique of activating pozzolanic activity of calcined kaolinitic clay in the form of metakaolin. The results showed the optimum heating temperature for the kaolin to be 750°C and the proportion of metakaolin in concrete to be 10% of the cement weight, while the strengths of concrete were 99% and 97% of normal concrete strength at 28 and 90 days respectively.

Maya et al [5]investigated the residual mechanical performance (compressive and splitting tensile strengths and elasticity modulus) of concrete made with recycled coarse aggregates after exposure to high temperatures. Specimens from all types of concrete, besides being tested at ambient temperature (about 20°C), were subjected to the following three temperatures for a period of 1h: 400°C, 600°C and 800°C at age 49 days and, after cooling down to ambient temperature, 4 days later they were finally subjected to mechanical testing. The results showed that in spite of higher porosity and the different thermal properties of the matrix-aggregate interface of recycled concrete coarse aggregate when compared to reference concrete, the incorporation of recycled aggregates does not influence the thermal response of the material. Accordingly, in terms of post-fire residual mechanical properties there are no limitations to the structural use of recycled aggregate concrete when compared with conventional concrete. These findings negate that of Katz [2] who proposed that the presence of the recycled aggregates influences fire resistance, hence, the residual strength.

The present strength work aims to highlight the differences in the behaviour of concrete with recycled coarse aggregates, subjected to ambient temperature.

2. Materials and experimental methodology

Cement used was Ordinary Portland Cement, with the corresponding properties. The aggregates were selected based on the limitation of BS 882.The fine aggregate used was river sand. Two different types of coarse aggregate were used in this study. One was natural coarse aggregate (crushed granite), collected from the quarry and the other was recycled aggregates, obtained from demolished concrete members at the site of the Pan Atlantic University building site, at IbejuLekki, Lagos. It consists of crushed concrete pavements, rejected structural precast elements and abandoned laboratory crushed concrete cubes. Prior to the mix design, the recycled coarse aggregate had to be sieved and graded to ensure quality concrete matrix. Previous work of Tam et al [6] suggests that the gradation of RCA be the same as that of the virgin aggregate. In order to reduce the RCA's water demand, all particles less than 4.75mm [No 4 sieve] were not used.

The coarse aggregates were soaked in water for 48 hours and then the excess water was drained off before they were mixed with the other ingredients in the production of the concrete.The water used for this work is clean, clear and fit for drinking (portable water) which satisfies the BS 5328standard for water requirements for mixing concrete. The mix ratio was 1:1.7:2.5 (cement: sand: aggregate recycled/virgin granite) with water cement ratio of 0.50. The granite normal concrete was substituted with 15% and 30% recycled aggregates. After the crushing of the parent concrete, the particles were sieved through the 25mm sieve size and any particle passing the sieve was discarded. The resulting aggregate particles were again passed through the 4.75mm sieve size and any particle passing was also discarded. A gradation test was performed on the sharp sand, natural and recycled coarse aggregate used and results are given in particle size curves.Various fresh trial concrete mixes were made to determine the ideal moisture content required for the experiment. As recommended byMatsushita et al [7],it is expected that concrete with RCA must have a minimum slump of 75mm.The test was carried out using a 300mm high concrete cone. The slump cone with the base plate below it was filled in three layers, and each layer tapped using a tapping rod to about 25 blows. The slump or reduction in height of the cone was taken to be the measure of workability.

The cube samples with 0%, 15% and 25% RCAwere heated in an electric furnace (carbolite GPC 12/65) at temperatures of 200°C, 400°C and 600°Ceach for 2hours at a heating rate of 2.5°C/min.The sample wasleft to cool for a few minutes before taking readings for strain and finally conducting the strength tests. The tests were also carried out on each sample at room temperature (i.e. about 24°C), as a basis for comparison.

3. Results and discussion

3.1. Particle size distribution of the coarse aggregates

The size distribution curves for sand, natural and recycled coarse aggregates are shown in figure 1. The obtained fineness modulus for the fine aggregate (sand) was 2.83 while the coefficient of uniformity (Cu) and coefficient of curvature (Cc) were2.83 and 1.41 respectively. The fineness modulus for fine aggregates should be in the range of 2.3 to 3.1. The fineness modulus is a measure of the fine aggregates' gradation, and is used primarily for Portland cement concrete mix design. It can also be used as a check in the production of concrete. The value of Cu is less than 6 for the fine aggregate which according to the Unified Soil Classification System infers a poorly graded soil.

For natural coarse aggregate (granite), the coefficient of uniformity (Cu) is 1.5 while that of recycled coarse aggregates is 2.2. Coeficient of curvature of both the RCA and NCA were similar with an average value of 5.5. Hence, according to the Unified Soil Classification System, both types of coarse aggregate (RCA) and (NCA) cannot be classified as well graded but similarity of the curves (figure 1) makes them suitable substitutes for each other. The results of the distribution curves show that they can hardly be differentiated in terms of the particle size.

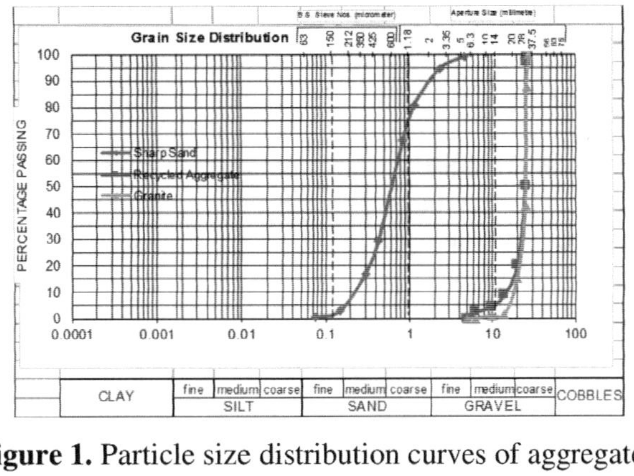

Figure 1. Particle size distribution curves of aggregates.

IOP Conf. Series: Materials Science and Engineering **96** (2015) 012078 doi:10.1088/1757-899X/96/1/012078

3.2. *Density and water absorption of all coarse aggregates*

Table 1 shows the results of density and water absorption of the coarse aggregates. The average particle densities of the natural coarse aggregate are 2895kg/m^3 and 2935 kg/m^3 for dry condition and saturated surface dry condition respectively, while the average particle densities of recycled aggregate are 2311kg/m^3 and 2446kg/m^3 for dry condition and saturated surface dry condition respectively. This means recycled coarse aggregate is lighter than natural coarse aggregate, and this can be a result of RCA also consisting of low density cement paste. When the particle size of recycle aggregate is increased, the volume percentage of residual mortar will increase; this invariably reduces the specific gravity and the density of the aggregate particles.

The average water absorption rate of the recycled coarse aggregate is 5.85%, while that of the natural coarse aggregate is only 1.4%that is, about 4 times. The results shown in table 1 indicate that more water was needed to be added when using the recycled aggregate in the concrete mixing to get an acceptable workability. This may be attributed to the fine aggregate and cement that have adhered to the recycled aggregate during its production. This result is in consonance with the results reported by Katz [2], that the total water absorption of concrete made with recycled aggregate was about 6.9-7.6%, and greater than absorption of the reference concrete of about 3.8-3.9%. The sources of recycled aggregates will determine its water absorption capacity.

Table 1. Density and water absorption capacity of recycled natural aggregates.

Types and Size of Aggregate	Particle Density (Dry), kg/m^3	Saturated Density (SSD)*, kg/m^3	Water Absorption,%
25mm natural aggregate	2885	2921	1.24
10mm natural aggregate	2895	2933	1.32
6.3mm natural aggregate	2906	2952	1.56
25mm recycled aggregate	2342	2473	5.54
14mm recycled aggregate	2281	2421	6.16

SSD*: Saturated surface dry condition

3.3. *Effect of temperature on the density and residual strength of RCA concrete*

The results of densities of concrete containing recycled aggregates and granite (virgin aggregates) with temperature increase are shown in table 2. The density of the concrete cubes for all the mix proportions increased as the curing period increased, irrespective of percentage replacement of the virgin coarse aggregate (granite) with the recycled ones. However, onthe 28th day, the normal concrete has the highest density and the density reduced as the percentage replacement of RCA increased. Density values, which tend to be smaller for concrete containing recycled aggregate than for natural aggregate concrete, can be attributed to a lower density and specific gravity of the recycled aggregates themselves and a weaker interface between old mortar and new mortar. The average density of all mix proportions reduced on heating by an average of 7% in each instance. With thermal treatment there was a reduction in the density of the cubes with the increase in the temperature. However, the residual

IOP Conf. Series: Materials Science and Engineering **96** (2015) 012078 doi:10.1088/1757-899X/96/1/012078

density is more than 95% of the initial density at 600°C heating. The obtained results of the residual compressive strength of thermally treated concrete cube samples 150mm x 150mm x 150mm at different temperatures after coolingareshown in table 3. In both 7 and 28 days, there was a notable increase in residual strength between 300°C and 400°C. The increase in the compressive strength may be due to the acceleration of the hydration of cement gel due to the increased rate of evaporation of free water. However, there is a general reduction in compressive strength of concrete with increasing replacement of natural coarse aggregates with the recycled coarse aggregates. The behaviour of the recycled aggregates concrete can be classified into three regimes with exposure to temperature as follows.

Table 2. Effects of temperature on the density of recycled aggregate concrete.

% RCA	Temp (°C)	Initial Initial W(kg)	Final Final W(kg)	Initial Density kg/m^3	Final Density kg/m^3
	25	8.25	8.25	2398	2398
	200	9.13	8.99	2654	2613
0%	400	8.19	7.68	2381	2232
	600	8.15	7.56	2369	2197
	25	8.16	8.16	2372	2372
15%	200	8.71	8.56	2532	2488
	400	8.13	7.59	2363	2206
	600	8.39	7.75	2439	2253
	25	8.15	8.15	2369	2369
30%	200	7.91	7.74	2299	2250
	400	8.85	8.17	2572	2375
	600	8.13	7.52	2363	2186

Regime I: *Temperature between 25°C and 200°C*: It was characterized with distinct pattern of strength loss. Control samples increased steadily up to about 8% (17.00N/mm^2 to 18.50N/mm^2), but samples with 15% RCA replacements displayed 8% (22.50N/mm^2 to 15.35N/mm^2) and 30% loss (21.03N/mm^2 to 14.70N/mm^2) and respectively reduction in strength. There was an initial drop in strength from 25°C through 100°C and 200°C which is suspected to be due to the relative weak interfacial bond between RCA and hardened paste within the concrete matrix

Regime II: *Temperature between 200°C and400°C*: In this regime, there was continuous gain in strength. A gradual increase in strength continued from 200°C to 400°C.Controlsamples increased steadily up to about 30% (18.5N/mm^2 to 26.70N/mm^2), but samples with 15% and 30% RCA replacements respectively displayed 36% (15.35N/mm^2 to 24.13N/mm^2) and 35%(14.70N/mm^2 to 22.70N/mm^2) increase in strength. It is suspected that there was thermal expansion and dehydration of concrete in this regime.

Regime III: *Temperature between 400°C and600°C*: There was a subsequent loss of concrete cube strength in this regime and steady drop occurred again as it approached 600°C. Control samples reduced steadily with about 13% (26.70N/mm^2 to 23.60N/mm^2), but samples with 15% and 30% RCA replacements displayed 5% (24.13N/mm^2 to 22.95N/mm^2) and 2% (22.70N/mm^2 to 22.20N/mm^2) reduction in strength. Surface cracks were observed in the cubes.

Table 3. Recycled aggregate concrete residual strength at different temperature and ages.

Temperature (°C)	Temperature of Thermal Treatment 25°C (Room Temp)		200°C		400°C		600°C	
% Recycled Aggregate	7 days	28 days	7 days	28 days	7 days	28 days	7 days	28 days
0	19.55	17.00	21.19	18.50	20.60	26.7	19.50	23.6
15	16.4	22.50	11.6	15.35	10.2	24.13	18.7	22.95
30	19.6	21.03	14.7	14.7	16.0	22.7	24.4	22.2

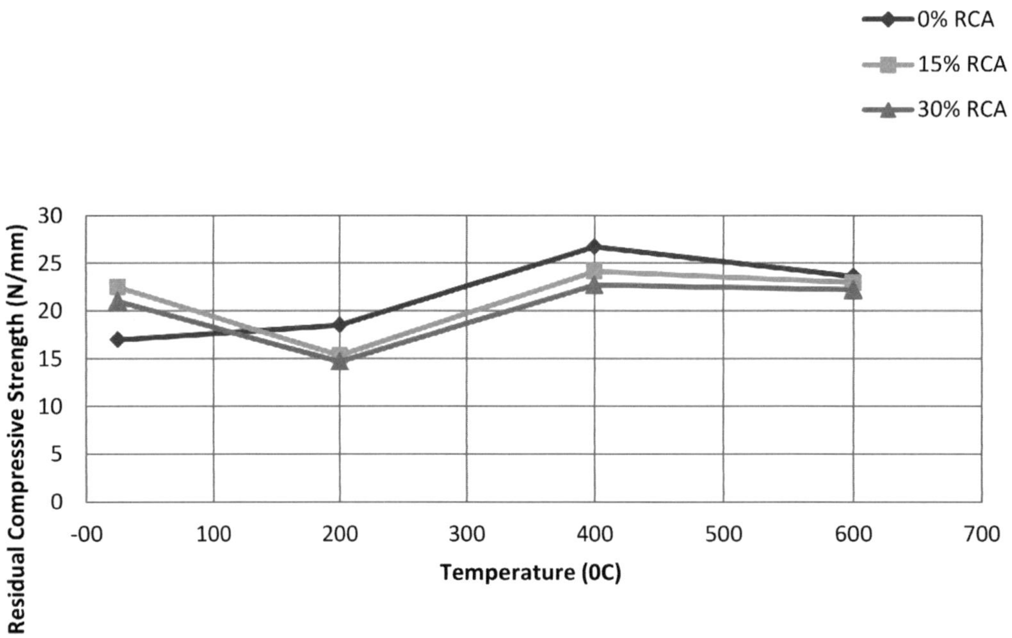

Figure 2. Effect of temperature on the residual strength of concrete (at 28 days) with different percentage of recycled aggregates.

3.4. Crack formation

The process leading to cracking is believed to be essentially the same as leading to spalling. Thermal expansion and dehydration of the concrete due to heating may lead to the formation of fissures in concrete in addition to spalling. Crack depths were not measured, but severities of the cracks were noticed as the temperatures approached 600°C. Figure 3 shows the cracks on 15% recycled aggregate cube on 28 days after heating for 2 hours at 400°C. Similar cracks were observed in the other cubes. However, the severity depends on the exposure to the thermal heating, especially after 400°C.

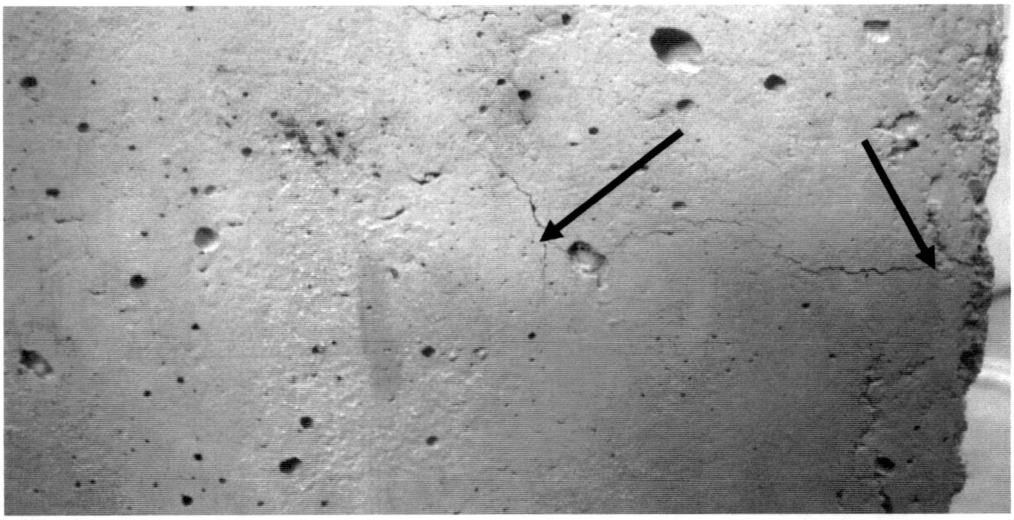

Figure 3. Arrows pointing to crack lines formed on 15% recycled concrete after tempering for 2 hours at 400°C.

4. Conclusions and recommendations

The most prominent physical feature of RAC that determines its behaviour just before it is subjected to high temperature is its porosity and density. This attribute has been seen to be due to mortar and cement pastes adhering to the aggregates. The density reduces with the increase in the percentage of recycled aggregate in concrete, while there is an increase in porosity with the increase in the recycled aggregates.

On heating the concrete samples, no noticeable changes took place until temperature was considerably above 100°C. There was an initial drop in strength from 100°C to 200°C which is suspected to be due to the relative weak interfacial bond between the RCA and hardened paste within the concrete matrix; a gradual increase in strength continued from 200°C to 450°C and steady drop occurred again after that up to 600°C. Also, with temperatures higher than 400°C the cubes showed further loss in weight and increased visibility and higher density of cracks.

Beyond 600°C, deterioration was relatively more severe and the mechanical properties of concrete (such as compressive strength) were largely affected by the temperature. In the case of a real fire, where temperatures may exceed 1000°C, the residual compressive strength values were feeble for the 2 hour exposure time with all mixes of recycled aggregate concrete; concrete structures need special attention in the repairing case if not totally demolished. For applications of concrete in places where temperatures shall not exceed 400°C concrete with the recycled aggregate with up to about 25% RCA content, in a mix proportion of 1:1.7:2.5 by weight and water-cement ratio of 0.50, can be recommended for use. It is recommended that further research should be conducted to find out the influence of longer durations of exposure such as 8hourswith various RCA content at lower temperature ranges (less than 400°C), which are more likely to be encountered in conditions like bakeries and large ovens. Further, cyclic heating and cooling of concrete made from various RCA content at lower temperature ranges less than 400°C could also considered.

5. References

[1] Fady M A 2002 *Using of Recycled Aggregate in producing Concrete Elements – A Thesis submitted in partial fulfillment of the requirement of Masters of Science in Civil Engineering-Design and rehabilitation of structures* (The Islamic University Gaza) p 1-5

[2] Katz A 2002 Properties of concrete made with recycled aggregate from partially hydrated old concrete *Cement and Concrete Research* **33** (5) 703 – 711

[3] Salau M A, Ikpowonsa E E and Olonode K A 2013 Effect of elevated temperature on Cassava peel ash concrete *American J. of Science* **3** (5) 142–148

[4] Salau M A and Oseafiana O J 2015 Effects of Temperature on the pozzolanic characteristics of metakaolin-concrete *Physical Science Int. J.* 1–13

[5] Maya T M, Nivin Philip and Job Thomas 2015 Review on Concrete Subjected to Elevated Temperature *Int. J. of Engin., Business and Enterprise Applications* 9163–69

[6] Tam V W Y and Tam C M 2006 Crushed aggregate production from centralized combined and individual waste sources in Hong Kong *Construction and Building Materials* **21** 1–8

[7] Matsushita H H, Tsuruta and Takao Chikada 2000 Experimental Studies on Compressive Strength and Drying Shrinkage of Recycled Aggregate Concrete *Int. Seminar on Recycled Concrete* Sponsored by Niigata University and Japan Concrete Institute, JCI69–78

Effect of Different Coarse Aggregate Sizes on the Strength Characteristics of Laterized Concrete

M A Salau[1,2] and A O Busari[1]
[1]University of Lagos, Department of Civil and Environmental Engineering, Lagos State, Nigeria

E-mail: ajibsalau@yahoo.co.uk

Abstract. The high cost of conventional concrete materials is a major factor affecting housing delivery in developing countries such as Nigeria. Since Nigeria is blessed with abundant locally available materials like laterite, researchers have conducted comprehensive studies on the use of laterite to replace river sand partially or fully in the concrete. However, the works did not consider the optimum use of coarse aggregate to possibly improve the strength of the laterized concrete, since it is normally lower than that of normal concrete. The results of the tests showed that workability, density and compressive strength at constant water-cement ratio increase with the increase in the coarse aggregate particle size and also with curing age. As the percentage of laterite increases, there was a reduction in all these characteristics even with the particle size of coarse aggregate reduction due to loss from the aggregate-paste interface zone. Also, when sand was replaced by 25% of laterite, the 19.5mm and 12.5mm coarse aggregate particle sizes gave satisfactory results in terms of workability and compressive strength respectively at 28 days of curing age, compared to normal concrete. However, in case of 50% up to 100% laterite contents, the workability and compressive strength values were very low.

1. Introduction

Due to increasing cost of producing concrete using conventional materials such as cement, river sand as fine aggregate and granite as coarse aggregate in Nigeria, researchers have been working on alternative, cheap and readily available materials that would serve perfect substitutes for such materials while still meeting the set requirements for concrete in the industry, see Udoeyo [1].

Abdullahi [2] established that aggregate type has effect on the compressive strength of normal concrete and concluded that concrete made from crushed quartzite demonstrates higher compressive strength at all ages compared to the concrete with granite as a coarse aggregate. However, Aginam et al [3] gave a contrary view with the higher compressive strength of the normal concrete with granite as a coarse aggregate when compared with the concrete made with washed and unwashed gravel as a coarse aggregate component.

[2] To whom any correspondence should be addressed.

IOP Conf. Series: Materials Science and Engineering **96** (2015) 012079 doi:10.1088/1757-899X/96/1/012079

Muhit et al [4] investigated the effects of "impact crushed" and "vertically shafted" coarse aggregates on the concrete blocks. They concluded that the slump values for the two types of crushed aggregates decreased moderately with water-cement ratio and concrete samples made with "vertically shafted" coarse aggregate having higher slump value than those with "impact crushed". However, the compressive strength of "impact crushed" aggregate concrete is higher compared to the "vertically shafted" aggregate with the decrease of water-cement ratio.

Jimoh and Awe [5] also investigated the size of aggregate that will improve the properties of the structural normal concrete and concluded that the use of quarry dust and 20mm granite improved the concrete strength by 34% over the strength produced by concrete with sand and gravel of 28mm maximum size. Also, as coarse aggregates' size increases, normal concrete strength decreases. The rate of fall in strength is highest with the concrete containing quarry dust as a fine aggregate. However, the case of laterized concrete was not studied. Xie et al [6] confirmed these earlier finding that the compressive strength decreases with the increase in the maximum coarse aggregates size. Bhiskshma and Florence [7] assessed the effect of size in higher grade concrete using high volume fly ash and found out that maximum size of coarse aggregates of 12.5mm gave the highest compressive, splitting and flexural strengths; hence the aggregate sizes influence strength.

Laterite has been used for wall construction around the world; it is a cheap, environmentally friendly and abundantly available building material in the tropical region. Lateritic soils, according to Osadebe and Nwakonobi [8], are widely used as construction material in Nigeria, other under-developed and developing countries of the world. However, they argue that laterite has not been extensively used as constructing medium in large-size building structures, probably because of the lack of adequate data needed in the analysis and design of structures built of lateritic soils. Joseph et al [9] investigated the structural characteristics of concrete using various combinations of lateritic sand and quarry dust as complete replacement for conventional river sand fine aggregate. The quantity of laterite was varied from 0% to 100% against quarry dust at intervals of 25%. Specifically compressive strength ranged from 17-34.2N/mm^2 for the mixes considered. These results compare favourably with those of conventional concrete. The concrete was found to be suitable for use as structural members for buildings and related structures, where laterite content did not exceed 50%.

Olubisi [10] ascertained the suitability of laterite as aggregate substitutes at 0, 10, 20, 30 and 40% of fine aggregate used in the construction industry. The results showed that the compressive strength of laterized concrete with laterite-fine aggregate ratio variation decreases when subjected to alternate wetting and drying and increases when subjected to magnesium sulphate (Mg_2SO_4). It was also observed that a laterized concrete with a laterite-fine aggregate ratio of 20% conditioned to a temperature range of 100°C attained optimum compressive strength of 12.9N/mm^2.

With all these research works on laterized concrete, no known attempt has been made to evaluate the effect of different coarse aggregate particle sizes on workability and compressive strength of laterized concrete, as in the case of normal concrete. However, it is well known that coarse aggregate of particular size is sometimes supplied and used on site. The aim of this study, therefore, is to determine the extent to which different sizes of coarse aggregate used for production would influence the workability and strength characteristics of laterized concrete.

2. Materials of experimental procedure

The coarse aggregates of different particle sizes viz; 19.5mm, 12.5mm, 9.5mm, 5.0mm and 2.36mm were obtained from Ojodu-Berger in Lagos State while the sharp river sand, used as a fine aggregate, was collected from Ogun River bed in Ogun state of Nigeria; the reddish-brown laterite, obtained from Sagamu in Ogun State, has maximum size 2.36mm sieve, just as the sharp sand; Ordinary Portland Cement with fineness value of 320m^2/kg, initial and final setting times of 85 and 218 minutes respectively was used. Portable water, free from any visible impurities was used for the experiments.

The mix proportion of the concrete used in this study is 1:2:4 (cement: sand/laterite: granite) with a constant water-cement ratio of 0.60. The sand was replaced at 25% and 50% with laterite while 0% was used for normal concrete with all the particle sizes of granite inclusive in the coarse aggregate.

Slump and Compacting Factor tests were carried out on the fresh concrete using granite (all particle sizes inclusive) at 0%, 25%, and 50% laterite contents, as well as for each of the coarse aggregate sizes (19.5mm, 12.5mm, 9.5mm, 5.0mm and 2.36mm) at 25% and 50% laterite contents to determine their workability according to ASTM C 143. Density and compressive strength tests were carried out on 150mm x 150mm x150mm cubes in a 2000kN capacity strength machine after curing for the relevant number of days in line with BS EN 12390 series. A total of 156 Nos. 150mm concrete cubes with different laterite contents were tested at different curing ages of 7, 14, 21 and 28 days.

3. Results and discussions

3.1. Preliminary tests on aggregates

The results of particle size distribution for the aggregates are shown in figure 1 while table 1gives the physical properties of the materials used in the laterized concrete. Table 1 shows that the coefficients of uniformity of the laterite, sand and granite are respectively, 2.35, 2.25 and 2.45 while those of curvature are 1.35, 1.38 and 1.24. In view of these, the aggregates are well graded and suitable for concrete production according to ASTM C33. The fineness moduli are within the specified limit of ASTM 125.

3.2. Tests on fresh concrete

The results of the slump test, indicating the workability of the concrete at 25% and 50% laterite contents with granite (all particle sizes inclusive) as controls, and using different coarse aggregate sizes (retained on 19.5mm, 12.5mm, 9.5mm, 5.0mm and 2.36mm sieve sizes) at 25% and 50% replacement of sand with laterite are shown in table 2.

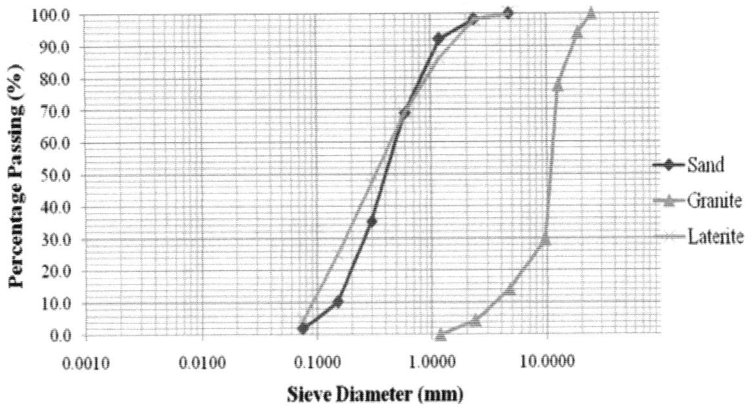

Figure 1. Particle size distribution curves for laterite, sand and granite.

For controls at 25% and 50% laterite contents using all granite particle sizes inclusive, the results show true slump all through but the degree of workability ranges from medium at 25% to a very low degree at 50% laterite contents due to high water absorption ability of laterite. In view of this, more water is needed as the percentage of laterite increases to achieve workable concrete. Moreover, the concrete slump decreases as coarse aggregate particle size decreases in the mixes from 80mm slump value for 19.5mm coarse aggregate particle size to zero for 2.36mm particle size at 25% laterite content. This is due to large surface area from the small particle sizes of coarse aggregate in the mix that requires more water to improve workability.

Table 1. Summary of the physical properties of the materials used.

Physical Properties	Cement	Laterite	Sand	Granite
Moisture Content (%)	-	0.72	3.56	1.21
Specific Gravity	3.15	2.54	2.62	2.65
Bulk Density (kg/m$^{3)}$	2298	2642	2668	2750
Dry Density (kg/m^3)	-	2638	2610	2729
Coefficient of Uniformity (cu)	-	2.35	2.25	2.45
Coefficient of Curvature (cc)	-	1.35	1.38	1.24
Fineness Modulus	-	1.75	1.98	4.50

Table 2. Slump value for laterized concrete with different coarse aggregate sizes at 25% and 50% laterite contents.

Particle Size (mm)	25% Replacement of Sand with Laterite			50% Replacement of Sand with Laterite		
	Slump Value (mm)	Type of Slump	Degree of Workability	Slump Value (mm)	Type of Slump	Degree of Workability
19.5	80	True Slump	Medium	35	True Slump	Low
12.5	65	True Slump	Medium	20	True Slump	Low
9.5	45	True Slump	Low	10	True Slump	Low
5.0	15	True Slump	Low	0	True Slump	Low
2.36	0	True Slump	Low	0	True Slump	Low
All Sizes of Coarse Aggregate	25	True Slump	Low	10	True Slump	Low

With 25% laterite content in the mix, using coarse aggregate sizes of 19.5mm and 12.5mm, medium workability (with slump of 80mm and 65mm) was achieved but its values were low for other remaining coarse aggregate sizes. This also implies that more water is needed for smaller coarse aggregate particle sizes to achieve medium workability because of large surface area. However, for 2.36mm coarse aggregate particle size at 25% laterite content and for both 5.0mm and 2.36mm coarse aggregate sizes at 50% laterite content values of slump were zero, showing a stiff consistency. This may be due to affinity of laterite for water and also large surface area caused by the small particle sizes of coarse aggregates.

The compaction factor test results for controls (all granite particle sizes inclusive) and with different coarse aggregate sizes (19.5mm, 12.5mm, 9.5mm, 5.0mm and 2.36mm sieve sizes) at 25% and 50% laterite contents show that the degree of workability ranges from medium at 25% laterite content to a very low degree at 50% laterite content for controls (all granite particle sizes inclusive) due to high water absorption ability of laterite. However, it was observed that the workability of the laterized concrete decreases as the coarse aggregate particle size decreases from 19.5mm particle size (0.938 to 0.804) to 2.36mm particle size (0.73 and 0.70) for both 25% and 50% laterite contents. This is in agreement with slump test results.

3.3. Tests on hardened concrete

3.3.1. Density.

The relationship between density of the laterized concrete for controls (all granite particle sizes inclusive) as well as with different coarse aggregate particle sizes at 25% and 50% laterite contents and curing age are shown in table 3.

Table 3. Variation of Density of Laterized Concrete with Different Coarse Aggregate Particle Sizesand Curing Age.

Curing Age (Days)	Density (kg/m^3) Coarse Aggregate Particle Sizes (mm) at 25% Laterite Content						Coarse Aggregate Particle Sizes (mm) at 50% Laterite Content					
	Control (All Granite Particle Sizes)	2.36	5.00	9.50	12.50	19.50	Control (All Granite Particle Sizes)	2.36	5.00	9.50	12.50	19.50
7	2565	2325	2391	2400	2435	2488	2477	2281	2311	2358	2468	2551
14	2577	2340	2400	2429	2506	2545	2500	2290	2325	2429	2497	2557
21	2651	2367	2414	2462	2518	2577	2610	2340	2370	2515	2533	2577
28	2740	2400	2429	2503	2548	2640	2651	2355	2456	2533	2595	2592

It was observed that the density of concrete produced using all granite particle sizes inclusive and different coarse aggregate particle sizes at 25% and 50% laterite contents generally increases with curing age and the increase in coarse aggregate particle size from the smallest to the biggest, ($2400kg/m^3$ for 2.36mm and $2640kg/m^3$ for 19.5mm at 25% laterite content and 28 days curing age), but reduces insignificantly with the increase in percentage of laterite content. All the laterized concrete studied can be considered as normal weight and suitable for structural purposes.

3.3.2. Compressive Strength.

The variation of compressive strength of concrete with curing age for controls (using all granite particle sizes inclusive) at 0%, 25% and 50% replacement of sand with laterite is shown in figure 2.

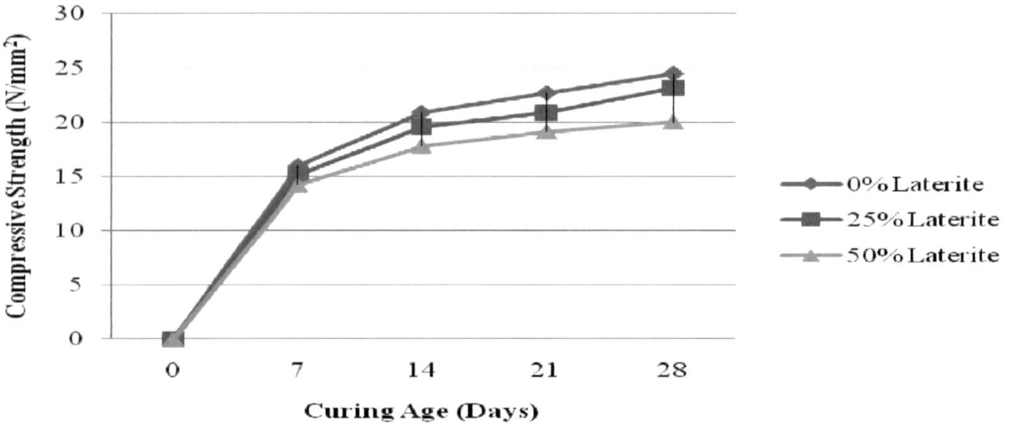

Figure 2. Variation of compressive strength with curing age for controls (all granite particle sizes) at 0%, 25% and 50% laterite contents.

Figure 2 shows that normal concrete with all granite particle sizes inclusive at 0% laterite content gives peak values of compressive strength at 7 days (16N/mm^2) and 28 days (24.44N/mm^2), while concrete with maximum laterite content (50%) gives minimum strength at 7 days (14.2N/mm^2) and 28 days (20.00N/mm^2). This shows that compressive strength decreases with increase in laterite content in concrete. This is in consonance with Udoeyo [1]. However, the development of strength in normal and laterized concrete is consistent.

The comparison of compressive strength for control (all granite particle sizes inclusive) and with different coarse aggregate particle sizes at 25% laterite content can be seen in figure 3.

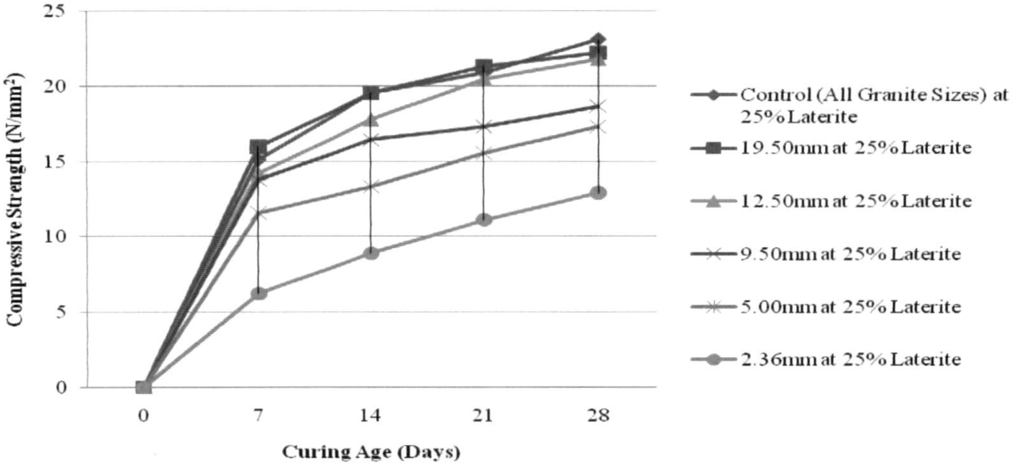

Figure 3. Comparison of compressive strengths of laterized concrete for control (all granite sizes) and with different coarse aggregate particle sizes at 25% laterite.

It can be observed from figure 3 that compressive strength for laterized concrete like normal concrete increases with curing age, {at 25% laterite content with 19.5mm coarse aggregate particle size (16.00N/mm^2 at 7days and 22.22N/mm^2 at 28days)}. Also, it increases with the increase in coarse aggregate particle size, (22.22N/mm^2 for 19.5mm and 12.89N/mm^2 for 2.36mm at 28 days) but for the two biggest particle sizes 19.5mm and 12.5mm, the values of compressive strength for both 25% and 50% laterite contents are almost the same with that of control at 28 days of curing.

Garrison [11] showed that concrete mixes with larger coarse aggregate at constant water-cement ratio give lower compressive strength due to the presence of a weak zone at the aggregate-paste interface where cracking will first occur. He further explained that with larger aggregates, larger cracks can form at the interface, and they can interact easily with paste cracks as well as other interfacial cracks. In contrast to this, the values of compressive strength at 28 days curing age for laterized concrete (25% laterite content) with the two biggest particle sizes of 19.5mm (22.22N/mm^2) and 12.5mm (21.78N/mm^2) from figure 3 are very close to that of normal concrete (24.44N/mm^2). This might be a result of the presence of laterite in the paste which improved the aggregate-paste interface zone to avoid any form of failure due to crack.

4. Conclusions and recommendations

From this study, the following conclusions can be arrived at:

1. The workability of laterized concrete at constant mix ratio of 1:2:4 and water-cement ratio of 0.6 increases as coarse aggregate particle size increases. However, it decreases with the increase in percentage of laterite content.
2. For the two biggest coarse aggregate particle sizes (19.5mm and 12.5mm), the workability of laterized concrete at 25% laterite content was almost the same (medium) compared to that of normal concrete but was low for 50% laterite content.
3. The density of the hardened concrete (using all granite particle sizes inclusive) increases from 0% (normal concrete) to 25% laterite content but later decreases from 25% to 50%.
4. The mean compressive strength at constant mix ratio of 1:2:4 and water-cement ratio of 0.6 increases with the increase in coarse aggregate particle sizes and curing age but decreases as percentage of laterite content increases. However, the 28th day compressive strengths of concrete for control (using all granite particle sizes inclusive) at 0%, 25% and 50% laterite contents were almost the same. Also, for the two biggest coarse aggregate particle sizes (19.5mm and 12.5mm), the 28th day mean compressive strengths of laterized concrete at 25% laterite content were almost the same with that of control but for 50% laterite content, it was low.

It is recommended that laterized concrete can be used in place of normal concrete provided the percentage of laterite content to replace sand does not exceed 50%, but 25% laterite content gives a good quality concrete in terms of workability and compressive strength even when compared to normal concrete.

Coarse aggregate with maximum particle size of 19.5mm should be used where high strength laterized concrete is required but it is significant to consider the dimensions of such concrete member, the size and spacing of reinforcement. However, when smaller particle sizes of coarse aggregate like 2.36mm, 5.00mm and 9.50mm are required to produce concrete, the cement paste would require high percentage of water to produce a concrete with good degree of workability.

5. References

[1] Udoeyo F F 2006 Strength performance of laterized concrete *Constr. and Buil. Mat.* 1057–62
[2] Abdulahhi M 2013 Effects of aggregate type on compressive strength of concrete *Int. J. of Civil and Structural Engin.* **2** (3) 791–800
[3] Aginam C H, Chidolue C A and Nwakire C 2013 Investigating the effects of coarse aggregates types on the compressive strength of concrete *Int. J. of Engineering Research and Applications* **3** (4) 1140–44
[4] Muhit B, Haque S and Rabiul Alam M D 2013 Influence of crushed aggregates on properties of concrete *American J. of Civil Engineering and Architecture* (Science and Education Publishing) **1** (5) 103–106
[5] Jimoh A A and Awe S S 2007 A study on the influence of aggregate size and type on the compressive strength of concrete *J. of Research Information in Civil Engin.* **4** (2) 157–168
[6] Xie W, Jin Y and Li S 2012 Experimental research on the influence of grain size of coarse aggregates on pebble concrete compressive strength *Applied Mechanics and Materials* **238** 133–137
[7] Bhikshma V and Florence G A Studies of effects of maximum size of aggregates types on the compressive strength of concrete *Int. J. of Engin. Research and Applications* **3** (4) 1140 – 44
[8] Osadebe N N and Nwakonobi T U 2007 Structural characteristics of laterized concrete at optimum mix proportion *Nigerian J. of Technology* (Nsukka, Nigeria) **26** (1) 12–17
[9] Joseph O U, Maurice E E and Godwin A A 2012 Compressive strength of concrete using lateritic sand and quarry dust as fine aggregate *ARPN J. of Engin. and Applied Sciences* **7** (1) 81–92
[10] Olubisi A I 2013 Performance of lateritic concrete under environmental harsh condition *Int. J. of Research in Engin. and Technology* **2** (8)
[11] Garrison 2007 The concrete mix design *Civil Engin. and Construction Mat.* **5** 55–106

IOP Conf. Series: Materials Science and Engineering **96** (2015) 012080 doi:10.1088/1757-899X/96/1/012080

Comparing the Environmental Impacts of Alkali Activated Mortar and Traditional Portland Cement Mortar using Life Cycle Assessment

P S Matheu[1,3], K Ellis[1] and B Varela[2]
[1]Department of Industrial and Systems Engineering, Rochester Institute of Technology, 1 Lomb Memorial Dr, Rochester NY, USA
[2]Department of Mechanical Engineering, Rochester Institute of Technology, 1 Lomb Memorial Dr, Rochester NY, USA

E-mail: pxs3517@rit.edu

Abstract. Since the year 1908 there has been research into the use alkali activated materials (AAM) in order to develop cementitious materials with similar properties to Ordinary Portland Cement. AAMs are considered green materials since their production and synthesis is not energy intensive. Even though AAMs have a high compressive strength, the average cost of production among other issues limits its feasibility. Previous research by the authors yielded a low cost AAM that uses mine tailings, wollastonite and ground granulated blast furnace slag (GGBFS). This mortar has an average compressive strength of 50MPa after 28 days of curing. In this paper the software SimaPro was used to create a product base cradle to gate Life Cycle Assessment (LCA). This compared the environmental impact of the AAM mortar to an Ordinary Portland Cement mortar (PCHM) with similar compressive strength. The main motivation for this research is the environmental impact of producing Ordinary Portland Cement as compared to alkali activated slag materials. The results of this LCA show that the Alkali Activated Material has a lower environmental impact than traditional Portland cement hydraulic mortar, in 10 out of 12 categories including Global Warming Potential, Ecotoxicity, and Smog. Areas of improvement and possible future work were also discovered with this analysis.

1. Introduction

The production of traditional Portland cement is harmful to the environment due to the high energy consumption and CO_2 emissions. On average 0.81kg CO_2 is released to the environment for every kilogram of Portland cement produced [1]. Most of the environmental impact is generated during the

[3] To whom any correspondence should be addressed.

clinker production. In order to manufacture clinker, a mixture of raw materials has to be slowly heated to approximately 1450°C then quickly cooled to a range of 100 to 200°C. This process requires a large amount of energy. In 2012, 3700 million cubic meters of Portland cement were produced around the world [2]. The demand for this product is growing, causing an increased risk to the environment.

Due to the environmental impacts of Portland cement it is important to find a viable, cost effective replacement for this product. A promising alternative to this issue are Alkali Activated Materials (AAMs). These types of materials are an emerging technology that utilizes industrial by-products such as fly ash or blast furnace slag to produce cementitious materials. The process involves the reaction of the by-product with an alkali solution at low temperature. The resulting binder has high early compressive strength, resistance to chemical attack and high thermal stability among other properties [3]. The energy requirement to produce these products is significantly lower than the requirement for the traditional Portland cement manufacture.

The cost of AAMs is normally higher than ordinary Portland cement, hindering their commercial application. However, previous research performed by the authors yielded a cost effective AAM. This design mix consisted of a mixture of Ground Granulated Blast Furnace Slag (GGBFS), mine tailings, and wollastonite activated with a potassium based alkali solution. Most of the components are by-products of different industrial processes [4]. The Alkali solution contains potassium hydroxide, amorphous silica, and water.

It was found that the current literature lacks quantitative research on the environmental impacts of these AAMs compared to those of Portland cement. This paper will focus on this gap by comparing the environmental impacts of a traditional Portland cement and the design mix AAM previously research by the authors [4]. A Life Cycle Assessment (LCA) using SimaPro was used in order to quantifiably compare the environmental effects of Portland cement and the design mix. According to the United States Environmental Protection Agency, an LCA is a methodology to assess the environmental impacts associated with a process, product, or service from cradle to grave [5].This methodology uses data to analyse all of the environmental impact of every production stage from raw material acquisition to disposal. This methodology helps to easily find opportunities to reduce the overall environmental impact. This will not only give a comparison of the total environmental impacts of both products but also the impact of each component, which will help to find future areas for improvements.

2. Analysis

The preparation of the AAM has been previously discussed by the authors [4]. This cementitious material was obtained by the use of a Potassium base solution which reacted with a mixture of blast furnace slag, mine tailings and Wollastonite. The resultant slurry was cured at room temperature, enhancing the workability. It is common for similar LCA analysis to not take into consideration any environmental impact caused by the by-products used. However, in this analysis the impact for grinding the GGBFS and the amorphous silica manufacture was included in the inventory.

In order to perform the LCA, it was necessary to determine the precise ingredients that go into a typical Portland cement hydraulic mortar. The ingredients for the analysis in this study were taken from a paper by Mahyuddin Ramli [6]. Both the AAM mix and the Portland cement mix had a compressive strength of 50 Mpa after 28 days of curing. The materials needed to produce one cubic meter of the AAM and the Portland cement hydraulic mortar were calculated and input into SimaPro to perform the analysis. The mix composition and percentage content can be seen in the tables below.

Table 1. AAM mix composition per cubic meter.

Material	Kg	%
45% KOH	83.95	4.37
Amorphous Silica	20.14	1.05
Water	251.85	13.10
Slag	711.91	37.04
20% NYAD G	142.38	7.41
Mining Tails	711.91	37.04

Table 2. Traditional portland cement hydraulic mortar mix composition per cubic meter.

Material	Kg	%
Lime (CaO)	387.84	17.32
Silica (SiO_2)	127.68	5.704
Alumina (Al_2O_3)	33.6	1.501
Iron Oxid (Fe_2O_3)	20.16	0.901
Magnesia (MgO)	12.36	0.552
Sulphur Trioxode (SO_3)	12.84	0.574
N_2O	0.3	0.013
Loss of Ignition	3.84	0.172
Water	240	10.72
Sand	1400	62.54

The LCA was performed as outlined by the ISO 14040 guidelines. The functional unit was a cubic meter of hydraulic mortar with an average compressive strength of 50 MPa after 28 days of curing. The scope of this LCA study focused on cradle to gate analysis. More specifically, the raw material acquisition, processing and manufacturing of both materials were studied. The packaging, use and disposal were not considered in this study. The AAM design mix has never been created in quantities equal to that of the functional unit, so the design mix materials were scaled up assuming a linear relationship.

Another major area of impact in the LCA is the transportation of materials. In this study Rochester Institute of Technology (RIT), Rochester, NY was used as the end point of all materials. When the exact starting location of materials was unknown, an average of all locations in the area was used. Given the location of the end point, it was decided to use the Building for Environmental and Economic Sustainability (BEES) method to compare the environmental impact of both mortars. This method was developed by the National Institute of Standards and Technology (NIST), an agency of the U.S. Department of Commerce. It was developed to measure the environmental impact of building materials specifically [7]. However, not all of the commonly used categories are calculated with this method. Due to this lack of categories and to verify the results, a sensitivity analysis using different methods was performed.

3. Results and discussion

The Life Cycle Assessment using the BEES method compares the impact of each material in thirteen different categories: Global Warming Potential, Acidification, Human Health: Cancer, Human Health: Non-Cancer, Human Health: Criteria Air Pollutants, Eutrophication, Ecotoxicity, Smog, Natural Resource Depletion, Indoor Air Quality, Habitat Alteration, Water Intake, and Ozone Depletion. As presented in table 3, in ten categories the AAM had less impact than Portland cement. In seven categories the Alkali-activated design mix had less than 40% of impact compared to the traditional Portland cement hydraulic mortar. The Indoor Air Quality category was excluded from the analysis because neither had a discernible impact. This was due to the fact that the use and disposal phase of the life cycle were not considered in this study. The two categories in which the traditional Portland cement mortar performed better, Eutrophication and Habitat Alteration, it had 97% and 38% of the impact of the AAM respectively. In an attempt to determine the cause of the higher AAM impact in the categories of Eutrophication and Habitat Alteration further analysis was done in these two categories.

Table 3. Comparing MT12 (AAM) with traditional portland cement hydraulic mortar BEES method.

Impact Category	Unit	MT12 (AAM)	PCHM	MT12/PCHM
Acidification	H+ moles eq	60609.15	143637.49	0.42
Ecotoxicity	g 2.4-D eq	500.53	2479.10	0.20
Eutrophication	g N eq	271.60	215.24	1.26
Global warming	g CO_2 eq	200309.50	790558.51	0.25
Habitat alteration	T&E count	8.00E-12	2.71E-12	2.95
HH cancer	g C_6H_6 eq	317.48	1226.27	0.26
HH criteria air pollutants	microDALYs	30.30	31.24	0.97
HH noncancer	g C_7H_7 eq	647311.80	2391701.70	0.27
Natural resource depletion	MJ surplus	266.25	686.17	0.39
Ozone depletion	g CFC-11 eq	0.002	0.011	0.22
Smog	g NOx eq	881.0962	3293.5687	0.27
Water intake	liters	588273.7	904118.08	0.65

The Eutrophication category refers to the amount of nutrients that a particular material emits into the surrounding waterways. Table 4 shows the main process contributions for the Eutrophication category. As can be seen, potassium hydroxide is the main contributor for the AAM design mix. For this analysis the potassium hydroxide is manufactured by the electrolysis of potassium chloride brine in an electrolytic cell, all information is based on industry data in the US.

Looking into the Habitat Alteration category, it is visible that the production of potassium hydroxide is again the largest contributor. However, as it can be seen in table 5 the total amount of T&E count, damage of Threatened and Endangered Species, is $8x10^-12$. This number shows that even though the Alkali Activated Material has a larger impact, it is so small that it is not a priority compared to other categories.

IOP Conf. Series: Materials Science and Engineering **96** (2015) 012080 doi:10.1088/1757-899X/96/1/012080

Table 4. Process contribution eutrophication BEES method.

Process	Unit	MT12	PCHM
Total of all processes	g N eq	271.60	215.24
Remaining processes	g N eq	1.83	3.69
Crude oil, at production/RNA	g N eq	1.47	6.58
Diesel, at refinery/US	g N eq	0.50	2.24
Electricity, high voltage, at grid/US S	g N eq	32.10	x
Electricity, low voltage, at grid/US S	g N eq	72.75	11.20
Operation, transoceanic freight ship/OCE S	g N eq	1.53	x
Portland cement, strength class Z 42.5, at plant/CH S	g N eq	x	105.47
Potassium hydroxide, at regional storage/RER S	g N eq	143.40	x
Sand, at mine/CH S	g N eq	x	5.45
Transport, combination truck, diesel powered/US	g N eq	16.59	74.25
Transport, ocean freighter, residual fuel oil powered/US	g N eq	1.42	6.36

Table 5. Habitat alteration comparing AAM to traditional portland cement hydraulic Mortar BEES Method.

Process	Unit	MT12	PCHM
Total of all processes	T&E count	8.00E-12	2.71E-12
Remaining processes	T&E count	2.51E-14	2.54E-12
Potassium hydroxide, at regional storage/RER S	T&E count	6.17E-12	x
Electricity, low voltage, at grid/US S	T&E count	1.12E-12	1.73E-13
Electricity, high voltage, at grid/US S	T&E count	5.99E-13	x
Operation, transoceanic freight ship/OCE S	T&E count	8.83E-14	x
Total of all processes	T&E count	8.00E-12	2.71E-12
Remaining processes	T&E count	2.51E-14	2.54E-12
Potassium hydroxide, at regional storage/RER S	T&E count	6.17E-12	x
Electricity, low voltage, at grid/US S	T&E count	1.12E-12	1.73E-13
Electricity, high voltage, at grid/US S	T&E count	5.99E-13	x
Operation, transoceanic freight ship/OCE S	T&E count	8.83E-14	x

4. Sensitivity analysis

ReCiPe model offers three different perspectives, the individualist perspective is based in a short term interest, and it assumes that the technological advances are going to help reduce the impact in the long run. This is contrary to Egalitarian which is the most precautionary perspective out of the three [8]. The hierarchist viewpoint is representative of the midpoint between the Individualist and Egalitarian, it is often considered to be the default model for ReCiPe. It was created with European normalisation in order to collaborate the results found by the BEES method, it was decided to compare both mortars using the ReCiPe midpoint hierarchist method.

IOP Conf. Series: Materials Science and Engineering **96** (2015) 012080 doi:10.1088/1757-899X/96/1/012080

Table 6. Comparing MT12 (AAM) with traditional portland cement hydraulic mortar ReCiPe method.

Impact Category	Unit	MT12 (AAM)	PCHM	MT12/PCHM
Climate change	kg CO2 eq	201.34	792.23	0.25
Ozone depletion	kg CFC-11 eq	0.00	0.00	0.55
Terrestrial acidification	kg SO2 eq	1.02	2.23	0.46
Freshwater eutrophication	kg P eq	0.01	0.00	3.00
Marine eutrophication	kg N eq	0.03	0.11	0.28
Human toxicity	kg 1,4-DB eq	41.98	162.83	0.26
Photochemical oxidant formation	kg NMVOC	0.84	3.05	0.27
Particulate matter formation	kg PM10 eq	0.38	0.83	0.46
Terrestrial ecotoxicity	kg 1,4-DB eq	0.01	0.01	1.26
Freshwater ecotoxicity	kg 1,4-DB eq	0.36	1.01	0.36
Marine ecotoxicity	kg 1,4-DB eq	0.35	1.06	0.33
Ionising radiation	kg U235 eq	12.09	16.50	0.73
Agricultural land occupation	m2a	2.29	2.61	0.88
Urban land occupation	m2a	1.28	1.47	0.88
Natural land transformation	m2	0.01	0.03	0.36
Water depletion	m3	1.07	3.83	0.28
Metal depletion	kg Fe eq	6.09	4.35	1.40
Fossil depletion	kg oil eq	58.87	131.59	0.45

As it can be seen in table 6, 18 categories are studied in this model. Similar to the other model, the AAM material has significantly lower environmental impact in most categories. The traditional Portland cement hydraulic mortar only has a lower impact in Freshwater Eutrophication, Terrestrial Ecotoxicity and Metal Depletion. Freshwater Eutrophication occurs when there is a discharge of nutrients into freshwater that create an increase in the nutrients level of the freshwater [9]. A simple definition for Terrestrial Ecotoxicity is the impact of chemicals on the terrestrial ecosystem. The last category with less impact created by traditional Portland cement, Metal Depletion, refers to the use of virgin metals.

After reviewing the main contributors to the environmental impacts of the AAM design mix, it was found that the main contributor for all of the categories in which the AAM had a higher impact was the potassium hydroxide production. However, it can be seen the effects from Freshwater Eutrophication and Terrestrial Ecotoxicity are minimal. For Metal Depletion, the kg FE equivalent for the potassium hydroxide is 4.56, the next contributor is the energy used for mixing the raw materials. The use of renewable energy sources, for example wind energy, will help to reduce the overall impact of the AAM making it an even more sustainable viable replacement.

5. Conclusions

Both methods used in this study demonstrate that the Alkali-Activated Material (MT12) has a significantly lower environmental impact in most categories. This suggests that AAMs would be a viable and much less impactful alternative to Portland cement hydraulic mortar. An important category for these materials is the Global Warming Potential, Climate Change in ReCiPe method, tables 3 and 6 show that the AAM has approximately 25% of the Portland cement hydraulic mortar impact. It was also demonstrated that the potassium hydroxide used in the activating solution is the main cause of environmental impact. Further work will search for a replacement to the potassium hydroxide activator as this would alleviate many of the environmental impacts of Alkali Activated Mortar. Another area for future research will be the possibility to reactivate or recycle this AAM making it an even more sustainable material.

6. References

[1] Hendriks C, Worrell E, Dejager D, Block K and Riemer P Emmision reduction of greenhouse gases from the cement industry [Available]: http://ieaghg.org/

[2] Oss H G 2013 Cement (Accessed Jan 2015)

[3] Provis J D J 2014 Alkali Activated Materials *RILEM state-of-the-art Reports* (Springer)

[4] Matheu P S, Waldron A and Varela B 2014 Optimization of Alkali Activation of Ground Granulated Slag with Mining Tails *NTCC2014: Int. Conf. on Non-Traditional Cement and Concrete (Brno, Czech Republic, 2014)*

[5] *Life Cycle Assessment (LCA) EPA* 2006 Environmental Protection Agency

[6] Ramli M and Dawood E T 2011 Comparative Study Between Flowable High Strength Mortar and Flowing High Strength Concrete *Concrete Research Letters* **2** (2) 249–261

[7] Barbara C Lippiatt 2007 *Building for Environmental and Economic Sustainability Technical Manual and User Guide* (EPA, National Institute of Standards and Technology)

[8] Goedkoop M, Heijungs R, Huijbregts M, Schryver An D and Struijs J 2009 *ReCiPe 2008 A Life Cycle Impact Assessment Method Which Comprises Harmonised Category Indicators at the Midpoint and the Endopoint Level* First edition

[9] Azevedo L B, Francesca V, Andrew D, Henderson, Rosalie Van Z, Jolliet O and Huijbregts M A J 2015 *Freshwater Eutrophication* LC-Impact

IOP Conf. Series: Materials Science and Engineering **96** (2015) 012081 doi:10.1088/1757-899X/96/1/012081

Statistical Analysis of Concrete Cover in New Highway Bridges

O S Ekolu[1,2] and F Solomon[1]
[1]University of Johannesburg, Department of Civil Engineering Science, Auckland Park 2006, Johannesburg, South Africa

E-mail: sekolu@uj.ac.za; sekolu@gmail.com

Abstract. Quality control is one of the important aspects of any major construction works, which is to be undertaken to ensure work execution according to design requirements. The work presented in this paper involved measurement of cover thickness in three newly constructed highway bridges. Testing was conducted to ensure that specified requirements were attained prior to commissioning of the structures; otherwise, the quality control survey would identify problem areas for consideration of corrective measures.
A total of 328 data sets were obtained during cover measurements. In this paper, the results obtained are discussed and evaluated. Data are characterised on the basis of statistical quantities.

1. Introduction

Concrete cover to steel reinforcement in reinforced concrete structures plays a significant role related to the service life of structures. This thin layer of concrete, typically 20 to 40 mm [1,2], depending on the severity of the exposure environment, is largely responsible for providing protection against corrosion of reinforcement. Worldwide, steel corrosion is the most predominant deterioration problem in reinforced concrete structures. The primary causes of corrosion are carbonation and chloride attack resulting from ingress of aggressive agents, CO_2 and chloride ions respectively. Concrete being non-metallic is, firstly, not subject to corrosion and secondly, the concrete matrix is highly alkaline with pH ranging from 12.6 to 13pH. At such a high pH, the steel embedded in concrete is protected and will not corrode. However, the ingress of CO_2/Cl^- ions causes loss of alkalinity, de-passivating steel and allowing corrosion to ensue. Clearly, the time taken for the aggressive agents to reach the level of steel is a defining point for service life of the structure.

It should also be considered that the concrete cover is usually a vulnerable layer of concrete during construction. This emanates from the fact that during construction, the cover is exposed to

[2] To whom any correspondence should be addressed.

environmental factors that lead to rapid loss of moisture to the atmosphere, while the interior concrete typically retains higher moisture levels. Accordingly, the permeability of cover concrete can be higher than that of the interior concrete. If the site curing applied is poor or inadequate, this effect can be exacerbated. Similarly, adequate compaction is required to ensure attainment of pore tortuosity towards lower permeability. Therefore the most important quality control considerations in cover placement are attainment of specified cover thickness, adequate curing and compaction [3]. Achieving the specified concrete cover during construction is usually quite difficult and leads to many failures [4]. Even when detailing is done correctly and spacers are properly placed, the loads applied during placement of fresh concretes and associated pressures may cause movement of the placed steel reinforcement. These effects from the construction process can cause lower cover and high variability of results in certain sections of elements.

Not only is concrete cover a major factor in the durability and service life of structures, its thickness has a significant structural effect on the capacity of elements to resist applied loads. The structural effects of concrete cover on members are two-fold: Firstly, a lower cover could reduce the moment capacity of a flexural member such as a slab or beam. Secondly, it would also rapidly lead to corrosion of reinforcement, which when combined with reduced compression area, can have the dual effect of causing a greater reduction of structural capacity. Figure 1 shows a suspended floor slab and a beam, experiencing severe delamination and corrosion of steel in the tension side. At this point, the slab has undergone all initial stages of deterioration from cracking, spalling to delamination; superimposed on these effects is the loss of steel area due to corrosion.

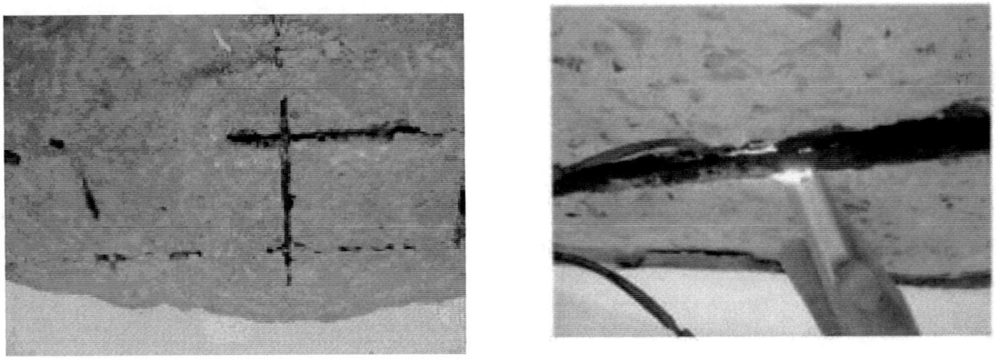

(a) (b)

Figure 1. Effects of thin concrete cover on corrosion of steel reinforcement; (a) slab soffits; (b) beam [5].

Tapan and Aboutaha [6] reported that the ratio of cover thickness to reinforcement bar size (C/D) has major influence on corrosion-induced cracking. The higher the C/D ratio, the greater the amount of steel corrosion required to cause cracking. For example, it was reported that 2.25% steel corrosion was required for cracking to occur under C/D =1, while 5.25% corrosion was required for cracking under C/D = 2.5. Yagaanbuyant and Bayar [7] found that a small change in cover thickness can lead to significant decrease in the moment capacity in floor slab. A reduction in cover thickness by 10 mm and 20 mm led to a moment capacity decrease of 12.5% and 21.6%, respectively. Investigations by Al-Kubaisy and Jumaat [8] found that by adding a layer of ferrocement cover on the tension side of

IOP Conf. Series: Materials Science and Engineering **96** (2015) 012081 doi:10.1088/1757-899X/96/1/012081

floor slabs, not only the capacity of the slab increased but the load or moment to first crack also increased.

Keeping these effects of concrete cover thickness in mind, it is therefore important to ensure that the construction processes not only achieve the specified cover thickness and quality during construction, but that quality control testing is conducted before commissioning of a new structure. Often, the project stakeholders, such as the contractor, consultant or owner of the structure would require quality control testing for cover thickness. This is done to determine the cover thickness in the structure as-built, against the specified requirements. Typically, testing for cover thickness may reveal areas or sections of the structure where the cover thickness may have fallen below the specified minimum thickness. Such findings provide an opportunity for corrective measures to be applied before commissioning of the structure. Taking remedial actions at this early stage would save future high costs of premature repairs that would result from early deterioration.

A standard covermeter is the most commonly used technique for conducting cover survey in massive structures such as bridges and buildings. Basically, it involves dividing the structure into a grid pattern, detecting the position of steel bars running in both directions and measuring the cover thickness over the bar location. Modern covermeters are quite sophisticated and measure a number of parameters including the location of bars and the bar size of steel reinforcement in addition to the cover thickness [9]. Most covermeters operate on the basis of electromagnetic fields; so care has to be exercised against factors that may influence the measurements. Another modern but quite complete technique that has been investigated for cover measurement is the Ground Penetrating Radar (GPR). GPR is often used for sub-surface imaging in investigation of pavements, concrete, or detection of subsurface objects such as metals in soils. Although not commonly used in cover measurement in the superstructure, some case study measurements have been conducted with satisfactory results [10].

2. Case studies

The purpose of this work was to determine and characterise the variability of cover in three newly constructed highway bridges, arbitrarily referred to as *1G*, *2R* and *3C*. The variability examined was expected to be a result of various natural site factors including materials used, site conditions and, more importantly, the construction process. The new bridges were among several highway structures that were newly constructed for the N17 highway extension. The highway is the main artery carrying eastbound traffic to and from Gauteng through Mpumalanga to neighbouring provinces and coastal cities. The three-span new bridges were approximately 40 to 50 m long and 10 to 12 m wide. Each bridge consisted of two piers and two abutments with the main (centre deck) spanning between the two piers, and two end jack spans running between the piers and abutments.

Cover survey on the bridges was conducted upon completion of the construction of the whole bridge superstructure. The survey was conducted based on different elements, consisting of the deck, piers and abutments. This was done in such a way as to survey the whole bridge structure. Each element was divided into a grid pattern that was used to select points of measurements. A view of one of the bridges can be seen in figure 2.

Figure 2. A newly completed highway bridge structure.

3. Results and discussion

Three bridges, *1G, 2R* and *3C*, were constructed consecutively and the survey was done once the superstructure of each bridge had been completed. A total of 111, 107 and 110 data points were measured for *1G, 2R* and *3C* bridges respectively. These measurements are distributed among the structural elements as shown in table 1. Measurements done on piers included pier heads. Since grid patterns were used, the distribution of the data points across the elements depended on the surface area of the element. Hence, the deck had more measurements than the piers and abutments. All deck cover measurements were made at the deck soffit, i.e. the tension side.

A cover depth of 40 mm was specified for the superstructure elements. As seen in table 1, the cover placed in the structure was in the range of 50 mm, which exceeded the minimum specified cover. It should be underscored, however, that the achievement of the minimum cover does not necessarily imply that all sections of the structural elements achieved the required criteria. Typically, certain sections may have very low cover while others may show excessively high cover. The scatter in data gives an indication of the uniformity of cover thickness as built. For this purpose, scatter diagrams of results have been plotted in figures 2, 3, and 4 for the bridges *1G, 2R* and *3C* respectively. It is evident, for example, in figures 3 and 4 that the deck soffits exhibit cover thickness values as low as 30 mm while others were as high as 67 mm. Similar scatter can be seen in figure 2 data for the abutments.

Histograms shown in figures 2, 3 and 4 show data which are characterised by a normal distribution curve, indicating normalcy of the data generation process. The histograms for bridge *1G* gave a much broader peak than the other two bridges 2R and 3C while the range of cover thickness remained similar for all the bridges. Table 1 also gives the standard deviation for the various elements and the overall coefficient of variation (CV) of the measurements. Of all the structural elements, only the bridge decks showed a consistent standard deviation of about 7.0. The piers had a relatively low standard deviation of 5.0 for bridges 1G and 3C but for some reason, the corresponding standard deviation was very high for bridge 2R. Of all the elements, abutments gave the highest standard deviation, ranging from 10 to 11 compared to 5 to 7 for the deck and pier. This observation may be related to higher quality control measures being applied to the deck and piers, compared to abutments, as the former elements are more crucial for load bearing than the latter. In all three bridges, however, the CV is consistent, falling in the range of 18 to 21%.

In order to investigate whether there may be some bias in data in relation to the type of structural element, statistical data were analysed according to the type of element. Data from all three bridges were combined according to the type of element i.e. decks, piers and abutments. Figure 5 gives a plot of histograms for the bridge decks, piers and abutments. It can be seen that a relatively large number of measurements were done on the bridge deck soffits compared to the piers and abutments. This was directly related to the relative surface areas of the elements, of which the bridge decks had larger areas. It was mentioned earlier that the selection of the measurement points was based on a grid spacing, which amounted to more data points being measured for elements of larger surface areas. Nonetheless, it is clear that all the elements gave normal distribution curves. Note that all the histograms in figure 5 are plotted on the same scale.

Table 1. Cover survey results for new highway bridges.

Bridge	Element	No. of data points	Average cover (mm)	Standard deviation	Coefficient of variation
1G	Deck	74	49	7.6	0.194
	Piers	11	50	5.2	
	Abutments	26	50	15.0	
	All elements	111	50	8.8	
2R	Deck	78	46	6.8	0.212
	Piers	11	62	18.3	
	Abutments	18	55	9.8	
	All elements	107	54	11.0	
3C	Deck	54	47	7.1	0.184
	Piers	20	49	4.8	
	Abutments	36	55	11.4	
	All elements	110	50	7.1	

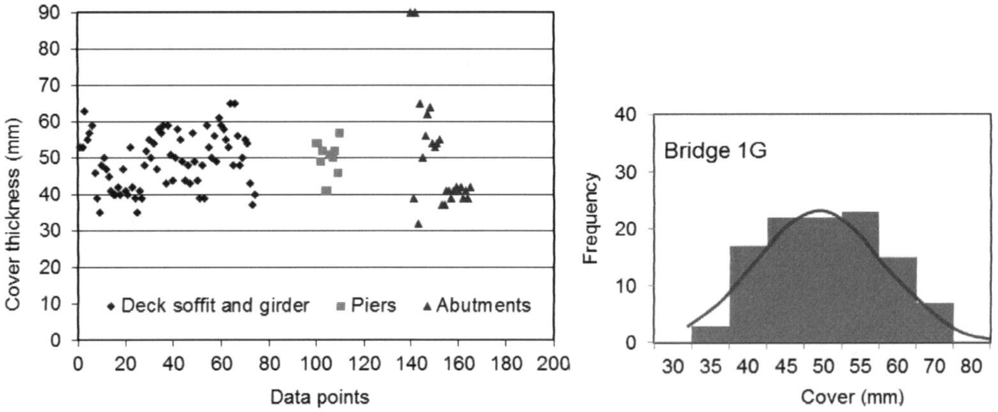

Figure 2. Scatter plot and histogram of cover for bridge *1G*.

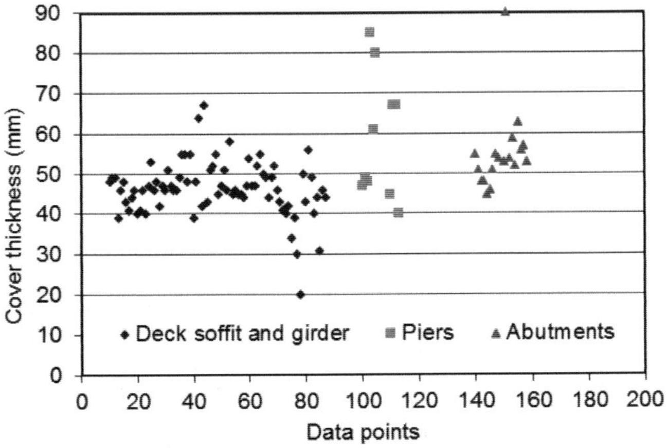

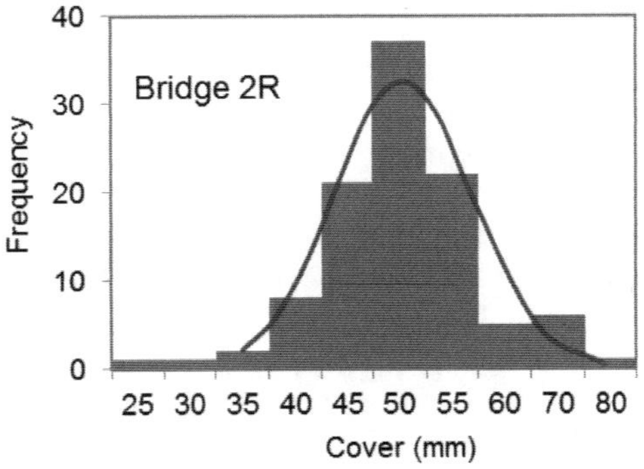

Figure 3. Scatter plot and histogram of cover for bridge *2R*.

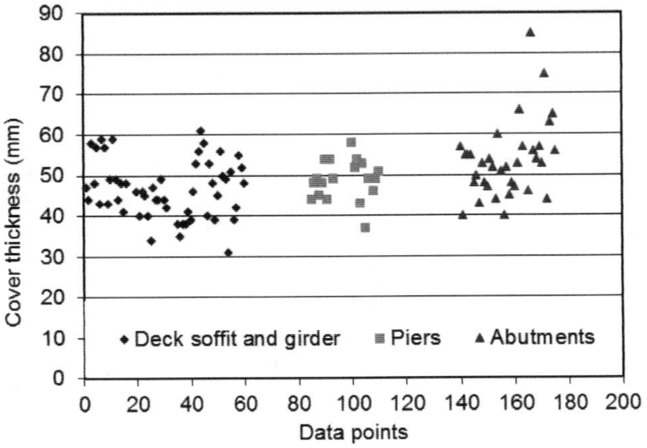

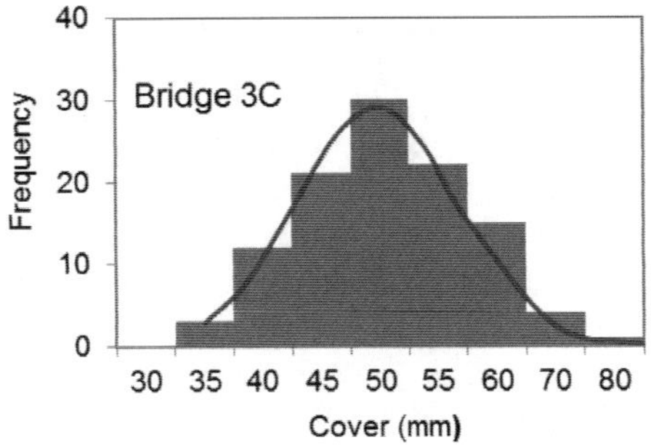

Figure 4. Scatter plot and histogram of cover for bridge *3C*.

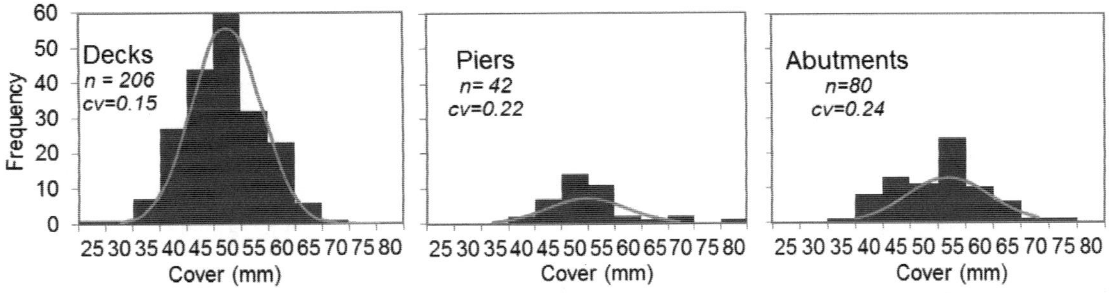

Figure 5. Histograms of cover for different structural elements (n – number of data points, CV – coefficient of variation).

In table 1 it can be seen that in all the bridges, decks had a cover thickness that was lower by about 2 to 3 mm, compared to the cover in piers and abutments. This may be attributed to the flexural behaviour of decks, in which case, steel reinforcement is vulnerable to deflection movement during concrete placement and compaction, unlike in compression members (piers and abutments) where no such problems may exist. Figure 5, however, shows that decks had the lowest CV of 0.15, much lower than CV = 0.22 for piers and CV = 0.24 for abutments. This pattern appears to indicate that much tighter quality control may have been placed on deck construction relative to the construction of other bridge elements.

4. Conclusions

In the foregone work, statistical analysis was conducted on the cover thickness data acquired from three new highway bridges. The inspection was done for the purpose of quality control. It was found that the specified cover depth of 40 mm was met with a margin of safety of 10 mm. The standard deviations of the deck and piers were generally in the range of 5 to 7, which indicates good quality control with respect to the material and construction execution. However, bridge 2R gave a particularly high standard deviation but this was due to the presence of higher cover while all values exceeded the minimum requirement. Similarly, abutments gave higher standard deviations but it was due to the presence of much higher values of cover rather than the opposite. The overall coefficient of variation was consistent in all bridges, giving values of coefficient of variation of 18 to 21%.

Bridge decks, which are flexural members, consistently gave a lower cover than compression members, consisting of piers and abutments.

5. References

[1] SANS 10100 2000, Code of Practice for the Structural use of Concrete, Part 1, & 2, Part 1: Design, Part 2: Materials. Pretoria, South African Bureau of Standards (2000) limited

[2] Kulkarni V R 2009 Exposure classes for designing durable concrete *The Indian Conc. J.* 23–43

[3] Merretz W and Smith G 2015 Achieving concrete cover in construction, Structural Concrete Industries (Aust) [Available]: http://www.sciaust.com.au/pdfs/CIAarticle.pdf

[4] Gjørv O E 2013 Durability design and quality assurance of major concrete infrastructure *Advances in Concrete Construction* **1** (1) 45–63

[5] Ekolu S O and Allan M 2012 Treatment of a stochastic service life prediction model to an evaluation of a distressed two-story RC building *Proc. 3th Int. Conf. on Concrete Repair, Rehabilitation and Retrofitting (ICCRRR), (South Africa, Cape Town, 3–5 September 2012)* p 451–457

[6] Tapan M and Aboutaha R S 2011 Effect of steel corrosion and loss of concrete cover on strength of deteriorated RC columns *Construction and Building Materials* **25** 2596–2603

[7] Yagaanbuyant D and Bayar J 2011 The influence of concrete cover to protect reinforcing bar on load carrying capacity of floor slab *The 12th East Asia-Pacific Conference on Structural Engineering and Construction, Procedia Engineering* **14** 2254–59

[8] Al-Kubaisy M A and Jumaat M Z 2000 Flexural behaviour of reinforced concrete slabs with ferrocement tension zone cover *Construction and Building Materials* **14** 245–252

[9] Barnes R and Zheng T 2008 Research on Factors Affecting Concrete Cover Measurement *NDT.net – The e-Journal of Nondestructive Testing* p 8

[10] Hasan Md I and Yazdani N 2014 Ground penetrating radar utilization in exploring inadequate concrete covers in a new bridge deck *Case Studies in Construction Materials* **1** 104–114

IOP Conf. Series: Materials Science and Engineering **96** (2015) 012082 doi:10.1088/1757-899X/96/1/012082

Usage of Alternative, Environmentally Acceptable Materials –Experience from Eastern Croatia

I Barišić[1], M Zagvozda[1,2] and S Dimter[1]
[1]Faculty of Civil Engineering, Drinska 16a, 31000 Osijek, Croatia

Email: mzagvozda@gfos.hr

Abstract. The concept of sustainability should be the main guiding principle in the construction industry today. It mandates conservation of natural resources and thus lower impact on the environment. In road construction, part of construction industry that consumes largest quantities of natural materials, sustainable building and maintenance of roads is possible trough application of secondary materials. Usage of industrial and construction waste presents energy, ecologically and financially effective alternative. Republic of Croatia, even as a new member of the European Union, still lags behind the well-established practices of the application of alternative materials in different European countries. The reasons for this can be found in the current legal and technical regulations for alternative materials. In this paper, the existing regulations for alternative materials and the impact they have on the application of these materials in practice in the region of eastern Croatia will be shown.

1. Introduction

The concepts of sustainability should be the norm in all aspects of human development today. In construction industry, this mandates conservation of natural resources and lower impact on the environment. Application of secondary "alternative materials" in road construction as one of the biggest exploiters of natural materials presents a more rational, economically and ecologically suitable solution. The term 'alternative material' covers all materials except natural virgin materials, and usually refers to construction and demolition waste and a variety of industrial by-products. In the countries of the European Union a wide range of such materials is widely used as a substitute for standard building materials for a number of years. For these materials to be applicable, they should meet certain criteria, show a satisfactory level of performance and be cost-effective when compared to traditional materials. As EU Waste Framework dictates the aim for 70% of all construction waste to be recycled, re-used or otherwise recovered by 2020 [1], many universities, institutes and government organizations have been involved in research and development of secondary or recycled materials. The

[2] To whom any correspondence should be addressed.

European Commission supported two such projects, ALT-MAT and COURAGE [2], which gave insight into the application of these materials and helped in creating the legal framework.

In Croatia, traditionally used natural materials still present the largest share of materials used in road construction. As a large proportion of the planned road network has been built, the amount of reconstruction and maintenance works is increasing rather than the construction of new roads. Therefore, recycling of existing pavement courses and construction waste should be a prevailing solution. According to Sherwood [3], encouragement towards application of alternative materials should be done through development of the new or supplement of the old standards and technical requirements, development of guidelines that should inform about the possibilities of application of those materials, various administrative measures (such as taxes on natural materials or waste disposal), as well as using domestic and foreign examples to educate and change attitude towards alternative materials.

In this paper, the current state of regulation concerning alternative materials in road construction is shown along with the research that has been done and the discussion of the state of application in road building practice of eastern Croatia.

2. Legal framework

The first step towards systematic waste management, reduction of waste amounts and the number of landfills, with the aim of protecting the environment and natural resources, which also implies application of alternative materials, was done by passing Waste Management Strategy for the Republic of Croatia [4]. It establishes a concept without landfills, which encompasses avoiding the generation of waste, reduction of waste amounts and waste recycling. In order to implement the Strategy, a legal framework is necessary that deals with creation and recycling of waste as well as technical regulations for the application of such recycled materials in construction projects.

Construction waste is defined by the Regulations on construction waste management [5] as waste generated during the construction, reconstruction, removal and maintenance of existing buildings, and waste generated from excavation material, which cannot be used for construction without prior recycling. It was common practice for this construction waste to be disposed of in landfills, but the Waste Management Plan [6] defines that those materials that can be reused have to be separated and sorted from such waste. The emphasis on recycling is specially given for the waste from road construction, such as crushed asphalt or concrete, as it is necessary to take it into asphalt mixing plants to be recycled. Waste Act [7] allows contractors to reuse their own waste as building materials on the building sites where it was generated, but only if it complies with the Technical regulations on construction products [8] and the given norms.

General technical requirements for road works [9] that have been in use since 2001 define the conditions for works in road construction and provide the requirements for materials. According to these requirements, the contractors can apply alternative materials if they:

a) possess the same quality or properties;

b) are of similar quality and properties in accordance with the recognized technical rules (norms);

c) meet the essential requirements for buildings according to Construction Act.

Table 1. Possibilities of application of recycled asphalt aggregate [10].

Recycled asphalt aggregate origin	Application of recycled asphalt aggregate in asphalt courses				
	Wearing course	Binder course	Base course	Base-wearing course	Poured asphalt
Wearing course	Yes	Yes	Yes	Yes	No
Wearing + binder	No	Yes	Yes	Yes	No
Binder course	No	Yes	Yes	Yes	No
Base course	No	No	Yes	No	No
Base-wearing course	No	No	No	Yes	No
Poured asphalt	No	No	No	No	Yes

General technical requirements do not give the possibility of application of recycled asphalt aggregate in production of new asphalt mixture, so for research purposes regulations for conventional materials had to be used. Therefore, from 2010 to 2012 new technical regulations for asphalt pavement layers [10] were being developed. Recycled asphalt aggregate, that is a product of milling of asphalt courses or crushing of asphalt slabs, is now allowed to be used as part of different bituminous mixes prepared by both hot and cold methods. Table 1 shows possibilities of application of recycled asphalt aggregate in different layers of pavement depending on their origin.

New regulations also define requirements for recycled asphalt aggregate to be used in new asphalt mixtures in terms of grading, proportion of bitumen and softening point of bitumen separated from the recycled aggregate. Before application, recycled aggregate has to go through testing defined by norms: sampling to assess homogeneity (HRN EN 932-1), bitumen extraction (HRN EN 12697-3 or HRN EN 12697-4), determination of bitumen softening point (HRN EN 1427), proportion of bitumen in reclaimed asphalt aggregate (HRN EN 12697-1), granulation of recycled asphalt aggregates (HRN EN 12697-2).

Beyond recycled asphalt aggregate, the possibility of application of fly ash from coal combustion as a filler and different types of slag (steelmaking, other metals) as an aggregate is mentioned for bituminous mixes.

3. Research into application of local alternative materials for road construction

Research on the application of alternative materials in Eastern Croatia is mostly related to Faculty of Civil Engineering Osijek, independently or in collaboration with few companies that are aware of benefits of such materials and want to improve their technology. A short overview of local research into the application of alternative materials for road construction is presented below.

3.1. Unbound aggregate base

The first instance of construction waste application in Croatia was done for unbound base courses in 2004 in Zagreb area [11]. Construction waste that was created in the course of demolition of existing buildings was crushed, sorted, sieved and then applied as unbound aggregate for transportation areas at the same location. Testing of compressibility modulus proved that this material is a sufficient substitute for natural aggregate both as a surface layer in construction stages and as a subgrade layer of the finished pavement.

3.2. Cement stabilized aggregate base

Research on cement stabilized base was done using alternative materials either as an aggregate or a binder. First research was done in 1996 with fly ash from the thermal power plant nearby Hungarian town Pecs. The research showed that cement stabilized mixture made with Drava River sand and 7 % of binder complies with Croatian regulations when maximally 25% of the cement is replaced with fly ash [12].

Researches of cement stabilized base with slag aggregate were done in order to define the optimum composition of the mixes that give great economical and resource savings [13] and to examine the suitability of 100% slag aggregate for base courses of pavement for heavy traffic [12]. First research showed that increase in slag content gives higher strength in mixes with low cement content and that mixes with 2 and 4% of cement fulfill strength class criterion, respectively. Second research confirmed that slag can be used for base courses of pavements for heavy traffic loads even though it requires more binder and water and develops lower compressive strength.

3.3. Asphalt courses

Research into application of alternative materials in asphalt layers in eastern Croatia is mostly done with recycled asphalt aggregate. Application was tested for asphalt base courses as partial replacement of both crushed stone aggregate and bitumen, as well as for the wearing course [14]. The first research justified application of 20, 25 and 30% of recycled aggregate from an economic and environmental point of view. While for the second solution a trial production with 10, 15 and 20% of recycled asphalt was carried out and trial road section was constructed to see the behavior of this mixes under traffic load and environment over time.

The use of recycled asphalt aggregate to a greater extent occurred during the Second program of reconstruction of national roads (Betterment II) [12]. Within this program, road sections were reconstructed with cold recycling methods by using cement, “geocrete” (commercial name for a special hydraulic binder) and water or foamed bitumen, cement and water as binders. Application of the second method served also for testing the technology and for defining the technical regulations.

4. Alternative materials in road building practice

During the CONWAS project [15], it was concluded that 80% of construction waste can be recycled and reused, but it was estimated that in 2006 only 7% of construction waste was recycled [4]. Through implementation of different regulations on waste management in the recent years both percentage of reported construction waste and the amount it has been recycled has risen (table 2). As it can be seen, in 2013 almost 72% of the reported construction waste was recycled [16]. Even though there is a positive trend, this is still a small amount considering the estimation that 2.3 million tones of construction waste is produced per year in the period from 2006 to 2015 [6].

Table 2. Amounts of reported and recycled construction waste [16].

Year	2010	2011	2012	2013
Reported construction waste [t]	220,911.16	186,960.98	317,310.84	374,407.36
Recycled construction waste [t]	67,237.43	46,848.23	115,477.98	269,427.67

That state of the application of recycled materials in road building practice is still poor; it testifies the fact that according to EAPA [17] in 2012 only 2% of production of new hot and warm asphalt mixes contained reclaimed materials. On top of that just 6 out of 55 asphalt plants in Croatia were fit for hot recycling.

Actual state in the application of alternative materials in road construction practice of eastern Croatia and reasons influencing such state were examined by a questionnaire survey. The survey was distributed to construction companies, which deal with road design, construction and maintenance [18]. It comprised 10 questions related to the experience in the application of alternative materials, reasons for usage of such materials and knowledge of technical regulations on this subject. In total 12 companies responded to the survey, 50% companies in the sample were engaged in construction, 22% in design and 28% in maintenance works.

When asked about the usage of alternative materials in their practice, 50% responded affirmatively, 33% do not use alternative materials and 17% of companies are still in research phases of possible application (figure 1). But the percentage of application of alternative materials in overall activity is quite low, because in almost 70% of companies alternative materials comprise up to only 15% of production (figure 2.). Other 23% of respondents that apply more than 15% of alternative materials in their work are in research only fields or that application is the usage of quarry waste in embankment/base layers.

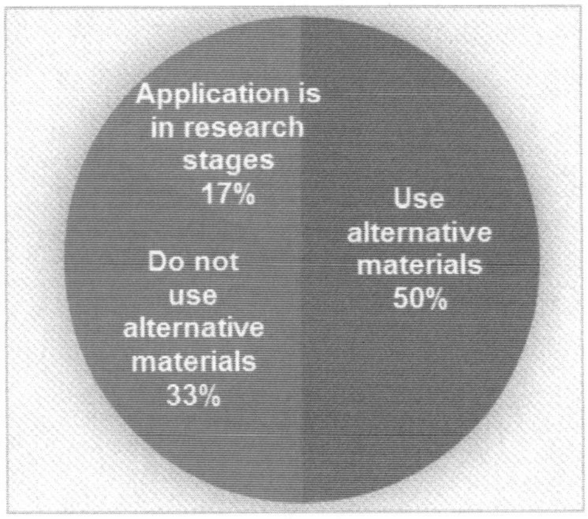

Figure 1. Application of alternative materials.

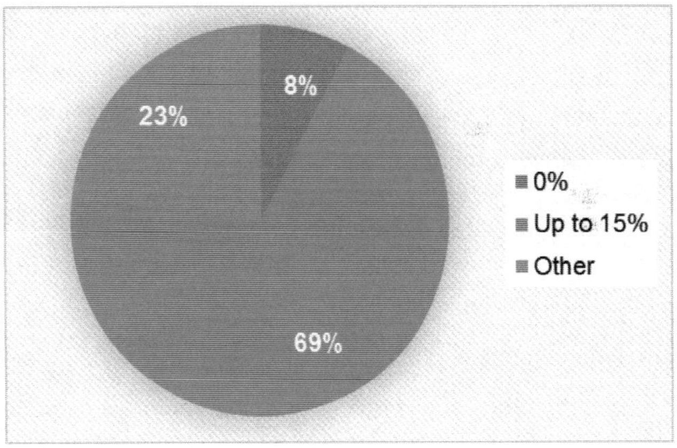

Figure 2. Representation of alternative materials in the overall production.

Out of possible materials to be used in road construction, the respondents mentioned quarry by-products and by-products of aggregate dust collection in asphalt plants, crushed concrete and brick as well as recycled asphalt as the ones they do apply in their practice.

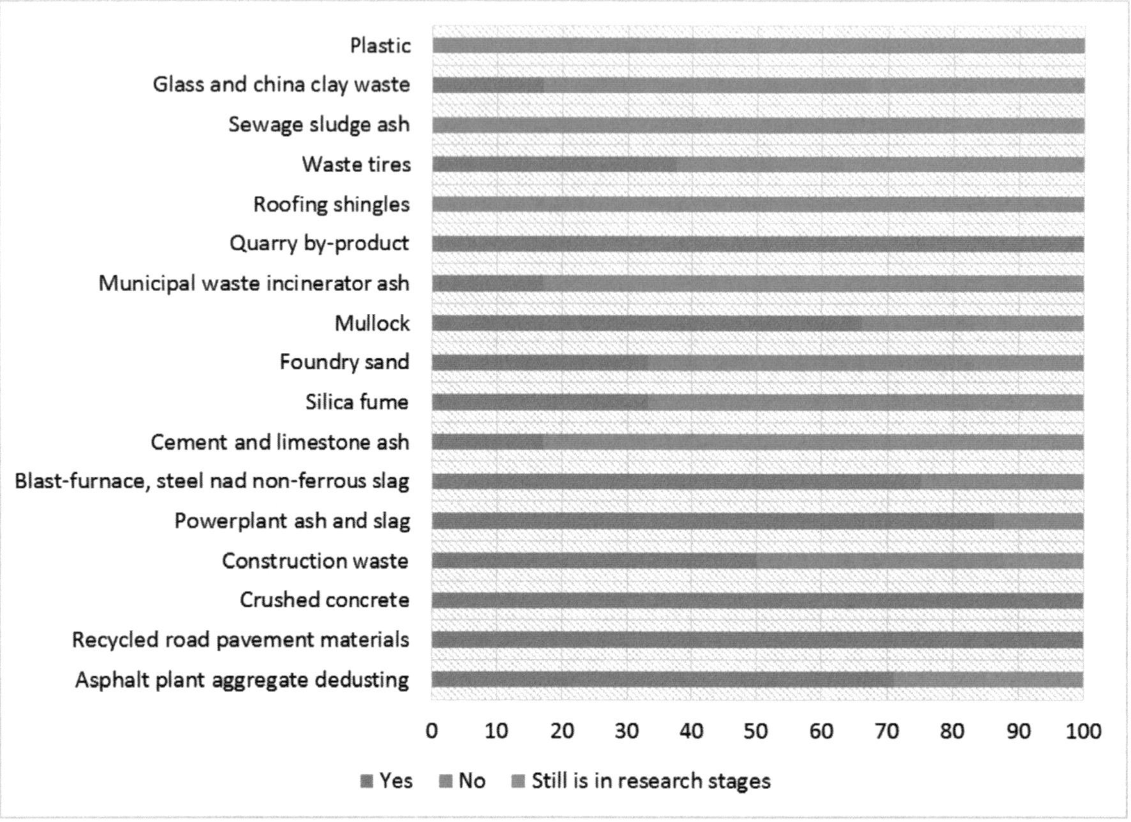

Figure 3. Potential of application of different alternative materials.

For a number of possible alternative materials, the respondents were asked to assess the possibility of whether those types of materials could be used in their field as is, whether more research should be done or whether their application is not possible. As is shown in figure 3, the greatest potential among all the materials already mentioned as being used is ascribed to different types of slag, construction waste and mullock, whereas shingles, plastics and sewage sludge ash do not have potential to be used or application is still in research stages.

Recycled asphalt is one of the most commonly used alternative materials in road construction that can be used as a wearing course, as aggregate in base and sub-base layers or as material for embankments. The survey showed that all of the respondents are familiar with different methods of asphalt recycling and 75% of them apply some method in their work (figure 4.).

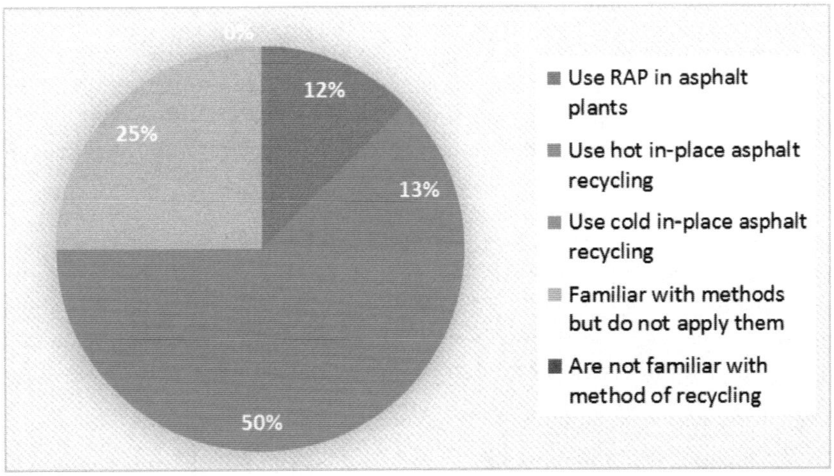

Figure 4. Application of recycled asphalt aggregate.

The biggest incentive for the usage of alternative materials can be found in ecological reasons and cost of these materials versus the cost of natural materials as the survey has shown (figure 5.). Respondents do not have good opinion with regard to the current state of law and regulations when application of alternative materials is concerned: only 38% are both familiar with the regulations and think it is sufficient for the field they are in, and the rest either are not familiar or do not think it is adequate.

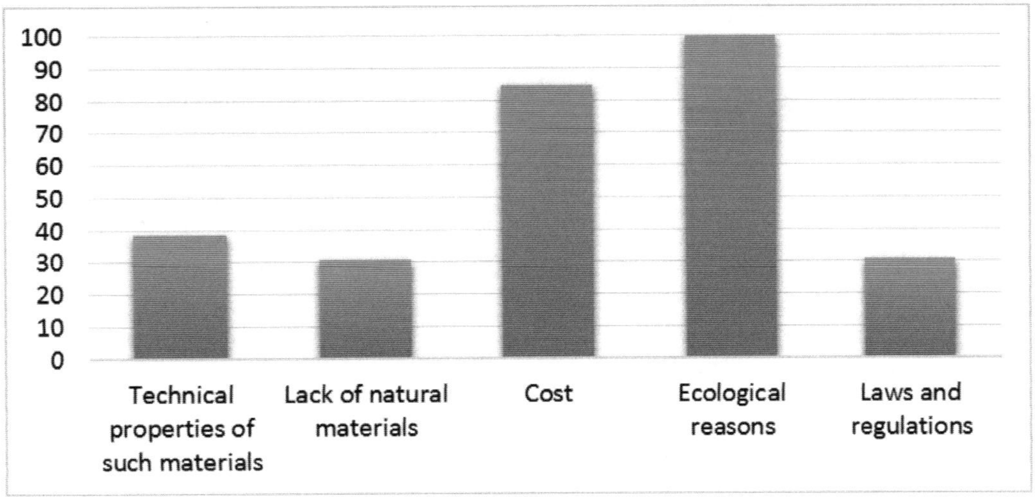

Figure 5. Reasons for the application of alternative materials.

When their own construction waste is concerned, even though it is allowed to reuse it as recycled material, only 27% do recycle, while others dispose their waste.

5. Conclusion

Even though Croatia recently became a member of the European Union, practice of application of more sustainable and ecological secondary materials still lags behind the well-established European practices. Some steps towards the improvement were made by adopting the Waste management strategy and its implementation. More construction waste is being reported and recycled every year and research is being done to support application of this materials. Research on recycled asphalt aggregate and its common application in other countries resulted in passing new technical regulations for asphalt courses. Such example of combining knowledge from alternative material research with experience from practice in other countries should be continued, since the main problem of application is presented in the lack of technical regulations, norms and criteria. The survey has shown that getting rid of that barrier should yield greater application of alternative materials as all respondents find them to be a better ecological solution and a possibility for financial savings.

6. References

[1] Directive 2008/98/EC on waste (Waste Framework Directive) [Available]: http://ec.europa.eu/environment/waste/framework/
[2] European Commissio 1999 *COURAGE*, Final report
[3] Sherwood P 2003 *Alternative Materials in Road Construction, A Guide to the Use of Recycled and Secondary Aggregates* Second edition (London: Thomas Telford Publishing)
[4] Waste Management Strategy for Republic of Croatia *Official Gazette* 130/05
[5] Regulations on construction waste management *Official Gazette* 38/08
[6] National Waste Management Implementation Plan for 2007-2015 *Official Gazette* 85/07, 126/10,31/11
[7] Waste Act, *Official Gazette* 178/04, 153/05, 111/06, 60/08, 87/09
[8] Technical regulation on construction products *Official Gazette* 33/10, 87/10, 146/10, 81/11, 130/12, 81/13, 136/14
[9] IGH 2001 *General Technical Conditions for Road Works, Book I: General Terms and Preparation Works* (Zagreb: Hrvatske ceste)
[10] *Razrada Tehničkih Svojstava i Zahtjeva za Građevne Proizvode za Proizvodnju Asfaltnih Mješavina i za Asfaltne Slojeve Kolnika* 2012 (Zagreb: Hrvatske ceste)
[11] Stanić A and Dimter S 2010 Application of recycled aggregates in pavement base courses *Proc. of 1st Int. Conf. on Road and Rail Infrastruction CETRA2010* ed Lakušić S pp 291–297
[12] Dimter S, Netinger I and Barišić I 2012 Hrvatska iskustva u primjeni recikliranih materijala u cestogradnji *Sabor Hrvatskih Graditelja 2012* Cavtat
[13] Barišić I, Dimter S and Rukavina T 2014 Strenght properties of steel slag stabilized mixes *Composites: Part B* **58** pp 386–391
[14] Dimter S and Androjić I 2014 Primjena recikliranog asfalta u proizvodnji asfaltnih mješavina *Dani Prometnica 2014-EU Fondovi i Projekti Prometne Infrastrukture* ed S Lakušić
[15] Primjena recikliranog agregata iz građevnog otpada [Available]: www.igh.hr/CONWAS/www/Primjena_HR.pdf
[16] Agencija za zaštitu okoliša 2010, 2011, 2012, 2013 *Izvješće o Podacima iz Registra o Onečišćavanja Okoliša*
[17] EAPA 2012, 2013 *Asphalt inFigures*
[18] Zagvozda M, Dimter S and Dolaček-Alduk Z 2012 Alternativni materijali – novi trend i izazov u graditeljstvu *Sabor Hrvatskih Graditelja 2012* Cavtat

IOP Conf. Series: Materials Science and Engineering **96** (2015) 012083 doi:10.1088/1757-899X/96/1/012083

Bearing Capacity Assessment on low Volume Roads

A Zariņš
Riga Technical University, Department of Roads and Bridges, Kalku Str. 1, Riga, LV–1658, Latvia

E-mail: atis.zarins@rtu.lv

Abstract. A large part of Latvian road network consists of low traffic volume roads and in particular of roads without hard pavement. Unbounded pavements shows serious problems in the form of rutting and other deformations, which finally lead to weak serviceability and damage of the road structure after intensive exploitation periods. Traditionally, these problems have been associated with heavy goods transport, overloaded vehicles and their impact. To find the specific damaging factors causing road pavement deformations and evaluate their prevention possibilities, and establish conditions that will allow doing it, the study was carried out. The tire pressure has been set as the main factor of load. Two different tire pressures have been used in tests and their impacts were compared. The comparison was done using deflection measurements with LWD together with dielectric constant measurements in a road structure using percometer. Measurements were taken in the upper pavement structure layers at different depths during full-scale loading and in different moisture/temperature conditions. Advisable load intensity and load factors for heavy traffic according to road conditions were set based on the study results.

1. Introduction

The most important road efficiency factor is the guaranteed road serviceability, which closely depends on the adequacy of actual traffic loads on particular road structure. Latvian road operation experience shows a number of problems associated with heavy transport and its impact on the road pavements. A large part of the paved road network has experienced strong rutting. It is mainly explained by overloaded cargo traffic loads. This leads to reducing of road capacity and shortens road service life. Regarding the roads with unbound pavement, they suffer from heavy loads during weak capability periods – winter and spring slush and during and after strong rainfall periods.

It is therefore important to find efficient possibilities to prevent road damage, while keeping necessary transportation volume.

The research is focused on low volume roads, typically with unbound pavement. Bearing capacity of such roads usually significantly decreases in slush seasons. Therefore responsible authorities close them for heavy transport during these periods. This results in economic disruption in the service area of the closed road.

To establish conditions that will allow using the road without interruptions and without damaging it, the study is carried out. The research methodology is based on road surface deflection measurements with low weight deflectometer (LWD) together with dielectric constant, temperature and moisture measurements in the road structure using percometer during full scale loading with maximum allowed axle load.

2. Comparative studies

According to Treube et al. [10], rutting in pavement surface layer decreases using lowered tire pressure, however, this phenomenon was less marked than expected. It has also been pointed out that low compaction and poor drainage contribute to significantly more rutting.

Bradley [1] reviewed research and user experience using variable tire pressures in the forest industry. In this review it is indicated that effect of reducing tire pressures on roads with thin pavements and low bearing capacity is essential in terms of road maintenance costs as well as construction costs and serviceability.

Studies of road structure using dielectric characteristics of pavement material in road structure are available from 1995 when percometer was used in tube suction test by Saarenketo and Scullion [8], [9]. Further, there were several studies with regard to the road pavement operational characteristics and dielectric properties of the road materials and soils: Saarenketo et al. [6] and [7]. All the studies mentioned were conducted under laboratory conditions.

Research on the road operating parameters considering their impact on the frost heave, during which dielectric properties of the medium are studied, is described by Saarenketo et al. [6] and Saarenketo and Aho [4]. The possibility to use the dielectric value of the medium under load as a parameter and characteristic of the loading impact on the road structure was also considered.

In the study of George [2], the elastic modulus calculations employing forward calculation equations (assuming an elastic half-space) obtained with deflectometer were regressed against laboratory resilient modulus. Among other it was concluded that the first sensor deflection exhibits a large variability. It was also detected that the first sensor modulus tends to be lower than each of the offset sensor modules when the surface material is softer than the underlying soil.

The research of Korkiala-Tanttu et al. [3] concentrates on the behavior of pavement with low bearing capacity under spring conditions and overload. Measurements were made in laboratory simulated conditions. The results led to conclusions that the majority of permanent deformations in all structures occurred in the subgrade sand, as well as that when the ground water was at the top of the subgrade (-0.5 m from the asphalt), a relatively larger share of permanent deformations occurred in the subbase gravel and the subgrade. When the ground water was lower (1.0 m), a greater share of displacements occurred higher up in the base course.

Zariņš [11] examined the feasibility to evaluate road pavement response under full scale loading in situ using the dielectric conductivity parameters of the road material. It was identified that evaluating change of dielectric parameters in unbound road material under loading it is possible to detect the structure condition and serviceability.

3. Dielectric behaviour of the road material

The dielectric permittivity of the granular material describes the degree of resistance of the media to the electric field. Dielectric permittivity of the subsoil material is determined by intensity of various processes in the subsoil structure. They include 1) the ionic and 2) dipolar polarisation and relaxation of the water molecules and 3) the atomic and 4) electronic resonances in the material media due to the

IOP Conf. Series: Materials Science and Engineering **96** (2015) 012083 doi:10.1088/1757-899X/96/1/012083

voltage applied to it. Dielectric permittivity $E_r(\omega)$ for soil or road material can be expressed with relationship:

$$E_r(\omega) = \frac{E(\omega)}{E_0} \tag{1}$$

Where, $E(\omega)$ –absolute dielectric permittivity of material, which depends on frequency and is a complex value.

E_0 – dielectric constant of material, or the relative permittivity of a material for a frequency of zero.

The character of each of these processes can be determined using different measurement frequency. The most important ones for the road structure are the first two values mentioned. The measurements to detect them must be carried out using the frequencies of about 2 kHz for detecting the ionic and domain polarization processes and of about 40-50 MHz for the molecular level polarization processes. However, the whole spectrum of the frequencies or at least both mentioned together can possibly raise interest in the subsoil material studies. The parameter measured with the percometer is the real part of the complex relative dielectric permittivity value E_r. Measurement E_r for the natural materials, including subsoil materials and other granular building materials, is relatively complicated. In order to avoid the impact of environmental heterogeneity on the conductivity parameters, it is important to choose the adequate frequency of measurement and the location of sensors. Also it is important to establish correct understanding about the water influence on the physical and chemical processes in the road material depending on actual chemical composition of the water. The conductivity of the granular material largely depends on its moisture, but much more – on the electrochemical properties of the water solution, i.e. salt content in the water as well as on the electrochemical properties of the medium particles. It is possible to use empirical relationships between conductivity E_r and the proportion of the water in the material – moisture W. In general, relation $E_r = f(W)$ at different moisture conditions and water electrical properties is non-linear and depends on water and material particle interactions. Most of the typical materials have fixed inherent values of W_v and E_r.

The main process, which can be analysed with respect to the road material response to the pressure, is the change of polarisation in the water solution in interaction with the soil particles and voids during the loading and consecutive volume change. Polarization reduces the electric field in the dielectric media because of the increase of polarization charge density on the particle surface. Consequently, the surface parameters of the media constituent material and their molecular structure must be taken into account while evaluating the measured dielectric data. The latter applies especially to the quantity of the water solution, density and its chemical structure. In order to compare the measurements made in different media, influence of all of the parameters must be precisely identified and evaluated. However, two different measurements taken at the same media in sufficiently similar conditions can be assumed as comparable, even without knowing those parameters, assuming that they are the same in both measurements compared.

The losses of material conductivity under pressure results from the inherent heat loss, the loss from the process of polarization of water molecules, and from the loss in the electrochemical processes between clay mineral particles. This phenomenon is the result of a complex frequency dependent dielectric permittivity. The real part of the dielectric permittivity for natural environments changes in the range of 1 for the air up to 81 for the water. Water is the component of the natural environment, including the unbound road materials, which to a great degree affect both the dielectric permeability and material mechanical properties, for example, the bearing capacity of the pavement structure. Note however that this effect also depends on other factors, for example, grading composition, etc. Road structure can contain several forms of water: adsorption (hygroscopic) water, capillary (matrix) water, and free water. Dielectric permeability of the water within soil or road material depends on the total soil particle surface area available for water molecules, on the polarization of the water molecules, as well as on the density of the soil. Thus, change of the material conductivity and hence the dielectric permittivity resulting from load or compression can be explained with the release of the colloidal particles and ions from the clay particle surface and their suspension in the free and matrix water. Others factors causing this effect are due to 1) suction and 2) pumping phenomenon in capillary material, as a result of cyclic loading. Loss of the dielectric capacity of the soil and the road material can also be explained with suction, when the free water from the upper layers is temporarily withdrawn to the lower layer as a result of the compression and subsequent release of the pavement structure.

Figure 1. The test vehicle.

4. Methodology and test site

The methodology of the research was based on the road bearing capacity evaluation using LWD together with dielectric permittivity measurements in the selected road structure using percometer. Measurements were made during full scale loading in different pavement structure layers at different depths and in different moisture/temperature conditions. The road structure performance was tested in two loading conditions while loading with standard six axle timber haulage vehicle (see figure 1): 1) with common tire pressure 0.8 MPa and 2) with the reduced tire pressure 0.35 MPa using CTI system installed on the same test car.

Figure 2. Test site.

Special infinity sign shape test site was prepared (see figure 2) in a way to simulate definite road conditions. Structure of the test road was made similar to most low volume roads under consideration. The structure of the road and sensor configuration is shown in figure 3. Water table during the first test session appeared at 0.65m below road surface. In the later sessions it was at the depth 0.70-0.80m.

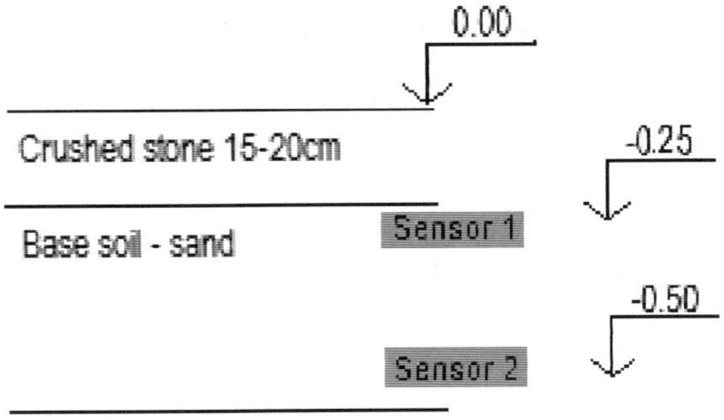

Figure 3. Configuration of sensors in the test station.

Loading was performed until direct and clear visible indications of pavement collapsing appeared. After each test session road structure was restored by profiling and compacting. Each test session contains test of road structure in the same conditions with both 0.35 and 0.8 MPa tire pressures in exactly that sequence.

Test equipment consists of the percometer equipped with two sensors, data logger, LWD and the portable weighting-machine for fixing determined axle loads in the particular test session.

The research methodology is based on the road structure dielectric constant measurements using a percometer during loading and LWD measurements before and after loading. Continuous E_r measurements were carried out at the depth of 0.25 and 0.50 m of the pavement structure during the test.

Project consists of four test sessions held in as different as possible road and moisture conditions – in late spring, before summer rain period, during and after summer rain period, during and after autumn rain period.

Figure 4. Installation of upper sensor.

The E_r evaluation of road pavement response is based on the consideration that the dielectric permittivity change of an unbound pavement material under the load correlates with the amount of the applied impact [11]. So the measurements of E_r in equal media conditions for different loads can be used for impact evaluation. Therefore it can be assumed that the induced parameter changes within a single measurement session (in equal conditions) are comparable.

Assuming that load impact influences the change of the dielectric parameters of the road material, and that to a great extent this change depends on matrix water, the measurements in the tests for detecting E_r were carried out using frequency 2 kHz. Samples of E_r were taken using sampling rate 100 Hz.

Duration between consecutive vehicle passes was set to approximately 60 seconds considering the speed of 20 km/h. Vehicle crosses the test section of the road in approx. 8-10 seconds. Dielectric permittivity measurements were taken using two sensors at 25 and 50 cm depth. These sensors were also used to determine the temperature and moisture in the structure.

Bearing capacity of road structure was tested using LWD equipped with set of three geophones. Deformation modulus E from the measured displacement was backcalculated according to Boussinesq equation. LWD measurements were done before, after and between all test sessions as well as few days prior and after. Measurements were located in wheel track and between wheel tracks. According to George [2], deflectometer measurements were done using 300 mm plate and interpreted considering displaced (second and third) geophones for road bearing capacity evaluation. The first geophone was located in the center of the loading plate (modulus signed as E(0)), the second and the third, respectively, at a distance 0.3 and 0.6 m from center of the loading plate (signed as E(0.3) and E(0.6)).

5. Test results

Overview of moisture dynamics during the research period is shown in figure 5. As shown, two of the performed test sessions were made in relatively dry conditions, while the other two – in wet. It was fixed that in all cases moisture level drops during the test, and to a higher degree it happens after passes using standard pressure – 8 bar tires. It can be also pointed out that higher drop appears at higher initial moisture. Such moisture drop can be explained by pumping action during load pass, which withdraws water from road structure to the surface. Higher tire pressure leads to deeper and more intensive action.

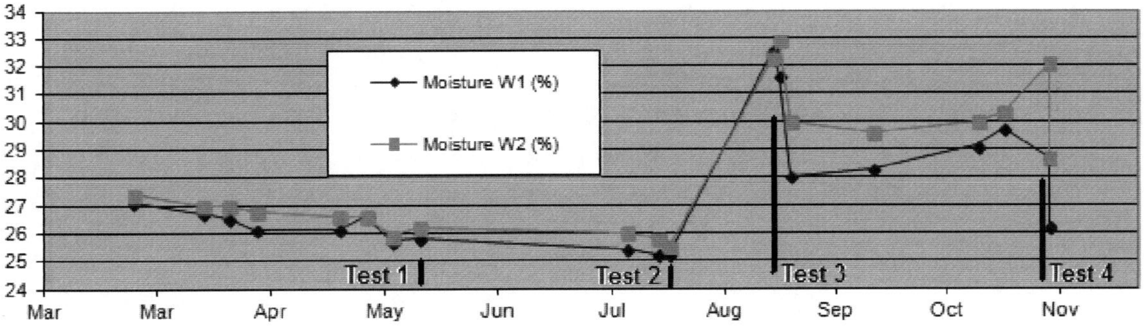

Figure 5. Moisture dynamics over test period (%).

Analysis of the correlation of bearing capacity with moisture does not show sufficiently reliable relationship (see figure 6). It can be explained by insufficient amount of data. However, it reflects the expected hypothesis that capacity is inversely proportional to moisture content. It is confirmed by measurements from all three geophones, and most strongly it appears in the upper part of pavement (see trendlines $E0/W_{up}$ and $E0.3/W_{up}$ in figure 6).

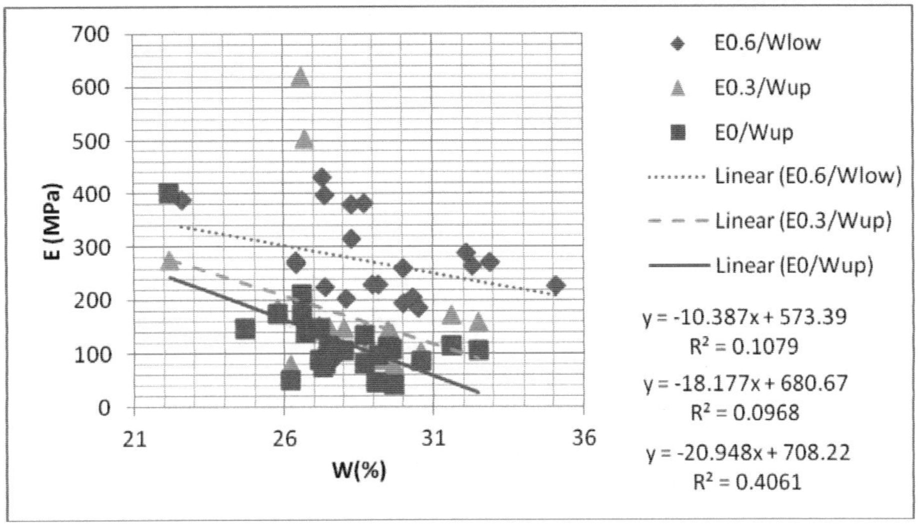

Figure 6. Relationship of moisture content to the bearing capacity.

IOP Conf. Series: Materials Science and Engineering **96** (2015) 012083 doi:10.1088/1757-899X/96/1/012083

The data obtained after test sessions indicate that using tires with lowered pressure can significantly increase service life of a road. The number of test vehicle passes with each type of tire pressure, until clear indications of pavement collapse was fixed, are collected in table 1. The indications of pavement collapsing are:

- Increasing dielectric permitivity under pavement (sensor 2). As shown in figure 5, until the 17th pass E_r in the lower sensor did not show any obvious response to the load. The 17th pass indicated clear response. This can be explained by loosening of the pavement bonds. The pavement does not adequately spread the load from wheels and this reflects on the deeper parts of road structure;
- After collapsing dielectric permittivity of the upper layer (sensor 1) decreases due to loosening of the pavement bonds and pavement material cracking in lateral displacement zone (see figure 7). It reflects as rising of dielectric resistance.
- Appearing of visible lateral cracks in the surface of pavement and rapid growth of lateral displacements along wheel tracks (see figure 8).

Table 1. Comparison of test vehicle passes until pavement collapse.

No of Test session	0.35 MPa (with CTI)	0.8MPa (standard pressure)
1	>22	17
2	>23	18
3	>26	17
4	=75	-

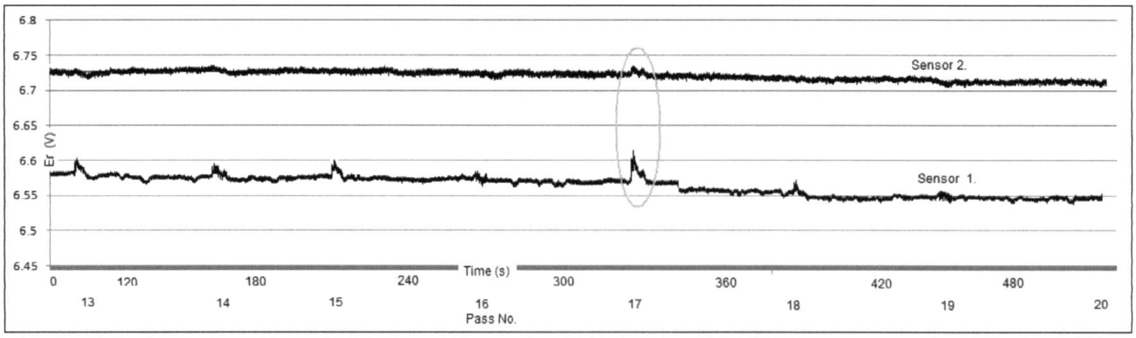

Figure 7. E_r response diagram during test drive with tire pressure 0.8 MPa.

Using lowered pressure (0.35 MPa) tires, collapsing conditions of the road structure were not observed after the designated count of passes.

Dielectric permittivity to a great extent depends on matrix water movement. In the first test session during regular test vehicle passes the pumping effect appears and it reflects in dielectric permittivity as from the lower sensor. E_r rises significantly when water reaches the sensor. In the upper sensor it appears after approximately 40 passes of the test vehicle in 3 hours. In this case it conforms to appearing of typical pavement collapse indications. However, in general any reasonable correlation between pavement moisture and bearing capacity during the observation period was not fixed, if considering those moisture levels typical for a particular structure. In some cases rising of moisture is

IOP Conf. Series: Materials Science and Engineering **96** (2015) 012083 doi:10.1088/1757-899X/96/1/012083

followed by rising of the bearing capacity. That can be explained by consolidation of matrix water or release of some molecular bound (hygroscopic) water during compaction of pavement material under load. This can also be justified by rising of temperature by 1- 2 degrees C in the upper sensor.

From the analysis of backcalculated modulus of deformation from all three LWD geophones (see figure 7) it can be concluded that the upper part or the layer of payment was compacted during loading while the lower part loses bearing either due to water pumping, or due to shear strain induced deformations in the base ground, or both together. It follows from the fact that during all test sessions bearing capacity backcalculated from the first geophone E(0) rose during loading with the lowered pressure tires, while other two E(0.3) and E(0.6) geophones gave falling capacities.

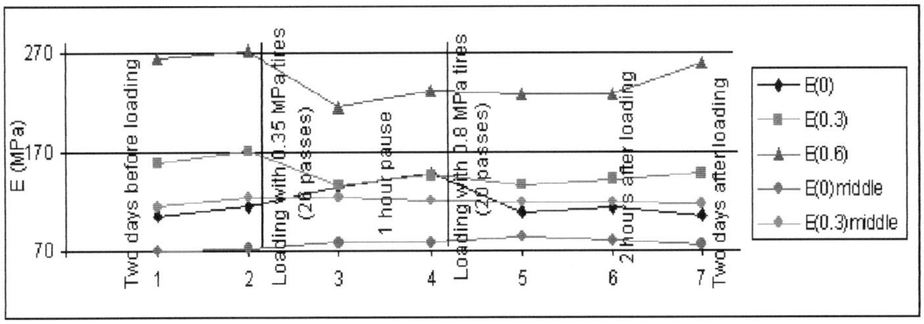

Figure 8. Dynamics of modulus of deformation during the test session.

The rising of bearing capacity E(0) during the use of lowered pressure tires once again points at the main advantage of CTI system – the upper layer compaction phenomenon.

Figure 9. Visible indications of pavement collapse.

IOP Conf. Series: Materials Science and Engineering **96** (2015) 012083 doi:10.1088/1757-899X/96/1/012083

After loading with full pressure tires, backcalculated capacities from all geophones showed falling bearing capacity. This to a greater extent relates to E(0), and confirms actual physical condition of the road after pavement collapse.

LWD measurements were performed in two zones of carriageway – in the wheel track and between the wheel tracks. Backcalculated modulus between wheel tracks typically was significantly lower and did not change during loading (see figure 6). This happens due to lower degree of compaction.

6. Conclusions

If managed in improper way, heavy traffic can seriously damage an unpaved road. If loading is intense, pavement material bonds are breaking down, shear stress in unbound pavement material and subsoil exceeds the threshold values and pavement collapses. As demonstrated in the current research, the structure of a typical low volume road can be destroyed by passing of less than twenty closely following full loaded timber haulage vehicles. To prevent this, considering of traffic intensity margins is necessary. These margins define the boundaries of pavement workability. Most of low volume roads with unbound pavement can sustain some ordinary heavy vehicle passes. Each of them breaks the pavement structure to some extent. Current research establishes that reasonable number of closely following heavy vehicle daily passes on a typical low volume road should not exceed 15-20. The exact number of close passes varies depending on subsoil, pavement, and drainage conditions. That number can be increased up to 50% by using lowered tire pressure. This prevents upper layer damage and supports pavement compaction. It was concluded that the reduced pressure in tires significantly reduces the impact on the upper layer of the pavement. However, in the depth of the base layer and subsoil this effect disappears. After intensive use, road structure must be profiled or repaired, if necessary, to reestablish loosened pavement bonds which to a great extent have hydraulic nature. Self recovering time after marginal loading lasts minimum 1-2 days.

It was also detected that pavement collapse is the function of passing frequency. Optimum frequency allows recovery of the pavement structure between consecutive passes. If pavement recovery is not complete, the next pass increases the previous damage. It is possible to put forward the hypothesis that optimum passing frequency is not a linear function. It depends on previous degree of damage and provided recovery time.

7. References

[1] Bradley H A 1997 *The Effect of Reduced Tire Inflation Pressures on Road Damage: A Literature Review* FERIC, Quebec, Canada

[2] George K P 2003 *Falling weight Deflectometer for Estimating Subgrade Resilient Moduli* Final report, FHWA, The University of Mississippi

[3] Korkiala-Tanttu L, Laaksonen R and Törnqvist J 2003 *Effect of the Spring and Overload to the Rutting of a low-volume Road* HVS-Nordic-research. Helsinki, Finnish Road Administration, Finnra Reports 22/2003

[4] Saarenketo T and Aho S 2005 *Managing Spring Thaw Weakening on low Volume Roads* Roadex II project, p 130 [Available]: http://www.roadex.org/services/knowledge-center/publications/permanent-deformation/

[5] Saarenketo T, Bell S, Berntsen G, Sundberg S and Vuontisjärvi E 2002 Roadex – benchmarking low traffic volume road condition management in EU northern periphery area *Proc. of the Sixth Int. Conf. on the Bearing Capacity of Roads, Railways, and Airfields (Portugal, Lisbon, 2002)*

[6] Saarenketo T, Kolisoja P, Luiro K, Maijala P and Vuorimies N 2002 Percostation for real-time monitoring of moisture variations, frost depth and spring thaw weakening *Transportation Research Board Meeting Proc.*

[7] Saarenketo T, Scullion T and Kolisoja P 1998 Moisture Susceptibility and Electrical Properties of Base Course Aggregates *Proc. of the Int. Conf. on the Bearing Capacity of Roads, Railways, and Airfields (Norway, Trondheim, 6-8 July 1998)* vol 3 pp 1401–10

[8] Scullion T and Saarenketo T 1995 Ground Penetrating radar technique in monitoring defects in roads and highways *Proc. of the Symposium on the Application of Geophysics to Engineering and Environmental Problems, SAGEEP (Florida, Orlando, 1995)*

[9] Scullion T and Saarenketo T 1997 *Using Suction and Dielectric Measurements as Performance Indicators for Aggregate Base Materials* Transportation Research Record vol 1577

[10] Treube M A, Evans G L and Bolander P 1995 Lowell test road: helping improve road surfacing design *Proc. of Sixth Int. Conf. on Low-Volume Roads, Mineapolis vol 2 (TRB, National Academy Press, Washington DC*

[11] Zariņš A 2011 The study of heavy load impact on road pavement *Proc. of 8th Int. Conf. on Environmental Engineering* (Vilnius, Lithuania: VGTU) vol 3 pp 1240–45

IOP Conf. Series: Materials Science and Engineering **96** (2015) 012084 doi:10.1088/1757-899X/96/1/012084

High Modulus Asphalt Concrete with Dolomite Aggregates

V Haritonovs[1], **J Tihonovs**[1,3] **and J Smirnovs**[2]
[1]Riga Technical University, Faculty of Building and Civil Engineering, Department of Roads and Bridges, Kalku Str. 1, Riga, LV-1658, Latvia
[2]Riga Technical University, Faculty of Building and Civil Engineering, Kalku Str. 1, Riga, LV-1658, Latvia

E-mail: janis.tihonovs@gmail.com

Abstract. Dolomite is one of the most widely available sedimentary rocks in the territory of Latvia. Dolomite quarries contain about 1,000 million tons of this material. However, according to Latvian Road Specifications, this dolomite cannot be used for average and high intensity roads because of its low quality, mainly, its LA index (The Los Angeles abrasion test). Therefore, mostly the imported magmatic rocks (granite, diabase, gabbro, basalt) or imported dolomite are used, which makes asphalt expensive. However, practical experience shows that even with these high quality materials roads exhibit rutting, fatigue, and thermal cracks. The aim of the research is to develop a high performance asphalt concrete for base and binder courses using only locally available aggregates. In order to achieve resistance against deformations at a high ambient temperature, a hard grade binder was used. Workability, fatigue and thermal cracking resistance, as well as sufficient water resistance is achieved by low porosity (3-5%) and higher binder content compared to traditional asphalt mixtures. The design of the asphalt includes a combination of empirical and performance based tests, which in laboratory circumstances allow simulating traffic and environmental loads. High performance AC 16 base asphalt concrete was created using local dolomite aggregate with polymer modified (PMB 10/40-65) and hard grade (B20/30) bitumen. The mixtures were specified based on fundamental properties in accordance with EN 13108-1 standard.

1. Introduction

If the local material does not fulfil the requirements, then one should seek the way for the improvement of its properties. If this is not possible, then one should seek the technological solution which will allow application of the weaker material [6]. One of the proper solutions might be the use of dolomite as a component of High Modulus Asphalt Concrete (HMAC). Knowing that the binder

[3] To whom any correspondence should be addressed.

courses situated between 5 and 12 cm below the road surface (figure 1) are subject to the highest stresses, high stiffness is probably the most important requirement for HMAC [2]. HMAC is a mixture of asphalt concrete designed for use in the base and binder courses of asphalt pavement. It has a closed structure with comparatively large content of bitumen. Hard road bitumen grades are applied, mainly 10/20, 15/25, 20/30 and polymer modified bitumen. Hard bitumen content ensures the mixture's resistance to rutting. Apart from workability, large content of bitumen assures fatigue durability and water resistance [7]. This type of an asphalt mixture is designed not only considering empirical properties but also performance based properties (rut test, stiffness modulus and fatigue tests) (SPENS-Sustainable Pavements for EU New Member States).

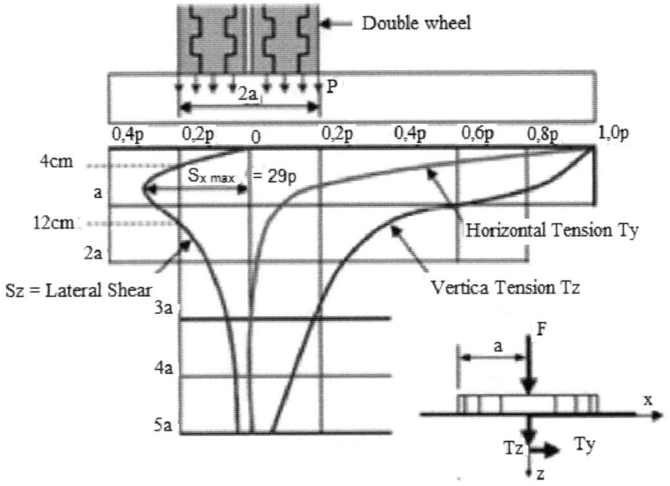

Figure 1. Lateral force diagram of heavy vehicle tire [4].

France was also one of the first countries in which mechanistic asphalt pavement design was introduced into the general practice (AFNOR – Organization of the French standardization system). In France, it is known under the acronym EM. In Poland, the acronym is AC WMS. Possible application of weaker mineral aggregate is one of the advantages of EM. Application of High Modulus Asphalt Concrete allowed for saving on asphalt pavement thickness thanks to higher stiffness modulus, which reduces tension strains in asphalt base layer. The aim of the paper is to develop high performance asphalt concrete for base and binder courses using only locally available aggregates – crushed dolomite. In order to achieve resistance to deformations at high ambient temperature hard grade binder was used.

2. Materials

The basic materials used in this study are fractionated crushed dolomite aggregate, unmodified hard grade bitumen B20/30 and SBS modified bitumen PMB 10/40-65 (B70/100 for reference mixture). Crushed dolomite aggregate was obtained from Pļaviņu DM ltd. (Latvia), and hard grade bitumen B20/30 – from Grupa LOTOS S.A (Poland).

2.1. Bitumen characteristics

The binder properties have been tested by means of conventional binder tests: needle penetration, softening point, aging and Fraas breaking point. The test results are listed in table 2.

2.2. Properties of dolomite aggregate

The test results of dolomite main properties show very low flakiness index – 5, high frost resistance with average MS (Magnesium sulphate test) value of 7 and low fines content – 0.6%. However, LA value is only 33. These aggregates are suitable for use as a component of High Modulus Asphalt Concrete, where LA value up to 40 (SPENS) is permitted. The properties of dolomite aggregate are shown in table 1.

Table 1. Physical and mechanical characteristics of the dolomite aggregate.

Physical and mechanical properties	Results	Requirement for HMAC-16	Requirement for reference mixture ACb-16	Standard
Los Angeles coefficient (LA), %	33	≤35	≤25	LVS EN 1097-2
Resistance to wear. Nordic test (AN), %	21	-	-	LVS EN 1097-9
Flakiness Index (FI), %	5	≤35	≤25	LVS EN 933-3
Water absorption, %	2	≤1	≤1	LVS EN 1097-6
Grain density, Mg/m3	2.80	-	-	LVS EN 1097-6
Fines content, %	1.0	-	-	LVS EN 933-1
Freeze/thawing (MS), %	7	≤25	≤25	LVS EN 1367-2

Table 2. Typical characteristics of the bitumens.

Bitumen				
Parameter	B 20/30	PMB 10/40-65	B70/100	Standard
Penetration at 25°C, dmm	25.3	35.0	80.0	LVS EN 1426
Softening point, °C	62.6	64.2	46.0	LVS EN 1427
Fraas temperature, °C	- 13.0	- 17.0	- 21.0	LVS EN 12593
Kinematic viscosity, mm2/s	1460	-	346	LVS EN 12595
Dynamic viscosity, Pa·s	3277	-	160	LVS EN 12596
Elastic recovery, %	-	85.0	-	LVS EN 13398
Ageing characteristics of bitumen under the influence of heat and air (RTFOT method)				
Loss in mass, %	-0.02	0.01	0.05	LVS EN 12607-1
Retained penetration, %	75.9	69.7	76.0	LVS EN 1426
Increase of a softening point, °C	6.9	5.5	5.4	LVS EN 1427
Fraas breaking point after aging, °C	-11.0	- 16.0	- 17.0	LVS EN 12593
Retained elastic recovery,%	-	> 50	-	LVS EN 13398

3. Asphalt mix design

HMAC-16 asphalt concrete mixtures have been designed by using unconventional raw materials (bitumen - B20/30, PMB 10/40-65 and dolomite aggregate LA > 30) (figure 2). The basic idea of HMAC is to design a mix with hard grade bitumen at high binder content [3]. The Marshall mix design procedure was used for the determination of the optimal bitumen content for the reference mixture, considering the mixture test results for Marshall stability and flow, as well as the volumetric values: air voids (V), voids in mineral aggregate (VMA) and voids filled with bitumen (VFB).

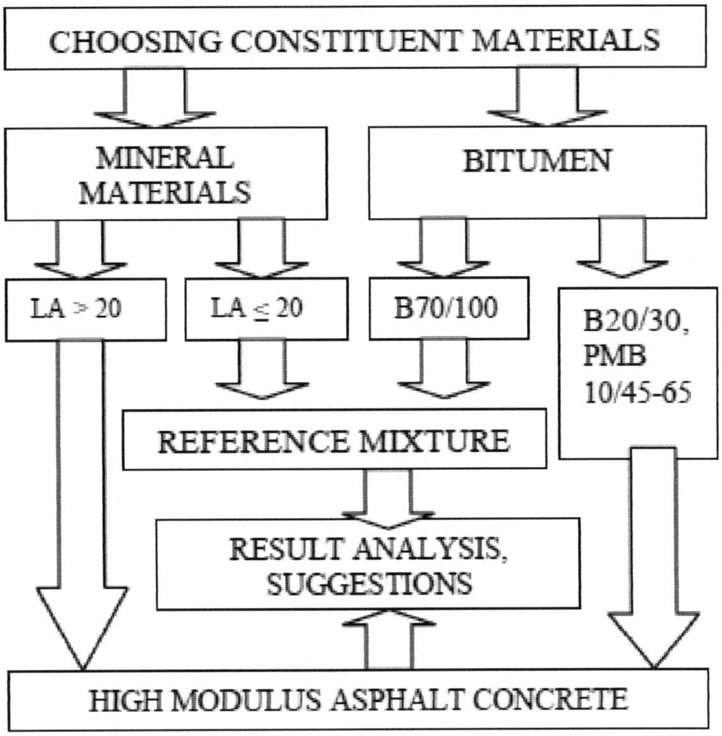

Figure 2. Experimental plan.

Table 3 shows all the designed and laboratory produced asphalt mixtures with different bitumens and dolomite aggregates.

Table 3. Types and components of HMAC mixes.

Bitumen	Mixture
PMB 10/45-65	HMAC-1/1
	HMAC-1/2
B 20/30	HMAC-2/1
	HMAC-2/2
	HMAC-2/3
B 70/100	Reference

Test specimens for Marshall Test were prepared in the laboratory by impact compactor according to LVS EN 12697-30 with 2×50 blows of hammer 140°C temperature (for mixes with conventional bitumen B70/100) and 150°C (for mixes with hard grade bitumen B20/30 and PMB 10/40-65). Mixing temperature for HMAC-16 asphalt concrete with conventional bitumen was 160°C and 170°C for HMAC-16 asphalt concrete with hard grade bitumen B20/30.

The following physical and mechanical characteristics of HMAC were determined:
1) Stiffness modulus, 10°C according to EN 12697-26 (4PB);
2) Resistance to rutting, 60°C and 10000 cycles, according to EN 12697-22;
3) Fatigue resistance 4PB-PR, 10°C and 10 Hz, according to12697-24;
4) Asphalt concrete mixture density, according to EN 12697-5;
5) Asphalt concrete mixture bulk density, according to EN 12697-6;
6) Air voids content, according to EN 12697-8;
7) Stability and flow, 60°C, according to EN 12697-34 (samples were prepared by 2×50 blows).

4. Results

4.1. Volumetric

Analysis of volumetric parameters of the different asphalt mixtures at different binder contents was performed. The results are presented in table 4. The binder content was optimized according to HMAC requirements developed in the SPENS programme (Sustainable Pavement for European New Member States).

Table 4. Volumetric properties of HMAC mixtures.

	PMB 10/45 -65			B20/30		B70/100
Parameter	HMAC-1/1	HMAC-1/2	HMAC-2/1	HMAC-2/2	HMAC-2/3	Reference
Bulk density, kg/m3	2455	2457	2430	2455	2457	2550
Maximum density, kg/m3	2555	2551	2586	2555	2551	2680
Voids content, %	3,9	3,7	6.0	3.9	3.7	4.85
VMA	17.8	18	18.3	17.8	18	17.6
VFB	78.2	79.6	67.2	78.2	79.6	72.4
Bitumen content, %	5.67	5.83	5.06	5.67	5.83	5.0

Table 5. Marshall test results.

Mixtures						
	PMB 10/45 -65		B20/30			B70/100
Parameter	HMAC-1/1	HMAC-1/2	HMAC-2/1	HMAC-2/2	HMAC-2/3	Reference
Bitumen content, %	5.67	5,83	5.06	5.67	5.83	5.0
Specimen height, mm	64.2	64.0	Not tested voids content > 5%	63.6	62.9	63.3
Stability at 60°C (kN)	16.5	15.8		16.6	15.4	12.0
Flow at 60°C (mm)	2.8	3.4		3.8	5.9	4.2

4.2. Marshall test

Table 5 contains Marshall test results at different binder contents for the mixtures that passed the requirement of having less than 5% air voids. The results show that HMAC mixtures have higher Marshall Stability compared to the reference mixture.

4.3. Wheel tracking test

A wheel tracking apparatus was used to simulate the effect of traffic and to measure the plastic deformations of the asphalt concrete samples. Tests were performed according to standard EN 12697-22 method B (wheel tracking test with small size device in air). This test method is designed to repeat the stress conditions observed in the field and therefore can be categorized as simulative. The resistance of asphalt mixture to permanent deformation is assessed by measuring the rut depth and its increments caused by repetitive cycles (26.5 cycles per minute) under constant temperature at 60°C. The rut depths are monitored by means of two linear variable displacement transducers (LVDTs), which measure the vertical displacements of each of the two wheel axles independently as rutting progresses. The obtained results after 20,000 wheel passes demonstrate the highest rut depth of 5.7mm for the HMAC 2/3 mixture with 5.83% bitumen content. However, the results show that HMAC 2/2 with a little lower bitumen content (5.70%) have higher rut resistance than HMAC 2/3. This can be explained by increasing of mixture stiffness due to decrease of bitumen content (see table 7). The reference mixture demonstrates similar result having 5.3mm rut depth. HMAC 1/1 and HMAC 1/2 mixture having 5.67% polimermodified bitumen PMB 10/45-65 content shows the highest rutting resistance with 3.8mm. This can be explained by high modified bitumen PMB 10/45-65 elastic properties when there is no difference between bitumen content 5.7% and 5.8%. Figure 3 and table 6 summarize the wheel tracking test results. Table 7 shows results compliance with SPENS requirements.

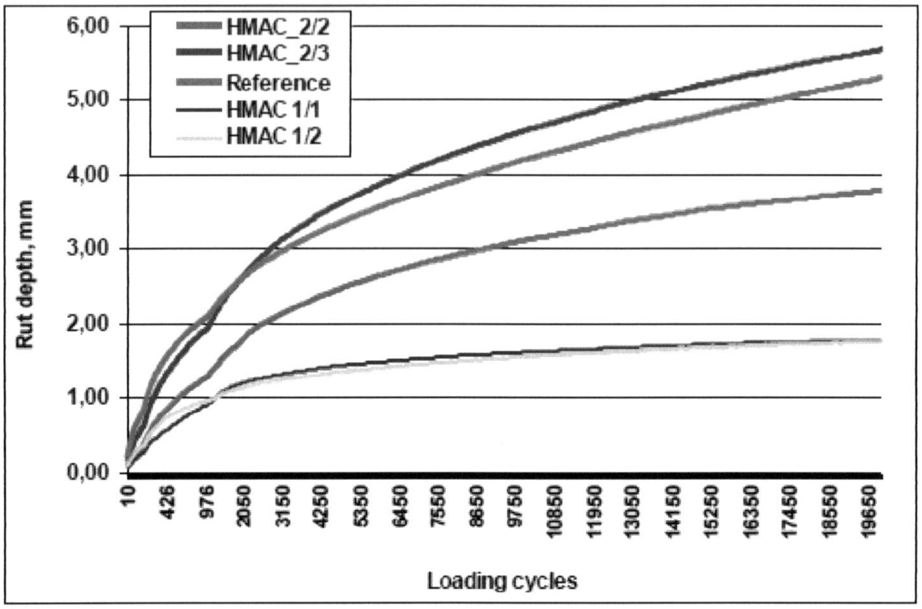

Figure 3. Wheel tracking test results.

Table 6. Numerical values of Wheel Tracking Test of HMAC mixtures with dolomite aggregate.

Parameter	Unit	Standard	Mixture				
			HMAC 1/1	HMAC 1/2	HMAC 2/1	HMAC 2/3	Ref
Bitumen content	%	EN 12697-1	5.7	5.8	5.7	5.8	5.0
Wheel tracking slope (WTS_{air})	mm/10^3 cycles	EN 12697-22 B method	0.04	0.04	0.14	0.22	0.20
Rut depth (RDair)	mm		1.8	1.8	3.8	5.7	5.3
Proport. rut depth (PRDair)	%		3.6	3.6	7.6	11.4	10.6

4.4. Fatigue

To determine the fatigue life of the prepared asphalt concrete mixes, a four point bending beam fatigue test was conducted. The test was run at 10°C, using 10Hz frequency at 130 μm/m strain level. The beams were compacted in the laboratory using roller compactor. They were saw cut to the required dimensions of 50mm width, 50mm height and 400mm length. The failure criterion used in the study is the traditional 50% reduction from initial stiffness. The obtained results indicate that HMAC mixtures have high resistance to fatigue, compared to the results of reference mixture made with conventional bitumen. HMAC mixes fatigue resistance corresponds to standard category ε6-130. HMAC mixes compliance with the SPENS requirements are given in table 5.

IOP Conf. Series: Materials Science and Engineering **96** (2015) 012084 doi:10.1088/1757-899X/96/1/012084

4.5. Water sensitivity

Water sensitivity test value ITSR give information about asphalt concrete durability with a respect to ingress of water. The water sensitivity test were performed according to EN 12697-12 standard method A on the base of indirect tensile strength of dry and wet Marshall specimens. HMAC mixes compliance with the SPENS requirements are given in table 5.

Table 7. Compliance with SPENS requirements.

	Mixtures				
	PMB 10/45 -65		B20/30		
Parameter	HMAC-1/1	HMAC-1/2	HMAC-2/2	HMAC-2/3	Requirement
Voids content, %	3,9	3,7	3.9	3.7	3.0 – 5.0
Rut resistance, mm/1000ciyles	0.04	0.04	0.14	0.22	0.03 – 0.25
Stiffness (10°C, 10Hz), MPa	16700	16100	17900	17100	Smin 14000
Fatigue (10°C, 10Hz, 130μmm/mm)	ε6-130	ε6-130	ε6-130	ε6-130	ε6-130 (≤ 50%)
Water sensitivity, ITSR, %	100	100	98	94	ITSR 80

5. Conclusions

Use of relatively weak dolomite aggregate in High Modulus Asphalt Concrete was evaluated. This mixture was designed to have less than 5% air voids using Marshall compactor, and had high hard (B20/30) and polymer modified (PMB 10/45-65) bitumen content. Testing was performed to compare this mix with AC16bin reference mixture that was produced using conventional bitumen B70/100 and granite aggregate. Test results demonstrated that with optimum mix design HMAC mixture can provide high rut und fatigue resistance. However, mixtures with hard grade bitumen B20/30 showed a little lower resistance to rutting. The reference mixture while having high rut resistance, proved that lower binder content results in shorter fatigue life. Rut resistance and stiffness of the reference mixtures do not meet SPEN project recommendations.

These results provide confidence that the weak Latvian dolomite may be applied in High Modulus Asphalt Concrete for base and binder courses. HMAC mixtures fulfil the HMAC asphalt concrete requirements in accordance with SPENS project recommendations (SPENS).

6. References

[1] AFNOR – Association Franēaise de Normalisation 1999 *Enrobēs Hydrocarbonēs: Couches d'assises: enrobēs a module ēlevē (EME): NF P 98-140* Paris

[2] Backer C, Visscher J, Glorie L, Vanelstraete A, Vansteenkiste S and Heleven L 2008 A comparative high – modulus experiment in Belgium *Proc. of Transport Research Arena Europe 2008 (TRA 2008) Int. Conf. (Slovenia, Ljubljana, 21–24 April 2008)*

[3] Rohde L, Ceratti J A P, Nūnez P V and Vitorello T 2008 Using APT and Laboratory testing to evaluate the performance of high modulus asphalt concrete for base courses in Brazil. *Proc. of the Third Int. Conf. on Accelerated Pavement Testing (APT '08) (Spain, Madrid, 1–3 October 2008)*

[4] Sivapatham P, Beckedahl H J and Jannsen S 2010 High stable asphalt for heavy loaded bus test lane sections *Proc. of the 11th Int. Conf. on Asphalt Pavements ISAP 2010 (Japan, Nagoya, Aichi, 1–6 August 2010)*

[5] Sustainable Pavements for European New Member States (SPENS) Document No. D8 [Available]: http://www.spens.fehrl.org

[6] Sybilski D, Bankowski W and Krajewski M 2010 High modulus asphalt concrete with limestone aggregate *Int. J. of Pavement Research and Technology* **3** (2) 96–101

[7] Sybilski D, Maliszewska D, Maliszewski M and Mularzuk R 2008 Experience with high modulus asphalt concrete in Warsaw street overlays *Proc. of Transport Research Arena Europe 2008 (TRA 2008) Internationa Conf. (Slovenia, Ljubljana, 21 – 24 April 2008)*

[8] Čygas D, Laurinavičius A, Vaitkus A, Perveneckas Z, Mo-tiejūnas A 2008 Research of asphalt pavement structures on Lithuanian roads (I) *The Baltic J. of Road and Bridge Engineering* **3** (2) 77–83

[9] Elliott R 2009 Implementing high modulus asphalt technology in The United Kingdom *The Internation Seminar Maintenance Techniques to Improve Pavement Performance* p 10

[10] Maupin G W and Diefenderfer B K 2006 *Design of High-Binder- High-Modulus Asphalt Mixture* Final Report VTRC 07- R15, Virginia Transportation Research Council p 28

[11] Vaitkus A, Laurinavičius A, Oginskas R, Motiejūnas A, Paliukaitė M, Barvidienė O 2012 The road of experimental pa-vement structures: experience of five years operation *The Baltic J. of Road and Bridge Engineering* **7** (3) 220–227

Acknowledgements
The research leading to these results has received the funding from Latvia State Research Programme under grant agreement "INNOVATIVE MATERIALS AND SMART TECHNOLOGIES FOR ENVIRONMENTAL SAFETY, IMATEH".